VISUALIZING

EVERYDAY CHEMISTRY

Michael Nichols/National Geographic Creative

VISUALIZING EVERYDAY CHEMISTRY

Douglas P. Heller
Senior Lecturer
University of Miami

Carl H. Snyder
Professor Emeritus
Department of Chemistry
University of Miami

WILEY

VICE PRESIDENT & PUBLISHER Petra Recter
SENIOR ACQUISITION EDITOR Nick Ferrari
DEVELOPMENT PROJECT MANAGER Charity Robey
DEVELOPMENT EDITOR David Chelton
EDITORIAL ASSISTANT Mallory Fryc
SENIOR MARKETING MANAGER Kristy Ruff
SENIOR PRODUCT DESIGNER Geraldine Osnato
SENIOR PROJECT EDITOR Jennifer Yee
DESIGN DIRECTOR Harold Nolan
INTERIOR DESIGN Wendy Lai
COVER DESIGN Thomas Nery
PHOTO EDITOR Elizabeth Blomster
SENIOR CONTENT MANAGER Kevin Holm
OUTSIDE PRODUCTION Jeanine Furino
SENIOR PRODUCTION EDITOR Elizabeth Swain

COVER PHOTOS Hand pumping gas: hxdbzxy/Shutterstock; boiling water: Steven Coling/Shutterstock; diving: T-Design/Shutterstock; oil: Angel Simon/Shutterstock; solar panels: R-J-Seymour/istockphoto

This book was set in 10.5/14 New Baskerville ITC Std. by codeMantra. The text/cover were printed and bound by Quad Graphics, Versailles. This book is printed on acid-free paper.

Founded in 1807, John Wiley & Sons, Inc. has been a valued source of knowledge and understanding for more than 200 years, helping people around the world meet their needs and fulfill their aspirations. Our company is built on a foundation of principles that include responsibility to the communities we serve and where we live and work. In 2008, we launched a Corporate Citizenship Initiative, a global effort to address the environmental, social, economic, and ethical challenges we face in our business. Among the issues we are addressing are carbon impact, paper specifications and procurement, ethical conduct within our business and among our vendors, and community and charitable support. For more information, please visit our Web site: www.wiley.com/go/citizenship.

ISBN: 978-0-470-62066-3
Binder version: 978-1-119-00306-9

Printed in the United States of America
10 9 8 7 6 5 4 3 2 1

About the Authors

Douglas Heller, Ph.D.
Senior Lecturer, Department of Chemistry
University of Miami

Douglas Heller received his Ph.D. in Organic Chemistry and MBA, both from the University of Chicago. He has worked in the field of technology transfer, helping to commercialize biomedical and physical science discoveries, and in the pharmaceutical industry, in project management and corporate communications. He then moved to academia with a visiting appointment in forensic science. Since 2005 he has been at the University of Miami, where he teaches organic chemistry, chemistry for nonscience students, and an innovative honor's general chemistry sequence integrating biology and calculus. In conjunction with the university's Writing Center, he also mentors science graduate students in writing and public speaking. He is particularly drawn to the art of communication, where first he found satisfaction and success in connecting the worlds of science and business, and now in bringing the world of chemistry to a variety of audiences. Since joining the University of Miami's Chemistry Department, he has assumed responsibility for the continuation and growth of its course for nonscience students, incorporating successful and popular innovations such as student response systems (clickers), collaborative projects, and novel demonstrations.

Carl Snyder, Ph.D.
Professor of Chemistry, Emeritus
University of Miami

Carl Snyder received his BS in Chemistry, summa cum laude, from the University of Pittsburgh and Ph.D. from the Ohio State University with specialization in organic chemistry. He then spent 1 year with Eastman Kodak Co. followed by 2 years of postdoctoral research on organoboranes at Purdue University with Prof. Herbert C. Brown. In 1961 he joined the University of Miami (FL) as Assistant Professor of Chemistry, retiring 45 years later as Professor of Chemistry Emeritus. During his tenure at the University of Miami he was author of two chemistry textbooks—*The Extraordinary Chemistry of Ordinary Things* (John Wiley & Sons), and *Introduction to Modern Organic Chemistry* (Harper and Row)—and author or coauthor of two dozen papers on chemical research and/or education. His interests lay principally in organic chemistry, the use of computers in chemical education, and chemistry for the nonscience student. He created the department's course for liberal arts students in 1976 as a 1-credit, 1-semester course and has led its growth to its current status as two 3-credit courses.

Dedication

To Jane, who gets me. To Lauren and Myles, who've been asking if I'm done with the book yet. Yes, and may your curiosities about the world continue to outpace the answers we have to explain it.

DPH

To the memory of Jean.

CHS

Brief Contents

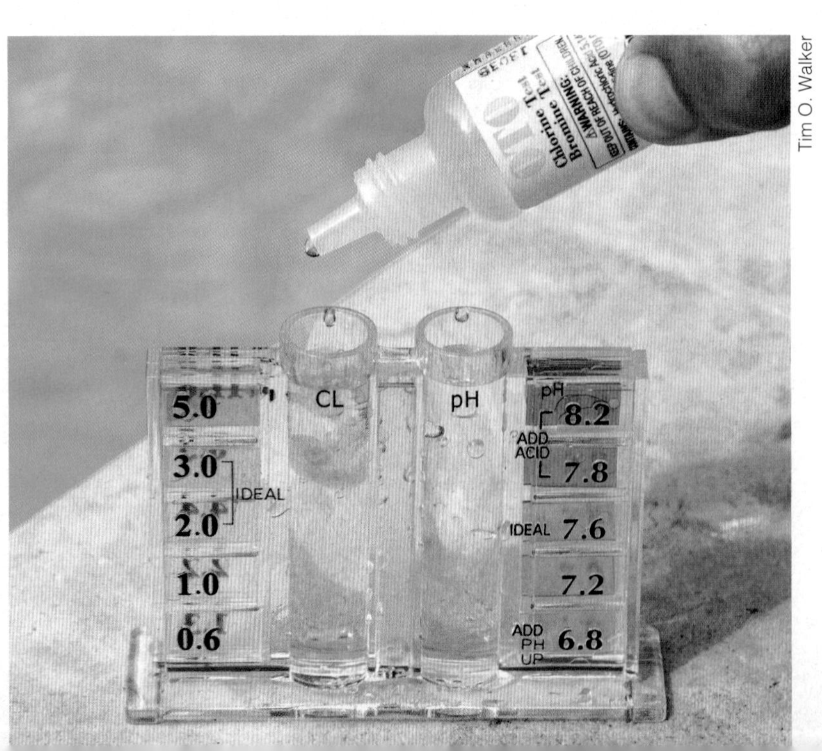

Tim O. Walker

Contents

Thomas Kcehler/Photothek via Getty Images

3D Sculptor/iStockphoto

edelmar/iStockphoto

© Godfrey Zwygart/Demotix/Corbis

Steve Raymer/National Geographic Creative

National Geographic Creative/Steve Raymer

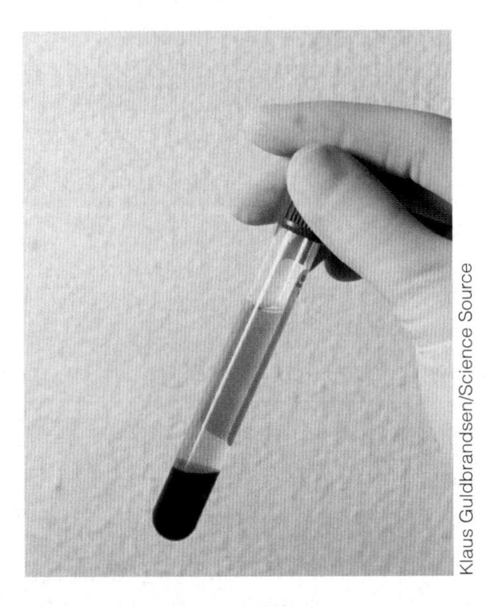
Klaus Guldbrandsen/Science Source

Tim O. Walker

RAYMOND GEHMAN/National Geographic Creative

Roberto Schmidt/Staff/AFP/Getty Images

Andreas Kuehn/Iconiaca/Getty Images

MCT via Getty Images

John Burcham/National Geographic/Getty Images

RedHelga/iStockphoto

Process Diagrams

A series or combination of figures and photos that describe and depict a complex process

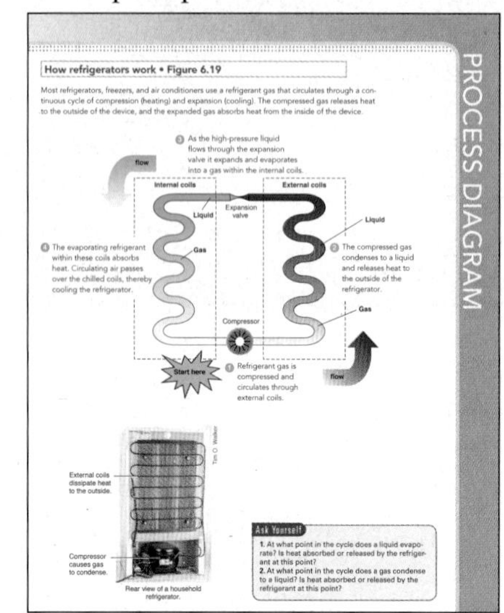

Chemistry InSight

A multi-part figure devoted to a major concept or topic in the chapter

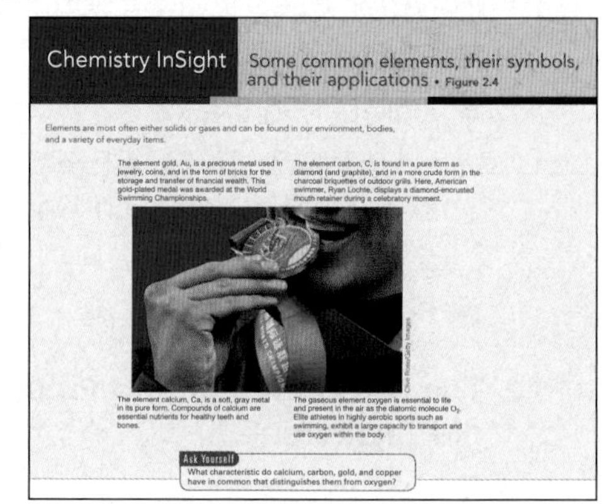

Preface

How Is Wiley Visualizing Different?

Wiley Visualizing is based on decades of research on the use of visuals in learning (Mayer, 2005).[1] The visuals teach key concepts and are pedagogically designed to **present**, **explain**, and **organize** new information. The figures are tightly integrated with accompanying text; the visuals are conceived with the text in ways that clarify and reinforce major concepts, while allowing students to understand the details. This commitment to distinctive and consistent visual pedagogy sets Wiley Visualizing apart from other textbooks.

The texts offer an array of remarkable photographs, maps, media, and film from photo collections around the world. Wiley Visualizing's images are not decorative; such images can be distracting to students. Instead, they are purposeful and the primary driver of the content. These authentic materials immerse the student in real-life issues and experiences and support thinking, comprehension, and application.

Together these elements deliver a level of rigor in ways that maximize student learning and involvement. Wiley Visualizing has proven to increase student learning through its unique combination of text, photographs, and illustrations, with online video, animations, simulations and assessments.

(1) Visual Pedagogy. Using the Cognitive Theory of Multimedia Learning, which is backed up by hundreds of empirical research studies, Wiley's authors create visualizations for their texts that specifically support students' thinking and learning—for example, the selection of relevant materials, the organization of the new information, or the integration of the new knowledge with prior knowledge.

(2) Authentic Situations and Problems. *Visualizing Everyday Chemistry* benefits from an array of remarkable photographs, maps, and media. These authentic materials immerse the student in real-life issues in chemistry, thereby enhancing motivation, learning, and retention (Donovan & Bransford, 2005).[2]

(3) Designed with Interactive Multimedia *Visualizing Everyday Chemistry* is tightly integrated with *WileyPLUS Learning Space*, our online learning environment that provides interactive multimedia activities in which learners can actively engage with the materials. The combination of textbook and *WileyPLUS Learning Space* provides learners with multiple entry points to the content, giving them greater opportunity to explore concepts and assess their understanding as they progress through the course. *WileyPLUS Learning Space* is a key component of the Wiley Visualizing learning and problem-solving experience, setting it apart from other textbooks whose online component is mere drill-and-practice.

Wiley Visualizing and the *WileyPLUS* Learning Environment are designed as natural extensions of how we learn

To understand why the Visualizing approach is effective, it is first helpful to understand how we learn.

1. Our brain processes information using two main channels: visual and verbal. Our *working memory* holds information that our minds process as we learn. This "mental workbench" helps us with decisions, problem solving, and making sense of words and pictures by building verbal and visual models of the information.
2. When the verbal and visual models of corresponding information are integrated in working memory, we form more comprehensive, lasting, mental models.
3. When we link these integrated mental models to our prior knowledge, stored in our *long-term memory*, we build even stronger mental models. When an integrated (visual plus verbal) mental model is formed and stored in long-term memory, real learning begins.

The effort our brains put forth to make sense of instructional information is called *cognitive load*. There are two kinds of cognitive load: productive cognitive load, such as when we're engaged in learning or exert positive effort to create mental models; and unproductive cognitive load, which occurs when the brain is trying to make sense of needlessly complex content or when information is not presented well. The learning process can be impaired when the information to be processed exceeds the capacity of working memory. Well-designed visuals and text with effective pedagogical guidance can reduce the unproductive cognitive load in our working memory.

---

[1]Mayer, R.E. (Ed) (2005). *The Cambridge Handbook of Multimedia Learning*. Cambridge University Press.

[2]Donovan, M.S., & Bransford, J. (Eds.) (2005). *How Students Learn: Science in the Classroom. The National Academy Press*. Available at http://www.nap.edu/openbook.php?record_id=11102&page=1

Wiley Visualizing is designed for engaging and effective learning

The visuals and text in *Visualizing Everyday Chemistry* are specially integrated to present complex processes in clear steps and with clear representations, organize related pieces of information, and integrate related information with one another. This approach minimizes unproductive cognitive load and helps students engage with the content. When students are engaged, they are reading and learning, which can lead to greater knowledge and academic success.

Research shows that well-designed visuals, integrated with comprehensive text, can improve the efficiency with which a learner processes information. In this regard, SEG Research, an independent research firm, conducted a national, multisite study evaluating the effectiveness of Wiley Visualizing. Its findings indicate that students using Wiley Visualizing products (both print and multimedia) were more engaged in the course, exhibited greater retention throughout the course, and made significantly greater gains in content area knowledge and skills, as compared to students in similar classes that did not use Wiley Visualizing.[3]

The use of *WileyPLUS* can also increase learning. According to a white paper titled "Leveraging Blended Learning for More Effective Course Management and Enhanced Student Outcomes" by Peggy Wyllie of Evince Market Research & Communications, studies show that effective use of online resources can increase learning outcomes. Pairing supportive online resources with face-to-face instruction can help students to learn and reflect on material, and deploying multimodal learning methods can help students to engage with the material and retain their acquired knowledge.

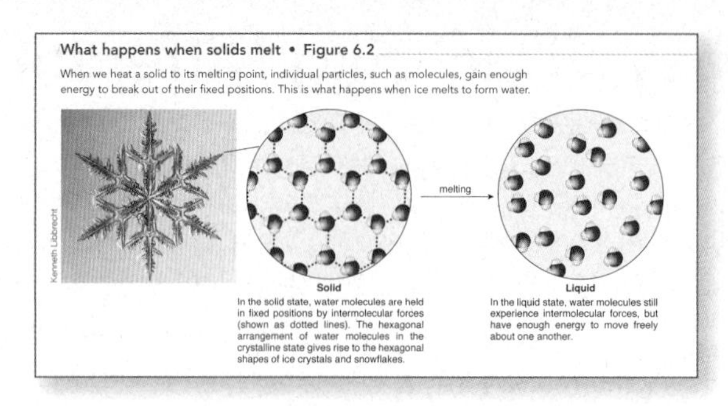

Figure 1: What happens when solids melt (Fig. 6.2) Figures show students what is happening at the molecular level, and how that relates to the observable world.

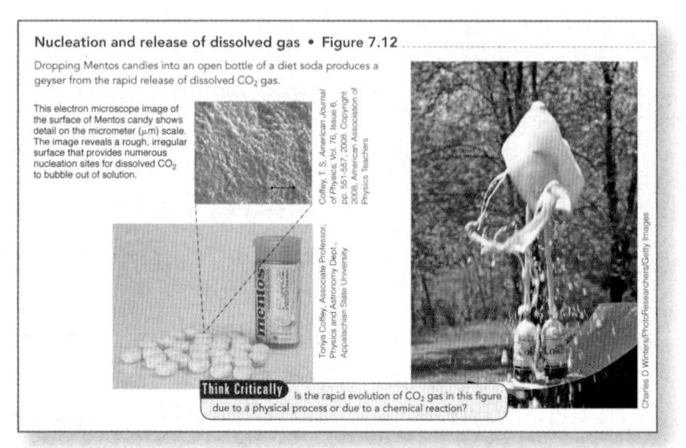

Figure 2: Nucleation and release of dissolved gas (Fig. 7.12) Microscopic and macroscopic images are paired so that students can compare and contrast them, thereby grasping the underlying concept. Adjacent captions eliminate split attention.

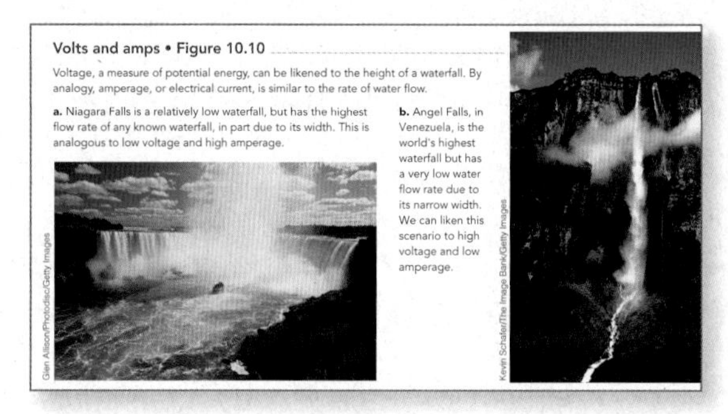

Figure 3: Volts and amps (Fig. 10.10) From abstraction to reality. Pairs of images explain abstract concepts visually.

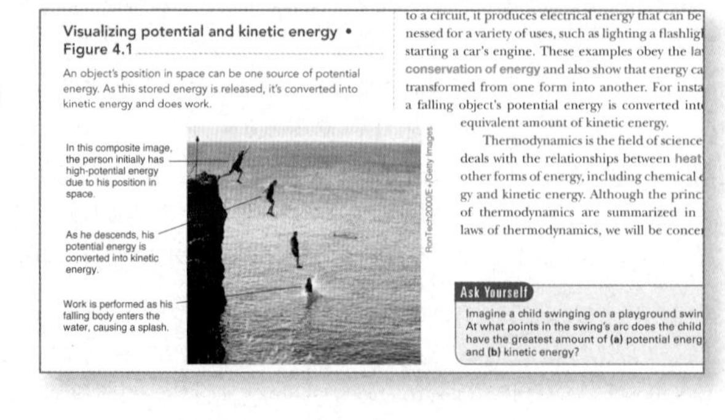

Visualizing potential and kinetic energy • Figure 4.1
An object's position in space can be one source of potential energy. As this stored energy is released, it's converted into kinetic energy and does work.
In this composite image, the person initially has high-potential energy due to his position in space.
As he descends, his potential energy is converted into kinetic energy.
Work is performed as his falling body enters the water, causing a splash.
RonTech2000/E+/Getty Images
to a circuit, it produces electrical energy that can be
nessed for a variety of uses, such as lighting a flashlig
starting a car's engine. These examples obey the la
conservation of energy and also show that energy ca
transformed from one form into another. For insta
a falling object's potential energy is converted int
equivalent amount of kinetic energy.
Thermodynamics is the field of science
deals with the relationships between heat
other forms of energy, including chemical
gy and kinetic energy. Although the princ
of thermodynamics are summarized in
laws of thermodynamics, we will be conce
Ask Yourself
Imagine a child swinging on a playground swin
At what points in the swing's arc does the child
have the greatest amount of (a) potential energ
and (b) kinetic energy?

Figure 4: Visualizing potential and kinetic energy (Fig. 4.1) Images like this composite photo teach concepts using visuals.

[3]SEG Research (2009). Improving Student-Learning with Graphically-Enhanced Textbooks: A Study of the Effectiveness of the Wiley Visualizing Series.

How Are the Wiley Visualizing Chapters Organized?

Student engagement is more than just exciting videos or interesting animations—engagement means keeping students motivated to keep going. It is easy to get bored or lose focus when presented with large amounts of information, and it is easy to lose motivation when the relevance of the information is unclear. The design of *WileyPLUS* is based on cognitive science, instructional design, and extensive research into user experience. It transforms learning into an interactive, engaging, and outcomes-oriented experience for students.

Each Wiley Visualizing chapter engages students from the start

Chapter opening text and visuals introduce the subject and connect the student with the material that follows.

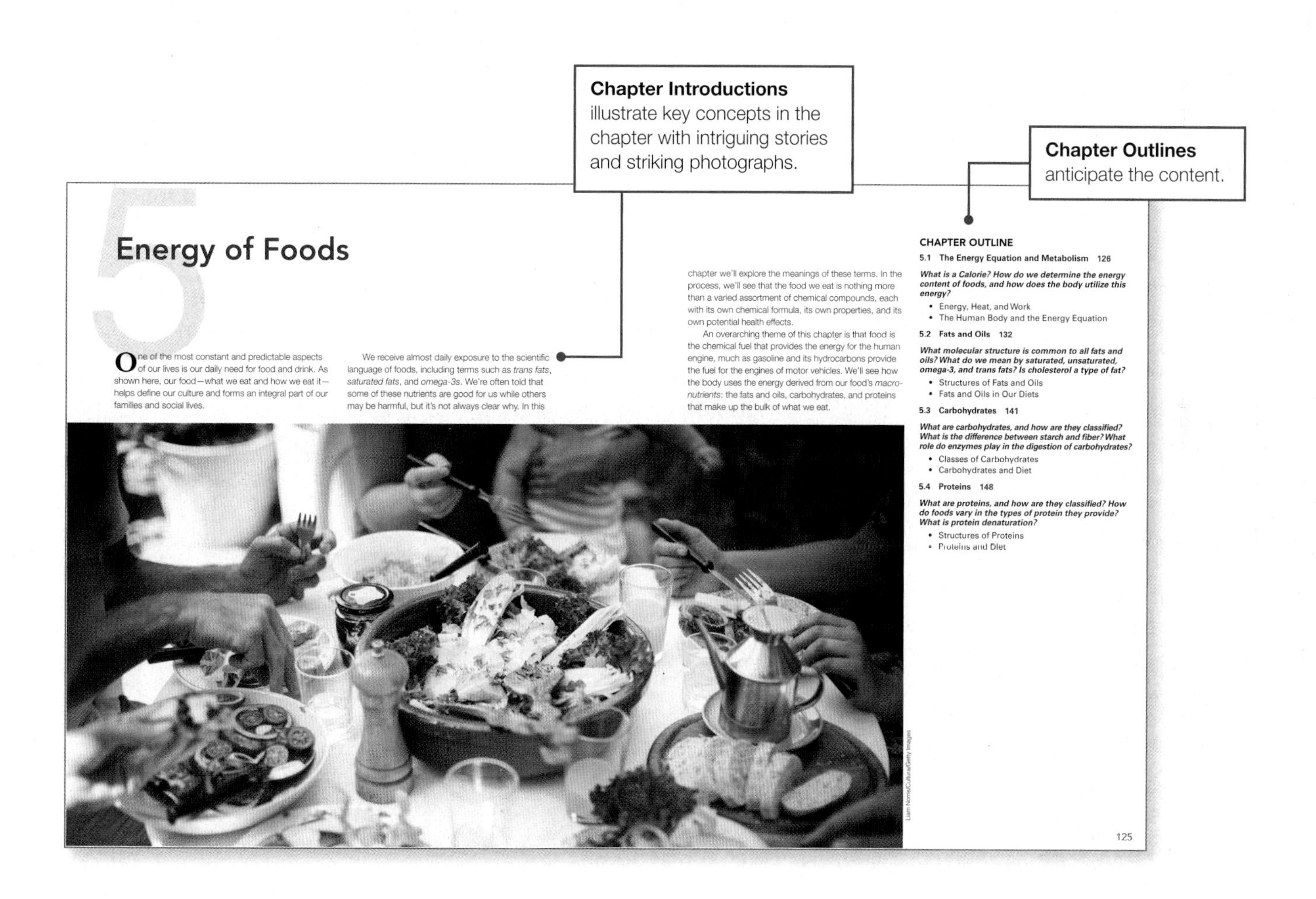
Chapter Introductions illustrate key concepts in the chapter with intriguing stories and striking photographs.

Chapter Outlines anticipate the content.

5

Energy of Foods

One of the most constant and predictable aspects of our lives is our daily need for food and drink. As shown here, our food—what we eat and how we eat it—helps define our culture and forms an integral part of our families and social lives.

We receive almost daily exposure to the scientific language of foods, including terms such as *trans fats*, *saturated fats*, and *omega-3s*. We're often told that some of these nutrients are good for us while others may be harmful, but it's not always clear why. In this chapter we'll explore the meanings of these terms. In the process, we'll see that the food we eat is nothing more than a varied assortment of chemical compounds, each with its own chemical formula, its own properties, and its own potential health effects.

An overarching theme of this chapter is that food is the chemical fuel that provides the energy for the human engine, much as gasoline and its hydrocarbons provide the fuel for the engines of motor vehicles. We'll see how the body uses the energy derived from our food's *macronutrients*: the fats and oils, carbohydrates, and proteins that make up the bulk of what we eat.

CHAPTER OUTLINE

125

WileyPLUS Experience the chapter through a *WileyPLUS* course.

Wiley Visualizing guides students through the chapter

The content of Wiley Visualizing gives students a variety of approaches—visuals, words, interactions, video, and assessments—that work together to provide a guided path through the content.

LEARNING OBJECTIVES at the start of each section indicate in behavioral terms the concepts that students are expected to master while reading the section.

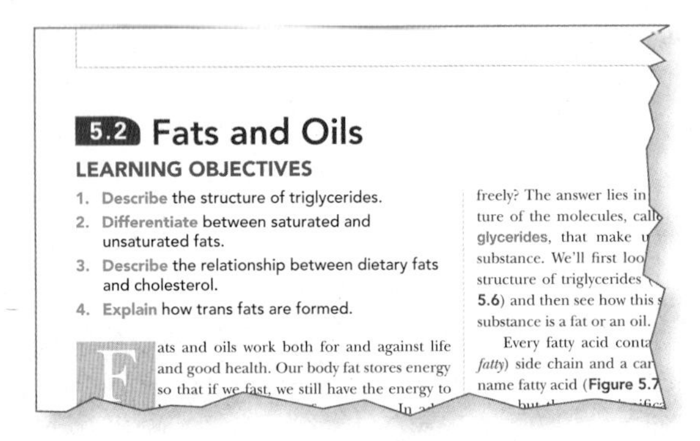

5.2 Fats and Oils

LEARNING OBJECTIVES

1. Describe the structure of triglycerides.
2. Differentiate between saturated and unsaturated fats.
3. Describe the relationship between dietary fats and cholesterol.
4. Explain how trans fats are formed.

Fats and oils work both for and against life and good health. Our body fat stores energy so that if we fast, we still have the energy to

freely? The answer lies in ture of the molecules, call glycerides, that make u substance. We'll first loo structure of triglycerides 5.6) and then see how this substance is a fat or an oil.

Every fatty acid cont *fatty*) side chain and a car name fatty acid (**Figure 5.7**

Chemistry InSight features are multipart visual sections that focus on a key concept or topic in the chapter, exploring it in detail or in broader context using a combination of photos, diagrams, maps, and data.

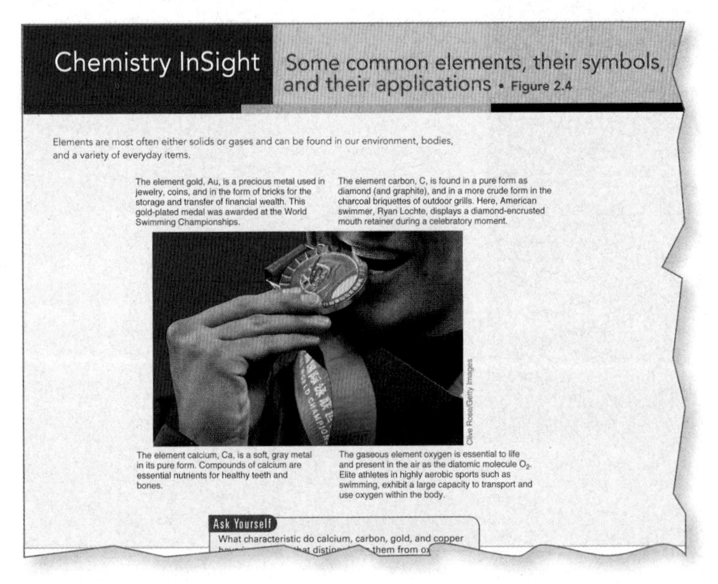

Chemistry InSight Some common elements, their symbols, and their applications • **Figure 2.4**

Elements are most often either solids or gases and can be found in our environment, bodies, and a variety of everyday items.

The element gold, Au, is a precious metal used in jewelry, coins, and in the form of bricks for the storage and transfer of financial wealth. This gold-plated medal was awarded at the World Swimming Championships.

The element carbon, C, is found in a pure form as diamond (and graphite), and in a more crude form in the charcoal briquettes of outdoor grills. Here, American swimmer, Ryan Lochte, displays a diamond-encrusted mouth retainer during a celebratory moment.

Clive Rose/Getty Images

The element calcium, Ca, is a soft, gray metal in its pure form. Compounds of calcium are essential nutrients for healthy teeth and bones.

The gaseous element oxygen is essential to life and present in the air as the diatomic molecule O_2. Elite athletes in highly aerobic sports such as swimming, exhibit a large capacity to transport and use oxygen within the body.

Ask Yourself

What characteristic do calcium, carbon, gold, and copper

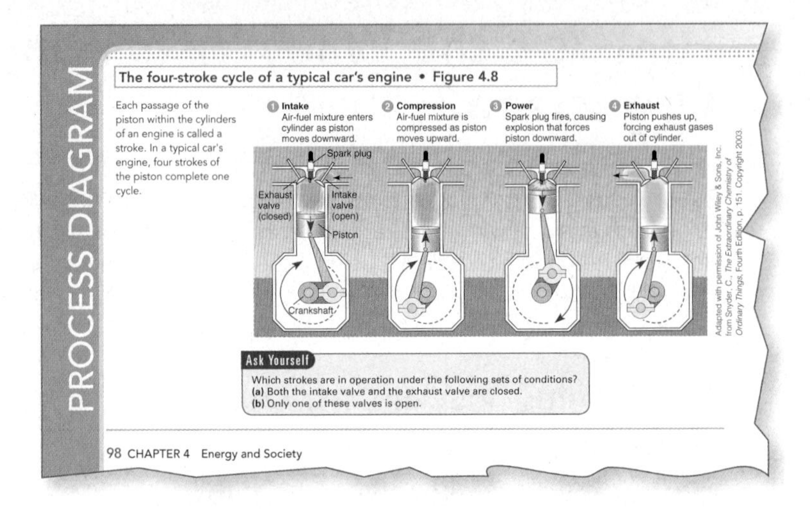

PROCESS DIAGRAM

The four-stroke cycle of a typical car's engine • Figure 4.8

Each passage of the piston within the cylinders of an engine is called a stroke. In a typical car's engine, four strokes of the piston complete one cycle.

1 Intake
Air-fuel mixture enters cylinder as piston moves downward.
2 Compression
Air-fuel mixture is compressed as piston moves upward.
3 Power
Spark plug fires, causing explosion that forces piston downward.
4 Exhaust
Piston pushes up, forcing exhaust gases out of cylinder.
Spark plug
Exhaust valve (closed)
Intake valve (open)
Piston
Crankshaft

Adapted with permission of John Wiley & Sons, Inc. from Snyder, C., *The Extraordinary Chemistry of Ordinary Things*, Fourth Edition, p. 151. Copyright 2003.

Ask Yourself

Which strokes are in operation under the following sets of conditions?
(a) Both the intake valve and the exhaust valve are closed.
(b) Only one of these valves is open.

98 CHAPTER 4 Energy and Society

Process Diagrams provide in-depth coverage of processes correlated with clear, step-by-step narrative, enabling students to grasp important topics with less effort.

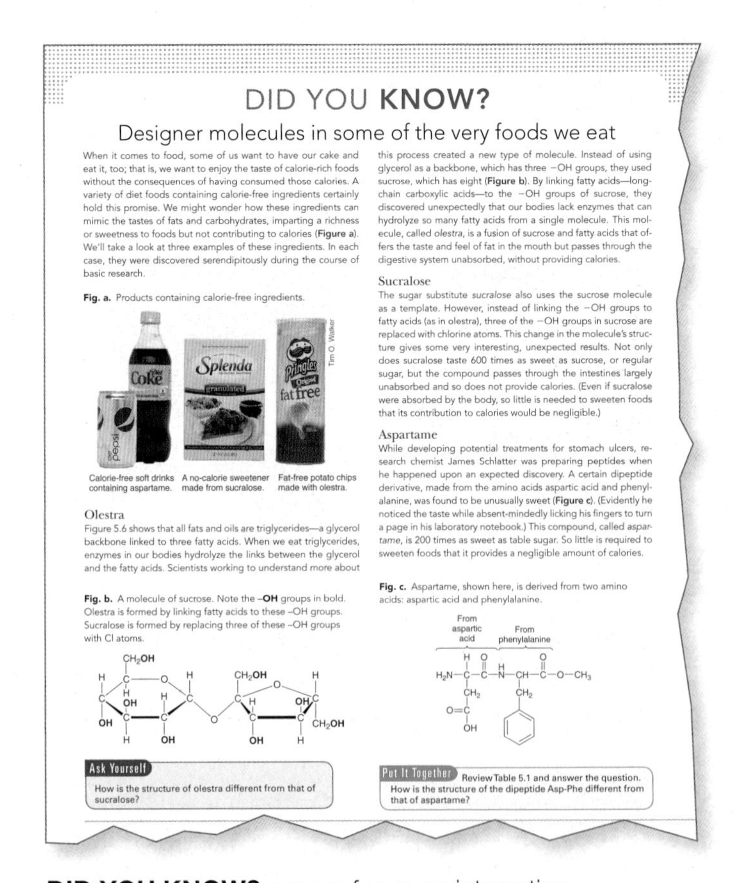

DID YOU KNOW?

Designer molecules in some of the very foods we eat

When it comes to food, some of us want to have our cake and eat it, too; that is, we want to enjoy the taste of calorie-rich foods without the consequences of having consumed those calories. A variety of diet foods containing calorie-free ingredients certainly hold this promise. We might wonder how these ingredients can mimic the tastes of fats and carbohydrates, imparting a richness or sweetness to foods but not contributing to calories (**Figure a**). We'll take a look at three examples of these ingredients. In each case, they were discovered serendipitously during the course of basic research.

Fig. a. Products containing calorie-free ingredients.

Calorie-free soft drinks containing aspartame. A no-calorie sweetener made from sucralose. Fat-free potato chips made with olestra.

Tim O. Walker

Olestra

Figure 5.6 shows that all fats and oils are triglycerides—a glycerol backbone linked to three fatty acids. When we eat triglycerides, enzymes in our bodies hydrolyze the links between the glycerol and the fatty acids. Scientists working to understand more about this process created a new type of molecule. Instead of using glycerol as a backbone, which has three –OH groups, they used sucrose, which has eight (**Figure b**). By linking fatty acids—long-chain carboxylic acids—to the –OH groups of sucrose, they discovered unexpectedly that our bodies lack enzymes that can hydrolyze so many fatty acids from a single molecule. This molecule, called *olestra*, is a fusion of sucrose and fatty acids that offers the taste and feel of fat in the mouth but passes through the digestive system unabsorbed, without providing calories.

Fig. b. A molecule of sucrose. Note the **–OH** groups in bold. Olestra is formed by linking fatty acids to these –OH groups. Sucralose is formed by replacing three of these –OH groups with Cl atoms.

Sucralose

The sugar substitute *sucralose* also uses the sucrose molecule as a template. However, instead of linking the –OH groups to fatty acids (as in olestra), three of the –OH groups in sucrose are replaced with chlorine atoms. This change in the molecule's structure gives some very interesting, unexpected results. Not only does sucralose taste 600 times as sweet as sucrose, or regular sugar, but the compound passes through the intestines largely unabsorbed and so does not provide calories. (Even if sucralose were absorbed by the body, so little is needed to sweeten foods that its contribution to calories would be negligible.)

Aspartame

While developing potential treatments for stomach ulcers, research chemist James Schlatter was preparing peptides when he happened upon an expected discovery. A certain dipeptide derivative, made from the amino acids aspartic acid and phenylalanine, was found to be unusually sweet (**Figure c**). (Evidently he noticed the taste while absent-mindedly licking his fingers to turn a page in his laboratory notebook.) This compound, called *aspartame*, is 200 times as sweet as table sugar. So little is required to sweeten foods that it provides a negligible amount of calories.

Fig. c. Aspartame, shown here, is derived from two amino acids: aspartic acid and phenylalanine.

From aspartic acid — From phenylalanine

Ask Yourself

How is the structure of olestra different from that of sucralose?

Put It Together Review Table 5.1 and answer the question.
How is the structure of the dipeptide Asp-Phe different from that of aspartame?

DID YOU KNOW? essays focus on interesting developments in contemporary chemistry, from forensic science to biofuels to designer macronutrients.

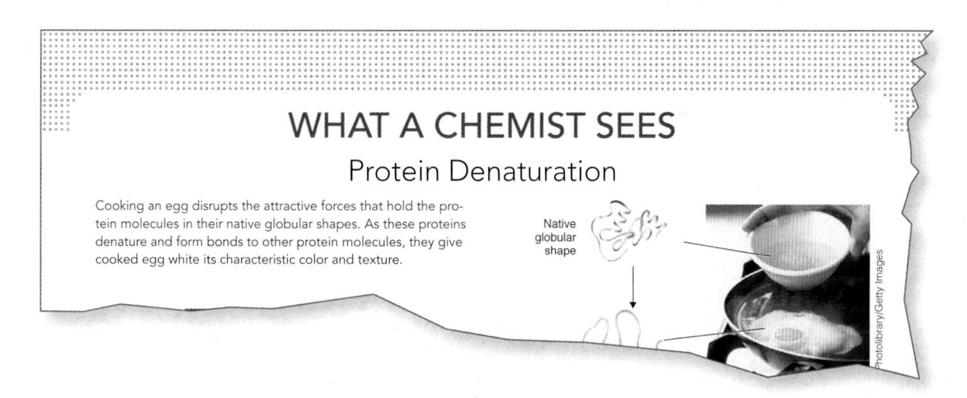

WHAT A CHEMIST SEES

Protein Denaturation

Cooking an egg disrupts the attractive forces that hold the protein molecules in their native globular shapes. As these proteins denature and form bonds to other protein molecules, they give cooked egg white its characteristic color and texture.

WHAT A CHEMIST SEES highlights a concept or phenomenon that would stand out to chemists. Photos and figures are used to improve students' understanding of the usefulness of a chemical perspective and to develop their observational skills.

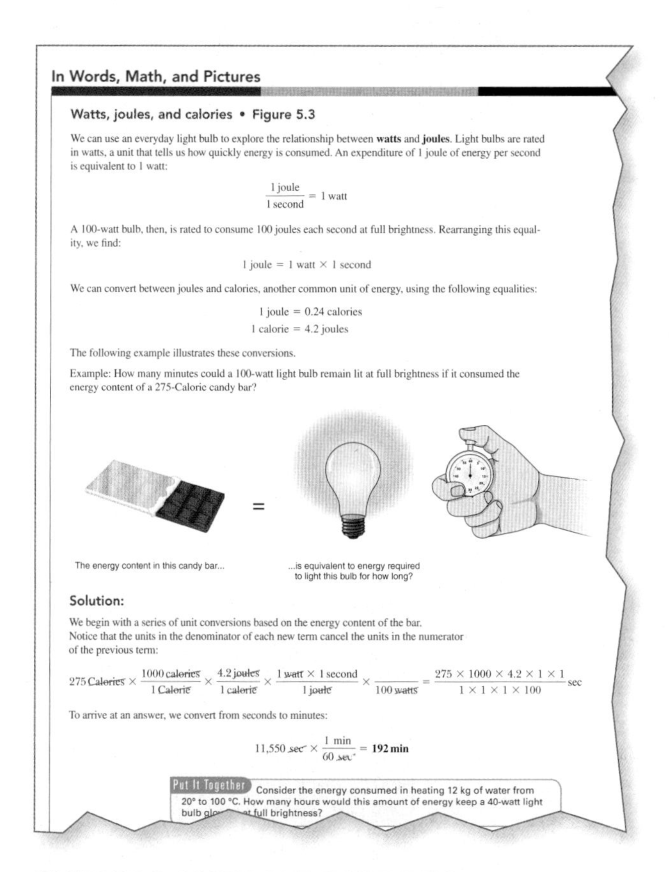

In Words, Math, and Pictures

Watts, joules, and calories • Figure 5.3

We can use an everyday light bulb to explore the relationship between **watts** and **joules**. Light bulbs are rated in watts, a unit that tells us how quickly energy is consumed. An expenditure of 1 joule of energy per second is equivalent to 1 watt:

$$\frac{1\text{ joule}}{1\text{ second}} = 1\text{ watt}$$

A 100-watt bulb, then, is rated to consume 100 joules each second at full brightness. Rearranging this equality, we find:

$$1\text{ joule} = 1\text{ watt} \times 1\text{ second}$$

We can convert between joules and calories, another common unit of energy, using the following equalities:

$$1\text{ joule} = 0.24\text{ calories}$$
$$1\text{ calorie} = 4.2\text{ joules}$$

The following example illustrates these conversions.

Example: How many minutes could a 100-watt light bulb remain lit at full brightness if it consumed the energy content of a 275-Calorie candy bar?

Solution:

We begin with a series of unit conversions based on the energy content of the bar. Notice that the units in the denominator of each new term cancel the units in the numerator of the previous term:

$$275\text{ Calories} \times \frac{1000\text{ calories}}{1\text{ Calorie}} \times \frac{4.2\text{ joules}}{1\text{ calorie}} \times \frac{1\text{ watt} \times 1\text{ second}}{1\text{ joule}} \times \frac{}{100\text{ watts}} = \frac{275 \times 1000 \times 4.2 \times 1 \times 1}{1 \times 1 \times 1 \times 100}\text{ sec}$$

To arrive at an answer, we convert from seconds to minutes:

$$11{,}550\text{ sec} \times \frac{1\text{ min}}{60\text{ sec}} = \mathbf{192\text{ min}}$$

Put It Together Consider the energy consumed in heating 12 kg of water from 20° to 100 °C. How many hours would this amount of energy keep a 40-watt light bulb glo... at full brightness?

IN WORDS, MATH AND PICTURES provides students with a detailed, worked-out example showing how the visual and verbal description of a problem, relate to the calculations used to solve it.

Concept Caching

A Deeper Look features conceptually advanced material, presented with a visual emphasis.

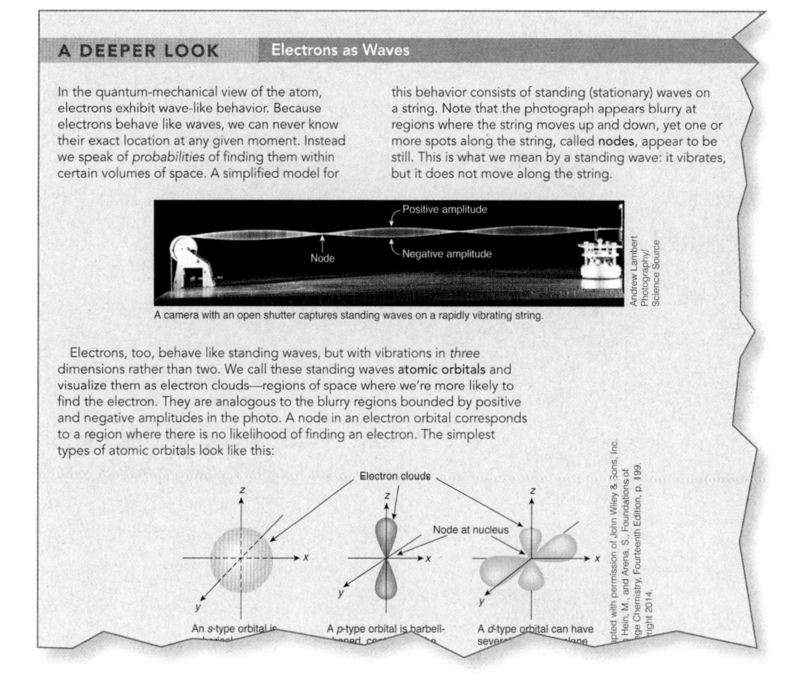

A DEEPER LOOK Electrons as Waves

In the quantum-mechanical view of the atom, electrons exhibit wave-like behavior. Because electrons behave like waves, we can never know their exact location at any given moment. Instead we speak of *probabilities* of finding them within certain volumes of space. A simplified model for this behavior consists of standing (stationary) waves on a string. Note that the photograph appears blurry at regions where the string moves up and down, yet one or more spots along the string, called **nodes**, appear to be still. This is what we mean by a standing wave: it vibrates, but it does not move along the string.

A camera with an open shutter captures standing waves on a rapidly vibrating string.

Electrons, too, behave like standing waves, but with vibrations in *three* dimensions rather than two. We call these standing waves **atomic orbitals** and visualize them as electron clouds—regions of space where we're more likely to find the electron. They are analogous to the blurry regions bounded by positive and negative amplitudes in the photo. A node in an electron orbital corresponds to a region where there is no likelihood of finding an electron. The simplest types of atomic orbitals look like this:

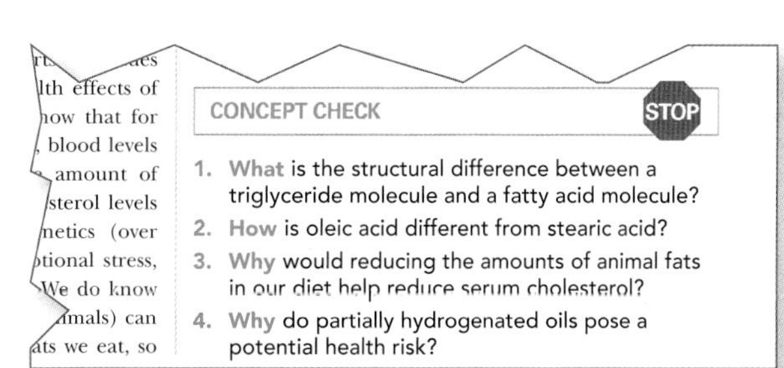

CONCEPT CHECK

1. What is the structural difference between a triglyceride molecule and a fatty acid molecule?
2. How is oleic acid different from stearic acid?
3. Why would reducing the amounts of animal fats in our diet help reduce serum cholesterol?
4. Why do partially hydrogenated oils pose a potential health risk?

Coordinated with the section-opening **Learning Objectives**, at the end of each section **Concept Check** questions allow students to test their comprehension of the learning objectives.

WileyPLUS Streaming videos are available to students in *WileyPLUS*.

Student understanding is assessed at different levels

Wiley Visualizing offers students lots of practice material for assessing their understanding of each study objective. Students know exactly what they are getting out of each study session through immediate feedback and coaching.

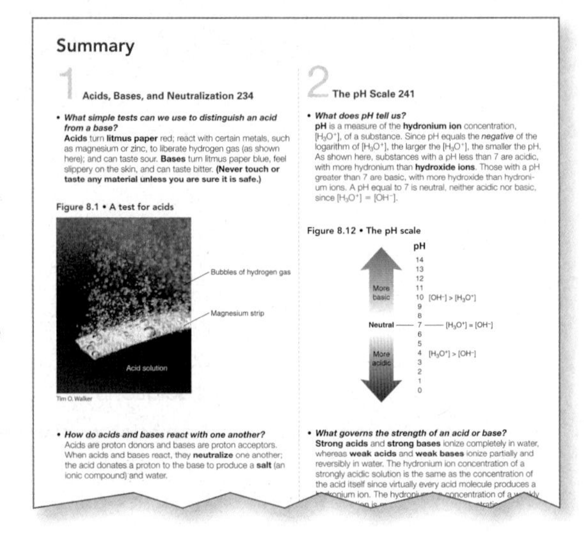
Summary

1 Acids, Bases, and Neutralization 234

- ***What simple tests can we use to distinguish an acid from a base?***
  Acids turn **litmus paper** red; react with certain metals, such as magnesium or zinc, to liberate hydrogen gas (as shown here); and can taste sour. **Bases** turn litmus paper blue, feel slippery on the skin, and can taste bitter. **(Never touch or taste any material unless you are sure it is safe.)**

Figure 8.1 • A test for acids

- ***How do acids and bases react with one another?***
  Acids are proton donors and bases are proton acceptors. When acids and bases react, they **neutralize** one another; the acid donates a proton to the base to produce a **salt** (an ionic compound) and water.

2 The pH Scale 241

- ***What does pH tell us?***
  pH is a measure of the **hydronium ion** concentration, $[H_3O^+]$, of a substance. Since pH equals the *negative* of the logarithm of $[H_3O^+]$, the larger the $[H_3O^+]$, the smaller the pH. As shown here, substances with a pH less than 7 are acidic, with more hydronium than **hydroxide ions**. Those with a pH greater than 7 are basic, with more hydroxide than hydronium ions. A pH equal to 7 is neutral, neither acidic nor basic, since $[H_3O^+] = [OH^-]$.

Figure 8.12 • The pH scale

- ***What governs the strength of an acid or base?***
  Strong acids and **strong bases** ionize completely in water, whereas **weak acids** and **weak bases** ionize partially and reversibly in water. The hydronium ion concentration of a strongly acidic solution is the same as the concentration of the acid itself since virtually every acid molecule produces a hydronium ion. The hydronium concentration of a weakly

The **Summary** revisits each major section, with informative images taken from the chapter. These visuals reinforce important concepts.

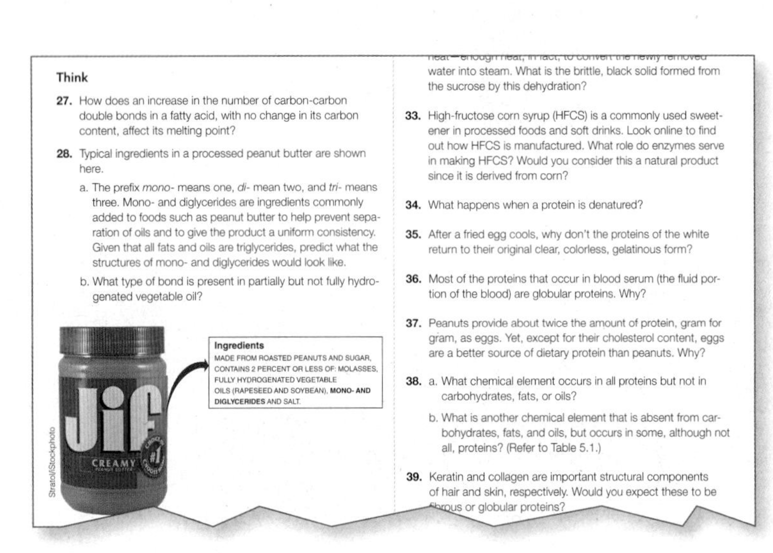
Think

27. How does an increase in the number of carbon-carbon double bonds in a fatty acid, with no change in its carbon content, affect its melting point?

28. Typical ingredients in a processed peanut butter are shown here.
    a. The prefix *mono-* means one, *di-* mean two, and *tri-* means three. Mono- and diglycerides are ingredients commonly added to foods such as peanut butter to help prevent separation of oils and to give the product a uniform consistency. Given that all fats and oils are triglycerides, predict what the structures of mono- and diglycerides would look like.
    b. What type of bond is present in partially but not fully hydrogenated vegetable oil?

water into steam. What is the brittle, black solid formed from the sucrose by this dehydration?

33. High-fructose corn syrup (HFCS) is a commonly used sweetener in processed foods and soft drinks. Look online to find out how HFCS is manufactured. What role do enzymes serve in making HFCS? Would you consider this a natural product since it is derived from corn?

34. What happens when a protein is denatured?

35. After a fried egg cools, why don't the proteins of the white return to their original clear, colorless, gelatinous form?

36. Most of the proteins that occur in blood serum (the fluid portion of the blood) are globular proteins. Why?

37. Peanuts provide about twice the amount of protein, gram for gram, as eggs. Yet, except for their cholesterol content, eggs are a better source of dietary protein than peanuts. Why?

38. a. What chemical element occurs in all proteins but not in carbohydrates, fats, or oils?
    b. What is another chemical element that is absent from carbohydrates, fats, and oils, but occurs in some, although not all, proteins? (Refer to Table 5.1.)

39. Keratin and collagen are important structural components of hair and skin, respectively. Would you expect these to be fibrous or globular proteins?

Think questions challenge students to think more broadly about chapter concepts. The level of these questions ranges from simple to advanced; they encourage students to think critically and develop an analytical understanding of the ideas discussed in the chapter.

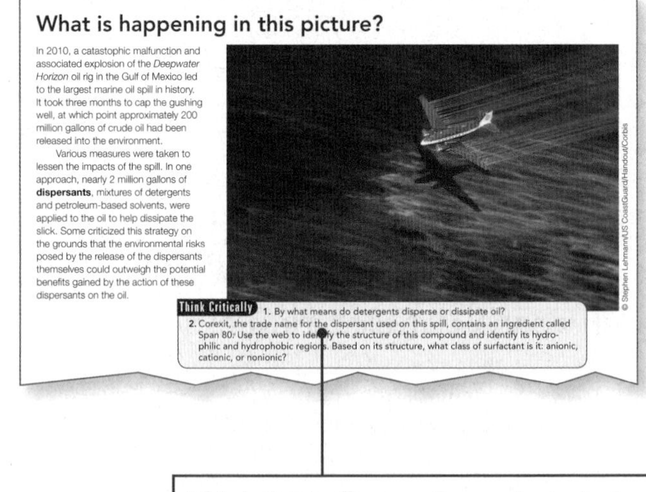
What is happening in this picture?

In 2010, a catastophic malfunction and associated explosion of the *Deepwater Horizon* oil rig in the Gulf of Mexico led to the largest marine oil spill in history. It took three months to cap the gushing well, at which point approximately 200 million gallons of crude oil had been released into the environment.

Various measures were taken to lessen the impacts of the spill. In one approach, nearly 2 million gallons of **dispersants**, mixtures of detergents and petroleum-based solvents, were applied to the oil to help dissipate the slick. Some criticized this strategy on the grounds that the environmental risks posed by the release of the dispersants themselves could outweigh the potential benefits gained by the action of these dispersants on the oil.

© Stephen Lehmann/US CoastGuard/Handout/Corbis

Think Critically
1. By what means do detergents disperse or dissipate oil?
2. Corexit, the trade name for the dispersant used on this spill, contains an ingredient called Span 80. Use the web to identify the structure of this compound and identify its hydrophilic and hydrophobic regions. Based on its structure, what class of surfactant is it: anionic, cationic, or nonionic?

What is happening in this picture? presents a photograph that is relevant to a chapter topic and illustrates a situation students are not likely to have encountered previously.

Think Critically questions ask students to apply what they have learned in order to interpret and explain what they observe in the image.

Exercises

Review

1. What do the terms a. *homogeneous* and b. *aqueous* mean as they pertain to solutions?
2. What dictates whether a solution is a gas, a liquid, or a solid?
3. Why is air considered to be a solution?
4. What term describes a solid solution consisting of two or more metals?
5. What is the primary element in steel? What other element is present in all steel?
6. The figure here depicts sodium chloride dissolving in water. With which ion does the a. oxygen atom of water; b. hydrogen atoms of water, associate? Why?

= Water
= Na^+
= Cl^-

Adapted with permission of John Wiley & Sons, Inc. from Hein, M., and Arena, S., *Foundations of College Chemistry*, Thirteenth Edition, p. 320. Copyright 2011.

7. Why are colloids and suspensions considered to be mixtures but not true solutions? What is the Tyndall effect?
8. In what aspects is blood considered
   a. a suspension?
   b. a solution?
9. Are nonpolar compounds hydrophobic or hydrophilic? Explain your answer.
10. Why is water an effective solvent for
    a. ionic compounds, such as sodium chloride,
    b. polar compounds, such as sugars and alcohols?
11. State Henry's Law and provide a common example of this law.
12. In addition to temperature, what is a factor that determines how much N_2 can dissolve in a given quantity of water? What is a factor that determines how fast the N_2 dissolves?
13. Why don't oil and water mix?
14. What characteristic of emulsifiers allows them to form attractions to both water and oils?
15. What unit of concentration is given by moles of solute per liter of solution?
16. What type of concentration unit is used most frequently in consumer products?

Visual end-of-chapter **Exercises** pose review questions that ask students to demonstrate their understanding of key concepts.

Why *Visualizing Everyday Chemistry*?

We created ***Visualizing Everyday Chemistry*** for a one-semester course designed to introduce chemistry to undergraduate nonscience majors. Our goal in writing this book is to show students that chemistry is important and relevant to their own lives, not because we say it is but because they see it is. We believe that regardless of one's academic interests, learning chemistry can not only be interesting and enjoyable, but rewarding as well—that familiarity with the language and concepts of chemistry can not only help in making better informed consumer choices, but can shed light on some of our most pressing environmental concerns as well as their potential solutions.

Many chemistry texts for the nonscience student provide a good survey of the major concepts, presenting the basics of chemistry in a standard pedagogical format. Others emphasize particular social and political issues and cover chemical concepts along the way. Although both kinds of books are well-illustrated, they do not fully exploit the learning potential of a more visual presentation of chemical concepts and applications.

Our book aims to take advantage of the highly visual learning style of today's students to accomplish the two most important goals of this course—to engage and motivate students about chemistry and to teach its basic concepts. Through the use of graphics to point out things that are not visible or not obvious to non-chemists (such as atoms, ions, and molecules), the visual approach of our book is intended to help students to see the world around them as chemists see it. In ***Visualizing Everyday Chemistry,*** graphics are not simply decorative. They present facts, concepts, processes, principles, and relationships. A visual text must be no less rigorous than a traditional text. ***Visualizing Everyday Chemistry*** incorporates mathematics and problem solving where appropriate, yet uses ample worked examples and visual tools to connect with students regardless of their background.

In our teaching of this course, we have found that encouraging students to share their experiences and curiosities about the material world creates a more approachable, dynamic learning experience. In writing this book, we have aimed for a more conversational tone, and throughout we encourage students to reflect on the relevance of chemistry to their lives, and to think about problems and weigh risks and benefits to themselves and to society.

These themes are woven throughout the text:

- Chemistry is not an abstract field of learning, but has practical, everyday applications that are important to all of us.
- Chemistry addresses a growing number of environmental concerns, from energy and resource consumption to pollution.
- A risk/benefit perspective of the role of chemistry in society and in our personal lives is central to developing informed opinions on policy and making better consumer choices.

To increase interest, our book examines contemporary topics such as:

- How chemists are working to make next-generation vehicles more fuel-efficient and less polluting.
- How consumer products are becoming greener through the use of biodegradable surfactants and other ingredients derived from renewable resources.
- How nano-scaled materials are being incorporated into consumer products.

Organization

- **Chapter 1** introduces the major themes of the book, including how chemistry impacts our everyday lives and society, how chemical technology offers solutions for a more sustainable future, and how a risk/benefit analysis helps inform our decisions about chemical use and exposure. The chapter also explores the science and methods of chemistry, including the use of scientific units.
- **Chapter 2** traces the development of our understanding of the atom and its structure from antiquity to the modern era. From this foundation, the chapter introduces the chemical elements, how these elements are organized within the periodic table, and how they are distributed in our environment and within our bodies.
- **Chapter 3** builds on the structure of atoms introduced in Chapter 2 to explain how atoms form bonds to one another. We explore how differences in bonding lead to different physical properties of the resulting chemical substances. We then learn how to identify chemical compounds through writing their formulas, drawing their structures, and naming them.
- **Chapter 4** is the first of various chapters in the book that explore energy. We begin by defining what we mean by energy and provide an overview of society's energy needs. We then show how society largely depends on the energy provided by fossil fuels and the impact of their use on the environment. The chapter closes by examining how alternative energies can provide solutions for future energy needs.

- **Chapter 5** explores the connection between food and energy. It explains how we define the energy content of foods and how our bodies utilize this energy. The chapter then examines the chemistry of the fats, carbohydrates, and proteins of our foods.
- **Chapter 6** shows the differences between physical and chemical changes and explores how these processes impact our daily lives. Quantitative concepts such as balancing chemical equations and determining the number of particles in a pure substance by weighing it are introduced and woven into the examples.
- **Chapter 7** defines chemical solutions and explores their myriad forms. We describe what makes water such an important solvent and show various ways of measuring the strengths or concentrations of solutions. We then learn about the composition of the water we drink and how society provides it to us.
- **Chapter 8** continues the discussion of water solutions by examining acids and bases. We first define the terms acid and base and then explore the meaning of the pH scale. The chapter concludes with an examination of the acids and bases in our bodies, foods, and consumer products, and in the environment itself.
- **Chapter 9** covers the chemistry that occurs within the nucleus of the atom. First we examine radioactivity. Then applications of nuclear processes are explored, including medical, safety, energy, and weaponry. A risk/benefit analysis of nuclear energy follows.
- **Chapter 10** continues the theme of energy by examining oxidation and reduction, chemical processes that underlie how fuels and batteries provide us with energy. We explore the chemistry of common types of batteries and see how oxidation and reduction underlie alternative energy sources such as fuel cells and solar cells.
- **Chapter 11** discusses cleaning and personal care agents. We first explore the chemistry of soaps and detergents and then examine cosmetics and the various products we use to care for our skin, hair, and teeth.
- **Chapter 12** explores the chemistry of medicines, drugs and genetics and the connections among them. We learn about common prescription and nonprescription medicines and how these products are developed, and then discuss drugs of abuse. The chapter then examines the molecular basis of heredity, the role of genetics in health and disease, and applications of genetic technologies.
- **Chapter 13** covers sustainability issues—first with a discussion of plastics and their uses, then with an examination of pollution and approaches to mitigate the release of pollutants into the environment.
- **Chapter 14** revisits the topic of chemistry and food by describing the vitamins, minerals, and various additives of our foods. We then ask what it means to label a chemical substance as "safe" and discuss food safety.

Given the limited time of a single term to cover a wide range of topics in a survey course such as this, instructors should be aware of the following rough guide to the chapter content:

Those that provide more foundational information include:

1 — Chemistry in Our World,
2 — Atoms and Elements,
3 — Chemical Compounds,
6 — Physical and Chemical Changes,
7 — Water, and Other Solutions, and
8 — Acid and Bases.

Those that emphasize energy and environmental themes include:

4 — Energy and Society,
9 — Nuclear Chemistry,
10 — Energy from Electron Transfer, and
13 — Plastics, Pollution, and Sustainability.

Those that address more health and personal care topics include:

5 — Food and Energy,
11 — Cleaning Agents, Personal Care, and Cosmetics,
12 — Genes, Medicines, and Drugs, and
14 — Micronutrients, Food Additives, and Food Safety

Some chapters do not fit neatly within any one of these three categories. For example, Chapter 7 (Water, and Other Solutions) and Chapter 8 (Acids and Bases) provide foundational information, but also cover environmental and health-related topics.

How Does Wiley Visualizing Support Instructors?

How Does Visualizing Everyday Chemistry Support Instructors?

The following resources are made available to all instructors and can be requested from your local Wiley sales representative:

Test Bank prepared by Jason Dunham, Ball State University, and Dan Stasko, University of Southern Maine, Lewiston-Auburn College. Includes over 1400 multiple choice, true/false, fill-in-the-blank, and open-ended questions and answers. A computerized version of the entire Test Bank is available with full editing features to help instructors customize tests.

PowerPoint Lecture Slides prepared by Don Fedie, Augsburg College, highlight key chapter concepts, contain numerous clicker questions, and include examples and illustrations that help reinforce and test students' grasp of essential topics.

PowerPoint Slides with Text Images Images, tables, and figures from the text are available in PPT format.

Digital Image Archive The text website includes downloadable files of text images in JPEG format. Instructors may use these images to customize their presentations and to provide additional visual support for quizzes and exams.

Personal Response Systems/"Clicker" Questions A bank of questions is available for anyone using personal response systems technology in their classroom.

How Has Wiley Visualizing Been Shaped by Contributors?

Wiley Visualizing and the *WileyPLUS* learning environment would not have come about without lots of people, each of whom played a part in sharing their research and contributing to this new approach.

Academic Research Consultants

Richard Mayer, Professor of Psychology, UC Santa Barbara. His Cognitive Theory of Multimedia Learning provided the basis on which we designed our program and provided guidance to our author and editorial teams on how to develop and implement strong, pedagogically effective visuals and use them in the classroom.

Jan L. Plass, Professor of Educational Communication and Technology in the Steinhardt School of Culture, Education, and Human Development at New York University. He co-directs the NYU Games for Learning Institute and is the founding director of the CREATE Consortium for Research and Evaluation of Advanced Technology in Education.

Matthew Leavitt, Instructional Design Consultant. He advised the Visualizing team on the effective design and use of visuals in instruction and has made virtual and live presentations to university faculty around the country regarding effective design and use of instructional visuals.

Independent Research Studies

SEG Research, an independent research and assessment firm, conducted a national, multisite effectiveness study of students enrolled in entry-level college Psychology and Geology courses. The study was designed to evaluate the effectiveness of Wiley Visualizing. You can view the full research paper at www.wiley.com/college/visualizing/huffman/efficacy.html.

Instructor and Student Contributions

Throughout the process of developing the concept of guided visual pedagogy for Wiley Visualizing, we benefited from the comments and constructive criticism provided by the instructors and colleagues listed below. We offer our sincere appreciation to these individuals for their helpful reviews and general feedback:

Visualizing Reviewers, Focus Group Participants, and Survey Respondents

James Abbott, *Temple University*
Melissa Acevedo, *Westchester Community College*
Shiva Achet, *Roosevelt University*
Denise Addorisio, *Westchester Community College*
Dave Alan, *University of Phoenix*
Sue Allen-Long, *Indiana University Purdue*
Robert Amey, *Bridgewater State College*
Nancy Bain, *Ohio University*
Corinne Balducci, *Westchester Community College*
Steve Barnhart, *Middlesex County Community College*
Stefan Becker, *University of Washington – Oshkosh*
Callan Bentley, *NVCC Annandale*
Valerie Bergeron, *Delaware Technical & Community College*
Andrew Berns, *Milwaukee Area Technical College*
Gregory Bishop, *Orange Coast College*
Rebecca Boger, *Brooklyn College*
Scott Brame, *Clemson University*
Joan Brandt, *Central Piedmont Community College*
Richard Brinn, *Florida International University*
Jim Bruno, *University of Phoenix*
William Chamberlin, *Fullerton College*
Oiyin Pauline Chow, *Harrisburg Area Community College*
Laurie Corey, *Westchester Community College*
Ozeas Costas, *Ohio State University at Mansfield*
Christopher Di Leonardo, *Foothill College*
Dani Ducharme, *Waubonsee Community College*
Mark Eastman, *Diablo Valley College*
Ben Elman, *Baruch College*
Staussa Ervin, *Tarrant County College*
Michael Farabee, *Estrella Mountain Community College*
Laurie Flaherty, *Eastern Washington University*
Susan Fuhr, *Maryville College*
Peter Galvin, *Indiana University at Southeast*
Andrew Getzfeld, *New Jersey City University*
Janet Gingold, *Prince George's Community College*
Donald Glassman, *Des Moines Area Community College*
Richard Goode, *Porterville College*
Peggy Green, *Broward Community College*
Stelian Grigoras, *Northwood University*
Paul Grogger, *University of Colorado*
Michael Hackett, *Westchester Community College*
Duane Hampton, *Western Michigan University*
Thomas Hancock, *Eastern Washington University*
Gregory Harris, *Polk State College*
John Haworth, *Chattanooga State Technical Community College*
James Hayes-Bohanan, *Bridgewater State College*
Peter Ingmire, *San Francisco State University*
Mark Jackson, *Central Connecticut State University*
Heather Jennings, *Mercer County Community College*
Eric Jerde, *Morehead State University*
Jennifer Johnson, *Ferris State University*
Richard Kandus, *Mt. San Jacinto College District*
Christopher Kent, *Spokane Community College*
Gerald Ketterling, *North Dakota State University*
Lynnel Kiely, *Harold Washington College*
Eryn Klosko, *Westchester Community College*
Cary T. Komoto, *University of Wisconsin – Barron County*
John Kupfer, *University of South Carolina*
Nicole Lafleur, *University of Phoenix*
Arthur Lee, *Roane State Community College*
Mary Lynam, *Margrove College*
Heidi Marcum, *Baylor University*
Beth Marshall, *Washington State University*
Dr. Theresa Martin, *Eastern Washington University*
Charles Mason, *Morehead State University*
Susan Massey, *Art Institute of Philadelphia*
Linda McCollum, *Eastern Washington University*
Mary L. Meiners, *San Diego Miramar College*
Shawn Mikulay, *Elgin Community College*
Cassandra Moe, *Century Community College*
Lynn Hanson Mooney, *Art Institute of Charlotte*
Kristy Moreno, *University of Phoenix*
Jacob Napieralski, *University of Michigan - Dearborn*
Gisele Nasar, *Brevard Community College, Cocoa Campus*
Daria Nikitina, *West Chester University*
Robin O'Quinn, *Eastern Washington University*
Richard Orndorff, *Eastern Washington University*
Sharen Orndorff, *Eastern Washington University*
Clair Ossian, *Tarrant County College*
Debra Parish, *North Harris Montgomery Community College District*
Linda Peters, *Holyoke Community College*
Robin Popp, *Chattanooga State Technical Community College*
Michael Priano, *Westchester Community College*
Alan "Paul" Price, *University of Wisconsin – Washington County*
Max Reams, *Olivet Nazarene University*
Mary Celeste Reese, *Mississippi State University*
Bruce Rengers, *Metropolitan State College of Denver*
Guillermo Rocha, *Brooklyn College*
Penny Sadler, *College of William and Mary*
Shamili Sandiford, *College of DuPage*
Thomas Sasek, *University of Louisiana at Monroe*
Donna Seagle, *Chattanooga State Technical Community College*
Diane Shakes, *College of William and Mary*
Jennie Silva, *Louisiana State University*
Michael Siola, *Chicago State University*
Morgan Slusher, *Community College of Baltimore County*
Julia Smith, *Eastern Washington University*
Darlene Smucny, *University of Maryland University College*
Jeff Snyder, *Bowling Green State University*
Alice Stefaniak, *St. Xavier University*

Alicia Steinhardt, *Hartnell Community College*
Kurt Stellwagen, *Eastern Washington University*
Charlotte Stromfors, *University of Phoenix*
Shane Strup, *University of Phoenix*
Donald Thieme, *Georgia Perimeter College*
Pamela Thinesen, *Century Community College*
Chad Thompson, *SUNY Westchester Community College*
Lensyl Urbano, *University of Memphis*
Gopal Venugopal, *Roosevelt University*
Daniel Vogt, *University of Washington – College of Forest Resources*
Dr. Laura J. Vosejpka, *Northwood University*
Brenda L. Walker, *Kirkwood Community College*
Stephen Wareham, *Cal State Fullerton*
Fred William Whitford, *Montana State University*
Katie Wiedman, *University of St. Francis*
Harry Williams, *University of North Texas*
Emily Williamson, *Mississippi State University*
Bridget Wyatt, *San Francisco State University*
Van Youngman, *Art Institute of Philadelphia*
Alexander Zemcov, *Westchester Community College*

Student Participants

Karl Beall, *Eastern Washington University*
Jessica Bryant, *Eastern Washington University*
Pia Chawla, *Westchester Community College*
Channel DeWitt, *Eastern Washington University*
Lucy DiAroscia, *Westchester Community College*
Heather Gregg, *Eastern Washington University*
Lindsey Harris, *Eastern Washington University*
Brenden Hayden, *Eastern Washington University*
Patty Hosner, *Eastern Washington University*
Tonya Karunartue, *Eastern Washington University*
Sydney Lindgren, *Eastern Washington University*
Michael Maczuga, *Westchester Community College*
Melissa Michael, *Eastern Washington University*
Estelle Rizzin, *Westchester Community College*
Andrew Rowley, *Eastern Washington University*
Eric Torres, *Westchester Community College*
Joshua Watson, *Eastern Washington University*

How Has *Visualizing Everyday Chemistry* Been Shaped by Contributors?

Our sincere appreciation goes to the following reviewers who were kind enough to read chapter manuscripts and give their professional comments:

Nicholas Alteri, *Community College of Rhode Island*
Mark Bausch, *Southern Illinois University—Carbondale*
Toni Bell, *Bloomsburg University*
Ruth Birch, *Saint Louis University*
Bill Blanken, *Reedley College*
Alison Bray, *Texas Lutheran University*
Bruce Burnham, *Rider University*
Andrew Burns, *Kent State University*
Kirsten Casey, *Anne Arundel Community College*
Susan Choi, *Camden Community College*
Roy Cohen, *Xavier University*
Milagros Delgado, *Florida International University*
Steven Desjardins, *Washington and Lee University*
Jason Dunham, *Ball State University*
Frank Dunnivant, *Whitman College*
Jeannine Eddleton, *Virginia Tech*
Ronald Fedie, *Augsburg College*
Ralph Gatrone, *Virginia State University*
Steven Gwaltney, *Mississippi State University*
Alton Hassell, *Baylor University*
Steven Higgins, *Wright State University*
Xiche Hu, *University of Toledo*
Richard Jarman, *College of DuPage*
Yasmin Jessa, *Miami University*
James Kilno, *SUNY Cobleskill*
Shari Litch Gray, *Chester College of New England*
Joseph Maloy, *Seton Hall University*
Nathan McElroy, *Indiana University of Pennsylvania*
Shaun Schmidt, *Washburn University*
James Schreck, *University of Northern Colorado*
Bradley Sieve, *Northern Kentucky University*
Anne Marie Sokol, *Buffalo State College*
Randy Sullivan, *University of Oregon*
Kenneth Traxler, *Bemidji State University*
Ken Unfried, *Sacred Heart University*
Ed Vitz, *Kutztown University*
Vidyullata Waghulde, *St. Louis Community College—Meramec*
Robert Widing, *University of Illinois at Chicago*

We also want to thank the following professors who participated in related surveys and offered valuable information on the course and market:

Michele Antico, *Farmingdale State College*
Chris Bahn, *Montana State University*
David Ball, *Cleveland State University*
John Bonte, *Clinton Community College*
Simon Bott, *University of Houston*
Tim Champion, *Johnson C. Smith University*
Douglas Cody, *Suffolk County Community College*
Jeannie Collins, *University of Southern Indiana*
Bettie Davis, *St. Vincent College*
Edward Delafuente, *Kennesaw State University*
Anthony Durante, *Bronx Community College*
Darlene Gandolfi, *Manhattanville College*
Marcia Gillette, *Indiana University—Kokomo*
Donna Gosnell, *Valdosta State University*
Karin Hassenrueck, *California State University—Northridge*
Shauna Hiley, *Missouri Western State University*
Adam Jacoby, *Southeast Missouri State University*
Mian Jiang, *University of Houston*
Subash Jonnalagadda, *Rowan University*
Amy Kabrhel, *University of Wisconsin—Manitowoc*
Joanne Kehlbeck, *Union College*
Angela King, *Wake Forest University*
John Kiser, *Western Piedmont Community College*
Terrence Lee, *Middle Tennessee State University*
Lisa Lindert, *California Polytechnic State University—San Luis Obispo*
Cynthia Maguire, *Texas Woman's University*
Garrett McGowan, *Alfred University*
Ricardo Morales, *University of La Verne*
R. John Muench, *Heartland Community College*
Aram Nersissian, *Occidental College*
Akinyele Oni, *Baltimore City Community College*
James Pazun, *Pfeiffer University*
Shaun Prince, *Lake Region State College*
Jeffrey Rahn, *Eastern Washington University*
Prafulla Raval, *Creighton University*
Ron Rolando, *LoneStar College System*
Scott Schlipp, *Milwaukee Area Technical College*
Jennifer Shanoski, *Merritt College*
Kim Simons, *Emporia State University*
Matthew Smith, *Walters State Community College*
Dan Stasko, *University of Southern Maine*
Gail Steehler, *Roanoke College*
Lou Sytsma, *Trinity Christian College*
Amy Vickers, *Cisco College*
Francisco Villa, *Northern Arizona University—Yuma*
Lane Whitesell, *University of Central Oklahoma*
Matthew Wise, *Concordia University Portland*
Joseph Wu, *University of Wisconsin—Platteville*
Regina Zibuck, *Wayne State University*

Special Thanks

Visualizing Everyday Chemistry could not have been created without the support and guidance of a skilled and accomplished team of professionals. The authors initially thank Nick Ferrari, Senior Acquisition Editor, our principal partner in the project. We are grateful to Nancy Perry, development manager for the Wiley Visualizing Imprint, for helping us hone our visual approach and for setting us on the path that culminated in what we see today. Our project editor Charity Robey and development editor David Chelton guided us successfully from initial ideas to completed manuscript with their insights and feedback. Jeanine Furino of Furino Productions has kept us efficiently on track in converting our manuscript into a finished book. Photo researchers Elizabeth Blomster and Hilary Newman and editorial assistant Mallory Fryc have been integral in helping us find and use the compelling images seen in this book. Senior Project Editor Jennifer Yee has been instrumental in gathering online resources for students and instructors. Senior Marketing Manager Kristine Ruff has provided strategy and enthusiastic support in bringing this book to its readers.

Special thanks to our families and friends, who supported us throughout and who graciously and patiently endured our absences over the course of the project. DPH owes a special gratitude to Dr. Tegan Eve, Senior Lecturer, and James Metcalf, both of the University of Miami Department of Chemistry.

WileyPLUS with Orion

Based on cognitive science, *WileyPLUS* with ORION provides students with a personal, adaptive learning experience so they can build their proficiency on topics and use their study time most effectively. ORION helps students learn by learning about them.

- Students **BEGIN** by taking a quick diagnostic for any chapter. This will determine their baseline proficiency on each topic in the chapter. A diagnostic report helps students decide what to do next.
- Students can either **STUDY** or **PRACTICE**. Study directs students to the specific topic they choose in *WileyPLUS*, where they can read from the e-textbook or use the variety of relevant resources. Student can also practice, using questions and feedback powered by ORION's adaptive learning engine.
- A number of reports and ongoing recommendations help students **MAINTAIN** their proficiency over time for each topic.

For more information, go to: www.wiley.com/college/sc/oriondemo

VISUALIZING

EVERYDAY CHEMISTRY

1 Chemistry in Our World

Studying chemistry has many practical benefits, from providing us with an understanding of the world around us to helping us make better informed consumer choices. Chemistry is also instrumental in addressing some of society's most pressing environmental concerns, from providing enough energy, food, and water for a growing world, to using resources more sustainably and responding to the challenges of climate change.

In your own life, the moment you start your day, you begin interacting with chemicals. Squeeze a strip of toothpaste onto your toothbrush and a mixture of chemicals (listed on the tube) helps you clean your teeth. As you dress, you cover your body with clothes made of chemicals, including cotton (a carbohydrate), wool (a protein), and/or synthetic fibers (polymers). Your foods are complex mixtures of chemicals—principally proteins, carbohydrates, and fats and oils. When you travel, you're likely to be carried along by energy derived from the hydrocarbons of petroleum. Your smartphone and other electronic devices run on chips of silicon. These and a variety of other chemicals form the sea of physical substances we're immersed in every day. Understanding them, how we use them, and how they affect us and the environment is what this book is about.

Blend Images /JGI,Tom Grill/Getty Images

CHAPTER OUTLINE

1.1 The Chemistry in Our Lives

LEARNING OBJECTIVES

1. **Define** chemistry.
2. **Describe** ways in which chemistry benefits both individuals and society as a whole.

Wherever we are in the world, **chemistry** is part of the picture. In essence, wherever there's matter, whether natural or made by humans, there's chemistry. To study chemistry is to engage in and to understand our material world—the kinds of things we can see, touch, feel, smell, and taste. Chemistry is much more as well. Understanding chemistry helps us turn the raw materials of our world—the components of our earth, water, and air—into useful products. It helps bring to reality the material objects that surround us and that we depend upon, whether as necessities or luxuries.

chemistry The study of matter—its composition, its properties, and the changes it undergoes.

We often start a typical day by turning on a light, showering, dressing, eating, and otherwise caring for ourselves and others. With this routine, we engage the world of chemistry in a multitude of ways. Since one of the goals of this book is to explore the chemistry of everyday things, we'll begin by looking at some of the chemistry involved in our use of energy, cleaning products, clothing, food, and medicine.

Energy

Our modern lives and prosperity depend on the widespread availability of reasonably priced energy. Whenever we turn on a light or plug a device into an electrical outlet,

Applications of chemical knowledge • Figure 1.1

The conversion of raw materials into valuable products often requires chemical knowledge.

a. Petroleum extraction and processing

Sarah Leen/National Geographic/GettyImages

Petroleum or crude oil—a natural resource—is pumped from the earth's crust and transported to a refinery.

Fred Froese/Photodisc/Getty Images

At the refinery, a combination of chemical, physical, and engineering processes separate the components of the petroleum and converts them into:

- fuels for transportation, heating, and the production of electricity,
- feedstocks for the production of plastics, agricultural chemicals, and pharmaceuticals,
- asphalt for paving roads.

b. Energy from sunlight and water

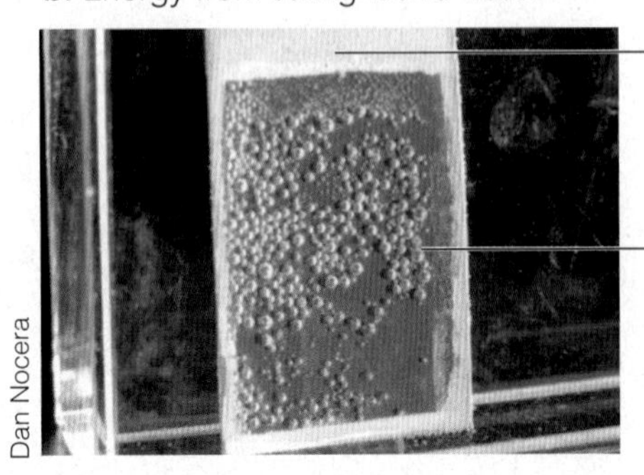

Thin, metallic wafer submerged in water.
Bubbles of oxygen gas form on the front face.
Bubbles of hydrogen gas form on the rear face.

Dan Nocera

Shining light on this "artificial leaf," developed by chemist Dr. Daniel Nocera of Harvard, converts water molecules into oxygen gas and hydrogen gas, a clean-burning fuel. Although still a prototype, this device points to a future in which sunlight and water, abundant natural resources, may be used to produce energy inexpensively and sustainably.

Ask Yourself

How does chemical expertise help turn petroleum (crude oil) into valuable products?

we're tapping into a vast energy network created by power companies. These utilities generate electricity through a variety of methods, but at present chiefly by burning fossil fuels—coal, natural gas, and oil. The energy produced by this combustion helps light our homes and streets, run our manufacturing plants, and keeps our homes, offices, and factories at comfortable temperatures and humidities.

Converting raw petroleum into useful fossil fuels takes chemical and engineering expertise. From the process we derive a host of valuable products (**Figure 1.1**). The fuels themselves—gasoline and diesel—help transport people and goods to where they need to go every day. Asphalt, another product of petroleum processing, helps pave our roads and highways.

physical change A transformation of matter that occurs without any change in chemical composition.

chemical change A process that produces substances with new chemical compositions.

In Chapter 4, we'll learn more about the **physical changes** that crude oil undergoes during its initial purification, a process called **distillation**. We'll also examine the **chemical changes** involved in the **combustion** of petroleum and other fossil fuels.

Cleaning

Chemistry brings us a variety of products to clean ourselves and the things around us (**Figure 1.2a** on the next page). Humans probably started using a crude form of soap about 4000 years ago. The active ingredients in soap are types of chemicals called **surfactants**. Chemists incorporate surfactants into various everyday cleansing products, including shampoos, shaving creams, laundry detergents, toothpastes, and even multipurpose contact lens solutions.

surfactant Shortened form of surface-active agent; a chemical that accumulates at a liquid's surface and changes the properties of that surface.

polymer A very large molecule formed by the repeated combination of much smaller molecules.

Clothing and Polymers

As we dress each morning, we cover our bodies with clothing made largely of **polymers**. Polymeric textiles include both natural fibers, principally cotton and wool, and synthetic fibers such as polyesters. Through the application of chemistry, synthetic fibers can be prepared with a wide range of properties. Whether the goal is imparting resistance to wrinkles or stains, or designing quick drying or antibacterial fabrics, the odds are that a textile chemist can find a technical solution to the problem at hand. We have workout clothes made from fabrics designed to wick away perspiration, backpacks made of lightweight but highly durable synthetic polymers, and fabrics with colorfast dyes (**Figure 1.2b**).

Even beyond the fabrics of clothing and containers, it's impossible to be far from a polymer at any moment. The plastics of our credit cards are polymers, as are cell phone and computer cases, plastic bottles, pens, and vinyl tabletops. Each of these comes from the work of chemistry.

Food

As we've seen, the hydrocarbons of our fuels—coal, natural gas, oil, and substances derived from them—supply the energy that sustains our society. Similarly, the **macronutrients** of our foods—their fats and oils, carbohydrates, and proteins—provide the energy that sustains our lives. These and the related **micronutrients**—the vitamins and minerals of our foods—come to our dining tables with the help of a variety of other natural and synthetic chemicals. These chemicals include fertilizers, herbicides, and pesticides used to grow food on a large scale, and the fuels needed to gather the food and transport it to local stores (**Figure 1.2c**). Throughout the entire process, food chemists address a variety of issues, including the safety, nutritional content, flavor, and appearance of what we eat.

macronutrients The major components of our foods that provide us with energy and the materials that form our bodies.

micronutrients Dietary substances needed in trace amounts for proper health.

Medicines

Medicines play a defining role in health care: They help cure the sick and enable those with chronic conditions to function normally (**Figure 1.2d**). Whether we might take a medication daily or only occasionally, we're likely to have first-hand knowledge of the many benefits a medicine can provide. Three classes of medicine that many of us may have benefited from are **analgesics** (pain reducers) and **antipyretics** (fever reducers), such as aspirin and ibuprofen, and **antibiotics** (antibacterial agents), which stem infection and have saved countless lives.

Chemistry InSight   Chemistry we see every day •

Figure 1.2

Chemical knowledge and innovation help bring to life a variety of everyday products, including our cleaning products, clothing, foods, and medicines.

a. Dry cleaning

The term *dry cleaning* is a little misleading; it refers to cleaning without water, but other liquids are used instead. Research chemists are constantly looking for dry-cleaning liquids that will remove dirt and stains from clothing without hurting the fabric, the workers, or the environment

Controls for regulating flow of dry-cleaning liquids

Kim Steele/Photodisc/Getty Images

b. Dyeing and printing textiles

Dyes have a long history; in the Middle Ages, colorful clothing was expensive and hard to obtain except by royalty. Today, chemists have developed dyes used in printers that can produce textiles with an unlimited range of colors and designs that do not fade or wash away.

Printing heads apply dyes to fabrics in precise patterns
Textiles can be made from natural of synthetic fibers

ROSENFELD IMAGESLTD/SCIENCE PHOTO LIBRARY

c. Agricultural chemistry

Growing large quantities of food often requires adding chemical fertilizer, as well as herbicides and pesticides.

Herbicides and pesticides applied to crops
Nutrients in soil

Andrew McLachlan/All CanadaPhotos/Getty Images

d. Developing modern medicine and medical equipment

Administering medication by intravenous therapy involves not only formulating the drug itself but also developing the plastics used in the container and in the valves that allow doctors to regulate the dosage.

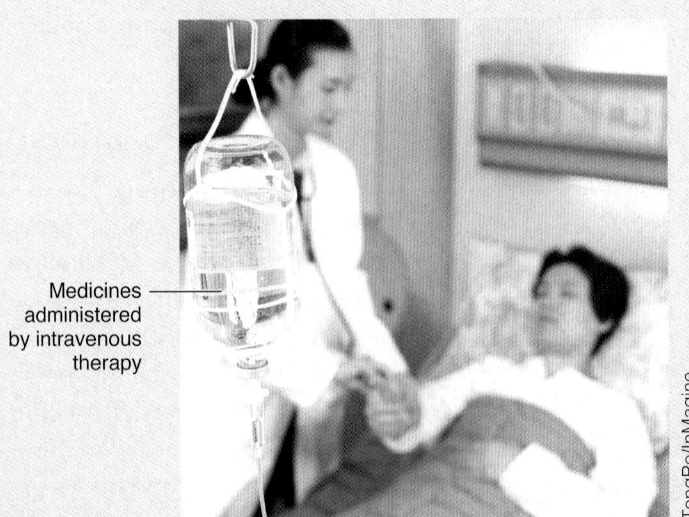

Medicines administered by intravenous therapy

TongRo/InMagine

WHAT A CHEMIST SEES

The Chemistry of Cars

When you step into a car, chances are you're not thinking about the polyurethane foam seats or the carbon black in the tires that improves durability, but these and a host of other chemicals are essential to the automobile being what it is today. Chemical know-how helps make cars more fuel efficient and less polluting. It's behind creating the materials that make up cars and the fuels, batteries, and lubricants that keep them running. For example, from petroleum we derive not only gasoline, motor oil, and lubricants, but also polymers—everything from rubber belts, tires, and wiper blades to plastic bumpers and fuel tanks.

Polymer components are used throughout the car, often replacing traditional materials such as steel, thereby reducing weight and improving fuel efficiency.

In the case of a front-end impact, air bags inflate within fractions of a second due to the rapid production of nitrogen gas from a chemical reaction.

Polymers make up a host of components in the interior— from the polyurethane foam in the seats to the materials that make up the paneling, carpeting, and upholstery. Some of these polymers are eco-plastics, derived from the cellulose of plants.

The windshield is made of safety glass, in which a thin film of the plastic, polyvinyl butyral, is sandwiched between two layers of glass so as to minimize shattering on impact.

© matsf/iStockphoto

Halogen headlights employ a tungsten filament encased in bromine, a halogen gas, and argon or xenon, inert gases.

A lead–acid storage battery uses chemical reactions to create the electricity needed for ignition and lighting. In addition, hybrid cars use a large bank of nickel–metal hydride or lithium batteries to help power the engine.

A catalytic converter, present in the exhaust system, reduces pollution. Within it, metals such as rhodium and platinum help turn unburned fuel and carbon monoxide into less hazardous gases.

The car's finish, a mixture of pigments, binding agents, and sealants, is designed to protect against the elements for years of use.

Conventional and hybrid cars rely on the combustion of the hydrocarbons of gasoline, which produces carbon dioxide and water vapor exhaust along with other trace pollutants.

Ask Yourself

In which parts of the car would you find chemical changes?

Since all medicines are by nature chemical compounds, chemical expertise is essential in developing and manufacturing them. For example, **statins**, a class of drugs used to lower cholesterol, were first developed by a biochemist who isolated these active compounds from certain fungi.

Clearly, whatever common product or process of our everyday lives we might examine, chemistry almost certainly plays a part in its origin, development, and current operation. Consider, for example, one of our most widely used consumer products—cars. The common automobile presents an almost unending story of chemistry behind our use of an everyday object. (See *What a Chemist Sees.*)

CONCEPT CHECK

1. **How** do we rely on chemistry to help keep our homes warm in the winter and cool in the summer?
2. **What** is a type of chemical that occurs in our (a) clothes; (b) cleaning materials; (c) foods; and (d) medicines?

1.2 Benefits Versus Risks

LEARNING OBJECTIVES

1. **Identify** Paracelsus and state how he distinguished between a poison and a remedy.
2. **Explain** the relationship between the benefits and the risks of the chemicals we take into our bodies as food and as medicine.
3. **Describe** how a synthetic chemical like BPA can enter our bodies from commercial products.

The word **chemical** has a bad reputation. Labeling anything as a "chemical" often generates caution, suspicion, fear, or worse, even though all the material substances in our everyday world—including the food we eat, the water we drink, and the air we breathe—consist of chemicals. Occasionally you will see a statement on a box of cereal or other processed foods assuring you (incorrectly) that the product contains no chemicals. Articles in the media often associate the word **chemical** with other words that give it a sinister aspect—words like **pollutant**, **additive**, or **toxin**. But the reality is far more complex than the simplistic notion that if something is a chemical it must be dangerous.

The reality is that any substance we take into our bodies to provide a benefit also carries with it the risk of harm, even death. This holds true whether the substance we ingest is produced by nature or made by humans and whether we take it as a medicine, a sedative, a stimulant, or as part of our normal food and drink. The answer to the question of which dominates—the benefits it provides or the risks it carries—depends on a variety of factors, including our own sensitivity to the substance, how it enters our bodies, and how much of it we absorb. The importance of this last factor, the dosage, has been recognized for hundreds of years. In the 16th century, a Swiss physician and alchemist known today as Paracelsus recognized that any substance that enters our bodies can harm us as well as benefit us, and that the amount of harm is usually proportional to the quantity. He proclaimed:

> Poison is in everything, and no thing is without poison. The dosage makes it either a poison or a remedy.

Three Common Chemicals with Known Benefits and Known Risks

Today we sum up Paracelsus's statement more simply as, "The poison is in the dosage." Caffeine is a good example. Caffeine occurs naturally in coffee beans, and many of us start our day with a lift from the caffeine in a cup or two of coffee. The caffeine in several cups of coffee provides a jolt to the central nervous system that many of us find pleasant and beneficial. Yet, in large amounts caffeine can be lethal, as studies with mice have shown. If we responded to caffeine as mice do, we would stand a 50% chance of dying if we drank about 70 cups of coffee in one sitting. Of course, drinking that much coffee or at one time would generate other, very urgent problems. In any case, Paracelsus was right: In small amounts caffeine can provide a benefit to many of us; in large amounts it's clearly a risk.

The risk–benefit balance can be more serious with aspirin. Known chemically as **acetylsalicylic acid**, aspirin has become one of our most widely used medications, with an annual worldwide production estimated at 45,000 tons and an annual consumption estimated at about 80 tablets per person. First synthesized in 1853, aspirin became accepted as an effective **analgesic** (pain reliever), **antipyretic** (fever reducer), and anti-inflammatory agent around the beginning of the 20th century. In addition to its capacity for reducing pain, fever, and inflammation, aspirin can inhibit the blood's ability to clot. This characteristic of aspirin provides a measure of protection against heart attacks resulting from internal blood clots. Small daily doses of aspirin are often prescribed to protect susceptible individuals from these kinds of heart attacks.

Aspirin's many benefits also come with associated risks. Its ability to reduce blood's clotting power, a benefit in small doses, becomes a risk with larger doses. Excessive use of aspirin can produce susceptibility to bruising and internal bleeding. Especially in children, large doses of aspirin can kill. The safety caps on bottles of aspirin and other medicines are designed to make it difficult for children to ingest their contents, perhaps mistaking the contents for candy.

Acetylsalicylic acid is a synthetic chemical; caffeine is a natural component of coffee beans. Thus chemicals produced by nature (caffeine) and chemicals manufactured by humans (acetylsalicylic acid) provide us with both risks and benefits. Still another chemical, one essential to our

Ingesting chemical substances • Figure 1.3

A variety of chemicals can enter our bodies through our intake of food and beverages.

	Deliberately Ingested	Unintentionally Ingested
Source	Substance	Mode of ingestion
Naturally grown coffee beans	Caffeine	BPA in food or beverage
Manufactured aspirin	Acetylsalicylic acid	BPA leaves plastic container or can liner and enters food
Environment	Water	BPA as a component of plastic bottles and food can liners

Ask Yourself

1. What are some advantages of BPA as a plastics additive?
2. Why might there be concerns about using BPA?

health and life, provides an even more striking example of Paracelsus's proclamation: water, a chemical combination of hydrogen and oxygen.

We must drink water to live. The benefits of drinking water are clear, yet drinking too much water, too intensely, over too short a time, can kill. The damage, known as **water intoxication**, results from a severe disturbance of the electrolyte balance in the body. Deaths have resulted from excessive water drinking in hazing incidents, water-drinking contests and during intensive exercise. Clearly, water is essential to life. Yet Paracelsus was right. Drunk in unreasonably large volumes, even plain water can kill.

BPA—A Chemical with Known Benefits but Uncertain Risks

Today we face types of risk–benefit situations that were unknown to Paracelsus and his contemporaries. Paracelsus was concerned with substances—medicines, for example—that we knowingly take into our bodies to produce certain beneficial effects. We've seen that taking large amounts of some common substances, such as caffeine, aspirin, and even water, over short periods of time can be harmful. But by now we've progressed beyond Paracelsus's idea that "the poison is in the dosage." We recognize that even very small amounts of some substances taken into our bodies over long periods of time may pose risks. The synthetic chemical known as **BPA**, provides an illustration of the subtleties and uncertainties that can be associated with a modern risk analysis.

BPA An abbreviation for bisphenol-A, a chemical added to plastics and resins that improves and strengthens their physical properties.

Early in the 1960s, BPA became an important ingredient in the manufacture of plastics and resins. By 2009, over 3 million tons were being manufactured annually, with a substantial portion used in polycarbonate plastics and epoxy resins. The properties of these two materials make them useful in products designed to contain food and drink, such as water bottles, infant formula bottles, and liners for food cans. Among the benefits provided by the added BPA are an increase in the resistance of the plastic to breaking or shattering and an ability of the resin to remain in contact with food for long periods of time without changing the food's flavor or consistency.

However, the use of BPA has risks as well, principally the risk that it may pass from the container or liner into the stored food or drink and then into our bodies as we eat or drink the contaminated contents. Animal studies indicate that exposure to BPA may present a risk to fetuses and newborns. **Figure 1.3** summarizes the routes by which BPA and other common but potentially hazardous chemicals can enter our bodies. In later chapters, we'll look again at the risks and benefits of chemicals in everyday use when we examine what we mean when we say that a chemical—or anything else—is *safe.*

CONCEPT CHECK STOP

1. **What** is the modern version of Paracelsus's concept of a poison?
2. **What** are the benefits of moderate amounts of (a) caffeine and (b) aspirin? What is the risk in ingesting excessively large amounts of each of these substances over a very short period of time?
3. **What** commercial products contain BPA, and how does BPA enter our bodies from these products?

1.3 Resources and Sustainability

LEARNING OBJECTIVES

1. **Explain** how consumption of resources is growing despite limited availability.
2. **Distinguish** between renewable and nonrenewable resources.
3. **Describe** how chemistry can provide solutions for using resources more sustainably.

In today's world, a combination of advertising and media messages reinforces the notion that the accumulation of material products can contribute to the good life. Consumers living in the more highly developed countries—about a fifth of the world's population—consume well over half the planet's energy and materials and generate about three-quarters of all pollution and waste products. For the United States, consumer spending alone represents over two-thirds of the nation's economy.

The continuing growth of Earth's population, improvements in standards of living, and steadily increasing worldwide consumption all contribute to concerns about expanding environmental pollution and about the sustainability of the world's natural resources. Chemistry plays an important role here.

Our Limited Resources

We can separate all our natural resources into two categories: renewable and nonrenewable. **Renewable resources** include fresh water (restored by rain), trees, food crops, and other things that grow naturally or renew readily. **Nonrenewable resources** can't be restored readily. Prime examples include fossil fuels, principally natural gas, coal, and petroleum. As we consume these nonrenewable resources, their natural supplies within Earth's crust shrink steadily, become more difficult to extract, and can become unavailable to us within the foreseeable future. Beyond this, even our renewable resources are susceptible to abuse and overuse. If we consume them faster than they are restored, we risk their depletion and loss even in the short term. Let's now look at some of the chemistry involved in using our resources wisely.

renewable resources Natural resources that can be renewed or replenished readily by natural processes.

nonrenewable resources Natural resources that are not replenished readily by natural processes and become depleted as they are used.

How Chemistry Improves Sustainability

Given the continuing growth in our consumption and the limitations of our resources, our challenge lies in learning how to use these resources in a sustainable manner. How, for example, do we accommodate increasing worldwide energy needs while simultaneously reducing carbon dioxide emissions that contribute to global warming? One promising approach would be to reduce our reliance on fossil fuels (which generate carbon dioxide as they burn) while simultaneously increasing our efforts toward energy conservation and our use of biologically derived fuels (biofuels) and of solar, wind, geothermal, hydroelectric, nuclear, and other minimally polluting forms of energy.

Chemistry plays a vital role here because of its focus on matter and energy. To help us expand our use of solar energy, for example, chemists are developing new materials for use in solar panels, which convert the energy of sunlight into electricity. The challenge here is to make solar energy more competitive with fossil fuels by making solar power more efficient and less expensive.

Chemists and other scientists are also seeking a cost-effective way to use the energy of sunlight to convert water into its components of hydrogen and oxygen. The hydrogen produced can be used as an energy-rich fuel that generates only water as a by-product when it burns. Many technical hurdles must be overcome to make this idea commercially feasible, but if and when it does come about, chemists will share much of the credit for the success.

In the realm of consumer products—food, toiletries, cosmetics, clothing, electronics, appliances, and cars—people have a growing awareness of sustainability and the environment. Increasingly, many household products offered to us in stores carry statements that they contain "all natural ingredients" or that they are free of "toxic chemicals," or more simply "chemical free" (**Figure 1.4**). One of the important topics we'll address in later chapters is what we mean by the term *chemical*. We'll examine whether any substance can indeed be free of chemicals; whether any chemical can be considered safe or nontoxic; and whether natural products are inherently safe or harmless.

Consumer products • Figure 1.4

Chemists help formulate useful products that serve their purpose without harming either the consumer or the environment.

STEVE WINTER/NationalGeographic Creative

Baby laundry detergent Plant-derived cleaning agents designed to remove stains with no added dyes or fragrances

Dish detergent Made with fruit and vegetable oils to dissolve grease without hurting your hands or leaving residue in the sink

Laundry wash Ingredients derived from plant and vegetable sources produce low suds and toxicity levels

Bleach cleaner Based on oxygen instead of chlorine to avoid harmful environmental waste

Drain opener Uses a combination of enzymes and bacteria instead of harsh lye to dissolve organic wastes

Another growing trend lies in the area of **life-cycle assessment** of consumer goods. Its goal is to reduce the environmental impact of manufacturing and using a product by considering everything from the raw materials and energy required to make and transport the product through its use by consumers to its being discarded or recycled at the end of its useful life (**Figure 1.5**). Laundry detergents, for example, now commonly come in concentrated form, which reduces packaging and transportation costs. Household paints feature low-VOC (low volatile organic compound) versions, which emit fewer fumes. Life-cycle assessments such as these require the application of chemistry to the design, production, use, and disposal of environmentally friendly consumer products.

Related to the concept of life-cycle assessment is the growing practice of **green chemistry**, a general set of manufacturing principles aimed at reducing the use and generation of hazardous substances and shifting toward greater use of renewable resources. Not only can green chemistry help protect the environment, but it can be useful economically, as it's often more productive to avoid generating hazardous substances in the first place than to remove them from the environment once they are created. As we progress in our examination of chemistry in the world about us, we'll continue to explore chemistry's important role in addressing sustainability issues.

life-cycle assessment Considering a product's full environmental impact, from the process of obtaining its raw materials to the disposal or recycling of the exhausted product.

green chemistry Chemical practices that aim to conserve resources and reduce the generation of waste and toxic substances.

Life-cycle assessment • Figure 1.5

Each step in the production, packaging, transportation, use, and disposal of a product has some effect on the environment. Sometimes a single step can be so damaging that the entire product needs to be redesigned for overall safety.

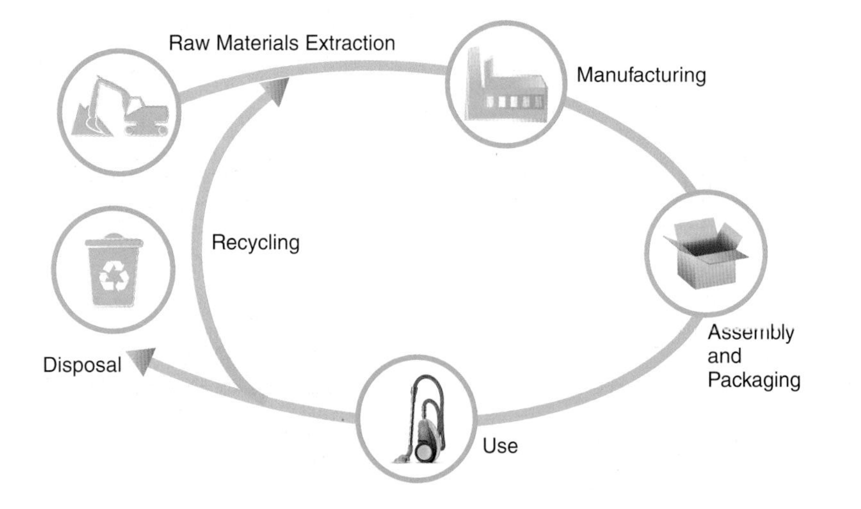

CONCEPT CHECK

1. **What** factors account for high levels of consumption of energy and goods on a worldwide basis?
2. **What** are three nonrenewable resources and three renewable resources?
3. **How** does chemistry provide solutions for sustainable energy? Give two examples.

1.4 The Science of Chemistry

LEARNING OBJECTIVES

1. **List** the types of jobs chemists perform and the fields in which they work.
2. **Describe** chemistry's role as a central science and the value of collaborative research.
3. **Explain** the general tenets of the scientific method.

The work chemists do is about as varied as the ways their activities affect our everyday lives. Most chemists work in the private or commercial sector; the remainder are in academic or governmental positions. They can spend their time in offices, in laboratories, in manufacturing plants, out in the field, or in any combination of these. They can work alone, in small groups, or in large teams. Their ability to work with others and to communicate can be as important as their technical and scientific skills. Chemists perform a variety of functions including:

- analyzing substances to determine their chemical compositions and properties,
- synthesizing new compounds,
- understanding and controlling chemical processes, and

Representative fields in which chemists work • Figure 1.6

Because chemistry touches on so many aspects of our lives, chemists work in a wide variety of fields, some of which are described here.

Pharmaceuticals/Cosmetics/Medical

Pharmaceutical chemists can help develop new drugs and drug formulations and ensure quality control of drug manufacturing. Cosmetic chemists can develop new creams and fragrances. Clinical chemists can ensure quality control of the equipment that analyzes bodily fluids.

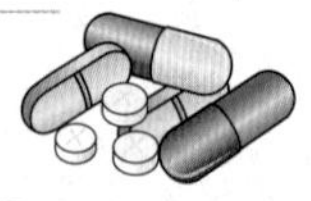

Energy/Materials

A large percentage of chemists work in energy or material related fields. Some help develop new battery or other energy technologies. Many help discover, produce, or test petrochemicals, plastics, paints, adhesives, or other valuable materials.

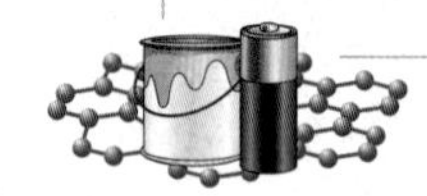

Chemistry

Agriculture/Food

Agricultural chemists can be involved in developing improved crop strains, or novel pesticides, herbicides, or fungicides. Food chemists can develop new flavorings and additives or work in food processing and safety.

Criminal Justice/Legal

Forensic chemists perform chemical analysis on drugs and other materials of criminal interest. Others in the legal profession often pursue an interest in patents.

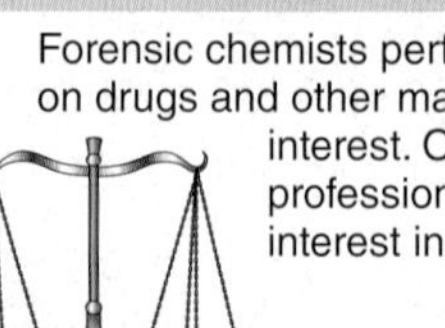

Environmental

Those in the environmental field often do chemical analysis of water, air, or biological samples to identify contaminants.

Research and development • Figure 1.7

High-throughput screening technology is used in drug discovery for the rapid evaluation of millions of compounds.

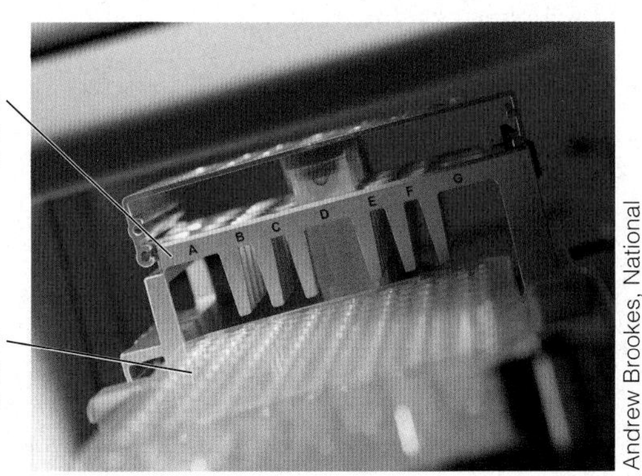

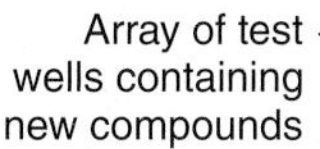

- formulating many types of new products, including medicines, cosmetics, foods, cleaning products, agricultural chemicals, and paints.

Figure 1.6 gives us a sense of what they do and how they do it.

Collaborative Research

Because chemistry deals with matter, its transformations, and its interactions with the world around us, chemistry is often called the central science. Much of the creative and productive work done in chemistry results from collaboration with scientists in other fields.

For example, chemists can gain insights from biology and medicine to develop strategies for finding new drugs. Using tools from computer science and robotics, chemists can screen millions of compounds to identify potential drug candidates—a process called **high-throughput screening**. Collaborative teams across the chemical, biological, computer science and engineering fields use this and related tools to aid in drug discovery. This is an example of a broader class of activities called **research and development** (**Figure 1.7**).

research and development Gathering knowledge with the goal of creating new products or improving existing ones.

basic research Fundamental research that increases our understanding of the world.

In addition, many scientists perform **basic research**, with the objective of adding to our fundamental knowledge of the universe. A chemist working in collaboration with astronomers, for example, could help identify the composition of other planets or distant stars (**Figure 1.8**). Although the goal of basic research is to increase our understanding of the world we live in, the insights it provides can help spawn the development of new and useful products and technologies, with important benefits to society. For example, medical imaging devices, such as magnetic resonance imaging (MRI) scanners, were developed

Basic research • Figure 1.8

Scientists involved in basic research explore fundamental questions with no specific application in mind. Planetary scientists, for example, may seek to understand the chemical composition of other planets and their moons. As they do, they often use tools of *spectroscopy,* analyzing the light reflected from these celestial objects to determine their compositions.

a. Saturn's moon Titan is the only moon in the solar system with a dense atmosphere.

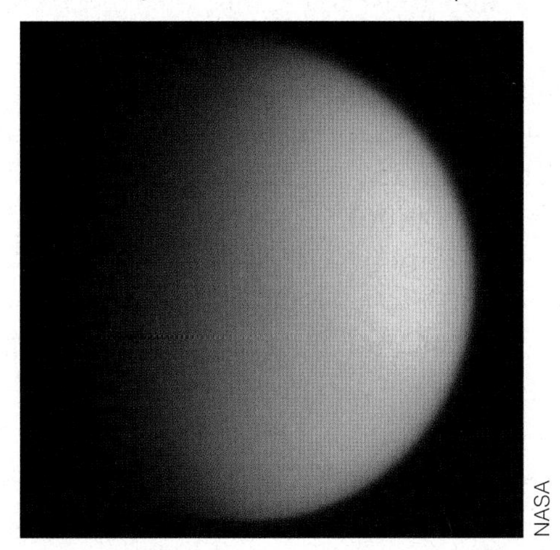

b. Chemists and astronomers have studied images of Titan (shown here in false color) and concluded that the dark-blue areas are lakes of liquid methane. This result has led to theories of a weather system on Titan based on methane instead of water, making Titan unique in the solar system.

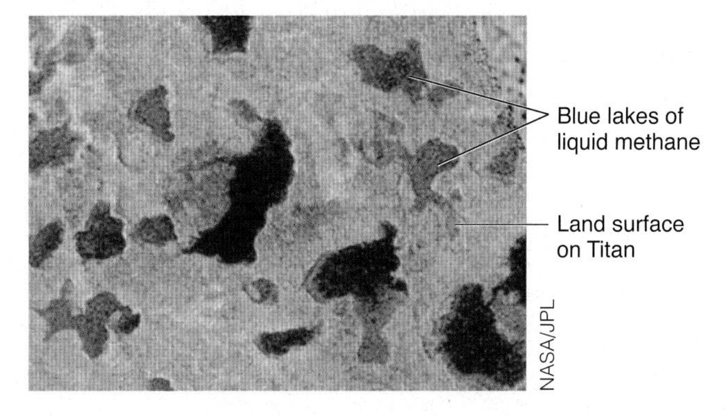

PROCESS DIAGRAM

Common steps in the scientific method • Figure 1.9

In practice, scientists often follow various procedures in carrying out research, such as trying several different hypotheses or experiments before proposing a theory.

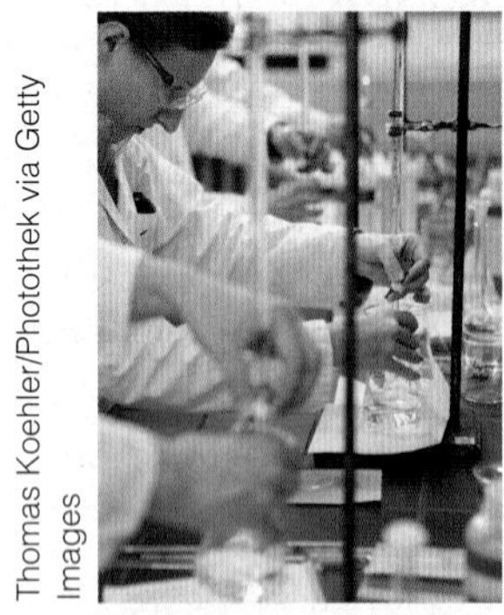
Thomas Koehler/Photothek via Getty Images

All of science is ultimately based on observation, asking questions, and experiment.

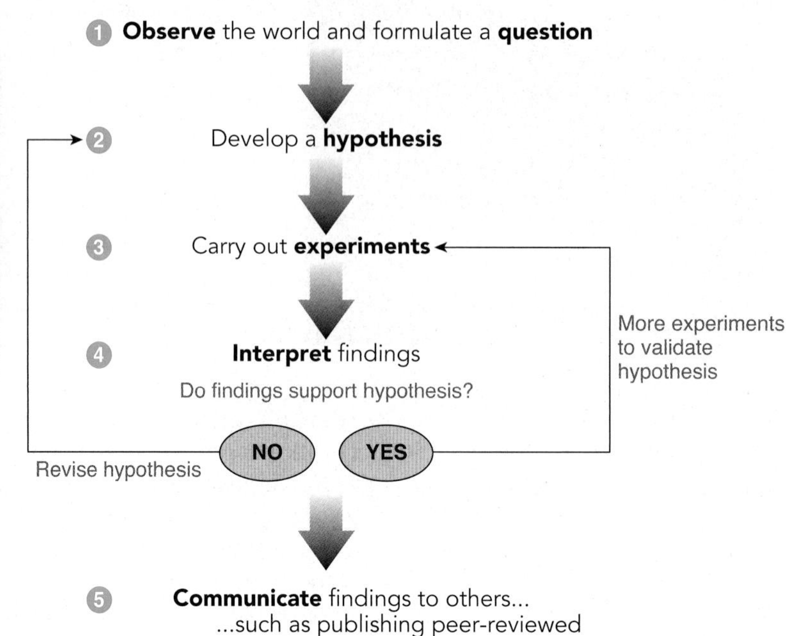
1 Observe the world and formulate a question
2 Develop a hypothesis
3 Carry out experiments
4 Interpret findings
Do findings support hypothesis?
NO
YES
Revise hypothesis
More experiments to validate hypothesis
5 Communicate findings to others...
...such as publishing peer-reviewed research papers and presenting results at scientific conferences.

Think Critically

1. Can a theory be proven to be correct? If *yes*, explain how. If *no*, explain why not.
2. In which steps of the scientific method is observation especially important?

from fundamental discoveries in the fields of physics and chemistry. The field of molecular biology and the development of a host of products to diagnose and treat medical conditions were built on the discovery of the structure of **DNA** (deoxyribonucleic acid) and an understanding of how it functions. The development of new drugs against diseases such as Alzheimer's, cancer, and others will likely benefit from fundamental insights into how protein molecules behave within the body. Funding for basic research is typically supported by government agencies.

The Scientific Method

Science, from the Latin *scire*, "to know," is a way of knowing and understanding the universe we live in. As scientists, chemists operate by a general set of principles known as the **scientific method**. This can take many forms, but it generally involves asking questions about the universe by means of experiments and tests; interpreting the resulting observations and findings; and communicating those results and interpretations to other scientists. The scientific method relies not only on evidence and observations gathered by a single scientist or group of scientists, but also on the validation of results and interpretations by a community of scientists who may accept or challenge them by suggesting alternative experiments or explanations.

scientific method The process by which science operates, involving the development of explanations for observations of the universe.

hypothesis A tentative explanation for a relatively small set of observations.

The scientific method involves the development of a tentative explanation called a **hypothesis**. Scientists perform experiments to test hypotheses. Over time, one or more substantiated, related hypotheses may mature into a **theory**. It is important to note, though, that even a widely accepted theory may have to be modified, revised, or even abandoned as the result of even a single new observation. A scientific theory gains footing not because it is proven per se, but rather because a large set of confirmed observations support it and none can disprove it. **Figure 1.9** describes the common steps in the scientific method.

theory A generally accepted principle based on a large set of confirmed observations.

CONCEPT CHECK

1. **What** is an example of chemical research in each of the five fields shown in Figure 1.6?
2. **Why** is high-throughput screening an example of how chemistry is a central science?
3. **What** are five general steps in the scientific method?

1.5 Working with Scientific Units

LEARNING OBJECTIVES

1. **List** the standard scientific units for distance, mass, and time.
2. **Explain** the use of prefixes in conjunction with scientific units.
3. **Convert** measurements from one unit to another using unit cancellation.

In our everyday lives, we regularly encounter and use units of measure: A speed limit may be so many **miles** per **hour**, referring to units of length and time; computer hard drives are rated in giga or terabytes, referring to units of data storage capacity; soft drinks commonly come in 2-**liter** bottles, referring to volume; and pretzels are often sold in 1-**pound** bags, referring to weight.

Long ago, probably well before the beginning of recorded history, humans started devising units of measure for use in agriculture, trade, and various other activities. Later, as commerce between societies and nations grew, people recognized the need for establishing common units of measurement.

SI Units and the Metric System

In the late 1700s, around the time of the French Revolution, the standardized system of units now used by most of the world—the metric system—was developed. A modern variant of the metric system, known as the International System in English and as the *Système International d'Unités* in French, is now the principal system of units used in science and in international trade and commerce. It is commonly referred to as the SI System, an abbreviation of its French title. Contained within this SI system are a set of **SI base units** for various measured quantities, such as length, mass, and time. The term **mass** is an inherent measure of the amount of matter in a substance. Mass and weight are often used interchangeably in everyday use but actually have different meanings, as we'll discuss in the next chapter.

SI base units Fundamental units of the SI system, such as the meter, kilogram, and second.

A meter extends slightly beyond 3 feet (a yard). For much greater lengths, perhaps the distance between two cities, we use a much larger unit, such as the **kilometer (km)**, which equals 1000 meters (m). Los Angeles, for example, is about 3900 km from New York. If, on the other hand, we were to describe something extremely small, say the width of a period on this page, a smaller unit makes more sense. A period on this page has a diameter of about 2 **millimeters (mm)**, or 2 one-thousandths of a meter.

Time standard • Figure 1.10

The cesium atomic clock at the National Institute of Standards and Technology sets the time standard for the United States. Atoms of the element cesium are suspended in a magnetic field within the vertical tube. A laser system counts the natural vibrations of each atom. Over 9 billion vibrations define a time interval of 1 second.

One of the great advantages of the SI or the metric system itself is that we can scale a reference unit either larger or smaller by using **SI prefixes** that refer to powers of ten. As we just saw, the prefix **kilo-** effectively multiplies a unit by 1000, while **milli-** divides it by 1000. In comparison, consider the older, English units of length: the inch, foot, yard, and mile. While these may be more familiar to most of us, the fact that 12 inches make a foot, 3 feet make a yard, and 5280 feet make a mile makes such conversions of English units to larger or smaller scales both more difficult and more cumbersome than in the metric system.

SI prefixes Prefixes that scale an SI unit either larger or smaller by some factor of ten.

reference standards Precise quantities upon which SI base units are defined.

The **kilogram (kg)**, the SI base unit of mass, is equivalent to 2.2 pounds (lb). From our everyday experience, this is what a liter of water weighs and is very nearly the weight of a quart of milk. A kilogram, as the name implies, equals 1000 grams (g). A gram, in turn, is just about the weight of a simple, large paper clip. A level teaspoon of sugar weighs roughly 5 g.

We blink, and a second of time passes. But have you ever stopped to wonder how we know exactly how long a second lasts? We could say that a second is 1/60th of a minute, but that's a *relative* answer, based on the length of a minute. It prompts us in turn to define just what a minute is. What we need is an *absolute* definition of a specific unit of time.

A solution does exist to the problem of absolute definitions of specific SI units. It's a solution based on **reference standards**, or definitions established through international bodies. The exact definition for the second is beyond the scope of our studies here, but it's based on properties of a substance called cesium. In the United States, the National Institute of Standards and Technology maintains what's known as a **cesium atomic clock**, a technical device that keeps exceedingly precise time (**Figure 1.10**). It's predicted to be accurate to within one second as it runs continuously over the next 60 million years! (A **meter**, in turn, is defined as the distance light travels in a vacuum during a well defined, minute fraction of a second.)

In the science of chemistry, as in our daily lives, we're often concerned with the volume of a substance, usually a liquid. In the U.S., for example, we usually buy milk by the quart and gasoline by the gallon. In science and international commerce, commonly used units of volume include the **liter (L)**, which runs just slightly larger than a quart, and the **milliliter (ml)**, which is one-thousandth of a liter. We'll have more to say about SI and other units as we proceed with our examination of chemistry.

SI Prefixes as Multipliers

As we saw earlier with the prefixes kilo- and milli-, the SI or metric system allows us to use prefixes to express larger or smaller variants of a unit easily and conveniently. Other useful prefixes include **giga-**, meaning one billion, **mega-**, one million, **centi-**, one-hundredth, and **micro-**, one-millionth. A microsecond, for example, is one-millionth of a second. **Figure 1.11** shows some examples of the use of both SI and English units in everyday life.

Chemistry InSight Units of measurement in everyday life • Figure 1.11

The SI or metric system uses prefixes, some of which are shown in **Figure a**. Almost all countries have adopted the metric system exclusively, but a few, such as the United States, commonly use both metric and English units.

a. SI Prefixes

giga-	Billion	1,000,000,000
mega-	Million	1,000,000
kilo-	Thousand	1,000
hecto-	Hundred	100
deca-	Ten	10
(none)	One	1
deci-	Tenth	0.1
centi-	Hundredth	0.01
milli-	Thousandth	0.001
micro-	Millionth	0.000 001
nano-	Billionth	0.000 000 001

b. Volume Bottled drink labels often show volumes in both English units (fluid ounces, fl oz) and metric units (milliliters, mL).

GIPhotoStock/Photo Researchers

c. Weight This scale measures weight in both pounds and kilograms.

© travelpixpro/iStockphoto

d. Distance A meter runs slightly longer than a yard.

Meterstick

80 81 82 83 84 85 86 87 88 89 90 91 92 93 94 95 96 97 98 99 1

Yardstick

32 33 34 35 36

Tim O. Walker

DID YOU **KNOW?**

Forensic Chemists are Unlikely to Set Foot in a Crime Scene as a Routine Part of their Work

The enormous popularity of crime scene investigation TV shows (**Figure a**) has shed a spotlight on forensic science, a multidisciplinary field that includes chemists, biologists, medical professionals, and others. In these programs, a single character may collect evidence at the crime scene, perform laboratory analysis, and interrogate suspects. But in the real world these various activities are carried out by separate individuals, each with specialized training. Forensic chemists, for example, routinely carry out their work in crime laboratories without any interaction in other aspects of a case, except occasionally testifying in court regarding their scientific findings.

Unlike the quick lab results presented on these shows, real forensic work can be quite detailed, complex, and time consuming. Take, for instance, the case of cocaine residues on money. It's widely reported that most U.S. paper currency is contaminated by minute amounts of cocaine, on the order of nanograms (billionths of a gram) per bill. The most reasonable explanation for this is that some bills become contaminated directly by drug traffickers or later by users who roll them up to inhale the drug. As these contaminated bills continue in general circulation, they cycle through banks, contaminating money-counting machines. The rollers on these machines, in turn, cross contaminate large numbers of "innocent" bills with minute particles of the drug.

With 90% of U.S. bills contaminated, as recent studies have suggested, would drug-detector dogs (**Figure b**) alert police to the presence of drugs on anyone carrying this currency? Defendants in drug-money confiscation cases have argued this very point: that a sniffer dog's alert should not constitute probable cause for authorities to search their property or premises. Controlled lab tests have shown that drug dogs respond to the presence of **methyl benzoate**, a volatile contaminant in cocaine, rather than to the cocaine itself. Furthermore, the minimum amount of methyl benzoate needed to trigger an alert is on the order of micrograms, or one-thousand times the level of nanograms. It follows that only currency carrying sufficient quantities of this signature odor—in other words, currency that was recently tainted by illicit cocaine—should trigger an alert, and makes it highly unlikely that "innocent" money would do the same.

a. Dramatic effects of dimly lit laboratories and procedures that produce rapid results are often the purview of TV forensic science (© CBS Corporation).

CBS via Getty Images

b. A drug-detector dog and handler during a training session. The dog is trained to discriminate between target odors (such as those from illicit drugs) and distractor scents (such as those from food or other animals.)

Mia Foster/PhotoEdit

Think Critically

1. Why would it be important for a crime lab chemist carrying out routine analyses to be ignorant of the other investigative aspects of a case?
2. The term *false positive* refers to an erroneous indication of a positive result. For example, evidence suggests that eating poppy seed bagels or pastries just prior to a drug test can lead to false positives for the presence of opiates. For which "diagnostic test" mentioned in this passage would it be helpful to understand false-positive rates?
3. Just as the smell of perfume or cologne on someone dissipates over time, the faint odor of methyl benzoate on cocaine dissipates with time. Why would it be useful to understand the rate at which it dissipates?

The *Did You Know?* feature shows how scientific principles and scientific units are brought to bear in solving real-world problems.

Converting Units Using Unit Cancellation

Applications occasionally require us to convert from one unit to another. A useful technique known as **unit cancellation** can help us considerably in problems of this sort. In track and field events, for example, a 100-yard dash is slightly shorter than a similar event called the 100-meter dash or sprint. Suppose, for example, we know that a certain athlete can run 100 yards in 12.0 seconds. Now we ask: "Assuming she ran at the same speed, how long would it take her to run 100 meters? We can solve this problem by converting the English unit of yards to the metric unit of meters, while keeping the unit of seconds unchanged. Here's how:

> **unit cancellation** A method of converting from one unit to another by multiplying by one or more equivalences.

- First, we want to know the time it takes her to run the course, so we'll set up a fraction with the unit of time, the second, in the numerator (the top part of the fraction).

$$\frac{12.0\ \text{seconds}}{100\ \text{yards}}$$

- In Appendix A, you can see that 1 meter is equivalent to 1.094 yard. That is, a piece of string that is 1 meter long is exactly the same length as another piece of string that is 1.094 yards long. So we can write:

$$1\ \text{meter} = 1.094\ \text{yard}$$

- Then, since 1 meter and 1.094 yard are identical, we know that each of the following fractions is equal to 1:

$$\frac{1\ \text{meter}}{1.094\ \text{yard}} = 1$$

and

$$\frac{1.094\ \text{yard}}{1\ \text{meter}} = 1$$

The following diagram shows this equivalence graphically:

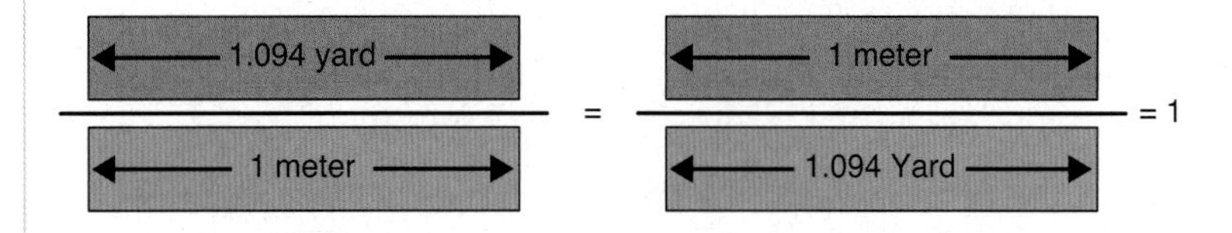

- Now we multiply the fraction that represents the runner's speed,

$$\frac{12.0\ \text{seconds}}{100\ \text{yards}}$$

by a fraction representing 1, in such a way that the unit *yards* that appears in the denominator (bottom part of the fraction) is canceled out by the same unit, *yards*, that appears in the numerator of the fraction containing both yards and meters:

$$\frac{12.0\ \text{seconds}}{100\ \cancel{\text{yards}}} \times \frac{1.094\ \cancel{\text{yard}}}{1\ \text{meter}} = \frac{(12.0 \times 1.094)\ \text{seconds}}{(100 \times 1)\ \text{meters}}$$

$$= \frac{13.13\ \text{seconds}}{100\ \text{meters}}$$

If the athlete ran at the same speed in the 100-meter sprint as she did in the 100-yard dash, it would take her 13.13 seconds to run the 100 meters.

In carrying out unit cancellations, be sure that the fractions representing 1—also known as equivalences—cancel all units except for those needed in the answer. In our example, the equivalence (1.094 yard/1 meter) canceled "yards", which was not needed in the answer, and replaced it with "meters". **Figure 1.12** on the next page, provides another example involving speed.

In Words, Math and Pictures

Unit conversions • Figure 1.12

A variety of chemistry and everyday problems require us to convert from one set of units to another. We can use the principles of unit cancellation just described to work through the following conversion.

3DDock/Shutterstock

If a commercial airplane's top speed is 600 miles per hour, how fast is this in meters per second?

Solution:

We'll need the following equivalences to answer this question:

1 mile = 5280 feet.
1 meter = 3.28 feet.

Step 1: Start with the initial value, expressed as a fraction:

$$\frac{600 \text{ miles}}{1 \text{ hour}}$$

Step 2: Think ahead to the units you'll need for the final answer:

$$\frac{600 \text{ miles}}{1 \text{ hour}} \Rightarrow \frac{\text{meters}}{\text{second}}$$

Step 3: Multiply the original fraction by a series of equivalences, canceling *units* (but not numbers) along the way, until you arrive at the desired units.

$$\frac{600 \text{ miles}}{1 \cancel{\text{hour}}} \times \frac{1 \cancel{\text{hour}}}{60 \text{ minutes}}$$

Notice we cancel the units but not the numbers. Continue on, one step at a time…

$$\frac{600 \text{ miles}}{1 \cancel{\text{hour}}} \times \frac{1 \cancel{\text{hour}}}{60 \cancel{\text{minutes}}} \times \frac{1 \cancel{\text{minute}}}{60 \text{ seconds}}$$

$$\frac{600 \cancel{\text{miles}}}{1 \cancel{\text{hour}}} \times \frac{1 \cancel{\text{hour}}}{60 \cancel{\text{minutes}}} \times \frac{1 \cancel{\text{minute}}}{60 \text{ seconds}} \times \frac{5280 \text{ feet}}{1 \cancel{\text{mile}}}$$

$$\frac{600 \cancel{\text{miles}}}{1 \cancel{\text{hour}}} \times \frac{1 \cancel{\text{hour}}}{60 \cancel{\text{minutes}}} \times \frac{1 \cancel{\text{minute}}}{60 \text{ seconds}} \times \frac{5280 \cancel{\text{feet}}}{1 \cancel{\text{mile}}} \times \frac{1 \text{ meter}}{3.28 \cancel{\text{feet}}}$$

… until you're left with the desired units, meters per second. Now use the numbers to arrive at the answer:

$$\frac{(600 \times 1 \times 1 \times 5280 \times 1) \textbf{ meters}}{(1 \times 60 \times 60 \times 1 \times 3.28) \textbf{ seconds}} = \frac{(600 \times 5280) \textbf{ meters}}{(60 \times 60 \times 3.28) \textbf{ seconds}} = 268 \text{ meters per second}$$

Ask Yourself

Calculate the answer to the same question, but this time, in Step 3, work with the *miles* first and then the *hours*. Do you arrive at the same answer?

KNOW BEFORE YOU GO

1. Identify the missing units in the figure.

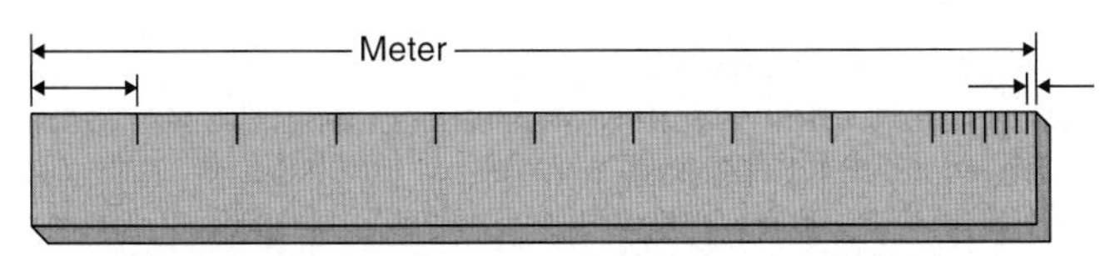

Solve the following unit conversion problems.

2. If pears sell for 2.80 euros (€) per kilogram (kg) in a market in France, what would be the equivalent price in U.S. dollars ($) per pound (lb)? For the purposes of this question, assume a currency exchange rate of 0.70 euros to the U.S. dollar. Also, 1 kg = 2.2 lb.

3. Your car gets 40 miles per gallon on the highway. What is the equivalent value expressed in kilometers (km) per liter (l)? Use the following equivalences to find your answer:

   1 mile = 1.6 km and 1 gallon = 3.8 l.

4. A pain reliever label states that adults can take up to 500 milligrams of the active ingredient every 4 hours. If the medicine comes in liquid form and each 15 ml supplies 500 mg of the active ingredient, how many milliliters can an adult receive over an 8-hour period?

CONCEPT CHECK

1. **What** *two* SI base units would you use to express an object's speed?
2. **How** are prefixes used in conjunction with scientific units?
3. **How** many millimeters are in a kilometer?

Summary

1 The Chemistry in Our Lives 4

- ***How does chemistry affect our lives and benefit society at large?***
  Chemistry is the study of matter, its composition, properties, and the changes it undergoes. Chemistry affects our everyday lives and society in multiple ways. Through combustion, fossil fuels provide most of the energy we need for transportation and for generating electricity. The cleaning power of soaps and detergents comes from chemicals called surface-active agents, or **surfactants**. Our clothing consists mostly of **polymers**, very large molecules formed by the repeated combination of much smaller ones. The energy content of our food is provided by its fats, carbohydrates, and proteins, known collectively as **macronutrients**. Our food also supplies us with **micronutrients**, the vitamins and minerals our bodies need to utilize the macronutrients effectively and to keep us in good health. All medicines are chemical compounds, such as the solution being administered to the patient shown here, and chemical expertise is essential in developing them.

Figure 1.2 • Chemistry We See Every Day

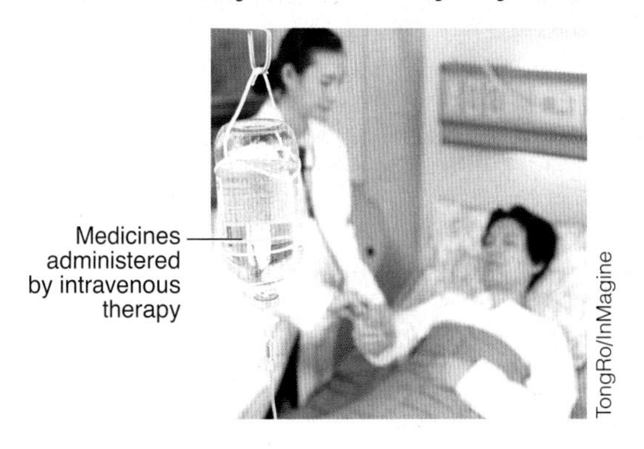

2 Benefits Versus Risks 8

- ***What are the trade-offs between the benefits and risks in our use of everyday chemical substances?***
  Any substance we take into our bodies that can benefit us in any way also carries with it the potential for harming us. The risk of harm depends on several factors, including the amount of the substance we ingest. Sufficiently high dosages

of the caffeine in coffee, of acetylsalicylic acid (the active ingredient in aspirin) and even pure water can be harmful and even toxic if taken over a short time span.

- **BPA**, an abbreviation for bisphenol-A, is an ingredient of plastics and resins that provides shatter resistance to beverage bottles made of plastic and prevents resin liners of food cans from affecting the flavor or consistency of the contained food. BPA appears to be able to leak from the plastic and resin of the bottles and can and enter our bodies as we consume their contents, as shown in the diagram. Concern about the potential harm this ingested BPA might cause to adults and children is leading to the abandonment of BPA as a component of food and beverage containers.

Figure 1.3 • Ingesting chemical substances

Deliberately ingested		Unintentionally ingested
Source	Substance	Mode of ingestion
Naturally grown coffee beans	Caffeine	BPA in food or beverage
Manufactured aspirin	Acetylsalicylic acid	BPA leaves plastic container or can liner and enters food
Environment	Water	BPA as a component of plastic bottles and food can liners

3 Resources and Sustainability 10

- ***How can chemistry address environmental concerns?*** With its focus on matter and energy, chemistry provides approaches to using resources more sustainably. Chemists are working to make solar energy, biofuels, and hydrogen fuels more cost-competitive with traditional fossil fuels. As shown in the diagram, **life-cycle assessment** aims to reduce the environmental impact of a product by considering everything from the raw materials used to produce it through to its disposal or recycling. Chemists address these issues in developing more environmentally friendly consumer products.

Life-cycle assessment • Figure 1.5

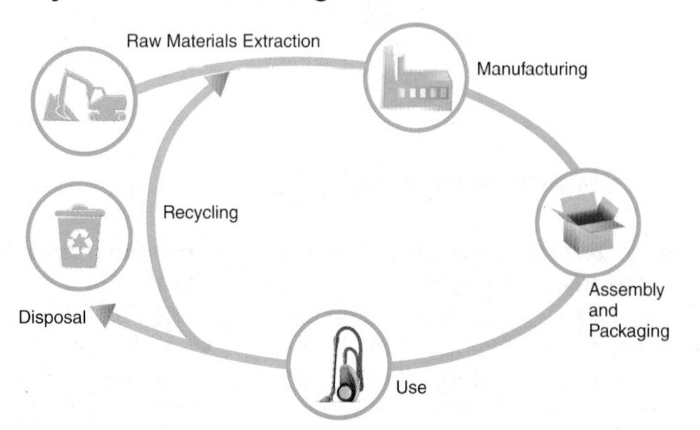

4 The Science of Chemistry 12

- ***What do chemists do?*** Chemists work in a variety of fields, including those shown below. Chemists perform a variety of functions, including synthesizing new compounds, analyzing samples to determine chemical composition, formulating new products, and understanding and controlling chemical processes.

Figure 1.6 • Representative fields in which chemists work

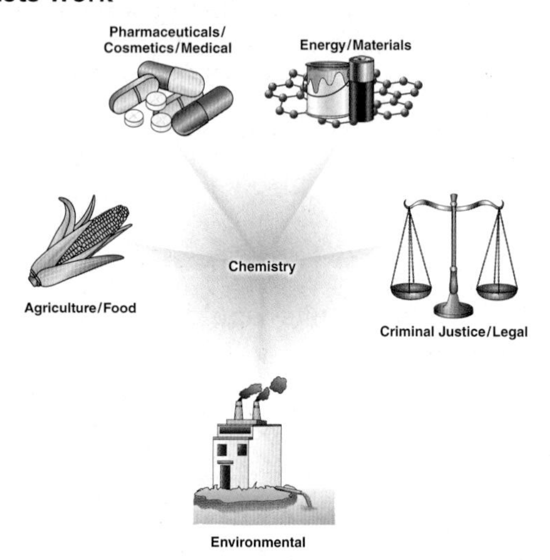

- ***What is the scientific method?*** The **scientific method** is a set of general guidelines under which scientists carry out their work. This generally involves asking questions about the universe by means of experiments and tests; objectively interpreting the resulting observations and findings; and communicating those results and interpretations to other scientists.

5 Working with Scientific Units 15

- ***What are scientific units?*** The SI system, a modern variant of the metric system, is the principal set of units currently used in science and in international trade and commerce. The **SI base unit** of length is the meter; for mass, the kilogram (as shown here);

Figure 1.11 • Units of measurement in everyday life

and for time, the second. A commonly used unit of volume is the liter. One advantage of the SI or metric system is the use of multiplier prefixes to scale a unit either larger or smaller by some factor of ten.

- ***How do you convert from one set of units to another?*** **Unit cancellation** is useful for converting units of one system to those of another. In this method, an initial value is multiplied by one or more equivalences, so that units appearing in both a numerator and a denominator (including the unit to be converted) cancel each other, with the newly desired unit(s) remaining uncanceled in the result.

Key Terms

- basic research 13
- BPA 9
- chemical change 5
- chemistry 4
- green chemistry 11
- hypothesis 14
- life-cycle assessment 11
- macronutrients 5
- micronutrients 5
- nonrenewable resources 10
- physical change 5
- polymer 5
- reference standards 16
- renewable resources 10
- research and development 13
- scientific method 14
- SI base units 15
- SI prefixes 16
- surfactant 5
- theory 15
- unit cancellation 19

What is happening in this picture?

In the photo, chlorine levels in a pool are being tested. Chlorine is a disinfectant widely used to help keep swimming pools sanitary and to treat municipal drinking water.
However, consider the warning label for chlorine found on storage tanks of this chemical.

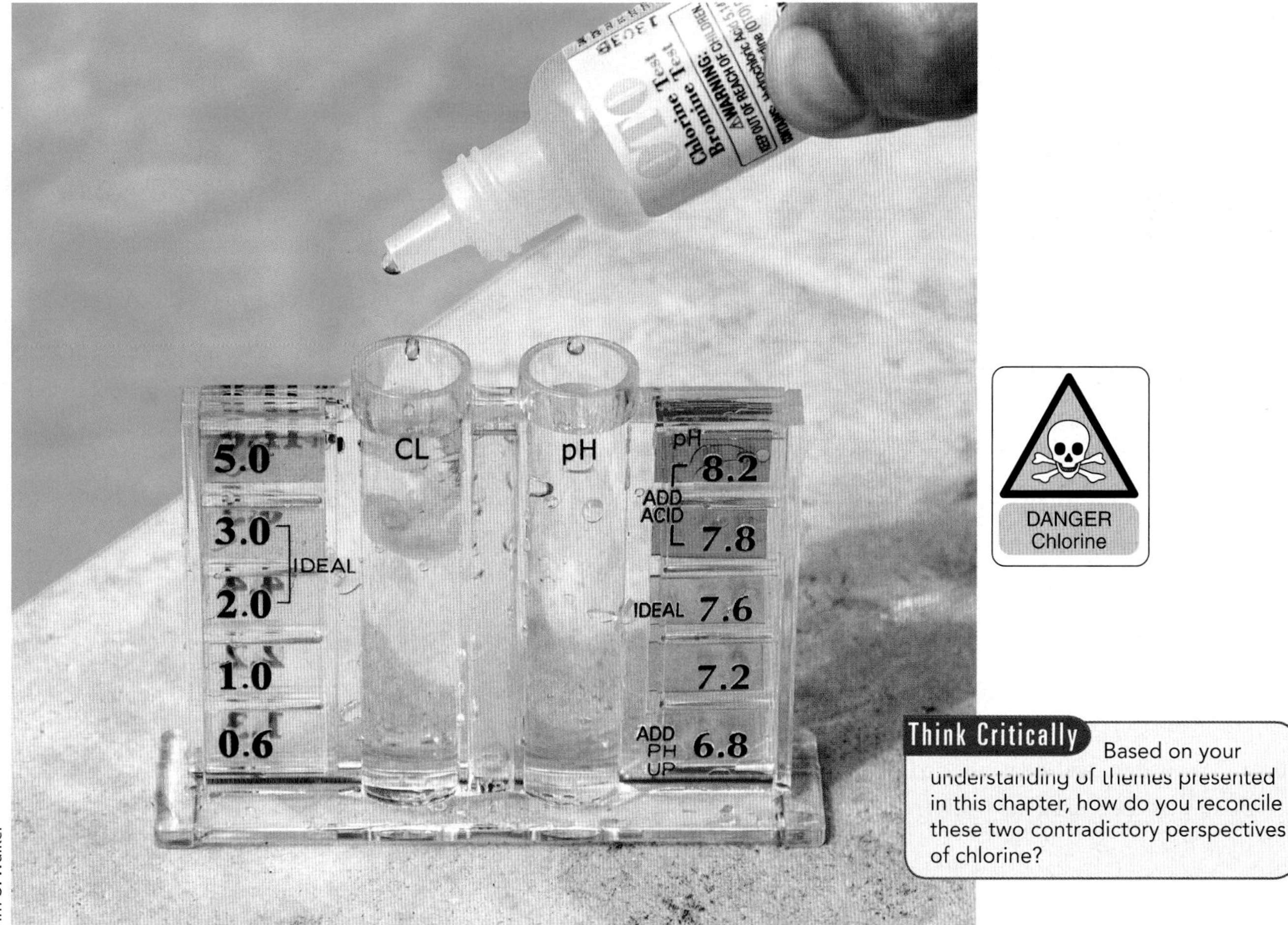

Tim O. Walker

Think Critically Based on your understanding of themes presented in this chapter, how do you reconcile these two contradictory perspectives of chlorine?

Exercises

Review

1. What are three fossil fuels that provide society with energy? Name or describe three sources of energy that we might use to replace fossil fuels as our major source of energy.

2. a. Name one common, commercial fuel that is derived from petroleum.

   b. Name one common fuel that is not derived from petroleum.

3. a. What common household product is most likely to contain chemicals known as surfactants?

   b. What function do these chemicals perform as you use this household product?

4. What common household product, often found in a medicine cabinet, contains an ingredient that serves as an analgesic and can also provide some protection against heart attacks? What is the chemical name of this ingredient? What function does this ingredient perform that allows it to reduce the risk of a heart attack?

5. Name or describe a substance or an object that you use every day that is composed entirely or largely of polymers.

6. Name the mathematical procedure used repeatedly in the process of unit cancellation.

7. Name one unit of the metric system and one unit of the English system used for measuring:

   a. length;

   b. volume;

   c. weight.

8. Identify the missing unit prefixes in the following chart.

	Million	1,000,000
	Thousand	1,000
hecto-	Hundred	100
deca-	Ten	10
(none)	One	1
deci-	Tenth	0.1
	Hundredth	0.01
	Thousandth	0.001
	Millionth	0.000 001

9. Describe who Paracelsus was and how he contributed to our understanding of what a poison or a toxin is.

10. a. What is BPA. and what is it used for?

    b. How can BPA enter our bodies?

11. a. Identify a renewable resource and explain how it is renewed.

    b. Identify a nonrenewable resource and explain why it is classified as nonrenewable.

Think

12. Burning a fossil fuel produces heat. Explain how burning a fossil fuel can help keep us cool rather than warm.

13. a. Identify the SI base units of mass, length, and time, respectively.

    b. Why are reference standards necessary in defining these units?

14. Which of the two systems—English and metric—do you think is more useful to us as individuals and as a society? Explain your reasoning.

15. What word would you use as the term for a single unit of mass that consists of exactly 1000 mg?

16. a. How does basic research differ from applied research and development?

    b. What arguments can be made for why funding for basic research is a worthy investment for society?

17. Describe the difference between

    a. a hypothesis and a theory and

    b. an observation and an experiment.

18. a. Describe the difference between the macronutrients and the micronutrients of our food. Give an example of each.

    b. Would a sample of "vitamin-fortified" bottled water that also contains small amounts of sugars supply macronutrients, micronutrients, or both?

19. Consider the life-cycle assessment of a common consumer product such as laundry detergent. For each of the stages shown in the diagram, how could the expertise of a chemist be helpful in reducing the environmental impact of the product?

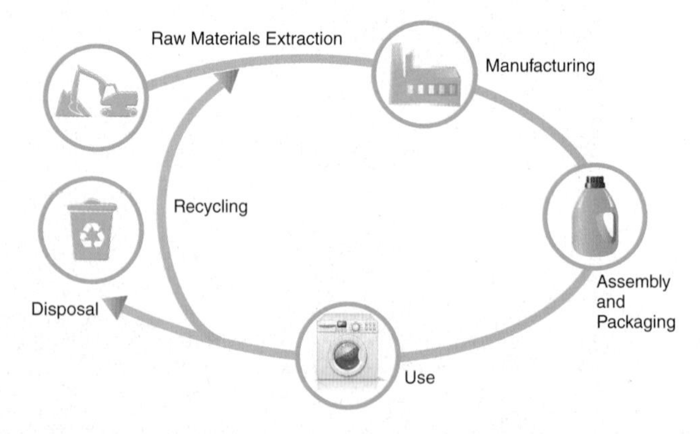

Calculate

20. Based on this image, the number of pounds in 1 kilogram is closest to:

a. 0.5 b. 1.0 c. 1.8 d. 2.2 e. 2.9

21. Calculate the number of

a. millimeters in a mile

b. milliseconds in a year

c. micrograms in a ton (2000 lb)

22. Calculate the number of kilograms in a gram.

23. If New York is approximately 4100 km from San Francisco, how far is this in centimeters?

24. Light travels at about 30 billion (30,000,000,000) centimeters per second. A light-year is defined as the distance that light travels in a vacuum in 1 year. What is the length of a light-year in

a. meters?

b. miles?

2 Atoms and Elements

To some of us, the terms **chemistry** and **chemicals** may sound remote and mysterious. Yet we all live our everyday lives immersed in a sea of chemicals, ranging from the simple to the incredibly complex, from the elemental **nitrogen** and **oxygen** in the air we breathe to the intricate molecular threads of DNA that carry our human characteristics to our children and their children. Without chemistry and chemicals, neither we nor the world we live in could exist.

We can begin to visualize our world of chemicals by looking at a common barbecue party. We might find a few **helium**-filled balloons. Someone might be drinking from an **aluminum** can or wearing a ring made of **silver** or **gold**. Cell phones run on semiconductor chips made from **silicon**. The grill itself might be loaded with charcoal briquettes, which are composed almost entirely of **carbon**. Each of these substances is a **chemical element**, composed of fundamental particles called **atoms**. In this chapter we explore the nature and meaning of these terms, including a valuable tool for categorizing them called the periodic table.

Tim O. Walker

CHAPTER OUTLINE

2.1 Atomic Structure

LEARNING OBJECTIVES

1. **Outline** the development of the modern atomic model.
2. **Describe** the Rutherford and Bohr models of the atom.
3. **Explain** why atoms are mostly empty space.

Our modern understanding of chemistry is based on the idea that all matter is made up of atoms. By understanding the nature of atoms, we can predict how observable matter behaves. In this section, we'll trace the development of our understanding of atoms to reveal the evidence underlying our current knowledge of chemistry.

Questioning the Nature of Matter

If you cut a piece of copper wire in half, you begin a process that puzzled some of the greatest thinkers at the beginnings of Western civilization. They asked, in essence, "What happens if you keep on cutting the wire this way, again and again and again (**Figure 2.1**)? How long could you keep cutting this way? Where does it end?"

Over 2000 years ago in ancient Greece, two schools of thought came to opposite conclusions. One, led by Aristotle, held that you could cut each new piece in half forever. You would get finer, narrower slices throughout all eternity, without end. In essence, Aristotle argued that all matter is completely continuous, without interruption from one surface to another.

Another group, gathered around Democritus, argued that eventually the slicing would have to come to an end. They held that all matter is composed of incredibly small, indivisible particles of various shapes, sizes, and textures. Cutting any object again and again would eventually bring you to these small, individual particles, the ultimate components of all matter. They named these components *atomos*, a Greek term for "indivisible."

Although Aristotle's theory of continuous matter dominated Western thought for about 2000 years, today we know that Democritus was right. **Atoms**, alone or joined to one another in molecules, form the chemical foundations of our everyday world (**Figure 2.2**).

atom A fundamental particle of matter.

Early Atomic Models

As mentioned earlier, Aristotle's vision of the structure of material objects—infinitely divisible substances, continuous in space from one surface to another—prevailed in Western thought for over two millennia. It was the product of nothing more than his own speculations about what the universe ought to be like, and it displaced Democritus's concept of atoms (which was also no more than speculation at that time). But as the centuries passed, the development of experimental science—with actual physical tests and investigations of real material substances under controlled conditions—changed all this.

In the first few years of the 19th century, John Dalton, an English teacher and tutor of mathematics and science (or, as it was then known, *natural philosophy*), reported the results of his careful, experimental study of various

Debating the existence of atoms • Figure 2.1

Can you keep cutting a copper wire into smaller and smaller pieces, forever?

Ask Yourself

Is the fine powder depicted in the final image still copper? What do you think?

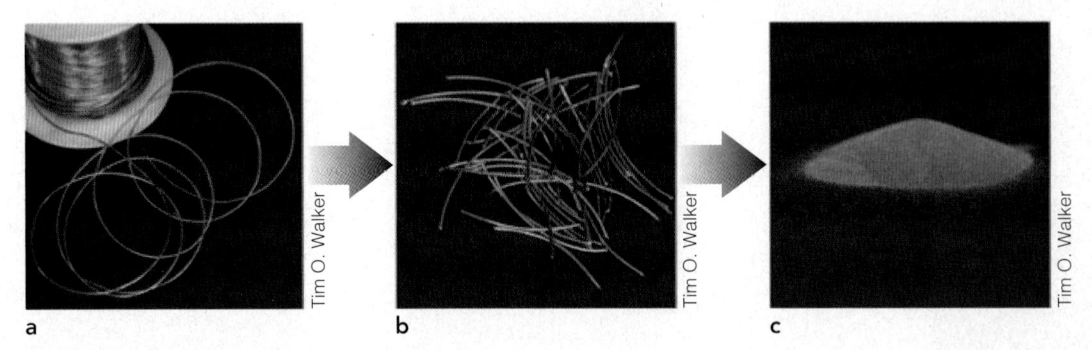

A modern image of atoms • Figure 2.2

This circle consists of individual cobalt atoms on a copper surface, in an image produced by modern visualizing technology at the National Institute of Standards and Technology (NIST).

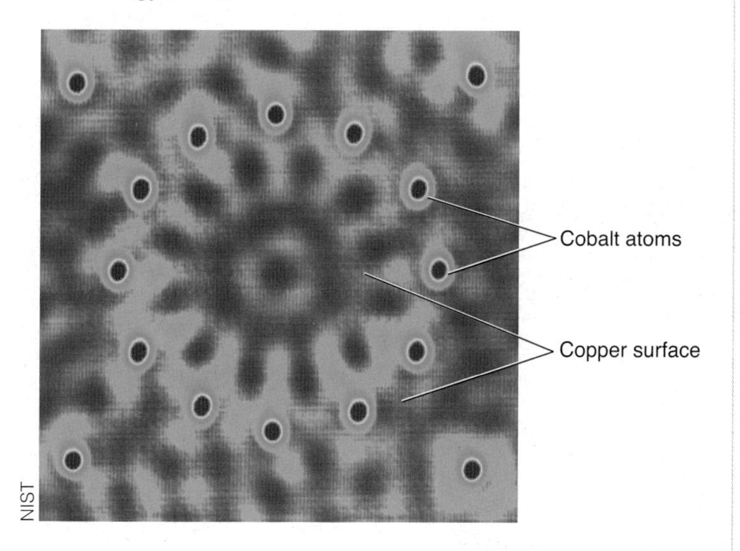

atmospheric gases. His experimental results demonstrated clearly that matter is not continuous (as Aristotle had claimed) and that Democritus's atoms do exist. Like Democritus, Dalton held that atoms are indivisible and indestructible. They cannot be split apart, sliced open, or broken into smaller pieces.

Dalton also found that different atoms can bond to one another in fixed proportions to form chemical **compounds**. We'll have more to say about compounds after we look more closely at the structures of the atoms themselves.

Figure 2.3 is a page from Dalton's publication, with his schematic drawings of atoms and the compounds formed as these atoms bond to one another. For example, Dalton depicts an atom of hydrogen as a simple circle with a dot at its center and an atom of carbon as a circle with a darker interior. Two atoms of oxygen bonded to the same atom of carbon form the compound we know as carbon dioxide. Today we represent carbon dioxide as CO_2, with C used as the symbol for a carbon atom, O for an oxygen atom, and the subscript $_2$ indicating the presence of two oxygen atoms.

The earliest symbols for atoms • Figure 2.3

An annotated page from Dalton's 1808 manuscript depicting both his early views of atom types, or *elements*, and bonded atoms of different types, or *compounds*.

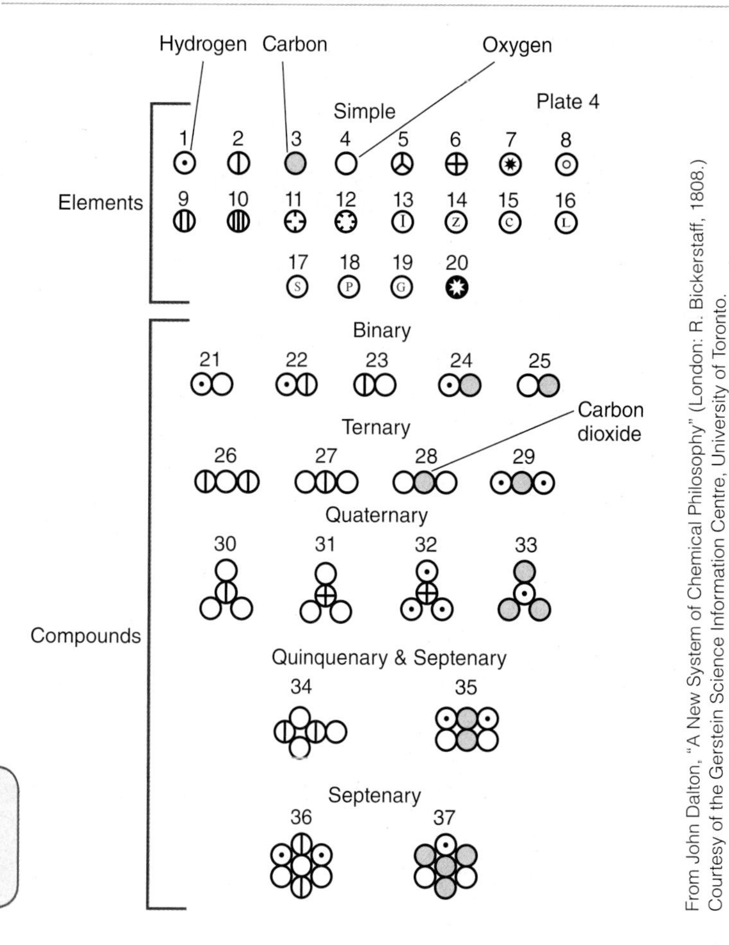

Ask Yourself

How might we represent the compound H_2O using Dalton's notation for atoms? Assume oxygen is bound to each hydrogen.

Chemistry InSight Some common elements, their symbols, and their applications • Figure 2.4

Elements are most often either solids or gases and can be found in our environment, bodies, and a variety of everyday items.

The element gold, Au, is a precious metal used in jewelry, coins, and in the form of bricks for the storage and transfer of financial wealth. This gold-plated medal was awarded at the World Swimming Championships.

The element carbon, C, is found in a pure form as diamond (and graphite), and in a more crude form in the charcoal briquettes of outdoor grills. Here, American swimmer, Ryan Lochte, displays a diamond-encrusted mouth retainer during a celebratory moment.

Clive Rose/Getty Images

The element calcium, Ca, is a soft, gray metal in its pure form. Compounds of calcium are essential nutrients for healthy teeth and bones.

The gaseous element oxygen is essential to life and present in the air as the diatomic molecule O_2. Elite athletes in highly aerobic sports such as swimming, exhibit a large capacity to transport and use oxygen within the body.

Ask Yourself

What characteristic do calcium, carbon, gold, and copper have in common that distinguishes them from oxygen?

We represent all elements, and their atoms as well, by capital letters or by combinations of a capital letter and a small letter. For example, H represents both the element hydrogen and an atom of hydrogen. The symbol He represents helium—a gas that's lighter than air and is used to inflate balloons—as well as an atom of helium. C stands for carbon, Ca is calcium (the element that gives strength to bones and teeth), and Cl is chlorine (naturally occurring as Cl_2), one of the two elements that form common table salt (sodium chloride). N represents nitrogen (naturally occurring as N_2), the major gas of our atmosphere, while Na is sodium, the other element in table salt (NaCl). Some common elements and their applications are discussed in **Figure 2.4**.

Nuclear Models of the Atom

As summarized in **Figure 2.5**, which presents the evolution of our knowledge of **atomic structure**, other scientists refined Dalton's view of the atom. The first few years of the 20th century in particular saw fundamental changes in how we picture the structure of an atom.

In 1904, British scientist J. J. Thomson proposed what became known as the "plum pudding" model. This was based on his 1897 discovery of **electrons** as

atomic structure The combination of all the particles that compose an atom, their relationships to one another, and their locations within the atom.

electrons Small, negatively charged particles located in shells surrounding an atom's nucleus.

Evolution of our knowledge of atomic structure up until the time of the Bohr model • Figure 2.5

The idea that all matter is made up of individual particles, or atoms, was first proposed in Ancient Greece, but models of atomic structure based on experimental evidence did not appear until early in the 20th century.

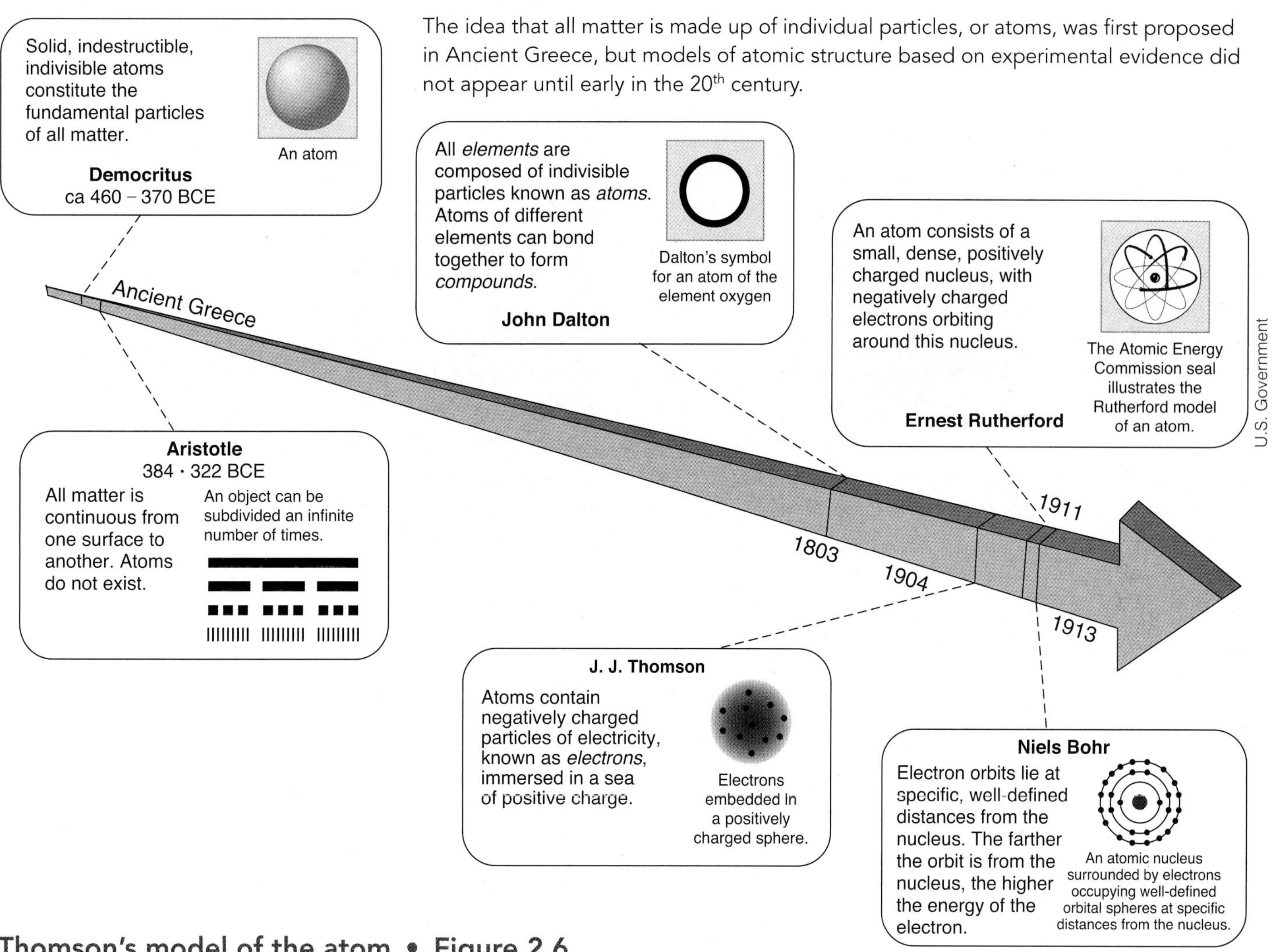

Solid, indestructible, indivisible atoms constitute the fundamental particles of all matter.
An atom
Democritus
ca 460 – 370 BCE
Ancient Greece
Aristotle
384 · 322 BCE
All matter is continuous from one surface to another. Atoms do not exist.
An object can be subdivided an infinite number of times.
All *elements* are composed of indivisible particles known as *atoms*. Atoms of different elements can bond together to form *compounds*.
Dalton's symbol for an atom of the element oxygen
John Dalton
1803
An atom consists of a small, dense, positively charged nucleus, with negatively charged electrons orbiting around this nucleus.
The Atomic Energy Commission seal illustrates the Rutherford model of an atom.
Ernest Rutherford
U.S. Government
1904
1911
1913
J. J. Thomson
Atoms contain negatively charged particles of electricity, known as *electrons*, immersed in a sea of positive charge.
Electrons embedded in a positively charged sphere.
Niels Bohr
Electron orbits lie at specific, well-defined distances from the nucleus. The farther the orbit is from the nucleus, the higher the energy of the electron.
An atomic nucleus surrounded by electrons occupying well-defined orbital spheres at specific distances from the nucleus.

Thomson's model of the atom • Figure 2.6

J.J. Thomson, who discovered electrons within atoms, proposed that an atom consists of a sphere of positive charge with smaller negative charges (the electrons) scattered about it. Though this so-called plum pudding model proved incorrect, Thomson properly recognized that electrons are negatively charged.

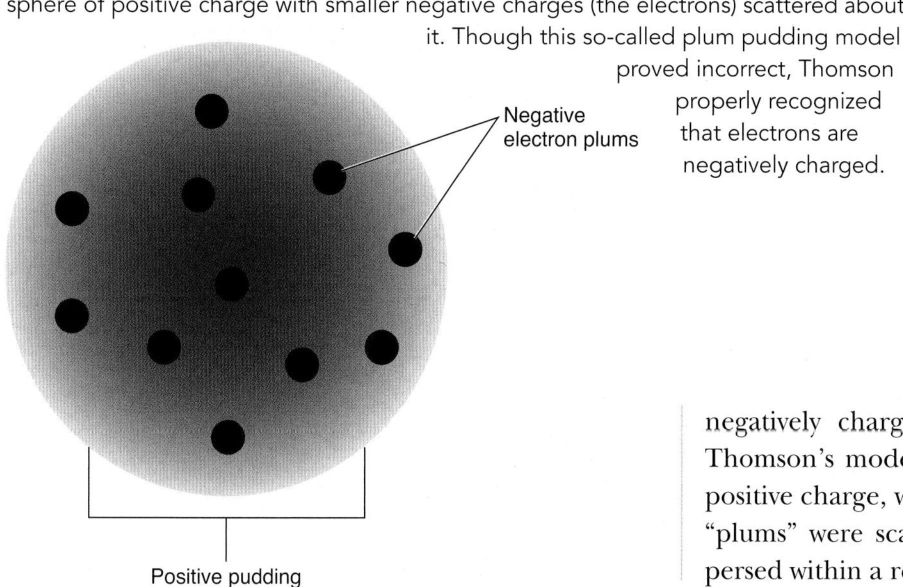

negatively charged particles contained within atoms. Thomson's model pictured an atom as a "pudding" of positive charge, within which the small negative electron "plums" were scattered, much as actual plums are dispersed within a real plum pudding (**Figure 2.6**).

Thomson's view of atomic structure held sway until Ernest Rutherford, born in New Zealand, began an experimental investigation of the more detailed structure of atoms. In a study designed by Rutherford, two of his assistants, Ernest Marsden and Hans Geiger—who later developed the Geiger counter—fired alpha particles at an extremely thin gold film, only a few atoms thick. (We now recognize alpha particles as helium nuclei, consisting of two protons and two neutrons [Sec. 9.1].) Any small deflections of these alpha particles as they passed through the thin gold film could give Rutherford valuable information about the detailed structure of Thomson's model.

Almost all the alpha particles behaved exactly as expected, passing through the thin gold film with only small deflections in their paths. However, to everyone's astonishment, a few particles did not pass through the film at all. Instead they hit "something," rebounded, and headed back toward their source (**Figure 2.7**). Rutherford later called this, "almost as incredible as if you had fired a 15-inch [artillery] shell at a piece of tissue paper and it came back and hit you." In 1911 Rutherford published these results and his conclusion: Atoms are not "plum puddings" at all. Instead each atom contains at its center a very small, highly massive positively charged **nucleus** (plural is "nuclei"), and the electrons exist in the space surrounding these positively charged nuclei.

Niels Bohr, a native of Denmark, refined the picture further. He proposed that electrons occupy well-defined orbits, now known as **quantum shells**. These shells are identified by their quantum numbers, n. For the shell closest to the nucleus, $n = 1$; for the next one, $n = 2$; then $n = 3$; and so on.

nucleus The extremely small, dense center of the atom in which the atom's positive charge is localized.

quantum shells The orbits of electrons at fixed distances from the nucleus, as described in the Bohr model of the atom.

Discovery of the nucleus • Figure 2.7

Rutherford's analysis of the Geiger–Marsden "gold-foil experiment" showed that Thomson's model of the atom could not be correct. Instead, the atom has a dense, positively charged nucleus in its center.

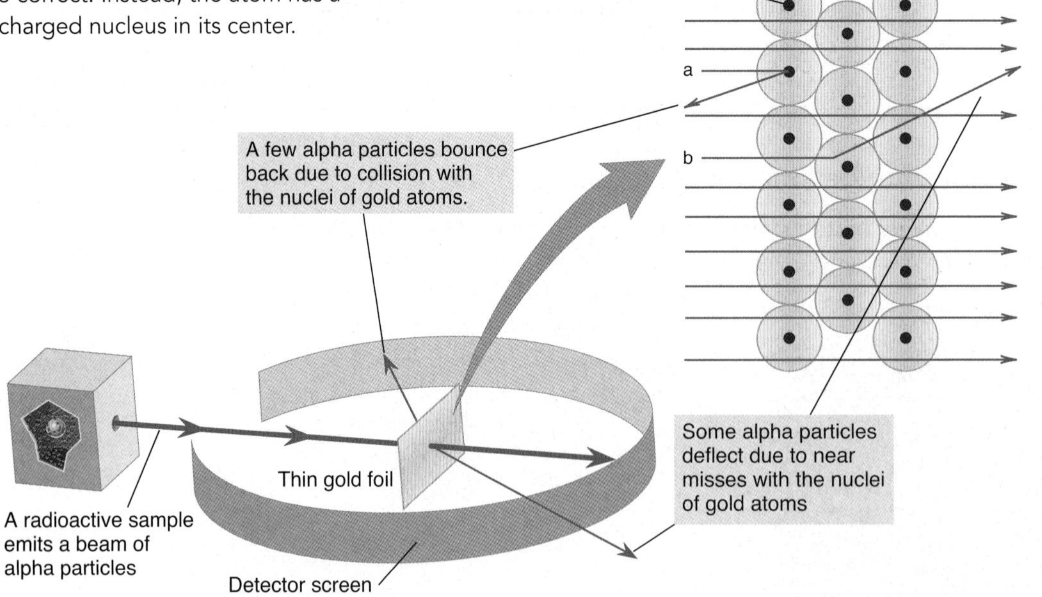
Gold atoms
Dense nucleus
a
b
A few alpha particles bounce back due to collision with the nuclei of gold atoms.
Some alpha particles deflect due to near misses with the nuclei of gold atoms
Thin gold foil
A radioactive sample emits a beam of alpha particles
Detector screen

Ask Yourself

As shown here, most of the alpha particles pass through the gold foil undeflected. How does this result support Rutherford's hypothesis that the volume of an atom's nucleus comprises a miniscule proportion of the volume of the entire atom.

Maximum number of electrons the first four shells can hold	Table 2.1
Quantum number, *n*, of shell	**Maximum number of electrons**
1	$2 \times (1)^2 = 2$ ••
2	$2 \times (2)^2 = 8$ ••••••••
3	$2 \times (3)^2 = 18$ ••••••••••••••••••
4	$2 \times (4)^2 = 32$ ••••••••••••••••••••••••••••••••

We should note that many of these investigations into the structure of the atom had unforeseen applications of great scientific and everyday importance. For example, Thomson discovered the electron by using a large vacuum tube with a coated screen that emitted light when struck by beams of electrons, known as "cathode rays" at that time. The "cathode ray tube" or CRT later became the basis of television, radar screens, and heart monitors. Rutherford's analysis of how particles rebounded from atoms became the basis of later x-ray analysis for determining the structure of medicines and molecules within our bodies. And as we shall see a little later in this section, Bohr's model of electron shells explained how atoms interact with light, leading to discoveries from neon lights to lasers.

In the Bohr model of the atom, two important conditions apply. First, electrons can exist only within these specified shells; they cannot occupy space that lies between two quantum shells. In addition, each shell can accommodate a maximum number of electrons, depending on the shell's quantum number. The maximum number of electrons any shell can hold is $2(n^2)$, where n is the shell's quantum number (**Table 2.1**).

All this—Dalton's concept of atoms as the smallest particles of elements, Thomson's discovery of the electron, Rutherford's discovery of the atomic nucleus, Bohr's elaboration of electron shells, and the data of Table 2.1—gives us two ways to picture an atom. The visualization based on Rutherford's contributions, shown in **Figure 2.8a**, is more popular and more attractive, but it doesn't provide much information about the atom. The other model, based on further insights by Bohr and shown in **Figure 2.8b**, is more schematic but gives chemists far more information. We'll see why as we examine atomic structure in more detail.

Visualizations of the Rutherford and Bohr models of the atom • Figure 2.8

Rutherford's contribution was focused more on the nucleus, whereas Bohr's was focused more on the electrons. Both of these models eventually proved to be oversimplifications.

a. Rutherford proposed that a minute, dense, positively charged nucleus exists at the center of the atom.

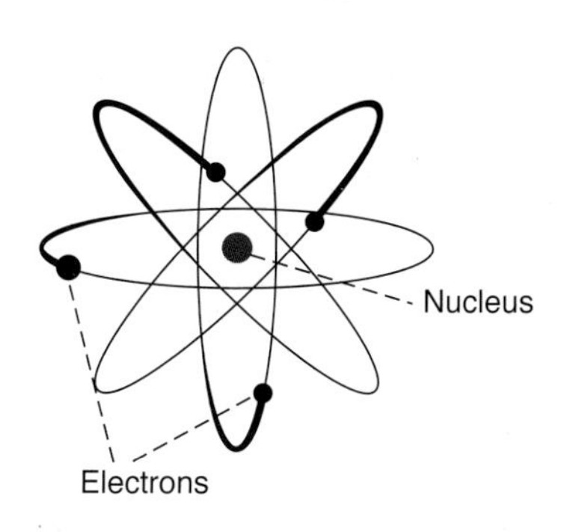

b. Bohr's model placed the atom's orbiting electrons in discrete shells, each with a specific energy and capable of holding a specific maximum number of electrons.

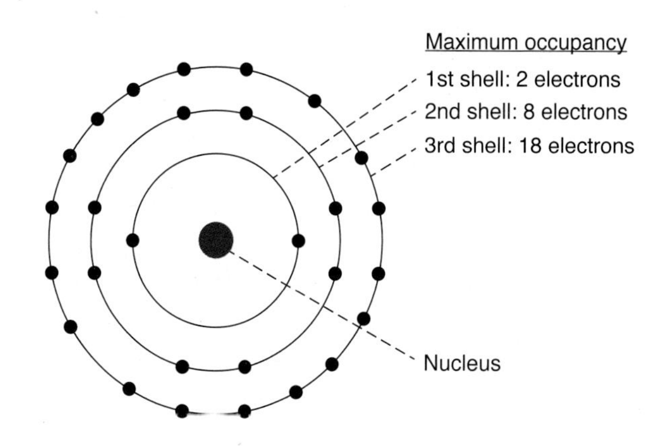

Ask Yourself

Which two components do the Rutherford and Bohr models of the atom have in common?

Bohr's atomic model and line spectra • Figure 2.9

The Bohr model explained why we see only certain colors of light in the hydrogen line spectrum.

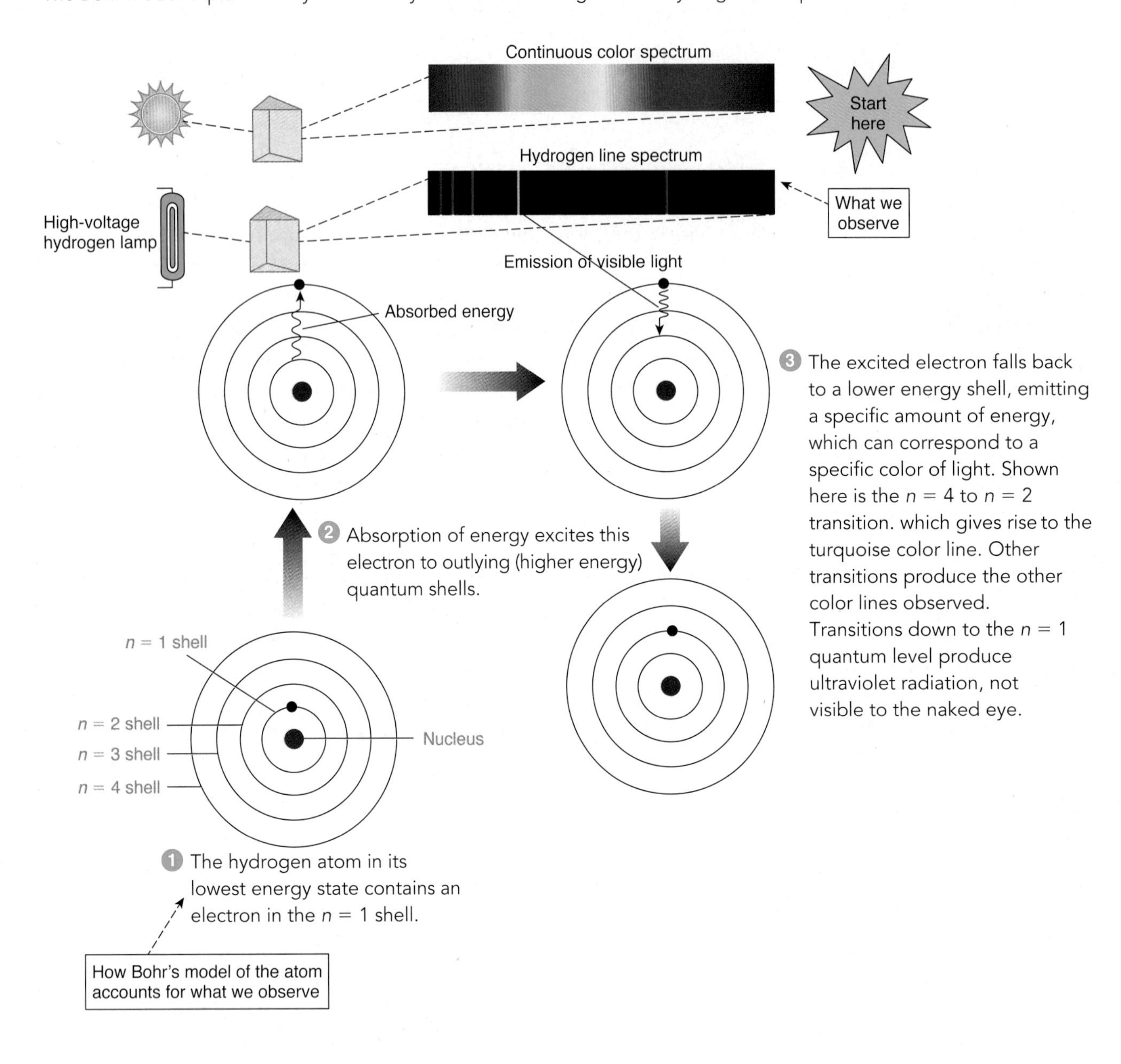
Continuous color spectrum
Start here
Hydrogen line spectrum
What we observe
High-voltage hydrogen lamp
Emission of visible light
Absorbed energy
3 The excited electron falls back to a lower energy shell, emitting a specific amount of energy, which can correspond to a specific color of light. Shown here is the n = 4 to n = 2 transition. which gives rise to the turquoise color line. Other transitions produce the other color lines observed. Transitions down to the n = 1 quantum level produce ultraviolet radiation, not visible to the naked eye.
2 Absorption of energy excites this electron to outlying (higher energy) quantum shells.
n = 1 shell
n = 2 shell
n = 3 shell
n = 4 shell
Nucleus
1 The hydrogen atom in its lowest energy state contains an electron in the n = 1 shell.
How Bohr's model of the atom accounts for what we observe

Interpret the Data

If all color lines observed in the hydrogen line spectrum are due to transitions down to the $n = 2$ level, and the red line arises from a smaller energy transition than that for the turquoise line, what produces the red line?

Bohr's model of the atom accounts for a phenomenon that had been observed by the early 1900s but was not well understood—the **emission spectrum** or **line spectrum** of hydrogen. To understand its significance, consider what we see when we place a prism in a sunlit window. The prism diffracts or separates the components of the sunlight into a continuous band of the colors of the rainbow—red, orange, yellow, green, blue, indigo, and violet.

However, suppose we do a different experiment by passing high-voltage electricity through a glass bulb containing hydrogen gas. The energized gas emits a different type of radiation. When this radiation passes through a prism, the result is a set of discrete, narrow lines of specific colors, rather than the continuous band of rainbow colors we get from sunlight. Bohr's model of the atom accounts nicely for this result, as shown in **Figure 2.9**.

Today we understand that every atom can exist at slightly different energy levels, corresponding to the types of electron shells shown in Figure 2.9. Also, we know that the energy differences between certain electron shells correspond to the absorption or emission of different colors of light. For example, Figure 2.9 shows that hydrogen emits a blue-green light when an electron in the $n = 4$ shell drops down to the $n = 2$ shell. Chemists have learned to utilize this property to create fireworks with different colors. For example, red fireworks can be produced from strontium compounds, green from barium compounds, and blue from copper compounds.

Chemists have also learned how to induce electrons to change energy levels in many atoms simultaneously so that when one atom emits radiation, many atoms emit the same radiation at the same time. The resulting light can be optically amplified, resulting in Light Amplification by Stimulated Emission of Radiation—a laser.

Later Atomic Models

Following Bohr's model, by 1926 other European physicists—notably Erwin Schrödinger, Max Born, and Werner Heisenberg—developed an even newer model for the behavior of electrons in an atom. This model, based on **quantum** or **wave** mechanics and now supported by ample experimental evidence, describes electrons as having both particle-like and wave-like characteristics (see **Figure 2.10**). Although this newer, quantum-mechanical model represents the current view of atomic theory, the Bohr model—emphasizing electrons as particles populating quantum shells—will serve us well in our discussions of chemistry.

The quantum-mechanical model of the atom made one very important advance over the Bohr model. This advance describes electrons as occupying shapes or regions of space around the nucleus called **atomic orbitals** (see *A Deeper Look* on the next page).

Note that almost all visualizations of atoms show a nucleus that's far too large. In all actual atoms, the nucleus is so very small in comparison to the rest of the atom that it's fair to say all atoms are almost entirely empty space! The diameter of a hydrogen atom, for example, is about 100,000 times the size of its nucleus. For comparison, if a hydrogen atom grew to be as large as Earth, the nucleus

Modern evidence supporting the quantum-mechanical model • Figure 2.10

This color-enhanced image using modern visualizing technology from IBM's Research Division shows concentric standing waves of electrons trapped in a "corral" of iron atoms on a copper surface. (Standing waves oscillate up and down but don't travel anywhere.)

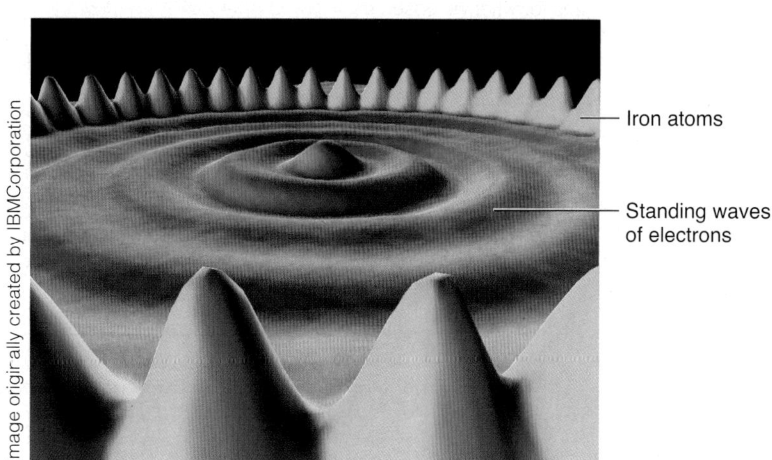

Image originally created by IBMCorporation

A DEEPER LOOK Electrons as Waves

In the quantum-mechanical view of the atom, electrons exhibit wave-like behavior. Because electrons behave like waves, we can never know their exact location at any given moment. Instead we speak of *probabilities* of finding them within certain volumes of space. A simplified model for this behavior consists of standing (stationary) waves on a string. Note that the photograph appears blurry at regions where the string moves up and down, yet one or more spots along the string, called **nodes**, appear to be still. This is what we mean by a standing wave: it vibrates, but it does not move along the string.

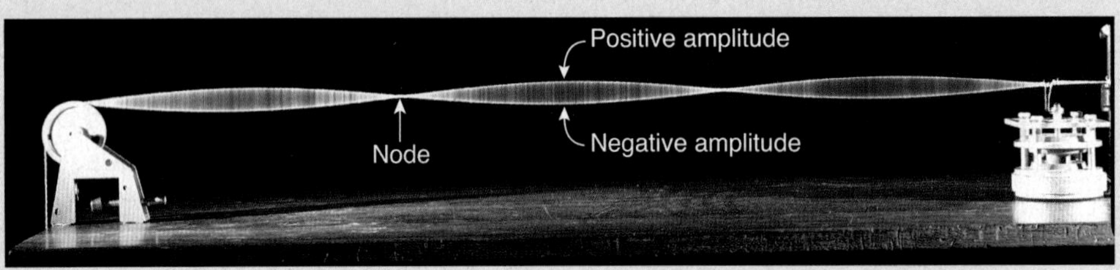

A camera with an open shutter captures standing waves on a rapidly vibrating string.

Andrew Lambert Photography/Science Source

Electrons, too, behave like standing waves, but with vibrations in *three* dimensions rather than two. We call these standing waves **atomic orbitals** and visualize them as electron clouds—regions of space where we're more likely to find the electron. They are analogous to the blurry regions bounded by positive and negative amplitudes in the photo. A node in an electron orbital corresponds to a region where there is no likelihood of finding an electron. The simplest types of atomic orbitals look like this:

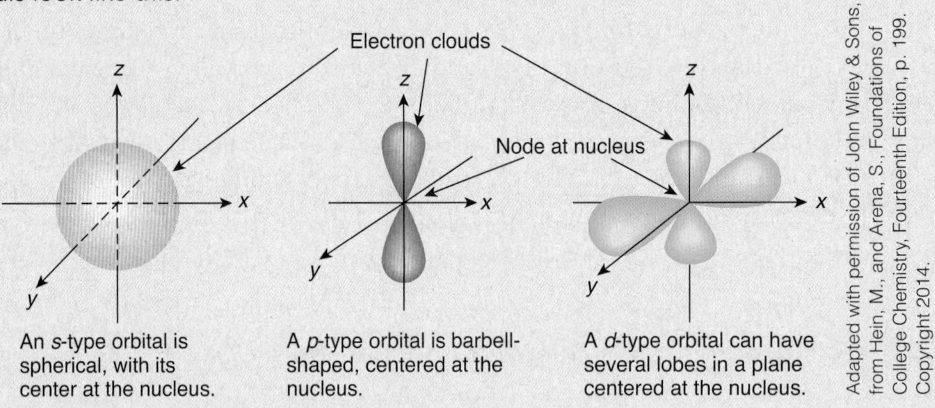

An *s*-type orbital is spherical, with its center at the nucleus.

A *p*-type orbital is barbell-shaped, centered at the nucleus.

A *d*-type orbital can have several lobes in a plane centered at the nucleus.

Each orbital can hold a maximum of two electrons. In addition, orbitals are organized into shells, as shown in the table, with the first shell closest to the nucleus and each additional shell progressively farther.

The following shell...	contains the following orbital(s)...	so can hold up to:
First	**1** *s*, for a total of **1 orbital**	2 electrons
Second	**1** *s* and **3** *p*, for a total of **4 orbitals**	8 electrons
Third	**1** *s*, **3** *p*, and **5** *d*, for a total of **9 orbitals**	18 electrons

Put It Together *Review Table 2.1 and answer this question.*

How many orbitals do you predict are in the fourth quantum shell?

a. 10 **b.** 12 **c.** 16 **d.** 24

Relative sizes of an atom and its nucleus • Figure 2.11

The nucleus is extremely small compared to the space occupied by electron shells.

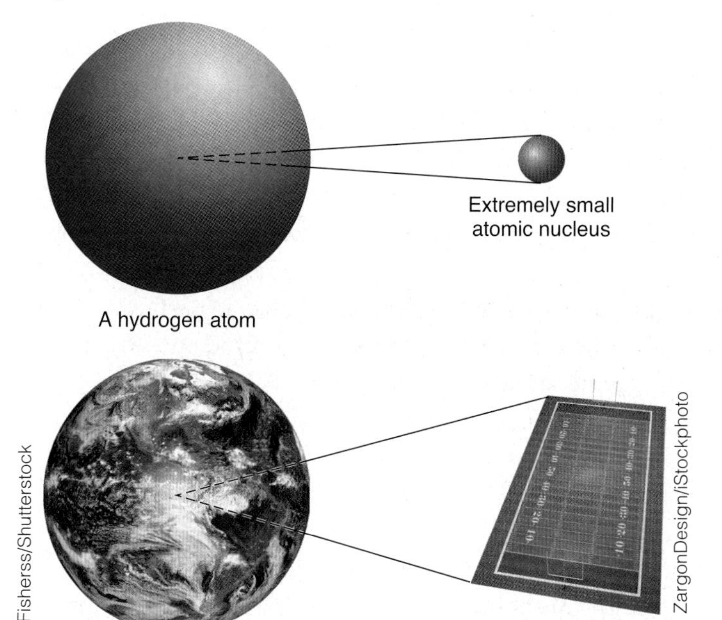

If a hydrogen atom were the size of Earth, almost 12,700 kilometers (or about 42,000,000 feet) in diameter, the atomic nucleus at its center would be a sphere with a diameter of only about 400 feet, able to hold a foodball field with just a little room to spare.

Ask Yourself

What is the ratio of the diameter of the hydrogen atom to the diameter of the nucleus at its center?

at its center would be just large enough to accommodate a football field, with room left over for a small band and some cheerleaders (**Figure 2.11**). Atoms are mostly voids.

CONCEPT CHECK

1. **How** did Dalton's work advance earlier notions of matter held by Democritus?
2. **How** did the Rutherford and Bohr models of the atom differ from the earlier model proposed by J. J. Thomson?
3. **What** is the relative size of a hydrogen atom as compared to its nucleus?

2.2 The Nucleus

LEARNING OBJECTIVES

1. **Compare** the electrical charges of a proton, a neutron, and an electron.
2. **Relate** atomic number to the definition of an element.
3. **Define** mass number.

Although the nucleus occupies only an extremely small portion of an atom's volume, it represents a unique part of an atom. With its own structure and composition, the nucleus defines the atom's identity. In this section we examine the nucleus, its structural components, and its relationship to the element the atom represents.

Atomic Number

The negative charge of the electrons in quantum shells surrounding the nucleus is balanced by the positive charge of the nucleus itself. This positive charge is provided by **protons**, with each proton carrying a positive charge exactly equal to the size of the negative charge

proton A positively charged particle found within the nucleus of an atom.

carried by every electron. The simplest atom, an atom of the most common form of the element hydrogen, contains a single proton as its nucleus and a single electron in its first quantum shell (**Figure 2.12**). With the negative charge of the electron exactly balancing the positive charge of the proton, the atom as a whole is electrically neutral. Similarly, all other electrically neutral atoms also carry exactly as many orbital electrons as nuclear protons.

The number of protons in an atom's nucleus gives us an important value for the atom, its **atomic number**. Thus the atomic number of a hydrogen atom is 1. Atomic numbers are particularly important because *all atoms of the same element have the same atomic number*. Stated another way, *all atoms of any specific atomic number are atoms of the same element*. This gives us a convenient definition of an **element**.

atomic number The number of protons in the nucleus of an atom.

element A substance, all of whose atoms have the same atomic number.

If we want to indicate an element's atomic number along with its symbol, we place the number as a subscript to the left of the symbol's capital letter. For hydrogen this gives us ${}_{1}H$. For the other elements discussed earlier—carbon, calcium, chlorine, nitrogen, sodium, and neon—attaching their atomic numbers to their symbols produces ${}_{6}C$, ${}_{20}Ca$, ${}_{17}Cl$, ${}_{7}N$, ${}_{11}Na$, and ${}_{10}Ne$. We'll see in Section 2.4 that the atomic numbers of all elements are listed in the periodic table.

Mass Number

In addition to the electron and the proton, a third kind of particle, the **neutron**, also contributes to the structure of atoms. Neutrons are found along with protons in the nuclei of all atoms—except the most common kind of hydrogen atom. Unlike protons and electrons, neutrons carry no electrical charge. Without either positive or negative charge, neutrons have no effect on atomic number. But they do have mass, so they do contribute to the weight of an atom. We'll have more to say about mass and weight in the next section. With no charge, neutrons can be written as n or n^0. Protons are represented as p or p^+, and electrons as e or e^-. These three particles—protons, neutrons, and electrons—constitute the building-blocks of all atoms (except the most common form of hydrogen) and are often known as subatomic particles.

neutron A nuclear particle that carries no electric charge.

We saw earlier that a subscript placed to the left of the atomic symbol shows the atomic number of the atom. The **mass number** is written as a superscript preceding the element symbol. The common hydrogen atom, with an atomic number of 1, contains no neutrons. Therefore it has a mass number of 1 and we can write

mass number The combined number of protons and neutrons in the nucleus of an atom.

$$\begin{array}{l}\text{Mass number} \longrightarrow \\ \text{Atomic number} \longrightarrow\end{array} {}^{1}_{1}H$$

Since subatomic particles are so very small, far smaller than the atoms that contain them, the measurements of length and mass we use in our everyday lives aren't suitable for describing their dimensions. It's more convenient to use exponential notation, sometimes known as **powers of ten** (**Figure 2.13**). Protons and neutrons are each about

Two views of a hydrogen atom • Figure 2.12

The most common form of the hydrogen atom contains one electron surrounding a nucleus containing one proton.

Ask Yourself

Why is the hydrogen atom electrically neutral?

a. In the **Bohr model**, we visualize the electron orbiting the nucleus at a fixed distance.

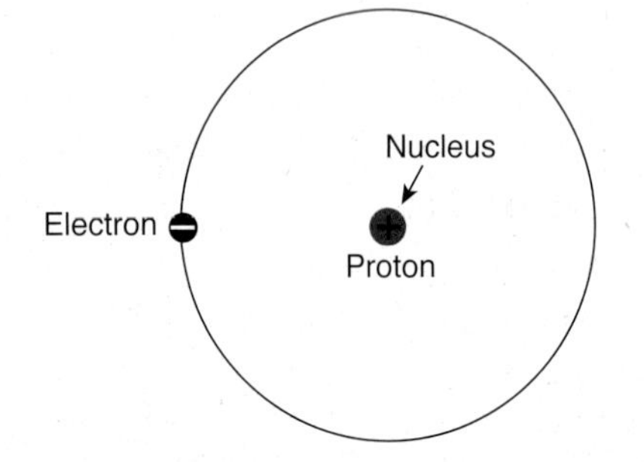

b. In the **quantum model**, we visualize the electron as a diffuse, spherical electron cloud with a proton (the hydrogen nucleus) at its center.

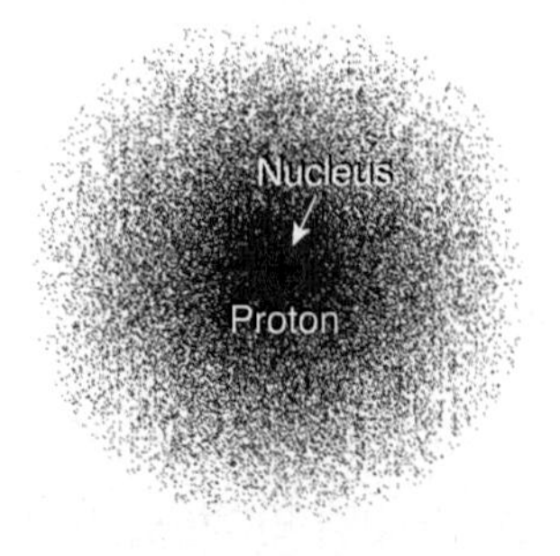

In Words, Math, and Pictures

Visualizing powers of ten • Figure 2.13

When we count or measure things, we sometimes use the same units for both relatively small numbers and extraordinarily large ones. For example, we might use the unit "year" to express both the age of a young child (3 years) and the age of Earth (over 4 billion years). We might use the unit "dollar" as the cost of a magazine (5 dollars) or the annual national budget of a large country (trillions of dollars).

In each case, the units are the same, but the numbers of these units vary enormously. For convenience in expressing wide ranges of numbers, we use **powers of ten**, which are simply multiples of the number 10 expressed as exponents of the number 10.

We can visualize powers of ten by looking at sets of 10 dots each:

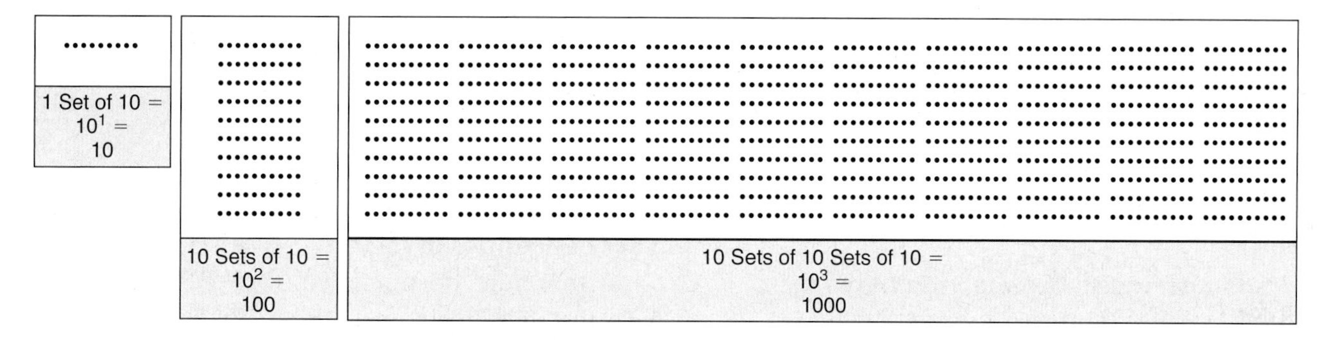

If we're dealing with 20 dots, we could view the set of 20 as two sets of 10 dots:

2 Sets of 10 = 2×10^1 = 20

But if we had, for example, 2,000,000,000 dots, we could more easily write this as 2×10^9 dots.

For numbers smaller than 1, we divide by 10 and use negative exponents of 10. The number 0.03, for example, is 3/100, or $3 \times 1/100$. To write this number, we use a negative exponent and write 3×10^{-2}.

A positive exponent, n, of 10 moves the decimal point to the right n places, and a negative exponent moves the decimal point to the left. For example, $5.3 \times 10^4 = 53{,}000$, and $5.3 \times 10^{-4} = 0.00053$.

Furthermore, when we multiply two numbers containing powers of ten, we *add* the exponents: $10^3 \times 10^2 = 10^5$

Example: Since the mass of an individual proton or neutron is 1.67×10^{-27} kg, an object that contains a combined total of 10^{30} of these nuclear particles must have a total mass of

$$(1.67 \times 10^{-27}\text{ kg}) \times 10^{30} =$$

$$1.67 \times (10^{-27} \times 10^{30})\text{ kg} =$$

$$1.67 \times 10^{(-27+30)}\text{ kg} =$$

$$1.67 \times 10^3\text{ kg} =$$

$$1670\text{ kg},$$

Stratol/iStockphoto

which is about the same as the mass of this car.

When we divide two numbers containing powers of ten, we *subtract* the exponents:

$$10^8 \div 10^5 = 10^3$$

Division by the subtraction of exponents is useful when we compare two numbers of different size.

Example: What is the relative size of Alaska compared to the size of Hawaii?

Solution: To find out, we divide the size of Alaska by the size of Hawaii.

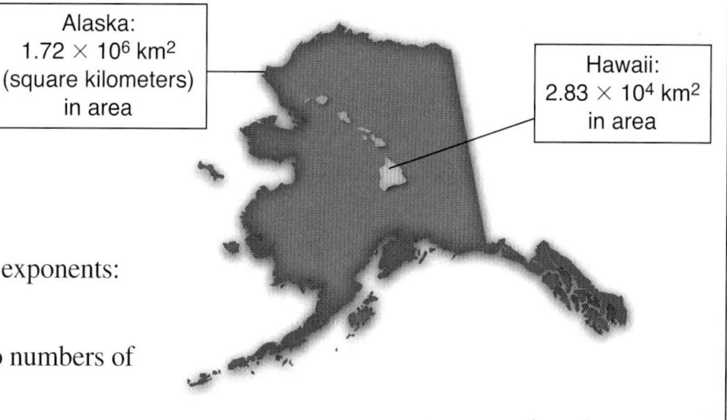

$$\frac{\text{Area Alaska}}{\text{Area Hawaii}} = \frac{1.72 \times 10^6\text{ km}^2}{2.83 \times 10^4\text{ km}^2} = \left(\frac{1.72}{2.83}\right) \times 10^{(6-4)}$$

$$= 0.608 \times 10^2 = 60.8\text{ times as large}$$

1×10^{-15} meters in diameter, or a millionth of a billionth of a meter. This is about 200 million times smaller than the smallest objects visible by high-power, light microscopes. Electrons are far smaller than even protons and neutrons.

CONCEPT CHECK STOP

1. **What** are the three subatomic particles? Which of these can be found in the nucleus?
2. **What** single property distinguishes a given element from any other element?
3. **How** can we determine the number of neutrons in an atom if we know its mass number and atomic number?

2.3 Isotopes and Atomic Mass

LEARNING OBJECTIVES

1. **Explain** what constitutes an isotope of a given element.
2. **Define** and distinguish between the terms *mass* and *weight*.
3. **Identify** the location within an atom where virtually all of the atom's mass lies.

We have seen that atoms with the same number of protons in their nuclei represent the same element. In this section, we will see how even atoms of the same element, each containing the same number of protons, can differ in significant ways.

Isotopes

Atoms of the same element must contain the same number of protons in their nuclei and must have the same atomic number. But atoms of the same element do not need to hold the same number of neutrons in their nuclei and therefore do not need to have the same mass number. Atoms such as these represent **isotopes**. The element hydrogen, for example, exists as three different isotopes (**Figure 2.14**).

isotopes Atoms of the same element that differ in the number of neutrons and therefore in mass number.

To help distinguish among these three isotopes of hydrogen, the isotope with a mass number of two—one proton and one neutron—$^{2}_{1}H$ is commonly known as **deuterium**, indicated by the symbol D. The isotope with a mass number of three—one proton and two neutrons—$^{3}_{1}H$ is **tritium**, indicated by the symbol T.

Naturally occurring hydrogen consists of all three isotopes, but the amount of the isotope with a mass number of 1 overwhelms the other two. It makes up about 99.985% of all the hydrogen in the universe. (Deuterium

The isotopes of hydrogen • Figure 2.14

Isotopes have the same numbers of protons and electrons, but different numbers of neutrons.

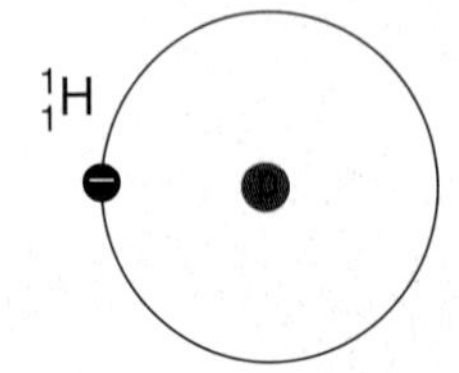

a. The most common isotope of hydrogen holds a single proton in its nucleus and no neutrons at all. This is the atom we saw earlier.

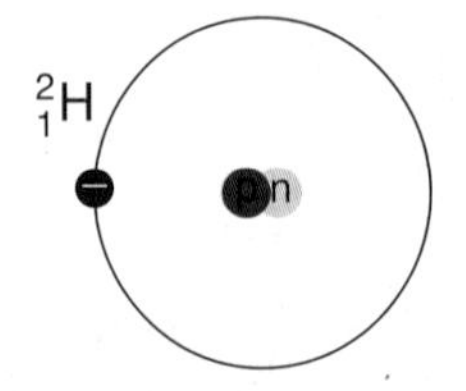

b. Deuterium, a much rarer isotope of hydrogen, contains *one* neutron in its nucleus along with the single proton that confirms it as an isotope of the element hydrogen.

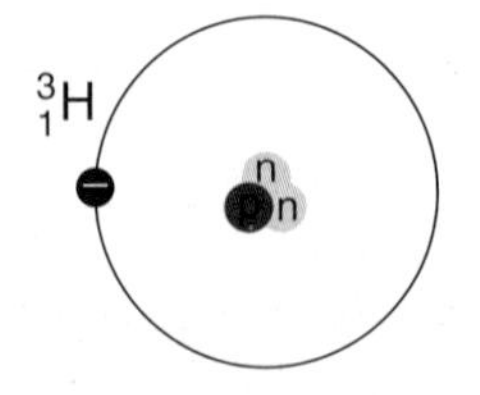

c. Tritium, an even rarer isotope of hydrogen, holds*two* neutrons in its nucleus along with the single proton.

Comparison of mass and weight • Figure 2.15

An object's mass does not change on Earth, on the moon, or in outer space, but its weight is different in each case.

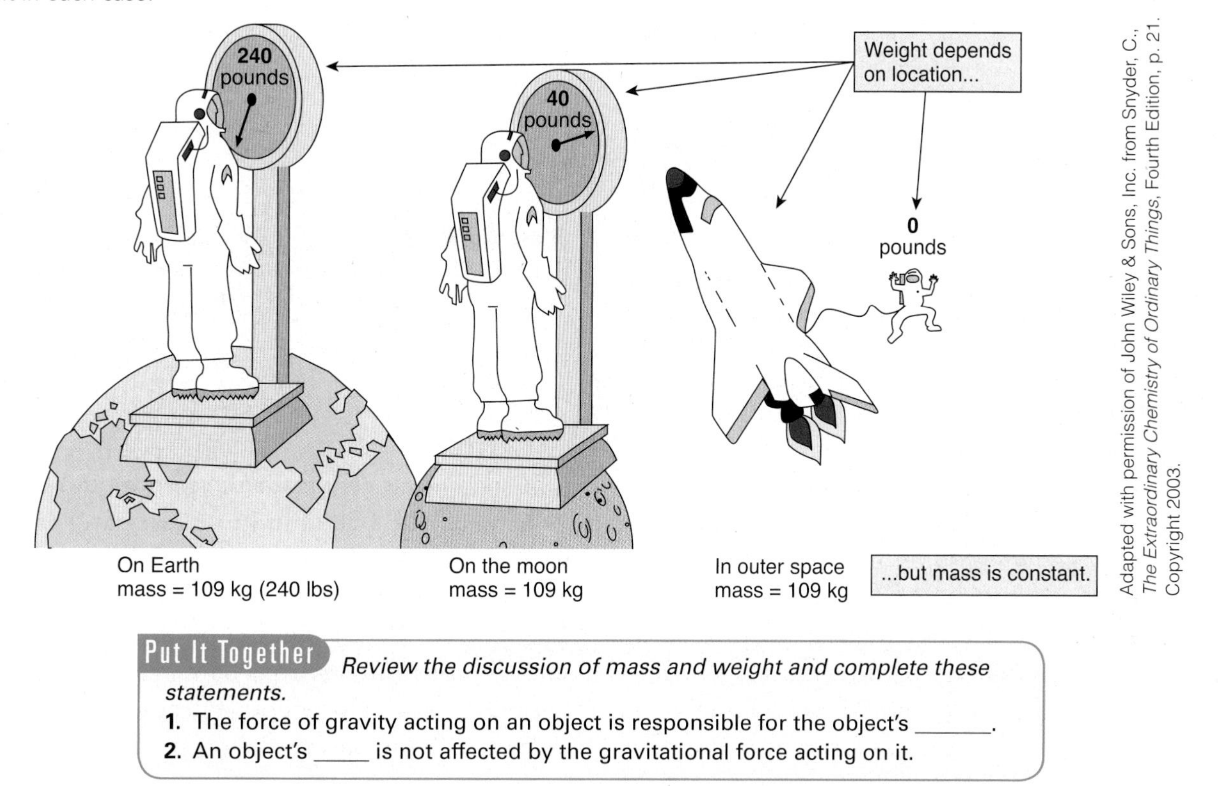

Put It Together *Review the discussion of mass and weight and complete these statements.*

1. The force of gravity acting on an object is responsible for the object's _______.
2. An object's _____ is not affected by the gravitational force acting on it.

forms almost all the rest, with tritium existing in only trace amounts.)

Other elements also exist as isotopes. With an atomic number of 6, all carbon atoms carry six protons in their nuclei. Like hydrogen, naturally occurring carbon also exists as three isotopes, with six, seven, and eight neutrons in their nuclei. Adding the protons and the neutrons gives us mass numbers of 12, 13, and 14:

$$^{12}_{6}\mathrm{C} \qquad ^{13}_{6}\mathrm{C} \qquad ^{14}_{6}\mathrm{C}$$

Instead of using different names for the various isotopes of all the elements beyond hydrogen, we can simply refer to each isotope by its symbol and its mass number joined by a hyphen: C-12, C-13, and C-14. Similarly, the three isotopes of hydrogen can be represented as H-1, H-2, and H-3.

Mass and Weight

In describing atoms and the particles that compose them, we've used two terms, mass and weight, that might seem identical or interchangeable. They're not.

Mass is the more fundamental of the two terms. It's an inherent property of any matter, any physical substance, anywhere in the universe. Your body has mass, a car has mass, a satellite orbiting in outer space has mass, and protons, neutrons, and electrons have mass. A body's **mass** reflects its resistance to acceleration or to any other change in its motion, wherever in the universe it might be. Mass is sometimes described more simply as the quantity of matter in a substance.

mass A measure of a body's resistance to acceleration.

weight A force due to the pull of gravity on an object.

Weight, on the other hand, is a product of the effect of gravity on a body. The weight of anything depends on where it is in the universe, and especially on what gravitational force is acting on it. If you weigh 60 kg (132 lb) on Earth, you would weigh only one-sixth of that—10 kg (22 lb)—on the moon because the moon's gravitational force is only one-sixth that of Earth. In an orbiting spaceship, you would weigh nothing at all since you would be essentially in free fall. Yet your body's mass would remain the same, 60 kg (132 lb) in all three places—on Earth, on the moon, and in the spaceship—because your mass is unaffected by gravity and by free fall. (**Figure 2.15**).

As you fill a grocery cart, both its mass and weight increase. It's easier to start pushing an empty cart because

Mass, force, and motion • Figure 2.16

To produce a given change in motion, you need to apply a large force to an object with large mass, but a small force to an object with small mass.

Adapted with permission of John Wiley & Sons, Inc. from Snyder, C., *The Extraordinary Chemistry of Ordinary Things*, Fourth Edition, p. 21. Copyright 2003.

Think Critically An empty shopping cart and a filled shopping cart are rolling down a hill. Does it take the same amount of force to stop each of them? If not, which one requires more force to bring it to a stop?

its mass is less than the mass of a full cart and it accelerates more readily (**Figure 2.16**). Similarly, pushing a real car is more difficult than pushing a small, toy car because of the much greater mass of the real car.

The mass of the substances of our everyday world has its foundation in the subatomic particles that form them—that is, their protons, neutrons, and electrons. Protons and neutrons of atomic nuclei have approximately the same mass: about 1.7×10^{-24} g each. Electrons are far lighter, with a mass of close to 9.1×10^{-28} g. For a more direct comparison with the mass of a proton and a neutron, we can rewrite the mass of an electron as 0.00091×10^{-24} g. A proton or neutron, then, is about $1.7 \div 0.00091$ or 1900 times as heavy as an electron, which is on the order of how much heavier this book is than a paper clip. For practical purposes, *virtually all the mass of an atom lies in its nucleus.*

To put these numbers into perspective, consider that the mass of a teaspoon of sugar is about 5 g. To match this mass we would need about 3×10^{24} protons or neutrons. That's about 3,000,000,000,000,000,000,000,000 (or 3 trillion *trillion*) of these particles.

Now let's simplify. Instead of writing out each of these complex numbers in units of grams, we can define a new unit, the **atomic mass unit** (abbreviated **u**). We'll define it in such a way that both the proton and the neutron (which are nearly identical in mass) have a mass of 1 u. To achieve this, we define the mass of an individual atom of the most common isotope of carbon, C-12, as exactly 12 u. Since C-12 contains 6 protons and 6 neutrons, and since protons and neutrons have nearly identical masses, this defines the mass of both a proton and a neutron as 1 u. **Table 2.2** summarizes the properties of the subatomic particles.

Atomic Mass

The mass of an individual atom or isotope can be expressed simply as its mass number in units of u. For example, an atom of H-3, with 1 proton and 2 neutrons, has a mass of 3 u. As you can see in **Figure 2.17**, we can use the mass numbers and the relative abundances of the naturally occurring isotopes of an element to calculate the **atomic mass** of that element. As you'll see in the next section, the names, symbols, atomic numbers, and atomic masses of the elements are vitally important information for everything you study in chemistry.

atomic mass The weighted average of the masses of all of the naturally occurring isotopes of an element.

Properties of subatomic particles Table 2.2

	Mass (g)	Mass (u)	Location in Atom	Charge	Symbol
Neutron	1.67×10^{-24}	1	Nucleus	0	n, n^0
Proton	1.67×10^{-24}	1	Nucleus	1+	p, p^+
Electron	0.0009×10^{-24}	0	Outside the nucleus	1−	e, e^-

In Words, Math, and Pictures

Calculating atomic mass • Figure 2.17

Here we show how to calculate the atomic mass of carbon.

Naturally occurring carbon exists as three isotopes: C-12 (which makes up about 98.9% of all the carbon in the universe), C-13 (which accounts for about 1.1% of all carbon), and C-14 (occurring in only trace amounts). Rounding these percentages gives us about 99 atoms of C-12 and about 1 atom of C-13 for every 100 carbon atoms in the universe.

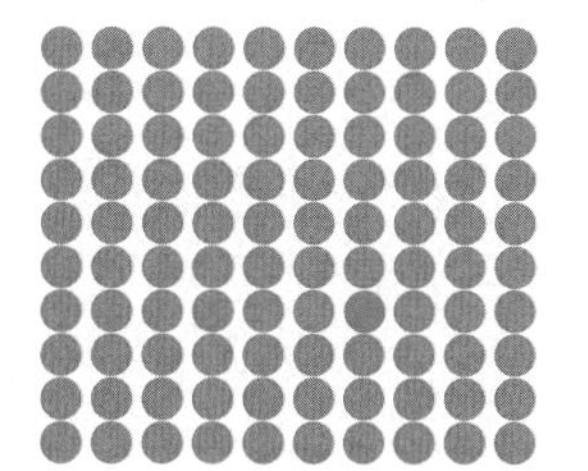

We can view this graphically by considering a set of 100 carbon atoms with a distribution of masses that's very nearly what we find in natural carbon throughout the universe. Here an orange dot represents a C-12 atom and a green dot represents an atom of the C-13 isotope.

We have 99 atoms of C-12 and one of C-13, which gives us 100 carbon atoms with a total mass of

$$(99 \times 12 \text{ u}) + (1 \times 13 \text{ u}) = 1201 \text{ u}.$$

The average mass for all 100 atoms is

$$\frac{1201 \text{ u}}{100} = 12.01 \text{ u}$$

(Total mass: 1201 u; Number of atoms: 100)

which represents the approximate atomic mass of naturally occurring carbon.

KNOW BEFORE YOU GO

1. The element, lithium, widely used in many types of rechargeable batteries, has two naturally occurring isotopes, Li-6 and Li-7. For every 100 naturally occurring lithium atoms, approximately 8 are Li-6 and 92 are Li-7. Use this information to calculate the atomic mass of lithium.

CONCEPT CHECK STOP

1. **How** do the isotopes of a given element differ from one another?
2. **Why** does the weight of an object depend on where it is in the universe?
3. **What** is the relative mass of a proton or neutron as compared to that of an electron?

2.4 The Periodic Table

LEARNING OBJECTIVES

1. **State** what makes one element different from the next element in the periodic table.
2. **Describe** in general terms the organizing principles of the periodic table.
3. **Explain** why elements in the same column of the periodic table often show similar behavior.

Today we know of more than 115 elements, both natural (about 90) and those made by humans. We can view some of their properties, their variety, and the relationships among them by following a trail of the first six, starting with hydrogen, atomic number 1

Atomic structure and properties of the first six elements • Figure 2.18

Each element has only one more proton (and one more electron and perhaps a different number of neutrons) than the preceding element. But notice that this small difference can produce a huge change in the properties of the element.

One electron in the first quantum shell

3D Sculptor/iStock photo

Hydrogen:
A highly flammable gas that has been used to lift space vehicles into orbit

Two electrons fill the first quantum shell

MARIA STENZEL/National Geographic Creative

Helium:
An unreactive gas that is less dense than air

With the first shell filled, the third electron enters the second shell

Ryan Mc Vay/ Media bakery

Lithium:
A light weight metal whose compounds are used in many types of batteries

NASA

Beryllium:
A light weight metal that has been used as a heat-shield for rocket reentry modules.

Bryan Myhr/E+/ Getty Images

Boron:
Commonly found combined with other elements in compounds. Boron carbide is a very hard, light weight material used in body armor.

gregstanfield/iStockphoto

Carbon:
An element central to life. Occurs in its elemental form in diamonds, graphite, and charcoal.

(see **Figure 2.18**). In each case we move from one element to the next one simply by adding one proton, one electron, and varying numbers of neutrons.

As we move beyond these first six elements and compare them with the remaining elements, we find some striking similarities. Helium (He, atomic number 2) and neon (Ne, atomic number 10) are both gases, quite inert in our everyday world. Lithium (Li, atomic number 3) and sodium (Na, atomic number 11) are both metallic solids that react vigorously with water to liberate and ignite hydrogen gas. Furthermore, the atomic numbers of He (2) and Ne (10) and the atomic numbers of Li (3) and Na (11) differ by the same value, eight.

Similarities of this sort continue and grow even more remarkable as we look further into the series of elements. Along with He and Ne, other ordinarily inert gases include argon (Ar, atomic number 18), krypton (Kr, 36), and xenon (Xe, 54). Also, as with Li and Na, other metallic solids that react similarly (and even more vigorously) with water include potassium (K, 19), rubidium (Rb, 37), and cesium (Cs, 55).

If we calculate the differences in atomic numbers for all the adjacent elements we have just described, we find these differences to be identical for each of the two series (**Figure 2.19**). Repetitions of this sort—repetitions in properties or characteristics that occur again and again over time or through members of a long series—are known as periodic repetitions. The brightness of the sky varies periodically each day as dawn leads to noon, then dusk and nightfall, and then to dawn again the next day. As the years go by, days grow longer and warmer periodically as spring leads to summer, then shorter and colder as autumn passes into winter. Daylight and temperature vary periodically as time passes. Flowers bloom periodically during the spring of each year, and farms are harvested periodically in the fall of each year.

In 1869 Dmitri Mendeleev, a Russian professor of chemistry, recognized periodicities of this sort in the properties of the 63 elements known at that time. Because atomic numbers were not identified until 1913, Mendeleev had to work with a sequence of atomic masses rather than atomic numbers. He created a table, now known as the **periodic table** of the elements, in which he placed the known elements into a series of horizontal rows so that their atomic masses increased regularly, with a few exceptions, from left to right. This generated a series of vertical columns that contained elements with similar properties. Today we call the table's rows its **periods** and its columns **groups**. The column that includes He and Ne is known as the **inert** or **noble gas** group because of their almost complete lack of chemical reactivity. The column including Li and Na is the **alkali metal** group.

periodic table An orderly arrangement of all chemical elements into rows (called periods) and columns (called groups or families).

periods Rows in the periodic table.

groups Columns in the periodic table.

Periodic repetitions • Figure 2.19

Similarities in the properties of elements often repeat themselves periodically throughout the periodic table. Within each column, adjacent atomic numbers also differ by the same numbers from one column to the next.

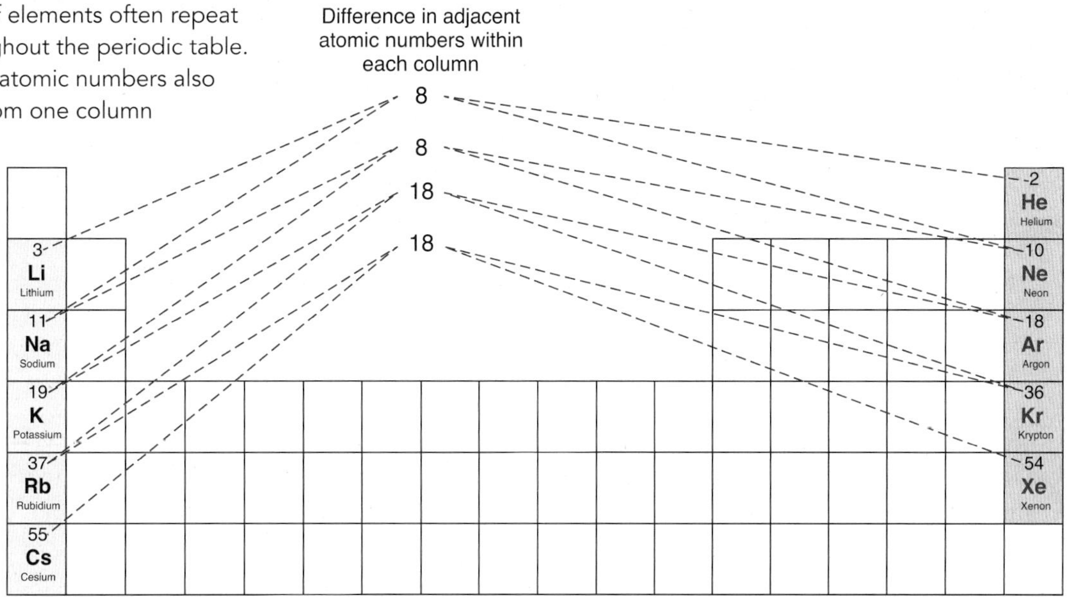

A version of the modern periodic table • Figure 2.20

The periodic table is the basic reference source in chemistry.

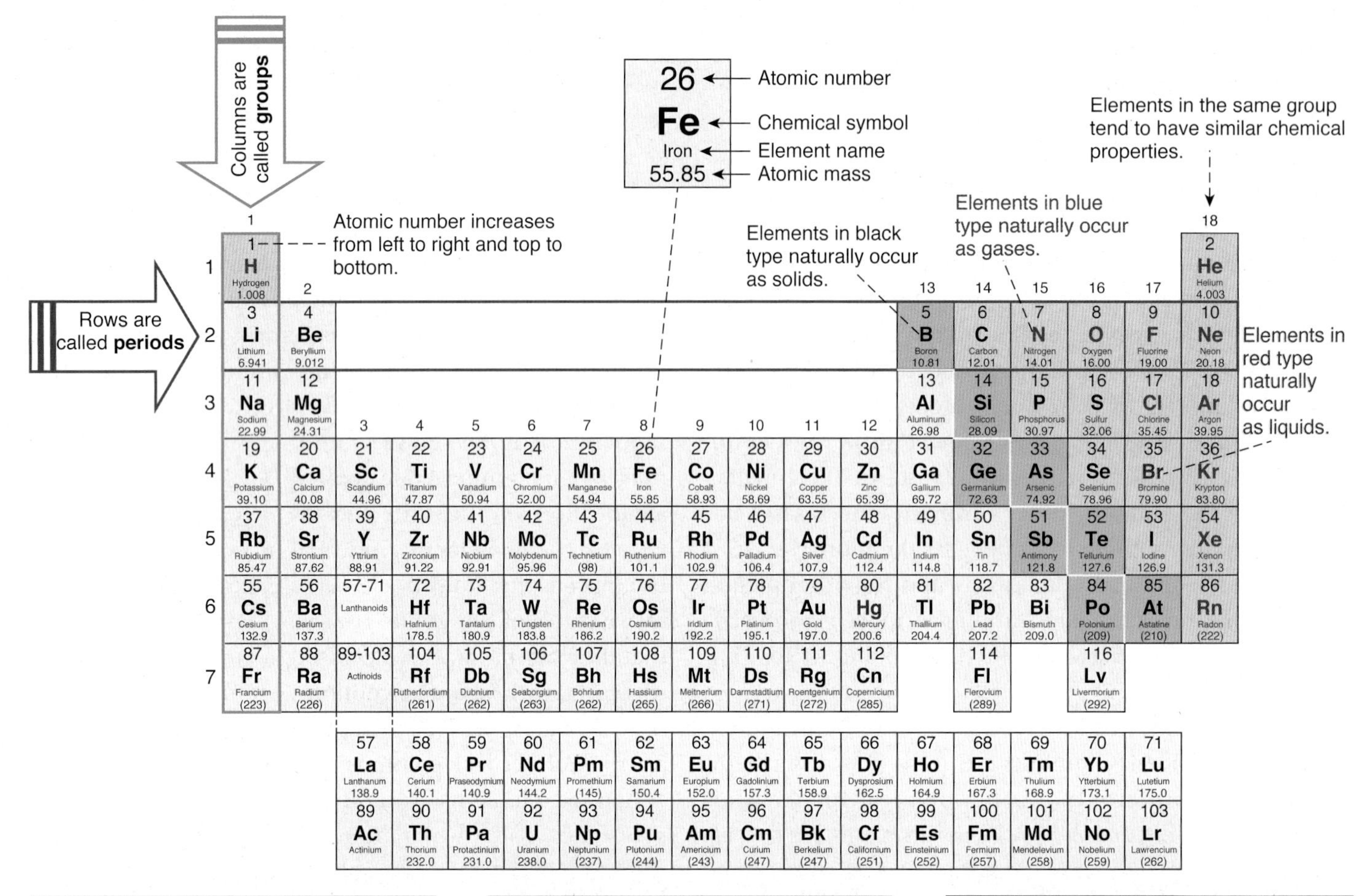
Columns are called **groups**
Rows are called **periods**
26 Fe Iron 55.85
Atomic number
Chemical symbol
Element name
Atomic mass
Atomic number increases from left to right and top to bottom.
Elements in black type naturally occur as solids.
Elements in blue type naturally occur as gases.
Elements in the same group tend to have similar chemical properties.
Elements in red type naturally occur as liquids.
Lanthanoids
Actinoids

Metals tend to be shiny solids and are good conductors of heat and electricity.

Gold vault at the Federal Reserve Bank of New York.

RANDY OLSON/ NationalGeographic Creative

Metalloids share properties of metals and nonmetals.

A silicon wafer. The *semiconductors*, silicon and germanium, are widely used in computers and electronic devices.

nicolas_/iStockphoto

Nonmetals are often gases or brittle solids and are poor conductors of heat and electricity.

A helium-filled weather balloon is launched over the Antarctic.

MARIA STENZEL/National Geographic Creative

Interpret the Data

1. Are most elements nonmetals or metals?
2. Identify the period containing the element potassium (K).
3. Identify the group containing the element phosphorus (P).
4. Identify the symbols for gold, silicon, germanium, and helium.
5. Which two elements naturally occur as liquids?

One atomic property—the number of electrons located in the outermost quantum shell of each atom—varies with the location of the column and is the same for all members of each of the groups in the periodic table. We call these outermost electrons **valence electrons**. In the next chapter we'll examine the chemical consequences of this characteristic. We will find that the number of an atom's valence electrons determines much of the chemical reactivity of the element.

Because hydrogen atoms share an important characteristic with other members of the alkali metal group—a single electron in its outermost quantum shell—hydrogen is placed at the top of the alkali metal column, even though it shares no other property with the other members of this family.

In the years since Mendeleev's work, many others have refined and expanded his periodic table. For example, to maintain a consistent set of properties within groups, Mendeleev had to leave some of the boxes empty. He predicted the discovery of elements that would fill these boxes and correctly described many of the properties they would exhibit. Later, with continuing work, new elements were discovered that fit perfectly within those empty boxes.

Also, to maintain periodicity Mendeleev had to reverse the atomic mass sequence occasionally within the table. With the later establishment of periodicity based on atomic numbers rather than masses, a steadily increasing series of atomic numbers now occurs throughout the table. Most modern versions of the periodic table contain an element's atomic number, chemical symbol, and atomic mass within each box. A full version of the table appears in **Figure 2.20**.

The Elements of Our Environment

Of the approximately 90 naturally occurring elements, most occur as part of chemical compounds, with atoms of different elements bonded to one another. Silicon, for example, is most commonly found in combination with oxygen in the form of the compound *silicon dioxide* (SiO_2). Moreover, the elements are not uniformly abundant in our physical environment. Just a few of the elements are far more represented than most, as we'll see shortly.

We find it convenient to categorize our physical environment in terms of three regions: Earth's crust, atmosphere, and hydrosphere. The **crust** is the relatively thin, surface layer of Earth—the familiar combination of rocks, sand, dirt, and all the other solid substances we walk on, pave over, build our buildings on, and plant our crops on. This crust holds the **hydrosphere**, the oceans, rivers, lakes, and groundwater of our environment. Above the crust and hydrosphere lies the **atmosphere**, the mixture of gases that surrounds Earth. **Figure 2.21** illustrates our physical environment and the most common elements found within it.

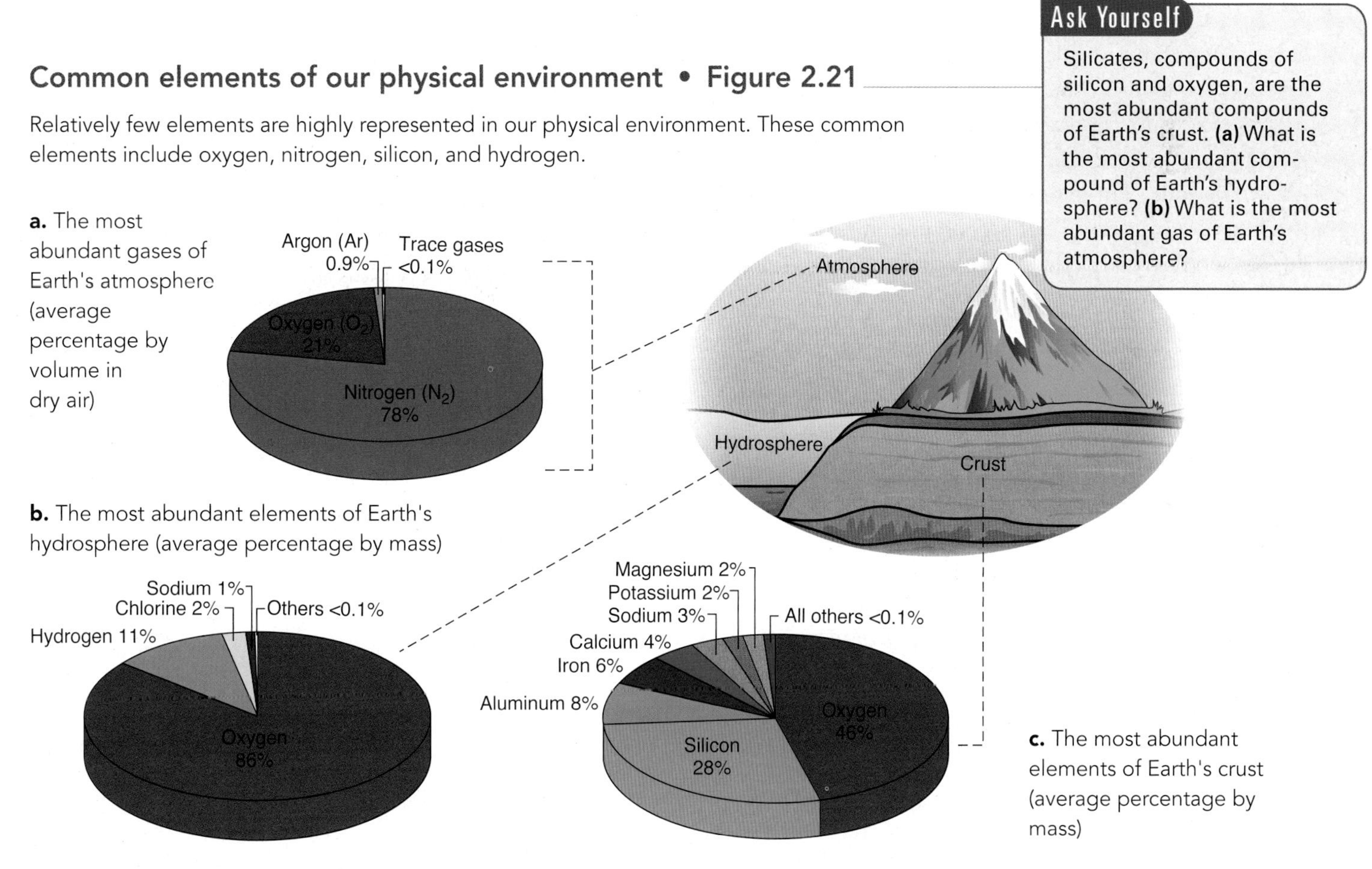

Common elements of our physical environment • Figure 2.21

Relatively few elements are highly represented in our physical environment. These common elements include oxygen, nitrogen, silicon, and hydrogen.

Ask Yourself

Silicates, compounds of silicon and oxygen, are the most abundant compounds of Earth's crust. **(a)** What is the most abundant compound of Earth's hydrosphere? **(b)** What is the most abundant gas of Earth's atmosphere?

Elemental composition of the human body • Figure 2.22

Elements within our bodies exist mostly as part of chemical compounds. For example, the compound water (H_2O) makes up nearly two-thirds of the mass of the body. Our soft tissues are rich in carbon-based compounds.

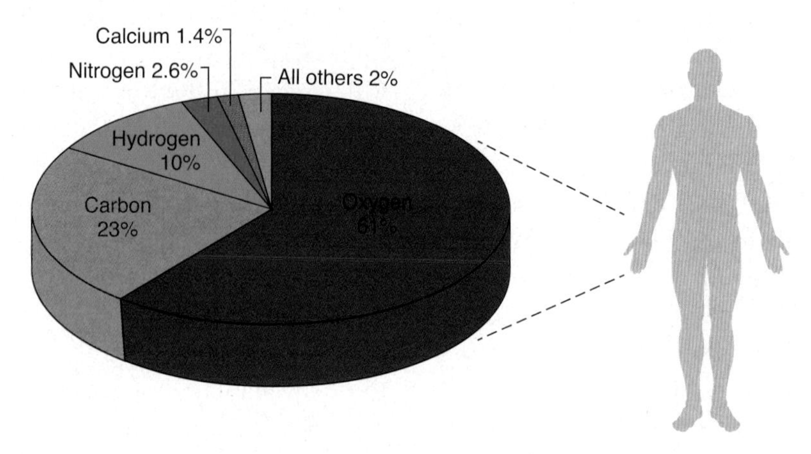

The most abundant elements of the human body (average percentage by mass)

The Elements of Living Things

All life as we know it, whether animal or plant, is predominantly based on four elements—oxygen, carbon, hydrogen, and nitrogen. These four elements comprise over 96% of the mass of our bodies (**Figure 2.22**).

In addition to the elements highlighted in Figure 2.22, other elements play critical roles in our health, but are present within our bodies at far lower levels. These include sodium, potassium, magnesium, iron, and zinc, which exist within the body as charged particles known as **ions**. The element phosphorus is also critical for life, forming part of the gene-encoding compound, DNA (deoxyribonucleic acid), and the energy-releasing agent, ATP (adenosine triphosphate), both found within our cells.

CONCEPT CHECK

1. **How** would you describe an element's nearest neighbors to the right or left in the periodic table?
2. **What** determines the vertical and horizontal location of an element in the periodic table?
3. **How** is a given element similar to the other elements in its column of the periodic table? Give examples.

Summary

1 Atomic Structure 28

- ***How have we come to understand that matter is composed of atoms?***
  Two ancient Greek philosophers, Aristotle and Democritus, proposed opposing theories of matter: Aristotle viewed matter as infinitely divisible, whereas Democritus conceived of matter as "atomistic." Aristotle's notion of matter dominated Western thought for two millennia. Then, in the early 1800s, Dalton began the development of our modern view by concluding that elements are, indeed, made up of **atoms** (as Democritus had proposed) and that these atoms can bond to each other to form compounds. Further investigations by Thomson, Rutherford, and Bohr refined this atomic model. The quantum mechanical model, even more recent, describes the behavior of electrons within the atom.

- ***What is the structure of the atom?***
  Thomson discovered that all atoms carry negatively charged **electrons**. Rutherford showed that atoms consist of a central, extremely small and highly dense, positively charged

nucleus, with (negatively charged) electrons orbiting around this nucleus. Bohr refined this model (depicted here) by proposing that electrons reside in **quantum shells** located outside the nucleus and at distinct distances from it.

Figure 2.8 • Variations of the Rutherford and Bohr models of the atom

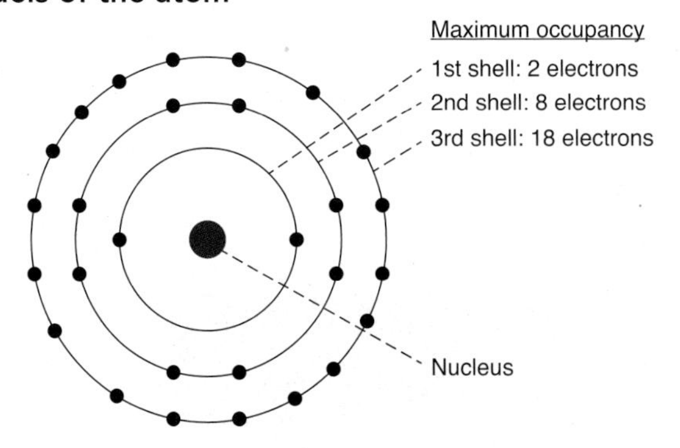

Visualizations of the atomic model often show the nucleus as too large compared to the surrounding electron shells, which define the outer dimensions of an atom. The diameter of an atom is about 100,000 times the diameter of its nucleus.

2 The Nucleus 37

- ***What defines an element?***
  All atoms of any given **element** contain the same number of **protons**, which are positively charged nuclear particles. The **atomic number**, which identifies an element and which appears as a subscript immediately preceding the element's symbol, represents the number of protons in the nucleus of each of the element's atoms.
- ***What is mass number, and how does it relate to an atom's nucleus?***
  Mass number is the combined number of protons and neutrons in an atomic nucleus. It appears as a superscript immediately preceding an element's symbol. A **neutron** resembles a proton except that the neutron carries no electrical charge. At least one neutron resides in all nuclei except in the nuclei of the simplest and most abundant isotope of hydrogen, symbolized as shown.

Mass number ⟶ $^{1}_{1}H$
Atomic number ⟶

3 Isotopes and Atomic Mass 40

- ***What are isotopes?***
  Isotopes are atoms of a given element that differ in the number of neutrons they contain and therefore in their mass numbers. Deuterium and tritium, depicted here, are isotopes of hydrogen. The mass of an isotope can be estimated based on the number of protons and neutrons in its nucleus.

Figure 2.14 • The isotopes of hydrogen

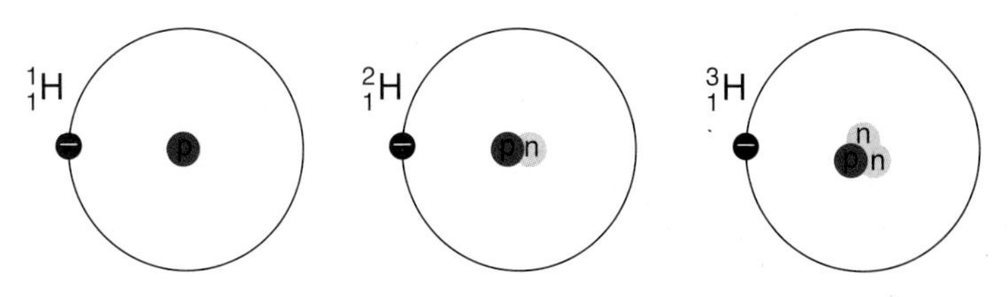

- ***How do we determine the atomic mass of an element?***
  The **atomic mass** of an element is the average of the masses of all the naturally occurring isotopes of that element, weighted for their individual abundances.

4 The Periodic Table 43

- ***How are elements organized in the periodic table?***
  The **periodic table**, a portion of which is depicted here, consists of rows and columns of all known elements. As we move from left to right within each row, the sequential elements within the row increase by one unit of atomic number. Each column consists of elements with similar chemical properties—except for hydrogen, which does not resemble any of the other elements in its column.

Figure 2.20 • A version of the modern periodic table

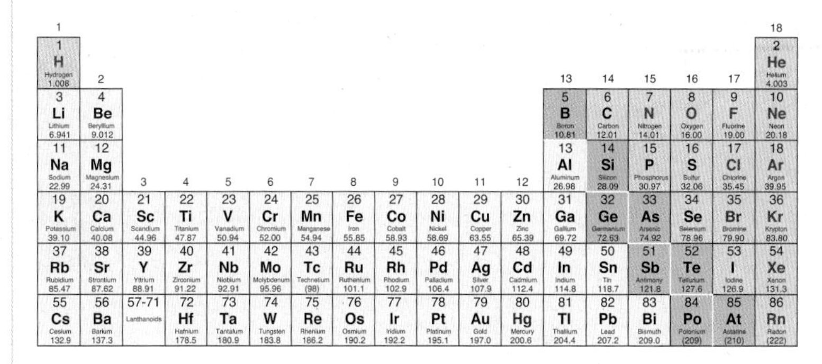

The term *periodic table* reflects the periodicity, or repeating pattern, of the elements as they appear in the rows and columns of this chart. For example, for helium, neon, argon, krypton and xenon, a series of gases with little or no reactivity, the difference in atomic numbers—8, 8, 18, 18—is mirrored by the difference in atomic numbers of lithium, sodium, potassium, rubidium and cesium, a series of metals that react similarly with water.

Key Terms

- atom 28
- atomic mass 42
- atomic number 38
- atomic structure 30
- electrons 30
- element 38
- groups 45
- isotopes 40
- mass 41
- mass number 38
- neutron 38
- nucleus 32
- periodic table 45
- periods 45
- proton 37
- quantum shells 32
- weight 41

What is happening in this picture?

The fabrication of computer chips, or integrated circuits, requires silicon with a purity exceeding 99.9999%. Semiconductor elements, such as silicon, are particularly suited for making computer chips because they can be easily modified to either conduct or block the flow of electricity.

Here, a clean-room technician inspects a cylindrical ingot of ultrapure silicon crystal. This cylinder will be sliced into very thin, circular cross sections or *wafers*, which will then be used in making integrated circuits.

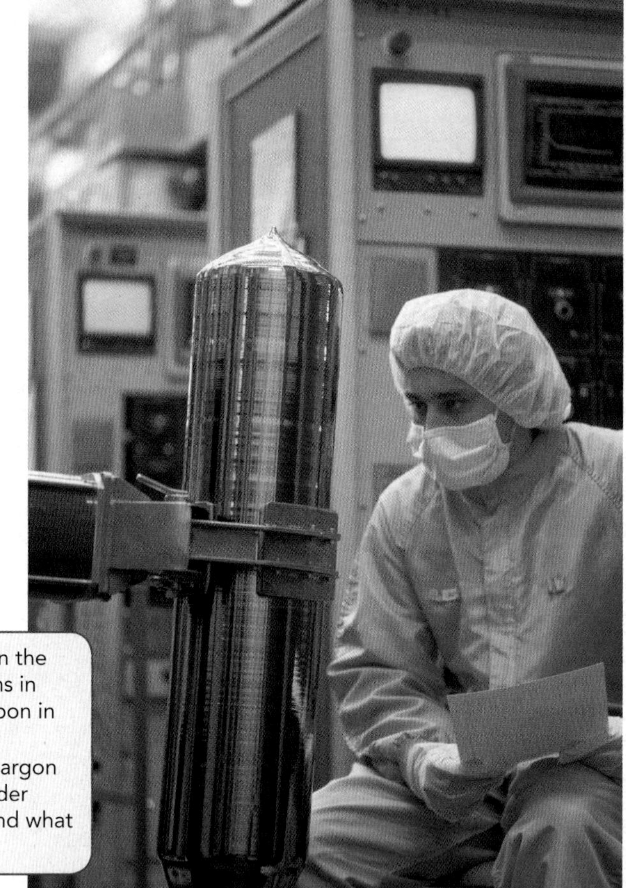

Michael Rosenfeld/Maximilian S

Think Critically 1. The arrangement of silicon atoms in the crystal form resembles the arrangement of carbon atoms in diamond, a pure form of carbon. Locate silicon and carbon in the period table and identify their group number.

2. Purified silicon is produced under an environment of argon gas to exclude oxygen, which can react with silicon. Under which group of the periodic table does argon belong and what property of argon makes it suitable for this use?

Exercises

Review

1. What was the major contribution of each of the following to our understanding of the composition and structure of the atom?

a. Ernest Rutherford

b. Niels Bohr

c. J. J. Thomson

2. The name *Aristotle* is not among those used in Question 1. Explain why not.

3. What contribution did each of the following make to our knowledge or our understanding of chemistry?

a. Dmitri Mendeleev

b. Erwin Schrödinger, Max Born, and Werner Heisenberg, as a group

4. What did Democritus and John Dalton have in common in their contributions to our understanding of modern chemistry?

5. Describe what each of the following represents or how it is determined.

a. an atom's atomic number

b. an atom's mass number

c. an element's atomic mass

6. Explain what each of the following represents.

a. the superscript that appears to the left of an elemental symbol

b. the subscript that appears to the left of an elemental symbol

7. What subatomic particles occupy an atom's quantum shells?

8. In what way or ways are protons and neutrons similar to each other? What is the most important difference between them?

9. Of all the subatomic particles that form an atom, which subatomic particle contributes *least* to the atom's mass?

10. What element is used to define the atomic mass unit (u)?

11. What distinguishes one isotope of an element from another isotope of the same element?

12. Write the chemical symbols for the three naturally occurring isotopes of hydrogen with appropriate superscripts and subscripts.

13. The nucleus of ${}^{1}_{1}H$ is different from the nuclei of all the atoms of all the other elements. In what way is the nucleus of ${}^{1}_{1}H$ different from all other nuclei?

14. Two elements are located next to each other within the same row of the periodic table. In what way *must* atoms of one of these two elements differ from atoms of the other element?

15. The nucleus of the atom shown here contains four protons, shown in red, and five neutrons, shown in gray. What is the atomic number of this atom? What is its mass number? What element does it represent? Write the elemental symbol, with the appropriate subscript and superscript.

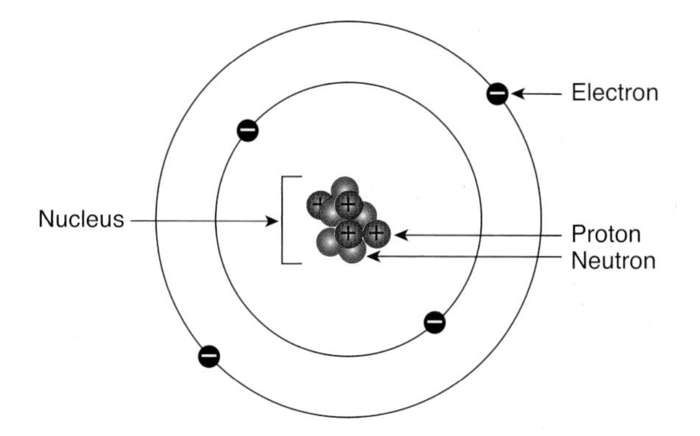

16. Explain the origin of the colored lines observed in the line spectrum emitted by hydrogen when high-voltage electricity is passed through this element.

17. What names are given to the *least* reactive group of elements of the periodic table?

18. What significance does the value of 2 (n^2) have in determining the structure of an atom of any particular element? What does the n of 2 (n^2) represent?

19. How do we define:

a. weight;

b. mass?

If an object weighs one pound on Earth and we transfer this object to the moon, which will change:

a. both its weight and its mass;

b. neither its weight nor its mass;

c. its weight but not its mass;

d. its mass but not its weight?

Explain your answer.

20. In what common commercial product would we find the element lithium?

21. a. Identify the group of the periodic table containing the element calcium (Ca) and list three other elements in this group.

b. How are the elements aluminum (Al), silicon (Si), and phosphorus (P) alike in terms of their location in the periodic table? How are these elements different in terms of metallic properties?

c. Identify the symbols for the elements nitrogen, neon, sodium, and nickel.

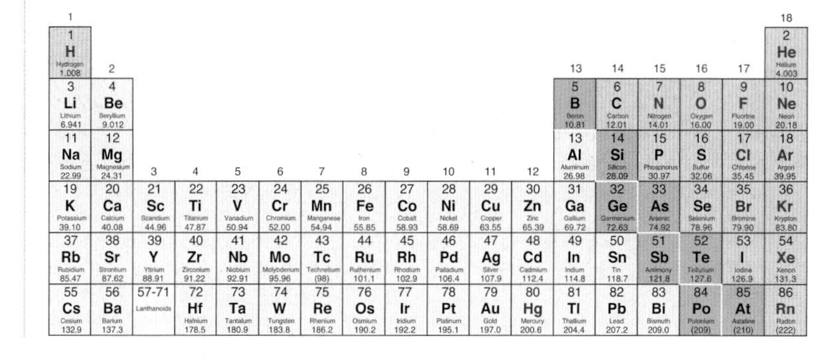

1	2	3	4	5	6	7	8	9	10	11	12	13	14	15	16	17	18
1 H Hydrogen 1.008																	2 He Helium 4.003
3 Li Lithium 6.941	4 Be Beryllium 9.012											5 B Boron 10.81	6 C Carbon 12.01	7 N Nitrogen 14.01	8 O Oxygen 16.00	9 F Fluorine 19.00	10 Ne Neon 20.18
11 Na Sodium 22.99	12 Mg Magnesium 24.31											13 Al Aluminum 26.98	14 Si Silicon 28.09	15 P Phosphorus 30.97	16 S Sulfur 32.06	17 Cl Chlorine 35.45	18 Ar Argon 39.95
19 K Potassium 39.10	20 Ca Calcium 40.08	21 Sc Scandium 44.96	22 Ti Titanium 47.87	23 V Vanadium 50.94	24 Cr Chromium 52.00	25 Mn Manganese 54.94	26 Fe Iron 55.85	27 Co Cobalt 58.93	28 Ni Nickel 58.69	29 Cu Copper 63.55	30 Zn Zinc 65.39	31 Ga Gallium 69.72	32 Ge Germanium 72.63	33 As Arsenic 74.92	34 Se Selenium 78.96	35 Br Bromine 79.90	36 Kr Krypton 83.80
37 Rb Rubidium 85.47	38 Sr Strontium 87.62	39 Y Yttrium 88.91	40 Zr Zirconium 91.22	41 Nb Niobium 92.91	42 Mo Molybdenum 95.96	43 Tc Technetium (98)	44 Ru Ruthenium 101.1	45 Rh Rhodium 102.9	46 Pd Palladium 106.4	47 Ag Silver 107.9	48 Cd Cadmium 112.4	49 In Indium 114.8	50 Sn Tin 118.7	51 Sb Antimony 121.8	52 Te Tellurium 127.6	53 I Iodine 126.9	54 Xe Xenon 131.3
55 Cs Cesium 132.9	56 Ba Barium 137.3	57-71 Lanthanoids	72 Hf Hafnium 178.5	73 Ta Tantalum 180.9	74 W Tungsten 183.8	75 Re Rhenium 186.2	76 Os Osmium 190.2	77 Ir Iridium 192.2	78 Pt Platinum 195.1	79 Au Gold 197.0	80 Hg Mercury 200.6	81 Tl Thallium 204.4	82 Pb Lead 207.2	83 Bi Bismuth 209.0	84 Po Polonium (209)	85 At Astatine (210)	86 Rn Radon (222)

Think

22. Can two atoms of two different elements be isotopes of each other? Explain.

23. Explain why atoms of two different elements *can* have the same mass number but *cannot* have the same atomic number.

24. All elements are referred to by their names. For example, we have carbon, hydrogen, nitrogen, oxygen, and so on. Another way of identifying the various elements would be by numbers. Explain how we could identify the various elements by numbers and what specific numbers we would use to identify carbon, hydrogen, nitrogen, and oxygen.

25. Hydrogen has few if any chemical or physical properties similar to any of the other elements the first column (or group) of the periodic table, the alkali metals that include lithium and sodium. Why, then, is hydrogen placed in the same column as these alkali metals?

26. How does Ernest Rutherford's concept of an atom differ from J. J. Thomson's? How does Niels Bohr's concept of an atom differ from J. J. Thomson's?

27. One of the principal rules of science is that particles with opposite electrical charges attract each other and those with the same electrical charges repel each other. Why, then, aren't all the negatively charged electrons that occupy the quantum shells of a neutral atom of any element pulled directly into the nucleus by the attraction of the positive charges of all the protons within the nucleus? Who developed the model of an atom that supports your explanation?

28. Suppose that the electron in the valence shell of an atom of $^{1}_{1}H$ was pulled into the nucleus and combined with the nuclear proton. What known subatomic particle would you expect to be produced? Explain your answer.

29. Suppose you have two atoms that represent isotopes of the same element. What characteristic(s) do they have in common? In what characteristic(s) do they differ from one another?

30. For an atom of tritium, write the symbol of the element it represents with the appropriate superscript and subscript to indicate its mass number and atomic number. Now suppose you are able to convert one of the neutrons in tritium's nucleus into a proton. Write the symbol of the new element with the appropriate superscript and subscript for the isotope you have generated.

31. Write the symbol, superscript and subscript for an atom of C-13. Now write the symbol, superscript, and subscript for the isotope of the element that would be generated if you converted one of the protons in the C-13 nucleus into a neutron.

32. If you smash together one deuterium atom and one tritium atom with enough energy, you can cause them to fuse together, forming a single new atom and ejecting one neutron in the process. The new atom formed in this reaction is no longer a hydrogen atom but is an atom of another, well-known element, generated by the combination of the deuterium and tritium nuclei and the simultaneous loss of the single neutron. Write the chemical symbol of this newly formed atom along with a subscript and a superscript to show its atomic number and its mass number. (The nuclear reaction described here is part of the process involved in the detonation of the hydrogen bomb.)

33. Suppose you have, here on the surface of Earth, a sphere of cork and a sphere of lead. Each sphere has a mass of exactly 1 kg.

a. Do the two spheres have the same size as well as the same mass? If your answer is Yes, explain why. If your answer is No, identify the larger sphere.

b. What is the weight of each?

c. Now you move them both to a location of the surface of the moon. What is the weight of each? What is the mass of each?

34. Examine the periodic table of Figure 2.16 and describe what is unusual about all four of the following pairs of elements: Ar and K (atomic numbers 18 and 19); Co and Ni (27 and 28); Te and I (52 and 53); Th and Pa (90 and 91). What explains this unusual characteristic?

35. When a baseball is hit by a swung bat, the baseball flies off into the baseball field. Given that the atoms of the materials that compose both the baseball and the bat consist mostly of empty space, suggest a reason why the bat doesn't simply pass harmlessly through the baseball.

36. Figure 2.7 shows a hydrogen line spectrum containing five clearly visible lines. Each line corresponds to energy emitted as an electron moves from one quantum shell to another. Yet the hydrogen atom contains only one electron, typically shown residing in the $n = 1$ quantum shell. How might this apparent contradiction be explained?

Calculate

37. The diameter of an atomic nucleus is on the order of 10^{-15} m, whereas the diameter of an atom itself is on the order of 10^{-10} m due to the size of quantum shells occupied by the atom's electrons. Suppose you wanted to build a scale model of a typical atom and used a beach-ball with a diameter of 1 m as the model's nucleus. How large would the entire atomic model have to be?

38. Provide the information missing from the following table. Assume that all atoms are electrically neutral.

Notation	$^{11}_{5}B$		$^{58}_{28}Ni$	
Atomic number				10
Mass number				
Number of neutrons		20		12
Number of electrons		20		

39. How many protons are there in 1 g of protons?

40. If the atomic mass of the element boron is 10.8 u, and there are only two naturally occurring boron isotopes, B-10 and B-11, what is the percent abundance of each isotope?

41. The following chart shows the elemental composition of the human body by mass.

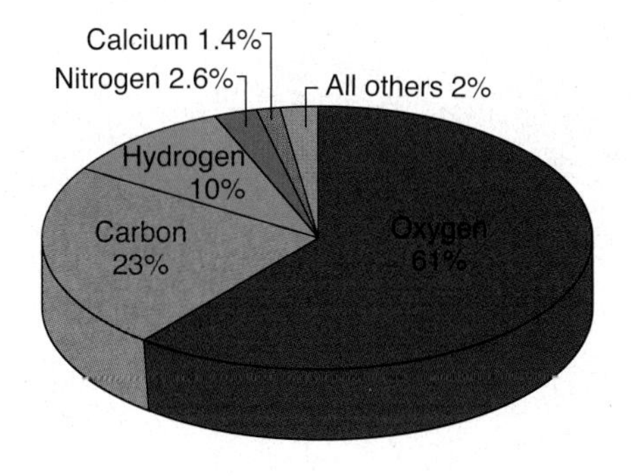

Use this information to answer the following questions:

a. Calculate the number of kilograms of oxygen, carbon, hydrogen, and nitrogen, respectively, in a person with a mass of 60 kg.

b. Calculate the number of carbon *atoms* in a 60 kg person. Note that 1.67×10^{-27} kg equals 1 u, and 12.01 u is the atomic mass of carbon, or the average mass of a carbon atom.

c. The element hydrogen makes up just 10% of our body by mass, but a much larger percentage in terms of the number of atoms present. Explain why.

42. Suppose you want to write out, in digits, the number represented by 10^{23}. Describe, in your own words, how you would write it. What numerals would you use, how many of each would you write, and in what order would you write them?

43. Since virtually all the mass of an atom lies in its nucleus, the mass of any object is effectively due to the nuclear particles (protons and neutrons) of the atoms that make up the object. How many combined protons and neutrons would you estimate to exist in a cell phone with a mass of 105.0 g?

3 Chemical Compounds

Most of the matter around us is made up of **chemical compounds**, substances in which atoms of different elements are bonded to one another. For instance, water (H_2O) is a compound composed of hydrogen and oxygen atoms bonded to each other.

We can explore compounds of everyday life through this image. The surfer's board and swimsuit are made up of polymers, chemical compounds formed by the repeated combination of smaller units. Polymers come in various types, including polyurethanes,

commonly used in surfboards; and polyesters, used in clothing and other materials. The deck of the surfboard has a coating of surf wax—a substance containing hydrocarbons—to help prevent slipping. Just below the surface of the water lie coral reefs, composed largely of *calcium carbonate* ($CaCO_3$). The saltiness of the ocean water results from the presence of *sodium chloride* (NaCl), along with minor amounts of other dissolved salts.

In this chapter we'll explore how atoms bond with one another to form compounds, and we'll examine the differences in bonding within substances such as water and sodium chloride. We also discuss bonding within compounds of carbon, known as organic compounds.

Paul Kennedy/Getty Images

CHAPTER OUTLINE

3.1 Periodicity and Valence Electrons

LEARNING OBJECTIVES

1. **Compare** electron configurations of elements in the same column of the periodic table.
2. **Define** valence electrons.
3. **Relate** valence electrons to Lewis structures of elements.

In Chapter 2, we saw that the organization of chemical elements within the periodic table is based on atomic numbers, which increase by one as we pass from one element to the next, from left to right, in each row. With each unit increase in atomic number, a corresponding increase of one electron occurs in the quantum shell being filled. In this section, we'll see that all atoms of elements located in the same column of the periodic table have the same number of electrons in their outermost shell, which has important consequences for chemical behavior.

Valence Electrons

electron configuration The arrangement of an atom's electrons in its quantum shells.

We can see changes in **electron configuration** as we look at those members of the first 18 elements—those with atomic numbers from 1 to 18—that occupy the first two and last two groups of the periodic table (**Figure 3.1**). Hydrogen— H, atomic number 1— has one proton in its nucleus and one electron in its first

PROCESS DIAGRAM

Electron configurations: Filling up quantum shells • Figure 3.1

Electron configurations of elements increase in complexity from left to right and from top to bottom of the periodic table.

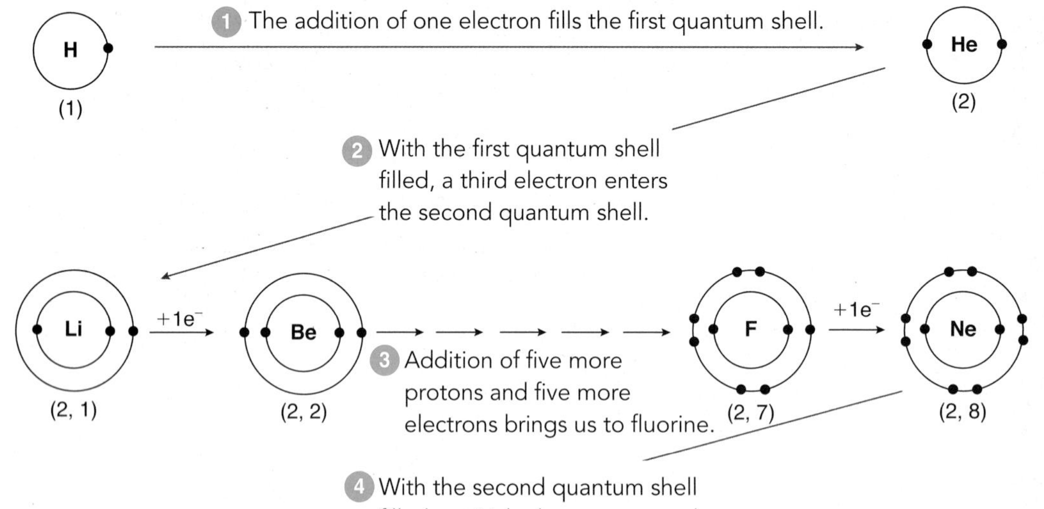

H
(1)
1 The addition of one electron fills the first quantum shell.
He
(2)
2 With the first quantum shell filled, a third electron enters the second quantum shell.
Li
(2, 1)
+1e⁻
Be
(2, 2)
3 Addition of five more protons and five more electrons brings us to fluorine.
F
(2, 7)
+1e⁻
Ne
(2, 8)
4 With the second quantum shell filled, an 11th electron enters the third quantum shell.

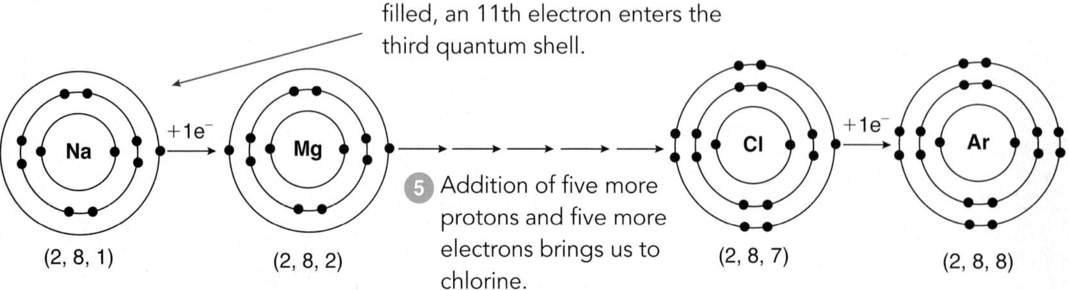

Na
(2, 8, 1)
+1e⁻
Mg
(2, 8, 2)
5 Addition of five more protons and five more electrons brings us to chlorine.
Cl
(2, 8, 7)
+1e⁻
Ar
(2, 8, 8)

Interpret the Data

1. From the figure, indicate the number of electrons in the outermost quantum shell of each element in the following sets of elements: (a) H, Li, Na; (b) Be, Mg; (c) F, Cl; (d) Ne, Ar.
2. From the figure, indicate the maximum electron occupancy of the: (a) 1st quantum shell; (b) 2nd quantum shell.

Alkali metals • Figure 3.2

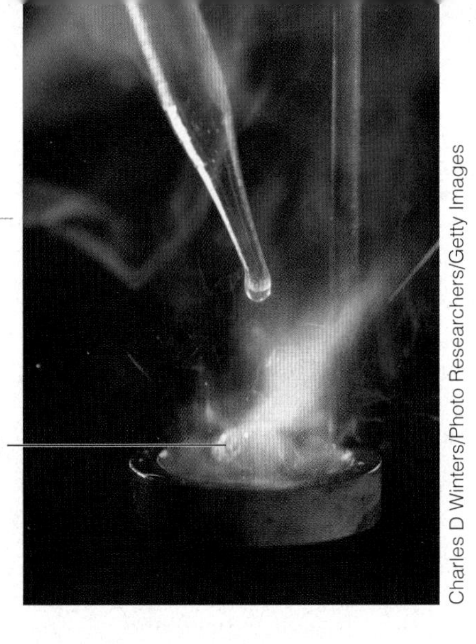

Lithium, sodium, and the other alkali metals react vigorously with water.

quantum shell. Helium— He, atomic number 2—has two protons within the nucleus, and two electrons, which fill the first quantum shell to its capacity. As we saw in the transition from helium to lithium in the *What a Chemist Sees* feature of Section 2.4, when a quantum shell is filled to its capacity with electrons, the next electron added to the atom enters the next higher quantum shell. For instance, the lithium atom has two electrons filling its inner quantum shell and one electron in its next higher shell. We can indicate this electron configuration as (2, 1).

With the exception of helium, not only do atoms of elements within the same column of the periodic table hold the same number of electrons in their outermost quantum shells (as we see in Figure 3.1), but the elements themselves (again, within the same column) often share similar chemical properties. Both lithium and sodium, for example, not only carry a single electron in their outermost quantum shells, but both are also metals that react vigorously (even explosively) with water to liberate hydrogen (**Figure 3.2**). These two and the remaining elements of their column in the periodic table, with the exception of hydrogen, are known as the **alkali metals**. Similarly, both fluorine and chlorine have seven electrons in their outermost quantum shells, and both are gases that react vigorously with a variety of other substances. Along with the remaining elements of their column, they are called the **halogens**.

alkali metals The elements in the first column of the periodic table, with the exception of hydrogen.

halogens Elements in column 17, the next to last column of the periodic table.

This connection between the electronic structure of outermost quantum shells and properties of elements continues to helium, neon, and argon. The atoms of these elements have outermost quantum shells with eight electrons (except helium, whose outermost shell is filled with two electrons), and all three elements are gases with little or no chemical reactivity at all under ordinary conditions. These and the remaining elements of their column are known as the **inert gases** or **noble gases**. The periodic table shown in **Figure 3.3** highlights the three **groups** or families of elements we've discussed.

inert or **noble gases** Elements in column 18, the last column of the periodic table.

Ask Yourself

Small amounts of bromine (Br) and iodine (I) are used in high-intensity lamps found in certain car headlights and other lighting applications. These ___________ bulbs take their name from the chemical group to which these two elements belong.

Selected groups in the periodic table • Figure 3.3

Elements within a given column of the periodic table are called a group or family and typically exhibit similar chemical properties. An exception is the gaseous element hydrogen, which is not a member of the alkali metals.

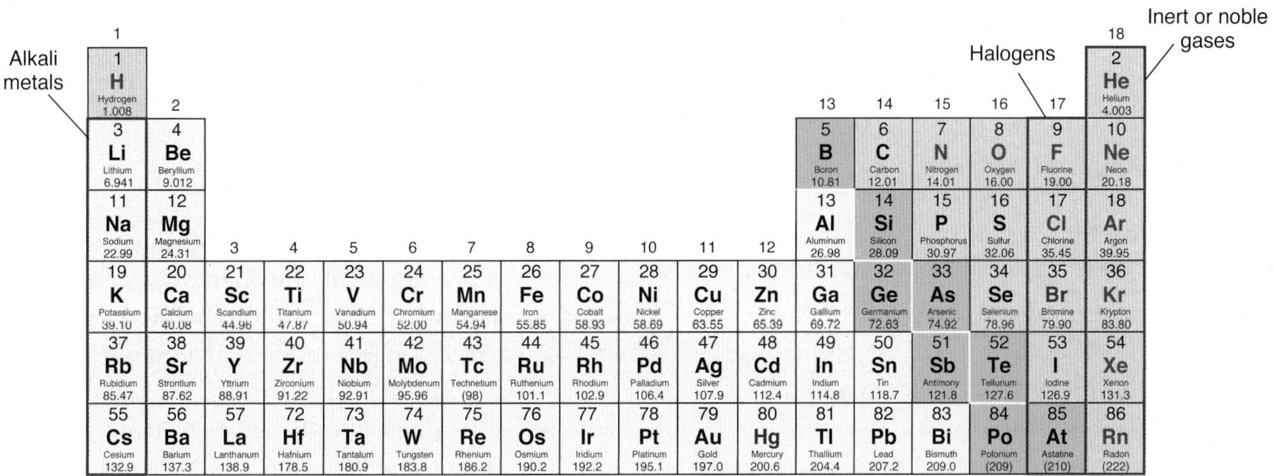

Lewis (or electron dot) structures of the first 20 elements Table 3.1

1	2											13	14	15	16	17	18
H·																	He:
Li·	·Be·											·Ḃ·	·Ċ̣·	·Ṇ̇:	:Ọ̇:	:F̣̈:	:N̤̈e:
Na·	·Mg·											·Ȧl·	·Ṩi·	·Ṗ̣:	:Ṩ:	:C̈̇l:	:Ä̤r:
K·	·Ca·																

Lewis Structures

Because of the importance of the number of electrons in an atom's outermost quantum shell, this shell is identified as an atom's **valence shell** and the electrons it contains are called the atom's **valence electrons**. We can use a simplified way of viewing the valence electron configurations of the atoms, known as **Lewis structures**, or **electron dot structures** (**Table 3.1**). With this table, we can clearly see that elements in the same column of the periodic table share the same number of valence electrons. Helium is the single exception, but like neon, it has a filled valence shell.

valence shell The outermost quantum shell of an atom.

valence electrons The electrons contained in an atom's valence shell.

Lewis or **electron dot structures** Notations that show only the valence electrons of an atom, arranged as dots around the element's symbol.

We can draw Lewis structures by:

- Starting with the symbol of the element;
- Placing dots around the symbol, one on each side at first, with the total equal to the number of valence electrons for that element. For elements in Groups 1 and 2, the number of valence electrons is identical to the group number. For elements in Groups 13–18, the number of valence electrons is simply the group number reduced by 10. (Helium, the lone exception, has two valence electrons.)
- Making sure that no more than two electrons are paired on any side of the symbol.

CONCEPT CHECK

1. **How** do we know that atoms of the elements beryllium and magnesium, with different total numbers of electrons, have the same number of valence electrons?
2. **How** many valence electrons would you predict iodine to have?
3. **Draw** the Lewis structure for selenium.

3.2 Ionic and Covalent Bonding

LEARNING OBJECTIVES

1. **State** the octet rule.
2. **Distinguish** between ionic and covalent bonds.
3. **Differentiate** between the types of covalent bonds.
4. **Describe** how the concept of polar covalent bonds is related to ionization.
5. **Explain** why ions are electrically conductive.

The connection between the filled electron valence shells of the inert gases and their very low or nonexistent chemical reactivity carries over to other elements as well. In this section, we'll see that atoms often react chemically in a way that gives them exactly eight electrons in their valence shells, mimicking the valence shells of the inert

Formation of sodium chloride by transfer of a valence electron • Figure 3.4

A sodium atom, with one valence electron, reacts with a chlorine atom, which has seven valence electrons, to form sodium chloride, NaCl. The process is the same for any alkali metal reacting with a halogen.

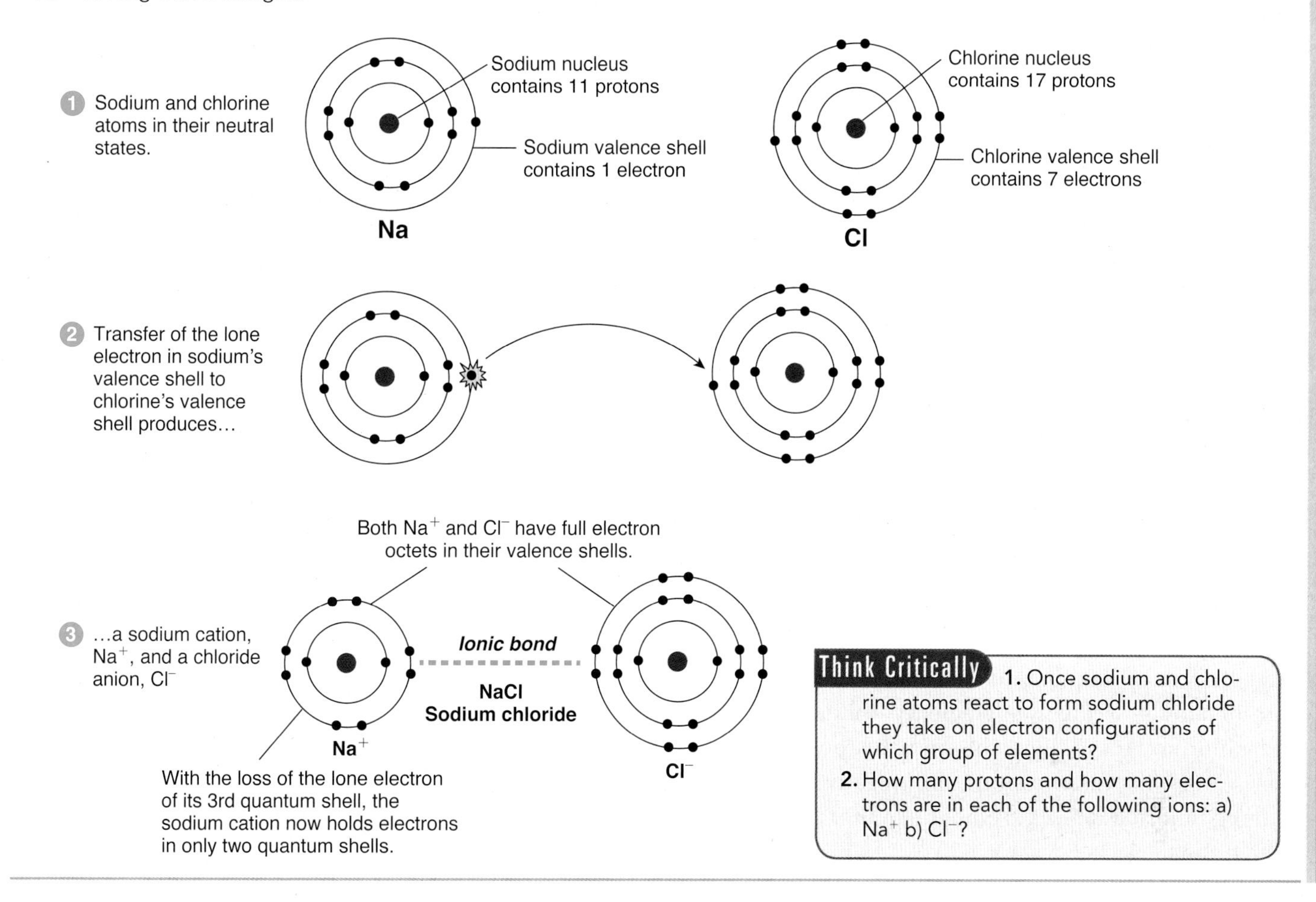

1 Sodium and chlorine atoms in their neutral states.
Sodium nucleus contains 11 protons
Sodium valence shell contains 1 electron
Na
Chlorine nucleus contains 17 protons
Chlorine valence shell contains 7 electrons
Cl
2 Transfer of the lone electron in sodium's valence shell to chlorine's valence shell produces...
Both Na+ and Cl− have full electron octets in their valence shells.
3 ...a sodium cation, Na+, and a chloride anion, Cl−
Ionic bond
NaCl
Sodium chloride
Na+
Cl−
With the loss of the lone electron of its 3rd quantum shell, the sodium cation now holds electrons in only two quantum shells.

Think Critically

1. Once sodium and chlorine atoms react to form sodium chloride they take on electron configurations of which group of elements?
2. How many protons and how many electrons are in each of the following ions: a) Na^+ b) Cl^-?

octet rule Atoms often react to obtain exactly eight electrons in their valence shell.

gases (except helium). This generalization, known as the **octet rule**, is a powerful framework for understanding why and how atoms bond to one another.

Ionic Bonds

Before exploring the octet rule, it's helpful to recall that an electrically neutral atom contains equal numbers of positively charged protons and negatively charged electrons. If a neutral atom were to *gain* an electron, it would acquire a *negative* charge because it would now hold more electrons than protons. Similarly, if it were to *lose* an electron, it would acquire a *positive* charge because it would now contain more protons than electrons. We call such positively or negatively charged atoms or groups of atoms **ions**. Those with a positive charge are called **cations** (CAT-ions) and those with a negative charge are known as **anions** (AN-ions). Furthermore, because they carry opposite electrical charges, cations and anions strongly attract one another, as we'll see shortly.

ion An atom or a group of atoms that carries an electrical charge.

cation A positively charged ion.

anion A negatively charged ion.

We can now see the octet rule in operation as sodium (a metal) reacts with chlorine (a nonmetal) to produce sodium chloride, NaCl (**Figure 3.4**). Here, an electrically neutral sodium atom (with equal numbers of protons and electrons) loses a valence electron and is transformed into a sodium cation: a positively charged sodium ion with an excess of one proton over its new total number of

A covalent bond between hydrogen atoms • Figure 3.5

Two hydrogen atoms form a covalent bond between them to produce the hydrogen molecule, H_2. By sharing valence electrons, each hydrogen atom of the H_2 molecule acquires a filled valence shell resembling that of the helium atom.

Here, the arrow pointing to the right represents a **chemical reaction**. The **reactants**—the H atoms—appear to the left of the arrow, and the **product**—the H_2 molecule—appears to the right.

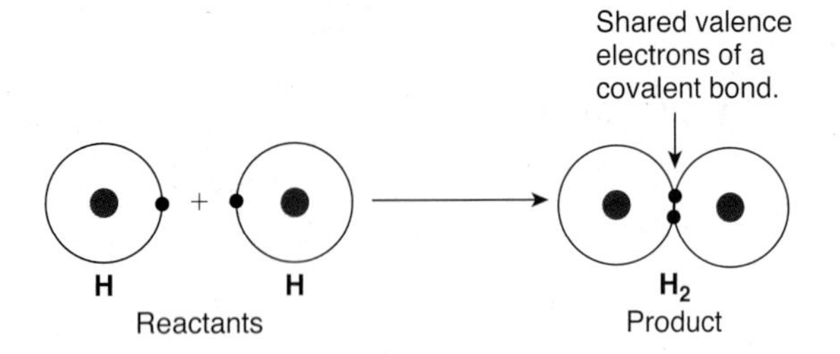

Put It Together Review the definition of the term **compound** and answer the question. Does the hydrogen molecule, H_2, represent a compound?

electrons. As the electrically neutral chlorine atom receives the transferred electron, the chlorine—with its new excess of one electron over its number of protons—is transformed into a negatively charged chloride anion. The result is the formation of NaCl, an ionic compound.

The **compound** sodium chloride forms by the chemical combination of equal numbers of atoms of the two **elements**, sodium and chlorine. In sodium chloride the positively charged sodium cations and negatively charged chloride anions are held together by **ionic bonds**.

Covalent Bonds

Another kind of chemical bond, a **covalent bond**, forms by the *sharing* of a pair of valence electrons between two atoms, rather than the complete *transfer* of a valence electron from one atom to another. The simplest example is the covalent bond formed by the chemical combination of two hydrogen atoms to form a **molecule** of hydrogen, H_2 (**Figure 3.5**). We call this a **diatomic** molecule because it is made up of *two* atoms (from the Greek prefix for two, *di-*).

By sharing valence electrons, covalently bonded atoms are able to achieve filled valence shells, as we saw in Figure 3.5 with the hydrogen atoms of the H_2 molecule. Atoms of other nonmetal elements (except for noble gases) readily form covalent bonds. For example, any halogen atom, such as chlorine, has seven valence electrons and so needs only a single electron to complete its octet. It can do so by forming a covalent bond, as shown in **Figure 3.6**, the formation of the Cl_2 molecule.

compound A pure substance formed by the chemical combination of two or more elements in a specific ratio.

ionic bond A chemical bond resulting from the mutual attraction of oppositely charged ions.

covalent bond A bond consisting of a pair of electrons shared by two atoms.

molecule An electrically neutral assembly of atoms held together by covalent bonds.

Formation of the Cl_2 molecule • Figure 3.6

Each chlorine atom shares one valence electron to form the covalent bond of the Cl_2 molecule.

$$:\ddot{\underset{\cdot\cdot}{Cl}}\cdot + \cdot\ddot{\underset{\cdot\cdot}{Cl}}: \longrightarrow :\ddot{\underset{\cdot\cdot}{Cl}}:\ddot{\underset{\cdot\cdot}{Cl}}:$$

Shared valence electrons of a covalent bond.

Lone pairs

Cl_2 molecule

Pairs of valence electrons that are not involved in bonding are called **lone pairs**. For instance, each chlorine atom in the Cl_2 molecule has three lone pairs. (The pair of valence electrons between the two chlorine atoms constitutes the covalent bond.) Lone pairs can play an important role in the shapes of molecules, as will be discussed shortly.

lone pairs A pair of valence electrons not involved in bonding.

The most common and most convenient way of showing a covalent bond between two atoms is through the use of a straight line joining them. Thus H_2 can be represented as H—H. Chemically, the two hydrogens of H—H are identical and have identical attractions for the shared pair of electrons. The same is true for the shared pair of electrons in a diatomic chlorine molecule, Cl—Cl. These are known as **nonpolar covalent bonds**. Since the two ends of the molecule are identical (H and H; and Cl and Cl), they have exactly the same electrical charge and are identical electronically.

nonpolar covalent bond A covalent bond in which the shared pair of electrons lies equidistant from both bonded atoms.

If atoms of two *different* elements form a covalent bond—one H and one Cl, for example—one of the atoms may have a slightly greater attraction for the shared pair of electrons. This attraction is determined by a property known as **electronegativity** (**Figure 3.7**). An atom with higher electronegativity may pull a bonding electron pair slightly closer to itself and farther from the other bonded atom. It happens that a chlorine atom's attraction for electrons is slightly greater than the attraction that a hydrogen atom exerts on the same electrons, so the bonded chlorine atom in HCl pulls the shared electrons toward itself by a small amount. As a result, the end of the molecule at the chlorine atom is slightly more negative than the end near the hydrogen. The resulting bond is called a **polar covalent bond**, and chlorine bears a partial negative charge. (The partial charges are indicated by the Greek letter *delta*, δ.) If the difference in the

electronegativity A measure of the ability of an atom to attract bonding electrons.

polar covalent bond A covalent bond in which the shared pair of electrons lies closer to one of the bonded atoms.

Electronegativity • Figure 3.7

This figure shows relative electronegativities of representative elements in the first four rows of the periodic table.

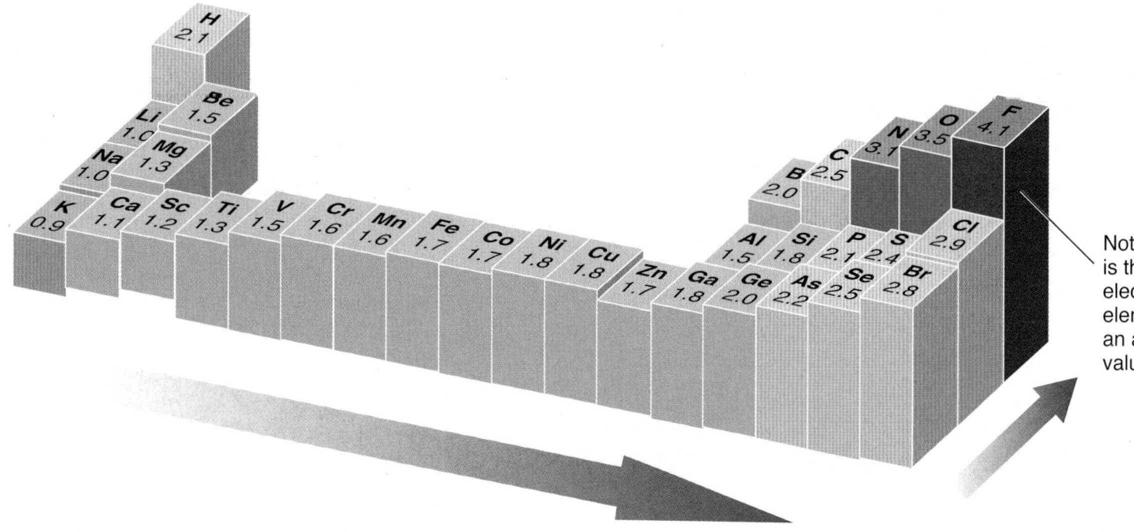

Adapted with permission of John Wiley & Sons, Inc. from Jespersen, N., Brady, J., and Hyslop, A., *Chemistry: The Molecular Nature of Matter*, Sixth Edition, p. 380. Copyright 2012.

Electronegativity and bond types • Figure 3.8

We can use the concept of electronegativity to help us understand bonding between two atoms, by imagining the bonded atoms competing for the bonding electrons.

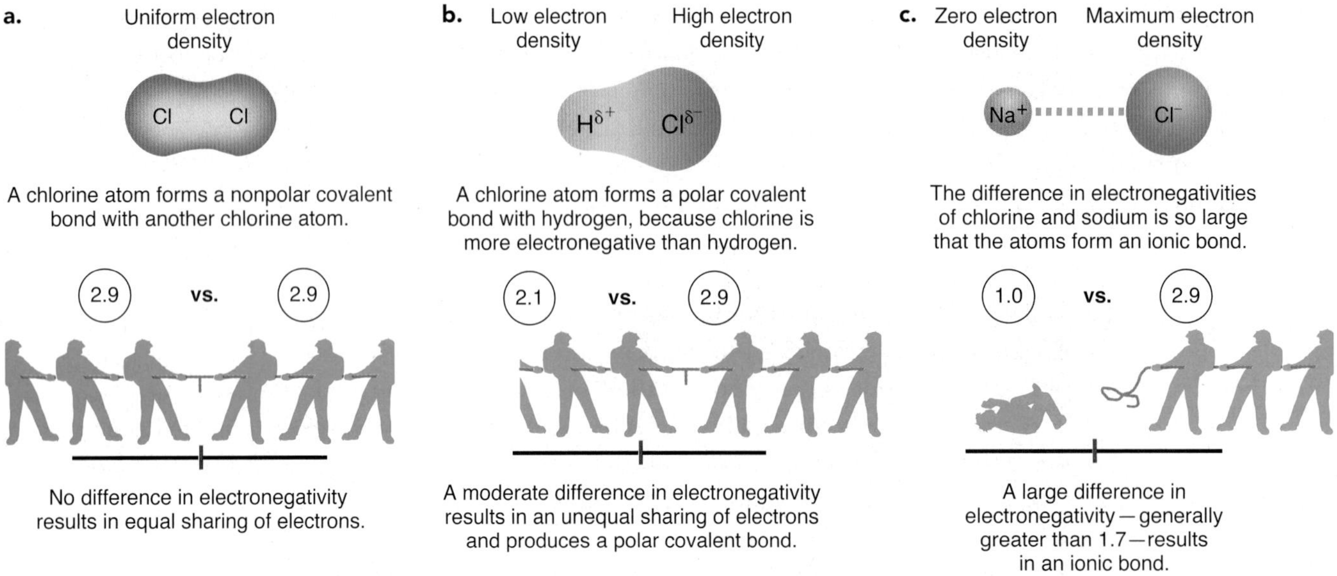

a. Uniform electron density

Cl Cl

A chlorine atom forms a nonpolar covalent bond with another chlorine atom.

2.9 vs. 2.9

No difference in electronegativity results in equal sharing of electrons.

b. Low electron density — High electron density

$H^{\delta+}$ $Cl^{\delta-}$

A chlorine atom forms a polar covalent bond with hydrogen, because chlorine is more electronegative than hydrogen.

2.1 vs. 2.9

A moderate difference in electronegativity results in an unequal sharing of electrons and produces a polar covalent bond.

c. Zero electron density — Maximum electron density

Na^{+} Cl^{-}

The difference in electronegativities of chlorine and sodium is so large that the atoms form an ionic bond.

1.0 vs. 2.9

A large difference in electronegativity—generally greater than 1.7—results in an ionic bond.

Adapted with permission of John Wiley & Sons, Inc. from Hein, M., and Arena, S., Foundations of College Chemistry, Fourteenth Edition, p. 227. Copyright 2014.

Ask Yourself

1. How many shared electrons constitute each bond represented in parts **a** and **b**?
2. How are the shared electrons in part **a** different from those in part **b?**
3. How is the bond represented in **c** different from that shown in **a** or **b**?

electronegativities of the two atoms is large enough, the result is an ionic bond, as in NaCl (**Figure 3.8**).

A large difference in the electronegativities of the atoms held together by a polar covalent bond can lead to **ionization** of the covalent compound into a cation and an anion. For example, under ordinary conditions, hydrogen chloride, HCl, exists as a gas, with its H and Cl atoms held together by a polar covalent bond. But when HCl is dissolved in water, the molecule ionizes to H^+ and Cl^- (**Figure 3.9**).

> **ionization** Conversion of a covalent molecule into ions.

Like HCl, water—the most common covalent compound in our everyday world—also contains polar covalent bonds. **Figure 3.10** presents an electron dot representation of two hydrogen atoms combining with one oxygen atom to form a water molecule, along with a more detailed picture of the water molecule itself. The resulting bonds are polar covalent, with the oxygen at the (slightly) negative end of the bonds and the hydrogens at the (again, slightly) more positive end. Notice that the two polar covalent bonds in the water molecule form an angle, which gives the overall molecule a polarity.

Ionization of a polar covalent compound • Figure 3.9

When HCl dissolves in water, it ionizes completely into H^+ and Cl^- ions.

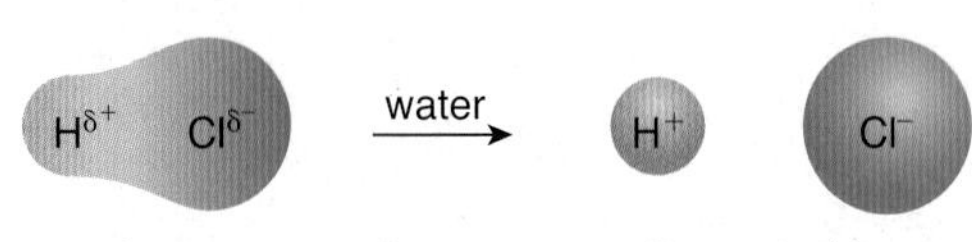

Adapted with permission of John Wiley & Sons, Inc. from Hein, M., and Arena, S., Foundations of College Chemistry, Fourteenth Edition, p. 227. Copyright 2014.

Water, a polar covalent compound • Figure 3.10

The polar bonds in water do not lie in a straight line, so the molecule has an overall polarity.

a. Electron-dot representation of formation of the water molecule

$H\cdot \; + \; \cdot\ddot{O}: \; + \; \cdot H \longrightarrow H:\ddot{O}: H$

b. Notice that the three atoms do not lie in a straight line. Instead, they produce an angular molecule with a 104.5° angle between bonds.

δ^{-} O, δ^{+} H, δ^{+} H, 104.5°

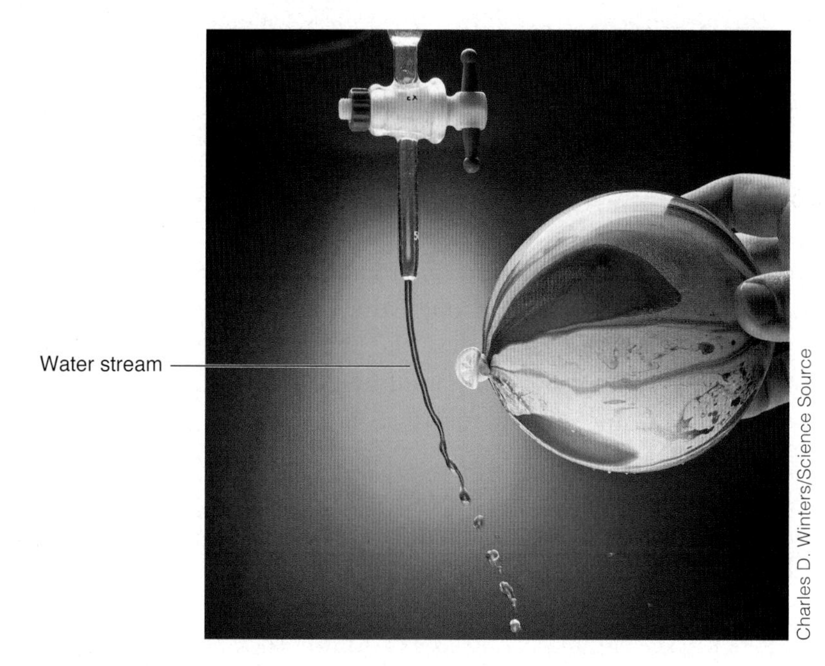
Water stream

Charles D. Winters/Science Source

c. In this photograph, a thin stream of water dropping straight down is bent to the right by the static electric field of the balloon, demonstrating the water's polarity.

The polar covalent O—H bond of water can ionize, but the ionization is minimal and reversible (**Figure 3.11**). Ionization is minimal because very few of the water molecules—roughly 1 out of every 100,000,000,000,000 or 1 out of every 10^{14}—are ionized at any given moment. It's reversible because the H^{+} and OH^{-} ions generated by the ionization readily recombine to form another covalent water molecule, H_2O.

The reversible ionization of water • Figure 3.11

Water ionizes and then reforms itself, but at any given moment the amount of ionized products is exceedingly small.

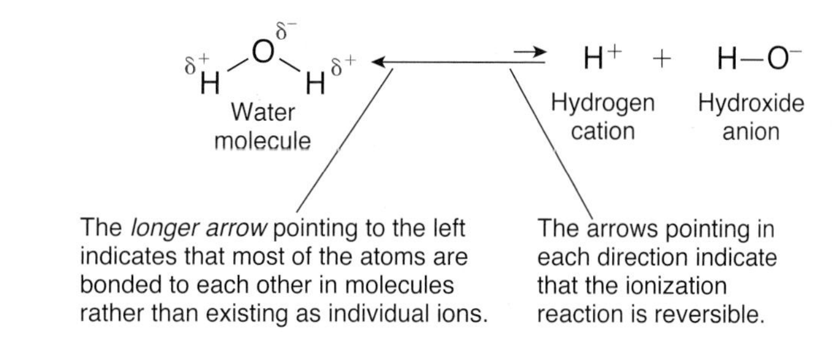
δ⁻ O
δ⁺ H
H δ⁺
Water molecule
H⁺ + H—O⁻
Hydrogen cation
Hydroxide anion
The *longer arrow* pointing to the left indicates that most of the atoms are bonded to each other in molecules rather than existing as individual ions.
The arrows pointing in each direction indicate that the ionization reaction is reversible.

Single, double, and triple covalent bonds • Figure 3.12

The bonds in Lewis structures consist of the electrons shared between the two atoms.

a. Each atom shares one valence electron to create a **single bond.**

:F· + ·F:

:F:F:

:F—F:

b. Each atom shares two valence electrons to create a **double bond.**

:O· + ·O:

:O::O:

:O=O:

c. Each atom shares three valence electrons to create a **triple bond.**

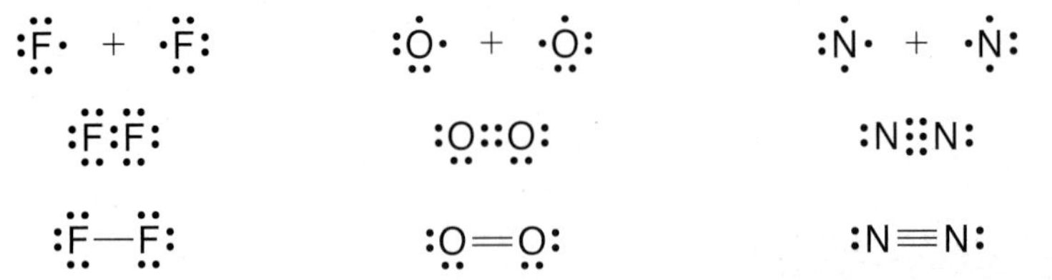

:N· + ·N:

:N⋮N:

:N≡N:

Ask Yourself

1. Are the covalent bonds represented in this figure polar or nonpolar?
2. How many shared electrons does each horizontal line in the figure represent?

single bond One pair of shared electrons serving as a covalent bond between two atoms.

double bond Two pairs of shared electrons serving as two covalent bonds between two atoms.

triple bond Three pairs of shared electrons serving as three covalent bonds between two atoms.

In addition to the categories of polar and nonpolar, covalent bonds can exist as **single, double,** or **triple bonds**, depending on the number of shared pairs of electrons holding two atoms together. A diatomic fluorine molecule, F_2, serves as an example of a molecule containing a single bond; a diatomic oxygen molecule, O_2, contains a double bond; and a diatomic nitrogen molecule, N_2, is an example of a molecule with a triple bond (**Figure. 3.12**). In each of these examples, notice that through the sharing of valence electrons, each atom experiences the equivalent of a filled octet.

Ions and Electrical Conductivity

With an understanding of the nature of ionic and covalent bonds, we can examine a fundamental and important difference between ionic and covalent compounds in terms of their abilities to conduct electricity. In **Figure 3.13** we see two views of a light bulb connected to an electrical circuit that's been split apart, with the ends of the break immersed in a beaker of water. For the bulb to illuminate, electricity must pass through the water and across the break.

The beaker on the left contains pure water. The bulb remains dark, indicating that pure water conducts electricity across the gap very poorly, if at all. But when we add a little table salt to the water, the bulb glows brightly. We conclude that the solution of salt in water conducts an electric current across the break in the circuit very well, far better than pure water does. Let's now see why.

Pure water is a covalent compound that ionizes only minimally, producing too few H^+ and OH^- ions to carry a current through the water—at least not of sufficient strength to cause the bulb to glow. Common table salt is an ionic compound that dissolves in water to produce large numbers of Na^+ and Cl^- ions, enough to carry the electrical current effectively through the salt-water solution and across the gap in the electrical circuit. We call a substance that produces large numbers of ions in solution, such as NaCl, an **electrolyte**.

electrolyte A substance that ionizes efficiently in solution and conducts electricity.

Temperature Effects on Water and Salt

A comparison of the effect of temperature on both water and salt provides another example of the differences between a covalent and an ionic compound. In ionic compounds, such as sodium chloride, all the anions and cations are held together tightly, in close contact to their oppositely charged neighbors, by the powerful electrical forces of ionic bonds. As a result, it takes a great deal of energy to disrupt the forces holding the ions of a salt crystal into a single, cohesive structure.

In covalent compounds, the picture is considerably different. The hydrogen and oxygen atoms of a single water molecule are held together tightly by the forces of their covalent bonds to form a stable water molecule, but the individual covalent water molecules cling weakly to their neighboring molecules in ice and in water, held together loosely by the relatively weak forces of **intermolecular**

The effect of table salt on water's ability to conduct electricity • Figure 3.13

The dissolved ions of NaCl allow a solution of common table salt in water to conduct an electric current. *Note*: The kinds of water we normally encounter in our everyday lives—for drinking, washing, swimming, or other uses—is only rarely pure. The small amounts of chemical impurities in our everyday water can convert the water into as good an electrical conductor as a bit of salt does. That's why it's always prudent to assume that any water you're dealing with conducts electricity very well and to be careful to avoid electrical shocks.

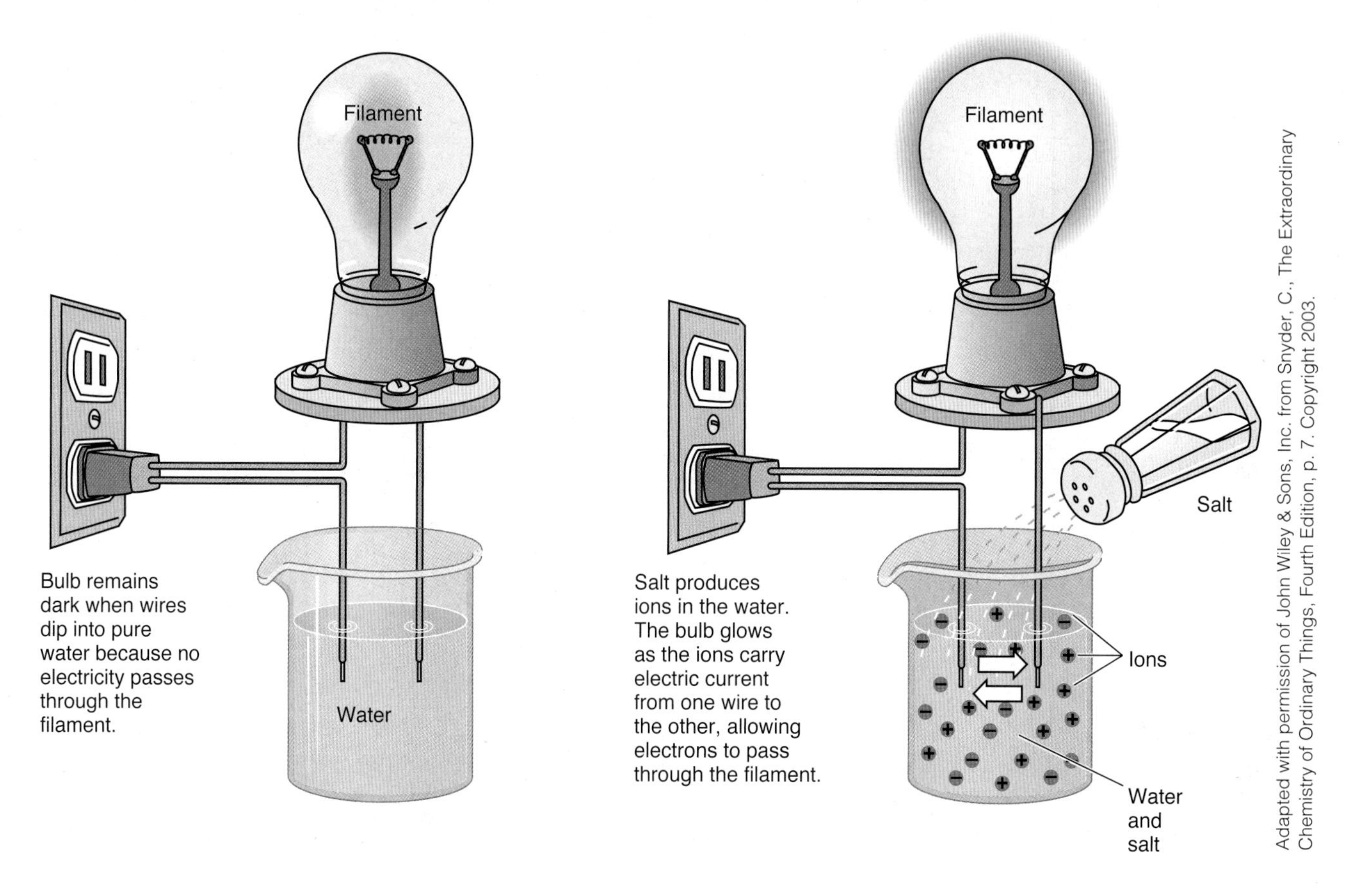

Filament
Bulb remains dark when wires dip into pure water because no electricity passes through the filament.
Water
Filament
Salt
Salt produces ions in the water. The bulb glows as the ions carry electric current from one wire to the other, allowing electrons to pass through the filament.
Ions
Water and salt

Adapted with permission of John Wiley & Sons, Inc. from Snyder, C., The Extraordinary Chemistry of Ordinary Things, Fourth Edition, p. 7. Copyright 2003.

attraction. When we heat both salt crystals and ice, we can see the effects of the powerful ionic forces that keep an ionic crystal intact as well as the effects of much weaker intermolecular forces of a covalent substance.

Well below room temperature, H_2O, a covalent molecule, exists as ice, the frozen form of water. As the temperature rises to 0°C (32°F), solid ice melts to liquid water, which remains a liquid (at normal atmospheric pressure) until it reaches a temperature of 100°C (212°F), where it boils and converts to steam, or water vapor.

Table salt, an ionic compound (NaCl), requires far higher temperatures for similar transformations. It remains a crystalline solid until it melts at the remarkably high temperature of 801°C (1474°F), and it doesn't vaporize until it reaches an astonishing 1413°C (2575°F) (**Figure 3.14** on the next page).

The major reason for these different responses to changes in temperature lies in the different attractive forces present in ionic and covalent compounds. In crystalline table salt, the sodium cations and the chloride anions are closely packed in an orderly arrangement that produces the **crystal lattice** of the salt we use with our food. In this sodium chloride lattice, the positively charged sodium cations and the negatively charged chloride anions lie close to each other in a repeating pattern, with each

crystal lattice Orderly, three-dimensional arrangement of the chemical particles that form a crystal.

The effect of temperature on covalent and ionic compounds • Figure 3.14

Because ionic bonds, formed by the mutual attraction of oppositely charge ions, are much stronger than the intermolecular attractions among covalent molecules, the ionic compound sodium chloride melts and vaporizes at much higher temperatures than does water, a covalent compound.

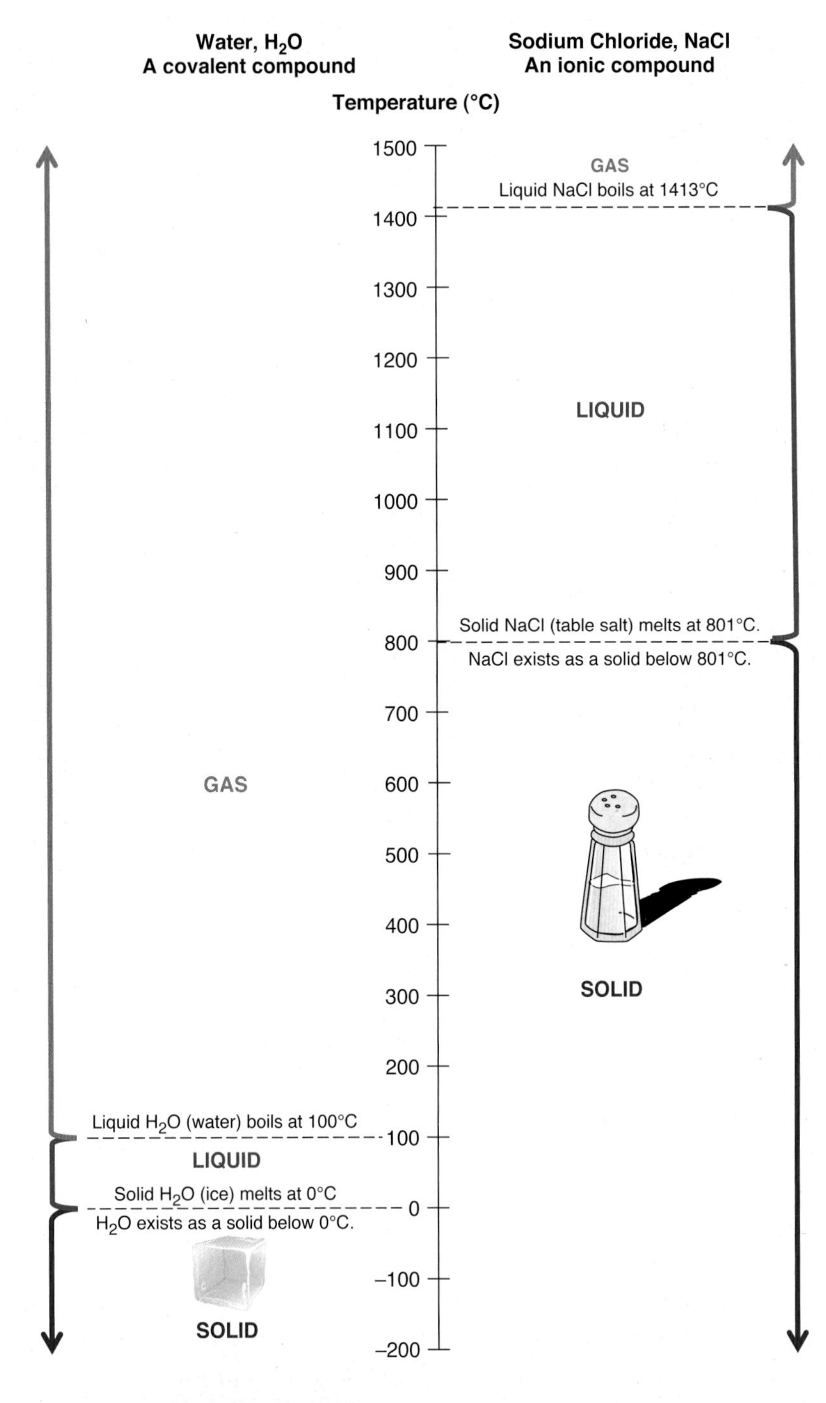

Interpret the Data

State whether each of these is a solid, liquid, or gas:

1. H_2O at 1412°C; H_2O at −5°C.

2. NaCl at 799°C; NaCl 1420°C.

Crystal lattice of NaCl • Figure 3.15

Each sodium ion is surrounded by chloride ions, and each chloride ion is surrounded by sodium ions.

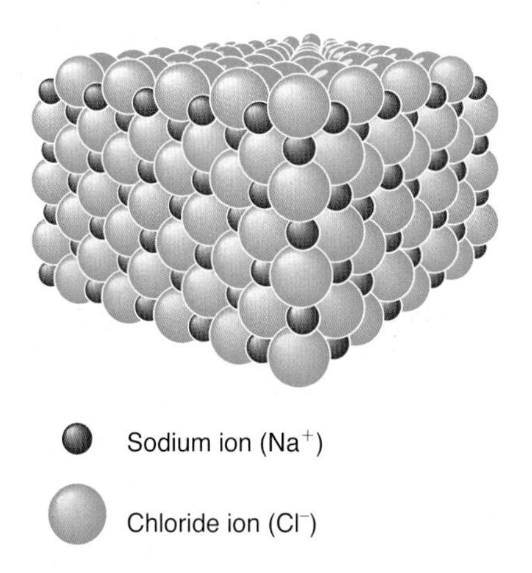

cation surrounded by anions and each anion surrounded by cations (**Figure 3.15**).

The closeness of the anions and the cations in this crystal, and the strong attractions of these oppositely charged ions for each other, give the crystal great stability. It takes a large amount of energy to pull the ions far enough apart from each other to cause the crystal to melt—thus the very high temperature that's required to convert the solid crystal into a liquid melt. Boiling the melted sodium chloride, thereby completely freeing the individual ions from all attractive forces, requires an enormous amount of energy. This explains the remarkably high boiling point of sodium chloride.

Water is a polar covalent compound, with more diffuse electrical charges than sodium or chloride ions. A water molecule has a much smaller electrical attraction for any of its neighbors than do sodium or chloride ions. Although the water molecules of solid ice are arranged in a more orderly manner than those of liquid water, the forces of attraction of the H_2O molecules in ice don't come anywhere near those of the cations and anions of the salt crystal (**Figure 3.16**). Thus much less energy and a much lower temperature are needed to melt ice than crystalline table salt. This same sort of reasoning applies to the comparison of the boiling point of water and the boiling point of liquid NaCl.

Weak and strong attractive forces • Figure 3.16

A water molecule has a relatively weak attraction for neighboring water molecules as compared to the very strong bond that attracts sodium cations and chloride anions.

a. The attractive force between water molecules is relatively weak because of partial electrical charges on hydrogen and oxygen atoms.

b. The attractive force between sodium cations and chloride anions is strong because of full electrical charges.

CONCEPT CHECK

STOP

1. **How** does the octet rule help explain why a sodium atom loses an electron and a chlorine atom gains an electron when they combine to form sodium chloride?
2. **Which** type of bond is characterized by mutually shared electrons, and which type results from electron transfer?
3. **Why** are bonding electrons not equally shared between the bonded atoms of a polar covalent bond?
4. **Why** would polar covalent bonds be more prone to ionize than nonpolar covalent bonds?
5. **Why** does salt water conduct electricity?

3.3 Ionic Compounds

LEARNING OBJECTIVES

1. **Describe** how ionic compounds are formed.
2. **Predict** chemical formulas of binary ionic compounds and learn how to name them.
3. **Define** formula mass.

A large majority of the atoms that make up the matter in our natural world exist bonded to atoms of other elements, as part of chemical compounds, rather than in their elemental states. As we've seen, a chemical compound is a pure substance made up of two or more different elements in a fixed ratio, in which the atoms are bound to each other by either ionic or covalent bonds. In this section, we'll further explore ionic compounds, starting with sodium chloride. In the following section (3.4), we'll look more closely at covalent compounds, beginning with water.

How Ionic Compounds Form

Ionic compounds often form via **electron transfer** reactions in which an atom of a metal transfers one or more valence electrons to an atom of a nonmetal. In the process, both the metal atom and nonmetal atom end up with an **octet** of electrons in their valence shells. We've seen this in operation earlier in this chapter when sodium, an alkali metal, reacts with chlorine, a halogen gas, to produce sodium chloride. The driving force for these types of reactions is the formation of a stable electron configuration in the valence shell, similar or identical to those of the noble gases. For example, Na^+ has the same electron configuration as Ne (two electrons in its first shell and eight in its second)—which we can indicate as (2, 8)—and Cl^- has the same electron configuration as Ar (2, 8, 8). Notice that the product, sodium chloride (table salt), has a different appearance and different properties than either of the reactants (**Figure 3.17**). This phenomenon occurs in many types of chemical reactions.

Chemistry InSight The formation of sodium chloride by chemical interactions • Figure 3.17

The final product has a different appearance and different properties than either of the original reactants.

What we observe

Sodium, a soft grey metal... added to...

Charles D Winters/Photo Researchers/Getty Images

chlorine, a pale green gas...

Photo Researchers/Getty Images

undergoes a fiery reaction...

© Richard Megna/Fundamental Photographs, NYC

producing sodium chloride (table salt), an edible white crystalline substance.

edelmar/iStockphoto

Atomic representation

Na atoms in a tightly packed three-dimensional arrangment

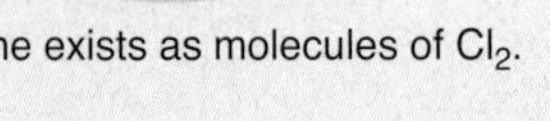

Chlorine exists as molecules of Cl_2.

Attracted to each other by ionic bonds, sodium and chloride ions form a highly ordered, three-dimensional crystal lattice.

Electron transfer reactions producing ionic compounds • Figure 3.18

The transfer of one or more electrons from an atom of a metal to an atom of a nonmetal yields an ionic compound.

a. Electron configuration view, showing lithium reacting with fluorine to form lithium fluoride, LiF.

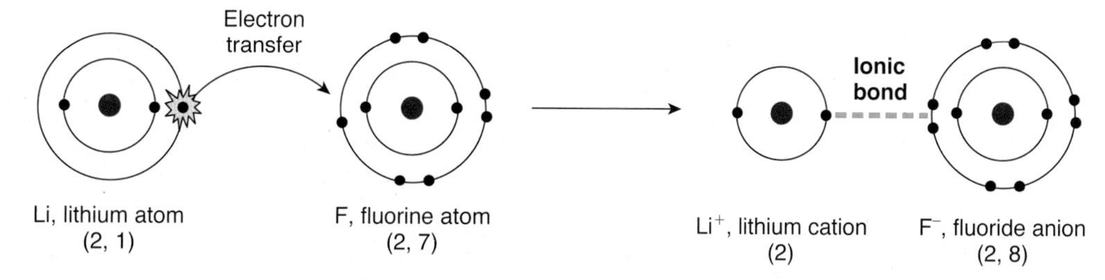

b. Lewis structures showing calcium reacting with chlorine to form calcium chloride, $CaCl_2$.

Electron transfer

$$:\ddot{\underset{..}{Cl}}\cdot \leftarrow \cdot Ca \cdot \rightarrow \cdot\ddot{\underset{..}{Cl}}: \longrightarrow\ :\ddot{\underset{..}{Cl}}:^- \ Ca^{2+}\ :\ddot{\underset{..}{Cl}}:^-$$

Ionic Compound Formulas and Names

The chemical formula of an ionic compound, or **compound formula** for short, gives us the identity and the ratio of its component elements. We've seen that in the formation of sodium chloride, sodium loses and chlorine gains one electron each, so they combine in a one-to-one ratio to form NaCl. When writing these formulas, we indicate the ratio of each element as a subscript following the elemental symbol. However, if the ratio for any element is 1, we omit the subscript. Therefore, we write sodium chloride as NaCl (rather than Na_1Cl_1). Combinations with this same ratio of cations to anions can occur between any alkali metal (lithium, sodium, potassium, etc.) and any halogen (fluorine, chlorine, bromine, etc.). In **Figure 3.18a**, for example, we see the electron configurations of lithium and fluorine as they react to form lithium fluoride, LiF, an ionic compound of lithium cations and fluoride anions in a one-to-one ratio. Here, both lithium and fluoride ions obtain the stability provided by filled valence shells.

compound formula A chemical formula showing the elements present in an ionic compound and the ratio of each.

Unlike sodium, calcium has two electrons in its valence shell. In calcium's reaction with chlorine, each calcium atom transfers two valence electrons to chlorine atoms. Since each chlorine atom needs only a single electron to complete its octet, one calcium atom combines with two chlorine atoms to form the ionic compound calcium chloride (**Figure 3.18b**). To simplify our representation of this electron transfer, we've shown Lewis structures, emphasizing valence electrons. By donating two electrons, each calcium atom becomes a Ca^{2+} ion, which has an electron configuration (2, 8, 8) and an octet in its valence shell. To show the association of one calcium cation with two chloride anions, we write the chemical formula of calcium chloride as $CaCl_2$.

Combinations with this ratio of cation and anion can occur between any member of the **alkaline earth** family, the second column of the periodic table, and any halogen. For example, a magnesium atom can combine with two iodine atoms to form magnesium iodide, MgI_2; and beryllium and fluorine can combine to form beryllium fluoride, BeF_2.

alkaline earth Any element in the second column of the periodic table.

The names of these sorts of **binary** ionic compounds consist of the name of the cation followed by the name of the anion, with an *–ide* suffix. Some examples are:

Formula	Name
LiF	Lithium fluoride
NaBr	Sodium bromide
MgI_2	Magnesium iodide
$CaCl_2$	Calcium chloride

As we examine the periodic table, we find that a variety of binary ionic compounds can form from pairing a metal with a nonmetal. The possibilities emerge as we become familiar with a few of the more common

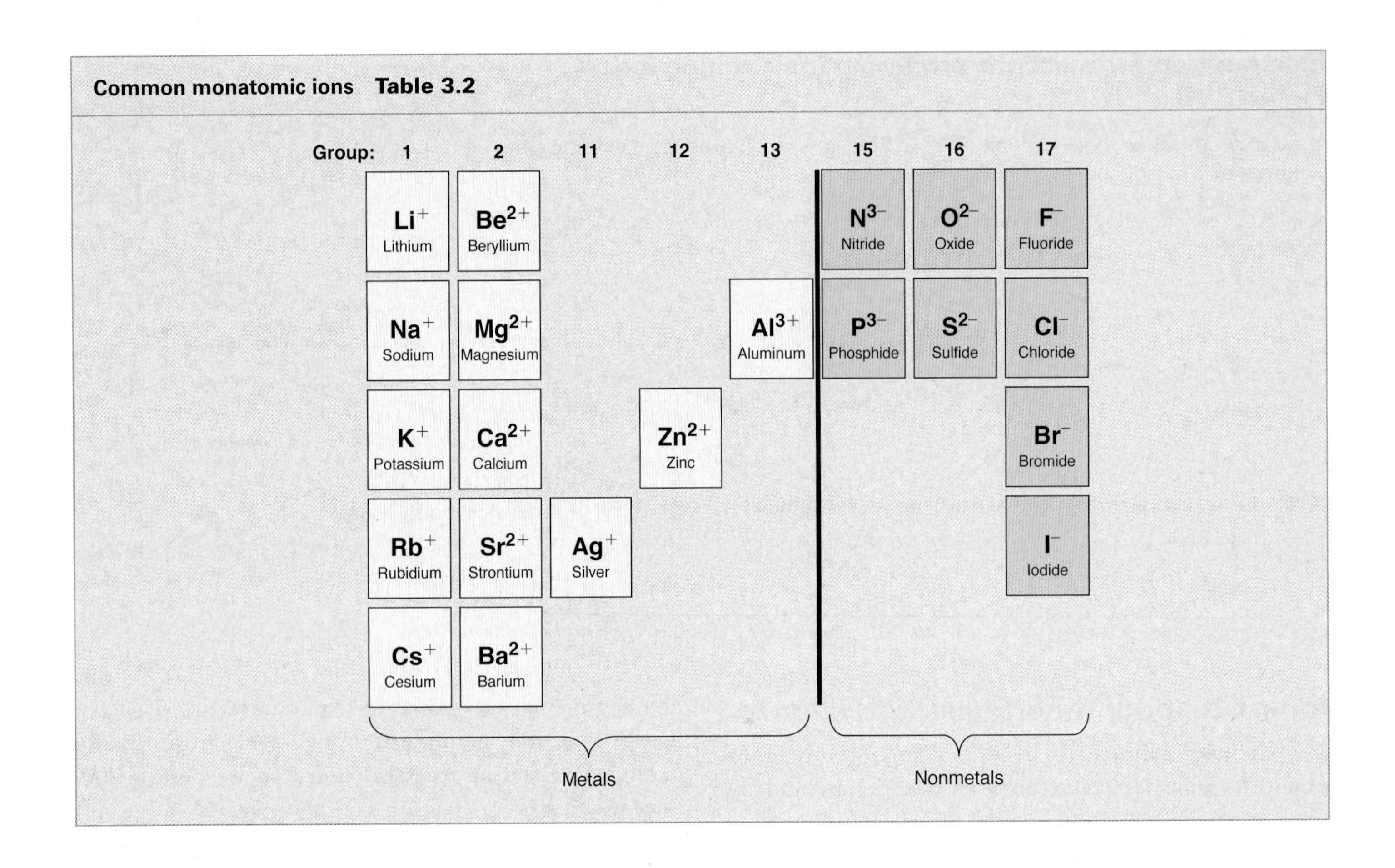

Common monatomic ions Table 3.2

monatomic, or *single*-atom ions (**Table 3.2**). Notice that the metals form positively charged ions and that the magnitude of the charge is either identical with the group number (for elements in columns 1 and 2) or equal to the group number reduced by 10 (for the remaining metals shown in Table 3.2). For example, aluminum, in group 13, forms a $(13 - 10) = +3$ ion, or Al^{3+}. When forming ionic compounds, nonmetals form negatively charged ions, with charges equal to the magnitude of the group number reduced by 18. Phosphorus, for example, a group 15 element, forms a $(15 - 18) = -3$ ion, or P^{3-}, called *phosphide*.

We can effectively combine any metal and nonmetal shown in Table 3.2 to produce a binary ionic compound. The ratio of the two elements in the binary compound requires that the net electrical charge of all the ions must be zero. In practice, this means that the total number of positive charges provided by the cation(s) must equal the total number of negative charges brought by the anion(s). There are two ways to arrive at this equality of charge:

1. If the magnitude of the charge on the cation is the same as that of the anion, the cation and anion form a one-to-one compound. For example, with Mg^{2+} and O^{2-}, the magnitudes of the charges are the same (two in this case), so they form the one-to-one compound MgO, magnesium oxide. Here, the 2^+ charge of each cation balances the 2^- charge of each anion (**Figure 3.19a**).
2. If the magnitude of the charge on the cation is *not* the same as that of the anion, the cation element's *subscript* is taken to be the same number as the anion's charge, and vice versa. For example, with aluminum and oxygen, the operation Al^{3+} ⤨ O^{2-} gives us, Al_2O_3, aluminum oxide.

This method ensures that the total number of positive charges brought by the cations equals the total number of negative charges brought by the anions. In aluminum oxide, two aluminum ions bring a charge of $(2 \times +3) = +6$, and three oxygen ions bring a charge of $(3 \times -2) = -6$ (**Figure 3.19b**).

Mixed ionic compounds • Figure 3.19

This figure shows the ratios of ions in two ionic compounds: MgO and Al_2O_3. Note that cations are generally smaller than anions.

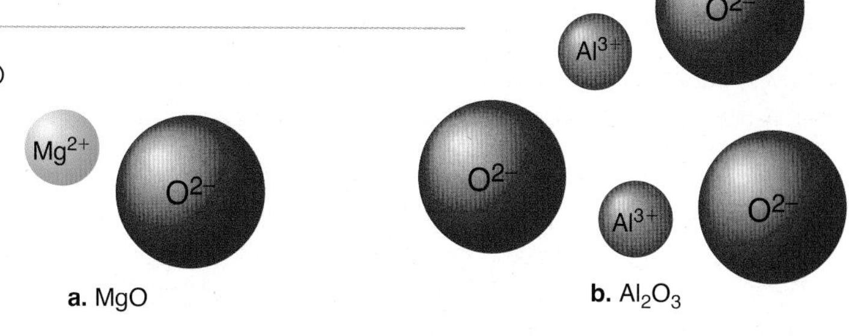

Formula Mass

formula mass Sum of the atomic masses of all the atoms in the formula of an ionic compound.

When working with ionic compounds, it is often helpful to know the **formula mass**. If we know a compound's formula, we can use atomic masses (in atomic mass units, u) to arrive at the formula mass. For example, aluminum oxide, Al_2O_3, has a formula mass of:

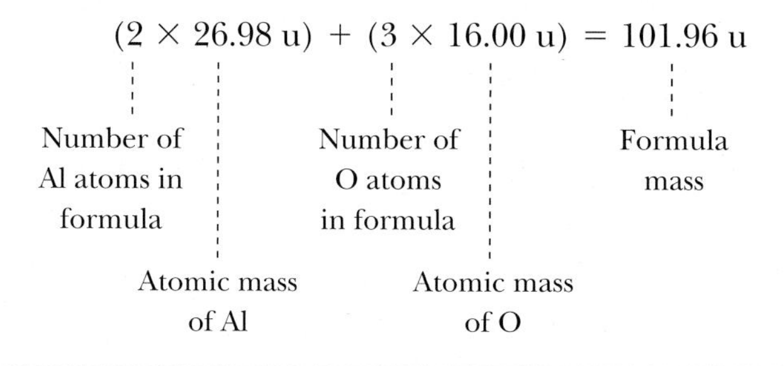

KNOW BEFORE YOU GO

1. Name and provide the formula mass of each of the following compounds:
   (a) MgI_2, (b) BeS, (c) K_2O.
2. Lead oxide, PbO_2, is an ionic compound used in the lead–acid batteries of automobiles. Knowing that each oxide anion, O^{2-}, has a 2− charge, and that there are two oxide anions for every lead cation, what is the charge on each lead cation?
3. What is the compound formula and name of the ionic compound made from: (a) silver and bromine, (b) strontium and sulfur, (c) potassium and nitrogen, (d) calcium and oxygen, (e) sodium and sulfur?

CONCEPT CHECK

1. **What** are the chemical formulas of potassium bromide, magnesium chloride, and calcium chloride?
2. **What** determines the ratio of ions in an ionic compound?
3. **How** can you determine the formula mass of an ionic compound?

3.4 Covalent Compounds

LEARNING OBJECTIVES

1. **Describe** how the structures of molecules are depicted.
2. **Show** how to write a molecular formula, calculate molecular mass, and name simple covalent compounds.
3. **Describe** how covalent compounds differ from ionic compounds.

When atoms share valence electrons to form covalent bonds, they generate a molecule. A molecule composed of atoms of different elements constitutes a **covalent compound**. Water, for example, consists of molecules formed by the covalent bonding of two hydrogen atoms to a single oxygen atom. These three atoms exist as a molecular unit of specific size and shape, no matter whether the water is in the form of ice, liquid water, or steam. To show the numbers of hydrogen and oxygen atoms in a water molecule, we write the chemical formula of water as H_2O. Here the subscript 2 indicates the presence of two hydrogen atoms and an implied subscript 1 after the oxygen indicates a single oxygen atom.

covalent compound A substance consisting of discrete molecules, each containing atoms of different elements held together by covalent bonds.

Model of HCl • Figure 3.20

Hydrogen chloride (HCl), also known as hydrochloric or muriatic acid, is an industrially important compound with varied uses.

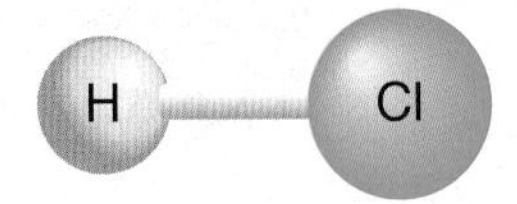

Visualizing Molecules

A **molecular structure** depicts the actual arrangement in space of the atoms of a molecule. In the case of a **diatomic**, or two-atom, molecule, the notion of molecular structure is simplified in that the bound atoms, by definition, form a straight line ●—●. Hydrogen chloride, HCl, is a linear molecule, shown in **Figure 3.20** in a ball and stick rendering.

> **molecular structure** The arrangement in space of the atoms that make up a molecule.

The chemical formula for water, H_2O, tells us what atoms are in the molecule and what their ratio is, but it doesn't convey anything about how the atoms are connected to each other or how they are oriented in space. For example, given that the molecular formula is H_2O, should we picture the connectivity of the atoms as H—H—O, or H—O—H? Furthermore, would these atoms all lie on the same straight line, ●—●—● or would they form an angular molecule ●^●^●? In the case of water, we find it is a bent molecule with an H—O—H atom connectivity. We can visualize this structure in several different ways (**Figure 3.21**).

You might wonder why HOH is an angular rather than a linear molecule—that is, why the two hydrogen atoms don't lie at the ends of a straight line with the oxygen atom at its center. Part of the answer is that the actual structure of the oxygen atom is more complex than the valence shell model suggests.

The valence shell model for water (**Figure 3.22a**) shows two hydrogen atoms connected to a single oxygen atom, with one hydrogen on each side of the oxygen. In this view, the nonbonding valence electrons (lone pairs) of the oxygen and the bonding electrons connecting the oxygen to each hydrogen are all separated from one another by 90° and thus exist as far apart as possible. But this simple view of H_2O shows the molecule in only two dimensions. A more accurate, three-dimensional model (**Figure 3.22b**) reveals that oxygen's lone pairs and bonding electrons all lie in **atomic orbitals**, extending outward to the corners of a tetrahedron. The two hydrogen nuclei reside in at the corners of this tetrahedron, with all groups separated by more than 90°—in fact, closer to 109.5°. This gives the water molecule its angular structure. In reality, the two lone pairs of the oxygen tend to repel each other a bit more strongly than the shared electrons of the OH bonds. As a result, the H—O—H angle is only 104.5°, whereas the two nonbonding electron pairs are separated by a little more than 109.5°.

> **atomic orbitals** Lobes or regions of space outside the atom's nucleus where there is a high likelihood of finding the atom's electrons. A maximum of two electrons may occupy each atomic orbital.

In the water molecule, the oxygen atom forms two bonds, one to each of the hydrogen atoms. In other molecules, we may find that atoms form bonds directly to more than two other atoms. For instance, in *methane* (CH_4), the single carbon atom is bonded to four hydrogen atoms (**Figure 3.23**). Methane is the principal component of natural gas, an important heating fuel.

Molecular models of water • Figure 3.21

Each representation has its own advantages. The model used should be the one most appropriate to the given topic of discussion.

a. Lewis structure showing geometry
Atom connectivity and bond angle can be seen clearly, but relative sizes of atoms cannot.

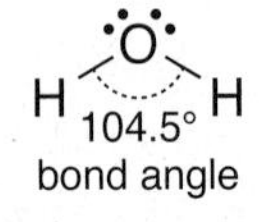

b. Ball-and-stick model
Atoms are shown as balls and bonds as sticks. Bond angle and relative sizes of atoms appear clearly, but this model overstates distances between atoms (bond lengths).

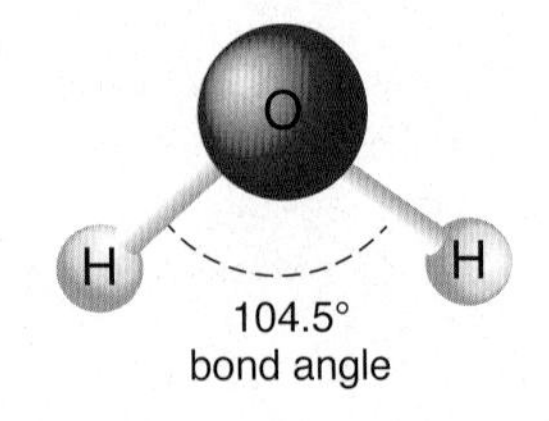

c. Space-filling model
The three-dimensional contour of the molecule can be seen clearly, but bonds and bond angles are vague.

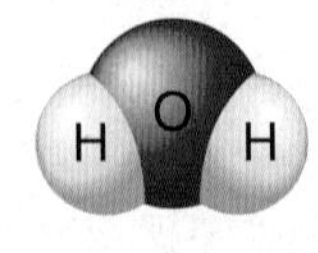

Visualizing why water is a nonlinear molecule • Figure 3.22

The angular shape of the water molecule is due partly to the arrangement of atomic orbitals in three-dimensional space.

a. Valence shell model.
If this simple valence shell structure were correct, then we might view the water molecule as a linear structure, with oxygen's two lone pairs on opposite sides of the oxygen atom (top and bottom) and the two bonded hydrogens also on opposite sides (left and right).

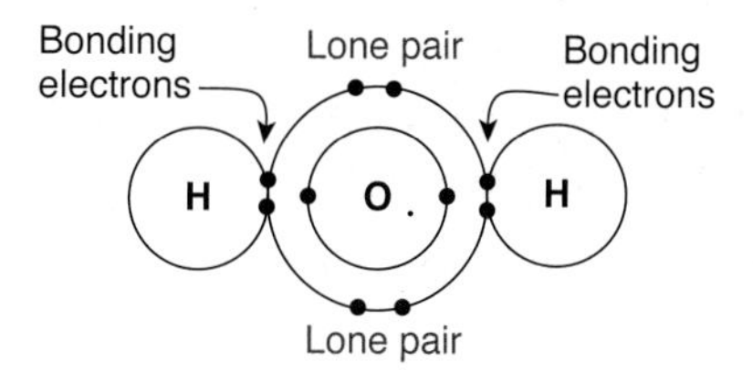

b. Atomic orbital model.
In a three-dimensional view of the water molecule, oxygen's lone pairs and bonding electrons all lie in atomic orbitals extending out from the central oxygen atom. Sets of repulsive forces tend to keep these electron orbitals as far apart as possible. This is achieved if the oxygen nucleus sits at the center of a *tetrahedron*, a four-sided figure with triangular faces, much like a pyramid.

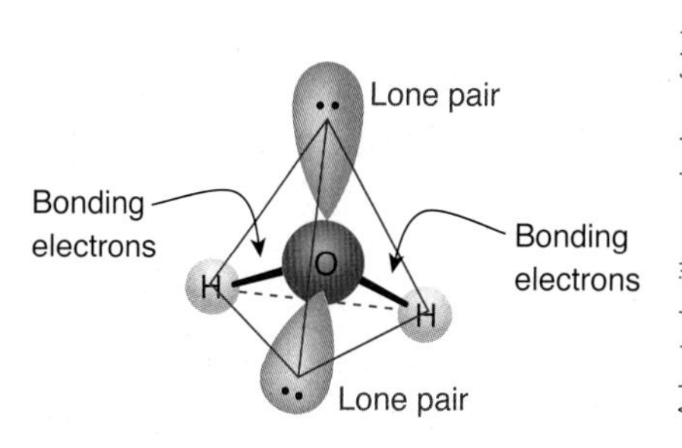

Put It Together Review Table 3.1 and answer the questions.

1. How many valence electrons do oxygen and hydrogen have, respectively?
2. (a) How many valence electrons in total are in the water molecule? (b) Of these, how many are involved in bonding? (c) Where do the remaining valence electrons reside?

The shape of the methane (CH_4) molecule • Figure 3.23

The carbon atom of methane forms four bonds.

a. In the methane molecule, covalent bonds point outward from the central carbon atom to the four corners of a tetrahedron, yielding four C-H bonds.

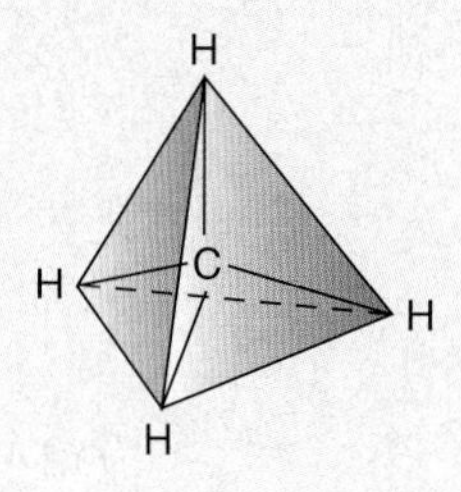

b. A ball-and-stick model of the methane molecule.

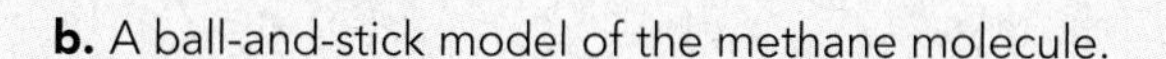

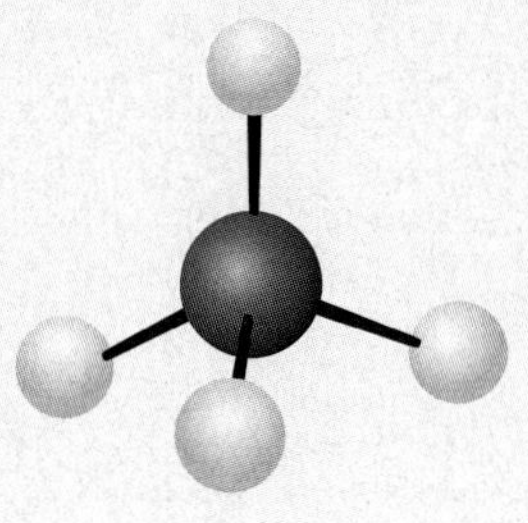

Think Critically

1. What is the total number of covalent bonds in the methane molecule?
2. What is the total number of electrons in these covalent bonds?
3. What is the total number of electrons in the entire methane molecule?

In **Figure 3.24**, we summarize the shapes of the water and methane molecules and present a third basic molecular form, the **pyramidal** shape. (We can find examples of other basic molecular shapes in addition to the ones we've already mentioned, but they are beyond the scope of our discussion.)

Molecular Formulas and Names

The **molecular formula** tells us how many atoms of each element are present in every molecule of a covalent compound, whether that compound exists as a solid, a liquid, or a gas. For example, the compound geraniol, a fragrant chemical derived from rose oil and used in perfumes, has the molecular formula $C_{10}H_{18}O$. This formula indicates that there are 10 carbon atoms, 18 hydrogen atoms, and 1 oxygen atom in each molecule of geraniol. Again, if the subscript for a given element is not shown, it is understood to be one. For molecules containing carbon, we write carbon first, hydrogen (if present) second, and then the other elements present in alphabetical order (**Figure 3.25**).

molecular formula The chemical formula of a covalent compound.

We can name covalent compounds in two different ways. **Systematic names** are based on an official nomenclature, or set of rules for naming. **Common names** have been carried down through history and are sometimes preferred for familiar substances such as water.

Examples of basic molecular shapes • Figure 3.24

The overall molecular shape depends on the arrangement of atoms within the molecule. This shape is influenced by sets of repulsive forces that tend to keep pairs of valence electrons and bonding electrons as far apart as possible. This may result in a tetrahedral arrangement of atomic orbitals, as shown in the examples here.

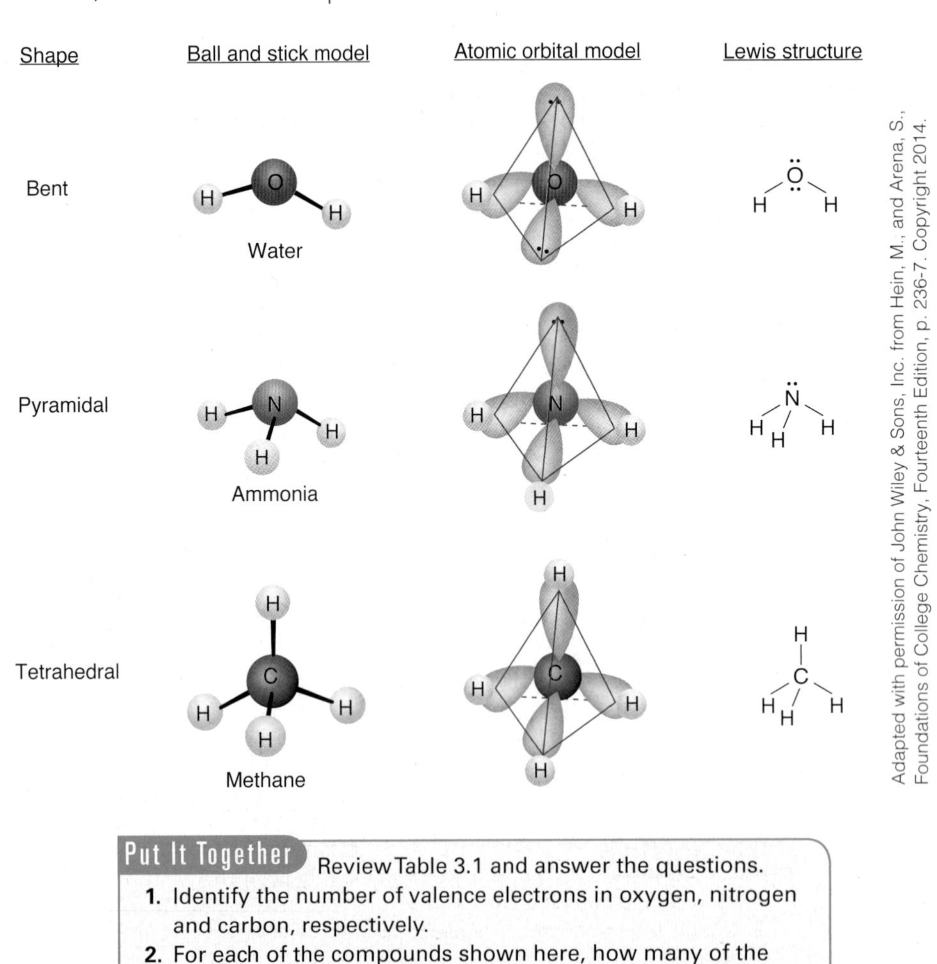

Put It Together Review Table 3.1 and answer the questions.

1. Identify the number of valence electrons in oxygen, nitrogen and carbon, respectively.
2. For each of the compounds shown here, how many of the central atom's valence electrons reside in: **(a)** lone pairs; **(b)** bonds?

Molecular formulas of covalent compounds • Figure 3.25

Every covalent compound has a molecular formula indicating the number of atoms of each element that are in every molecule of the compound.

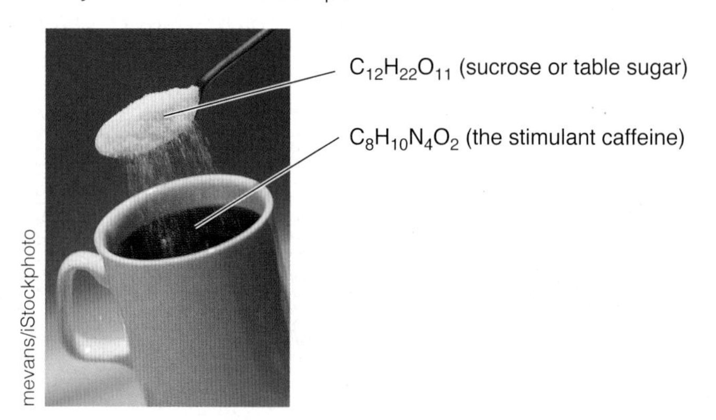

Systematic names of binary covalent compounds—compounds made from two nonmetals—simply tell us the names of the atoms within the molecule and the number of each of these atoms the molecule contains. For example, CO_2, with one carbon and two oxygens in each molecule, is known simply as *carbon dioxide.* With no prefix, the word "carbon" implies only a single carbon atom in each molecule. The prefix *di-* in *dioxide* tells us that there are two oxygen atoms attached to each carbon atom. Similarly, the covalent compound NO_2 is named *nitrogen dioxide* while SO_2 is *sulfur dioxide.* Other prefixes give us the names of more complex covalent molecules. *Sulfur trioxide* tells us there are one sulfur atom and three oxygen atoms in SO_3, while *dinitrogen pentoxide* identifies the two nitrogens and five oxygens in the molecule N_2O_5.

For molecules containing only a single oxygen atom and one or more atoms of some other element, the prefix *mon-* (a shortened form of *mono-*) is usually attached to the oxygen atom. With this convention we have *carbon monoxide* for CO and *dinitrogen monoxide* for N_2O. **Table 3.3** lists prefixes commonly used in the names of binary covalent compounds.

Common names tell us nothing at all about the number or kinds of atoms within the molecules. Instead these often ancient names reflect the various characteristics, origins, or histories of the substances involved. The word *water* (H_2O), for example, came into English long ago, from a variety of other languages. Its origin is obscure. Ammonia, the name of the compound NH_3, originated in 1782 as the name of a gas emitted by a substance known as *sal ammoniac.* This substance was itself named for its source, which was close to a Libyan temple named for the Egyptian God *Amun.*

Molecular mass is the sum of the atomic masses of all the atoms in a covalent compound or molecule. For example, water, a covalent compound with formula H_2O, has a molecular mass of:

> **molecular mass** Sum of the atomic masses of the atoms in a molecule.

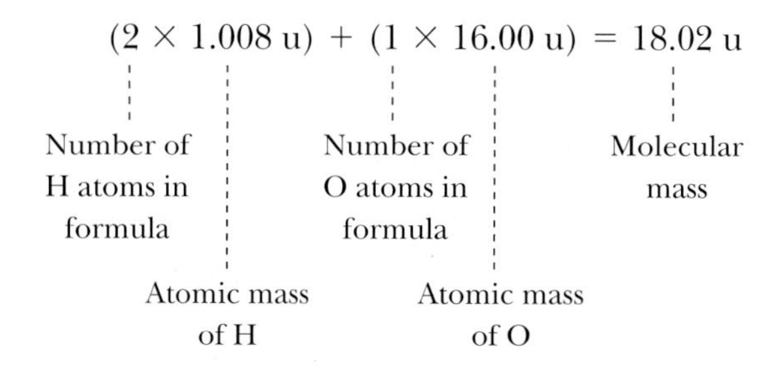

In another example, testosterone, a covalent compound with formula $C_{19}H_{28}O_2$, has a molecular mass of:

$$(19 \times 12.01\ u) + (28 \times 1.008\ u) + (2 \times 16.00\ u) = 288.4\ u$$

Prefixes indicating the number of atoms of each element in a binary covalent compound Table 3.3

Prefix	Number of atoms
mono-*	1
di-	2
tri-	3
tetra-	4
penta-	5

* *mono-*, or its shortened version, *mon-*, is only used with the second element, not the first, For example, CO is carbon monoxide, not monocarbon monoxide.

Covalent Versus Ionic Compounds

Let's review how covalent compounds differ from ionic compounds:

1. Covalent compounds exist as discrete molecular units, as opposed to ionic compounds, which exist as a regularly ordered array (crystal lattice) of ions that extends out in space.
2. The atoms in the molecules of covalent compounds are held together by covalent bonds. In ionic compounds, the ions of the crystal lattice are held together by ionic bonds.

WHAT A CHEMIST SEES

Everyday Products Containing Polyatomic Ions

Many common household products contain ionic compounds made up of polyatomic ions. These products may have applications as medicines, cleaning agents, or any number of other purposes. The one thing they have in common is that they contain ions consisting of two or more atoms covalently bonded together.

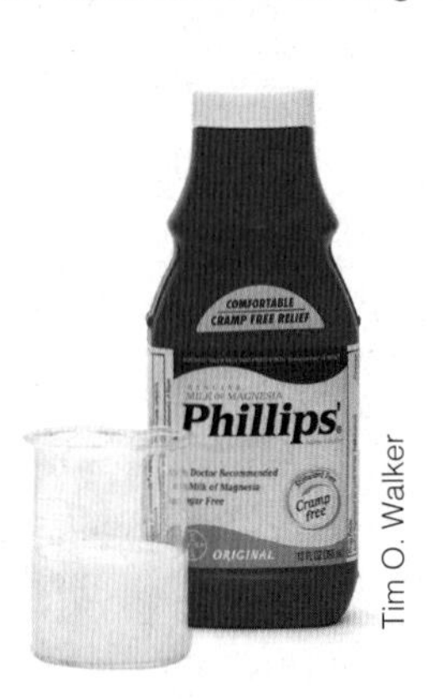

Tim O. Walker

Milk of magnesia – a suspension of *magnesium hydroxide*, $Mg(OH)_2$, used to treat indigestion and constipation

Mg^{2+} $[O—H]^{-}_{2}$

Hydroxide ion

Tim O. Walker

Bleach – a solution of *sodium hypochlorite*, NaOCl.

Na^{+} $[O—Cl]^{-}$

Hypochlorite ion

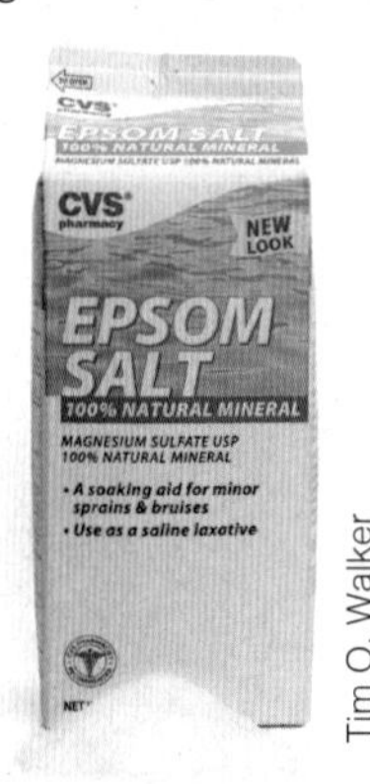

Tim O. Walker

Epsom salt–used for therapeutic baths.

Also known as *magnesium sulfate*, $MgSO_4$.

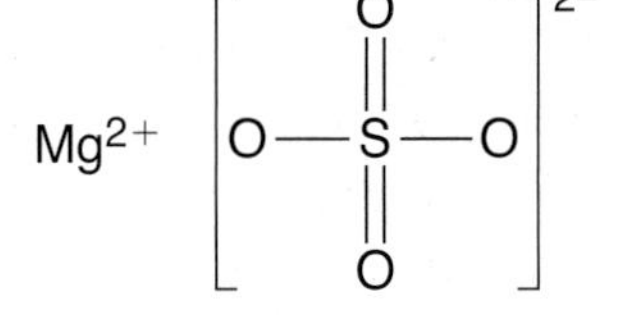

Sulfate ion

Tim O. Walker

Baking soda – used in cooking, cleaning, and a variety of other purposes.

Also known as *sodium bicarbonate*, $NaHCO_3$.

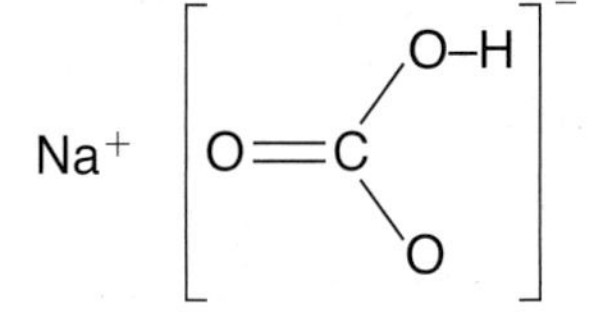

Bicarbonate ion

Common polyatomic ions Table 3.4

Name	Formula
Ammonium	NH_4^{+}
Bicarbonate	HCO_3^{-}
Carbonate	CO_3^{2-}
Hydroxide	OH^{-}
Nitrate	NO_3^{-}
Nitrite	NO_2^{-}
Phosphate	PO_4^{3-}
Sulfate	SO_4^{2-}

Looking back at the ionic compounds we've studied, we see that they were all of the binary type, combining a metal with a nonmetal in some fixed proportion. Some ionic compounds contain **polyatomic ions**. The atoms forming a polyatomic ion are bound together by covalent bonds, with the cluster as a whole bearing a positive or negative electric charge. In the sulfate ion, SO_4^{2-}, for example, sulfur and oxygen atoms are connected by covalent bonds, with the overall cluster of atoms carrying a negative charge. A variety of polyatomic ions occurs in everyday chemical substances (see **Table 3.4** and *What a Chemist Sees*).

polyatomic ion An ion made up of more than one atom.

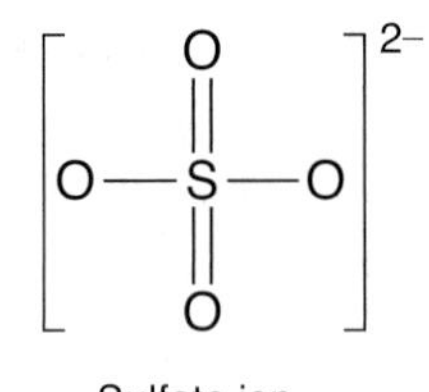

Sulfate ion

CONCEPT CHECK STOP

1. **How** is the covalent compound H_2O structurally different from the ionic compound Na_2O?
2. **What** are the molecular formula and molecular mass of *cortisone*, a covalent compound with 21 carbon atoms, 28 hydrogen, and 5 oxygen atoms in each molecule?
3. **Why** does sodium chloride not exist as a molecule?

3.5 Introduction to Organic Chemistry

LEARNING OBJECTIVES

1. **Define** organic chemistry.
2. **Recognize** the structure of alkanes and describe how to name them.

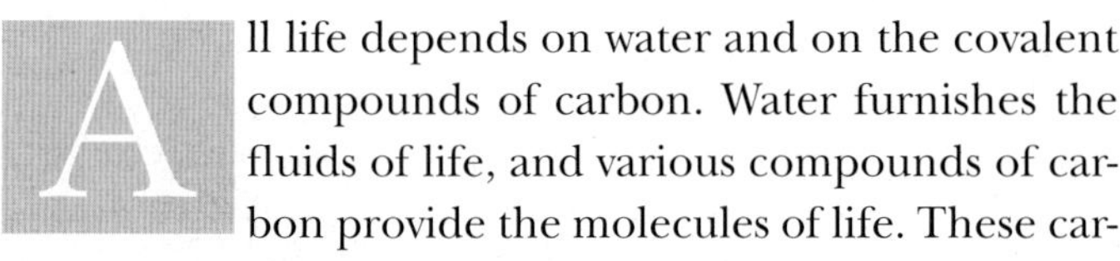

All life depends on water and on the covalent compounds of carbon. Water furnishes the fluids of life, and various compounds of carbon provide the molecules of life. These carbon compounds occur in all living things. No life exists without them.

Defining Organic Chemistry

The chemistry of carbon compounds is **organic chemistry**, a term coined by the earliest chemists. Until about 150 years ago, organic matter—substances obtained from things that are alive or were once alive—served as the only source of these compounds. Ethyl alcohol, for example, an **organic compound** whose intoxicating effects have been known from antiquity, has long been obtained from the fermentation of grains, which gives it the alternative name, grain alcohol. Soap, another organic substance, has been made for centuries from fats (also organic) rendered from slaughtered animals. Organic dyes and drugs of all sorts have been extracted for centuries from a great variety of plants. The red-orange dye henna, used since antiquity for coloring hair and leather, comes from the lawsonia shrub; quinine, the first effective treatment for malaria, was originally isolated from the bark of South American trees. If you had lived as late as the first third of the 19th century and you had wanted one of these compounds, or any other compound of carbon, you would have had to isolate it from organic matter. There were simply no other sources.

> **organic compound** A covalent compound containing carbon.

The "organic" mystique was so powerful in those years that it engendered a belief in a "vital force," which was supposedly possessed by all living things and was thought to be uniquely capable of producing the carbon compounds they contain. Urea illustrates the idea nicely. It's through urea, with its molecular formula CH_4N_2O, that almost all mammals excrete the unused nitrogen of proteins in their foods. Urea makes up 2–5% of human urine and was considered "organic" in the sense that it was believed to be generated only through the action of the mysterious vital force that exists only within living bodies. Substances like the urea that comes from living things were supposedly different in a very mysterious sense from those obtained from nonliving sources—water, for example—which were called "inorganic."

In 1828, the German chemist Friedrich Wöhler changed all this. He prepared urea in his laboratory from two chemicals, proving that the idea of a "vital force" unique to living things was irrelevant to the development of chemistry. The urea from Wöhler's laboratory was shown to be identical in every way to the urea formed in the bodies of mammals. Indeed, since the time of Wöhler's work, no difference has ever been demonstrated between the structure, properties, or behavior of a pure substance isolated from a living or once living thing and that very same pure substance prepared in a chemist's laboratory.

By now the term **organic compound** has lost its mysterious aura and, with only a very few exceptions, is simply the category of the compounds of carbon. In general use, however, the term **organic** has another meaning, describing foods produced without the use of synthetic fertilizers or pesticides, or through the use of genetic engineering.

Naming Organic Compounds

Organic compounds have structures and properties as varied as life itself. More than 20 million organic compounds are now known, each with its unique molecular structure, name, and chemical and physical properties.

To bring a sense of order to this enormous number and variety of carbon compounds, chemists have organized them into families, each of which consists of compounds of similar composition and properties. One of the largest of these is the family of **hydrocarbons**, which are compounds composed exclusively of carbon and hydrogen. It is convenient to organize the large hydrocarbon family even further into smaller families, such as the **alkanes**, which we'll discuss here.

> **alkane** A hydrocarbon molecule containing only single bonds.

Recall that a carbon atom possesses four valence electrons, $\cdot\dot{\underset{\cdot}{\mathrm{C}}}\cdot$, so it can form covalent bonds with as many as four other atoms. For example, in methane (CH_4), the carbon atom forms four single bonds, one to each of the hydrogen atoms. Methane not only represents the simplest of all the hydrocarbons, but the simplest of the alkanes as well.

Introduction to alkane structure • Figure 3.26

The carbon atoms within alkanes form single bonds only. Note the presence of carbon–carbon bonds in ethane and propane.

Name	Molecular Formula	Lewis Structure	Condensed Structural Formula	Space-filling Model
Methane	CH_4	H–C(H)(H)–H	CH_4	
Ethane	C_2H_6	H–C(H)(H)–C(H)(H)–H	CH_3CH_3	
Propane	C_3H_8	H–C(H)(H)–C(H)(H)–C(H)(H)–H	$CH_3CH_2CH_3$	

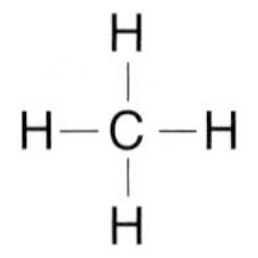

Adapted with permission of John Wiley & Sons, Inc. from Hein, M., and Arena, S., Foundations of College Chemistry, Fourteenth Edition, p. 447. Copyright 2014.

Ask Yourself

For each of the molecules shown, identify the number of C-H and C-C bonds.

The next larger alkane molecule is *ethane*. The molecular formula of ethane, C_2H_6, tells us that each of its molecules is made up of two carbon atoms and six hydrogen atoms. The **Lewis structure**, though, shows explicitly all the covalent bonds within the molecule. The **condensed structural formula** of ethane, CH_3CH_3, tells us that the molecule is made up of two -CH_3 units, also known as *methyl* groups, linked to one another. In **Figure 3.26**, we summarize these ways of representing molecules and present a comparison of the first three alkanes: methane (CH_4), ethane (C_2H_6), and propane (C_3H_8).

In naming alkanes, we use the suffix, *–ane.* Following methane, we can form a series of **linear** or **straight-chain alkanes**, of increasing carbon-chain length (**Table 3.5**). If you were to trace a path along the carbon chain of each molecule in this series, you would find that all carbon atoms lie in a single, unbranched chain. In other words, each carbon chain is linear.

Unlike the linear alkanes shown in Figure 3.26, many alkanes have branch points along their carbon chains. For example, consider a compound with the molecular formula C_4H_{10}. We know this represents an alkane,

The first 10 straight-chain alkanes Table 3.5

Alkanes have the general molecular formula, C_nH_{2n+2}, where n is equal to the number of carbon atoms in the molecule.

Name	Molecular formula C_nH_{2n+2}	Condensed structural formula
Methane	CH_4	CH_4
Ethane	C_2H_6	CH_3CH_3
Propane	C_3H_8	$CH_3CH_2CH_3$
Butane	C_4H_{10}	$CH_3CH_2CH_2CH_3$
Pentane	C_5H_{12}	$CH_3CH_2CH_2CH_2CH_3$
Hexane	C_6H_{14}	$CH_3CH_2CH_2CH_2CH_2CH_3$
Heptane	C_7H_{16}	$CH_3CH_2CH_2CH_2CH_2CH_2CH_3$
Octane	C_8H_{18}	$CH_3CH_2CH_2CH_2CH_2CH_2CH_2CH_3$
Nonane	C_9H_{20}	$CH_3CH_2CH_2CH_2CH_2CH_2CH_2CH_2CH_3$
Decane	$C_{10}H_{22}$	$CH_3CH_2CH_2CH_2CH_2CH_2CH_2CH_2CH_2CH_3$

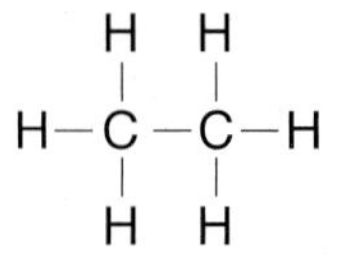

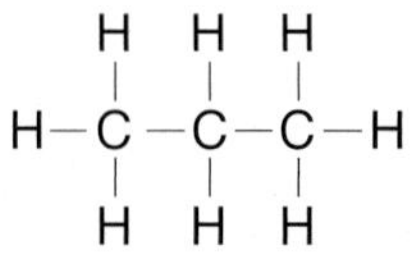

Ask Yourself

1. What is the name of the straight-chain alkane with six carbons?
2. What is the molecular formula of an alkane with 16 carbon atoms?

Alkane isomers • Figure 3.27

These two compounds have the same molecular formula, C_4H_{10}, but have different structures and are named differently.

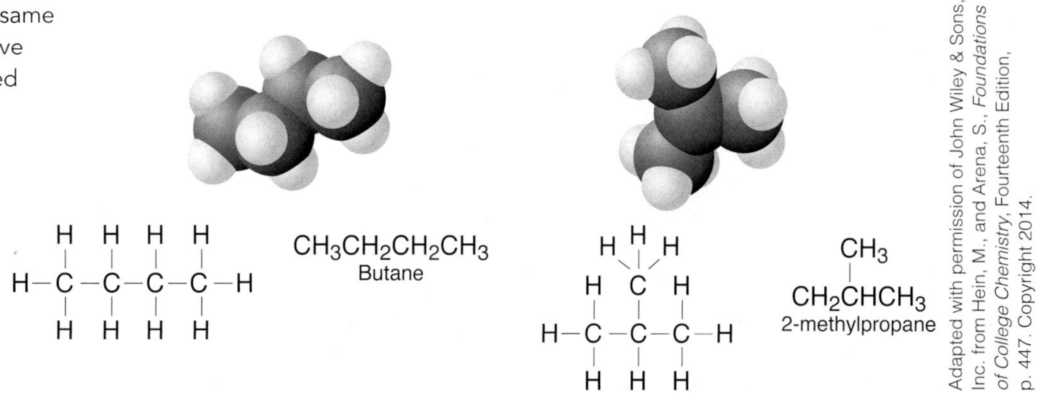

because it follows the general formula, C_nH_{2n+2}, where $n = 4$ in this case. However, two distinct alkanes share this same formula: One has a linear structure and the other has a branch point in its structure (**Figure 3.27**). We call compounds with the same molecular formula but different structures **isomers**.

isomers Two or more compounds that share the same molecular formula but differ in structure.

As the number of atoms in a molecule increases, so does the capacity for isomerism, and so does the number of names required to assign a unique name to each of these molecules. In the case of an alkane with four carbons, we can form two isomers: *butane* and *2-methylpropane*. For an alkane with five carbons, we can form three isomers. For alkanes with greater than five carbons, the number of possible isomers increases dramatically. For example, we can form 75 possible isomers for an alkane with 10 carbons. Since each organic compound (including all isomers) requires a unique name, chemists have developed a systematic method of naming organic compounds. This set of rules is known as the **IUPAC system** (**Figure 3.28**), which is the acronym for the chemistry organization that developed them, The **I**nternational **U**nion of **P**ure and **A**pplied **C**hemistry.

The IUPAC system • Figure 3.28

The IUPAC system is a set of rules for naming organic compounds. Here, we show how these rules apply to the naming of alkanes.

1. Identify the longest continuous chain of carbon atoms within the molecule. This gives the **parent** or **root name**.
2. Identify any groups attached to this chain—a *methyl* group, for example. Number the carbons of the chain starting from the end that gives the position(s) of the attached group(s) the lowest number(s).
3. Indicate the total number of each set of mutually identical groups with a prefix. For example, use *dimethyl* for two methyl groups, *trimethyl* for three methyl groups, and so on.
4. Give each group present on the chain its individual location number.
5. Separate numbers by commas, and separate numbers from letters by a hyphen.

For example, the molecule shown in **a.** is 2-methylpropane. Its root name is propane because there are three carbons in the longest continuous carbon chain. The number 2 places the single methyl group at the second of the three-carbon (propane) chain.

The molecule shown in **b.** is 2,2,4-trimethylpentane, a compound used in determining octane ratings of gasoline. The compound's name indicates that there are three methyl groups, located at the 2-, 2-, and 4- positions, respectively, along the parent (pentane) chain.

a. 2-methylpropane

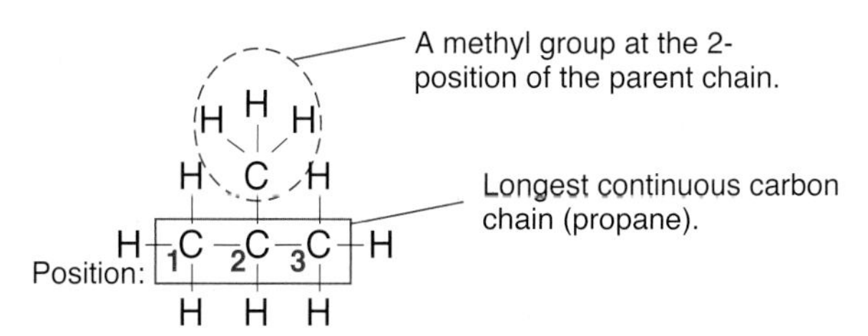

b. 2,2,4-trimethylpentane

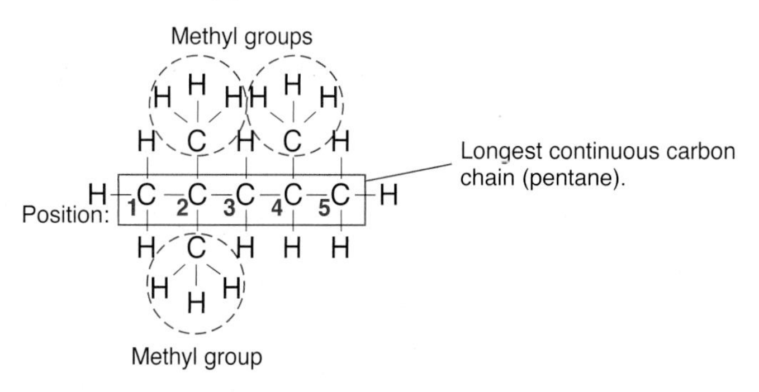

Geraniol: An alcohol • Figure 3.29

The scent of a rose is due to a variety of organic compounds, one of which is geraniol, $C_{10}H_{18}O$, a type of alcohol.

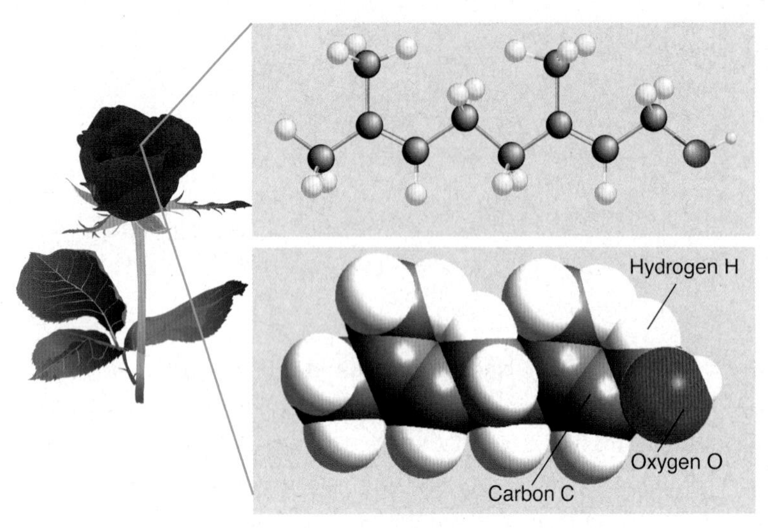

a. Ball and stick model of geraniol.

b. Space-filling model of geraniol.

Think Critically

1. Identify the location of the hydroxyl group in the geraniol molecule shown in **Figure a.**
2. Examine Figure a and identify the number of carbon atoms directly bonded to each of the following:
   (a) no hydrogen atoms;
   (b) one hydrogen atom;
   (c) two hydrogen atoms;
   (d) three hydrogen atoms.

KNOW BEFORE YOU GO

4. (a) How many carbon atoms are in 2,2,4-trimethylpentane? (b) What is its molecular formula?
5. Name the straight-chain alkane that is an isomer of 2,2,4-trimethylpentane.
6. Draw the Lewis structures of the three C_5H_{12} isomers: pentane, 2-methylbutane, and 2,2-dimethylpropane.

To summarize, we've just explored identifying and naming alkanes, one of the subfamilies within the larger family of hydrocarbons. Since carbon atoms form bonds to atoms of other elements besides hydrogen and carbon, a wide variety of other organic families exist in addition to the hydrocarbons. For example, compounds containing a bond from carbon to a *hydroxyl* group, -OH, can belong to the **alcohol** family (**Figure 3.29**). Throughout the book, we'll explore these and other families of organic compounds.

CONCEPT CHECK

STOP

1. **What** do chemists mean by the term *organic* and what alternate meaning does this word have as it pertains to foods?
2. **What** do you call compounds, such as pentane and 2-methylbutane, that have the same molecular formula but different structures?

Summary

1 Periodicity and Valence Electrons 56

- ***Why do elements in the same column of the periodic table show similar chemical reactivity?***
  Elements in the same column of the periodic table have the same number of valence electrons (as shown in the table). As a result, they exhibit similar chemical behavior (with rare exceptions).

Table 3.1 • Lewis (or electron-dot structures) structures of the first 20 elements

H·																	He:
Li·	·Be·											·B·	·C·	·N:	:O:	:F:	:Ne:
Na·	·Mg·											·Al·	·Si·	·P:	:S:	:Cl:	:Ar:
K·	·Ca·																

2 Ionic and Covalent Bonding 58

- ***How and why do atoms bond to one another?***
  Atoms often react so as to obtain filled valence shells. When atoms of identical nonmetals share two valence electrons—one electron supplied by each atom—they form a nonpolar covalent bond, such as the Cl— Cl bond (shown below).

  A polar covalent bond forms when valence electrons are shared by atoms of nonmetals of differing electronegativity, as in the H— Cl bond.

  When atoms of a metal react with atoms of a nonmetal, electrons are transferred from the metal to the nonmetal to form cations and anions, respectively. For example, when sodium and chlorine atoms react, an electron is transferred from sodium to chlorine, resulting in an ionic bond between a sodium cation and chloride anion.

Figure 3.8 • Electronegativity and bond types

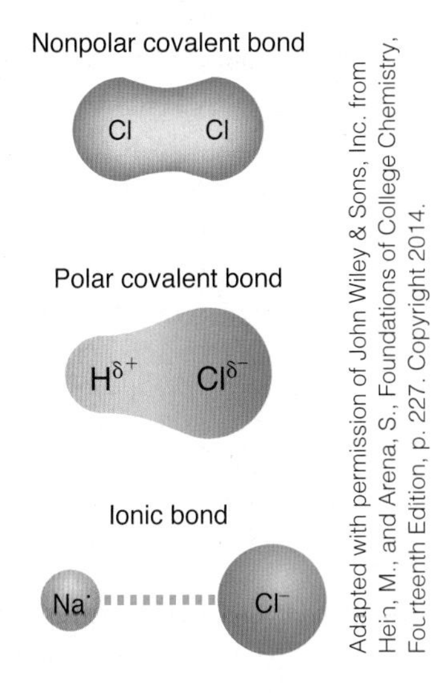

- ***Why does salt water conduct electricity?***
  Salt water conducts electricity because the dissolved sodium chloride dissociates into sodium cations and chloride anions, which can carry an electric current as they travel through the aqueous solvent.

3 Ionic Compounds 68

- ***When two elements interact to form an ionic compound, in what ratio do they combine?***
  Metals react with nonmetals to form ionic compounds. The ratio of cations to anions in an ionic compound is such that the net electrical charge of all the ions in the compound must be zero. For example, sodium chloride consists of equal numbers of sodium cations, Na^+, and chloride anions, Cl^-, producing the chemical formula NaCl and the crystal structure shown here. Calcium chloride, $CaCl_2$, requires two Cl^- ions for each Ca^{2+} ion present, with each 2+ charge of the Ca^{2+} ions balanced by two −1 charges of the two Cl^- ions.

Figure 3.17 • The formation of sodium chloride by chemical interactions

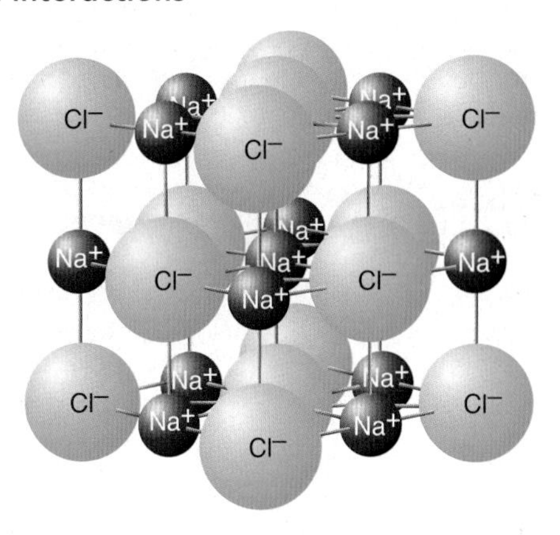

4 Covalent Compounds 71

- ***How do we represent the structure of molecules?***
  One common representation of covalent compounds employs ball-and-stick molecular models, shown here for the molecule, methane, with balls representing atoms and sticks representing covalent bonds. The molecular formula for methane is CH_4, which indicates the presence of one carbon atom and four hydrogen atoms in every molecule of methane.

Figure 3.23 • The shape of the methane moelcule (CH_4)

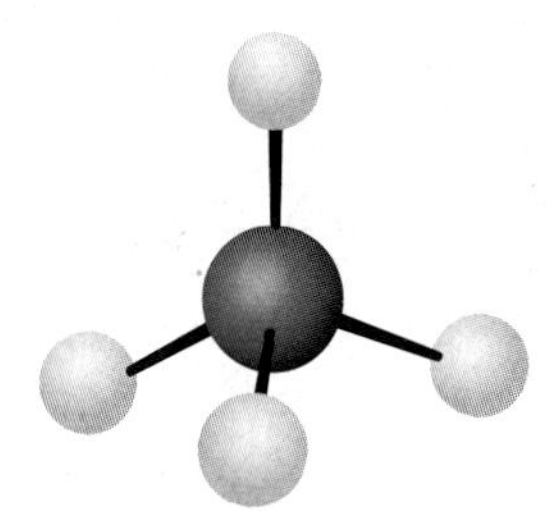

- ***How do covalent and ionic compounds differ?***
  Covalent compounds exist as discrete molecular units, as opposed to ionic compounds, which exist as a regularly ordered array (crystal lattice) of ions. The atoms in the molecules of covalent compounds are held together by covalent bonds. In ionic compounds, the ions of the crystal lattice are held together by ionic bonds.

5 Introduction to Organic Chemistry 77

- ***What are organic compounds?***
  Organic compounds are covalent compounds containing carbon. Chemists organize organic compounds into families, with similar composition and properties. One of the largest of these families is the hydrocarbons, compounds composed exclusively of carbon and hydrogen. Alkanes are a subtype of hydrocarbon, containing only single bonds. Alcohols are organic compounds containing the C— OH bond, such as shown here for geraniol.

Figure 3.29 • Gernanol: An alcohol

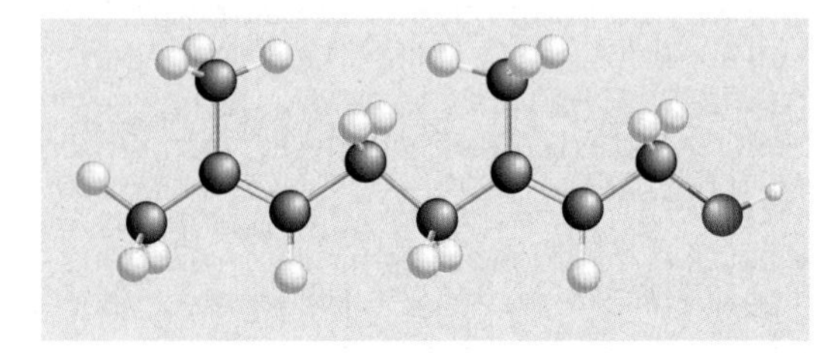

Key Terms

- alkali metals 57
- alkaline earth 69
- alkane 77
- anion 59
- atomic orbital 72
- cation 59
- compound 60
- compound formula 69
- covalent bond 60
- covalent compound 71
- crystal lattice 65
- double bond 64
- electrolyte 64
- electron configuration 56
- electronegativity 61
- formula mass 71
- halogens 57
- inert or noble gases 57
- ion 59
- ionic bond 60
- ionization 62
- isomer 79
- Lewis or electron dot structures 58
- lone pair 61
- molecule 60
- molecular formula 74
- molecular mass 75
- molecular structure 72
- nonpolar covalent bond 61
- octet rule 59
- organic compound 77
- polar covalent bond 61
- polyatomic ion 76
- single bond 64
- triple bond 64
- valence electron 58
- valence shell 58

What is happening in this picture?

AFP/Getty Images

An observer surveys the damage left in the wake of an industrial accident in 2010 that flooded several towns in western Hungary with a red sludge from a nearby aluminum refining plant. Approximately 1 million cubic meters of this waste material was released into the environment when a reservoir at the factory ruptured, ultimately resulting in nine deaths and significant property and environmental damage.

This red material owes its color to the presence of iron (III) oxide, Fe_2O_3, an ionic compound containing Fe^{3+} cations and O^{2-} anions. Other compounds present in this sludge included aluminum oxide, calcium oxide, and sodium oxide. Much of the biological damage from this accident was due to the strong basicity of the material, which can cause chemical burns of the skin. (We'll discuss acids and bases in Chapter 8.) Because of these risks, rescue workers responding to this accident were advised to wear protective gear to prevent skin damage.

Think Critically

1. What is the chemical formula for each of the following: **(a)** aluminum oxide; **(b)** calcium oxide; **(c)** sodium oxide? [You may find it helpful to review Table 3.2 in answering the question.]
2. Are these ionic or covalent compounds?
3. Many cations of transition metals are colored. What cation is responsible for the red color shown here?

Exercises

Review

1. Which would you expect to be a better conductor of electricity: water taken from the ocean or water taken from an inland lake? Explain your choice.
2. How can you identify an atom's valence shell?
3. What is the name of the group of elements consisting of fluorine, chlorine, bromine and iodine?
4. What, specifically, do the dots of a Lewis structure represent?
5. What is the maximum number of electrons that can occupy the valence shell of a hydrogen atom?
6. Atoms of what elements have valence shells that are filled by only two electrons? In which columns of the periodic table are these elements located?
7. What is the name of the group of elements that occupy the column of the periodic table headed by lithium?
8. How many electrons are shared between the two oxygen atoms in a molecule of O_2?
9. a. Explain the octet rule.
   b. Show how the octet rule applies to the ionic compound sodium chloride and to the covalent compound water.
   c. Which atoms within the water molecule do not obey the octet rule?
10. a. How many valence electrons does the oxygen atom contain?
    b. How many of oxygen's valence electrons are used to form bonds in the water molecule shown here? Are these bonds covalent or ionic?
    c. What do the gray lobes and pairs of dots in this figure represent?

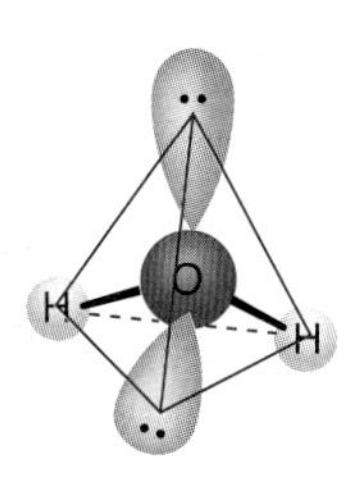

Adapted with permission of John Wiley & Sons, Inc. from Hein, M. and Arena, S., Foundations of College Chemistry, Fourteenth Edition, p. 237. Copyright 2014.

11. When either lithium or sodium is added to water, what gas is produced?
12. Describe the physical and chemical characteristics of those elements composed of electrically neutral atoms with filled valence shells.
13. a. Which represents a polar covalent bond? Which represents a nonpolar covalent bond?

    b. What structural characteristic distinguishes these two types of bonds?
    c. What property measures the ability of a bonded atom to attract electrons?
14. What does a double arrow $\rightleftharpoons$ indicate about a chemical reaction?
15. Identify a chemical reaction that involves an ionization.
16. What kind of chemical bond is formed when two atoms share a pair of electrons?
17. What *two* pieces of information does the chemical formula of a compound give us?
18. Give an example of a molecule in which two atoms share two pairs of electrons.
19. What term do we use to indicate that two bonded atoms share a total of six electrons?
20. Refer to Table 3.2 and the periodic table in answering the following questions. Consider the series of elements from lithium to magnesium (i.e., with atomic numbers 3–12). Which can form:
    a. cations containing eight valence electrons?
    b. anions containing eight valence electrons?
    c. an electrically neutral atom with eight valence electrons?

21. Identify an element among the first eight elements of the periodic table that exists as diatomic molecules, with each molecule containing a

a. single bond

b. double bond

c. triple bond

22. Refer to Table 3.2 in answering the following questions. Consider the series of elements from lithium to fluorine (i.e., with atomic numbers 3–9). Of these, give the chemical formulas and names of the ionic compounds that have general formulas of the following types, where X is a metal and Y is a nonmetal.

a. XY (*Hint*: Two compounds within this series of elements satisfy this formula.)

b. X_2Y

c. XY_2

d. X_3Y

23. What is the most important factor that governs the ratio of cations to anions in an ionic compound?

24. Given the molecular formula of a covalent compound, how do we calculate the compound's molecular weight?

25. What are the advantages and disadvantages of each of the following two types of molecular models?

a. ball and stick

b. space filling

26. Carbon dioxide, CO_2, is depicted here using two different molecular models.

a. What does the ball-and-stick model reveal about the nature of the carbon–oxygen bonds in this molecule that is not apparent in the space-filling model?

b. What is the molecular shape: linear, bent, pyramidal, or tetrahedral?

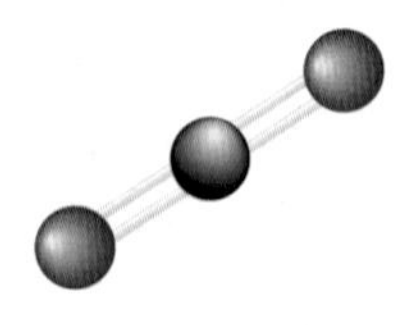

Ball and stick

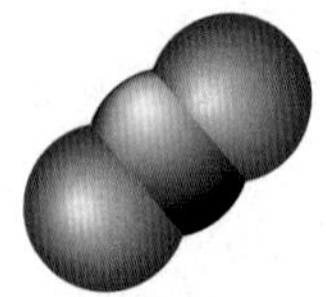

Space filling

27. a. What is a polyatomic ion?

b. Identify four different commercial products, often found in the home, that contain polyatomic ions.

c. What are the name and chemical formula of the principal polyatomic ion in each product?

28. Give the name and the chemical formula of the compound formed by the electron transfer between atoms A and B, shown here.

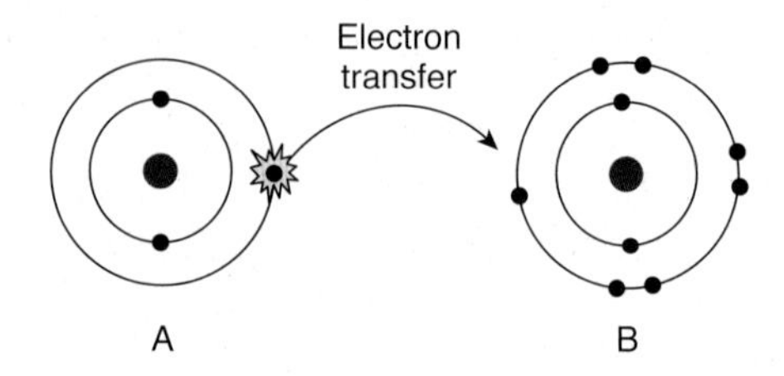

29. Explain why ionic compounds generally have higher melting points than covalent compounds.

30. Heating a thin strip of magnesium metal in the air causes the magnesium to burn with an intense white flame as the magnesium combines with the oxygen of the air to form an ionic compound. What are the name and the chemical formula of this compound?

31. When a small piece of sodium metal, Na, is dropped into a beaker of water, it reacts vigorously with the water, producing sodium hydroxide, NaOH, and hydrogen gas, H_2. The flame observed is due to the released hydrogen gas immediately reacting with oxygen, O_2, to produce water vapor, H_2O.

a. In this process, which polyatomic ion is produced?

b. Knowing that elements in the same group of the periodic table show similar chemical behavior, what products would we expect from the reaction of water with i) potassium? ii) rubidium?

32. a. Provide the names and molecular formulas of the following alkanes.

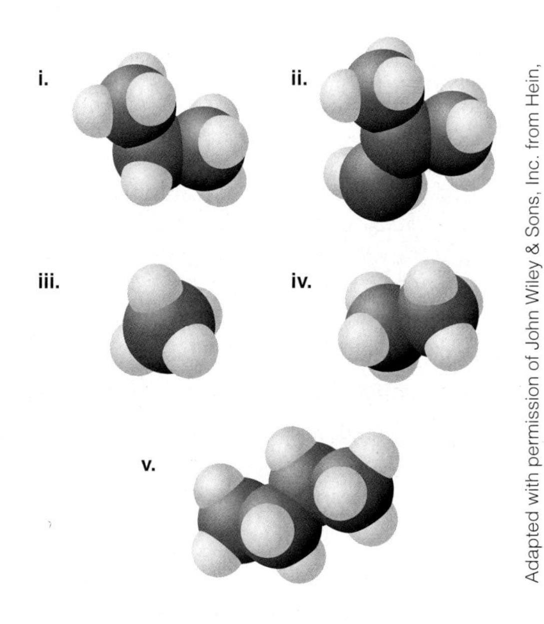

Adapted with permission of John Wiley & Sons, Inc. from Hein, M., and Arena, S., Foundations of College Chemistry, Fourteenth Edition, p. 447. Copyright 2014.

b. Which compounds shown represent isomers?

33. Consider the compounds 3-methylpentane, 2,2-dimethylbutane, and 2,3-dimethylbutane.

a. Provide the Lewis structure and molecular formula of each.

b. Which of these compounds represent isomers?

34. In the year 1800, what characteristic was used to define a chemical as an organic chemical? What characteristic is used today to define a chemical as an organic chemical?

Think

35. How does the electronic configuration of noble gases help explain why these gases are unreactive under ordinary conditions?

36. a. In what way are the electron configurations of lithium and sodium atoms different?

b. In what way are they similar?

37. While ionic and covalent bonds are fundamentally different, the concept of a filled octet comes into play when describing each kind of bond. Why?

38. In the reaction of two different elements with each other—which we will identify as X and Y—two electrons transfer from the valence shell of an atom of X to the valence shell of an atom of Y. What, if anything, can we conclude about the compound formed from the reaction of X and Y?

39. The approximate relative sizes of some common atoms and ions illustrate the generality that monatomic cations are smaller than their corresponding neutral atoms, while monatomic anions are larger than their neutral atoms. Why would removing one or more electrons from an atom make it smaller, while transferring one or more electrons to an atom make it larger?

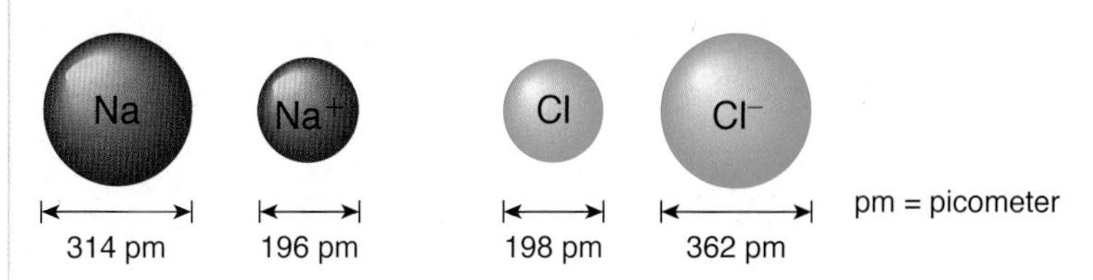

40. Compound formulas, such as $CaCl_2$, and molecular formulas, such as CO_2, look similar in that they each show elemental symbols with subscripts (or an implied subscript for$_1$). But a compound formula expresses a somewhat different concept than a molecular formula. What is the difference?

41. A molecule of propane, which is the fuel of liquid propane gas supplied by tanks to outdoor grills, trailers, and many homes, has the structure shown here.

```
    H  H  H
    |  |  |
 H—C—C—C—H
    |  |  |
    H  H  H
```

a. What is the total number of covalent bonds in propane?

b. What is the total number of shared electrons in propane?

c. What is the molecular formula of propane?

42. We can simulate the formation of the covalent bond in H_2 by imagining what happens as two separated hydrogen atoms (Step 1), move closer to each other (Step 2), until they form a bond (Step 3).

If we were to force the atoms even closer after the bond is formed, they would repel one another as shown in Step 4. Why?

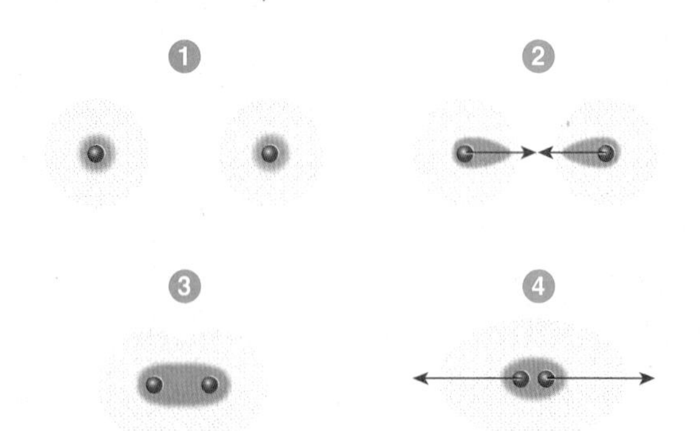

Adapted with permission of John Wiley & Sons, Inc. from Jespersen, N., Brady, J., and Hyslop, A., *Chemistry: The Molecular Nature of Matter*, Sixth Edition, p. 369. Copyright 2012.

43. Pure hydrogen chloride, HCl, is a gas under ordinary conditions. Molecules of pure, gaseous HCl consist of a hydrogen atom covalently bonded to a chlorine atom. When HCl gas dissolves in water, the resulting solution easily conducts an electric current, even though the gaseous HCl is a covalent compound. Explain why.

44. The covalent compound acetylene, which is the fuel of the oxyacetylene torch used by welders, has the molecular formula C_2H_2. The covalent compound benzene, a commercial solvent, has the molecular formula C_6H_6. Each of these covalent compounds contains carbon and hydrogen atoms in a one-to-one ratio. Would it be correct to write the chemical formulas of each as CH? Explain.

45. When we say that a certain food is high in sodium or that bananas are a good source of potassium, we are referring to the *ions*, Na^+ and K^+, and not to the *metals*, Na and K. These ions are highly water soluble and important in human nutrition. The metals, Na and K, on the other hand, are highly reactive in water and completely unsafe for consumption. Explain these different properties by comparing the electronic structures of Na^+ and K^+, with Na and K, respectively.

46. Would you expect a sample of calcium phosphide, Ca_3P_2, to contain molecules? Why or why not?

47. In Figure 3.18 we see that a lithium atom reacts with a fluorine atom to form LiF. Suppose that the maximum number of electrons that can occupy the first electron shell remains at two, but that the second quantum shell could hold a maximum of *nine* electrons (instead of the eight electrons that actually do fill the shell). In that hypothetical case, what would be the formula of lithium fluoride?

48. All the elements of column 2 of the periodic table, beginning with beryllium and magnesium, hold two electrons in their valence shells. Why is helium, which also contains two electrons in its valence shell, located in column 18 rather than in column 2?

49. Are all molecules necessarily covalent compounds? Explain, with examples.

50. Both lithium and sodium react with chlorine, a halogen, to form respectively, lithium chloride and sodium chloride. Both lithium and sodium can also react with hydrogen to form, respectively, lithium *hydride*, LiH (composed of Li^+ and H^- ions), and sodium *hydride*, NaH (composed of Na^+ and H^- ions).

a. Explain why hydrogen can sometimes act as if it were a member of the halogen family of elements.

b. Draw the electron dot representation of the hydride ion.

51. Nitric oxide, NO, is a covalent compound that serves several important roles in the body. Draw an electron dot representation of the NO molecule, showing a double bond between the atoms.

52. Each of the following represents a binary compound. For each structure, describe the bond type, determine the chemical formula, and calculate the molecular or formula mass, as appropriate.

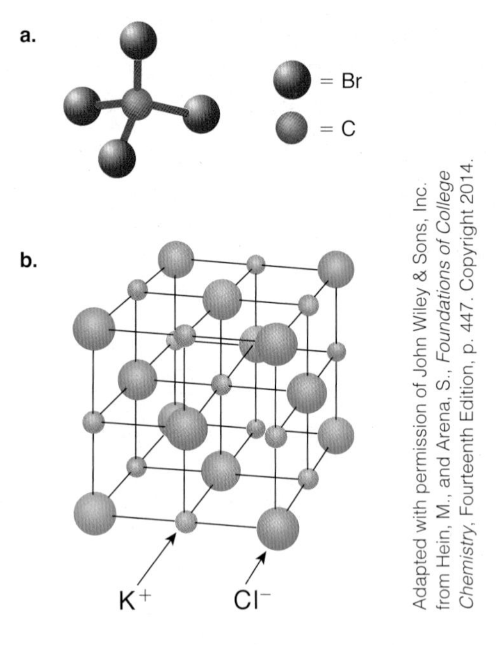

Adapted with permission of John Wiley & Sons, Inc. from Hein, M., and Arena, S., *Foundations of College Chemistry*, Fourteenth Edition, p. 447. Copyright 2014.

53. One atom of one element (element A) reacts with one atom of another element (element B) to form a new, binary compound, AB. Describe in detail what determines whether one or more electrons are transferred from A to B to form an ionic AB, or whether two (or more) electrons are shared between A and B to form a covalent compound.

Calculate

54. How many protons and how many electrons are in each of the following ions:

a. K^+ b. S^{2-} c. Al^{3+} d. Fe^{2+}

55. What is the formula mass of each of the following compounds? Which has the largest formula mass? Which has the smallest?

a. CdSe b. Al_2O_3 c. Na_2S d. Li_3N

56. Which has the larger molecular mass: a molecule of vitamin C ($C_6H_8O_6$), or a molecule of vitamin B_6 ($C_8H_{11}NO_3$)?

57. To help prevent cavities, certain brands of toothpaste contain *stannous fluoride*, an ionic compound consisting of Sn^{2+} ions and fluoride ions.

a. What is the compound formula of stannous fluoride?

b. What is its formula mass?

58. Glucose, or blood sugar, shown here, is a key source of energy for the brain and body. Use the structure of glucose to determine its molecular mass.

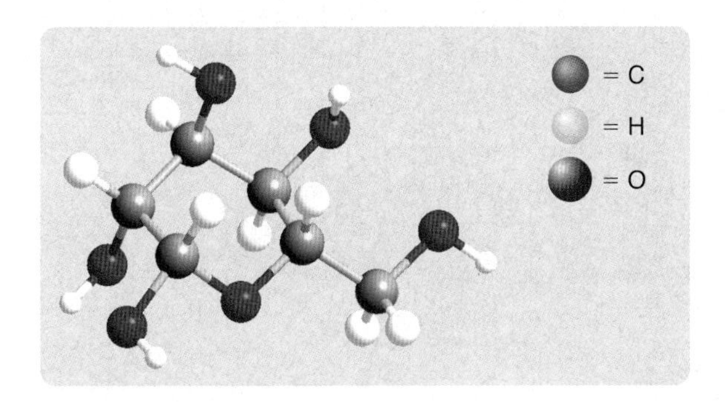

59. The caffeine content of an 8 oz cup of coffee varies, depending on several factors. If we assume an average value of 90 mg of caffeine per cup, how many caffeine molecules are in the average 8 oz cup of coffee? The molecular formula of caffeine is $C_8H_{10}N_4O_2$. Also, 1 atomic mass unit (u) $= 1.66 \times 10^{-21}$ mg.

60. a. Given that the atomic mass of sodium is 22.99 and the atomic mass of chlorine is 35.45, what is the percentage, by mass, of sodium in sodium chloride?

b. The U.S. Department of Agriculture Dietary Guidelines (2005) recommend a dietary intake of no more than 2300 mg (2.3 g) of sodium each day for the average, healthy American. Use your answer to part a. to calculate the mass of sodium chloride that would contain this recommended maximum amount of sodium.

c. Assuming that one teaspoon of common table salt has a mass of 5 g, how many teaspoons of table salt would it take to reach this maximum recommended daily intake of sodium?

4

Energy and Society

Throughout the world, fossil fuels provide the energy that supports our modern lives. We use these fuels for transporting people and goods, as well as for providing electricity and heat to our homes, buildings, and factories. Even the food we eat would be scarce if not for the fossil fuels that power the farm equipment, factories, and trucks we use to produce it, process it, and bring it to market. It shouldn't surprise us then, that a nation's consumption of fossil fuels reflects its level of economic activity.

However, the burning of fossil fuels has an effect on the environment, releasing carbon dioxide gas into Earth's atmosphere. The accumulation of this gas is affecting our climate, creating changes on a global scale. Scientists today are working to develop alternatives to fossil fuels to preserve our modern society without leading to catastrophic climate changes.

In this chapter we explore how **hydrocarbons**—covalent compounds of carbon and hydrogen that constitute fossil fuels—provide the energy of combustion and in the process generate carbon dioxide gas. We examine the problems created by the presence of this gas within the atmosphere, and we also look at alternative energy sources that may allow vehicles throughout the world, including those shown here, to operate without releasing harmful emissions.

CHAPTER OUTLINE

4.1 Energy and Its Uses

LEARNING OBJECTIVES

1. **Define** types of energy.
2. **Describe** how energy is used by society.

We've all experienced what it's like to feel energetic or to be lacking in energy, but what do we mean when we speak of energy? In this case we're referring to biological energy or an emotional state. Chemists, though, think of energy in ways that can be more clearly defined and measured. In this section we'll define energy and examine how it's used by society.

Defining Energy

Energy itself is simply the capacity to do work. When you do work, you use energy; and conversely, when you use energy, you do work. The more energy you have, the more work you can do. In physical terms, work is a force acting through a distance, such as throwing a ball or pushing a shopping cart. We can describe the energy of an object as the sum of its **potential energy**, which we can think of as stored energy, and **kinetic energy**, which is the energy due to the object's motion (**Figure 4.1**).

energy The capacity to perform work.

Some materials, such as fuels, also possess potential energy, which can be converted into kinetic energy and perform work. Consider gasoline, for instance. The combustion of gasoline within the cylinders of a car's engine transforms potential energy into kinetic energy of the pistons, causing them to do work that moves the car. The nutrients of food represent a vital source of potential energy for all animals, including ourselves. The metabolism of these nutrients enables us to perform work, whether simply breathing, walking, climbing stairs, or engaging in more vigorous activities. Other materials can possess potential energy as well, such as the chemicals within batteries. When a battery is connected to a circuit, it produces electrical energy that can be harnessed for a variety of uses, such as lighting a flashlight or starting a car's engine. These examples obey the **law of conservation of energy** and also show that energy can be transformed from one form into another. For instance, a falling object's potential energy is converted into an equivalent amount of kinetic energy.

potential energy Stored energy due to an object's position in space or due to the composition of a substance.

kinetic energy The energy of motion.

law of conservation of energy A natural law recognizing that energy can neither be created nor destroyed but can only be converted from one form to another.

heat The energy that flows from a warmer body to a cooler one.

Thermodynamics is the field of science that deals with the relationships between **heat** and other forms of energy, including chemical energy and kinetic energy. Although the principles of thermodynamics are summarized in four laws of thermodynamics, we will be concerned

Visualizing potential and kinetic energy • Figure 4.1

An object's position in space can be one source of potential energy. As this stored energy is released, it's converted into kinetic energy and does work.

In this composite image, the person initially has high-potential energy due to his position in space.

As he descends, his potential energy is converted into kinetic energy.

Work is performed as his falling body enters the water, causing a splash.

RonTech2000/E+/Getty Images

Ask Yourself

Imagine a child swinging on a playground swing. At what points in the swing's arc does the child have the greatest amount of **(a)** potential energy and **(b)** kinetic energy?

Entropy and the second law of thermodynamics • Figure 4.2

An increase in entropy increases the disorder or randomness of a collection of atoms, ions, or molecules.

a. Visualization of an increase in entropy.
Food dye added to water disperses to form a uniform mixture. Here, entropy increases because as the dye particles distribute throughout the liquid, their specific positions at any moment become more uncertain.

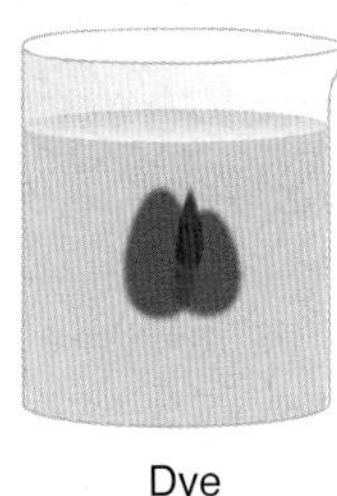

Ask Yourself

Which of the following laws of thermodynamics— the first or the second— best applies to the following statements?

(a) The energy supplied by foods is conserved, though it may be converted into work and heat.

(b) We cannot fully utilize the energy of foods, since some of this energy dissipates from the body as heat.

b. How increases in entropy relate to energy use.
Machines and living organisms are energy-conversion devices, utilizing energy and performing work. But during these conversions some of this energy invariably dissipates as heat. This dispersion of energy increases entropy.

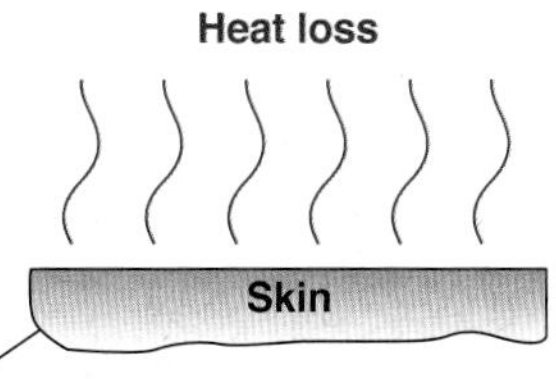

The body utilizes the stored energy of the nutrients of food to perform work – from keeping our vital organs functioning to physical activities, such as this human truck-pull. However, more than half of the energy provided by foods is lost as heat from the body.

with only two of these laws, the **first and second laws of thermodynamics**.

The first law of thermodynamics is what we just stated as the law of conservation of energy. It recognizes that although energy itself can neither be created nor destroyed in any isolated system, the energy that does exist within the system can be converted from one form to another. For example, the chemical energy of the components of a battery is converted into electrical energy when the battery is connected to a circuit. Similarly, the chemical energy of a car's gasoline is converted into the kinetic energy that moves the car along a road as the gasoline burns in the car's engine.

Though energy can be converted from one form to another, these energy conversions are inefficient. A portion of the energy undergoing conversion is invariably dissipated as heat and becomes unavailable for work. We detect this when we touch a battery that is emitting large amounts of electricity and find that it has become warm. Also, as a car's engine operates, only about 20% of the energy released by the gasoline's combustion produces the kinetic energy that moves the car. The remainder of the gasoline's energy is dissipated to the car's surroundings as heat. According to the second law of thermodynamics, this dispersal of energy is associated with an increase in a property called **entropy**, which is a measure of disorder in a physical system (**Figure 4.2**).

entropy A measure of the disorder or randomness of the positions of a collection of atoms, ions, or molecules.

Consumption of energy in the United States by sector • Figure 4.3

Four broad sectors of society account for most of our energy consumption: residential; commercial/municipal; transportation; and industrial.

In total, the United States consumes roughly 100 *quads* of energy each year. (One quad, shorthand for quadrillion British thermal units [Btu], is equivalent to 1×10^{18} joules. This is about as much energy as 5.5 million U.S. homes use annually, or about as much energy as that produced by the winds of a medium-sized hurricane over the course of a day).

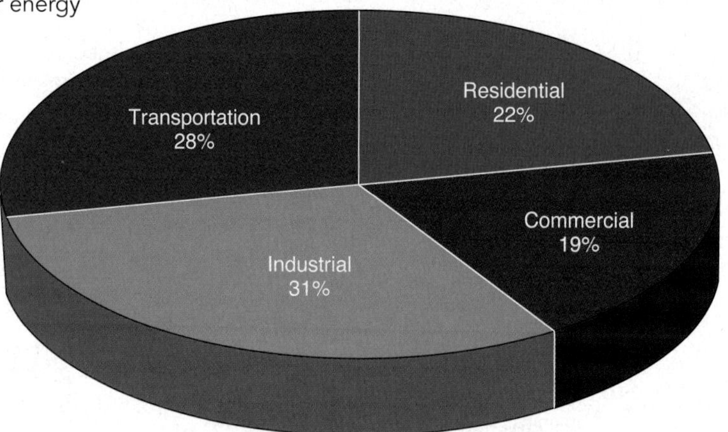

U.S. Energy Information Administration/Annual Energy Review 2011.

Energy exists in many forms—chemical, electrical, biological, thermal, and more—but energy itself is usually described most simply and conveniently in terms of its equivalent in heat. An early scientific unit used to describe energy, and still in use today, is the **calorie**, from the Latin, *calor*, for heat. Note the use of the small "c" for calorie and its abbreviation as cal. One thousand calories is equal to one kilocalorie (kcal), and a kilocalorie is equal to one **Calorie** (capitalized "C"), a unit used to describe the energy content of foods.

1000 calories (cal) = 1 kilocalorie (kcal) = 1 Calorie (Cal)

A unit of energy favored for most scientific work and used in the metric system is the **joule** (J) (rhymes with *tool*), named in honor of the 19th-century English physicist James Joule, who investigated the relationship between work and heat. One joule is equivalent to 0.24 calories. In the chapter on energy and food, we'll further discuss how these units of energy are defined and used.

1 joule (J) = 0.24 calories

Uses of Energy

Consider how you use electricity, fuels, and other sources of energy during a typical day. In our homes we use energy to keep the rooms at a comfortable temperature, to heat water, to provide light, and to operate a variety of appliances. Commercial buildings, schools, stores, office buildings, and hospitals require energy for many of these same purposes. Municipalities use energy to light streets, operate traffic signals, provide water, and for other public services. Modern transportation consumes enormous amounts of energy, from the fuels that power cars, trucks, airplanes, and ships (including those used for military purposes), and from the electricity that powers many types of trains. Society's largest expenditure of energy, though, is for industrial production. Our factories and other industrial operations use vast amounts of energy to create a variety of useful commercial products, including steel and other metals, paper and other wood products, petroleum fuels and chemicals, textiles, foods, electronics, and a myriad of other goods. **Figure 4.3** summarizes these uses of energy.

Worldwide annual energy consumption is about 500 quads and is expected to grow to about 700 quads by 2030, according to estimates by the U.S. Energy Information Agency. How does the world meet these enormous energy needs? Currently, the vast majority of the world's energy is supplied by fossil fuels, a topic we address in the next section. Because fossil fuels constitute a limited resource, alternatives must be developed to meet future energy requirements.

CONCEPT CHECK

1. **What** unit of energy is used for most scientific work?
2. **What** four main sectors of society account for our energy use?

4.2 Fossil Fuels

LEARNING OBJECTIVES

1. **Identify** the types of fossil fuels and their uses.
2. **Explain** the degree to which world energy needs are met by fossil fuels.
3. **Describe** the products of fossil fuel combustion.

The development of industrial societies throughout the world has been made possible to a large extent by the energy of fossil fuels. In our own daily lives, we rely on the energy provided by fossil fuels for most of our private and public transportation, for the generation of most of our commercial electricity, for the production and distribution of much of our food supply, and for many other necessities and conveniences of our daily lives. In this section we will examine fossil fuels and their many uses.

Types of Fossil Fuels

Most of the world's energy comes from the burning of **fossil fuels**: natural gas, petroleum (oil), and coal. These carbon-rich fuels derive primarily from the plant matter that existed in Earth's oceans, swamps, and forests some 280–360 million years ago, well before the age of dinosaurs. Over the course of perhaps tens of millions of years, this organic material was compressed under sedimentation and rock, and, through the effects of pressure and heat, formed today's fossil fuels. Chemically, fossil fuels consist largely of **hydrocarbons**. The common heating fuel, propane, provides a good introduction to this class of compounds (**Figure 4.4**).

fossil fuels Fuels such as natural gas, petroleum (oil), and coal, derived from decaying plant and animal matter.

hydrocarbons A class of chemical compounds composed of carbon and hydrogen. The general molecular formula C_xH_y is used to represent any hydrocarbon molecule, where *x* and *y* denote the number of carbon and hydrogen atoms, respectively.

Propane normally exists as a gas but easily liquefies under pressure. Propane is commonly stored in pressurized tanks.

Natural gas Like other fossil fuels, **natural gas** forms from the decay of organic matter compressed under Earth's surface for millions of years. Its major component is methane (CH_4), but crude natural gas can also contain other hydrocarbon gases, as well as impurities such as carbon dioxide, which must be removed through purification. Russia and the United States are the world's largest producers of natural gas. In the United States, about half of all homes are heated by natural gas, which is the cleanest burning fossil fuel. Because methane is odorless, utility companies blend minute amounts of sulfur-based odorants into natural gas to enable its easy detection.

natural gas A mixture of hydrocarbon gases, chiefly methane, often found associated with petroleum deposits.

Propane, a hydrocarbon • Figure 4.4

Like all hydrocarbons, propane contains only carbon and hydrogen atoms. This important fuel is found in both petroleum and natural gas deposits.

Comstock.Stockbyte/Getty Images

DANGER–FLAMMABLE GAS UNDER PRESSURE

A ball-and-stick model of a molecule of propane, C_3H_8. Note the "backbone" of covalently bonded carbon atoms in black.

Propane normally exists as a gas but easily liquefies under pressure. Propane is commonly stored in pressurized tanks.

Many natural gas deposits exist within deep underground rock formations. A growing proportion of the natural gas mined in the United States is produced through hydraulic fracturing, or **fracking** for short. Fracking involves injecting a high-pressure mixture of water, sand, and chemical additives deep underground to produce fractures in the rock formations, thus releasing the natural gas. This technique allows access to vast quantities of natural gas reserves that were previously uneconomical to mine, but also yields large volumes of wastewater that can contaminate groundwater. As a result, several communities have raised opposition to this practice. However, according to the U.S. Energy Information Agency, from 2011 to 2040 annual U.S. natural gas production is projected to grow 44%, and almost all of this growth is anticipated to be due to new wells that utilize fracking.

Petroleum The origin of the word **petroleum** as a combination of two Latin terms: *petra*, meaning "rock," and *oleum*, "oil," reflects petroleum's source as an oil extracted from rocks that lie just below Earth's surface. This oil, known as "crude oil," provides almost 40% of all the energy consumed in the United States.

petroleum An oily, usually dark, flammable liquid, consisting of a complex mixture of hundreds of hydrocarbons and other minor components.

coal A solid, carbon-rich fuel with a widely varying composition that depends on its source. This relatively plentiful fuel is used chiefly for generating electricity.

Through the process of petroleum refining, a variety of energy-rich fuels are obtained from crude oil, including gasoline, kerosene, diesel, heating oil, and jet fuel. In the next section we'll examine how crude petroleum is converted into refined gasoline, how the internal combustion engines of our cars convert the chemical energy of gasoline's hydrocarbons into mechanical energy, and how gasoline is modified to make our cars run more efficiently. Note that although crude oil is used largely for making fuels for transportation and heating, a smaller portion serves as feedstock for the production of plastics, agricultural chemicals, pharmaceuticals, and other products.

Coal Of the three fossil fuels, **coal** is the most plentiful, the most widely distributed across the world, and the least expensive per unit of energy it provides. Because coal is a solid, not a liquid like petroleum or a gas like natural gas, it is used primarily for large-scale applications, such as power plants. The United States has the largest proven reserves of coal, but China is the world's largest producer and consumer of this fuel. According to the National Academy of Sciences, coal production and use presents several environmental concerns:

- Surface mining of coal disturbs the landscape, and mining runoff can harm water quality.
- Combustion of coal is responsible for several environmental contaminants, including sulfur dioxide and nitrogen oxide gases, which are associated with acid rain (Chapter 8); heavy metals such as mercury; and particulate matter.
- Coal produces more carbon dioxide per unit of heat or electricity generated than other fossil fuels.

However, newer technologies are being developed to reduce emissions and improve efficiencies of coal-fired power plants.

Fossil fuel consumption Fossil fuels are considered to be **nonrenewable resources**, since they are being consumed far more rapidly than they are being generated. This is in contrast with **renewable resources** such as wind power, solar power, or biofuels, which are replenished relatively quickly over time. Having formed over millions of years, fossil fuels are being consumed, worldwide, as if in an instant on a historical time scale. Estimates vary considerably as to how much longer reliable supplies of each of the fossil fuels will be available, but there is general consensus that crude oil will be the first to be depleted, followed by natural gas and then coal. At current consumption rates, recoverable oil reserves may last only until the middle part of this century, while relatively plentiful coal reserves could last well into the next century.

For the time being, however, fossil fuels are still relatively abundant. According to the International Energy Agency, about 80% of the world's energy is currently supplied by these fuels (**Figure 4.5**). This figure may not change significantly for the foreseeable future, barring international agreements restricting carbon emissions.

In terms of total energy usage, China has eclipsed the United States as the largest consumer of energy in the world and by 2035, is projected to be using 70% more energy than the United States. However, on a per capita basis (energy use per person), the United States still consumes more energy than most other countries. The United States has 5% of the world's population, yet accounts for nearly 20% of the world's energy consumption. Fossil fuels provide the vast majority of the energy consumed in the United States (**Figure 4.5b**).

World and U.S. energy demand by type of fuel • Figure 4.5

Fossil fuels provide most of our present and projected energy needs. **a.** Global energy demand is expected to grow over 35% from 2010 to 2040, with fossil fuels continuing to account for the vast majority of usage. The fast-growing economies of China and India are projected to account for half of the growth in energy demand over the time frame shown here. *Source:* International Energy Agency. International Energy Outlook 2013. **b.** Fossil fuels provide the vast majority of present U.S. energy needs. *Source:* U.S. Energy Information Agency, Annual Energy Review 2009.

a. World energy consumption trends by fuel type

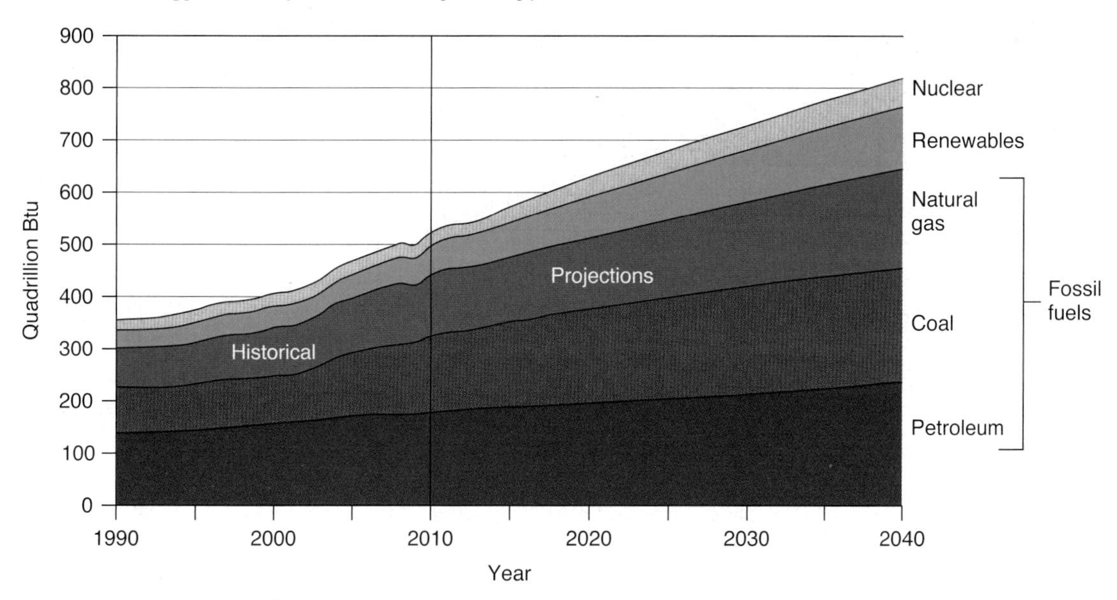

b. U.S. energy consumption by fuel type

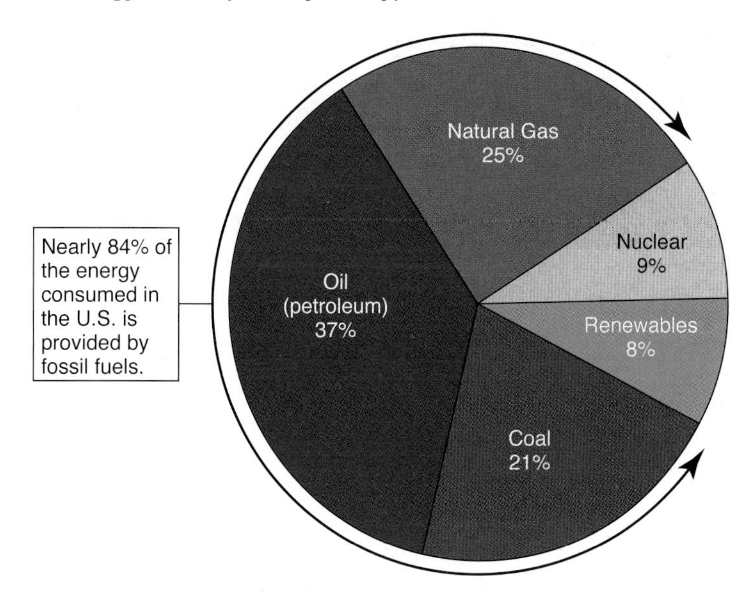

Ask Yourself

What does the term *renewables* in this figure refer to?
Provide three examples of these renewables.

Interpret the Data

Which two types of fossil fuels:

a. provide the greatest proportion of the world's current energy needs?

b. are expected to provide the greatest proportion of the world's energy needs in the year 2040, according to the projections shown in part a?

U.S. petroleum consumption and production • Figure 4.6

The United States consumes more petroleum than any other country. Nearly half of our petroleum needs are met by imports. The remainder is produced domestically, with Texas, North Dakota, Alaska, the Gulf of Mexico, and the Pacific coast as the major oil-producing regions.

a. U.S. daily petroleum consumption
The United States consumes approximately 20 million barrels of petroleum every *day*. This volume is equivalent to a column rising from the boundaries of a football field (including end zones) to a height of 2000 feet.

1 barrel = 42 US gallons

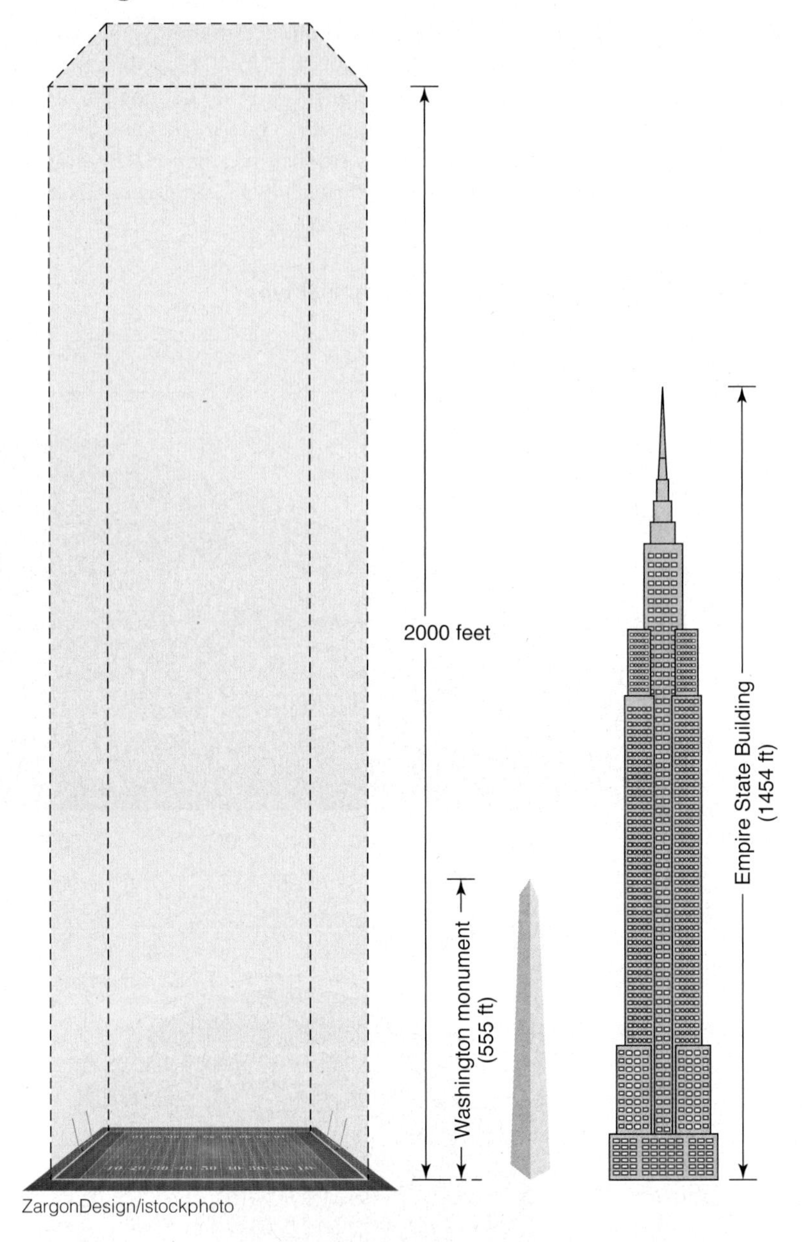

2000 feet
Washington monument (555 ft)
Empire State Building (1454 ft)

ZargonDesign/istockphoto

b. Consumption vs. production of petroleum in the United States.
Source: U.S. Energy Information Administration. Annual Energy Outlook 2010.

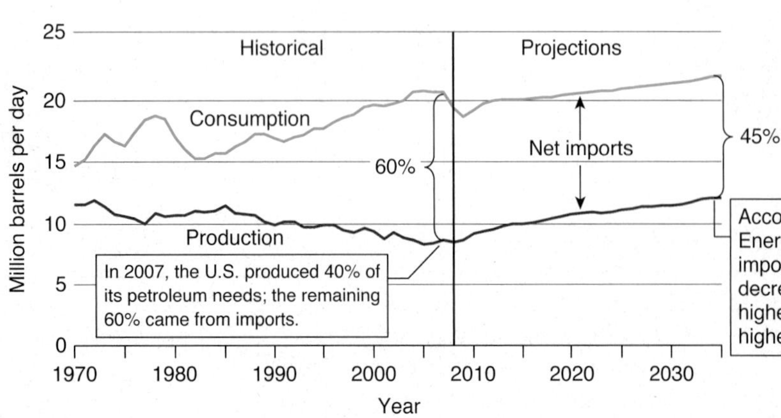

Historical
Projections
Consumption
Net imports
60%
45%
Production
In 2007, the U.S. produced 40% of its petroleum needs; the remaining 60% came from imports.
According to estimates by the U.S. Energy Information Association, net imports are projected to slowly decrease due to a combination of higher domestic oil production and higher fuel economy standards.
Million barrels per day
0
5
10
15
20
25
1970
1980
1990
2000
2010
2020
2030
Year

Interpret the Data

Approximately how many gallons of petroleum are consumed each day in the United States? Roughly what proportion of this is produced domestically?

Fossil fuel combustion • Figure 4.7

The compositions of fossil fuels vary, but each of these fuels is rich in carbon. When fossil fuels combust, carbon dioxide gas and water vapor are produced and heat is released. The following equation describes a general chemical reaction for the combustion of hydrocarbons.

$$C_xH_y + O_2 \longrightarrow CO_2 + H_2O + \text{heat}$$

C_xH_y: Hydrocarbons from natural gas, petroleum-based fuels, or coal; O_2: Oxygen; CO_2: Carbon dioxide; H_2O: Water

O=C=O

In the CO_2 molecule, carbon is bonded to 2 oxygen atoms.

We can also write equations for the combustion of specific hydrocarbons. Methane, CH_4, the major component of natural gas, reacts with oxygen as shown here:

$$CH_4 + 2\,O_2 \longrightarrow CO_2 + 2\,H_2O + \text{heat}$$

Each molecule of methane.... ...reacts with two molecules of oxygen.... ...to produce... ...one molecule of carbon dioxide... ...two molecules of water... ...and release heat.

This **balanced chemical equation** provides the molecular ratios of reactants and products. We'll have more to say in Chapter 6 about working with these types of equations.

Due to a large dependence on transportation fuels in the United States, the greatest share of our energy needs is supplied by petroleum. The United States consumes far more petroleum than it can produce on its own, with the difference consisting of imports (**Figure 4.6**).

Combustion of Fossil Fuels

Fossil fuels burn in a reaction known as **combustion**, consuming oxygen and producing carbon dioxide and water along with varying amounts of pollutants. This reaction is **exothermic** (heat-releasing) and is an example of a process called **oxidation**. (If a reaction needs energy to take place, it is **endothermic**, or heat-absorbing.) As fossil fuels burn, their carbon atoms combine with oxygen (O_2) to form carbon dioxide (CO_2)—hence the name oxidation (**Figure 4.7**).

combustion The reaction of a fuel with oxygen to produce heat and light.

exothermic Any process that releases heat.

endothermic Any process that absorbs heat.

Like all chemical reactions, the combustion of fossil fuels obeys the law of conservation of energy. The reactants (hydrocarbons and oxygen) collectively contain more potential energy within their chemical bonds than exists within the chemical bonds of the products (carbon dioxide and water). As fossil fuels burn, this difference in energy is released as heat.

For a given amount of heat generated, fossil fuels vary in the amount of CO_2 emitted. Natural gas emits the least CO_2 per unit of heat released—29% less than petroleum and 43% less than coal. Though the use of coal presents the most concerns in terms of CO_2 emissions, the use of gasoline, a petroleum-based fuel, is also problematic due to the enormous volumes of this fuel consumed globally. For every gallon of gasoline burned, almost 20 pounds of CO_2 are released into the atmosphere.

CONCEPT CHECK

1. **What** are major uses of natural gas, petroleum and coal?
2. **What** percentage of the world's energy needs is supplied by fossil fuels?
3. **What** are the two primary combustion products of fossil fuels?

4.3 Petroleum Refining and Gasoline

LEARNING OBJECTIVES

1. **Describe** how octane rating relates to engine knocking.
2. **Describe** how crude oil is turned into gasoline.
3. **Explain** the purpose of catalytic converters and fuel oxygenates.

When you turn the ignition key of a car, you begin a sequence of chemical reactions. A stream of electrons leaves the battery to provide the electric current that turns the starting motor; the starting motor activates the engine; and the engine begins burning a mixture of gasoline and air. As spark plugs fire, the combustion reaction of hydrocarbons (from gasoline) with oxygen (from air) keeps the car running. In this section, we'll examine how gasoline combustion provides the mechanical energy to power a car and how petroleum is refined to produce gasoline and a variety of other products.

The Internal Combustion Engine

Almost all gasoline engines consist of four, six, or eight cylinders. Each cylinder consists of a hollow tube, only a few centimeters in diameter, containing a close-fitting piston that rides smoothly up, nearly to the top of the cylinder, and down to the bottom. Each full passage of the piston, upward or downward, is a stroke. A cleverly designed arrangement of rods and gears converts the linear motion of the piston's strokes into the rotary motion of the wheels. In the simplest engines, two valves at the top of the cylinder open and close in rhythm with the movements of the piston, and a spark plug ignites the gasoline–air mixture at precisely the right moment. The entire cycle takes four strokes of the piston to accomplish, hence the name "four-stroke internal combustion engine" (**Figure 4.8**). When all this operates properly, the engine converts the energy released by the burning gasoline into the energy of motion, and the car runs smoothly.

PROCESS DIAGRAM

The four-stroke cycle of a typical car's engine • Figure 4.8

Each passage of the piston within the cylinders of an engine is called a stroke. In a typical car's engine, four strokes of the piston complete one cycle.

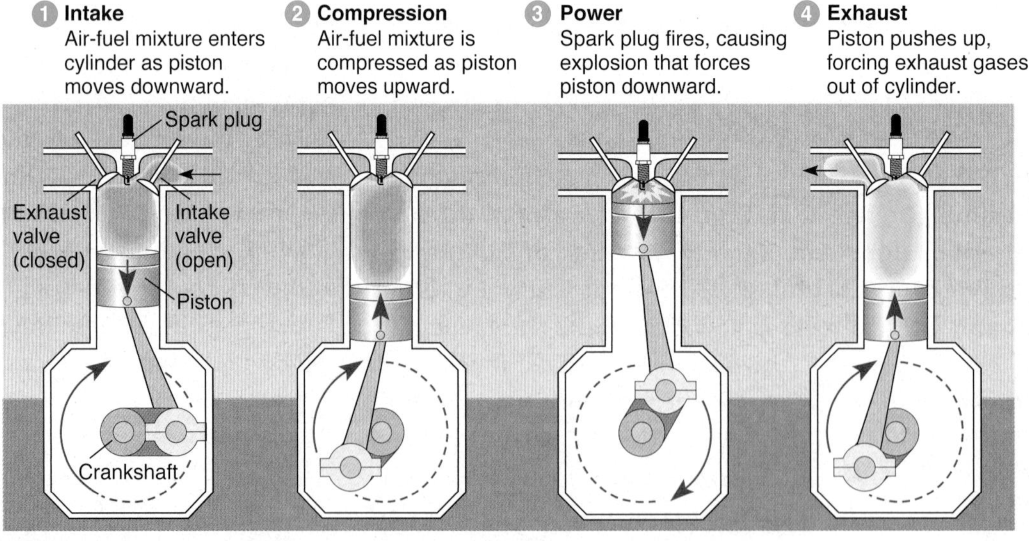
1 Intake
Air-fuel mixture enters cylinder as piston moves downward.
2 Compression
Air-fuel mixture is compressed as piston moves upward.
3 Power
Spark plug fires, causing explosion that forces piston downward.
4 Exhaust
Piston pushes up, forcing exhaust gases out of cylinder.
Spark plug
Exhaust valve (closed)
Intake valve (open)
Piston
Crankshaft

Adapted with permission of John Wiley & Sons, Inc. from Snyder, C., *The Extraordinary Chemistry of Ordinary Things*, Fourth Edition, p. 151. Copyright 2003.

Ask Yourself

Which strokes are in operation under the following sets of conditions?
(a) Both the intake valve and the exhaust valve are closed.
(b) Only one of these valves is open.

Knocking, an analogy • Figure 4.9

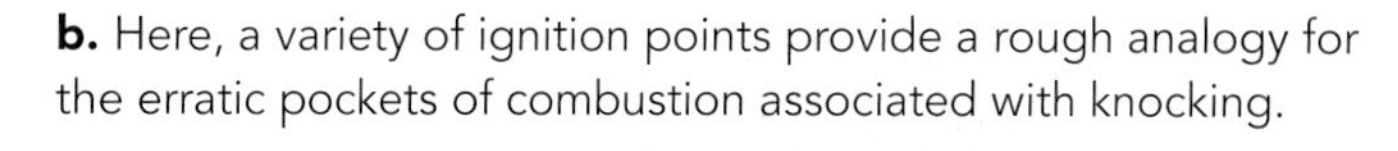

Ideally, each time a spark plug fires, the compressed gasoline–air mixture within a cylinder ignites from a single point in a controlled manner. Engine knocking results when multiple ignition points occur instead of the more effective single point.

a. Here, one combustion point appears to radiate outward, analogous to a smooth wave of combustion emanating from the spark plug.

SondraP/iStockphoto

b. Here, a variety of ignition points provide a rough analogy for the erratic pockets of combustion associated with knocking.

TakashiTAKEUCHI/iStockphoto

When the spark plug fires, expansion of the hot gases generated as a result of combustion provides the mechanical force to drive the piston downward. (Most of the heat generated from this gasoline combustion dissipates from the engine block as lost energy.) The exhaust gases, which leave the car's tailpipe, consist of carbon dioxide and water vapor, as well as a variety of other gases. Inefficient combustion adds unburned hydrocarbons and carbon monoxide to the exhaust gases. Often, these gases include small quantities of nitrogen- and sulfur-containing impurities that carry over from the crude petroleum to the commercial gasoline. These impurities contribute oxides of nitrogen and sulfur to the exhaust gases, which add to the problem of acid rain and other forms of pollution (Chapter 8).

To squeeze as much energy out of the burning gasoline as possible, an engine must be designed with a high **compression ratio**. The higher the compression ratio, the greater the compression of the gasoline–air mixture when it is ignited, and the more powerful the thrust it delivers to the piston on its way down in the power stroke. The compression ratio is measured as the ratio of the gasoline–air volume at the beginning of the compression stroke to the gasoline–air volume at the moment the spark plug fires.

Automobile engines of the 1920s had a compression ratio of about 4, meaning the gasoline–air mixture was compressed to one-fourth of its original volume before it was ignited. Today's compression ratios are 10 or even more, meaning the mixture can be compressed to as little as one-tenth (or even less) of its original volume. As a result, today's engines can get much more energy out of hydrocarbon combustion.

> **compression ratio** A measure of the extent to which the fuel–air mixture is compressed during the compression stroke.

However, there is a limit to all this compression: **knocking**. Knocking is a rapid pinging or knocking sound that comes from an engine when it is pushed to produce a lot of power quickly. Knocking can occur when a car is accelerating, especially while going uphill or when pulling a trailer. With any particular grade of gasoline, the higher a car's engine compression ratio, the greater the likelihood of knocking (**Figure 4.9**).

> **knocking** A metallic pinging sound sometimes heard from automobile engines when the air–fuel mixture combusts erratically in pockets, instead of in one smooth wave emanating from the spark plug.

Knocking can cause loss of power, inefficient and uneconomical use of fuel, and, in severe cases, it can produce pits or fractures in the top surface of the piston. Fortunately, the high octane ratings of gasoline (and other advances in engine design) have greatly curtailed knocking in today's cars. We'll define a gasoline's octane rating and examine how it relates to knocking.

Petroleum Refining

In the United States alone, more than 250 million motor vehicles of all kinds, including cars, vans, trucks, buses, and motorcycles, travel an estimated 3,000,000,000,000 (3 trillion) miles each year. To power all this transportation, in 2012 we consumed some 3 billion barrels (roughly 125 billion gallons) of gasoline, which in turn was derived from 7 billion barrels of petroleum.

volatility A measure of how readily a substance evaporates. Highly volatile compounds have low boiling points.

petroleum refining A process by which petroleum is separated into its different components.

fractional distillation A process by which a liquid mixture of compounds is separated into fractions based on boiling points.

Gasoline is a complex mixture of hydrocarbons. The familiar smell of gasoline is due to the relative ease of evaporation, or **volatility**, of these compounds. We often encounter other types of volatile compounds in our everyday lives (**Figure 4.10**).

Petroleum is a crude natural product, containing hundreds of hydrocarbons with a wide range of volatility. The process by which petroleum is converted into gasoline (and a host of other useful products) is called **petroleum refining**. This purification process begins with **fractional distillation**, in which the mixture of hydrocarbons in petroleum is separated into groups or *fractions* based on boiling points. These groups range from highly volatile gaseous fractions to those containing nonvolatile waxy or tarry solids (**Figure 4.11**).

Distillation alone isn't sufficient to transform petroleum into gasoline for today's engines. One reason is that there just aren't enough of the right kinds of hydrocarbons in petroleum to provide the modern world with all the gasoline it demands. Although the composition of crude oil varies considerably, depending on the region of the world it comes from, generally not much more than about 20% of any barrel of crude oil consists of hydrocarbons that can be blended into a gasoline mixture.

catalyst A substance, often a specialized metal, that speeds up the rate of a chemical reaction without itself being consumed.

catalytic cracking A petroleum refining process that uses a catalyst to break down (crack) higher-boiling, higher-molecular-weight hydrocarbon molecules into lighter molecules.

One solution to this problem is to use **catalysts** to convert certain hydrocarbons present within petroleum into structures more suitable for gasoline. (A catalyst is a substance that is not consumed in a reaction, but rather increases the speed and efficiency of the reaction. Catalysts are very important in both industry and biology, as we will see in later chapters.) **Catalytic cracking** uses a catalyst to facilitate the cracking, or breaking apart, of large covalent molecules into smaller ones. This process yields two or more smaller molecules from a larger parent compound. For example, cracking a 12-carbon-atom molecule of the

Volatile compounds in everyday life • Figure 4.10

Cologne and perfume contain a mixture of volatile compounds. As these compounds evaporate, they dissipate into the air and produce the fragrances we smell. The warmth of our skin helps speed up evaporation of these compounds.

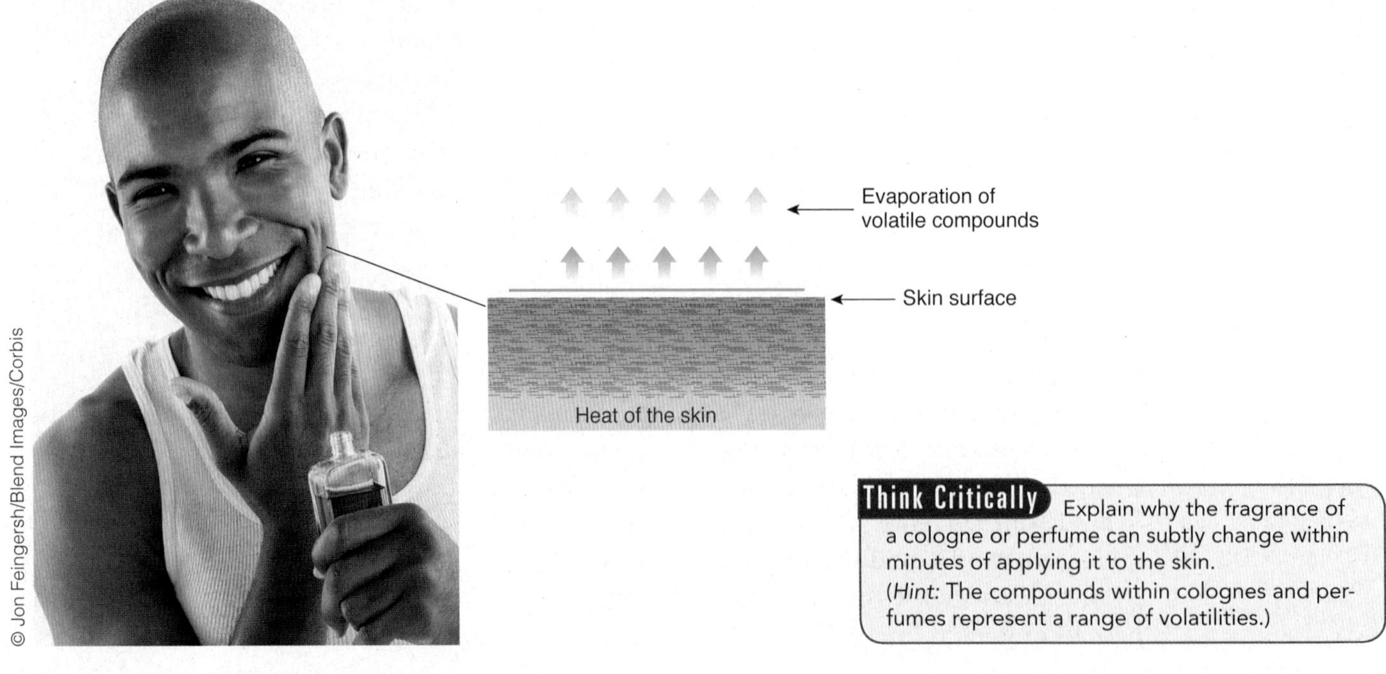

Think Critically Explain why the fragrance of a cologne or perfume can subtly change within minutes of applying it to the skin.
(*Hint:* The compounds within colognes and perfumes represent a range of volatilities.)

Fractional distillation of petroleum • Figure 4.11

The mixture of compounds in crude oil is separated into fractions based on boiling points of the hydrocarbon components.

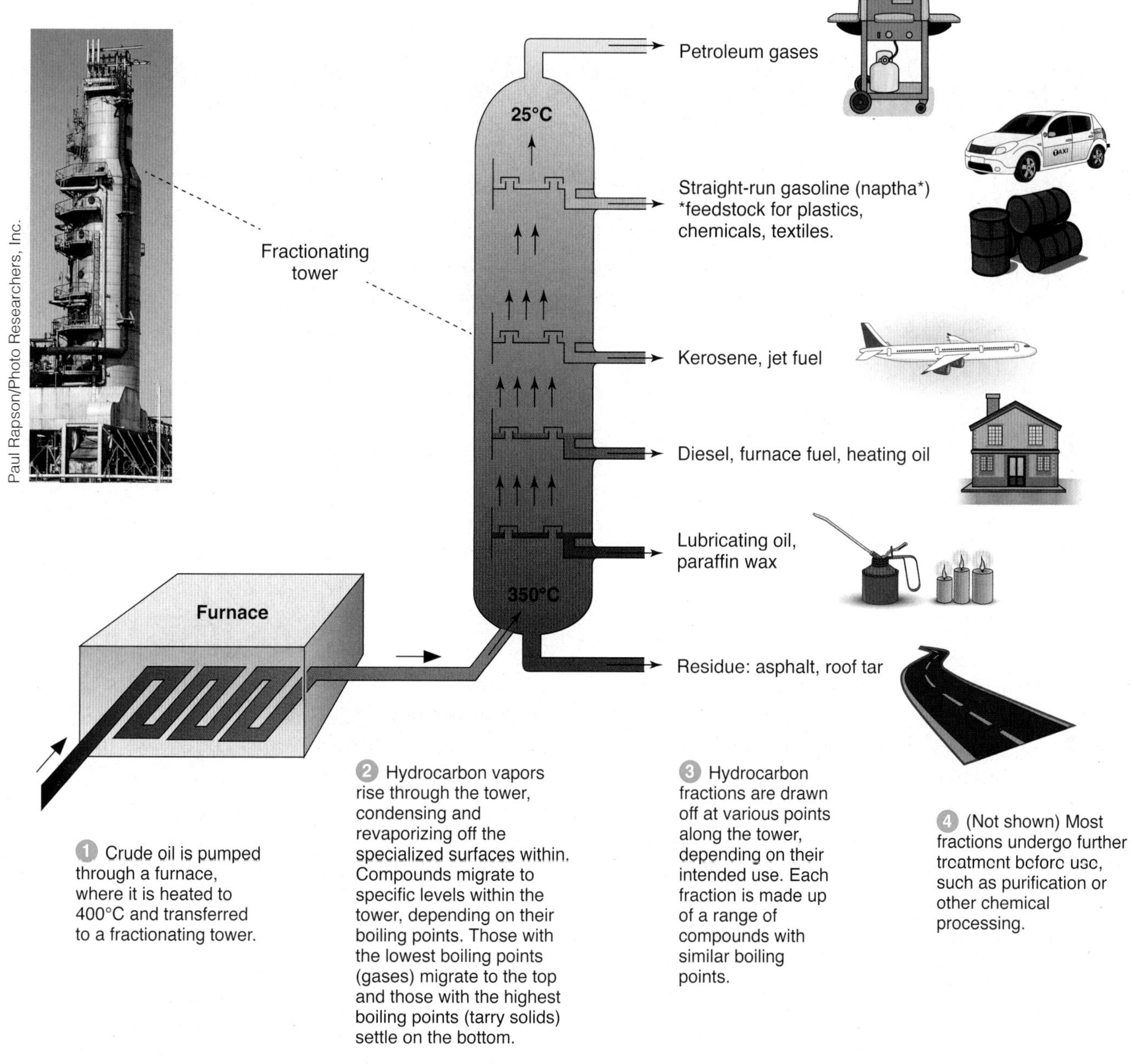

1. Crude oil is pumped through a furnace, where it is heated to 400°C and transferred to a fractionating tower.

2. Hydrocarbon vapors rise through the tower, condensing and revaporizing off the specialized surfaces within. Compounds migrate to specific levels within the tower, depending on their boiling points. Those with the lowest boiling points (gases) migrate to the top and those with the highest boiling points (tarry solids) settle on the bottom.

3. Hydrocarbon fractions are drawn off at various points along the tower, depending on their intended use. Each fraction is made up of a range of compounds with similar boiling points.

4. (Not shown) Most fractions undergo further treatment before use, such as purification or other chemical processing.

Adapted with permission of John Wiley & Sons, Inc. from Snyder, C., *The Extraordinary Chemistry of Ordinary Things*, Fourth Edition, p. 162. Copyright 2003.

Think Critically Where do **(a)** the most volatile compounds and **(b)** the least volatile compounds leave the fractionating tower?

Catalytic processes • Figure 4.12

Catalysts are widely used in petroleum processing to speed up chemical transformations. (For clarity, hydrogen atoms are not shown in the figure.)

a. Catalytic cracking.
Here, a hydrocarbon with a higher molecular weight is "cracked" or broken down into smaller molecules.

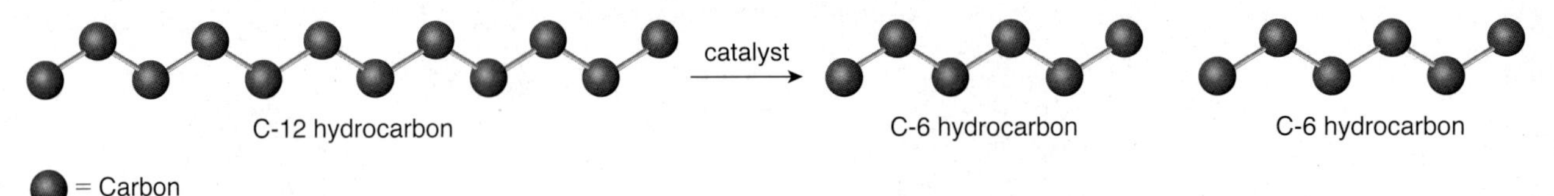

b. Catalytic reforming.
Here, the carbon skeleton of an unbranched hydrocarbon is catalytically reorganized into a branched structure. This process can also turn unbranched hydrocarbons into cyclic structures that can be useful for industrial solvents and a host of other products.

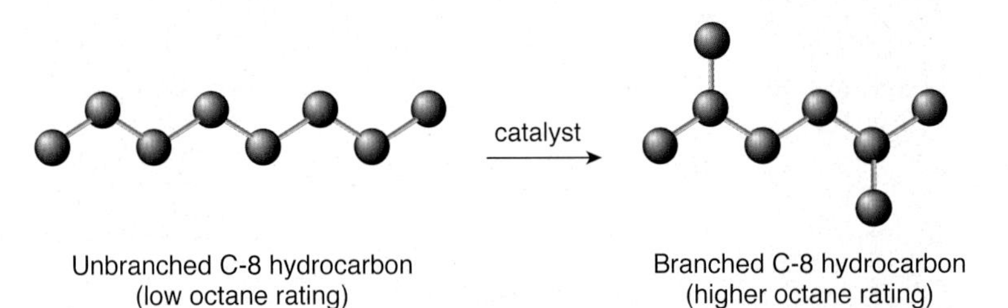

Ask Yourself

How does catalytic cracking help increase the yield of gasoline from a given quantity of petroleum?

kerosene fraction (abbreviated C-12) could yield two C-6 hydrocarbon molecules (**Figure 4.12a**). A C-6 hydrocarbon is more useful than a C-12 hydrocarbon in blending a gasoline mixture because of its greater volatility. In any case, cracking can increase the amount of low-molecular-weight hydrocarbons obtained through petroleum refining, *but only at the expense of the higher-molecular-weight hydrocarbons.* By using a catalytic cracking process, the yield of gasoline from a barrel of petroleum can be increased from 20% to about 45%, but at the expense of the higher-boiling hydrocarbons, including kerosene and home heating oil.

> **catalytic reforming** A petroleum refining process in which a catalyst converts low-octane-rated compounds into those more suitable for gasoline.

Petroleum refineries not only separate the various petroleum fractions through distillation and convert large hydrocarbon molecules into several smaller ones through catalytic cracking, but they also reorganize the molecular shapes of molecules through **catalytic reforming**. This refining process reorganizes the carbon skeletons of unbranched or only slightly branched hydrocarbons into branched or much more highly branched molecules (**Figure 4.12b**). This increases the octane rating of the hydrocarbons and makes them far more useful for gasoline.

As long as supplies of crude oil are ample, the fact that catalytic cracking and reforming deplete certain hydrocarbons so that we may have more of others is not generally an issue. But a cold winter, for example, coupled with restrictions in petroleum imports, could potentially threaten our ability to maintain ample supplies of kerosene/heating oil and gasoline, for example. This could establish a difficult choice between creating enough gasoline to satisfy everyone (but at the expense of cold homes) and producing enough heating fuels to generate winter warmth for all (but at the expense of gasoline shortages). To mitigate potential shortages, the U.S. federal government has established the **Strategic Petroleum Reserve (SPR)**, an emergency inventory of some 700 million barrels of crude oil, stored in underground caverns in Texas and Louisiana, situated near petroleum refineries. The SPR can be tapped only by presidential order.

Gasoline

Because of its intricate design, the internal combustion engine requires a carefully regulated blend of hydrocarbons. Gasoline is a mixture of over a hundred different hydrocarbons, trace contaminants, and additives. On the one hand, the gasoline must be rich in volatile hydrocarbons to ensure that enough of their vapors mix with the air in the cylinders to start the engine on a cold morning. On the other hand, an overabundance of volatile hydrocarbons would result in a fuel–air mixture that's too rich in fuel and too poor in oxygen. In this case there wouldn't be enough oxygen to oxidize the vaporized gasoline effectively. The results would be poor ignition, roughness, poor fuel economy, and even vapor lock, a condition in which a bubble of gasoline vapor blocks the flow of fuel to the engine. Since gasoline must be equally efficient in Miami in July and in Chicago in January, manufacturers blend their gasolines to match the season and the locale of use.

To avoid knocking, gasoline must burn smoothly, even under the rigorous conditions of a modern high-compression engine. Studies of the tendencies of various hydrocarbons to knock in test engines reveal one consistent trend: The more highly branched the hydrocarbon, the greater its tendency to burn smoothly and to resist knocking. For example, a highly branched C-8 hydrocarbon, named 2,2,4-trimethylpentane, shows very little tendency to knock. In connection with gasoline, this compound is sometimes (and mistakenly) called "octane." On the other hand, an unbranched C-7 hydrocarbon, named heptane, knocks readily, even under mild conditions. "Octane," because of its ability to burn smoothly and to resist knocking, is assigned an **octane rating** of 100. Heptane, with its great tendency to knock, receives an octane rating of 0. These two compounds serve as references in assigning octane ratings to gasoline (**Figure 4.13**).

octane rating A measure of the antiknock properties of a fuel—a gasoline with a higher octane number is less prone to knocking than a gasoline with a lower octane number.

Cars designed to run on higher grades of gasoline generally have higher compression ratios, so they benefit from the antiknocking properties of these fuels. For cars designed to run on regular grade gasoline (with an octane rating of 87), the use of higher-octane fuel has not been shown to deliver noticeably better fuel economy or greater power.

The chemistry of gasoline and gasoline additives has an interesting history. In the 1920s, a simple and inexpensive method for increasing octane ratings was developed. The addition of a very small amount (roughly one teaspoon per gallon) of a compound called tetraethyllead to gasoline was found to increase the octane number of the gasoline by 10–15 points. This ushered in the era of leaded gasoline, which permitted the production of more powerful engines that would be free of knocking.

Assigning octane ratings to gasoline • Figure 4.13

To determine the octane rating of any particular blend of gasoline, we identify a specific ratio of "octane" and heptane that has the same knocking tendencies as the particular blend.

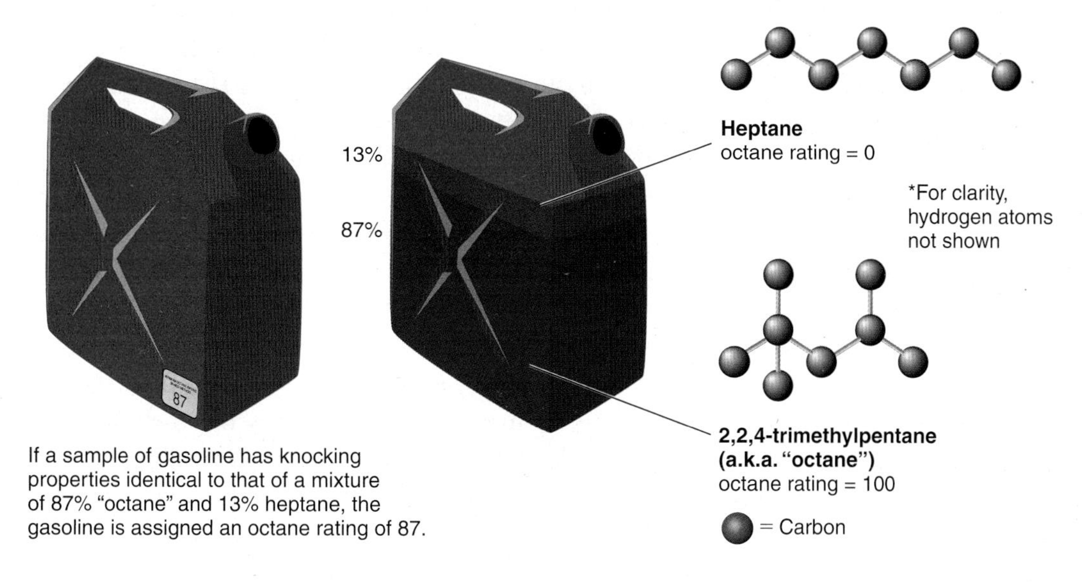

If a sample of gasoline has knocking properties identical to that of a mixture of 87% "octane" and 13% heptane, the gasoline is assigned an octane rating of 87.

Ask Yourself

1. Is gasoline with an octane rating of 87 composed of 87% "octane" and 13% heptane? Explain.
2. What mixture of "octane" and heptane exhibits the same knocking tendency as premium (octane 93) gasoline?

Catalytic converter • Figure 4.14

As exhaust gases from the engine pass through the catalytic converter:

- Unburned hydrocarbons (C_xH_y) and carbon monoxide (CO) **oxidize** into carbon dioxide (CO_2) and water vapor (H_2O).
- Nitrogen oxides (NO_x) convert to harmless nitrogen gas (N_2) through chemical **reduction**, the opposite of **oxidation**.

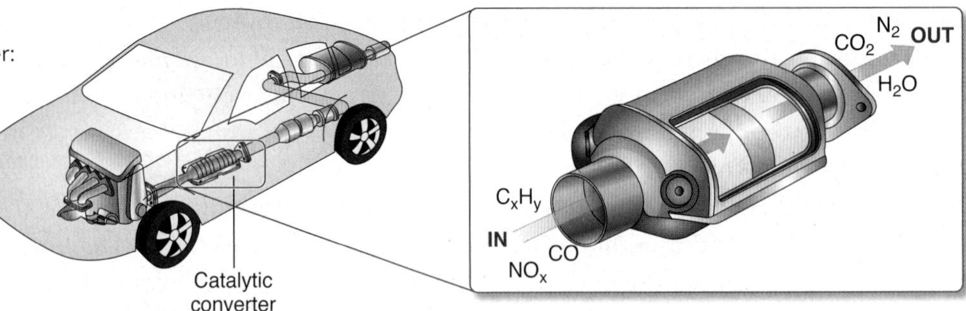

Think Critically
1. Do catalytic converters decrease carbon dioxide emissions?
2. Which gas reacts with the unburned hydrocarbons that enter the catalytic converter?

By the 1950s, increasing numbers of automobiles were emitting a variety of noxious exhaust gases. Ideally, the hydrocarbons of gasoline burn completely to yield only carbon dioxide, water, and energy. Under real operating conditions, however, exhaust gases can also contain unburned hydrocarbons; carbon monoxide (from incomplete oxidation of hydrocarbons); and sulfur dioxide and various oxides of nitrogen, resulting from oxidation of impurities within the gasoline. [As a note of caution, burning any fossil fuel or wood can produce carbon monoxide gas as a result of incomplete combustion. Any devices burning these fuels should be properly vented to the outside to reduce the risk of inhaling this gas, which is potentially lethal.]

catalytic converter A device built into an automobile's exhaust system that uses catalysts to reduce the levels of hydrocarbons, carbon monoxide, and other pollutants emitted.

oxygenate An oxygen-containing compound added to gasoline to improve oxidation of fuel and decrease harmful emissions.

In 1970, spurred by concern over increasing air pollution, Congress enacted the Clean Air Act, which set specific limits on tailpipe emissions (among other pollutants). For a chemical solution to a chemical problem, auto manufacturers turned to the **catalytic converter**, a device mounted between the engine's exhaust system and the tailpipe. The converter contains finely divided platinum, palladium, or other substances that serve as catalysts to oxidize hydrocarbons contained in the exhaust gases and carry out other reactions (**Figure 4.14**). Since 1975, virtually all automobiles built or imported into the United States have been equipped with this device.

The catalysts themselves are sensitive to a form of deterioration called **catalytic poisoning**, in which another chemical coats their surfaces and renders them ineffective. Lead, in particular, can poison catalysts. For this reason, Congress prohibited the use of leaded gasoline in cars equipped with catalytic converters, and today leaded gasoline is no longer available. Even if Congress hadn't acted, it's likely that lead would no longer be used as a gasoline additive, since lead and its compounds are poisonous not only to catalysts but to humans as well.

By the early 1980s, a lead-free octane enhancer, methyl *tert*-butyl ether (MTBE), was introduced as a gasoline additive. MTBE is an **ether**, a compound in which an oxygen atom is covalently bonded to two carbon groups (**Figure 4.15a**). In addition to acting as an octane enhancer, MTBE also serves as an **oxygenate**, which is an additive that improves the efficiency of hydrocarbon combustion by introducing additional oxygen into the combustion process. Inefficient combustion in an engine can contribute to emissions of carbon monoxide and unburned hydrocarbons. To decrease levels of these pollutants, the U.S. Congress amended the Clean Air Act in 1990 to require that oxygenates be added to gasoline sold in certain metropolitan areas that fall below federal air quality standards. In effect, the addition of oxygenates causes this reformulated gasoline, or RFG, to carry some of the needed oxygen along with its own hydrocarbons. As a result of this legislation, MTBE's popularity as an additive grew enormously.

Although MTBE is an effective octane enhancer and oxygenate, by the mid-1990s, it was becoming clear that MTBE carried its own environmental risks. The tendency for underground gasoline storage tanks to leak occasionally, combined with MTBE's solubility in water, led to

Oxygenates and octane enhancers • Figure 4.15

Ethanol has now replaced methyl tert-butyl ether (MTBE) as a gasoline additive used to reduce harmful emissions and improve octane ratings.

Ask Yourself

How are the structures of MTBE and ethanol **(a)** similar and **(b)** different?

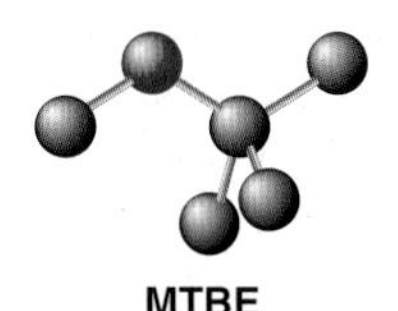

MTBE

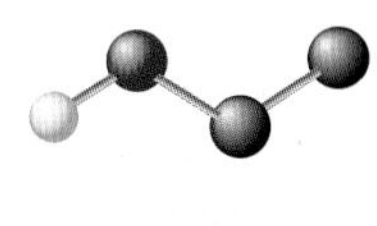

Ethanol

● = C
● = H
● = O

[Hydrogen atoms on carbon not shown, to highlight the framework of each molecule.]

the intrusion of small amounts of MTBE into groundwater and drinking water at various locations. As a result, by 2005 MTBE was banned by almost all states. Ethanol (**Figure 4.15b**) is now widely used as its replacement, in large part due to federal laws encouraging greater use of **biofuels**. In the United States, ethanol is produced almost exclusively from the fermentation of corn. In 2010, about 30% of the U.S. corn crop went to the production of ethanol. Labels on gasoline pumps not only show the ethanol content of the gasoline but reflect a variety of federal standards for gasoline formulation (see *What a Chemist Sees*).

biofuels Fuels derived from living matter or biomass, a renewable resource.

WHAT A CHEMIST SEES

At the Pump

A typical gasoline pump carries considerable information on octane ratings, additives, contaminants, and other characteristics of the fuel it dispenses. A chemist can interpret this information and apply it to vehicle performance and the environmental effects of the fuel. Here is the significance of some of the terms found on a typical gasoline pump.

Tim O. Walker

ULTRA-LOW SULFUR HIGHWAY DIESEL FUEL
(15 ppm Sulfur Maximum)
Required for use in all highway diesel vehicles and engines.
Recommended for use in all diesel vehicles and engines.

CONTAINS UP TO 10% ETHANOL

Maximum contaminant levels
Sulfur is a natural contaminate present in petroleum. Combustion of sulfur-containing fuels produces sulfur dioxide, SO_2, a component of acid rain. By law, petroleum refiners are required to reduce sulfur levels in these fuels. For instance, all diesel fuel sold in the U.S. must contain no more than 15 parts per million (ppm) sulfur.

Octane Rating
In the U.S., two different test methods are used to determine octane ratings. The R, or "research" octane rating, measures the ability of a gasoline to burn smoothly under relatively moderate operating conditions. The M, or "motor" octane rating, is more appropriate for conditions of acceleration or heavier loads. The rating posted at the pump is an average of the two, hence (R + M) / 2.

Additives
Gasoline now commonly contains up to 10% ethanol as an octane enhancer and oxygenate.

All gasoline, regardless of brand or octane rating, also contains small amounts of detergent, or "deposit control" additives, which help prevent the build-up of harmful deposits in the fuel system.

CONCEPT CHECK

1. **What** property of gasoline helps to reduce engine knocking?
2. **Why** is crude oil heated in order to undergo fractional distillation?
3. **How** do catalytic converters and fuel oxygenates help reduce emissions of unburned hydrocarbons?

4.4 Fossil Fuels and the Carbon Cycle

LEARNING OBJECTIVES

1. **Describe** the carbon cycle.
2. **Explain** how the use of fossil fuels affects the carbon cycle.

Life as we know it would not exist without a source of energy some 90 million miles away—the Sun. A small fraction of the energy emitted by the Sun travels to Earth in the form of a combination of visible light and, ultraviolet and infrared radiation. However, this small fraction supplies most of the energy available to all life on the planet.

Photosynthesis and Cellular Respiration

Through the process of **photosynthesis**, plants convert a portion of the Sun's radiant energy into chemical energy stored within the molecules the plant produces. In its overall chemistry, plant photosynthesis transforms water and atmospheric carbon dioxide into **organic compounds**, or carbon-based compounds, such as carbohydrates (**Figure 4.16**).

Animals that eat plants generate their own energy by a biological reversal of the photosynthetic process used by plants. These animals consume the organic compounds of plants and, with chemical reactions using atmospheric oxygen as a reactant, convert these compounds back to carbon dioxide and water. This process, known as **cellular respiration**, releases the energy stored in the organic molecules of foods.

The Carbon Cycle

In harnessing solar energy through photosynthesis, plants consume carbon dioxide and release oxygen. In producing life-sustaining energy, animals consume oxygen and release carbon dioxide. The shuttling of carbon back and forth between atmospheric carbon dioxide and the molecules of living things—through the complementary processes of plant photosynthesis and animal cellular activity—forms part of a larger process called the **carbon cycle**. Within this system, carbon cycles among various storage locations, or **sinks**, including the **atmosphere**, the **hydrosphere** (oceans, lakes, and rivers), and the **biosphere** (living matter).

carbon cycle A global cycle in which carbon is exchanged among the atmosphere, the oceans, geological systems, and living things.

Photosynthesis and cellular respiration • Figure 4.16

Photosynthesis and cellular respiration are complementary processes. Photosynthesis uses energy to synthesize biomolecules, whereas respiration breaks down biomolecules to obtain energy.

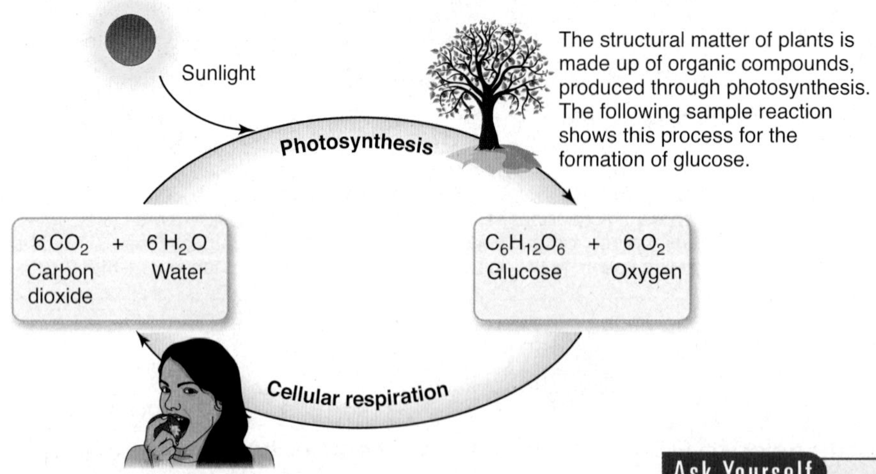

Sunlight
Photosynthesis
The structural matter of plants is made up of organic compounds, produced through photosynthesis. The following sample reaction shows this process for the formation of glucose.
$6CO_2$ + $6H_2O$
Carbon dioxide
Water
$C_6H_{12}O_6$ + $6O_2$
Glucose
Oxygen
Cellular respiration
The energy we derive from food and the carbon dioxide we produce and exhale is the result of cellular respiration.

Ask Yourself

Which compound shown in this figure supplies the carbon atoms found in glucose?

The carbon cycle and fossil fuels • Figure 4.17

Carbon and its compounds flow among four carbon sinks—the atmosphere, biosphere, hydrosphere, and lithosphere. Normally, the amounts of carbon within each of these sinks remains relatively constant as carbon flows among them. However, the combustion of fossil fuels has disrupted the carbon cycle, leading to an increase in the CO_2 present within the atmosphere and oceans. This increase is measured in gigatons (or billions of tons) of carbon per year (GtC/yr).

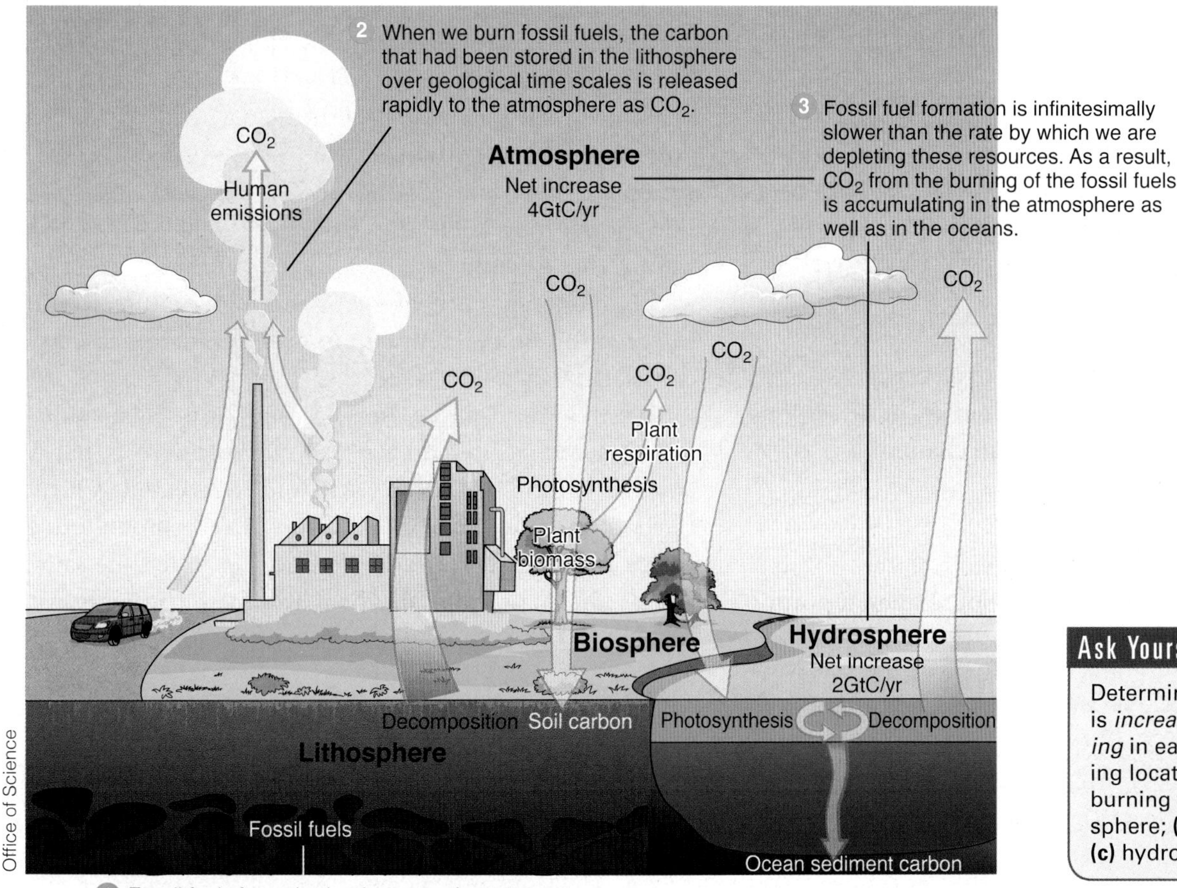

Source: Office of Biological and Environmental Research of the U.S. Department of Energy Office of Science

Ask Yourself

Determine whether carbon is *increasing* or *decreasing* in each of the following locations as a result of burning fossil fuels: **(a)** lithosphere; **(b)** atmosphere; **(c)** hydrosphere.

The **lithosphere**, which is the portion of Earth's crust that contains fossil fuels, represents yet another important storage location for carbon. Petroleum, for example, is thought to have been formed when plankton (microscopic organisms that float near the surface of bodies of water) died and settled to the ocean bottom long ago, then mixed with fine marine sediments and were buried. Over tens of millions of years, subjected to heat, pressure, and chemical activity, this organic matter converted into crude oil and natural gas. Coal is believed to have been formed in a similar manner, but from trees and similar plants, growing in prehistoric swampland.

When we burn fossil fuels—organic matter that has been stored in the lithosphere for eons—we release carbon dioxide to the atmosphere. Currently, we release on the order of 30 billion metric tons of CO_2 per year worldwide though burning fossil fuels. Because the processes by which fossil fuels form are so slow, this newly released CO_2 represents a disruption in the carbon cycle (**Figure 4.17**). Increasing levels of carbon dioxide in the atmosphere are a cause of significant global concern, as CO_2 plays an important role in regulating surface temperatures of Earth. We'll examine this topic further in the next section.

CONCEPT CHECK STOP

1. **How** are photosynthesis and cellular respiration complementary processes?
2. **Why** does the carbon dioxide released from burning fossil fuels produce a net addition of carbon to the atmosphere?

4.5 Greenhouse Gases and Climate Change

LEARNING OBJECTIVES

1. **Define** greenhouse gases.
2. **Explain** how greenhouse gases affect Earth's climate.

An accumulating body of evidence indicates that an increasing concentration of carbon dioxide within the atmosphere is changing the world's climate through an increased **greenhouse effect**. In this section we'll learn about the basis for this effect and examine responses to the challenges posed by climate change.

greenhouse effect A process by which infrared radiation is trapped by certain atmospheric gases, thereby warming Earth's surface and lower atmosphere.

The Greenhouse Effect

The glass walls and roof of a greenhouse allow a portion of the sun's radiant energy to penetrate into the enclosed space. There, inside the greenhouse, the energy is absorbed and partly transformed into **infrared radiation**, which we feel as heat. Since the glass panels of the greenhouse don't transmit the generated heat back to the outside easily, they allow the space within the greenhouse to grow warm. This trapped heat keeps the temperature within the greenhouse high enough to grow temperature-sensitive plants in a cold climate.

infrared radiation A type of radiation that we sense as heat.

greenhouse gases Gases that trap infrared radiation in the atmosphere.

Earth's atmosphere behaves similarly. Solar radiation that penetrates through the atmosphere is partly absorbed at the surface of Earth and re-radiated as heat. Carbon dioxide and other atmospheric **greenhouse gases** trap this newly generated heat in much the same way as the glass traps the heat within the greenhouse (**Figure 4.18**).

How the greenhouse effect works • Figure 4.18

The glass walls of a greenhouse and the greenhouse gases within the atmosphere both trap infrared radiation.

a. Warming of a greenhouse
2 A portion of the IR radiation is trapped by the glass walls, thereby warming the greenhouse.
Greenhouse
Escaping infrared radiation
1 The sun's radiant energy passes through the greenhouse walls and is absorbed and re-radiated as infrared (IR) radiation or heat.
b. Warming of Earth's surface
2 Some of the IR radiation escapes to space, but a large portion is absorbed by greenhouse gas molecules and re-radiated. This traps heat in the lower atmosphere and warms Earth's surface.
Escaping infrared radiation
CH4
CO2
Troposphere
Earth's surface
Trapped infrared radiation
1 A portion of the sun's radiant energy that reaches Earth's surface is absorbed and then re-radiated as IR radiation.

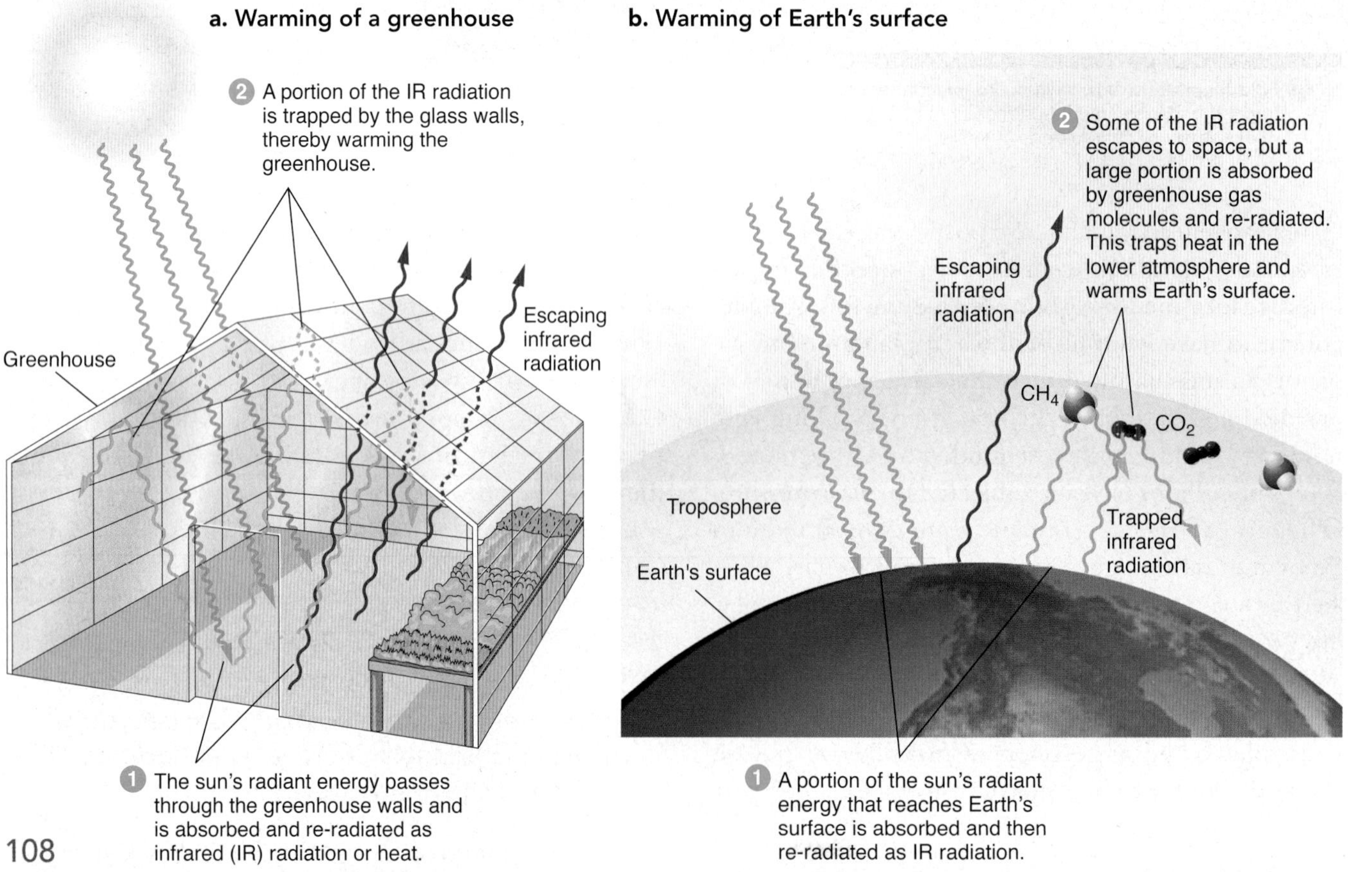

The greenhouse effect and planetary surface temperatures • Figure 4.19

The greater the quantities of carbon dioxide present in a planet's atmosphere, the greater the heat-trapping capacity of its atmosphere.

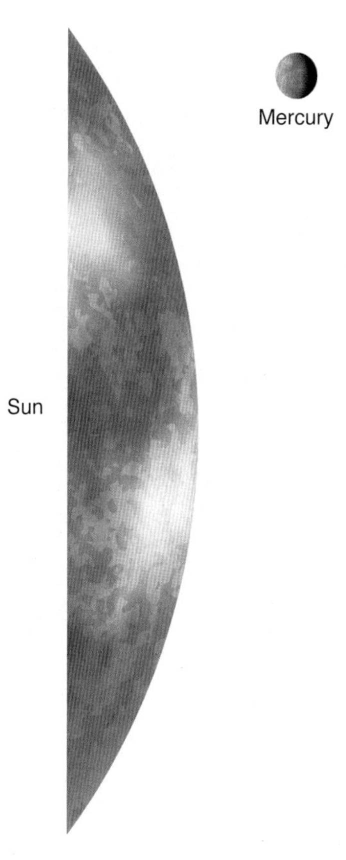

Mercury has practically no atmosphere and no greenhouse effect. Although it's the closest planet to the sun, its average surface temperature is 167°C, far cooler than that of its nearest neighbor, Venus.

Venus's atmosphere is about 95% carbon dioxide and is far denser than that of Earth, giving rise to a potent greenhouse effect. Venus receives only 25% as much solar energy as Mercury, since it's farther from the sun, yet its surface temperature is far hotter, averaging 465°C, well above the melting point of lead.

Earth's atmosphere contains just four-hundredths of 1% (0.04%) carbon dioxide and overall is about 1% as dense as that of Venus. With a modest greenhouse effect, Earth has just the right temperature range for life as we know it. Global surface temperatures average 15°C.

The greenhouse effect, in and of itself, is not harmful; in fact, it's necessary. Life as we know it could not be sustained on Earth if it were not for the presence of just the right amount of carbon dioxide and other greenhouse gases in our atmosphere. By contrast, our neighboring planet, Venus, has a thick atmosphere composed almost entirely of greenhouse gases. Largely as a result, Venus has furnace-like surface temperatures, far hotter even than the planet Mercury, which lies closest to the sun (**Figure 4.19**).

Climate Change

Throughout history, humans have harnessed various sources of energy to do our work. In earlier times we used animals to pull plows, wind to move ships, falling water to grind grain, and wood to feed fires and provide heat. But the middle of the 18th century marked a new era in energy usage. In Great Britain, the development of an efficient steam engine and the expanded mining of coal led to the transformation of a primarily agricultural society into a newly industrial society. With coal readily available and the steam engine able to convert coal's energy into industrial work, the Industrial Revolution had begun.

As the Industrial Revolution spread from Great Britain to what we now call the industrialized world, the use of fossil fuels expanded and the CO_2 content of the atmosphere began to creep upward. By 1958, atmospheric concentrations of carbon dioxide were estimated to have risen by about 12% from their levels in the 1750s. Today they are roughly 40% higher than preindustrial levels and are growing at an estimated 0.5% each year (**Figure 4.20** on the next page). Many observations indicate that Earth's temperature is increasing and that this warming trend is due at least in part to the increasing use of fossil fuels as sources of energy.

Carbon dioxide is not the only greenhouse gas produced by human activity (often referred to as **anthropogenic** activity, from the Greek *anthropos*, "man," and Latin *genus*, "born of"). The *Chemistry InSight* on page 111, describes some of the more important anthropogenic greenhouse gases and their relative contributions to the greenhouse effect.

Atmospheric carbon dioxide levels • Figure 4.20

Atmospheric CO_2 levels are growing rapidly compared to historical norms. Virtually the entire increase in these levels is attributed to human activities, largely through the burning of fossil fuels for transportation, heating, and industrial production. Worldwide, fossil fuel combustion releases an estimated 26 billion tons of CO_2 annually. Roughly half of this remains in the atmosphere; the rest is taken up by carbon sinks, such as the oceans and living plants.

a. Recent atmospheric CO_2 levels.
The data below were recorded at the Mauna Loa Observatory in Hawaii, a remote location, far from any urban and industrial sources of CO_2. In 2013, CO_2 levels reached 400 parts per million by volume, equivalent to a concentration of 0.04%.

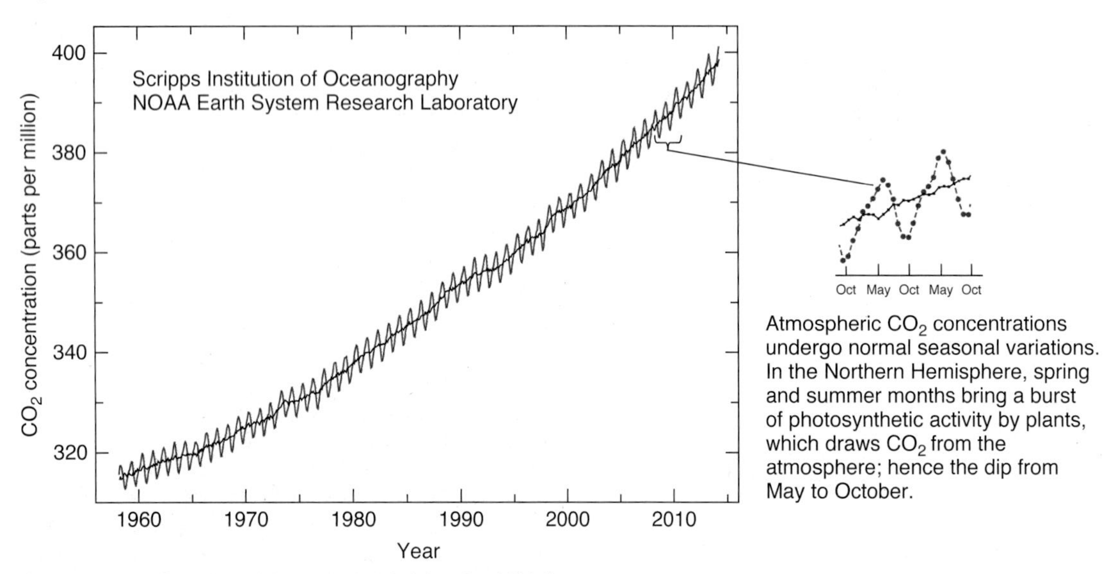

Atmospheric CO_2 concentrations undergo normal seasonal variations. In the Northern Hemisphere, spring and summer months bring a burst of photosynthetic activity by plants, which draws CO_2 from the atmosphere; hence the dip from May to October.

Source: National Oceanic and Atmospheric Administration (NOAA)

b. Prehistoric atmospheric CO_2 levels.
The data below were obtained from ice core samples taken in Antarctica. Chemical analyses of air bubbles trapped within this ice show that atmospheric levels of CO_2 have varied over the last few hundred millennia, but are well below current values.

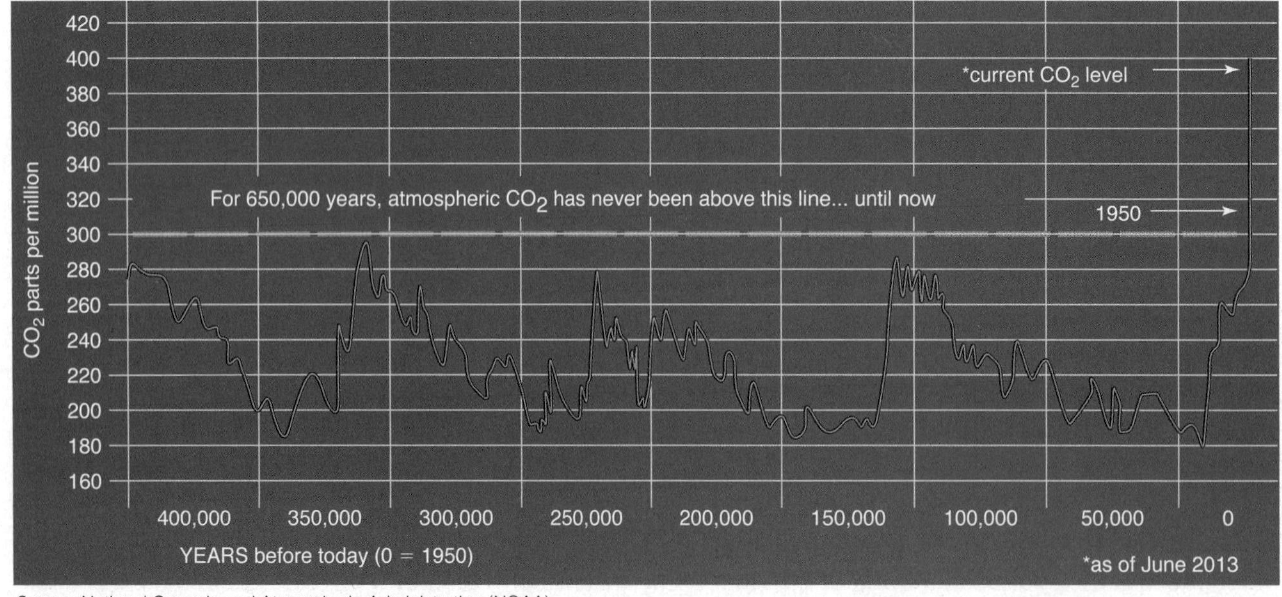

Source: National Oceanic and Atmospheric Administration (NOAA)

Chemistry InSight Anthropogenic greenhouse gases •

Figure 4.21

Carbon dioxide is the most significant greenhouse gas produced by human activity.

a. Important greenhouse gases produced by human activity and their sources

Carbon dioxide (CO_2)

Burning fossil fuels and plant material (wood), deforestation

Methane (CH_4)

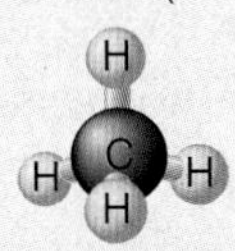

Biomass decomposition in solid waste landfills, cattle farming, rice farming, coal mining, etc.

Nitrous oxide (N_2O)

Use of nitrogen-containing fertilizers in commercial agriculture

b. U.S. emissions of anthropogenic greenhouse gases (GHGs). *Source:* U.S. Energy Information Agency, Annual Greenhouse Gas Emissions (2007).

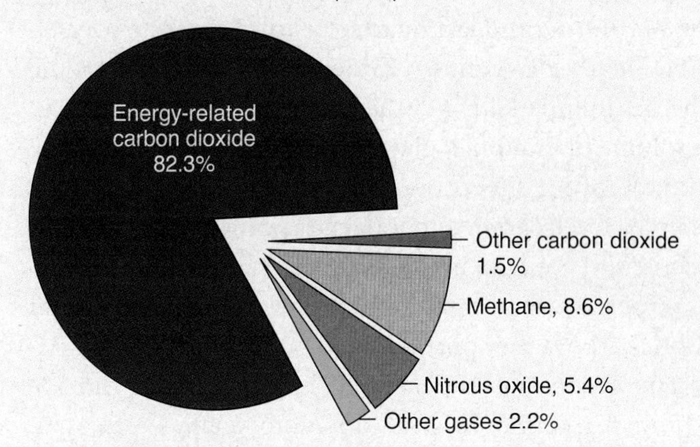

GHGs vary in how efficiently they trap infrared radiation. For example, methane (CH_4) and nitrous oxide (N_2O) trap heat more readily than CO_2. The values in this chart are based on *carbon dioxide equivalents*, which take into account these differences in heat-trapping strength. In spite of its weaker heat-trapping effect, CO_2 still represents the most significant anthropogenic GHG because of the quantities emitted by human activity.

c. How greenhouse gases absorb infrared radiation

The terms *stretching* and *bending* may sound like exercise routines, but they're also used to describe the vibrational motions greenhouse gas molecules undergo when they absorb infrared (IR) radiation. In some sense, the bonds that connect the atoms of a molecule behave like springs. As a greenhouse gas molecule absorbs IR radiation, its bonds undergo rapid vibrations. For example, CO_2, a linear molecule in its resting state, can undergo IR-induced bending vibrations, as modeled below. These and other molecular vibrations absorb and re-radiate IR energy, providing a basis for the greenhouse effect.

Along with the increasing concentrations of atmospheric CO_2 and other greenhouse gases, the average temperature of Earth's surface appears to be rising as well (**Figure 4.22**). The period 2000–2009 was the warmest decade on record, and in the United States, 2012 was the warmest year on record. Moreover, the temperature increase seems to be accelerating. During the 20th century as a whole, the temperature rose at a rate of about 0.6°C/century. But during the last quarter century, the rate accelerated to nearly 2°C/century. Evidence of a general warming trend lies in the melting of the Arctic ice cap; a decrease in the volume of mountain glaciers throughout the world; the shrinking of ice sheets over Greenland and Antarctica; a rise in sea level; earlier and longer growing seasons in many regions; and a spread in the territorial range of several species of plants and animals to higher elevations and higher latitudes, which they once avoided because of the cold.

The greenhouse effect itself is real. Carbon dioxide and other greenhouse gases do act to retain Earth's surface heat. But although a large portion of the increase in Earth's surface temperature results directly from human activities, part of the warming trend may also reflect normal variations in the planet's climate. These kinds of variations brought the latest ice age, some 20,000 years ago, and the subsequent warming to today's moderate climate. Both of these climate changes—the cooling and the warming—took place long before humans dominated the planet.

The Intergovernmental Panel on Climate Change (IPPC) has been asked to assess the contributions of both human activity and nonhuman influences on Earth's climate. In 2013 the IPPC reported that[1]:

- "Warming of the climate system is unequivocal, and since the 1950s, many of the observed changes are unprecedented over decades to millennia. The atmosphere and ocean have warmed, the amounts of snow and ice have diminished, sea level has risen, and the concentrations of greenhouse gases have increased."
- "It is *extremely likely* that human influence has been the dominant cause of the observed warming since the mid-20th century and that more than half of the observed increase in global average surface temperature from 1951 to 2010 was caused by the anthropogenic increase in greenhouse gas concentrations and other anthropogenic forcings together."
- "Continued emissions of greenhouse gases will cause further warming and changes in all components of the climate system. Limiting climate change will require substantial and sustained reductions of greenhouse gas emissions."

Regarding the future, an increase of an anticipated two to four degrees Celsius in Earth's average surface temperature, spread over the next century, may not seem like much. However, global warming is causing sea levels to rise, with potentially serious consequences for coastal communities and water supplies. Also, however gradual it might be, a persistent global warming could also generate public health concerns through an increase in the level of tropical diseases, heat-related illnesses, and malnutrition.

Climate change is a long-term, gradual phenomenon that for most of us has not had a discernible effect on our daily lives. As a result, climate change can be hard to perceive as a pressing issue. Not surprisingly then, apart from the scientific community, some of the earliest voices calling attention to the issue have been those

Global mean temperatures, 1900–2010 • Figure 4.22

Average global temperatures are nearly 1°C higher than they were 100 years ago, and the warming trend appears to be accelerating.

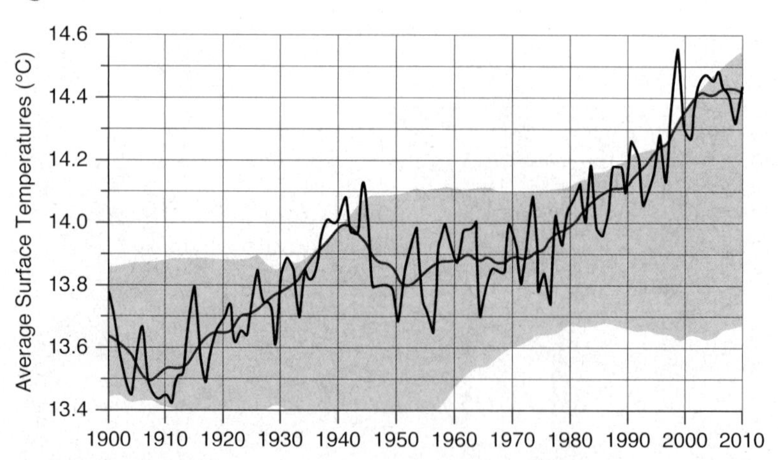

Graph drawn by Hanno Sandvik, using data made available by the Met Office Hadley Centre for Climate Change, U.K. and made public via Wikipedia Commons

Evidence of climate change • Figure 4.23

The Arctic region has been particularly susceptible to the impacts of climate change. As the Arctic ice shelf recedes, the warming trend is accelerating, partly because the relatively darker surfaces of ocean water absorb more of the sun's energy than the white surfaces of ice and snow.

1980

NASA/Goddard Scientific Visualization Studio

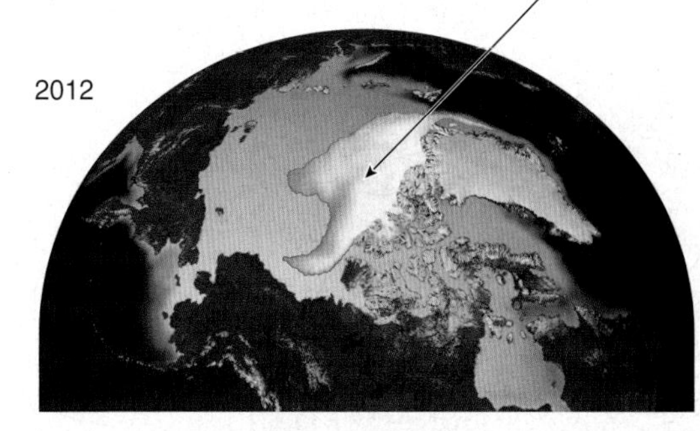

NASA/Goddard Scientific Visualization Studio

This visualization from satellite data shows wintertime Arctic sea ice in 1980 and 2012. Goddard Scientific Visualization Studio

carbon taxes An alternative to carbon cap and trade, whereby a tax is levied on the use of fossil fuels.

cap and trade A system whereby total CO_2 emissions are limited or capped by a government body. Participants in the system, such as electric utilities, buy permits allowing them to emit defined amounts of CO_2. Those who emit less than their quota can sell their excess allowances to others in the system.

whose ways of life and livelihoods are likely to be threatened by climate change. For example, Sheila Watt-Cloutier, a Canadian Inuit activist, has argued that the human rights of the Inuit—an indigenous people of the Arctic, in Russia, Canada, Alaska, and Greenland—are threatened by human-induced climate change. Summertime Arctic sea ice has declined dramatically over the last 30 years, threatening polar bear, seal, and other populations (**Figure 4.23**).

If the extent of future global temperature rise depends largely on the amount of greenhouse gases we release into the atmosphere, what are some of the ways we might mitigate those emissions? Many have suggested economic incentives to curb the use of fossil fuels, such as **carbon taxes** or **cap and trade** systems. One stumbling block in addressing climate change, however, has been getting the support of those countries with the largest or fastest-growing GHG emissions levels to agree on national goals for reducing those levels and enacting policies to achieve these goals. The **Kyoto Protocol**, an international treaty introduced in 1997 (in Kyoto, Japan), was created with the intent of committing industrialized nations to set such targets for themselves. Whether the Kyoto Protocol or other international treaties help deliver significant reductions in global GHG emissions remains to be seen.

CONCEPT CHECK

1. **How** do greenhouse gases trap heat in the atmosphere?
2. Predict **what** Earth's average surface temperature will reach in the year 2100 if it keeps increasing at the same rate as it did in the last quarter of the 20th century.

4.6 Energy for the Future

LEARNING OBJECTIVES

1. **Describe** the benefits as well as limitations of renewable energies.
2. **Discuss** strategies for reducing the environmental impacts of fossil fuels.

According to the International Energy Agency, worldwide demand for energy is expected to grow 40% between 2012 and 2040, yet the proportion of this energy supplied by fossil

Direct solar energies • Figure 4.24

Sunlight—a free, essentially limitless source of energy—can be harnessed to heat water and produce electricity.

a. A solar hot water heater. Photothermal systems such as this, which can also be mounted on the roof of a home, are price competitive with conventional water heaters that use electricity or natural gas.

Mark Boulton/Photo Researchers/ Getty Images

b. A large-scale photovoltaic (PV) installation for generating electricity. Because of the relatively low efficiency in converting sunlight into electricity, large areas of land are needed for these facilities. Chemists, material scientists, and others continue to develop novel materials to improve the efficiencies of PV cells.

John Burcham/National Geographic/Getty Images

fuels—about 80%—is expected to remain roughly consistent over this time frame, assuming current energy consumption trends continue. Since fossil fuels are a nonrenewable resource, our continued reliance on them is not sustainable. In addition, serious environmental concerns associated with their use point to a potential future in which alternative energies play a greater role and ultimately supplant fossil fuels. In this section we explore alternative energies, such as solar and nuclear power, as well as strategies for reducing the environmental impacts of fossil fuels.

Alternative Energies

A broad range of strategies will likely be needed to reduce our reliance on fossil fuels. For example, we can attempt to curb energy use through energy conservation and improved efficiency, which can help buy us time to develop alternative energies. Energy conservation means reducing energy use and waste through behavioral choices we can collectively make, such as by turning off lights and appliances when they're not needed, setting thermostats to energy-saving temperatures, biking instead of driving, taking public transportation, carpooling, and driving at slower speeds. Energy efficiency refers to using less energy to accomplish a given task, usually through technological improvements. For instance, the designs of new cars continue to offer improved fuel efficiencies, which reduces gasoline consumption. Also, compact fluorescent and light-emitting diode (LED) light bulbs consume less power than traditional (incandescent) bulbs, which waste more than 90% of their energy as heat.

Most alternative energies rely either directly or indirectly on sunlight—a free, essentially limitless source of energy. Capturing this solar energy efficiently is the key. Direct solar technologies include **photothermal** systems, which convert sunlight to heat, and **photovoltaics**, which convert sunlight to electricity (**Figure 4.24**). Indirect solar energies include **wind energy**, which results from the sun's warming of the atmosphere; **biomass** (plant or animal materials used as fuel), which grows as a result of photosynthesis, driven by the sun's energy; and **hydropower**, which results from the sun's role in driving Earth's water cycle. In this cycle, water that evaporates from oceans and other bodies of water falls as precipitation onto higher elevations. As the runoff from this precipitation flows into rivers, the currents generated can be used to drive turbines, which in turn produce electricity.

Nonsolar alternative energies include nuclear and geothermal energy. Nuclear energy (discussed in Chapter 9) results from **nuclear fission**, in which the nuclei of certain heavy elements, such as uranium, split into smaller fragments under specific conditions. The energy released during this process is used to boil water, and the steam thus generated can power turbines that produce electricity. Geothermal energy systems capture the heat produced naturally within Earth and can be used for heating homes and generating electricity.

Currently about 5% of the world's energy needs are met by nuclear power and 12% are met by renewable energies, such as the burning of biomass, as well as solar, wind, hydroelectric, and geothermal power. [We discuss technical aspects of nuclear and solar power in Chapters 9 and 10, respectively.] Nuclear power does not contribute greenhouse gas emissions, but a combination of cost issues and safety concerns—including accidents and the storage of the long-lived radioactive waste produced by these plants—has limited growth in the industry. Of the total energy consumed in the United States, nuclear power provides 8.5% and renewables supply about 7.5%. Of this energy supplied by renewables, two-thirds comes from hydroelectric sources, and about one-fifth comes from wind power.

Various factors such as cost and convenience influence the adoption of alternative energies. Regarding costs, according to the International Energy Agency, governmental subsidies are essential to the growth in renewable energy, since some renewables are still more expensive than fossil fuels. For example, photovoltaic (PV) cells represent a promising technology, yet costs have limited their growth. Continued improvements in both the design of PV cells as well as in their installation and maintenance offer hope that they may become more cost-competitive with conventional energies.

Transportation accounts for nearly 30% of our energy needs and is the cause of significant air pollution, so reducing gasoline consumption offers various benefits. Hybrid, electric, and fuel cell vehicles (discussed in greater detail in Chapter 10) offer promise in reducing our need for gasoline.

- Hybrid vehicles, currently the most popular alternative to conventional, gasoline-powered vehicles, can operate on either gasoline or battery power. Batteries are **electrochemical** devices, meaning that chemical reactions that occur within them generate a flow of electricity. Hybrids utilize regenerative braking, in which an electric generator slows the moving car by converting its energy of motion into electricity rather than heat. This regenerated electricity is fed to the battery to help maintain its charge.
- Electric vehicles run solely on batteries and thus are zero emission vehicles (ZEVs). Although electric vehicles also employ regenerative braking, their batteries are recharged primarily by plugging into electrical sources. In locations where this electricity is supplied wholly or in part by fossil fuels, the use of electric vehicles contributes, albeit indirectly, to greenhouse gas emissions.
- Fuel cell vehicles, like those powered by batteries, use chemical reactions to generate electricity, but they require fuel supplied from an external source. In a common form of a fuel cell, hydrogen (H_2) combines with oxygen (O_2) on the surface of the cell's catalysts to form water and to produce electricity (**Figure 4.25**). Hydrogen fuel is currently produced from fossil fuels, which limits its environmental benefits. But since hydrogen can also be produced directly from water through **electrolysis** (Chapter 10), chemists and others are seeking cost-effective ways to accomplish this. One promising approach is through **photocatalysis**, in which sunlight shining on a metal catalyst submerged in water helps to convert this water directly into hydrogen (and oxygen) gases.

Fuel cells • Figure 4.25

Within a hydrogen fuel cell, hydrogen reacts with oxygen to produce water vapor. The energy released in this process yields electricity, which can be used to power electric motors and in other applications.

a. The chemical reaction that occurs within a hydrogen fuel cell and the essential components of this device.

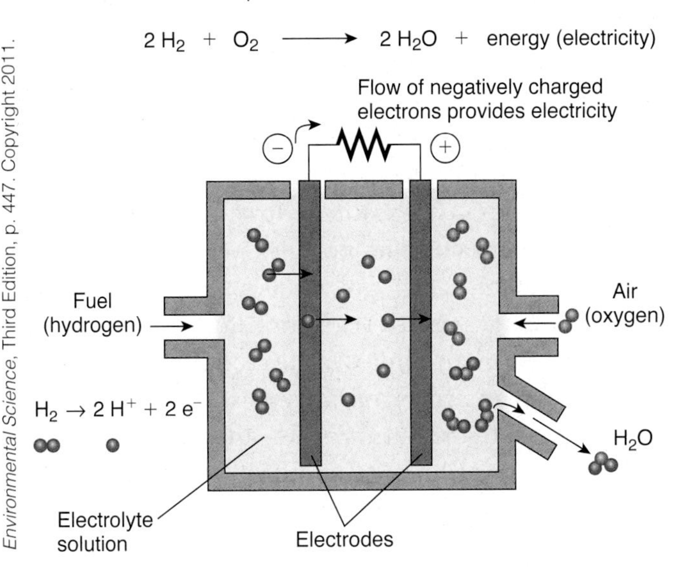

b. A hydrogen fuel cell bus. A combination of technical and cost issues has thus far limited wider adoption of fuel cell vehicles. For instance, unlike gasoline, hydrogen fuel is not widely available.

Ask Yourself

Would you agree that the bus shown here produces "zero emissions"? Explain.

Carbon-Based Fuels

Fossil fuels contain carbon, so burning them releases carbon dioxide. However, the environmental impacts of combusting these fuels vary. The use of coal raises the most concern because it releases more carbon dioxide and other pollutants for a given amount of heat generated than natural gas or petroleum-based fuels. In addition, because coal is the least expensive and most plentiful fossil fuel, its worldwide use is growing. China already consumes more coal than the United States, India, and Russia combined, and coal use in China as well as in other developing economies is accelerating.

Several approaches can help decrease the environmental impacts of coal use. Many newer coal-fired power plants are designed with improved efficiencies so that more power is generated for the same amount of CO_2 emitted (Chapter 13). In addition, **carbon capture and sequestration** (CCS) is a method of capturing CO_2 gas from stationary sources of these emissions, such as coal-fired power plants, and storing this gas long-term in rock formations deep underground. Although some large-scale CCS projects are in operation worldwide, CCS is still an emerging technology and it remains to be seen how widely it may be deployed.

Since natural gas is the cleanest-burning fossil fuel, a shift from coal to natural gas can help lower CO_2 emissions. According to the U.S. Energy Information Agency, from 2010 to 2040, natural gas is projected to be the fastest-growing fossil fuel and will therefore play a greater role in meeting future energy demand. However, a greater reliance on natural gas may inadvertently serve to postpone the inevitable shift from fossil fuels to more environmentally friendly energy alternatives, and as a result present a greater risk for climate change.

The United States is the world's largest consumer of petroleum, and overall, two-thirds of all the petroleum it uses goes, in one form or another, into transportation of all kinds. We add more pollutants to our environment through the use of transportation fuels than we do with any other single activity. Auto exhaust contributes about a quarter of all the CO_2 discharged into the air above the United States, as well as almost all of the carbon monoxide in and above our cities.

Increases in fuel-efficiency standards for new cars and trucks—known as **CAFE standards** in the United States (for Corporate Average Fuel Economy)—can help reduce gasoline consumption and air pollution. According to estimates by the Environmental Protection Agency and the National Highway Traffic Safety Administration, currently proposed CAFE standards for cars and light-duty trucks sold in 2017–2025 are projected to save approximately 4 billion barrels of oil and to reduce greenhouse gas emissions by 2 billion metric tons.

Cleaner-burning alternatives to gasoline exist as well. These include liquefied petroleum gas (LPG), compressed natural gas (CNG), and ethanol.

Liquefied petroleum gas (LPG), often referred to simply (and erroneously) as propane, is a liquefied mixture of hydrocarbons that normally exist as gases. Its composition varies, but it's largely propane with the isomeric butanes and smaller amounts of methane and ethane. LPG is currently the most widely used alternative fuel. Only gasoline and diesel fuel surpass LPG as a fuel for transportation. It's a clean-burning fuel, but it is still a fossil fuel, with all the disadvantages that implies.

Compressed natural gas (CNG) is composed principally of methane, with small amounts of other alkanes. Plentiful, clean-burning, and often at a cost not far from gasoline's, it ranks with LPG as a cleaner-burning alternative to gasoline. Because CNG is compressed, this fuel must be stored in heavy, sturdy tanks. Its disadvantages are the short driving range it provides—about 100 miles—the heavy and awkward fuel tanks it requires, and the complexity of refueling. It is more suitable to fleet vehicles than to private automobiles.

Ethanol (ethyl alcohol) belongs to a class of fuels known as **biofuels** (see *Did You Know?*). Unlike fuels obtained from petroleum, natural gas, or coal, ethanol comes principally from the fermentation of grain and other crops, including corn and sugarcane, and is thus a renewable resource. Add to this ethanol's octane rating of 105, its lower potential impact on the environment, and its relatively low toxicity, and ethanol becomes a potentially attractive replacement for hydrocarbon fuels. Among its disadvantages are its lower energy content than gasoline (meaning fewer miles per gallon) and its lower availability.

Since plants absorb carbon dioxide from the atmosphere through photosynthesis, land-use changes that remove plant cover on a large scale can contribute to higher atmospheric carbon dioxide levels. For example, forests still cover about 30% of the world's land mass, but are being lost through deforestation.

DID YOU **KNOW?**

Biofuels are not a new technology

Given our association of cars with gasoline, you might be surprised to learn that some of the earliest internal combustion engines ran on biofuels. For example, in the 1890s, Rudolf Diesel, the inventor of the motor that bears his name, ran some of his first engines on vegetable oil. A decade later, Henry Ford's famous Model T automobile was capable of running on ethanol, derived by the fermentation of grains. By the 1920s, however, petroleum became widely available and cheap, making gasoline and diesel the transportation fuels of choice.

Interest in biofuels reemerged in response to the 1973 oil embargo, a period remembered for high gas prices and long lines at gas stations. Today, concerns about dependence on oil imports and impacts on the environment continue to underscore the need for alternatives to gasoline. A major attraction of biofuels is that they are based on plant materials, a renewable resource that can be grown domestically. Furthermore, when we burn these fuels, the CO_2 emissions are offset by the CO_2 that had been previously removed from the atmosphere by the plants as they grew, as shown in the following carbon cycle for corn-based ethanol.

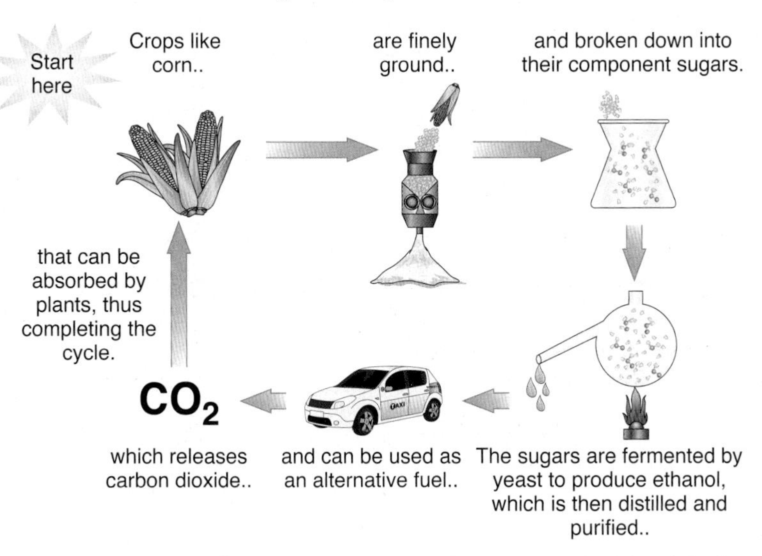

Start here
Crops like corn..
are finely ground..
and broken down into their component sugars.
The sugars are fermented by yeast to produce ethanol, which is then distilled and purified..
and can be used as an alternative fuel..
which releases carbon dioxide..
CO2
that can be absorbed by plants, thus completing the cycle.

Adapted from U.S. Energy Information Administration.

What this figure doesn't reveal, however, is that fossil fuels themselves are required to run the farm equipment and factories needed to produce biofuels. Also, forests or grasslands, which can serve as rich carbon sinks, are sometimes cleared in the planting of biofuel crops. Since these activities contribute to greenhouse gases, critics have argued that the **carbon footprint** of biofuels, such as corn-based ethanol, is not as small as many perceive it to be. (Carbon footprint is a measure of how much carbon dioxide is released into the atmosphere by a particular product or activity.)

More efficient ways of producing biofuels are on the horizon. Cellulose, derived from the woody or fibrous parts of plants, is a large molecule that can be broken down into simple sugars. These sugars can be fermented to produce ethanol. The challenge has been in decomposing cellulose efficiently into these smaller sugars. Switchgrass, a robust prairie plant that can be grown in much higher yields per acre than corn, is being explored as a source of cellulose in order to produce ethanol. Specialized algae show promise for producing biofuels (**Figure a**), as they do not require agricultural land and do not pose the "food versus fuel" trade-offs associated with corn. [As we divert an increasing portion of our corn crop toward ethanol production, a concern is that the price of corn used for food will rise as a result of decreased supply.]

Today, **E10** (gasoline formulated with 10% ethanol) is widely used in the United States. Cars and light-duty trucks that normally operate on regular gasoline can run on E10 without needing any modification. Flexible fuels vehicles (FFVs), on the other hand, have fuel systems that can accommodate **E85,** a far richer ethanol blend of up to 85% ethanol and the remainder unleaded gasoline. FFVs often bear a small "FLEXFUEL" nameplate to alert drivers of these vehicles of the option to use E85. When operating on E85, a vehicle gets somewhat fewer miles per gallon because ethanol provides less energy than the hydrocarbons of conventional gasoline.

Greater use of biofuels, such as ethanol, may be a step in the right direction, but so far it has not markedly reduced our dependence on petroleum. One of the challenges has been in scaling up ethanol production. For example, converting all the grain grown in the United States each year into ethanol would still supply only about a quarter of the transportation fuel needed. Clearly, the use of biofuels represents but one of many alternatives to the use of fossil fuels.

Bloomberg via Getty Images

a. Different types of algae are being explored as sources of biofuels. With sunlight, water, and nutrients, these organisms can potentially turn CO_2 directly into oils or other fuel components. The challenge lies in scaling up these processes to produce fuels on a cost-competitive basis.

Ask Yourself

What are three advantages and three disadvantages of the use of corn-based ethanol as a transportation fuel?

As trees are cut down, photosynthetic activity within them not only stops, but the burning of these plant materials or allowing them to decay naturally releases CO_2 to the atmosphere. Reforestation on a large scale represents one potential strategy for mitigating the rise in atmospheric CO_2 levels.

CONCEPT CHECK 

1. What are two factors limiting the greater adoption of photovoltaic technology?
2. How might the use of biofuels help in curbing greenhouse gas emissions?

Summary

1 Energy and Its Uses 90

- ***What is energy?***
  Energy is the capacity to do work. The **joule** is a standard scientific unit of both energy in general and the specific form of energy we recognize as heat. **Potential energy**, such as that present in fuels, is stored energy, whereas **kinetic energy** is the energy of motion (as shown here).

Figure 4.1 • Visualizing potential and kinetic energy

RonTech2000/E+/Getty Images

- ***How is energy used by society?***
  Four broad sectors of society account for most of our energy consumption: residential, transportation, commercial (such as schools and offices), and industrial (such as factories). This energy is provided largely in the form of electricity and in fuels for transportation and heating.

2 Fossil Fuels 93

- ***What are fossil fuels, and why are they vital to society?***
  Fossil fuels are **natural gas**, **petroleum** (oil), and **coal**—carbon-rich fuels that slowly formed over millions of years from decaying organic matter. Found in Earth's crust, these fuels consist of hydrocarbons, which are molecules containing hydrogen and carbon (as shown here for propane), and are widely used to provide **energy** for heating, transportation, and the generation of electricity. **Fossil fuels** are also used as feedstocks for the manufacture of plastics, textiles, chemicals, and many other products.

Propane, a hydrocarbon • Figure 4.4

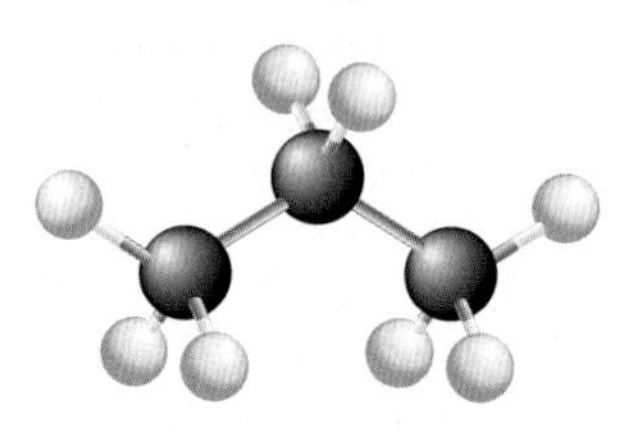

3 Petroleum Refining and Gasoline 98

- ***What is petroleum refining?***
  Petroleum refining is a process by which crude oil, a complex mixture of compounds, undergoes separation into its various components, including petroleum gases, transportation and heating fuels, and lubricants. Separation is accomplished through **fractional distillation** in the type of equipment shown here.

Figure 4.11 • Fractional distillation of petroleum

Paul Rapson / Photo Researchers, Inc.

- ***What is gasoline made of, and what does its octane rating represent?***
  Derived from petroleum, gasoline is a mixture of over a hundred different **hydrocarbons**, trace contaminants, and additives. To burn smoothly in a car's engine, gasoline is formulated to meet certain **octane** ratings, a measure of how well the gasoline avoids **knocking**, or erratic **combustion**. Additives such as ethanol are blended into gasoline to increase the efficiency of combustion and improve octane ratings.

4 Fossil Fuels and the Carbon Cycle 106

- ***What is the carbon cycle?***
  The **carbon cycle** describes how carbon is exchanged between living things, the atmosphere, the oceans, and geological systems as shown here. In one phase of this cycle, plants convert CO_2 from the atmosphere into organic compounds through photosynthesis. In another phase, animals convert the organic compounds of food into CO_2 through cellular respiration.

Figure 4.17 • The carbon cycle and fossil fuels

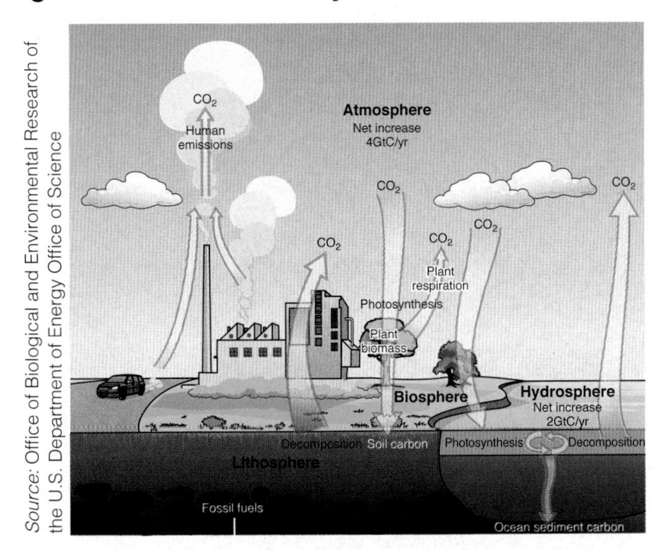

Source: Office of Biological and Environmental Research of the U.S. Department of Energy Office of Science

- ***How does burning fossil fuels impact the carbon cycle?***
  Burning fossil fuels disrupts the carbon cycle because fossil fuels form so slowly compared to the speed with which the carbon contained within fossil fuels is released into the atmosphere.

5 Greenhouse Gases and Climate Change 108

- ***What are greenhouse gases, how do they affect the environment, and why are they a cause for concern?***
  Greenhouse gases help warm Earth's surface by allowing the sun's radiant energy to reach the surface, but preventing the resulting **heat** from leaving (as shown here). Levels of CO_2, an important greenhouse gas, have been steadily rising in the atmosphere as a result of the burning of fossil fuels. There are widespread concerns about climate change associated with increased levels of **greenhouse gases**.

Figure 4.18 • How the greenhouse effect works

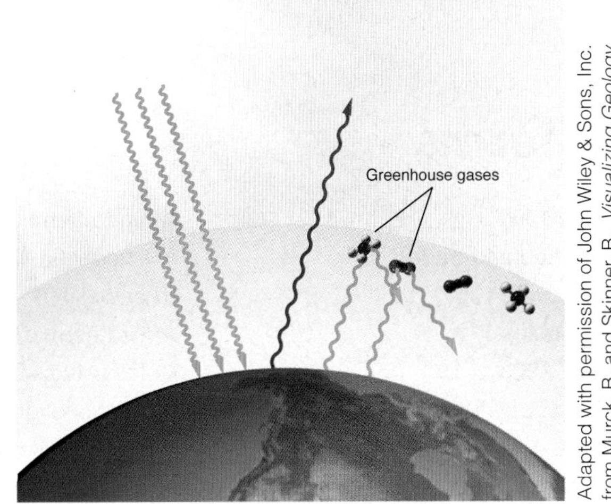

Adapted with permission of John Wiley & Sons, Inc. from Murck, B. and Skinner, B., *Visualizing Geology*, Third Edition, p. 369. Copyright 2012.

6 Energy for the Future 113

- ***How can alternative energies meet future energy demand?***
  The sun's energy—used either directly or indirectly—enables most alternative energies, with the notable exceptions of nuclear and geothermal energy. Safety, cost, and convenience factor into the market growth of alternative energies. Nuclear power provides a reliable source of electricity and does not emit greenhouse gases, but safety and cost issues have hampered its growth. Photovoltaics, like other alternative energies, will likely gain wider acceptance as they become more cost-competitive with fossil fuels. Alternative-powered vehicles, such as hybrids, electrics, and those using hydrogen fuel cells (shown here), are also helping to reduce fossil fuel consumption.

Figure 4.25 • Fuel cells

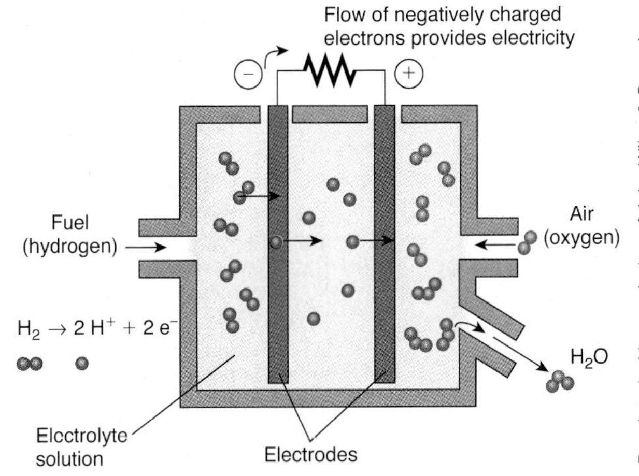

Reprinted with permission of John Wiley & Sons, Inc. from Berg, L. and Hager, M.C., and Hassenzahl, D., *Visualizing Environmental Science*, Third Edition, p. 447. Copyright 2011.

- ***How can we reduce the environmental impacts of carbon-based fuels?***
  Since the combustion of coal produces more air pollutants than the burning of **natural gas**, a shift from coal to natural gas can help reduce these emissions. Cleaner-burning alternatives to gasoline, including liquefied petroleum gas, compressed natural gas, and ethanol (which is a **biofuel**), can also help reduce pollution. Raising vehicle fuel-efficiency standards can reduce gasoline consumption and pollution as well. Aside from shifting to alternative energies, strategies for slowing growth of atmospheric carbon dioxide levels include carbon capture and sequestration, limiting deforestation, and the large-scale planting of trees.

Key Terms

- biofuels 105
- carbon cap and trade 113
- carbon cycle 106
- carbon taxes 113
- catalyst 100
- catalytic converter 104
- catalytic cracking 100
- catalytic reforming 102
- coal 94
- combustion 97
- compression ratio 99
- endothermic 97
- energy 90
- entropy 91
- exothermic 97
- fossil fuels 93
- fractional distillation 100
- greenhouse effect 108
- greenhouse gases 108
- heat 90
- hydrocarbons 93
- infrared radiation 108
- kinetic energy 90
- knocking 99
- law of conservation of energy 90
- natural gas 93
- octane rating 103
- oxygenate 104
- petroleum 94
- petroleum refining 100
- potential energy 90
- volatility 100

What is happening in this picture?

For many years, the Brazilian government provided incentives for cattle ranchers to burn and clear land in the Brazilian rainforest, converting it to cattle pasture. Today the incentives have ended but the significant deforestation has altered the landscape.

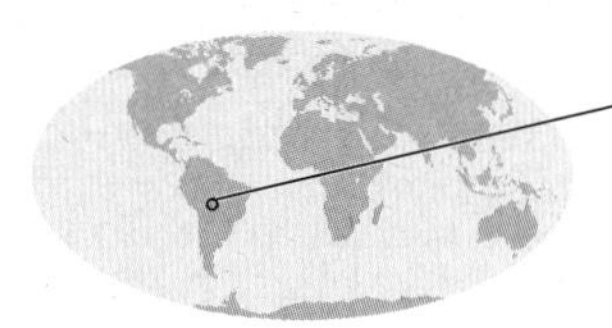

Global Locator
Rondonia State, Brazil

Michael Nichols/National Geographic Creative

Think Critically

1. Does burning plant material in and of itself provide a net addition of carbon dioxide to the atmosphere?
2. If burning plant material is accompanied by land-use changes, as in the clearing of tracts of rainforest, how does this impact the balance of atmospheric greenhouse gases?
3. What hydrocarbon greenhouse gas is associated with cattle farming? (Refer to Figure 4.21.)

Exercises

Review

1. Fossil fuels supply what percentage of the energy requirements of

a. the world;

b. the United States?

2. a. Identify three fossil fuels.

b. Which is more commonly used as a source of transportation fuels?

c. Which two are more commonly used to produce electricity (among other uses)?

3. Rank the fossil fuels in decreasing order in the amount of CO_2 given off for a given amount of heat produced.

4. a. Recoverable supplies of which fossil fuel are expected to be depleted the soonest?

b. Which is expected to last the longest?

5. a. What two gases result from the complete oxidation of a hydrocarbon?

b. What substance is produced on incomplete oxidation of a hydrocarbon?

6. a. What is a major advantage of increasing the compression ratio of an engine?

b. What characteristic of a gasoline must be changed as the compression ratio increases?

7. To what does the *volatility* of a liquid refer, and how does this relate to petroleum refining?

8. In the fractionating column at right, identify where the following substances would be withdrawn:

a. petroleum gases, such as propane;

b. liquids suitable for use in gasoline;

c. tarry solids used in road paving and roofing.

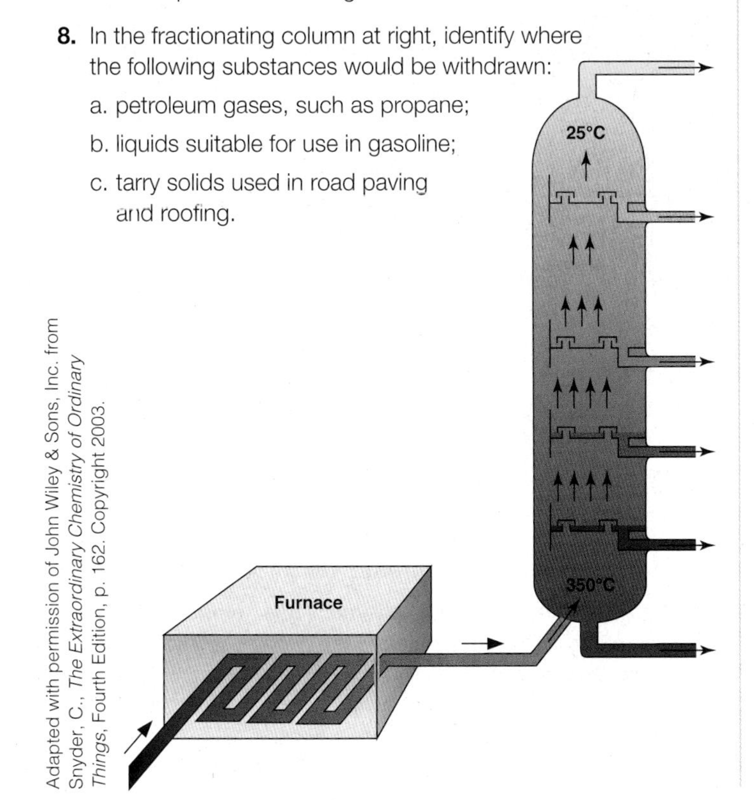

Adapted with permission of John Wiley & Sons, Inc. from Snyder, C., *The Extraordinary Chemistry of Ordinary Things*, Fourth Edition, p. 162. Copyright 2003.

9. a. Why was tetraethyllead originally added to gasoline?

b. What initially prompted the removal of tetraethyllead from gasoline?

c. What additional benefit comes from removing this additive?

d. Name another chemical additive that can serve the same function as tetraethyllead.

e. What refining process produces the same characteristic in gasoline as that obtained by adding tetraethyllead?

10. a. In this figure of a catalytic converter, what do the formulas CO, C_xH_y, and NO_x represent?

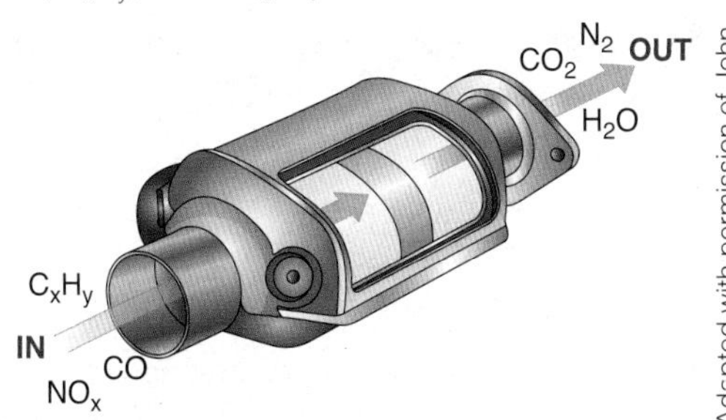

Adapted with permission of John Wiley & Sons, Inc. from Sclar, D., *Auto Repair for Dummies*, Second Edition, p. 155. Copyright 2008.

b. What causes the generation of CO and C_xH_y?

c. What compound(s) are converted within this device to produce: (i) CO_2; (ii) N_2?

11. What automotive emissions are reduced by the use of catalytic converters? What emissions are slightly increased? What types of elements are responsible for the catalytic activity of these devices?

12. a. What is the difference between the function of an *octane enhancer* and an *oxygenate*?

b. Can a single compound provide both functions? If so, provide an example.

13. How are photosynthesis and cellular respiration related? How are they different?

14. Name two other gases, in addition to carbon dioxide, that are classified as greenhouse gases and describe their principal sources.

15. What does the term *anthropogenic* mean with regard to greenhouse gases?

16. a. What type of radiation is absorbed by greenhouse gases?

b. By what mechanism do greenhouse gases absorb and reradiate this energy?

17. Venus is farther from the sun than Mercury, yet its atmosphere is far hotter than that of Mercury. Why is this so?

18. a. In what way does the use of gasoline contribute to the greenhouse effect?

b. What arguments can be made for why the use of corn-derived ethanol as a fuel contributes comparatively less to the greenhouse effect?

19. a. What type of additive is used to increase the rate of each of the following reactions of hydrocarbons? (In the figure, hydrogen atoms have been omitted for clarity.)

b. Which reaction helps increase the amount of gasoline that can be produced from a given quantity of petroleum?

c. Which reaction produces material that increases the octane rating of gasoline? Explain.

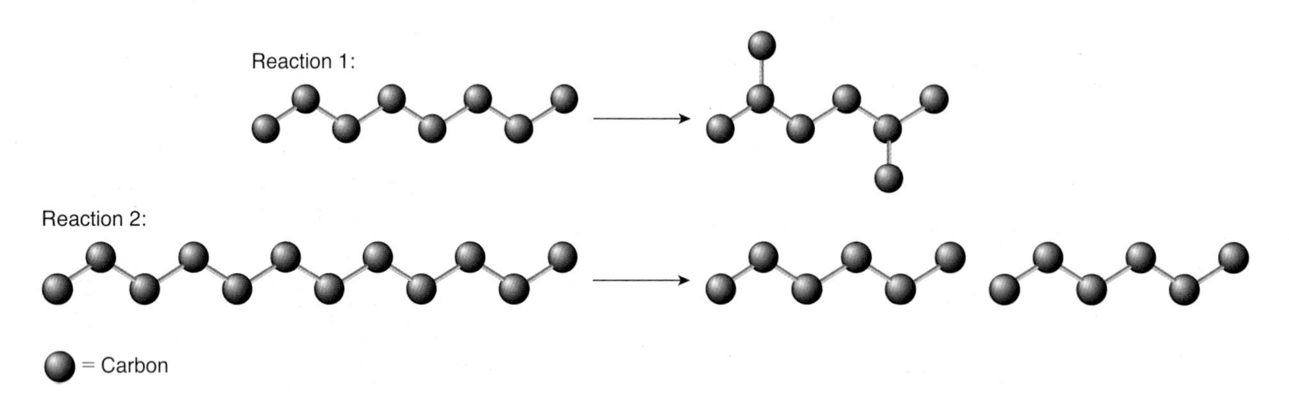

20. a. What is being drawn into the cylinder in Step 1?

b. How would you determine the compression ratio of the engine? During which step is there a potential for knocking?

c. What two major gases are being expelled in Step 4?

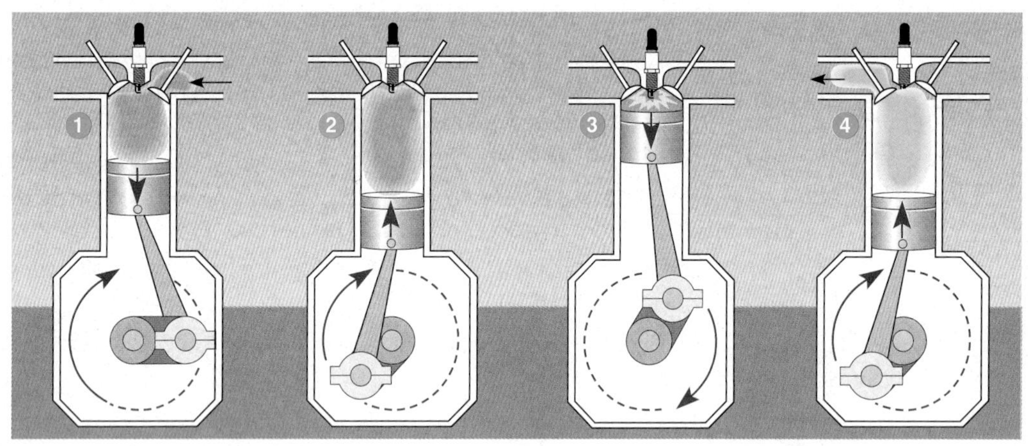

Think

21. Even if the internal combustion engine were so efficient in burning gasoline that no catalytic converters were needed to protect the environment, it's likely that "leaded" gasoline would no longer be available. Why?

22. Suppose we discovered a way to modify the combustion products of fossil fuels by converting CO_2 emissions into CH_4 (methane) emissions through the chemical reaction:

$$CO_2 + 2\,H_2O \xrightarrow{\text{catalyst}} CH_4 + 2\,O_2$$

What effect, if any, would this have on the greenhouse effect?

23. Planting 100 million trees has been suggested as a means for decreasing the greenhouse effect of our atmosphere. Describe the reasoning behind this proposal.

24. What would have been the fate of the carbon of plankton if the plankton had been completely metabolized by other organisms instead of being trapped, only partially decomposed, at the bottom of bodies of water and eventually converted into fossil fuels?

25. You have a mixture of two different liquid compounds and you attempt to separate them from each other by fractional distillation. You find, however, that no matter how carefully you distill them and no matter how complex and intricate a distillation apparatus you use, you cannot separate the two components. What can you conclude about these two liquids?

26. Why does the production and use of biofuels, such as corn-derived ethanol, still contribute to the greenhouse effect, even though the carbon atoms in the corn (and hence in the ethanol) originate in atmospheric CO_2?

27. Electric cars do not give off tailpipe emissions, but they still contribute to the greenhouse effect, although significantly less per mile driven than gasoline-powered cars. How do electric cars contribute to the greenhouse effect?

28. a. What does the term *hydraulic fracturing*, or "fracking," mean, and what primary type of fossil fuel is recovered using this process?
    b. Use the Internet to identify the potential benefits and environmental risks associated with this activity.

29. In 2010, an explosion on the *Deepwater Horizon* oil rig in the Gulf of Mexico led to the largest oil spill in U.S. history. Use the Internet to identify how this event compared in magnitude to that of the *Exxon Valdez* oil spill in 1989. What are the benefits and risks associated with domestic oil production?

30. Is it possible for one country to emit more CO_2 into the atmosphere than another from burning fossil fuels, yet consume less energy through fossil fuels? Explain.

31. Burning a gallon of gasoline releases nearly 20 pounds of CO_2 into the air, even though a gallon of gasoline weighs only about 6 pounds. How is it possible that this emitted CO_2 weighs more than the gallon of gasoline?

Calculate

32. What would be the octane rating of a particular blend of gasoline if it showed knocking tendencies identical to that of a mixture of 7% heptane and 93% 2,2,4-trimethylpentane under standard test conditions?

33. Atmospheric CO_2 concentrations increased from 325 parts per million (ppm) in 1970 to 400 ppm in 2013. What percentage increase does this represent?

34. According to the U.S. Department of Transportation, cars, vans, and light-duty trucks are driven an average of about 13,000 miles per year in the United States and consume and an annual average of roughly 700 gallons of gasoline.
    a. What is the average fuel economy, in miles per gallon (mpg), obtained by these vehicles?
    b. How many gallons of gasoline would the average vehicle save each year if average fuel economy increased by 5 mpg?
    c. Assuming that 19.5 pounds of CO_2 are emitted for every gallon of gasoline consumed, how many fewer pounds of CO_2 would the average vehicle emit annually with a 5 mpg increase in fuel economy?

35. If petroleum consumption in the United States is 20 million barrels per day, how many barrels are consumed on an annual basis? If the U.S. population is approximately 300 million, what is the per capita annual consumption of petroleum, expressed in barrels? How many gallons does this represent?

36. Petroleum refineries can produce about 19 gallons of gasoline for every barrel (42 gallons) of petroleum processed. What percentage of the volume of a barrel of oil does this represent? If petroleum consumption in the United States is 20 million barrels per day, approximately how many gallons of gasoline can be produced domestically each day?

37. A gallon of gasoline supplies roughly 1.3×10^8 J of energy. How many (i) calories; (ii) kilocalories; does this represent? If a Big Mac supplies 540 kilocalories of energy, how many Big Macs would represent the same energy as that supplied by a gallon of gasoline?

5

Energy of Foods

One of the most constant and predictable aspects of our lives is our daily need for food and drink. As shown here, our food—what we eat and how we eat it—helps define our culture and forms an integral part of our families and social lives.

We receive almost daily exposure to the scientific language of foods, including terms such as *trans fats*, *saturated fats*, and *omega-3s*. We're often told that some of these nutrients are good for us while others may be harmful, but it's not always clear why. In this

chapter we'll explore the meanings of these terms. In the process, we'll see that the food we eat is nothing more than a varied assortment of chemical compounds, each with its own chemical formula, its own properties, and its own potential health effects.

An overarching theme of this chapter is that food is the chemical fuel that provides the energy for the human engine, much as gasoline and its hydrocarbons provide the fuel for the engines of motor vehicles. We'll see how the body uses the energy derived from our food's *macronutrients*: the fats and oils, carbohydrates, and proteins that make up the bulk of what we eat.

Liam Norris/Cultura/Getty Images

CHAPTER OUTLINE

5.1 The Energy Equation and Metabolism

LEARNING OBJECTIVES

1. **Differentiate** between a chemical calorie and a dietary Calorie.
2. **Determine** how many Calories are in a food through an analysis of the food's macronutrient content.
3. **Describe** how the body uses food energy.

Different as they obviously are, your human body and a car's engine have this in common: Both use energy stored in chemical structures to do work. In the case of the car, hydrocarbon molecules of gasoline react with oxygen to produce gaseous byproducts and heat (energy), a process called **combustion**. In the case of your body, nutrients in food supply the chemical energy to do work and sustain life, part of a complex process called **metabolism**. In this section, we first examine the interrelationships of energy, work, and heat, and then we look at how the nutrients of food supply the energy our bodies need to function and do work.

Energy, Heat, and Work

In Section 4.1, we found that energy is simply the capacity to do work and that energy is typically measured in terms of its equivalent in heat. A common unit used in quantifying heat or energy is the **calorie** (lowercase *c*), from the Latin *calor*, for "heat" (**Figure 5.1**). One thousand calories (cal) is equal to 1 kilocalorie (kcal), and 1 kilocalorie is equal to 1 **Calorie** (uppercase *C*), abbreviated Cal, and used principally in the field of nutrition:

calorie The amount of heat (or energy) needed to raise the temperature of 1 gram of water 1°C.

Calorie A dietary unit of energy defined as 1000 calories, or 1 kilocalorie.

$$1000 \text{ cal} = 1 \text{ kcal} = 1 \text{ Cal}$$

Units of heat or energy • Figure 5.1

Visualizations of the calorie (a) and Calorie (b).

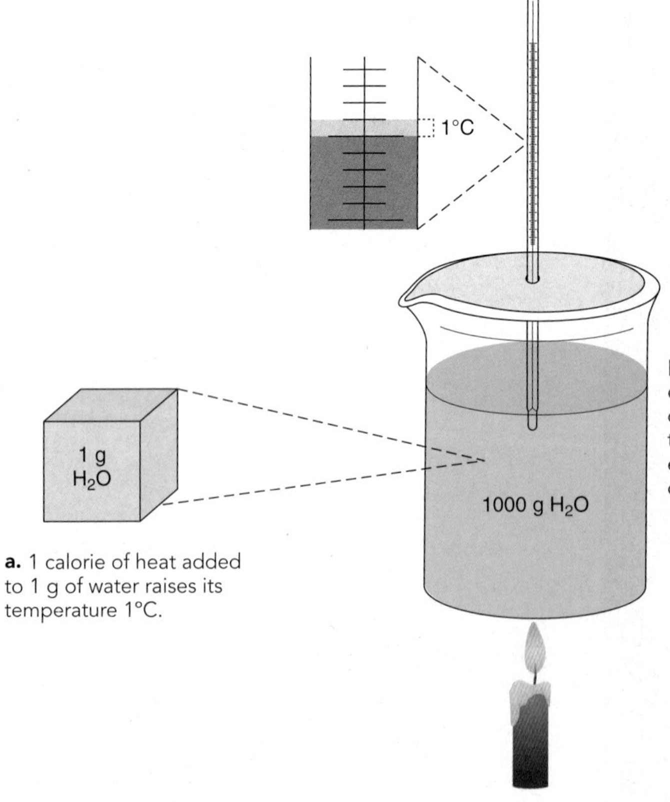

a. 1 calorie of heat added to 1 g of water raises its temperature 1°C.

b. Similarly, 1000 calories of heat added to 1000 g of water increases its temperature by 1°C. This is equivalent to 1 kilocalorie or 1 Calorie.

Ask Yourself

1. How many calories are in 0.1 Calorie?
2. Could a Calorie or kcal also be defined as the amount of heat required to raise the temperature of 1 g of water by 1000°C? Why or why not?

Calories and joules • Figure 5.2

The energy content of foods can be described in terms of kilocalories, kilojoules, or both, depending on conventions in a given country.

Here nutritional information on a food label is reported in both kilojoules (kJ) and kilocalories (kcal).

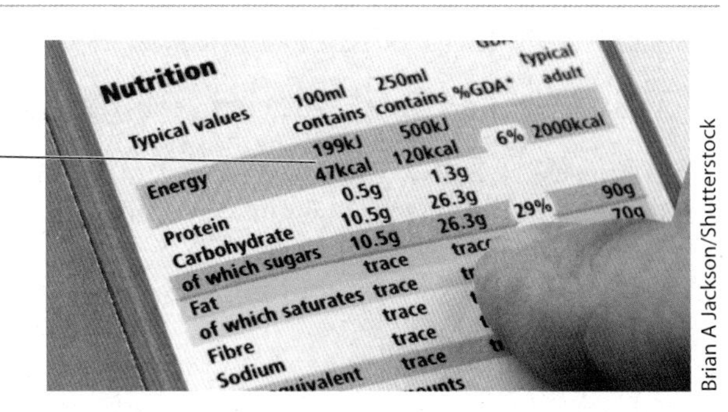

Brian A Jackson/Shutterstock

Interpret the Data

Based on information provided in the figure,

1. How many kilojoules are equal to 1 kilocalorie?
2. Which is a larger unit of energy: the kilojoule or the kilocalorie?

The technique of measuring the amount of heat that is equivalent to a specific amount of energy or work is called **calorimetry**. The device used in carrying out this measurement is a **calorimeter**. You can carry out a very simple bit of calorimetry with the materials shown in Figure 5.1. Measure the temperature of the kilogram of water and then heat the water for a short time and measure its temperature again. The increase in temperature, in degrees Celsius, gives you the number of kilocalories absorbed by the kilogram of water in the beaker. Very roughly, heating for 1 minute with a common household candle adds about 1 kilocalorie of energy to the water. The exact value depends, of course, on the specific conditions you use. If you replaced the candle in this apparatus, with, say, a few peanuts, burning them instead, you would have a very crude way of estimating the caloric content of the nuts. Originally, the energy content of foods was determined in this general way, using a more sophisticated, highly insulated calorimeter to prevent heat losses. Today we can determine the number of Calories in a given food by knowing the masses of the fats, carbohydrates, and proteins the food contains, a topic we'll explore shortly.

calorimetry The measurement of heat released or absorbed by a chemical or physical process.

joule The standard unit of energy in the metric system, abbreviated J, equivalent to 0.24 calories. One kilojoule (kJ) is 1000 joules.

One of the people who helped refine the connection between work and heat was the 19th-century English physicist James Joule. He established that work of any kind—whether mechanical, electrical, or chemical—has an exact equivalent in heat. The more work done, the greater the amount of heat produced. In Joule's honor, we now have the **joule** as a unit of work or energy, in addition to the calorie. Although both the joule and the calorie provide us with convenient units for measuring work and energy, the joule is now favored for most scientific work and is used in the metric system (**Figure 5.2**). The **watt** is a related unit used in measuring the *rate* at which energy (in joules) is consumed. **Figure 5.3** explores the relationships among these terms.

KNOW BEFORE YOU GO

1. Suppose you heat 2.5 kg of water and find that the temperature of the water rises by 7°C. How many kilocalories have you added to the water?

   Answer: Increasing the temperature of 1.0 kg water by 7 C requires:

   $$(1.0 \times 7) = 7 \text{ kilocalories.}$$

   So raising the temperature of 2.5 kg of water by the same amount requires:

   $$(2.5 \times 7) = 17.5 \text{ kilocalories.}$$

2. How many kilocalories of work is accomplished by heating 12 kg of water from 20° to 100°C?

The Human Body and the Energy Equation

We can view the human body as an incredibly complex engine that operates in the physical world like any other engine, using energy to perform work and

In Words, Math, and Pictures

Watts, joules, and calories • Figure 5.3

We can use an everyday light bulb to explore the relationship between **watts** and **joules**. Light bulbs are rated in watts, a unit that tells us how quickly energy is consumed. An expenditure of 1 joule of energy per second is equivalent to 1 watt:

$$\frac{1 \text{ joule}}{1 \text{ second}} = 1 \text{ watt}$$

A 100-watt bulb, then, is rated to consume 100 joules each second at full brightness. Rearranging this equality, we find:

$$1 \text{ joule} = 1 \text{ watt} \times 1 \text{ second}$$

We can convert between joules and calories, another common unit of energy, using the following equalities:

$$1 \text{ joule} = 0.24 \text{ calories}$$
$$1 \text{ calorie} = 4.2 \text{ joules}$$

The following example illustrates these conversions.

Example: How many minutes could a 100-watt light bulb remain lit at full brightness if it consumed the energy content of a 275-Calorie candy bar?

The energy content in this candy bar...

...is equivalent to energy required to light this bulb for how long?

Solution:

We begin with a series of unit conversions based on the energy content of the bar. Notice that the units in the denominator of each new term cancel the units in the numerator of the previous term:

$$275\,\cancel{\text{Calories}} \times \frac{1000\,\cancel{\text{calories}}}{1\,\cancel{\text{Calorie}}} \times \frac{4.2\,\cancel{\text{joules}}}{1\,\cancel{\text{calorie}}} \times \frac{1\,\cancel{\text{watt}} \times 1 \text{ second}}{1\,\cancel{\text{joule}}} \times \frac{}{100\,\cancel{\text{watts}}} = \frac{275 \times 1000 \times 4.2 \times 1 \times 1}{1 \times 1 \times 1 \times 100} \text{ sec}$$

To arrive at an answer, we convert from seconds to minutes:

$$11{,}550\,\cancel{\text{sec}} \times \frac{1 \text{ min}}{60\,\cancel{\text{sec}}} = \mathbf{192\,min}$$

Put It Together Consider the energy consumed in heating 12 kg of water from 20° to 100 °C. How many hours would this amount of energy keep a 40-watt light bulb glowing at full brightness?

functioning under the laws of the physical universe. Just as hydrocarbons provide the fuel for mechanical engines, the nutrient molecules of food furnish the fuel for the biological engines of our bodies. Simply put, gasoline fuels our cars; food fuels our bodies. However, although we may look for alternative fuels for our cars, as we described in Chapter 4, our bodies have no alternative to food.

As we examine food as fuel for our bodies, we'll use the term **metabolism** to describe the conversion of food into energy and the physical substance of our bodies. The body's metabolic chemistry releases the potential energy of food by processes that are similar to the way combustion within an engine releases the energy of hydrocarbons. That is, the chemical reactions of metabolism form products—largely carbon dioxide and water—that contain less total energy than the reactants. In a typical example, a series of metabolic reactions converts the carbohydrate sucrose (table sugar) and oxygen into carbon dioxide, water, and biological energy:

> **metabolism** All the chemical reactions that take place in a living organism to support life.

$$\underset{\text{sucrose}}{C_{12}H_{22}O_{11}} + 12\ O_2 \rightarrow 12\ CO_2 + 11\ H_2O + \text{energy}$$

Although metabolism is similar to combustion in its results, metabolism differs from combustion in two important ways: Metabolism occurs more slowly than combustion and under much milder conditions. Metabolism neither requires nor generates the very high temperatures that accompany combustion and therefore does not endanger body tissue.

Like any other physical process, metabolism operates under the Law of Conservation of Energy, which states that in a chemical reaction, energy can neither be created nor destroyed. Because of this conservation of energy, all the potential energy that comes into our bodies with food must be accounted for; it must be either used or stored. We can state this requirement through an **energy equation** (**Figure 5.4**).

Energy In The **macronutrients** of our food supply biological energy. We classify macronutrients as one of three types:

- Fats and oils
- Carbohydrates
- Proteins

> **macronutrients** The major components of our foods that provide us with energy and the materials that form our bodies.

Micronutrients, on the other hand, are nutrients we need in very small quantities, such as vitamins and minerals. We'll have more to say about them in a later chapter.

Through the complex chemical reactions that constitute human metabolism, the chemical energy of macronutrients becomes transformed into the human energy of our lives. Before we examine the structures of macronutrients, let's look more closely at the amounts of energy they furnish.

The actual quantity of energy provided by each class of macronutrients is well known. Fats and oils yield 9 Cal/g, and carbohydrates and proteins each yield 4 Cal/g. The amount of energy each of these chemicals delivers to our bodies is independent of both the kind of food in which it occurs and the presence or absence of

The energy equation • Figure 5.4

We can put the whole, sometimes complex, relation between human nutrition and body metabolism into these simple terms: Either we burn up the energy of the food we eat or we store it as fat.

Energy In = Energy Out + Energy Stored (as fat)

To maintain weight, "Energy In" should equal "Energy Out."

To gain weight, "Energy In" should be larger than "Energy out." This means eating more and/or exercising less.

Energy In

Energy Out

To lose weight, "Energy Out" should be larger than "Energy In." This means exercising more and/or eating less.

In Words, Math, and Pictures

Calories from macronutrients • Figure 5.5

The number of calories provided in a meal or specific food can be determined from the masses (in grams) of the protein, carbohydrate, and fat it contains. Carbohydrates and protein each provide about 4 Cal/g, while fats provide 9 Cal/g.

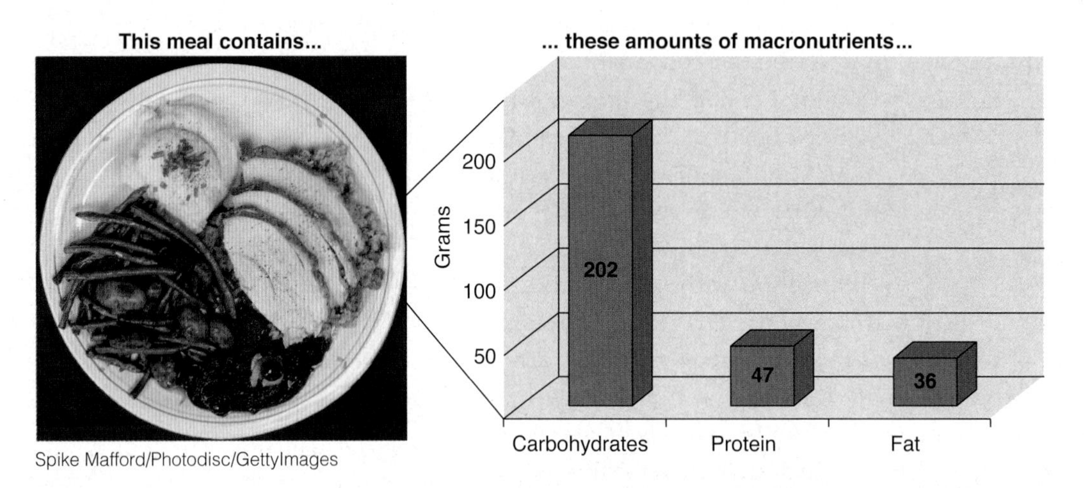

Spike Mafford/Photodisc/GettyImages

...which provide the following number of Calories:

$$202 \text{ g carbohydrates} \times \frac{4 \text{ Cal}}{\text{g carbohydrates}} = 808 \text{ Cal}$$

$$47 \text{ g protein} \times \frac{4 \text{ Cal}}{\text{g protein}} = 188 \text{ Cal}$$

$$36 \text{ g fat} \times \frac{9 \text{ Cal}}{\text{g fat}} = 324 \text{ Cal}$$

1320 Cal total

Ask Yourself

Suppose we added gravy to the meal, increasing carbohydrates by 5 g, protein by 6 g, and fat by 20 g. How many Calories would the meal now have?

other macronutrients. One gram of fat or oil provides 9 Calories, whether it comes from, say, butter, beef, nuts, or corn, and 1 g of carbohydrate provides 4 Calories, whether its source is bread, potatoes, or sugar. **Figure 5.5** illustrates how we calculate the Calories in a food or meal based on the macronutrient content.

The macronutrients fats and oils provide us with the most concentrated form of food energy available. At 9 Cal/g, this is 2.25 times as much as the energy content of carbohydrates and proteins. The high energy density of fats and oils makes fat an ideal storage medium for reserve supplies of energy in animal bodies. For comparison, ethanol, the alcohol that some of us consume along with food, provides 7 Cal/g. In its energy content, this alcohol stands a bit closer to fats and oils than to carbohydrates and proteins.

Energy Out The human body uses energy through exercise and several forms of metabolism. We'll now briefly examine the expenditure of energy through exercise and through two forms of metabolism: the thermic effect of food and basal metabolism.

Exercise, the energy path we know best, includes more than the standard jogging, weightlifting, team sports, and the like. In the sense we're using here, *exercise* is the composite of all the physical work we do with our bodies. The rate of energy expenditure through exercise ranges from the enormous output demanded by the most vigorous physical activities, such as competitive swimming, to the minimal output required by the most restful activities, such as sitting quietly. (Even while sitting, we're exercising in this particular sense. We're burning up a minute amount of energy as we use muscles to keep our balance and to keep our head up and eyes open.)

The **thermic effect of food (TEF)** accounts for the energy consumed in digesting and metabolizing food. It's the price we pay, in energy, for extracting energy from food. In this sense, it's analogous to the energy cost of refining petroleum into gasoline, heating oil, and other products through distillation and refining (Chapter 4).

thermic effect of food (TEF) Energy the body expends in digesting and metabolizing food.

You can sometimes sense TEF in operation as you grow a little warmer and find your heart beating a bit faster after a meal. These are both signs that your body is hard at work digesting food and using up energy through TEF. (Drowsiness after a large meal comes from the diversion of blood from the brain to the digestive system. The body is working hard to process all the fuel that's just been brought in and—first things first—digestion takes momentary precedence over mental alertness.)

The amount of energy lost through TEF depends on what you're eating. For fat, it's about 4% of the calories provided; for carbohydrates, roughly 6%; for protein, around 30%. As a very rough estimate, about 10% of the energy we get from a meal is spent in TEF. That is, something like 10% of food energy goes toward the very act of extracting and using the remaining 90%.

Basal metabolism accounts for all the work that goes on inside our bodies just to keep us alive. It's the energy we use to keep our heart beating, our lungs expanding and contracting, and all other organs working to maintain life.

basal metabolism The body's energy expenditure while at rest.

The standard technique for measuring basal metabolism is to determine our total energy output in the absence of both exercise and TEF. In this measurement, a person lies at rest after 12 hours of fasting. Lying at rest eliminates virtually all energy expenditure by exercise. The added 12-hour fast precludes energy expenditure through TEF. In a healthy adult, the basal metabolism rate amounts to roughly 1 Calorie per hour per kilogram of body weight. Basal metabolism increases with any bodily stress, such as illness or pregnancy.

Energy Stored Excess body fat poses several health risks, but having adequate body fat is necessary for proper health. If we didn't have body fat as a means for storing the unused chemical energy of our food—and if we couldn't then convert this fat into energy later when we need it—none of us would be alive today. Without body fat, any temporary, even minor, shortage of food experienced by our cave-dwelling ancestors would have been catastrophic; it would have wiped out the human race.

KNOW BEFORE YOU GO

3. The population of the entire world is estimated at 7 billion people.

a. Assuming that the average person spends 1200 Calories per day through exercise, calculate the annual energy output, through exercise alone, of the world's population.

b. Estimate the annual use of energy through basal metabolism alone, again by all the people in the world. Assume an average body weight of 50 kg.

c. Neglecting the effects of TEF, what is the estimated total annual biological energy expenditure of all people on Earth?

Converting an excess of a macronutrient into body fat is a particularly effective means of long-term energy storage. (Short-term energy storage, usually from one meal to another, occurs through a starch-like substance called glycogen, which we'll examine shortly.) The high energy density of fat—its ability to store energy compactly in relatively little space and with relatively little mass—allows us to carry our stored energy with us efficiently. It gives us and other animals the mobility and freedom necessary for survival in a world that was, and can still be, unforgiving to those poorly equipped for survival during periods when food isn't readily available.

The body stores fat in **adipose tissue**. One pound of adipose tissue stores (and provides when needed) roughly 3500 Calories of energy.

KNOW BEFORE YOU GO

4. Suppose someone is 15 pounds above what he or she would consider ideal weight and that this extra weight is held purely in adipose tissue. How many excess stored Calories (from this fat) would the person be carrying at all times?

CONCEPT CHECK

STOP

1. **How** many (a) kilocalories, (b) Calories, and (c) joules are equivalent to 1000 calories?
2. **Why** are fats considered energy-dense macronutrients?
3. **How** does the body process food energy that is not expended through exercise, TEF, or basal metabolism?

5.2 Fats and Oils

LEARNING OBJECTIVES

1. **Describe** the structure of triglycerides.
2. **Differentiate** between saturated and unsaturated fats.
3. **Describe** the relationship between dietary fats and cholesterol.
4. **Explain** how trans fats are formed.

Fats and oils work both for and against life and good health. Our body fat stores energy so that if we fast, we still have the energy to keep our vital organs functioning. In addition, our fat tissue helps insulate our bodies heat loss and also forms a protective cushioning around major organs. Although fats have no flavors of their own, many of the substances that do add flavor and enjoyment to eating are far more soluble in fats than in the more watery substances of food. The fat in meat, for example, carries the flavor we associate with meat. Without fat, meats would be tasteless.

Clearly fats are beneficial; but in excess, they can be dangerous. Obesity, for instance, is linked to various diseases. In this section we'll explore the chemistry of fats and oils and their role in our diets and health.

Structures of Fats and Oils

The difference between fats and oils is a practical one: At ordinary temperatures, fats, such as butter, are solids, whereas oils, such as olive oil, are liquids. What makes fats hold their shape at room temperature while oils flow freely? The answer lies in the nature of the molecules, called **triglycerides**, that make up each substance. We'll first look at the structure of triglycerides (**Figure 5.6**) and then see how this structure relates to whether a substance is a fat or an oil.

triglycerides Molecules composed of glycerol and three fatty acid chains.

Every fatty acid contains a long hydrocarbon (i.e., *fatty*) side chain and a carboxylic *acid* head—hence the name fatty acid (**Figure 5.7**). A wide variety of fatty acids exist, but the most significant difference among them, lies in the structures of their linear hydrocarbon chains. These chains can vary in:

- The number of carbons they contain and hence their lengths
- The number and nature of carbon-carbon double bonds in their side chains

A **saturated fatty acid** is a molecule in which the carbons of the hydrocarbon chain are saturated with hydrogen atoms. In other words, each carbon in the chain is bonded to the maximum number of hydrogens it can hold. A quick way to tell if a fatty acid is saturated is simply to look for carbon-carbon double bonds within the hydrocarbon chain. If there are none, it's saturated. An **unsaturated fatty acid** has one or more carbon-carbon double bonds.

saturated fatty acid A fatty acid with no carbon-carbon double bonds.

unsaturated fatty acid A fatty acid with one or more carbon-carbon double bonds.

The structures of triglycerides • Figure 5.6

All fats and oils are triglycerides—molecules composed of a glycerol backbone connected to three fatty acids by ester linkages.

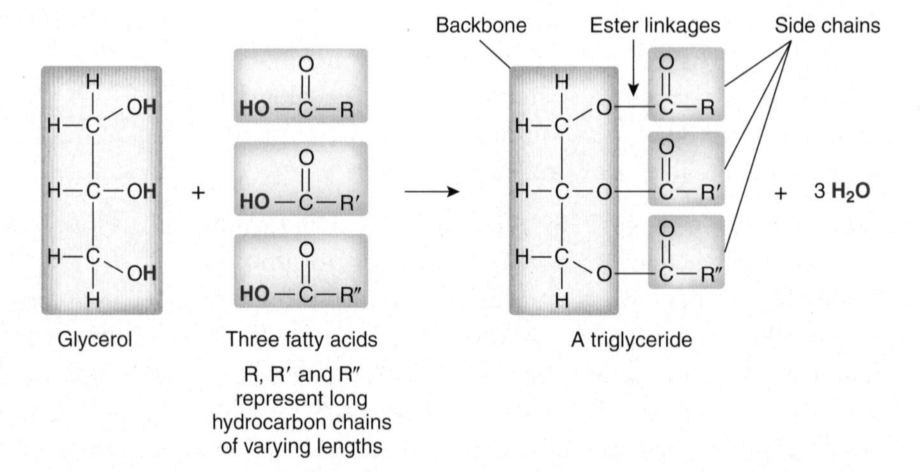

The structures of fatty acids • Figure 5.7

Fatty acids vary in the nature of their hydrocarbon chains. Saturated fatty acids, such as stearic acid, have no carbon-carbon double bonds.

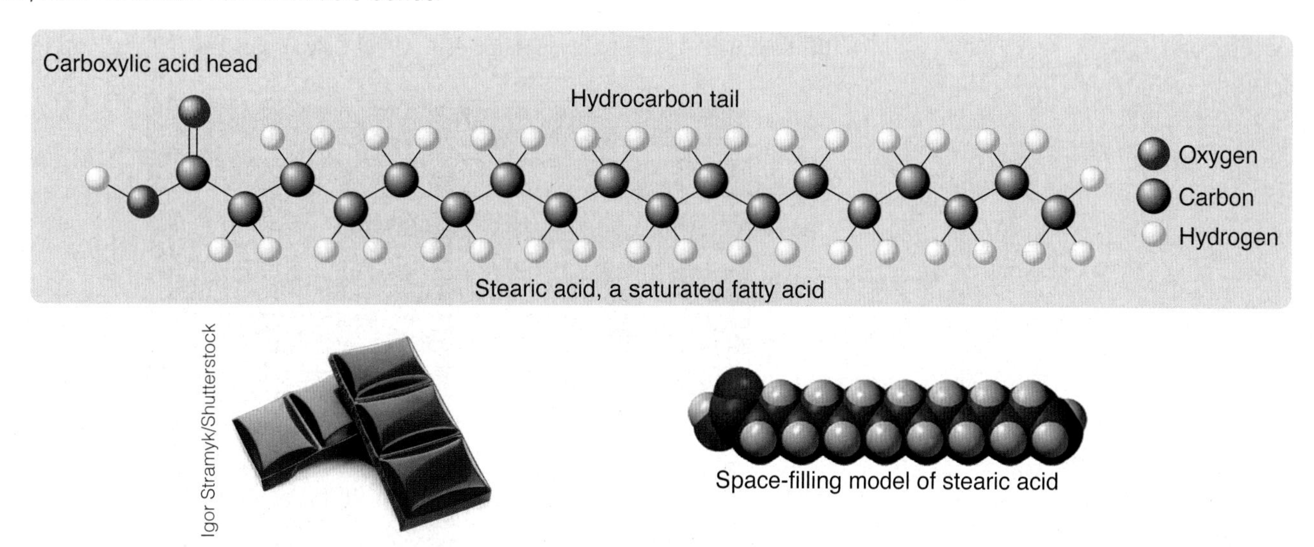

Chocolate's rich, creamy, melt-on-the-tongue texture derives largely from the nature of its fats. Stearic acid makes up about one-third of the fatty acids in the triglycerides of chocolate.

Unsaturated fatty acids, such as oleic acid, have at least one carbon-carbon double bond.

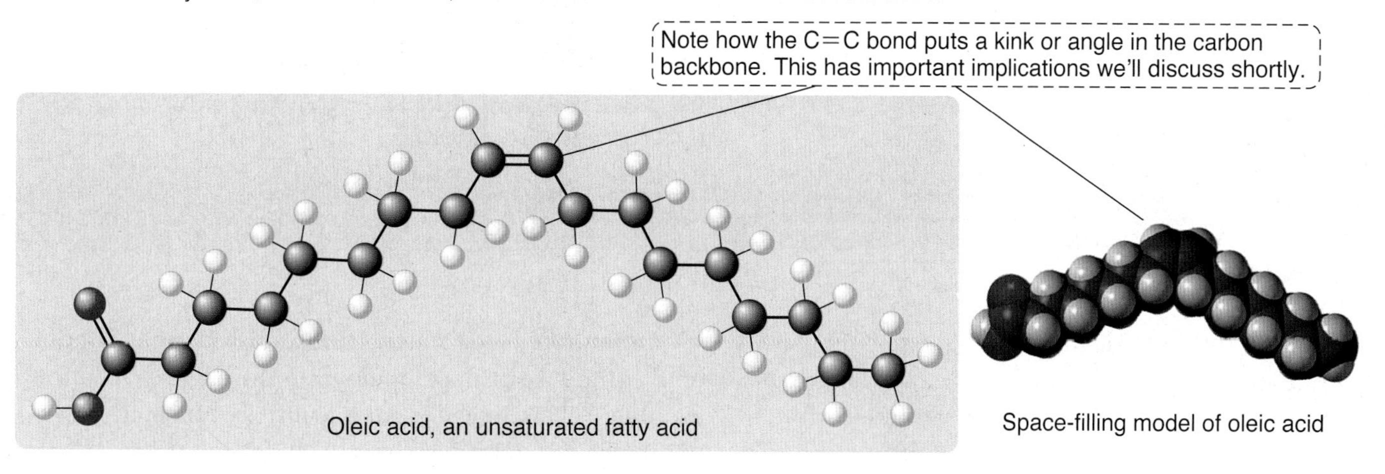

Unsaturated fatty acids, can be further classified by the extent of unsaturation in their chains.

- Monounsaturated fatty acids have one C=C bond.
- Polyunsaturated fatty acids have more than one C=C bond.

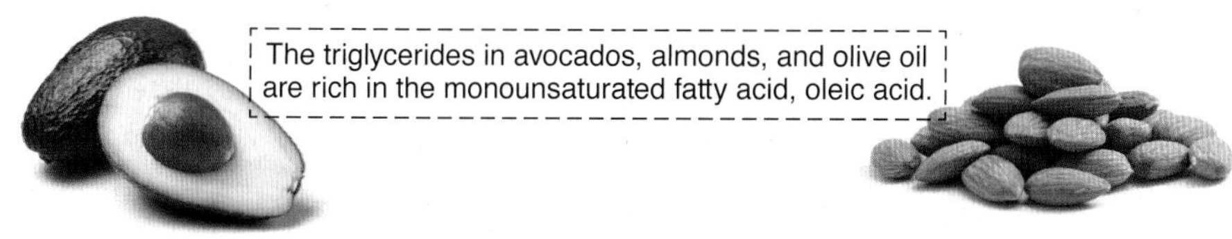

The triglycerides in avocados, almonds, and olive oil are rich in the monounsaturated fatty acid, oleic acid.

Oliver Hoffmann/istockphoto

Alasdair Thomson/istockphoto

Ask Yourself

Naturally occurring fatty acids typically contain an even number of carbon atoms and generally carry a total of about 8 to 24 carbons. Determine the number of carbons in stearic and oleic acids to learn whether each conforms to this observation.

Composition of fats and oils • Figure 5.8

A comparison of the fatty acid composition of fats and oils reveals a general trend. Fats contain more triglycerides with saturated side chains.

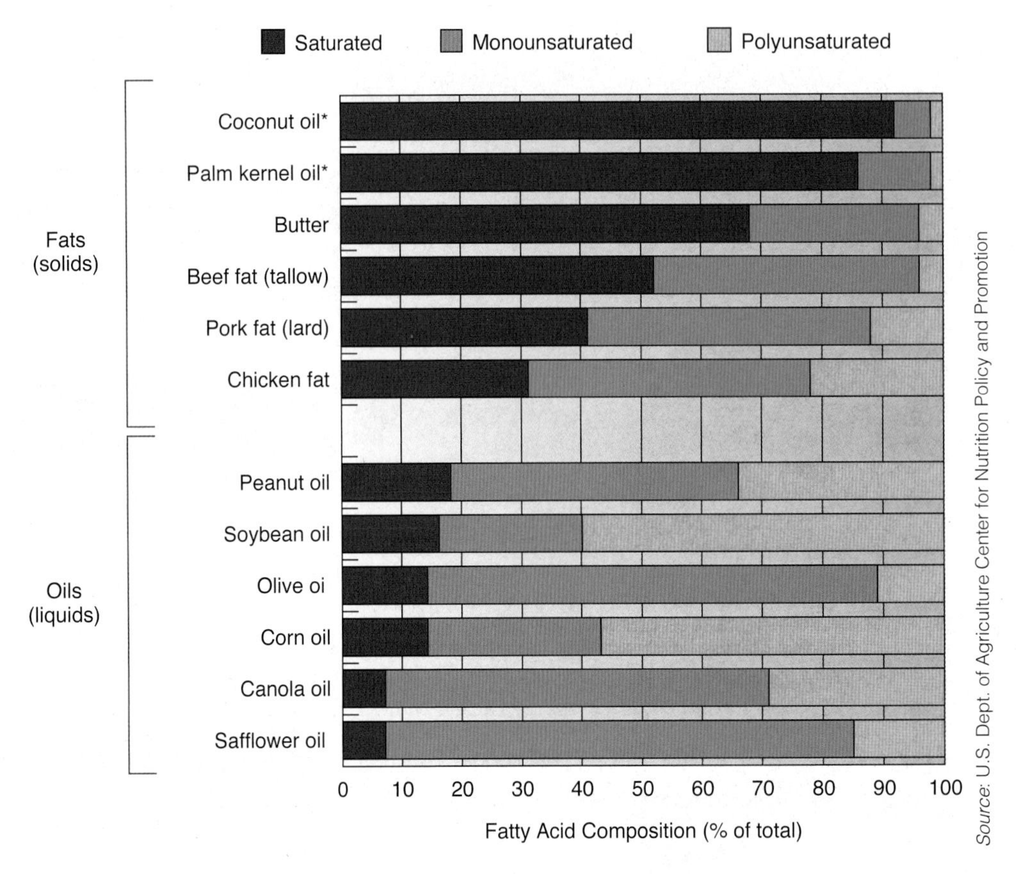

*Coconut oil and palm kernel oil are commonly known as oils because they derive from plants; however, like fats, they are semi-solid at room temperature, due to their high content of saturated fatty acids.

Generally, a triglyceride with a high melting temperature and a tendency to exist as a fat is characterized by long side chains and a high degree of saturation (fewer C=C bonds). A triglyceride with a low melting temperature and a tendency to exist as an oil is associated with shorter side chains and a greater degree of unsaturation (plenty of C=C bonds). **Figure 5.8** illustrates these relationships.

Why, then, do fats tend to be solids at room temperature? Saturated side chains (those with no double bonds) more readily adopt a linear orientation (see stearic acid in Figure 5.7, for example). This regularity allows for stronger associations between triglyceride molecules. Only at temperatures greater than room temperature is there enough heat energy to disrupt these **intermolecular associations** (attractions between molecules) so that the triglyceride molecules can move more freely about one another, thereby producing melting and the liquid state.

Unsaturated side chains (those with double bonds) tend to be bent (see oleic acid in Figure 5.7, for example). This weakens interactions between triglyceride molecules, allowing them to slide past one another more easily, so less heat energy (i.e., lower temperature) is required for melting (**Figure 5.9**).

As the graph in Figure 5.8 suggests, nature isn't so neat and simple as to form animal fats exclusively from highly saturated fatty acids and to generate plant oils exclusively from unsaturated acids with plenty of double bonds. Instead, the fats and oils of our diets are complex mixtures of triglycerides containing a considerable variety of side chains: some long, some short, some fully saturated, some monounsaturated, some polyunsaturated. Although fats are generally richer in saturated side chains, as we've already noted, no fat consists entirely of fully saturated side chains, and no oil is free of them.

Why fats are solids and oils are liquids at room temperature • Figure 5.9

The difference between fats and oils lies in their *melting temperatures* (a).

Maximilian StockLtd./Photolibrary/Getty Images

Fats, such as butter, exist as solids at room temperature, which means their melting points lie above room temperature.

Oils exist in the liquid state at room temperature, which means their melting points lie *below* room temperature.

a. Fats and oils

Triglycerides with saturated side chains (b) tend to be more abundant in fats, while triglycerides with unsaturated side chains (c) tend to be dominant in oils.

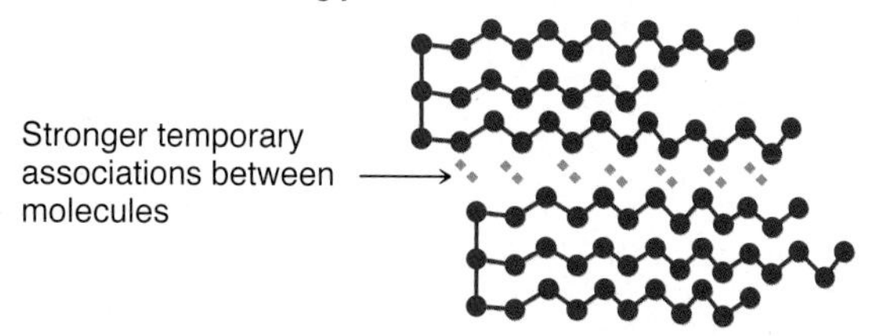

b. Saturated side chains tend to follow more linear patterns, allowing for stronger intermolecular interactions. This results in a higher melting temperature.

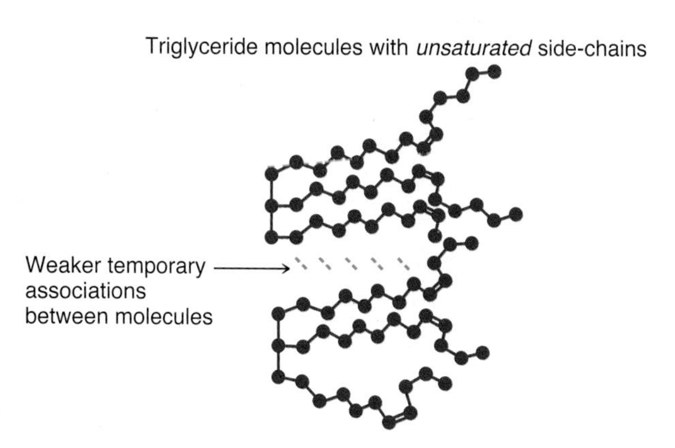

c. Unsaturated side chains tend to have kinks, which prevents close, orderly associations between molecules. This gives rise to lower melting temperatures.

Ask Yourself

What would you observe if you put a cup of oil in the freezer overnight (at a temperature below its melting temperature) and then removed it and let it warm to room temperature?

Determining the extent of unsaturation of a fat or an oil • Figure 5.10

When you mix iodine (I_2), a dark purple diatomic element, with a fat or an oil, a reaction takes place, in which iodine adds to C=C bonds in the side chains to produce a colorless product. After the fat or oil has absorbed all the iodine it can, the continued addition of iodine to the mixture produces a violet-purple color in the solution.

You can liken the C=C bonds in fats and oils to hungry chicks with open mouths in a nest. The greater the number of chicks, the greater the capacity to consume food. Similarly, the greater the number of double bonds in a given amount of fat or oil, the more I_2 can be taken up, and hence the greater the iodine number.

I—I + HC=CH ⟶ I—CH—CH—I

Iodine (I_2)
Dark purple

C=C bond(s) on the side chains of triglycerides

Colored mixture

Colorless product

Put It Together Refer to Figure 5.8 to predict which has a larger iodine number: palm kernel oil or corn oil.

See *WileyPLUS* for more on iodine number, including a demonstration you can carry out yourself using tincture of iodine (a common antiseptic) and some familiar cooking oils.

Chemistry InSight Fat, cholesterol, and heart disease • Figure 5.11

The body requires some cholesterol to aid in digestion as well as to produce vitamin D and hormones, such as estrogen and testosterone. But high levels of cholesterol are closely associated with atherosclerosis—the buildup of fatty deposits on the inner lining of arteries (a). This buildup reduces blood flow and can lead to high blood pressure and heart disease.

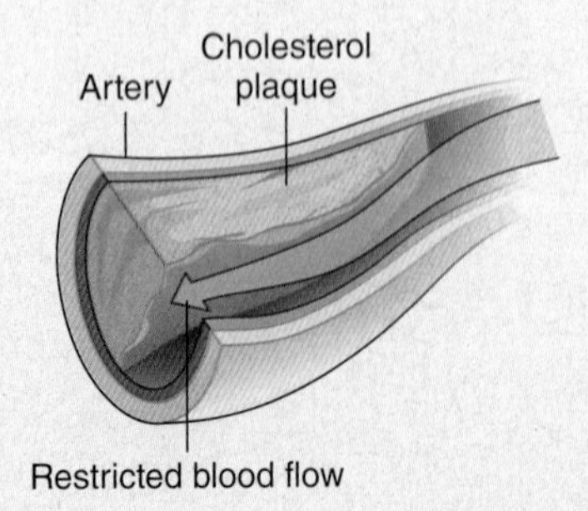

a. In atherosclerosis, plaque builds un on the wall of the artery, thus restricting blood flow. This plaque is composed of cholesterol, fat, and other blood components.

b. Cholesterol is a **hydrophobic** molecule—from the Greek *hydro* for "water" and *phobic* for "fearing", meaning it has very little tendency to dissolve in water. Cholesterol is classified as a **steroid** because of its core molecular structure of four rings (highlighted in color). (Hydrogen atoms on the ring carbons have been omitted for clarity.)

Because of the importance of unsaturation in the chemistry of triglycerides, it's useful to have a simple, convenient measure of the extent of unsaturation. We have such a measure in a value known as the **iodine number**, which is the number of grams of iodine that can add to 100 g of a triglyceride. When we compare fats and oils, those with a *higher* iodine number are *more* unsaturated, whereas those with a *lower* iodine number are *less* unsaturated (more fully saturated). **Figure 5.10** explores this relationship.

Fats and Oils in Our Diet

Both fats and oils are essential to our diets, but there are important health differences between them. The triglycerides that are rich in mono- and polyunsaturated fatty acids are considered healthier, lowering the risk of heart disease. Moreover, those rich in saturated fatty acids or containing trans fatty acids are considered potentially harmful. Cholesterol (which is not a triglyceride) is closely related to the story of dietary fats. You can see the relationships among these substances in **Figure 5.11**.

trans fatty acid An unsaturated fatty acid containing one or more C=C bonds with a *trans* orientation.

catalytic hydrogenation A process by which hydrogen is added to unsaturated chemical compounds in the presence of a catalyst.

Trans fatty acids (also known as *trans fats*) have received much attention from public health officials. Consumption of these fats appears to lower levels of beneficial HDLs and raise levels of potentially harmful LDLs. In 2006, the U.S. Food and Drug Agency began requiring food manufacturers to list levels of trans fats on food labels. In that same year, New York City became the first major city in the United States to ban the use of trans fats in restaurants. Other municipalities have since followed suit. Trans fats are an unintended byproduct of the **catalytic hydrogenation** of vegetable oils (**Figure 5.12** on the next page).

To transport cholesterol through the blood, our bodies wrap packets of cholesterol in a sheath of proteins and triglycerides, presenting a more **hydrophilic** (water-loving) surface to the blood. The resulting cholesterol-transporting bundle is a type of **lipoprotein**. The lipoproteins are classified by their density: HDLs are high-density lipoproteins and LDLs are low-density lipoproteins (c).

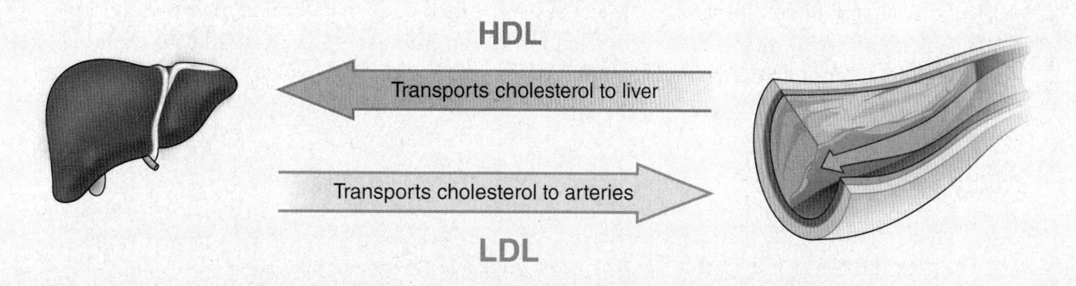

c. HDLs are beneficial in that they remove cholesterol deposits from the walls of arteries and transport them to the liver for disposal. LDLs, on the other hand, deposit cholesterol on arterial walls.

Catalytic hydrogenation and trans fats • Figure 5.12

Catalytic hydrogenation converts carbon-carbon double bonds into single bonds (a).

a. Catalytic hydrogenation

$$H{-}H \; + \; \underset{}{HC{=}CH} \xrightarrow{\text{Catalyst}} H{-}\overset{H}{C}{-}\overset{H}{C}{-}H$$

Hydrogen gas

C=C bond(s) on the side chains of triglycerides

This process is used in the food industry to harden vegetable oils in order to convert them into semi-solids that are useful in making margarines and vegetable shortening. Since complete hydrogenation of all the C=C bonds of a typical vegetable oil would produce a hard, waxy substance, partial hydrogenation is often preferred because it yields a softer material.

Carbon-carbon double bonds can exist in either of two orientations, *cis* or *trans*. Under normal circumstances, these two forms do not interconvert (**b**).

b. *Cis* and *trans* double bonds

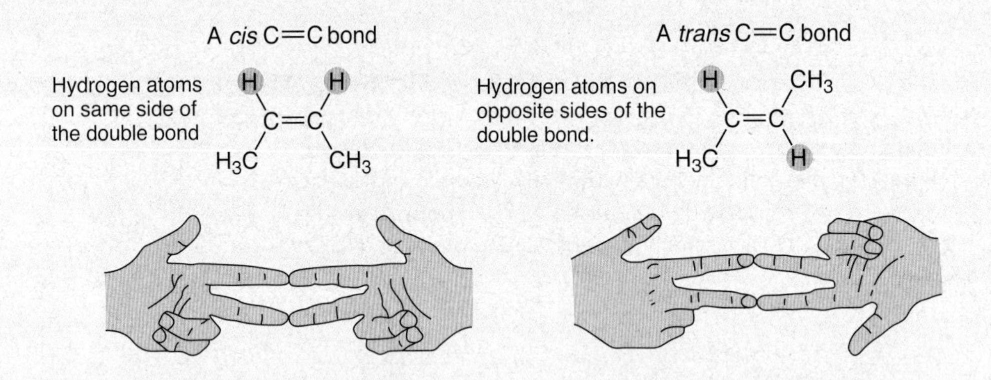

The C=C bonds we find in vegetable oils are typically in the *cis* configuration (**c**). Partial hydrogenation of vegetable oils causes some of these C=C bonds to change their orientation from the *cis* form to the *trans* form (**d**). These trans fats are an unintended consequence of partial hydrogenation.

c. A *cis* fatty acid

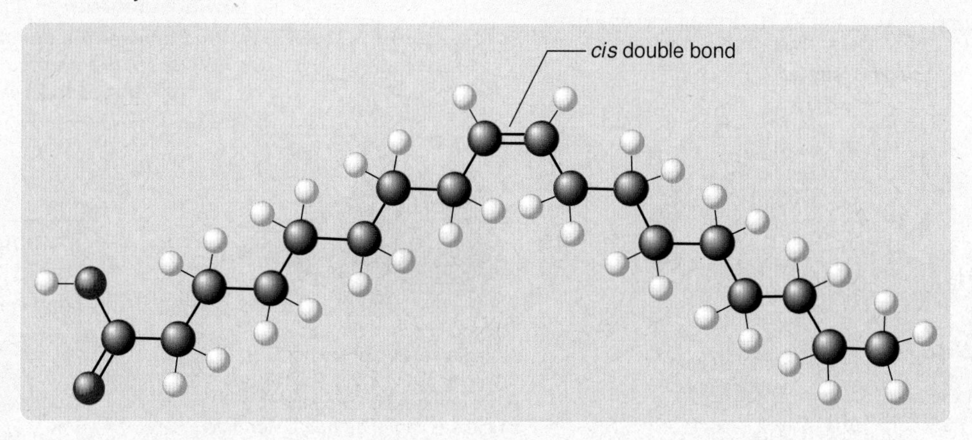

d. A *trans* fatty acid

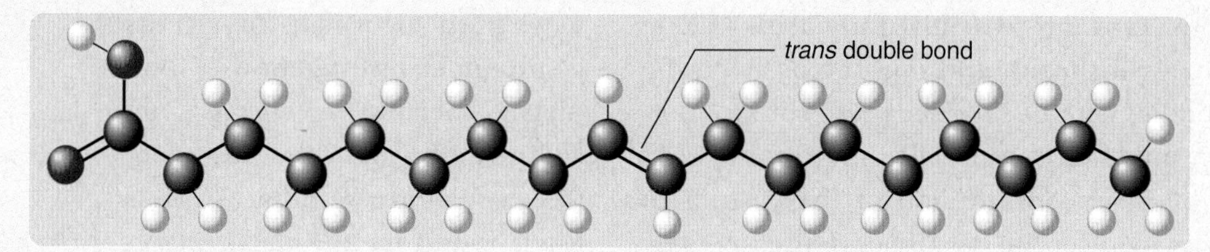

Food packaging labels reflect a growing consumer awareness of trans fats. Since the FDA began requiring the reporting of trans fat content in foods, levels of these fats have generally fallen as manufacturers have found alternative ways of formulating their products (e).

e. Labeling of trans fat content in foods

Tim O. Walker

Think Critically

1. In catalytic hydrogenation, hydrogen (H_2) adds to a C═C bond to produce a C—C bond. What other reagent discussed previously in this chapter converts a C═C bond to a C—C bond?
2. Does partial hydrogenation raise or lower the melting point of vegetable oils? Why?
3. Describe the types of carbon-carbon bonds in the side chains of the triglycerides of partially hydrogenated vegetable oils.

Omega-3 fatty acids • Figure 5.13

Omega 3s are unsaturated fatty acids containing a carbon-carbon double bond starting at carbon 3 (counting from the $-CH_3$ end of the chain).

DHA, an omega-3 fatty acid

Start counting carbons here.

Salmon is rich in omega 3s. A single serving of salmon provides about 2.0 g of DHA and other omega-3s. By contrast, a single serving of omega-3 -fortified eggs supplies only 1/10th the level of these nutrients.

Put It Together Review Figure 5.12 and answer the question. What are the configurations of the double bonds in DHA?

In contrast to trans fats, **essential fatty acids (EFAs)** are necessary for good health and must come from the diet, since our bodies cannot produce these compounds. **Omega-3 fatty acids** are especially important EFAs (**Figure 5.13**) because consumption of these kinds of fatty acids appears to lower the risk of heart disease. (The mechanism by which this occurs is still under investigation.) Good dietary sources of omega-3s include salmon and other fatty fish, flaxseed, and walnuts. Various foods found in the marketplace,

omega-3 fatty acid An unsaturated fatty acid with a carbon-carbon double bond occurring between the third and fourth carbons from the end of the chain. (The last carbon of the fatty-acid chain is designated as the omega carbon. Omega is the last letter of the Greek alphabet.)

Cholesterol and fatty acid compositions of common foods • Figure 5.14

For most of us, the majority of saturated fats in our diet comes from dairy and meat. Here, milligrams of cholesterol (left) and grams of fat (right) are shown per 100-g serving* of each type of food.

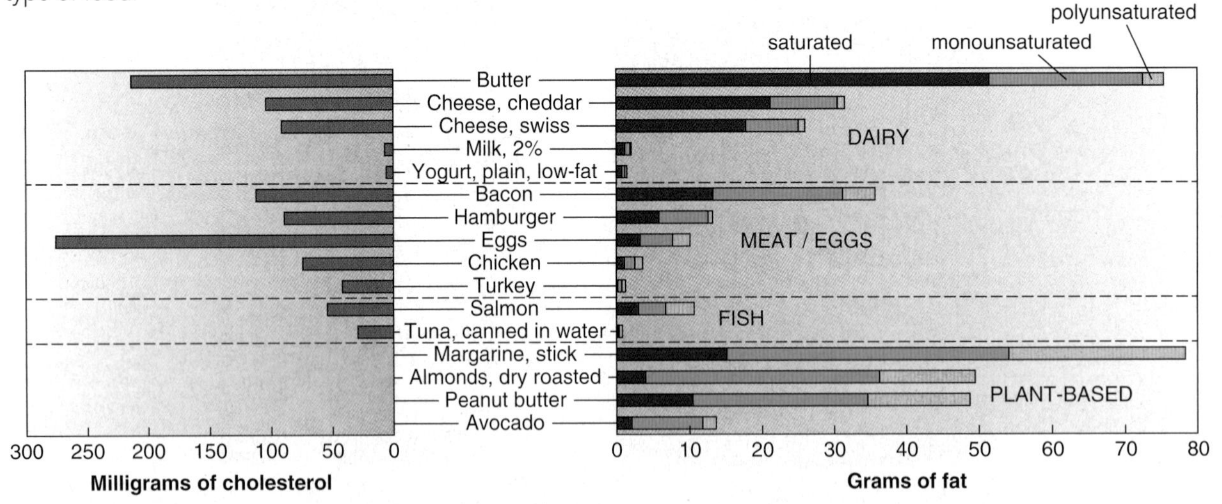

*Note that a 100-g serving gives us a basis of comparison across food types, but does not necessarily represent a typical serving size of each. Foods higher in water content, such as milk and yogurt, appear lower in fat in this type of analysis.

Interpret the Data

1. What types of foods have zero cholesterol content?
2. Which has a greater percentage of saturated fat per serving: dry roasted almonds or peanut butter?

including milk, eggs, and margarines, are now fortified with omega-3s. However, the levels of these compounds present in these fortified foods are still typically far less than those found in foods naturally rich in omega-3s.

Nutritional advice from health experts continues to evolve as we learn more about the health effects of the food we eat. For instance, we now know that for most people, serum cholesterol levels (i.e., blood levels of cholesterol) only modestly reflect the amount of cholesterol in a person's diet. Serum cholesterol levels depend on several factors, including genetics (over which we have no control), exercise, emotional stress, and the types of fiber and fat in the diet. We do know that our own livers (and those of all other animals) can generate cholesterol from the saturated fats we eat, so limiting our consumption of saturated fats is recommended. **Figure 5.14** compares the fat and cholesterol contents of some common foods.

CONCEPT CHECK STOP

1. **What** is the structural difference between a triglyceride molecule and a fatty acid molecule?
2. **How** is oleic acid different from stearic acid?
3. **Why** would reducing the amounts of animal fats in our diet help reduce serum cholesterol?
4. **Why** do partially hydrogenated oils pose a potential health risk?

5.3 Carbohydrates

LEARNING OBJECTIVES

1. **Distinguish** between mono-, di-, and polysaccharides.
2. **Describe** how enzymes help us digest carbohydrates.
3. **Explain** why we can digest starch but not cellulose.

Bite into a juicy apple or dig into a bowl of spaghetti or other noodles, and you're consuming **carbohydrates**, a class of nutrients known for the quick energy they provide. If it were not for this class of nutrients, you wouldn't be able to interpret the words on this page since the fuel that supplies energy to your brain and nervous system is the carbohydrate **glucose**, also known as dextrose or blood sugar. It's quite literally our food for thought. Glucose also supplies the energy that keeps our bodies at a constant temperature, flexes our muscles, and keeps our digestive and respiratory systems running. In addition to glucose, a variety of other carbohydrates are critical to our daily lives, as we'll see in this section.

carbohydrates A class of macronutrients including sugars and starches that is an important source of energy for organisms.

glucose A simple carbohydrate that serves as the main source of energy for the functioning of our bodies.

Classes of Carbohydrates

We have seen that all fats and oils are esters of glycerol and fatty acids, and they differ from one another entirely through the lengths and the degrees of unsaturation of their side chains. Unlike the fats and oils, carbohydrates come in a great variety of shapes and sizes, as shown in **Figure 5.15**. However, all carbohydrates share certain key molecular characteristics:

- Carbohydrates are composed exclusively of three elements: carbon, hydrogen, and oxygen.
- The ratio of hydrogen to oxygen atoms in all carbohydrate molecules is almost always exactly 2:1, as it is in water (H_2O).

This 2:1 ratio suggested to early investigators that these molecules were *hydrates* of carbon—that is, clusters of carbon atoms bound to water molecules. For example, the molecular formula of glucose, $C_6H_{12}O_6$, can be written as though the molecule were composed of six carbon atoms bonded to one another and also to six water molecules, as in $C_6(H_2O)_6$. Although this interpretation of the molecular structure of carbohydrates proved to be incorrect, the name *carbohydrate* has stuck.

Types of carbohydrates • Figure 5.15

Carbohydrates are classified according to the number of connected sugar molecules, called **saccharides**, that form their molecules. The smallest, simplest carbohydrates are the **monosaccharides**, meaning one sugar molecule per carbohydrate molecule.

Monosaccharides

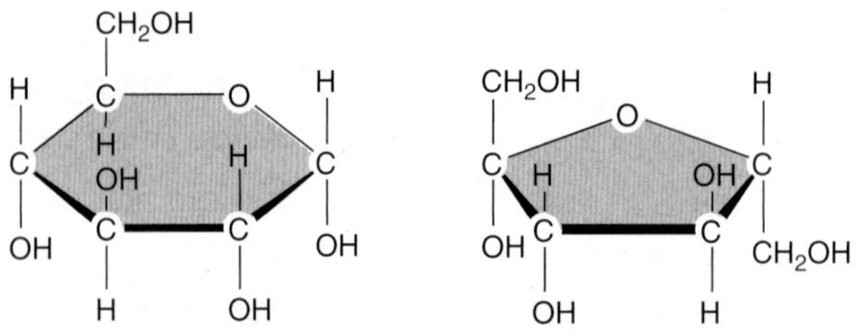

Glucose, or *blood sugar*

Fructose or *fruit sugar*, common in honey and fruits

Glucose and fructose each has the same molecular formula, $C_6H_{12}O_6$, but they have different molecular structures, so they are **isomers** of one another.

If you remove an –OH group from one of these molecules and an –H from the other, then join the two remaining structures together by a covalent bond, you get a **disaccharide**.

Disaccharides

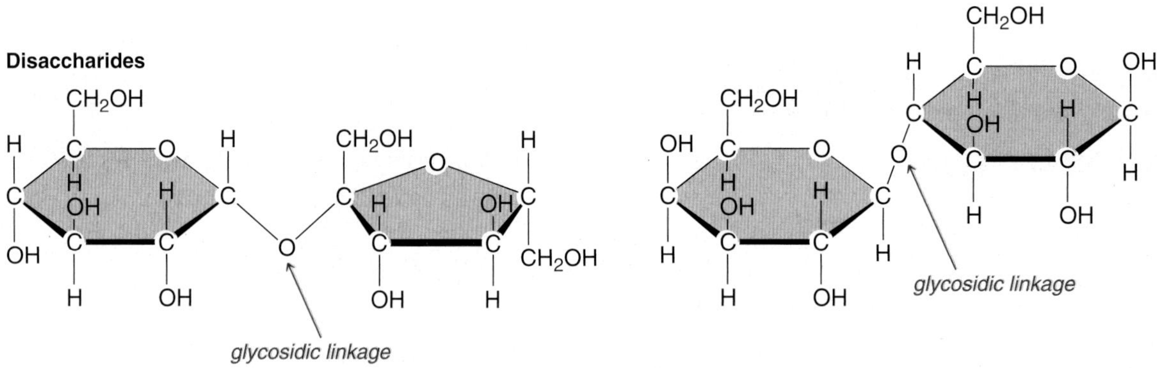

glycosidic linkage

Sucrose ($C_{12}H_{22}O_{11}$), or common *table sugar*, is formed by the combination of glucose with fructose.

glycosidic linkage

Lactose ($C_{12}H_{22}O_{11}$), or *milk sugar*, is formed by the combination of glucose and another monosaccharic *galactose*

Three monosaccharides joined in this way by covalent bonds, with the loss of two water molecules, produce a trisaccharide; four give a **tetrasaccharide**; and so on. Linking very large numbers of monosaccharides into very long chains gives **polysaccharides**, such as starch and cellulose, important compounds we'll discuss shortly. In all these cases, from disaccharides on up, the individual monosaccharides are linked to each other through an oxygen atom, called a **glycosidic linkage**.

Interpret the Data

1. When glucose and fructose (each $C_6H_{12}O_6$) combine to form sucrose ($C_{12}H_{22}O_{11}$), what small molecule is lost in the process?
2. Are sucrose and lactose isomers? Why or why not?

Carbohydrates and global nutrition • Figure 5.16

The majority of the world's people get the bulk of their caloric intake through carbohydrates. Carbohydrate-rich dietary staples such as rice, corn, and wheat provide two-thirds of the world's caloric needs. Vietnamese workers with sacks of rice, a carbohydrate-rich component of many of the world's diets.

Steve Raymer/National Geographic Creative

Carbohydrates and Diet

Carbohydrates form an essential part of our diet, and indeed of the world's diet (**Figure 5.16**). Yet in the industrialized world, where food is relatively plentiful, carbohydrates are sometimes believed to be the principal source of weight gain. The popularity of low-carbohydrate diets in recent years has undoubtedly fueled this perception. But as we saw earlier with fats, some carbohydrates can be considered healthy components of our foods, while others may pose risks to our health.

At issue is how quickly the body digests carbohydrates. Those that are digested and absorbed quickly can produce a rapid increase in blood sugar. Consuming large amounts of foods rich in these types of carbohydrates, known as *high-glycemic index* foods, is associated with increased risk of diabetes and obesity. The culprits in this case tend to be the refined sugars in soft drinks and the flours in white bread, both of which are digested quickly in the body. *Low-glycemic* foods, on the other hand, such as whole grains, whole fruits, beans, and most vegetables, contain carbohydrates that are digested and absorbed more slowly, resulting in more controlled changes in blood sugar levels. Let's take a look at the chemistry behind these differences.

Although any of the monosaccharides—glucose or fructose, for example—easily penetrate the intestinal wall to enter the bloodstream, none of the larger carbohydrates, from disaccharides on up, normally pass through the intestinal barrier. They are simply too large to penetrate the walls of the intestine and enter the bloodstream. To assimilate these larger carbohydrates, we must first separate them into their component monosaccharides, which are small enough to pass through the intestinal wall and into the bloodstream. This process, known as **hydrolysis**, involves the cleavage of each glycosidic link by the addition of a water molecule (**Figure 5.17**).

hydrolysis The process of breaking down a chemical compound through reaction with water.

Hydrolysis of carbohydrates • Figure 5.17

Here, a segment of a polysaccharide undergoes **hydrolysis**. The word **hydrolysis** derives from the Greek *hydro* for "water" and *lysis*, meaning "to split apart, loosen, or set free."

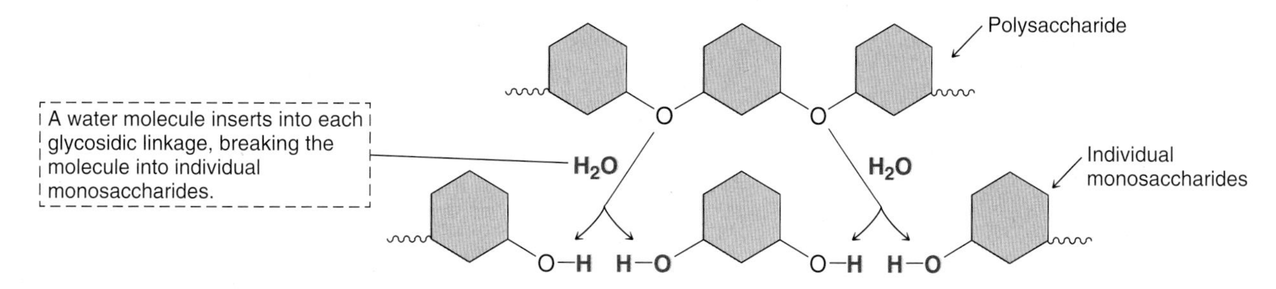

PROCESS DIAGRAM

How enzymes work • Figure 5.18

Enzymes are large molecules with intricate and well-defined shapes. For an enzyme to function, the molecule on which it acts, called the **substrate**, must fit into the convolutions or shape of the enzyme, much as a key fits into a lock. When the key's shape exactly matches the contours of the lock, the lock opens. Similarly, when the substrate fits precisely into the **active site** of the enzyme, a reaction proceeds. This then frees up the enzyme to rapidly repeat the process, forming a cycle. We can see how this cycle operates in the hydrolysis of sucrose.

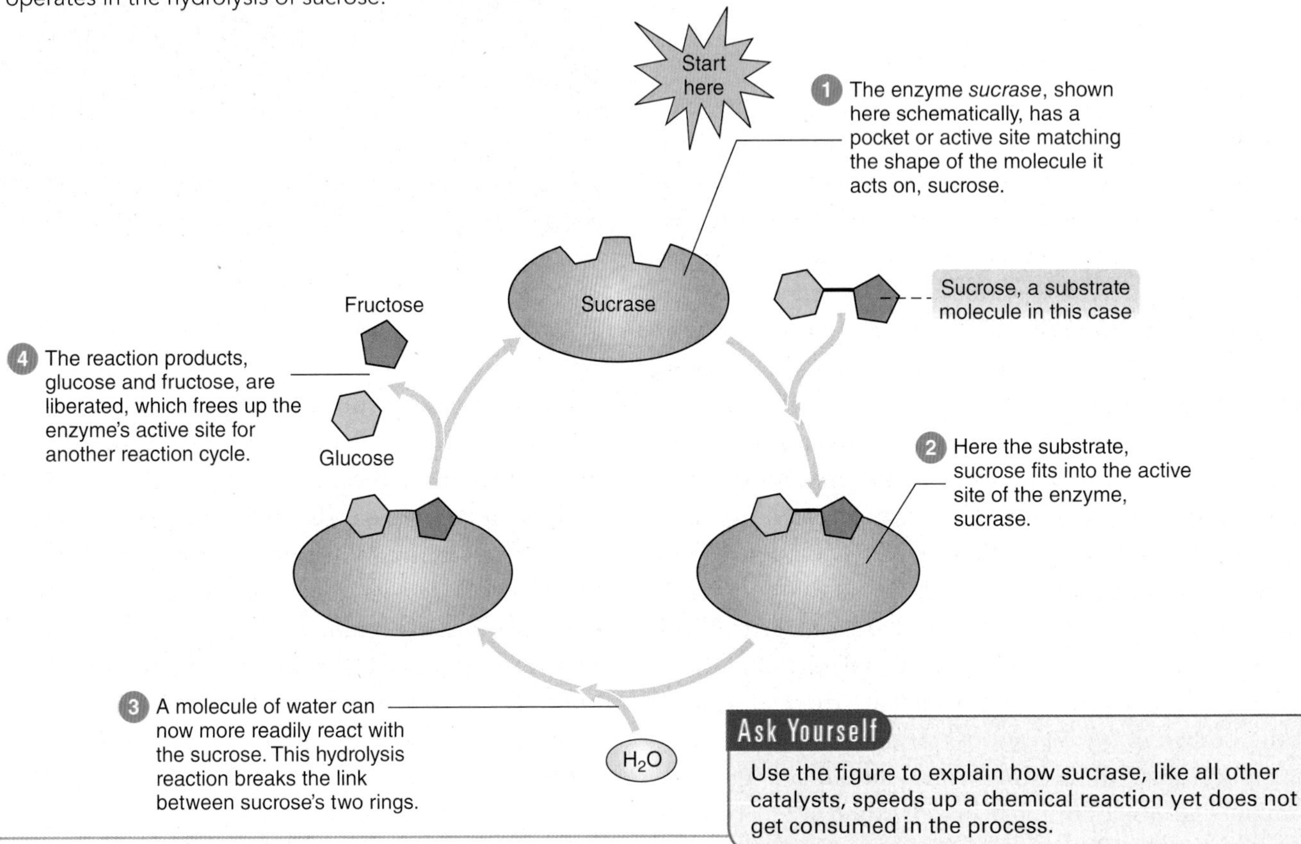

Start here
1 The enzyme *sucrase*, shown here schematically, has a pocket or active site matching the shape of the molecule it acts on, sucrose.
Sucrase
Sucrose, a substrate molecule in this case
2 Here the substrate, sucrose fits into the active site of the enzyme, sucrase.
3 A molecule of water can now more readily react with the sucrose. This hydrolysis reaction breaks the link between sucrose's two rings.
H_2O
4 The reaction products, glucose and fructose, are liberated, which frees up the enzyme's active site for another reaction cycle.
Fructose
Glucose
Ask Yourself
Use the figure to explain how sucrase, like all other catalysts, speeds up a chemical reaction yet does not get consumed in the process.

Our bodies split apart carbohydrate molecules with the help of digestive **enzymes**. Enzymes act as biological catalysts that allow chemical reactions to take place more rapidly or under milder conditions than they might without the enzymes. Like all other catalysts, enzymes are not consumed themselves during chemical reactions. Enzymes allow us to fracture nutrient molecules quickly and efficiently in the relatively mild conditions of our digestive systems. **Figure 5.18** shows how enzymes work.

enzyme A biological molecule, typically a protein, that acts as a catalyst.

Enzymes account for our ability to digest certain carbohydrates but not others. As a prime example, consider the carbohydrates **starch** and **cellulose**. The two are structurally similar in that they are very long polysaccharide molecules made up exclusively of glucose rings. Yet although we digest and absorb the starch of our foods easily, the cellulose, or fiber, passes through our bodies unchanged and unabsorbed (**Figure 5.19**). For further details on the different orientations of glucose structures in starch and cellulose, see *A Deeper Look* (on the next page).

Although we depend on plants that provide starch for much of our energy needs, there is probably more cellulose on this planet than any other single organic (carbon-based) compound. Growing plants produce close to 100 billion tons of cellulose worldwide each year. This polysaccharide, which doesn't occur at all in animals, is the principal structural component of plant cells. Cellulose is the main constituent of a variety of plant-derived products, such as paper, books, cardboard, and cotton and linen fabrics. Depending on the source, cellulose molecules vary in length from a few hundred to several thousand consecutive β-glucose rings.

Chemistry InSight — Why we can digest starch but not cellulose • Figure 5.19

The starch contained in the seeds or roots of such foods as wheat, corn, rice, and potatoes serves as a macronutrient. However, the *cellulose* in the more fibrous structures of plants—such as bran or seed husks of wheat and oats; celery stalks; and lettuce leaves—provides us with dietary fiber, sometimes called roughage. Why is it we can digest starch but not cellulose? The answer lies in the molecular structure of each **(a)**.

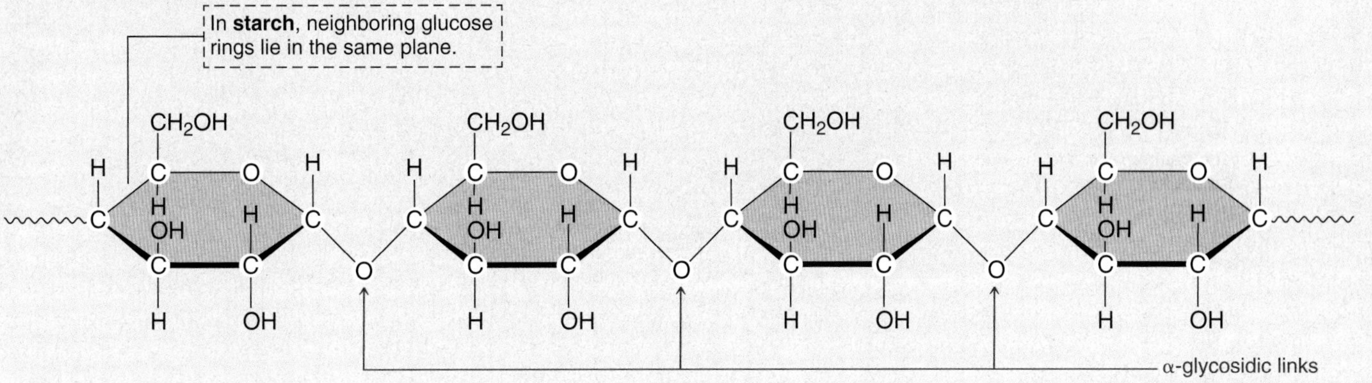

In **starch**, neighboring glucose rings lie in the same plane.

α-glycosidic links

In **cellulose**, neighboring glucose rings lie in different planes, somewhat like stair steps.

β-glycosidic links

a. Both starch and cellulose are polysaccharides composed of glucose rings. (We've shown just a small segment of each molecule to illustrate the repeating theme.) In starch, the rings are joined by α-links, whereas in cellulose they are joined by β-links.

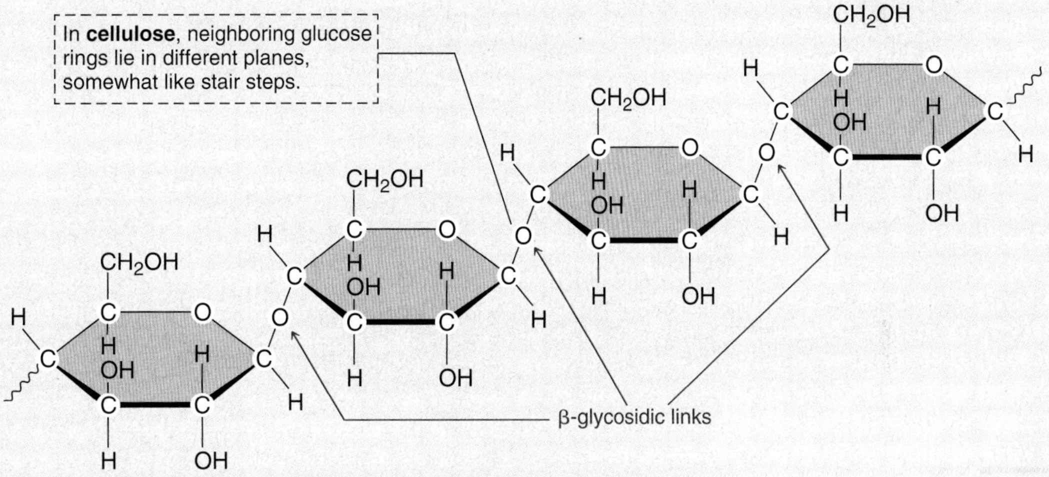

Our bodies have enzymes that recognize the shape of the α-linkages of starch, so we can hydrolyze these links and digest starch. However, we don't have enzymes that recognize the shape of the β-linkages of cellulose, so dietary fiber, which is largely cellulose, simply passes through our digestive system unabsorbed.

b. The sight, smell, or even thought of a savory meal can stimulate the production of saliva. Enzymes present in saliva help begin the process of digestion. One of these, *amylase*, helps hydrolyze starch into smaller segments, chiefly the disaccharide *maltose*.

This dish is rich in the polysaccharide, starch.

johnfoto18/Shutterstock

Think Critically What monosaccharide is released when the *enzyme maltase*, present in our intestines, helps hydrolyze maltose?

A DEEPER LOOK Alpha and Beta Glucose

We have shown glucose represented in either of two ring forms, α− and β−, but glucose also can exist in an open-chain form. As the open-chain form of glucose closes on itself, or *cyclizes*, it forms either of the two ring forms, α− or β−.

The carbon backbone in the open chain form of glucose is flexible, allowing it to bend around...

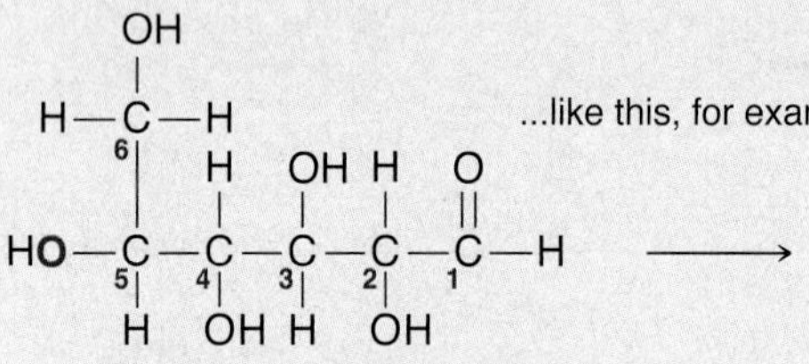

Open-chain glucose

...like this, for example.

The oxygen (in red) is now close enough to carbon #1 to form a new bond.

Open-chain glucose has an *aldehyde* (AL-de-hide) functional group at carbon #1. The reactivity of this group helps the ring form in this case. We'll have more to say about aldehydes in Chapter 11.

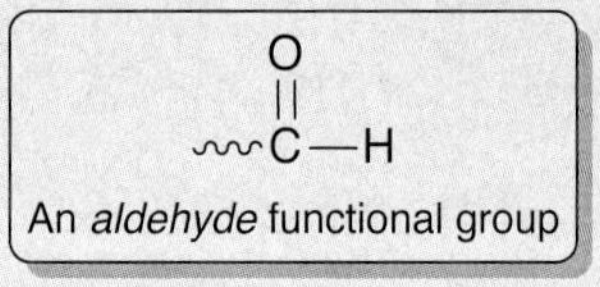

An *aldehyde* functional group

In this representation of β-glucose, note how the hydroxyl group (–OH) attached to carbon #1 points **up**.

This gives either β-glucose (above) or α-glucose (below)

Tail above nose

Tail below nose

In this representation of α-glucose, note how the hydroxyl group points **down**.

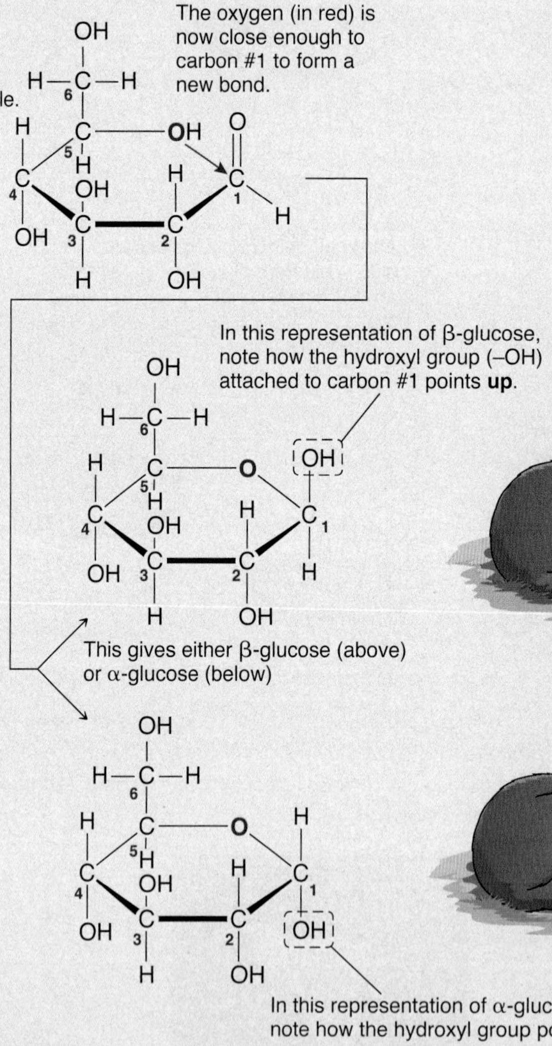

We can liken this *cyclization* to a fox curling up, with its tail either above or below its nose. As the ring forms, an –OH group at carbon 1 rests above the ring (in β-glucose) or below the ring (in α-glucose). The simple phrase, "Beta is for the birds, and alpha is for the ants," may help in remembering which form is which. In β-glucose, the highlighted –OH points up, as if to the sky, while in α-glucose, this –OH group points down, as if to the ground.

Interpret the Data

As the ring forms, what happens to the aldehyde functional group at carbon 1?

Another polysaccharide of interest is **glycogen**, which serves as a source of quick energy in animals (**Figure 5.20**). Glycogen resembles starch in that it is made up exclusively of α-glucose rings, except that it is more highly branched and contains fewer glucose units. In our bodies, glycogen is deposited largely in the liver and muscles, where it can be cleaved quickly to yield individual glucose molecules, supplying the body with a ready source of energy. We cannot store much glycogen, though; the liver's supply of glycogen is generally used up each night as we sleep and must be replenished the next day.

Enzymes that act on a given carbohydrate are often given a name that resembles the carbohydrate's name. For example, in most cases, replacing the *-ose* ending of a sugar with *-ase* gives us the name of the enzyme that hydrolyzes the sugar. *Sucrase* helps hydrolyze the links of *sucrose*, for instance. *Cellobiase* helps hydrolyze *cellobiose*, a disaccharide made up of β-linked glucose units. Since we do not have cellobiase in our bodies, we cannot cleave cellobiose, nor other saccharides made up of β-linked glucose units, such as cellulose. Grass, leaves, and other plant material, all of which are rich in cellulose, are indigestible in our own intestines. This same plant material, though, is a source of energy for cows, goats, sheep, and termites simply because the digestive systems of these organisms harbor microorganisms that produce the needed cellobiase.

Our dependence on enzymes to hydrolyze sugars can lead to problems for some of us. Lactose (from the Latin *lac*, for "milk"), the principal carbohydrate in milk, offers a good example. Lactose is a disaccharide made up of a glucose ring joined to a galactose ring through a β-link. (The monosaccharide galactose differs only subtly from glucose.) Since virtually no disaccharide of any kind penetrates the intestinal wall and enters the bloodstream, we need the enzyme *lactase* to hydrolyze lactose into its components, glucose and galactose (which can pass through the intestinal wall and provide us with nourishment). The more milk and milk products we consume, the more lactase we need.

Normally, the digestive systems of infants and children have plenty of lactase. That's fortunate because nursing babies get about 40% of their calories from the lactose of their mother's milk. Older children generally have more than enough lactase to digest all the lactose of the milk, cheese, ice cream, and other dairy products they eat. But gradually, as we grow from infancy to adolescence, most of us lose the ability to produce lactase in large quantities, so we're largely lactase-deficient as adults. In this condition, we produce very small amounts of lactase, too little to handle more than a glass or two of milk at a time. It's been estimated that about two out of three adults throughout the world have very low levels of lactase in their digestive systems. Estimates for the percentage of lactase-deficient adults within a given population range widely, though, from as low as 3% for Scandinavians and others of northern European descent to 80% and higher for Asian populations.

Cellulose and glycogen • Figure 5.20

Both cellulose and glycogen are polysaccharides composed of many glucose rings. Plants produce cellulose as a structural component. Animals utilize glycogen as an energy source.

Ask Yourself

Which compound more closely resembles starch: cellulose or glycogen? Why?

Cellulose is the main component of various plant-derived products, such as this paper.

Glycogen provides a major source of energy for exercise. For especially strenuous activities, such as sprinting and powerlifting, stores of glycogen can be depleted within as little as 15 minutes.

Adam Gault/Digital Vision/GettyImages

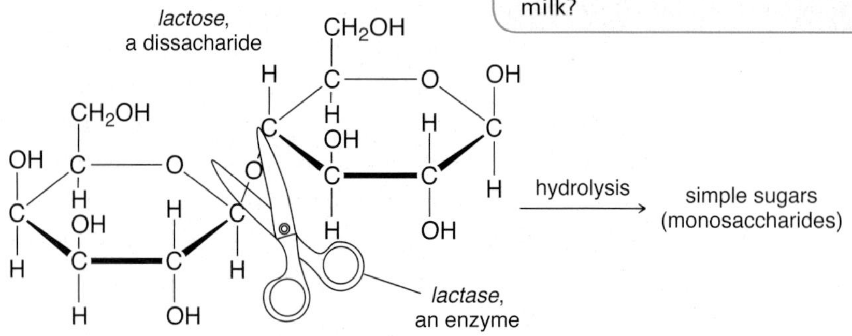

Put It Together Review the structure of lactose in Figure 5.16 and answer the question.
What sugar molecules would you expect to be present in lactose-free milk?

Lactose-free dairy products • Figure 5.21

The enzyme lactase helps hydrolyze the carbohydrate lactose.

100% LACTOSE FREE

© The Photo Works/Alamy

In lactose-free dairy products, lactase has been added in order to hydrolyze the lactose naturally present in milk.

lactose, a dissacharide

hydrolysis → simple sugars (monosaccharides)

lactase, an enzyme

Without enough lactase in digestive fluids, the lactose of milk and milk products isn't hydrolyzed effectively. Therefore, much of it passes unabsorbed along the intestinal path to the lower intestines. There it undergoes fermentation to produce the bowel irritant *lactic acid*, as well as gases such as carbon dioxide and hydrogen. The combination easily produces gastric distress. Today, consumers can choose from a variety of lactose-free dairy products (**Figure 5.21**).

CONCEPT CHECK

STOP

1. **Rank** the following carbohydrates in size, from largest to smallest: cellulose, fructose, maltose.
2. **What** small molecule reacts with glycosidic linkages in the enzyme-mediated cleavage of carbohydrates?
3. **Why** can't humans live on grass and hay?

5.4 Proteins

LEARNING OBJECTIVES

1. **Recognize** the general structure of amino acids and how they link to each other to form peptides.
2. **Distinguish** between primary and higher-order structures of proteins.
3. **Describe** the role of essential amino acids in the diet.
4. **Explain** how protein denaturation takes place.

We would not have survived as a species without the compact storehouse of energy that fat provides, but although our brain, nerves, and muscles depend on the carbohydrate glucose for energy, it is **protein** that gives the very shape to our bodies (**Figure 5.22**). Proteins are also the molecules of our enzymes, those catalysts that help us digest food and keep a myriad of chemical reactions operating within our bodies. Without proteins, we could not function.

protein A polymer made up of a long sequence of amino acids.

Proteins and the body • Figure 5.22

Proteins are important components of our skin, hair, nails, muscles, and organs.

Our hair and nails are composed of the protein **keratin**.

Our muscles contain the proteins **actin** and **myosin**.

Our skin contains the protein **collagen**.

Rita Jacobs/E+/Getty Images

Amino acids and proteins • Figure 5.23

Some 20 different amino acids constitute the building blocks of proteins. All naturally occurring amino acids share the same general structure, but each has a different side group (R).

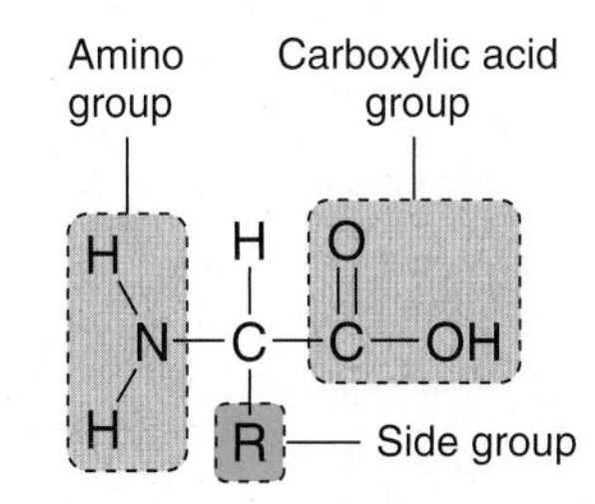

General structure of amino acids

Structures of Proteins

To appreciate the critical role protein plays in nutrition and in giving shape to our bodies, it's helpful to first understand protein structure. Like the polysaccharides starch and cellulose, proteins exist as long, threadlike molecules (polymers) formed by linked smaller units (monomers). But while each unit of starch and cellulose is a cyclic glucose molecule, the units that make up protein molecules consist of a variety of **amino acids** (**Figure 5.23**).

Amino acids carry both **amino** ($-NH_2$) and **carboxylic acid** ($-CO_2H$) functional groups. (We'll have more to say about carboxylic acids in Chapter 8.) The only structural difference among the amino acids that form proteins lies in the side group (R) attached to the carbon atom bonded to the CO_2H. For example, in the simplest amino acid, *glycine*, the side group is $-H$, while in *alanine*, the side group is $-CH_3$ (**Table 5.1** on the next page). Proteins used as nutritional supplements vary in the composition of their amino acids, depending on the source of the protein (**Figure 5.24**).

amino acid A small organic molecule containing both **amino** and **carboxylic acid** functional groups that serves as a structural unit for proteins.

Amino acids link together linearly, like beads on a chain, to form **peptides**. Linking any two amino acids, identical or different, gives a **dipeptide**; joining any three yields a **tripeptide**. A large number of amino acids linked into a chain produces a **polypeptide**. A protein consists of one or more polypeptide chains, each of which can contain hundreds of amino acids.

peptide A compound made up of linked amino acids.

Protein supplements • Figure 5.24

Whey protein (derived from milk) and soy protein (obtained from soybeans) are common types of protein supplements. The amino acid content of a single serving of one brand of whey protein is shown below here.

Tim O. Walker

Typical Amino Acid Profile (milligrams per 41 g scoop*)**

Essential Amino Acids		Nonessential Amino Acids	
Histidine	339 mg	Alanine	892 mg
Isoleucine	1,150 mg	Arginine	458 mg
Leucine	1,993 mg	Aspartic Acid	2,029 mg
Lysine	1,657 mg	Cysteine	419 mg
Methionine	408 mg	Glutamic Acid	3,011 mg
Phenylalanine	613 mg	Glycine	4,348 mg
Threonine	2,279 mg	Proline	1,111 mg
Tryptophan**	325 mg	Serine	926 mg
Valine	1,056 mg	Tyrosine	567 mg

**L-Tryptophan is naturally occurring, not added.
***approximate values

Naturally occurring amino acids   Table 5.1

Of the 20 amino acids that occur naturally in proteins, our bodies lack the ability to synthesize about half, shown in red. We call these **essential amino acids** because we must get them from the proteins of our foods. **Nonessential amino acids,** shown in black, can be produced by our own bodies as needed.[1]

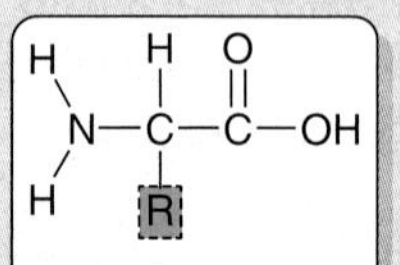

Name	Abbreviation	Side group. R
Alanine	Ala	$-CH_3$
Arginine	Arg	$-CH_2-CH_2-CH_2-NH-C(=NH)-NH_2$
Asparagine	Asn	$-CH_2-C(=O)-NH_2$
Aspartic Acid	Asp	$-CH_2-C(=O)-OH$
Cysteine	Cys	$-CH_2-SH$
Glutamic Acid	Glu	$-CH_2-CH_2-C(=O)-OH$
Glutamine	Gln	$-CH_2-CH_2-C(=O)-NH_2$
Glycine	Gly	$-H$
Histidine	His	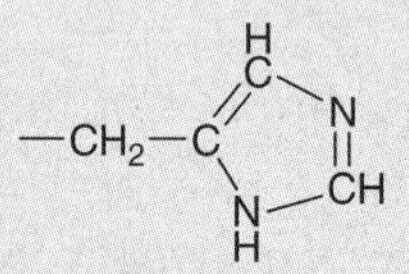 $-CH_2-C$ (imidazole ring: HC=, N, CH, N–H)
Isoleucine	Ile	$-CH(CH_3)-CH_2-CH_3$
Leucine	Leu	$-CH_2-CH(CH_3)-CH_3$
Lysine	Lys	$-CH_2-CH_2-CH_2-CH_2-NH_2$
Methionine	Met	$-CH_2-CH_2-S-CH_3$
Phenylalanine	Phe	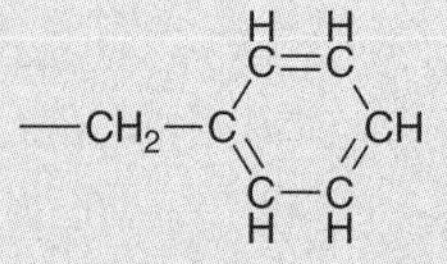 $-CH_2-C_6H_5$ (benzene ring)
Proline	Pro	*see note
Serine	Ser	$-CH_2-OH$
Threonine	Thr	$-CH(OH)-CH_3$
Tryptophan	Trp	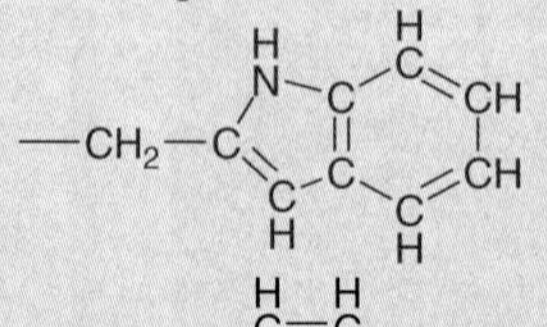 $-CH_2-C$ (indole ring)
Tyrosine	Tyr	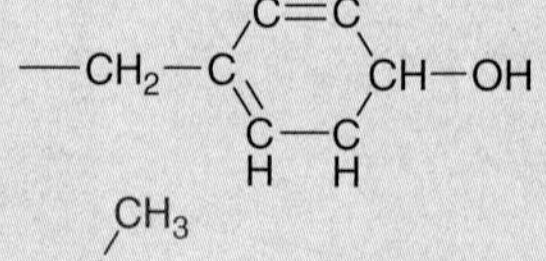 $-CH_2-C_6H_4-OH$ (ring with CH–OH)
Valine	Val	$-CH(CH_3)-CH_3$

general structure of amino acids: $H_2N-C(H)(R)-C(=O)-OH$

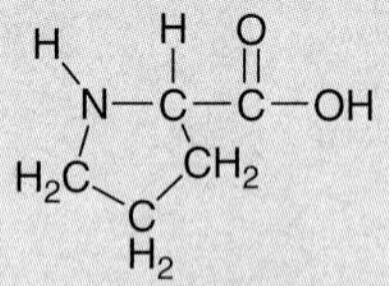

*In proline, shown here, the side chain forms a ring with the amino nitrogen

[1]The classification of amino acids as either essential or nonessential isn't always compeletely clear. Although nine of the amino acids appear to be essential for everyone, three more—arginine, cysteine, and histidine—probably can be generated in suffcient supply by an adult but not by a young child.

Interpret the Data

1. Which amino acids have purely hydrocarbon side chains?
2. Which amino acids have an alcohol functional group within their side chains?

The structures of proteins • Figure 5.25

The order in which amino acids are linked together to make any particular protein is called that protein's **primary structure**. We can liken primary structure to the sequence of beads on a chain. As this chain forms, it twists and folds into higher-order structures, the shapes of which are critical to the protein's function.

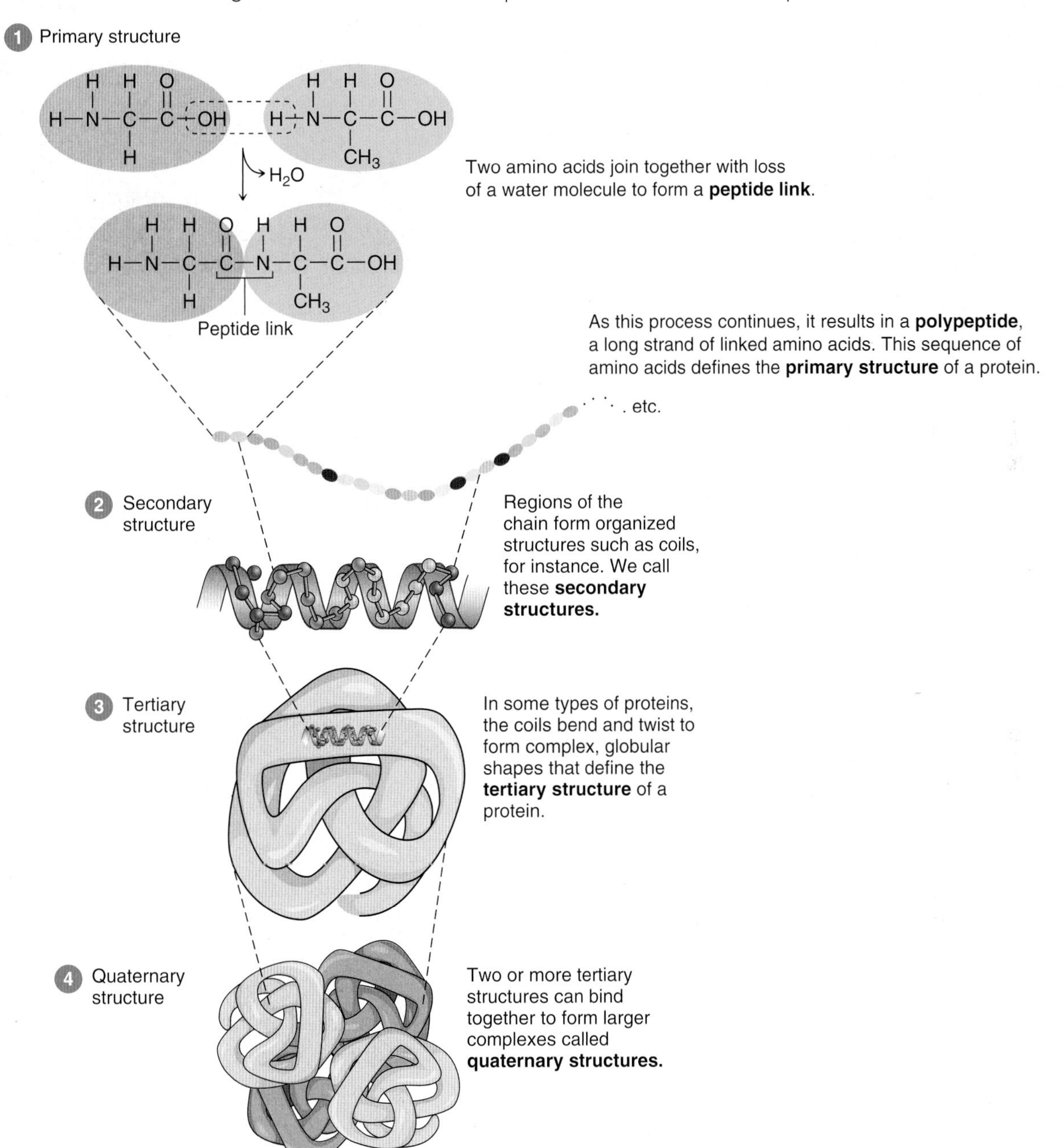

1 Primary structure
H₂O
Two amino acids join together with loss of a water molecule to form a **peptide link**.
Peptide link
As this process continues, it results in a **polypeptide**, a long strand of linked amino acids. This sequence of amino acids defines the **primary structure** of a protein.
. . . etc.
2 Secondary structure
Regions of the chain form organized structures such as coils, for instance. We call these **secondary structures.**
3 Tertiary structure
In some types of proteins, the coils bend and twist to form complex, globular shapes that define the **tertiary structure** of a protein.
4 Quaternary structure
Two or more tertiary structures can bind together to form larger complexes called **quaternary structures.**

Since our bodies construct their peptide chains from a set of 20 different amino acids, and since it's possible to produce a peptide chain of any length from one or only a few different amino acids, in principle it might be possible for our bodies to generate any number of different polypeptides. Yet our bodies are designed by our genetic heritage to produce only specific peptides. Each of these must serve a rational, chemical purpose, performing a specific and important function in sustaining our lives. The reason for this is that the sequence of amino acids in a protein, called its **primary structure**, determines the higher-order structures—the coils and bends—that define the overall shape and function of the protein molecule (**Figure 5.25**).

Fibrous and globular proteins • Figure 5.26

The native shapes adopted by a protein's peptide chains strongly influence the protein's function.

GORDON WILTSIE/National Geographic Creative

a. Our hair, nails, and muscles are tough and strong because of their **fibrous** proteins. Their strength comes from the organization of these protein molecules, which lie in parallel strands. entwining much like strands of fiber twisted into strong rope.

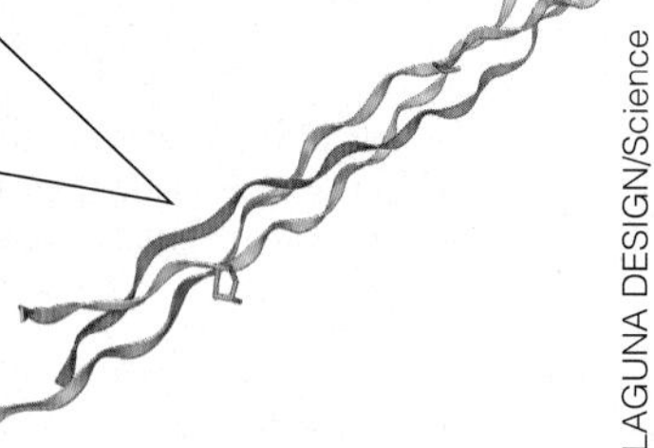

Visualization of a segment of *collagen*, a protein that makes up skin.

LAGUNA DESIGN/Science Photo Library/Getty Images

Joff Lee/Photolibrary/Getty Images

b. The peptides of **globular** proteins bend and twist back into themselves, forming small spheres. As a result, they move around easily in the blood and other fluids and are available to do their chemical work where they are needed. Enzymes are globular proteins, as are the proteins of egg white.

Visualization of a globular protein, such as *albumin*, a protein found in egg white.

The shape of a protein's higher structure dictates its function. Two types of proteins, **fibrous** and **globular**, have shapes that suit them particularly well to the functions they perform (**Figure 5.26**).

Raw egg white consists of about 90% water, with the remainder chiefly globular proteins. The long strands of these proteins maintain their globular forms by attractive forces within each protein molecule. Heating an egg white overcomes these attractive forces and allows the molecules to unravel, while still keeping their original primary structure. As a molecule unravels—a process called **denaturation**—it comes in contact with other protein molecules, which have also lost their globular shapes (see *What a Chemist Sees*). These proteins then gather together into shapes more like those of fibrous proteins and form the familiar tough white of a cooked egg. In the chapter on medicines and drugs, we'll have more to say about a protein's shape and how its shape influences its function, but for now we turn our attention to proteins in our diets.

denaturation A process by which a protein's higher-order structure is disrupted due to heat or exposure to chemical compounds.

Proteins and Diet

Protein is unique among the three macronutrients in that it's the only one that contains nitrogen. Our bodies regularly excrete nitrogen compounds, mostly through urine, and we need to consume adequate amounts of protein daily to maintain a **nitrogen balance**. But how much protein is enough to maintain this balance?

nitrogen balance The balance between the amount of nitrogen consumed in the diet (in the form of protein) and the amount excreted.

With the popularity of protein-rich diets, protein supplements, and nutrition bars, you might think we could benefit from more protein in our diets. Yet the vast majority of us living in industrialized countries get more than adequate protein in our normal diets, without having to resort to specialized diets or supplements. Each day, a typical adult needs 0.8 g of protein per kilogram of

WHAT A CHEMIST SEES

Protein Denaturation

Cooking an egg disrupts the attractive forces that hold the protein molecules in their native globular shapes. As these proteins denature and form bonds to other protein molecules, they give cooked egg white its characteristic color and texture.

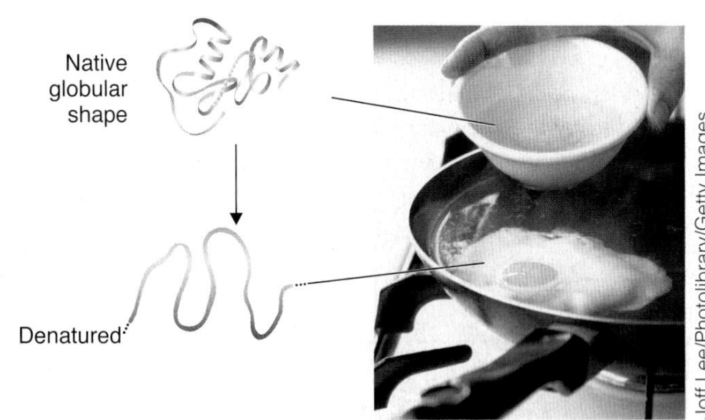

Think Critically Would you expect the denaturation of the proteins in egg white to be a reversible process or an irreversible process?

body weight, which for a 68-kg (150-lb) person amounts to 55 grams of protein per day. This **recommended daily allowance (RDA)** value can be a bit higher for those with special needs, such as children, pregnant women, those recovering from serious illness, or athletes. Endurance athletes, for instance, need up to 1.2–1.4 g of protein per kilogram of body weight daily. For bodybuilders, the value may be as high as 1.7 g/kg, but levels beyond this have not been shown to improve athletic performance or increase muscle mass. Most of us can meet dietary protein needs through a variety of foods known to be rich in protein (**Figure 5.27**).

recommended daily allowance (RDA) The level of daily consumption of a given macronutrient recommended for maintaining proper health.

Protein-rich foods • Figure 5.27

Dairy, eggs, meat, poultry, fish, beans, and nuts contain relatively high levels of protein (**a**). The graph shows the amount of protein provided by recommended servings of various foods (**b**).

a. High-protein foods

b. Protein content of various high-protein foods

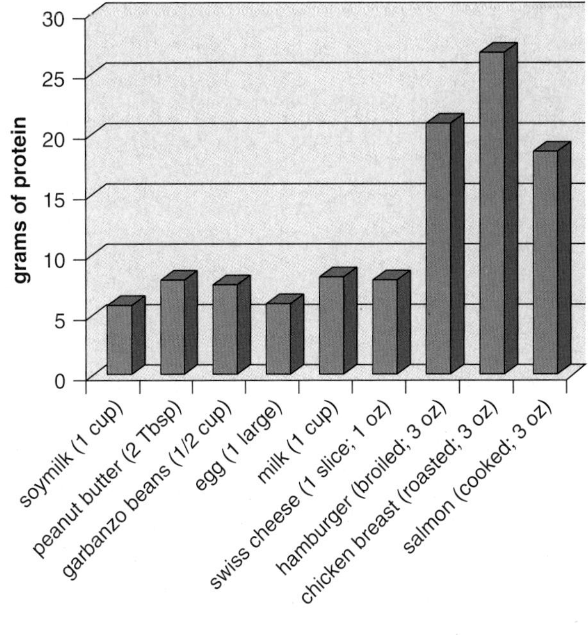

Meeting protein needs through a vegetarian diet • Figure 5.28

Protein complementation refers to eating proteins from different plant sources so that collectively they provide adequate amounts of essential amino acids (EAAs). Many vegetarian dishes, for example, pair a legume, such as lentils, beans, peanuts, or tofu (soybean curd), with a grain, such as rice or corn. EAAs that may be lacking in the legume can be provided by the grain and vice versa. Dietitians now recognize that complementary proteins do not need to be eaten at the same meal, so any well-varied vegetarian diet supplying sufficient calories can adequately meet protein needs.[2]

Beans, such as these garbanzos, are generally rich in lysine but poor in methionine.
Grains, such as rice, are generally poor in lysine, but rich in methionine.

@lexandra panella/Flickr Open/GettyImages

Ask Yourself

Would you expect a peanut butter sandwich to provide complementary proteins? Why or why not?

Although it's important to consume *enough* protein, the quality of the protein in our diets is equally important. By "protein quality," we mean how well a dietary protein provides the essential amino acids. Since we can't store significant amounts of essential amino acids in our bodies—as we store fat, for example—all the essential amino acids should be present in daily diets so that our bodies can generate the necessary proteins. Dietary proteins that provide substantial amounts of all the essential amino acids, and in about the same ratio as they occur in our own proteins, are considered complete proteins, or **high-quality proteins**. Dietary proteins deficient in one or more of the essential amino acids are considered **incomplete proteins**.

Foods of animal origin, such as meat, poultry, fish, dairy products, and eggs, provide high-quality protein. On the other hand, plant-based foods, such as grains, legumes, and nuts, are generally deficient in one or more of the essential amino acids. One exception is soy protein, which offers adequate amounts of all the essential amino acids. Although you might savor the taste of a juicy burger or enjoy the taste of cheese, it is by no means necessary to eat animal-derived foods in order to meet protein needs. A vegetarian diet based on a range of foods can more than adequately do so (**Figure 5.28**). Plant-based diets also tend to be lower in saturated fat and can be associated with a variety of health benefits, such as decreased risk of heart disease, stroke, and diabetes.

Failure to secure a good dietary supply of the essential amino acids, in the proper proportions, can produce devastating results, especially among young people. *Kwashiorkor* (kwa-shee-OR-kor), a potentially fatal condition most common in Africa, develops from a protein deficiency due to the virtually complete absence of essential amino acids from the diet. At the other extreme, habitual overconsumption of protein can add unnecessary calories to the diet and pose other health risks. Concerns about overconsumption of calorie-rich foods in general may lead some to seek low-calorie alternatives (see *Did You Know?*).

When we digest protein, we break it down into its constituent amino acids, which our bodies reassemble into other proteins for use as muscle tissue, antibodies, or hair, for example. If there's no demand for muscle tissue, antibodies, or some other specialized molecules that only the amino acids can form, we either use them for energy or store them by converting them to fat.

CONCEPT CHECK STOP

1. **What** small molecule is eliminated when amino acids link together to form peptides?
2. **What** information does the primary structure of a protein provide?
3. **Why** is the *quality* of protein you consume as important as or even more important than the *quantity* of protein in your diet?
4. **Why** do proteins denature when subject to high temperatures?

DID YOU KNOW?

Designer molecules in some of the very foods we eat

When it comes to food, some of us want to have our cake and eat it, too; that is, we want to enjoy the taste of calorie-rich foods without the consequences of having consumed those calories. A variety of diet foods containing calorie-free ingredients certainly hold this promise. We might wonder how these ingredients can mimic the tastes of fats and carbohydrates, imparting a richness or sweetness to foods but not contributing to calories (**Figure a**). We'll take a look at three examples of these ingredients. In each case, they were discovered serendipitously during the course of basic research.

Fig. a. Products containing calorie-free ingredients.

Calorie-free soft drinks containing aspartame. A no-calorie sweetener made from sucralose. Fat-free potato chips made with olestra.

Olestra

Figure 5.6 shows that all fats and oils are triglycerides—a glycerol backbone linked to three fatty acids. When we eat triglycerides, enzymes in our bodies hydrolyze the links between the glycerol and the fatty acids. Scientists working to understand more about this process created a new type of molecule. Instead of using glycerol as a backbone, which has three –OH groups, they used sucrose, which has eight (**Figure b**). By linking fatty acids—long-chain carboxylic acids—to the –OH groups of sucrose, they discovered unexpectedly that our bodies lack enzymes that can hydrolyze so many fatty acids from a single molecule. This molecule, called *olestra*, is a fusion of sucrose and fatty acids that offers the taste and feel of fat in the mouth but passes through the digestive system unabsorbed, without providing calories.

Sucralose

The sugar substitute *sucralose* also uses the sucrose molecule as a template. However, instead of linking the –OH groups to fatty acids (as in olestra), three of the –OH groups in sucrose are replaced with chlorine atoms. This change in the molecule's structure gives some very interesting, unexpected results. Not only does sucralose taste 600 times as sweet as sucrose, or regular sugar, but the compound passes through the intestines largely unabsorbed and so does not provide calories. (Even if sucralose were absorbed by the body, so little is needed to sweeten foods that its contribution to calories would be negligible.)

Aspartame

While developing potential treatments for stomach ulcers, research chemist James Schlatter was preparing peptides when he happened upon an expected discovery. A certain dipeptide derivative, made from the amino acids aspartic acid and phenylalanine, was found to be unusually sweet (**Figure c**). (Evidently he noticed the taste while absent-mindedly licking his fingers to turn a page in his laboratory notebook.) This compound, called *aspartame*, is 200 times as sweet as table sugar. So little is required to sweeten foods that it provides a negligible amount of calories.

Fig. b. A molecule of sucrose. Note the **–OH** groups in bold. Olestra is formed by linking fatty acids to these –OH groups. Sucralose is formed by replacing three of these –OH groups with Cl atoms.

Fig. c. Aspartame, shown here, is derived from two amino acids: aspartic acid and phenylalanine.

From aspartic acid

From phenylalanine

Ask Yourself

How is the structure of olestra different from that of sucralose?

Put It Together Review Table 5.1 and answer the question.
How is the structure of the dipeptide Asp-Phe different from that of aspartame?

Summary

1 The Energy Equation and Metabolism 126

- ***What is a Calorie?***
  A Calorie (Cal), a unit of energy used in nutrition, is defined as the amount of energy needed to raise the temperature of 1000 g of water by 1°C as shown here. It's equal to 1000 calories.

Figure 5.1 • Units of heat or energy

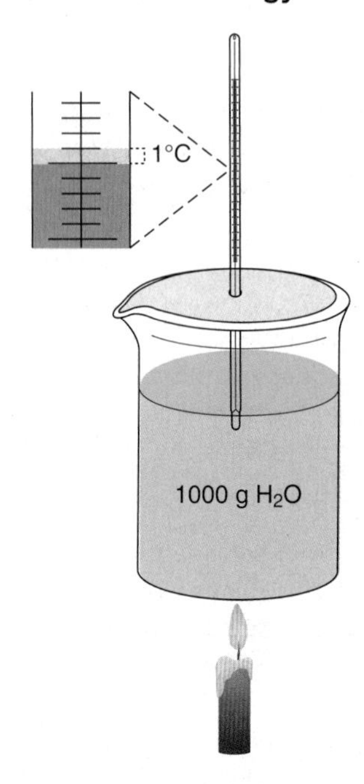

- ***How do we determine the energy content of foods, and how does the body utilize this energy?***
  The number of Calories in a given serving of food is determined from the masses of the individual macronutrients—fats, carbohydrates, and protein—in the portion. Per gram, fats provide 9 Calories, while carbohydrates and protein each provide 4 Calories. The body uses this energy to digest food, to support basal metabolism, and in exercise and physical work. Excess energy is stored as adipose, or fat, tissue.

2 Fats and Oils 132

- ***What molecular structure is common to all fats and oils?***
  Chemically, all fats and oils are triglycerides, molecules made up of glycerol linked to three fatty-acid side chains. Triglycerides vary in the structures of these side chains.

- ***What do we mean by* saturated, unsaturated, omega-3, *and* trans *fats?***
  These are all types of fatty acids. Saturated fats have no C=C bonds. Unsaturated fats contain one or more C=C bonds as shown here and are generally healthier than saturated fats. Omega-3s are a type of essential fatty acids, in which a C=C bond begins three carbons in from the $-CH_3$ end of the chain. Trans fats can result from partial hydrogenation of unsaturated fats.

Figure 5.7 • The structure of a fatty acid

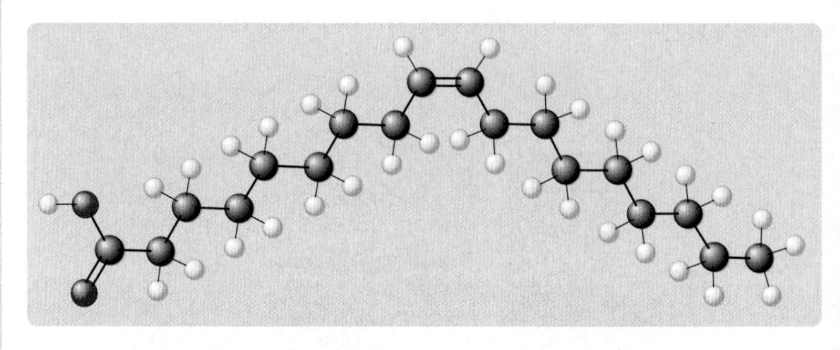

- ***Is cholesterol a type of fat?***
  Cholesterol is not a triglyceride, so it is not a fat. It is produced only in animals so is present only in animal-derived foods.

3 Carbohydrates 141

- ***What are carbohydrates, and how are they classified?***
  Carbohydrates are common in starchy foods. These molecules exist largely as single-ring structures or as their combinations and are classified by the number of joined rings. Monosaccharides, such as glucose, have one ring; disaccharides have two, as shown here; and so on.

Figure 5.15 • Types of carbohydrates

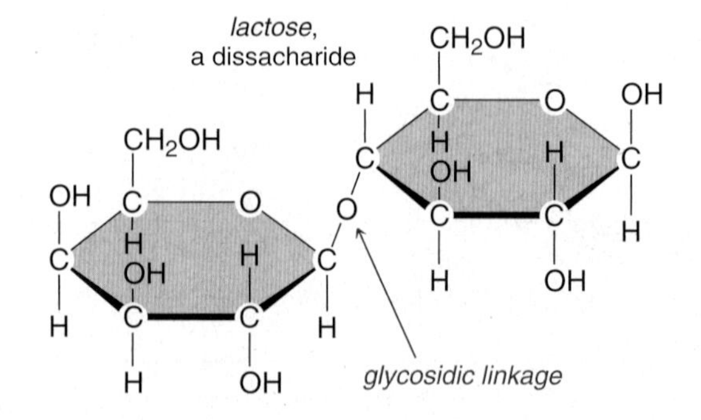

- ***What is the difference between starch and fiber?***
  Each is a polymer of glucose, but starch, made of α-glucose, is digestible by humans, although fiber, made of β-glucose, is not.
- ***What role do enzymes play in the digestion of carbohydrates?***
  Enzymes hydrolyze or cleave (by the action of water molecules) the linkages between carbohydrate rings. We can digest carbohydrates for which we have the appropriate enzymes. For example, the enzyme maltase helps break down the linkages in starch molecules.

Figure 5.26 • Fibrous and globular proteins

Joff Lee/Photolibrary/Getty Images

A globular protein

4 Proteins 148

- ***What are proteins, and how are they classified?***
  Proteins are polypeptides, or polymeric chains of amino acids. The primary structure of a protein describes the sequence of its amino acids within the polypeptide chain. Fibrous proteins are more strand-like in structure and interconnected; globular proteins, as shown here, are more spherical and stand alone.
- ***How do foods vary in the types of protein they provide?***
  Complete protein offers adequate amounts of all the essential amino acids, while incomplete protein may be poor in one or more essential amino acids. Animal-based foods provide complete protein, but protein needs can be met through a well-varied vegetarian diet.
- ***What is protein denaturation?***
  Protein denaturation refers to the unraveling of higher-order structures of proteins upon exposure to heat or other agents.

Key Terms

- amino acid 149
- basal metabolism 131
- calorie 126
- Calorie 126
- calorimetry 127
- carbohydrates 141
- catalytic hydrogenation 137
- denaturation 152
- enzyme 144
- glucose 141
- hydrolysis 143
- joule 127
- macronutrient 129
- metabolism 129
- nitrogen balance 152
- omega-3 fatty acid 140
- peptide 149
- protein 148
- recommended daily allowance (RDA) 153
- saturated fatty acid 132
- thermic effect of food (TEF) 131
- triglycerides 132
- trans fatty acid 137
- unsaturated fatty acid 132

What is happening in this picture?

The origins of the wonderful aromas, colors, and textures of a freshly baked pizza can be understood with a bit of chemical insight. For instance, the browning we see in baked pizza, toasted bread, and other cooked foods results as heat initiates a chemical reaction between protein and carbohydrate molecules naturally present in the food. Specifically, amino groups ($-NH_2$) on the side chains of amino acids, such as lysine, form a bond with certain carbons in carbohydrate molecules. (This *Maillard* reaction is named after the French chemist who first described it.) After subsequent chemical reactions, the foods form *melanoidins*, which are brown-colored compounds associated with the rich flavors of cooked food.

Richard Nowitz/National Geographic/Getty Images

Think Critically (You may find it helpful to refer to Table 5.1 when answering both questions.)

1. Arginine can participate in a Maillard reaction but methionine cannot. Why?
2. When the lysine in food undergoes a Maillard reaction, this can lead to some loss in nutritional value. However, the same cannot be said for arginine. Why?

Exercises

Review

1. a. Which is a larger unit: the calorie or the joule? b. How many calories are in a (nutritional) Calorie?
2. Which macronutrient contributes the greatest amount of energy per gram to the body?
3. How do our bodies store unused energy?
4. Given a sample of a triglyceride, how would you determine whether it is a fat or an oil?
5. Vegetable oils provide the same number of calories, per gram, as animal fats, but replacing animal fats with vegetable oils usually decreases the dietary levels of two undesirable substances. What are they?
6. What do you call types of fatty acids that the body requires but cannot synthesize from other substances, so these fatty acids must come from diet?
7. An important group of fatty acids that our bodies can synthesize has chains that end in the structure $CH_3-CH_2-CH_2-CH_2-CH_2-CH_2-CH_2-CH_2-CH{=}CH-$. Using the omega system of naming fatty acids, what designation would you give to this group of acids?
8. What major class or category of foods contains no cholesterol whatever?
9. What two changes in the molecular structure of a triglyceride occur as a result of partial hydrogenation?

10. Give the chemical names of two of the following:
   a. monosaccharides
   b. disaccharides
   c. polysaccharides

11. Classify each of the following as a monosaccharide, a disaccharide, or a polysaccharide:
   a. glycogen
   b. glucose
   c. lactose
   d. sucrose
   e. cellulose
   f. galactose
   g. cellobiose
   h. fructose

12. a. What process is depicted in this figure?
   b. What class of compounds facilitates this reaction within the human body?

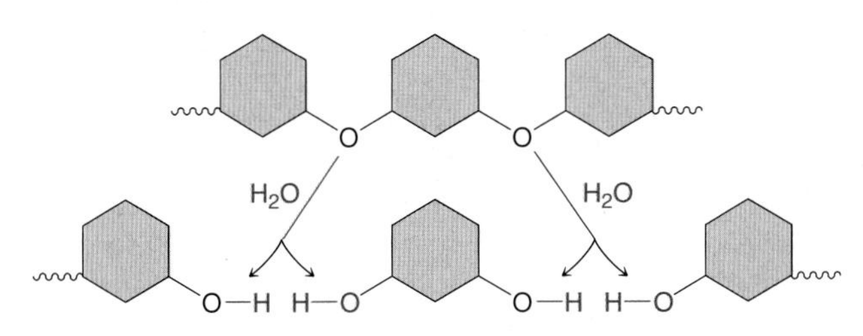

13. What monosaccharide(s) do you get from the hydrolysis of:
   a. sucrose
   b. lactose
   c. cellulose
   d. starch

14. Which of the following terms can be applied to Figure a? To Figure b?
   i. cellulose ii. starch
   iii. α-glucose units iv. β-glucose units

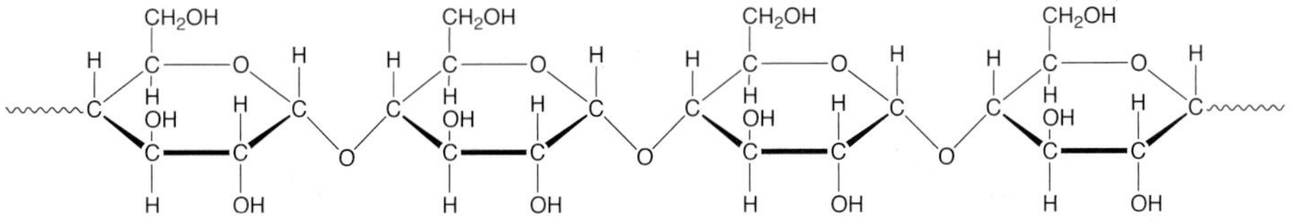

15. Of what value is glycogen to us?

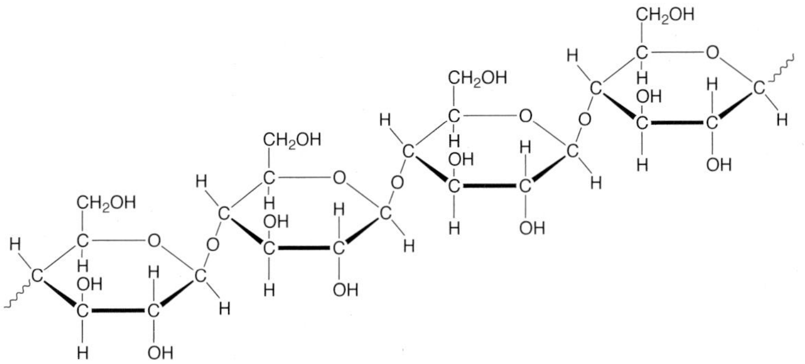

16. Fats and oils are chemical combinations of glycerol and long-chain fatty acids. Polysaccharides consist of chemically linked chains of monosaccharides. What chemical units form the chains of proteins?

17. a. Refer to Table 5.1 to determine which two compounds combine in the reaction shown here.
   b. What is the name of the link formed between the two units?
   c. Do reactions of this type form the primary (1°), secondary (2°), tertiary (3°), or quaternary (4°) structures of proteins?

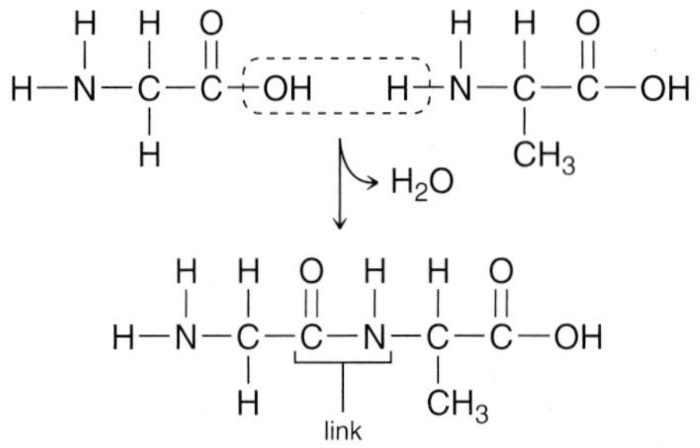

18. What is the difference between an essential amino acid and a nonessential amino acid? Name two amino acids in each category.

19. In what way are cysteine and methionine different from the other amino acids listed in Table 5.1?

20. Which, if any, of the following are essential amino acids for adults? (Refer to Table 5.1.)
   a. tryptophan
   b. valine
   c. isoleucine
   d. serine
   e. leucine

21. Which amino acids are essential for children but not for adults?

22. Name three foods that provide large amounts of high-quality protein.

23. Why do animal products generally provide better-quality protein than plant products?

24. What's the difference between a globular protein and a fibrous protein?

25. a. Name the structures that carry cholesterol through the blood.

b. Name the form of starch stored in the muscles for quick bursts of energy.

26. Classify each of the following structures as a(n): amino acid, carbohydrate, dipeptide, fatty acid, or triglyceride.

a.

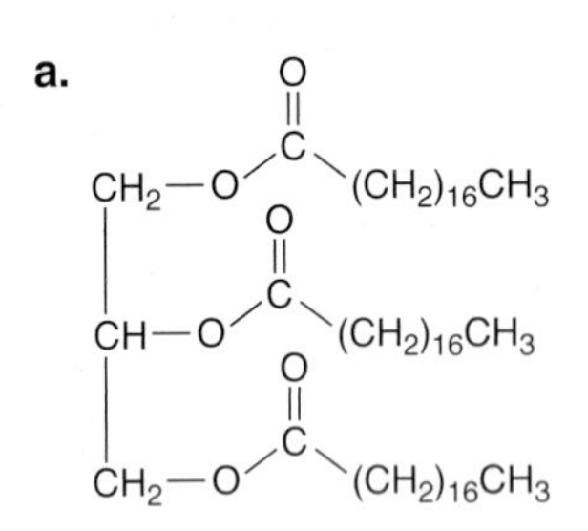

b. $NH_2—CH(CH_3)—C(=O)—NH—CH(CH_3)—C(=O)—OH$

c.

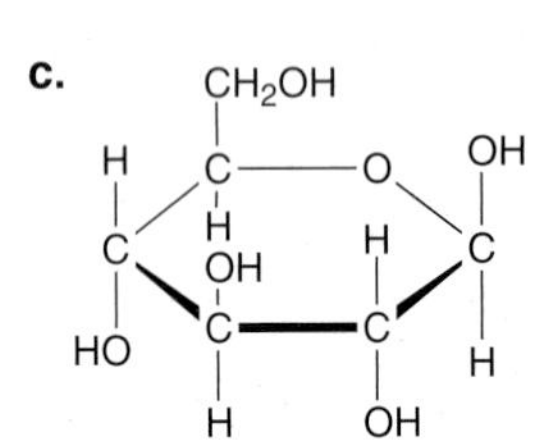

d. $H_2N—CH(CH_3)—C(=O)—OH$

e. $HO—C(=O)—(CH_2)_{16}CH_3$

Think

27. How does an increase in the number of carbon-carbon double bonds in a fatty acid, with no change in its carbon content, affect its melting point?

28. Typical ingredients in a processed peanut butter are shown here.

a. The prefix *mono-* means one, *di-* mean two, and *tri-* means three. Mono- and diglycerides are ingredients commonly added to foods such as peanut butter to help prevent separation of oils and to give the product a uniform consistency. Given that all fats and oils are triglycerides, predict what the structures of mono- and diglycerides would look like.

b. What type of bond is present in partially but not fully hydrogenated vegetable oil?

Stratol/iStockphoto

Ingredients
MADE FROM ROASTED PEANUTS AND SUGAR, CONTAINS 2 PERCENT OR LESS OF: MOLASSES, FULLY HYDROGENATED VEGETABLE OILS (RAPESEED AND SOYBEAN), **MONO- AND DIGLYCERIDES** AND SALT.

29. What effect, if any, does catalytic hydrogenation have on iodine numbers?

30. Why might a baby who is born lactase-deficient not be able to obtain sufficient nourishment through nursing alone, without supplemental nutrition?

31. Certain foods, including legumes (beans, lentils, soy, etc.) are known to cause gas because the body lacks enzymes that help hydrolyze certain carbohydrates, such as *raffinose* and *stachyose*, that are present in these foods. These undigested carbohydrates pass through the digestive system to the lower intestine, where microorganisms help degrade them. In doing so, they produce gaseous compounds, such as hydrogen, carbon dioxide, and methane. The active ingredient in *Beano*, a dietary additive intended to reduce gas, is *alpha-galactosidase*.

a. Based on the suffix of this name, what type of compound is this?

b. Also based on this name, predict what monosaccharide is liberated when *raffinose* and *stachyose* are broken down.

32. Adding concentrated sulfuric acid, a very strong acid and a very powerful dehydrating agent, to powdered table sugar effectively and dramatically dehydrates the sucrose, forming water and a black, brittle substance resembling badly charred wood. The reaction is vigorous and evolves considerable heat—enough heat, in fact, to convert the newly removed water into steam. What is the brittle, black solid formed from the sucrose by this dehydration?

33. High-fructose corn syrup (HFCS) is a commonly used sweetener in processed foods and soft drinks. Look online to find out how HFCS is manufactured. What role do enzymes serve in making HFCS? Would you consider this a natural product since it is derived from corn?

34. What happens when a protein is denatured?

35. After a fried egg cools, why don't the proteins of the white return to their original clear, colorless, gelatinous form?

36. Most of the proteins that occur in blood serum (the fluid portion of the blood) are globular proteins. Why?

37. Peanuts provide about twice the amount of protein, gram for gram, as eggs. Yet, except for their cholesterol content, eggs are a better source of dietary protein than peanuts. Why?

38. a. What chemical element occurs in all proteins but not in carbohydrates, fats, or oils?

b. What is another chemical element that is absent from carbohydrates, fats, and oils, but occurs in some, although not all, proteins? (Refer to Table 5.1.)

39. Keratin and collagen are important structural components of hair and skin, respectively. Would you expect these to be fibrous or globular proteins?

Calculate

40. a. Use the following table to determine which has a greater number of Calories.

b. Calculate the percentage of Calories from fat for each.

	Fat (g)	Carbohydrates (g)	Protein (g)
24-fluid oz. sweetened coffee drink with whipped cream	4	70	5
Blueberry muffin	11	58	7

41. Use the total fat, total cholesterol, and protein values in the Nutrition Facts label to estimate the number of Calories per serving for this soup.

Chicken Noodle Soup

Nutrition Facts

Serving Size 1/2 cup (120 ml) condensed soup
Servings Per Container about 2.5

Amount Per Serving	% Daily Value
Total Fat 1.5g	2%
Saturated Fat 0.5g	3%
Trans Fat 0g	
Cholesterol 15mg	
Sodium 890mg	37%
Total Carbohydrate 8g	3%
Dietary Fiber 1g	4%
Sugars 1g	
Protein 3g	

42. A person in normal health and weighing 165 pounds (75 kg) fasts for 36 hours while lying still in bed. What is the total number of Calories lost during the last 24 hours of the fast?

43. a. Based on the structures of stearic acid and oleic acid in Figure 5.7, determine the molecular formula of each.

b. How does the number of hydrogen atoms in a fatty acid change as the molecule becomes less unsaturated (more highly saturated)?

44. Which has a greater percentage of unsaturated fats: peanut oil, soybean oil, olive oil, or canola oil? (Refer to Figure 5.8.)

45. Many food labels now proclaim that they are fat-free to a specified percentage. If a label claims that a processed lunch meat is "95% fat-free," what percentage of the calories of that lunch meat comes from fat? (Assume that if the meat is 95% fat-free, the macronutrient content of the meat is 5% fat by mass and the rest of it is a combination of carbohydrate and protein.)

46. The carbohydrate sucrose (table sugar) has the molecular formula $C_{12}H_{22}O_{11}$. Rewrite this molecular formula as it would appear for a true hydrate of the same carbon content.

6 Physical and Chemical Changes

Have you ever wondered how the kinds of dramatic stage effects shown here are created, such as low-lying fog and flashes of pyrotechnics?

Stage fog is typically produced by machines that heat up a solution of water and glycerol or propylene glycol (see inset) to generate a vapor or a gas. As the vapor leaves the machine, it condenses in the cooler external air to produce a fog—a dispersion of tiny liquid droplets suspended in the air. You can also make a thick, low-lying fog by immersing dry ice, which is solid carbon dioxide, in warm water.

Fireworks displays are initiated by chemical reactions between *oxidizing* and *reducing agents*, which are substances that undergo transfers of electrons. The rapidly

expanding gases generated by these reactions create a loud bang, and the heat released causes metal ions or pure metal powders within the fireworks to glow, emitting specific colors of light.

The compounds involved in making fog undergo physical changes—changes in form but not chemical composition. The combustion of fireworks, on the other hand, involves chemical changes as new compounds are formed. In this chapter we'll explore how matter undergoes both physical and chemical changes and examine the enormous importance of these processes to society and to our everyday lives.

Propylene glycol, $C_3H_8O_2$, a common component of stage fog.

Steven J. Messina

CHAPTER OUTLINE

6.1 States of Matter and Their Changes

LEARNING OBJECTIVES

1. **Describe** what happens at the molecular level when a substance changes from one state to another.
2. **Compare** densities of common substances and explain how density can change with temperature.

Water in its three physical states—solid, liquid, and gas—is a familiar part of our everyday lives. For example, on a summer's day you might use ice (solid water) to cool a drink (made from liquid water) or feel the humidity (water vapor) in the air. In this section, we'll see how the same molecular substance, H_2O, can take on such different forms. We also see how the density and other properties of a substance can change as the material undergoes a change of physical state.

States of Matter

Anything we contact in our daily lives is a solid, liquid, or gas, depending on the way it retains its shape and volume. Solids, such as ice cubes, occupy distinct volumes of space and have well-defined shapes. Liquids, such as water, also occupy fixed volumes, but have no shape of their own. Except for small droplets that tend to become spheres, liquids always take the shape of whatever container they're held in. Gases, such as water vapor, do not retain their own shapes or volumes, but instead expand into their surroundings (**Figure 6.1**). Solids, liquids, and gases exhibit these different characteristics largely as a result of the way their chemical particles are arranged or move about in space (see *What a Chemist Sees*).

Whether an element or a compound exists as a solid, a liquid, or a gas depends principally on its temperature. We can melt ice and boil water through heating. Since heat is a form of energy, heating anything increases the kinetic energy of its chemical particles. A molecule, for instance, can possess kinetic energy in any of three ways:

- By **rotating** about its center of mass, like a spinning figure skater.
- By **vibrating**, like an oscillating spring.
- By **moving** about from one location to another, with what we call **translational** energy.

In what follows, we'll be concerned largely with translational energy.

Solids keep their own shapes because the forces of attraction that hold the particles close to one another and relatively static are much larger than the energies of the individual particles. In ice crystals, for instance, the attractive forces between H_2O molecules—their **intermolecular forces** of attraction—are larger than the energies of the individual molecules. As a result, these molecules remain firmly fixed next to one another in the crystalline lattice, unable to move about within the bulk of the material.

intermolecular forces Attractive forces that exist between molecules in close proximity.

When you heat a solid, though, its chemical particles absorb energy. When the temperature of the solid reaches

Distinguishing states of matter • Figure 6.1

All common objects are either a solid, a liquid, or a gas. Some substances, like water, can be in any state, depending on factors such as temperature.

a. Solid
- fixed volume
- retains own shape

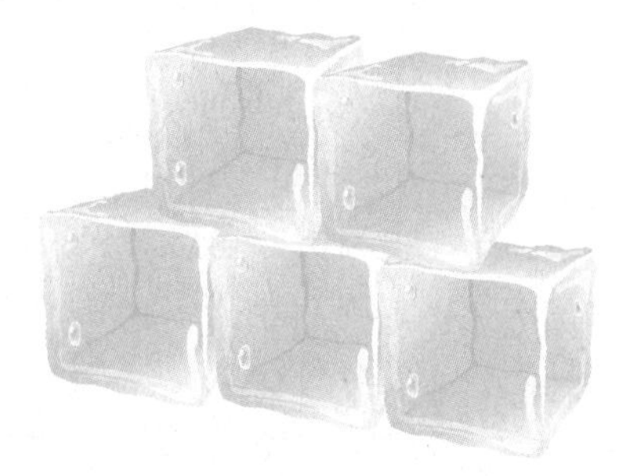

b. Liquid
- fixed volume
- adopts shape of container

c. Gas
- adopts volume and shape of container

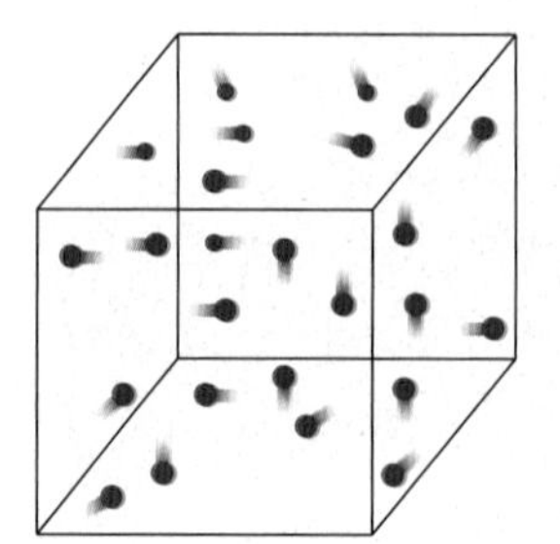

Ask Yourself

Which of the three common states of matter—solids, liquids, or gases—**(a)** maintain their own shapes, no matter what container holds them? **(b)** maintain their own volumes, no matter what container holds them?

WHAT A CHEMIST SEES

States of Matter at the Molecular Level

Water is found in its three physical states in this image of geothermal activity in Yellowstone National Park. In each state of matter, water molecules differ in their arrangements and movements.

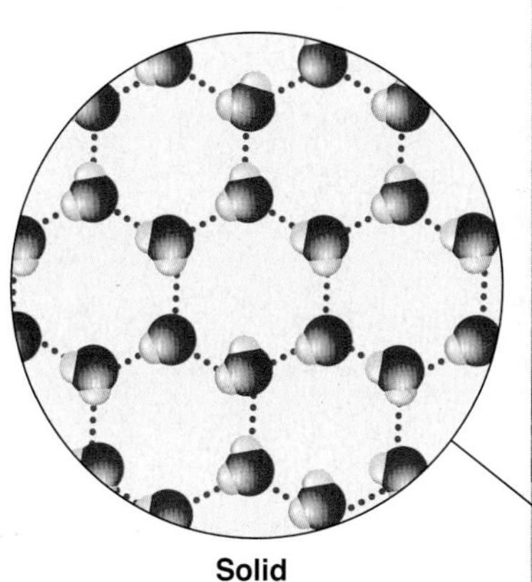

Solid
In ice and snow, water molecules are packed together in a static, organized array.

Media Bakery

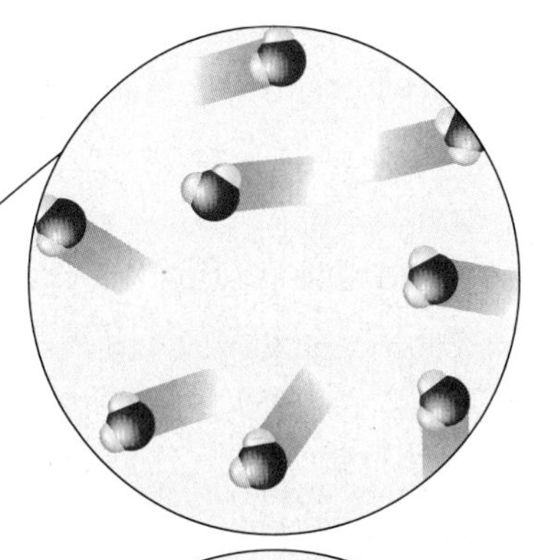

Gas
In water vapor, water molecules move about at high speeds and at relatively large distances from one another. Water vapor is invisible, but as it cools in the air the vapor condenses into tiny water droplets that we observe as clouds, fog, or steam.

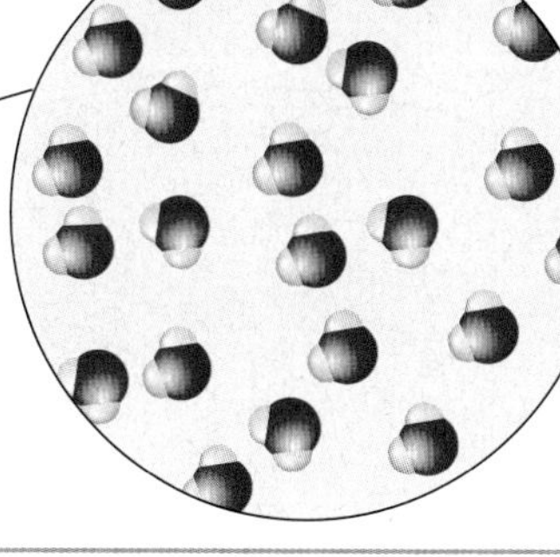

Liquid
In liquid water (whether moving or still), water molecules move rapidly about within close proximity to one another.

melting point The temperature at which a solid is transformed into a liquid.

its **melting point**, the movements of its particles become sufficiently vigorous to tear them away from their neighbors and out of their fixed positions. As they begin to move about freely within the bulk of the material, we observe that the solid melts to a liquid (**Figure 6.2**). Melting and other changes in state are examples of a **physical change**.

physical change A transformation of matter that occurs without any change in chemical composition.

What happens when solids melt • Figure 6.2

When we heat a solid to its melting point, individual particles, such as molecules, gain enough energy to break out of their fixed positions. This is what happens when ice melts to form water.

Kenneth Libbrecht

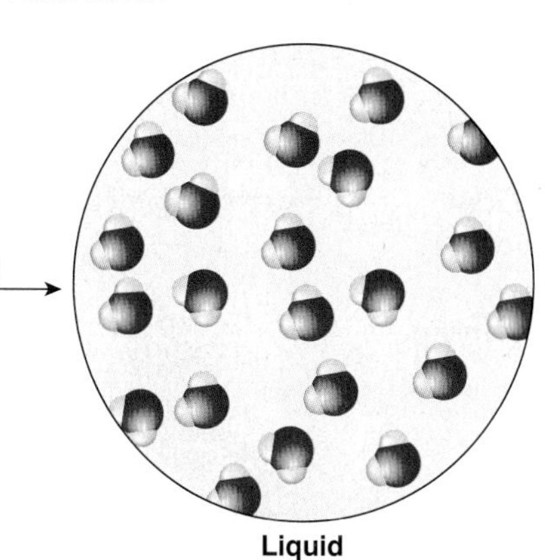

Solid
In the solid state, water molecules are held in fixed positions by intermolecular forces (shown as dotted lines). The hexagonal arrangement of water molecules in the crystalline state gives rise to the hexagonal shapes of ice crystals and snowflakes.

Liquid
In the liquid state, water molecules still experience intermolecular forces, but have enough energy to move freely about one another.

In the liquid state, although the particles are now moving about freely, the intermolecular forces are still strong enough to keep the particles trapped, for the most part, in the form of a liquid. We say "for the most part" because a portion of these particles have high enough translational energies to overcome these attractive forces and escape into the vapor state. We call this process **evaporation** (**Figure 6.3**).

When the temperature of the entire liquid reaches the **boiling point**, all the chemical species

evaporation A process by which a portion of a liquid at its surface turns into gas.

boiling point The temperature at which a liquid is transformed into a gas.

How liquids evaporate • Figure 6.3

Chemical particles with enough translational energy can escape from the liquid state into the gaseous state.

a. Liquids evaporate as they absorb heat from their surroundings, making evaporation an **endothermic** process. Liquids evaporate more quickly at warmer temperatures because a greater portion of the particles gain enough translational energy to escape from the surface than at colder temperatures.

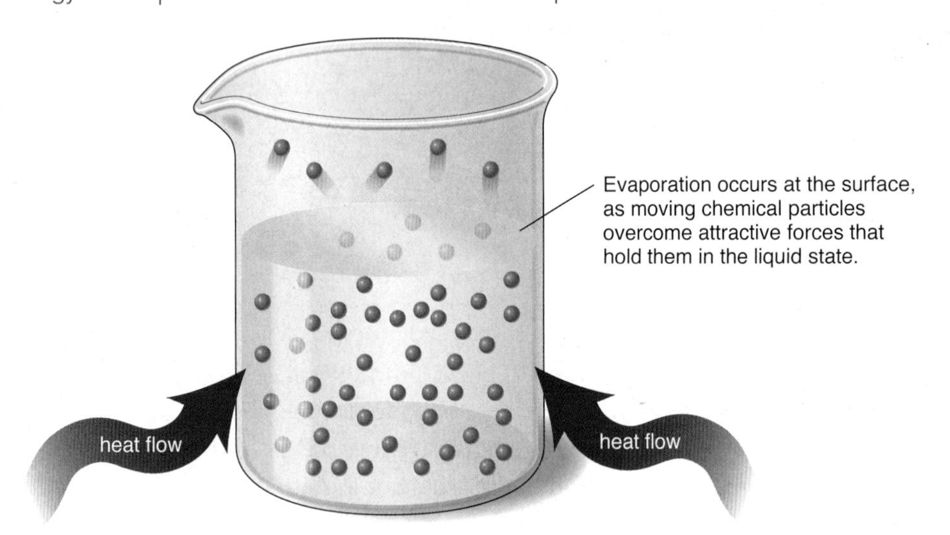
Evaporation occurs at the surface, as moving chemical particles overcome attractive forces that hold them in the liquid state.
heat flow
heat flow

Think Critically When you apply cologne or perfume to your skin, why do you detect the scent so rapidly?

b. Sweating cools the body. When you sweat, the perspiration on your skin absorbs excess heat from your body as the perspiration evaporates. This is an example of **evaporative cooling**.

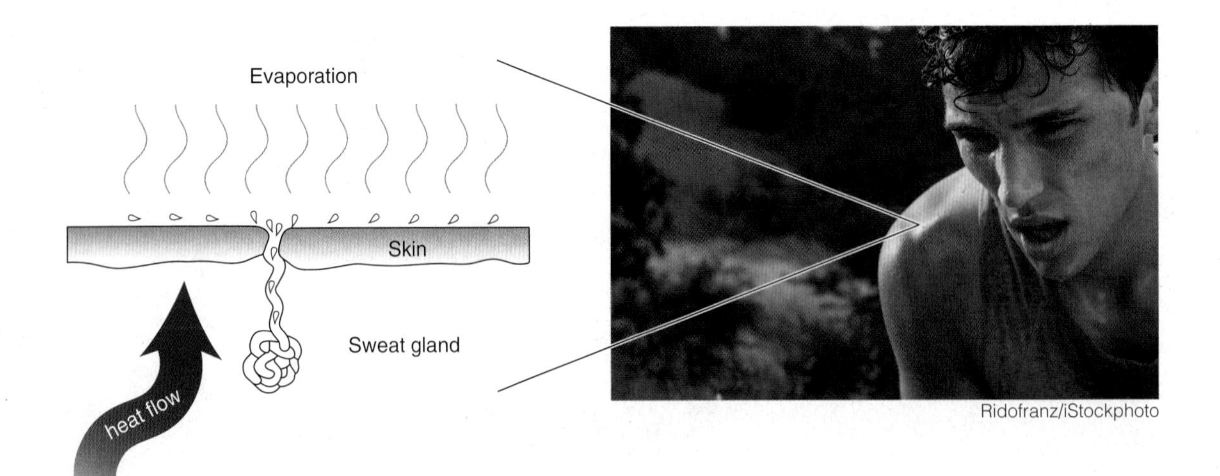
Evaporation
Skin
Sweat gland
heat flow

Ridofranz/iStockphoto

acquire translational energies high enough to escape the pull of their neighbors and enter into the vapor phase. As this happens throughout the entire bulk of the liquid, we see it boil. For instance, as water boils, the bubbles seen escaping are water vapor. Chemical particles in the vapor or gaseous state have translational energies far higher than the forces that would bind them to their neighbors, so they move about freely, virtually independent of one another.

A substance with greater attractive forces among its chemical particles tends to melt and boil at higher temperatures than a substance with relatively weaker intermolecular forces. For instance, water (H_2O) boils at a much higher temperature than propane (C_3H_8)—100°C for H_2O as compared to −42°C for C_3H_8—because the intermolecular forces among H_2O molecules are much stronger than those among C_3H_8 molecules. Water is a **polar** compound influenced by attractive intermolecular forces called **hydrogen bonds** (**Figure 6.4**).

polar Having a separation of charge due to differences of electronegativities of component atoms or ions.

hydrogen bond An attractive force between an oxygen, nitrogen, or fluorine atom and a nearby hydrogen atom bonded to an oxygen, nitrogen, or fluorine atom.

Hydrogen bonding in water • Figure 6.4

The water molecule contains polar O— H bonds. As a result, water molecules can experience intermolecular attractions called hydrogen bonds.

a. The polarity of water. The angular shape of the water molecule and the difference in electronegativity between oxygen and hydrogen make water a polar compound. Since oxygen has a greater affinity for bonding electrons than hydrogen, this makes...

...the oxygen atom electron rich (as indicated by a partial negative charge, δ^-).

....the hydrogen atoms electron poor (as indicated by a partial positive charge, δ^+).

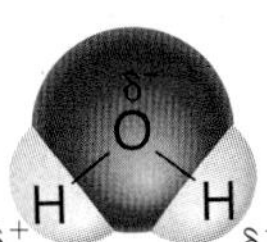

b. Hydrogen bonding in water. The polarity of water causes H_2O molecules in both the liquid and solid states to experience attractive forces called **hydrogen bonds**. These attractions can occur between the electron-rich oxygen atom within one H_2O molecule and the electron-poor hydrogen atoms of nearby H_2O molecules.

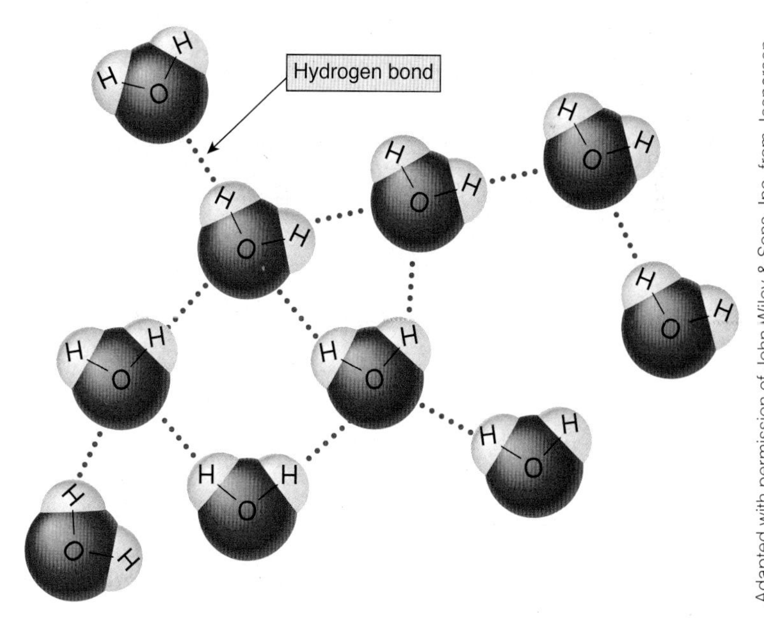

Think Critically Ethylene glycol, a component of automobile antifreeze, boils at 197°C. Glycerine, used in drug and cosmetic formulations, boils at 290°C. How does hydrogen bonding help account for the much higher boiling temperature of glycerine?

```
   OH OH              OH OH OH
   |  |               |  |  |
H—C—C—H            H—C—C—C—H
   |  |               |  |  |
   H  H               H  H  H
Ethylene glycol       Glycerine
```

Dispersion forces • Figure 6.5

Nonpolar compounds such as propane can experience weak intermolecular attractions called dispersion forces. These dispersion forces are generally much weaker than hydrogen bonds.

a. Propane (C_3H_8), a nonpolar compound. Carbon and hydrogen atoms have nearly the same affinities forbonding electrons. As a result, hydrocarbons, such as propane, are nonpolar. This means that *on average*, valence electrons are evenly distributed within these molecules.

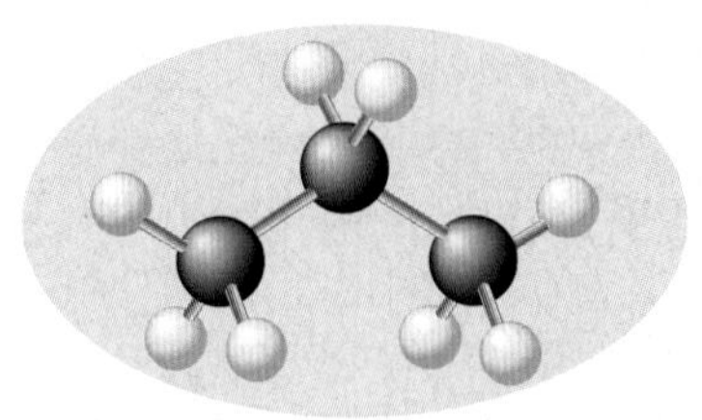

The propane molecule lacks permanent regions of positive and negative electrical charge.

b. Dispersion forces in propane. Because electrons within a molecule are in constant motion, at any given moment one end of a nonpolar compound may experience a slightly higher concentration of electrons than the other end. We can visualize this as an electron cloud...

... in which one end of the molecule experiences a transient partial negative charge...

... and the other end experiences a transient partial positive charge.

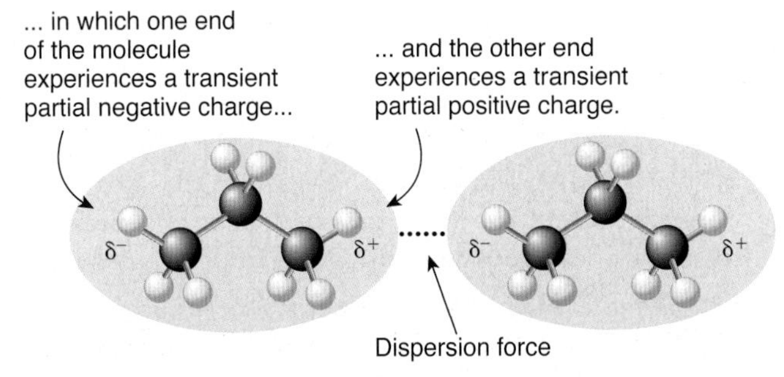

This polarization or charge separation within one molecule can induce a transient polarization of the electron cloud within a nearby molecule. The result is a momentary attraction between the two molecules called a **dispersion force**.

Dispersion forces tend to increase with the number of atoms in a molecule. For instance, octane molecules (C_8H_{18}) experience greater dispersion forces than smaller hydrocarbons, such as pentane (C_5H_{12}). As a result, the melting (and boiling) temperatures of hydrocarbons generally increase with molecular size. For example, pentane boils at 36°C and octane boils at 125°C.

Think Critically

1. Which compound would you expect to boil at a higher temperature, ethane (CH_3CH_3) or ethanol (CH_3CH_2OH)? Explain your reasoning.
2. Both mineral oil and petroleum jelly (such as Vaseline®) are composed of mixtures of hydrocarbons. But mineral oil exists as a liquid at room temperature and petroleum jelly is a semisolid. What does this suggest about the average size of the hydrocarbon molecules in mineral oil as compared to those of petroleum jelly?

nonpolar Lacking an overall separation of charge within a molecule.

dispersion force A weak attractive force between nonpolar molecules in close proximity.

Propane is a **nonpolar** compound influenced by weaker intermolecular attractions called **dispersion forces** (**Figure 6.5**).

Boiling points are particularly sensitive to changes in the external pressure, so we usually report them as temperatures at normal atmospheric pressure. **Table 6.1** lists the melting points and boiling points of some common substances.

The physical changes we've just described are reversible processes. **Freezing**, for example, is the reverse of **melting**; and **condensation** is the reverse of **evaporation**. Some substances do not exist as liquids under normal conditions but instead pass directly from the solid to a gaseous state. This transition is known as **sublimation** (**Figure 6.6**).

sublimation Changing directly from a solid to a gas, without passing through a liquid state.

Melting and boiling points of common substances Table 6.1

A substance with greater attractive forces between its chemical particles tends to melt (and boil) at a higher temperature than a substance with relatively weaker forces. For instance, sodium chloride (NaCl) melts at a very high temperature because of very strong attractive forces between Na^+ and Cl^- ions. Water melts at a much lower temperature because the attractive forces between H_2O molecules are relatively weak by comparison.

Substance	Common Source or Use	Melting Point (°C)	Boiling Point (°C)
Oxygen	Atmosphere	−218	−183
Nitrogen	Atmosphere	−210	−196
Propane	Gas grills	−190	−42
Ammonia	Household cleaner	−78	−33
Ethanol	Alcoholic drinks	−117	78
Acetone	Nail polish remover	−94	56
Water	Water	0	100
Acetic acid	Vinegar	17	118
Sucrose	Table sugar	185	(decomposes)
Sodium chloride	Table salt	801	1413
Gold	Jewelry, coins	1064	3080

Interpret the Data

1. At room temperature (about 22°C), which of the substances in the table are: (a) gases, (b) liquids, (c) solids?
2. From the data in the table, what can you infer about the strength of the intermolecular forces in liquid propane as compared to those in liquid water?

KNOW BEFORE YOU GO

1. Why does a cold drink often accumulate beads of condensation on the outside of its container?

In Figure 6.4, we saw that partial electrical charges on the (slightly negative) oxygen of one water molecule and a (slightly positive) hydrogen of a neighboring water molecule produce attractive forces between adjacent water molecules. Because of intermolecular forces such as these, water molecules not only attract one another but can associate with a variety of other kinds of nearby molecules as well (see *Did You Know?* on the next page).

Density

When a substance converts from one physical state—solid, liquid, or vapor—to another, not only does its appearance change, but so do many of its other properties, such as its **density**. Generally (but with some important

density A measure of how much mass is in a given volume of a substance.

Sublimation of dry ice • Figure 6.6

At normal atmospheric pressure, solid carbon dioxide, also known as *dry ice*, sublimes at temperatures at or above −78°C (−109°F) to produce gaseous carbon dioxide.

This fog effect can be created by placing dry ice (solid CO_2) in water. The CO_2 gas given off is invisible but is so cold that it condenses the water vapor in the surrounding air, thus creating a fog.

© RASimon/Istockphoto

DID YOU **KNOW?**

Smarter clothes depend on intermolecular forces

Clothes designed for exercise and fitness activities are commonly made from synthetic fabrics designed to wick away moisture. In contrast, many athletes find that cotton, while highly comfortable for everyday wear, is less desirable for vigorous exercise because of its high capacity for retaining moisture. The key to this different behavior—what chemists refer to as **hydrophobic** (water-fearing) versus **hydrophilic** (water-loving) properties—lies in the nature of the molecules that make up each of these materials.

A common synthetic fabric used in workout clothing is polyester, a polymer molecule made up of long chains of repeating units (monomers) of the molecule shown in **Figure a**. Cotton is made up of a natural polymer called cellulose. The monomer in cellulose, **Figure b**, is based on glucose, a common carbohydrate.

The hydrophilic nature of cotton is due to the presence of O—H units, called **hydroxyl** groups (**Figures b** and **c**), which have the same polar, covalent bonds present in the water molecule. Hydroxyl groups in cellulose are capable of **hydrogen bonding** to water molecules. A common type of hydrogen bond (**Figure d**) is an attractive force between the hydrogen of a hydroxyl group on one molecule and the oxygen of a hydroxyl group on another molecule. A hydrogen bond is far weaker than a conventional covalent bond. Nonetheless, hydrogen bonds can be quite important. Cotton's ability to retain water through hydrogen bonding makes it a preferred fabric for towels, for example. Polyester, on the hand, is hydrophobic; it does not contain hydroxyl groups and has mostly nonpolar covalent bonds, which repel water.

Moof/Cultura/Getty Images

....to another ... to another....

b. The monomer that makes up the cellulose in cotton.

δ^- δ^+ O—H

c. The hydroxyl group, O—H, has a highly polar covalent bond.

....to another ... to another....

a. The monomer that makes up a common type of polyester used in workout clothes.

Hydrogen bond

Water molecule

Hydroxyl group from cellulose

d. In one type of hydrogen bond, electron-poor (δ^+) hydrogen atoms of hydroxyl groups on cellulose experience an attraction to electron-rich (δ^-)oxygen atoms of water molecules.

To enhance the wicking effect, the synthetic fabric used in some types of fitness clothing is treated with a fine hydrophilic coating that helps draw moisture away from the skin to speed up evaporation.

Think Critically Is there another way to draw a hydrogen bond between a water molecule and a hydroxyl group on cellulose? (*Hint:* Can the oxygen atom in the hydroxyl group on cellulose hydrogen bond to water?)

Density and states of matter • Figure 6.7

In the solid or liquid state, chemical particles are relatively close to one another, so density is high. In the gaseous state, they are much farther apart and density is much lower.

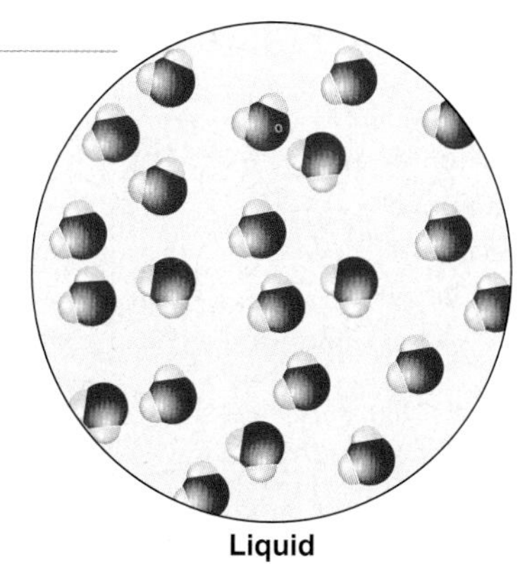

Liquid

Higher density. Many chemical particles per unit volume.

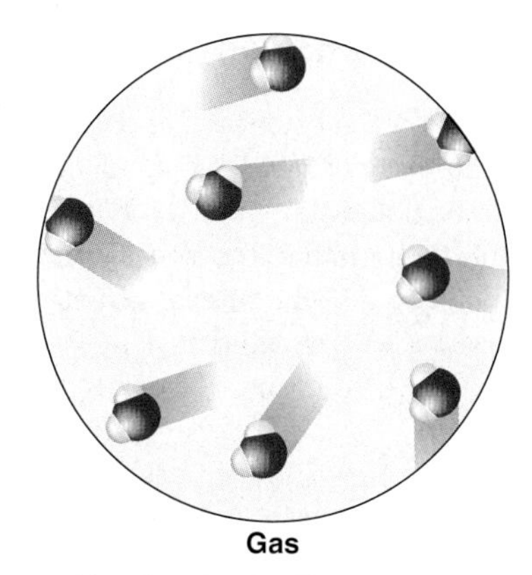

Gas

Very low density. Few chemical particles per unit volume.

exceptions), a substance is most dense in its solid state and slightly less dense in its liquid state. The gaseous state of a substance is far less dense than either its liquid or solid states (**Figure 6.7**).

Densities of solids and liquids are typically measured in units of grams per milliliter (g/mL) or grams per cubic centimeter (g/cm^3). Since a milliliter is the same as a cubic centimeter, these two measures of density are equivalent. (An easy way to determine the volume of an object is to immerse it in water and determine the volume of water it displaces. Naturally, if the object can be damaged by water, we must use some other technique).

It's important to recognize the significant difference between the mass of an object and its density. We sometimes think of lead, for example, as a heavy metal. But what we really mean is that it's dense. A single lead fishing sinker has a mass of only a few grams and is small enough to hold in your hand. By comparison, a huge tree certainly has much more mass than the sinker, but nonetheless the tree is far less dense. The key here is that one cubic centimeter of the sinker (lead) has a greater mass than the same volume—one cubic centimeter—of the tree (wood). **Table 6.2** lists the average densities of some typical solid substances.

Fats and oils float on water because they're intrinsically less dense than water. Boats float because the air trapped inside their hulls gives them an average density less than that of water. Some brands of bar soap (Ivory, for example) float because of an enormous number of microscopic bubbles of air that are whipped into them as they're manufactured, giving the bar of soap a smaller density than water.

We've already seen that as materials become warmer, their chemical particles possess greater kinetic energy. As particles move about with greater energy, the average distance between the particles increases. This causes the

Average densities of common solids Table 6.2

Water, which has a density of about 1.0 g/mL, is a convenient reference for density. Substances that are less dense than water—most woods, for example—float in water; substances that are more dense than water—like lead—sink in water.

Substance	Density (g/mL or g/cm^3)
Cork	0.2
Oak wood	0.8
Waxes	≈0.9
Ice	0.9
Aluminum	2.7
Diamond	3.5
Lead	11.4
Gold	19.3

These floating candles are less dense than water.
These sunken stones are more dense than water.

Tim O.Walker

A substance is typically most dense in its solid state and slightly less dense in its liquid state. Water is unusual in that the opposite is true—solid water (ice) is less dense than liquid water. Since ice is less dense than water, it floats.

Think Critically Consider the densities of ice and liquid water. What can you predict about the average distance between H_2O molecules in ice as compared to their distances in liquid water?

Thermal expansion of liquid water • Figure 6.8

Most pure substances are densest in their solid state. Water is unusual in that it reaches a maximum density in the liquid state (at 4°C). When warmed above this temperature, the density of water decreases, meaning a given mass of water occupies a larger volume when heated.

A given mass of water occupies 3% greater volume at 90°C as compared to 4°C.

110 mL
100 mL
4°C
Thermal expansion
110 mL
100 mL
90°C

material to expand slightly, a phenomenon called **thermal expansion**. Since the same mass of material occupies a greater volume, this means that the density of the material—its mass per unit of volume—typically decreases with temperature (**Figure 6.8**).

The thermal expansion of water has important implications for the world's oceans. In the last century alone, average sea levels rose about 20 cm (8 in.). As the Earth becomes warmer, this rate is anticipated to accelerate. Although the melting of land-based ice sheets certainly contributes to a rise in sea level, the major cause is thermal expansion of the warming oceans. In the United States, nearly 4 million people live within a few feet of high-tide levels, putting them at risk. In addition, populated coastal communities around the world, especially in Asia, are at particular risk from the projected impacts of sea-level rise.

Since water reaches a maximum density at 4°C, the density of water not only decreases above this temperature, but below this temperature as well, causing water to expand as it freezes. In winter, water trapped in pipes can freeze, causing pipes to burst because of the pressure created by the expanding mass of ice.

KNOW BEFORE YOU GO

2. The density of "950 platinum," an alloy of platinum used to make jewelry, is 20.1 g/cm^3. What is the volume of a solid chunk of this material if its mass is 75 g? Answer: Use unit cancellation as follows:

$$75\ \cancel{g} \times \frac{1\ \text{cm}^3}{20.1\ \cancel{g}} = \mathbf{3.7\ cm^3}$$

3. A solid block of aluminum has a volume of 10 cm^3, and a solid block of lead has a volume of 2 cm^3. Which has a greater mass: the block of aluminum or the block of lead? (See Table 6.2 for densities.)

CONCEPT CHECK

1. **How** does the kinetic energy of chemical particles change as a substance is warmed?
2. **Why** is a substance far less dense in its gaseous state than in its liquid or solid states?

6.2 Gas Behavior

LEARNING OBJECTIVES

1. **Identify** the gases in our atmosphere and the basis for atmospheric pressure.
2. **Describe** how the volume of a gas changes with changes in pressure and temperature.
3. **Explain** how the expansion of gases can be used for cooling purposes.

Most chemical reactions take place in substances that are in the gaseous or liquid state rather than in solids because atoms and molecules can move around, collide, and interact more readily in gases or liquids than they can in solids. We will examine some properties of gases in this section and study liquid solutions in the next chapter.

Composition of Earth's lower atmosphere • Figure 6.9

At sea level, nitrogen and oxygen, together, make up about 99% of the volume of dry air. Argon accounts for almost all of the remainder.

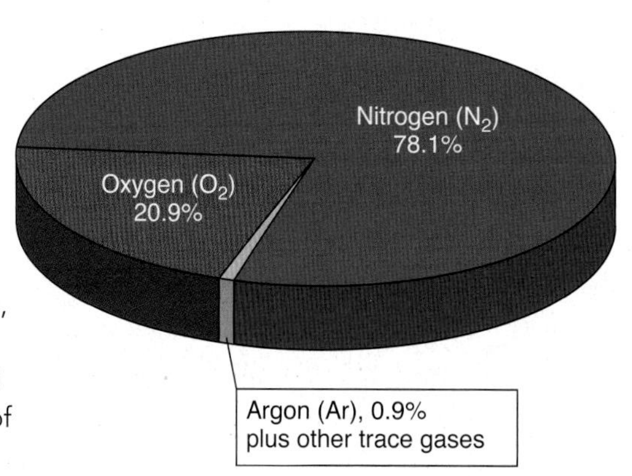

The trace remainder of gases, in order of decreasing proportions, consists of carbon dioxide (CO_2), neon (Ne), helium (He), methane (CH_4), and yet other gases. Humid or wet air also contains varying amounts of water vapor, which lowers the percentages of all other gases in the total combination. Polluted air carries with it noxious fumes, or suspensions of solids or liquids in the form of smoke or haze.

From pumping up a tire, to dispensing shaving cream from a pressurized can, to the very act of breathing, gases are a familiar and essential part of our everyday lives. In this section we'll explore the basis for how gases behave. In the process, we'll see how gas behavior underlies some important modern conveniences, such as air-conditioning and refrigeration.

The Gas We Live In

A good starting place for exploring the world of gases is simply the gas we live in—our atmosphere. This gas enters our lungs as a complex mixture of simple substances (**Figure 6.9**). The combined weight of all the gases above any particular point, either on Earth's surface or anywhere within the atmosphere itself, generates the atmospheric pressure at that spot. At sea level, the atmosphere exerts a pressure, on average, of 14.7 pounds per square inch (**Figure 6.10**).

Atmospheric pressure, also known as barometric pressure, is measured with a **barometer**, from the Greek words *baros*, for "pressure," and *metros*, for "measure." Some of the earliest barometers measured atmospheric pressure in terms of the height of a column of mercury

Atmospheric pressure • Figure 6.10

Our atmosphere consists of a vast ocean of gases covering the surface of Earth and reaching up to an altitude of over 60 miles. At sea level, we are situated at the bottom of this mass of gases. This atmosphere acts as a fluid, exerting pressure on objects within it, much as a body of water exerts pressure on submerged objects.

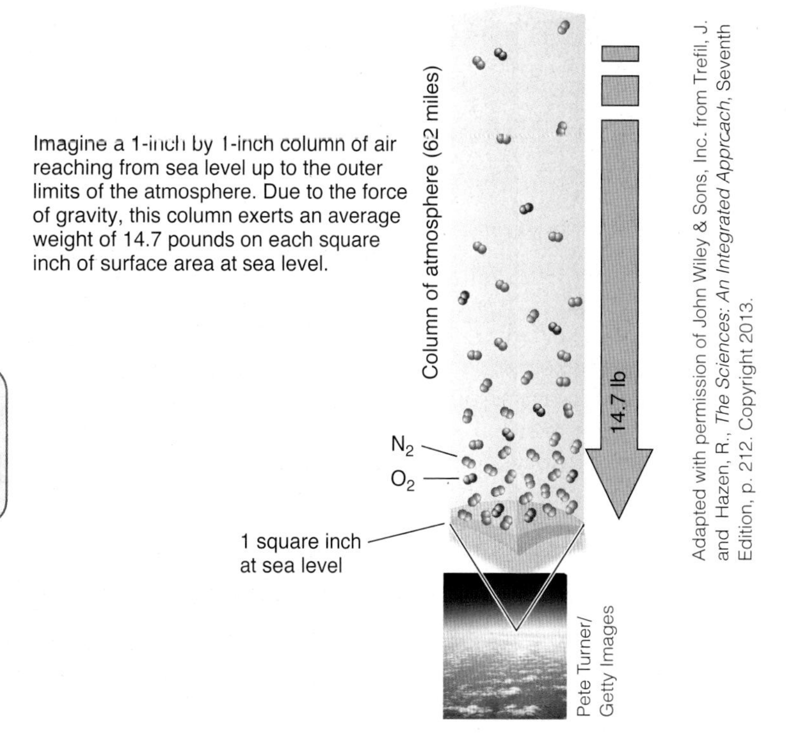

Interpret the Data

From the figure, what would you predict about the relative density of air at sea level as compared to its density at very high altitudes?

supported by the atmosphere itself, a practice still used today (**Figure 6.11**), along with yet newer methods.

Atmospheric pressure varies with both altitude and weather conditions. This pressure drops rapidly with altitude, decreasing to half the sea-level value at about 5500 m (18,000 ft). Changes in the density of the atmospheric gases above Earth's surface produce regions of high atmospheric pressure and low atmospheric pressure—the "highs" and "lows" of our weather forecasts. A rising atmospheric pressure usually forecasts a change to clear, sunny skies, whereas a decreasing atmospheric pressure indicates cloudy, rainy weather ahead. Some of the lowest barometric pressure readings ever recorded at sea level are associated with hurricanes and typhoons.

Gas Laws

Gases differ markedly from solids and liquids in that gases are readily compressed, whereas solids and liquids are not. You can feel this ease of compression when you manually inflate a bicycle tire, especially at the start of the process. A compressed gas can also pose safety hazards. Many aerosol cans carry a warning label that their contents are under pressure and the can must not be incinerated. The reason is that the pressure of a confined gas increases with temperature, and at a high enough pressure the can could actually explode. You can observe this same relationship between gas pressure and gas temperature if you check the pressure of a car's tires when they are cold,

Measuring pressure • Figure 6.11

The earliest type of barometer was developed by Evangelista Torricelli in 1643. Torricelli was the first to suggest that the atmosphere has mass and that it exerts a pressure on everything within it.

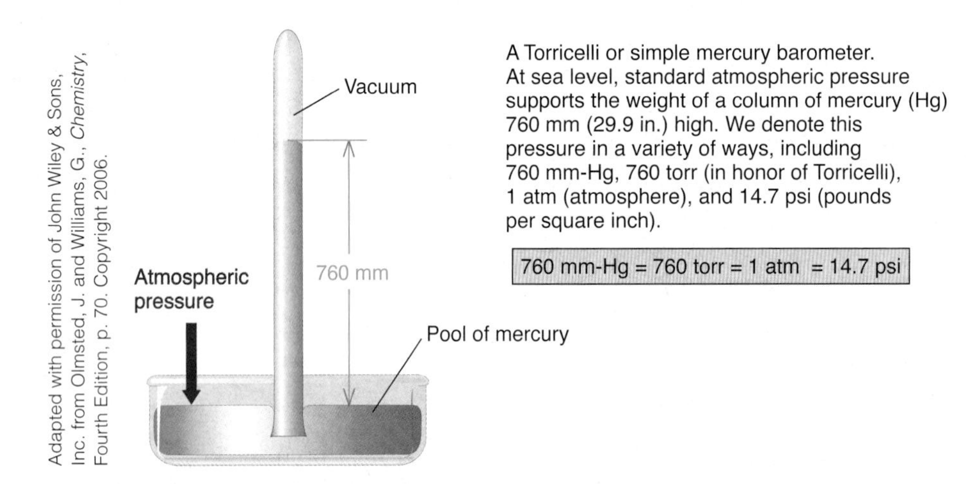

A pressure gauge differs from a barometer in that the pressure gauge measures the *difference* in pressure between the inside of a hollow object, such as a tire, and the surrounding air.

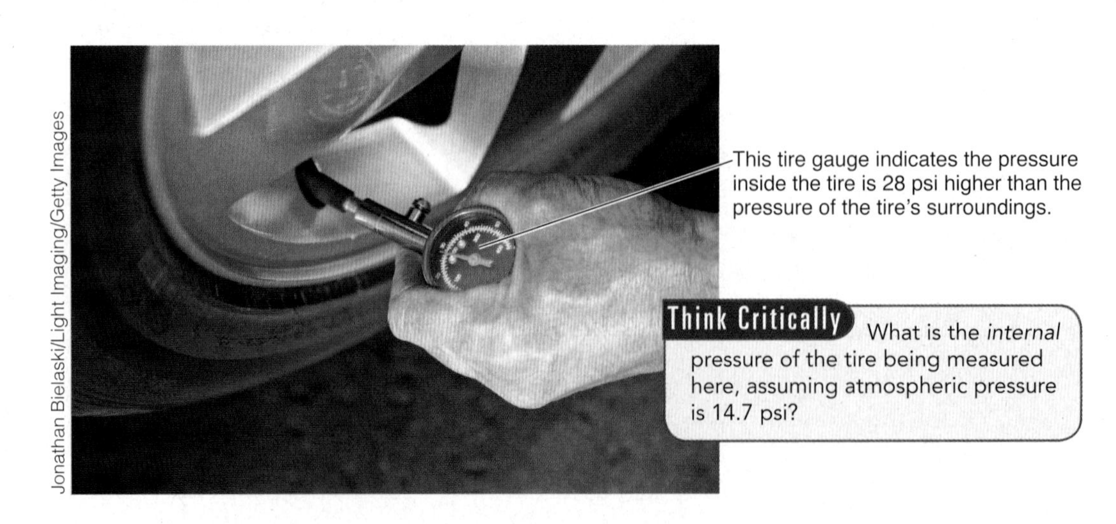

Jonathan Bielaski/Light Imaging/Getty Images
This tire gauge indicates the pressure inside the tire is 28 psi higher than the pressure of the tire's surroundings.
Think Critically What is the *internal* pressure of the tire being measured here, assuming atmospheric pressure is 14.7 psi?

Motions of gas particles • Figure 6.12

The word "gas" is derived from the Greek word *chaos*. As the temperature of a gas increases, the average kinetic energies and hence the velocities, of the gas particles increase.

Container wall

Gas particles are in constant, chaotic motion, colliding off one another and the walls of any container or surface.

Gas molecule =

Ask Yourself

Outside on a perfectly still day or inside a room without any drafts, are the molecules that make up the air moving about in constant motion?

before the car is driven, and then again when they grow warmer after the car has been driven for a few miles. Pressure is higher in the warmer tires.

We can account for these and other simple observations of the behavior of gases with a general theory called the **kinetic-molecular theory** of gases. According to this model, gases act—ideally, at least—as though their individual, component molecules or atoms are perfectly resilient, infinitesimal particles, taking up no space whatever. They behave like point-sized spheres that move about continuously, bouncing off each other and off the walls of their containers, losing no energy in the process and exerting absolutely no attraction for one another. Any gas made up of chemical particles that behave in exactly this way is called an **ideal gas**.

ideal gas A gas that behaves as if it were made up of molecules or atoms that have no volumes of their own and bounce off each other without losing any energy.

According to this view, the physical properties of any gas—or at least any gas that behaves as if it were an ideal gas—depend only on the kinetic energy (the energy of motion) of its molecules (or its individual atoms if it's a monatomic gas). This is why we call it the "kinetic-molecular" theory. What's more, the kinetic energy of the molecules depends entirely on the temperature of the gas. The higher the temperature, the greater the molecular energy of the gas. The pressure of the gas in an enclosed container results from repeated collisions of the particles of the gas with the walls of the container (**Figure 6.12**).

Atoms and molecules of real gases do occupy an extremely small but finite amount of space, and they do exert small but measurable attractions for one another. Otherwise, no gas would ever condense into a liquid; there would be, for example, no such substance as liquid water. Nevertheless, this idea of point-sized ping-pong balls careening about with an energy that depends entirely on their temperature does give us a very good description of the ideal behavior of a gas, particularly at high temperatures and low pressures. Under these conditions, all gases follow a pattern of relationships easily described by simple mathematical equations, known as the **ideal gas laws**. These laws come to us from the observations of early scientists who studied the ways gases respond to changes in pressure, temperature, and volume.

Gas pressure and volume: Boyle's Law Robert Boyle, an Irishman who studied many natural phenomena, was the first of the great gas law investigators. In 1657, Boyle learned of a newly devised air pump. With some ingenuity, he improved its original design, built a model, and began studying the effects of pressure on the volume occupied by a quantity of air. Soon he described what we now know as **Boyle's Law**: With the temperature and quantity of a gas sample held constant, the volume of the gas varies inversely with its pressure.

We can express Boyle's Law mathematically as

$$V = k \times \frac{1}{P}$$

Here V is volume, P is pressure, and k is a proportionality constant. Any change in pressure produces a change in volume that is the exact inverse of the pressure change. For example, if we triple the pressure, the volume becomes one-third its original size.

Kinetic-molecular theory helps explain Boyle's Law. The pressure the gas exerts on the walls of its container results from the force of its molecules colliding with the walls. The greater the number of collisions at any instant, the greater the pressure on the container. Squeeze the

Boyle's Law • Figure 6.13

For a sample of gas at constant temperature, as the pressure on the gas increases, the volume of the gas decreases proportionally. We can observe this behavior in this piston held at constant temperature (**Figure a**).

a. Piston showing the effect of pressure on gas volume. Plotting a large number of pressures and their corresponding volumes gives the curve shown in Figure b.

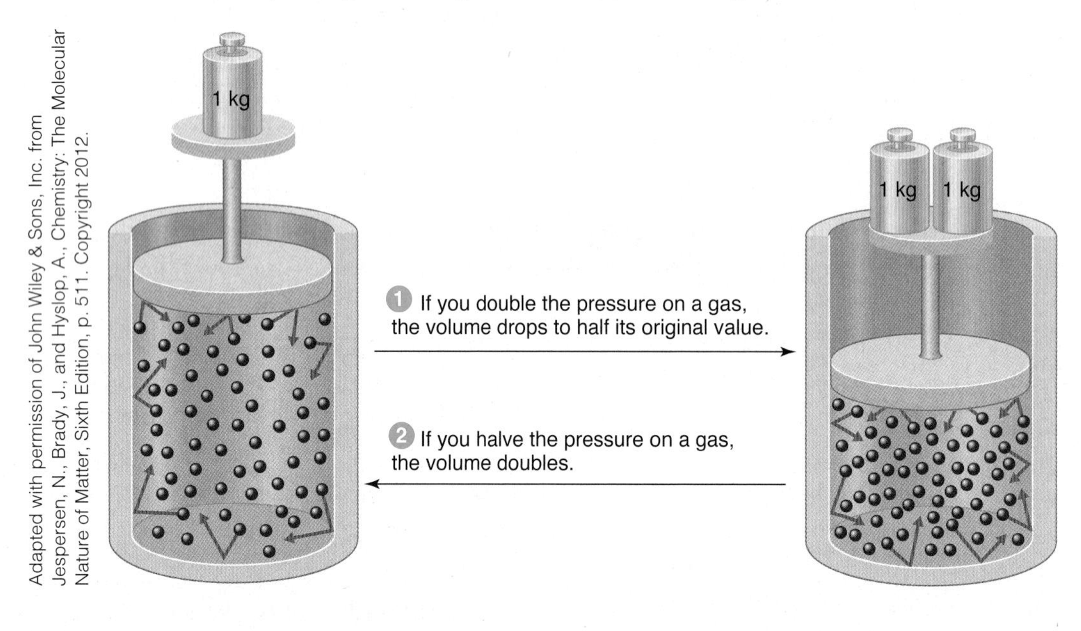

1 kg
1 kg
1 kg
1 If you double the pressure on a gas, the volume drops to half its original value.
2 If you halve the pressure on a gas, the volume doubles.

b. The volume of a gas varies inversely with its pressure.

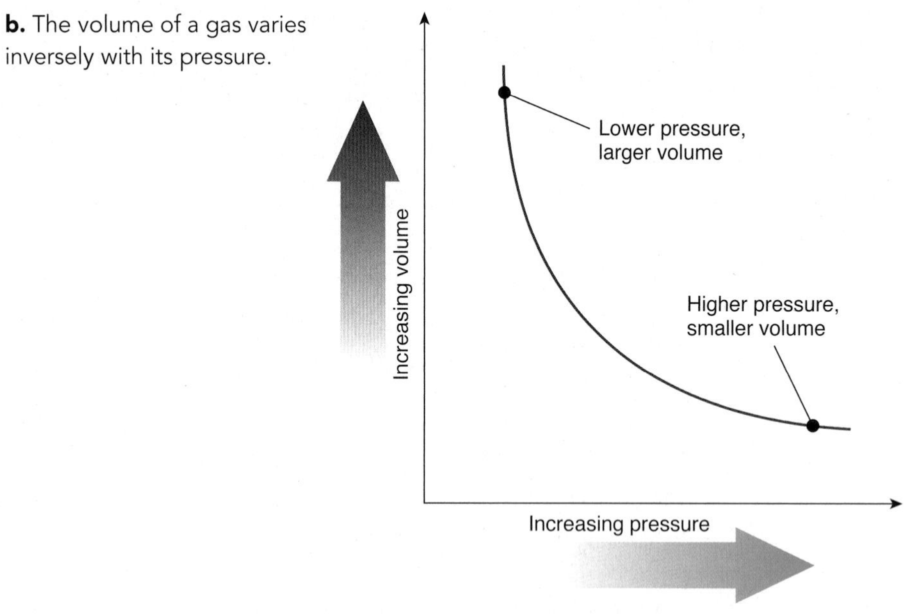

gas into half its original volume and the rate at which the molecules bounce off the walls doubles. So does the pressure (**Figure 6.13**).

Gas temperature and volume: Charles's Law Born nearly a century after Boyle first read of the newly devised air pump, Jacques Charles turned to science from a career in the French Ministry of Finance. In the 1780s, he became the first person to inflate a balloon with hydrogen gas (which is less dense than air) and the first to use a balloon to ascend some 3 km into the atmosphere. Charles discovered that a quantity of gas kept at a constant pressure expands as it warms and contracts as it cools (**Figure 6.14**).

Charles's Law • Figure 6.14

A sample of gas held at constant pressure contracts as it is cooled and expands as it is warmed.

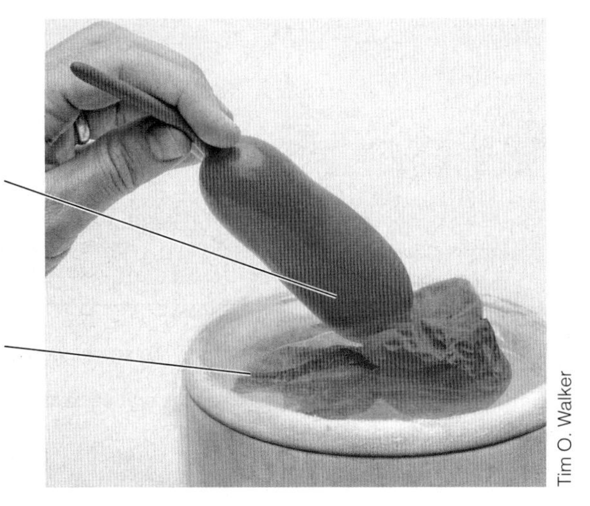

The volume of air in this balloon contracts as the balloon is submerged in the exceedingly cold bath of liquid nitrogen.
Insulated flask of liquid nitrogen (−196 °C)
Tim O. Walker

Ask Yourself
The balloon seen here was fully inflated at room temperature prior to being cooled in liquid nitrogen. What would you expect to observe as the balloon is removed from the cold bath?

With the pressure and the quantity of gas held constant, the volume of the gas varies directly with its temperature. Mathematically, this amounts to

$$V = k \times T$$

Here k represents a proportionality constant and T is the temperature.

The temperature scale used in this equation was devised by the British physicist William Thomson, who was given the title of Lord Kelvin, partly in recognition of his contributions to the design and implementation of the first trans-Atlantic telegraph cable. The **Kelvin or absolute temperature scale** begins 273 degrees below zero degrees Celsius, at **absolute zero**, and moves upward with units the same size as the degrees of the Celsius scale (**Figure 6.15**).

absolute zero The coldest possible temperature, corresponding to zero Kelvin or −273°C.

The Kelvin temperature scale • Figure 6.15

Nothing can become colder than zero on the Kelvin scale, 0 K, also known as **absolute zero**. (The degree symbol, °, is not used in expressing temperatures on the Kelvin scale.)

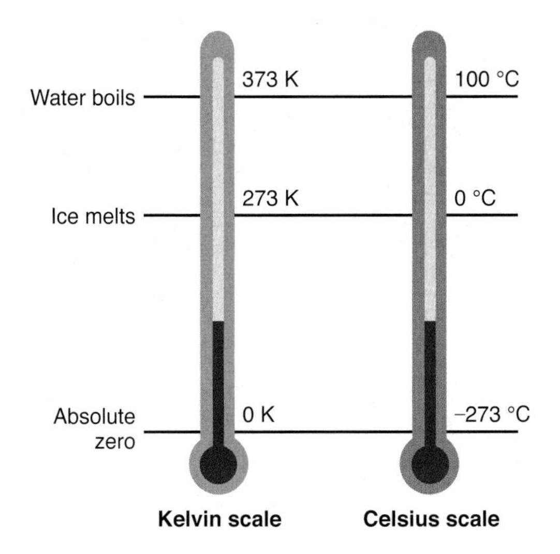

To convert from Celsius to Kelvin, add 273 to the Celsius temperature.
To convert from Kelvin to Celsius, subtract 273 from the Kelvin temperature.

Ask Yourself
Which is warmer, 40°C or 310 K?

Liquefied gases • Figure 6.16

Charles's Law predicts that the volume of a sample of gas should continue to shrink as the temperature drops. However, when cooled sufficiently, real gases eventually condense into liquids.

Liquefied gases, such as **cryogenic** (extremely low temperature) nitrogen (boiling point, 77 K) and helium (boiling point, 4 K), are used in specialized applications, such as cooling the powerful magnets used in magnetic resonance imaging (MRI) devices.

Interpret the Data

1. What does the dashed line in this figure represent?
2. Under normal atmospheric pressure, at what Kelvin temperature does helium gas condense into a liquid? What Celsius temperature does this correspond to?

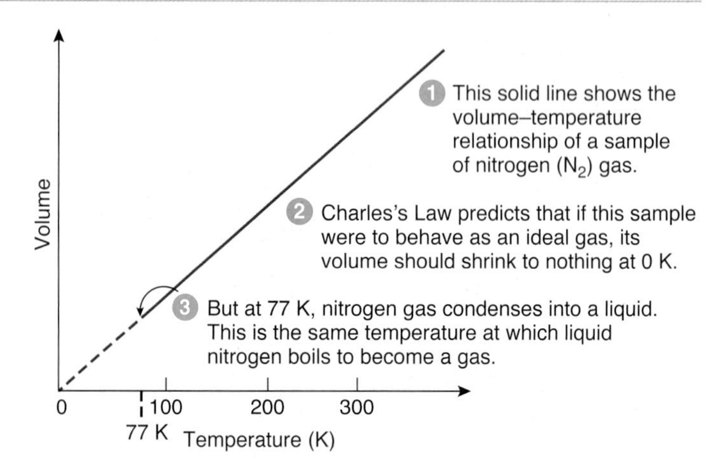

1 This solid line shows the volume–temperature relationship of a sample of nitrogen (N_2) gas.
2 Charles's Law predicts that if this sample were to behave as an ideal gas, its volume should shrink to nothing at 0 K.
3 But at 77 K, nitrogen gas condenses into a liquid. This is the same temperature at which liquid nitrogen boils to become a gas.
Volume
0
100
200
300
77 K
Temperature (K)

Volume–temperature relationship of a gas.

Charles's Law implies that an ideal gas, with its zero volume of chemical particles, would vanish completely at absolute zero (0 K). Mathematically,

$$V = k \times T, \text{ so if } T = 0, \text{ then } V = 0.$$

However, gases show ideal behavior only at high temperatures (and low pressures), not at temperatures anywhere near absolute zero. The real volumes of gas particles and the real attractive forces they exert on each other cause all real gases to liquefy as temperatures drop and pressures rise. The ideal world of bouncing, point-sized ping-pong balls fades into the real world of material substances, of actual atoms and molecules. At a pressure of 1 atmosphere (1 atm), for example, nitrogen liquefies at −196°C, or 77 K (**Figure 6.16**).

As with Boyle's Law, the kinetic-molecular theory explains things nicely here, too. As the temperature of a gas drops, so does the kinetic energy of its molecules. At lower temperatures, molecules don't bounce off one another or off the walls of their container with quite as much energy or as often as they do at higher temperatures. To maintain a constant pressure on the walls of the container, the volume of a gas must shrink. On the other hand, as the temperature rises, the energy and frequency of collisions also increase, and so does the volume if the pressure is to remain constant. We can illustrate Charles's Law with the following example (**Figure 6.17**).

In Words, Math, and Pictures

Charles's Law in action • Figure 6.17

Suppose we inflate a balloon with nitrogen gas to volume of 1.0 L at 27°C, then place it in a freezer kept at −23°C. What is the volume of the balloon when the gas cools to the freezer's temperature?

Solution:

We designate V_1 as the original volume, T_1 as the original temperature, V_2 as the final volume, and T_2 as the final temperature. Since $V_1 = k \times T_1$ and $V_2 = k \times T_2$, the ratio of these two volumes is

$$\frac{V_1}{V_2} = \frac{k \times T_1}{k \times T_2} \text{ ...which simplifies to } \frac{V_1}{V_2} = \frac{T_1}{T_2}$$

$$\text{...which can be arranged to } V_2 = \frac{V_1 \times T_2}{T_1}$$

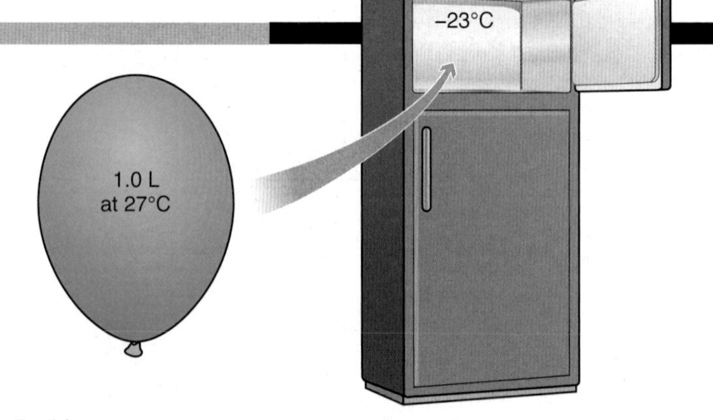

In this case:

$V_1 = 1.0 \text{ L}$

$T_1 = (27 + 273) = 300 \text{ K}$

$T_2 = (-23 + 273) = 250 \text{ K}$

$$\text{so } V_2 = \frac{1.0 \text{ L} \times 250 \cancel{\text{K}}}{300 \cancel{\text{K}}} = 0.83 \text{ L}$$

KNOW BEFORE YOU GO

4. A gas kept at constant pressure in a balloon has a volume of 1.00 L at 25°C. At what Celsius temperature would the gas have a volume of 1.25 L? (Assume the balloon expands to this volume without bursting and can tolerate the temperature change.)

The Combined Gas Law So far we have seen two laws that describe the behavior of gases:

- Boyle's Law, which relates pressure and volume for a fixed mass of gas held at a constant temperature, and
- Charles's Law, which relates volume and temperature for a fixed mass of gas held at a constant pressure.

Instead of treating these as two distinct laws, independent of each other, we can combine them into a single law involving the pressure, volume, and temperature of a fixed mass of gas. We can write this combination of Boyle's Law and Charles's Law in the following convenient form:

$$\frac{P_1 V_1}{T_1} = \frac{P_2 V_2}{T_2}$$

(The subscripts $_1$ and $_2$ refer to the initial and final conditions, respectively.) If we know any five of these six quantities, we can solve for the unknown variable, as in the following example (**Figure 6.18**).

In Words, Math, and Pictures

Applying the Combined Gas Law • Figure 6.18

A helium-filled balloon has a volume of 1.0 L at sea level, where the pressure is 760 mm-Hg and the temperature is 27°C. If the balloon escapes and ascends to 5000 m, where the pressure is 420 mm-Hg and the temperature is −18°C, what is its volume under these new conditions?

Solution:

The balloon is subjected to two contrasting effects during its ascent. The drop in atmospheric pressure tends to cause the balloon to expand, while the decrease in temperature tends to make it contract. We can use the combined gas law as follows:

$$\frac{P_1 V_1}{T_1} = \frac{P_2 V_2}{T_2}$$

We can rewrite the equation as

$$V_2 = \frac{P_1 V_1 T_2}{T_1 P_2}$$

Now we solve for the final volume, V_2, using these values:

$$P_1 = 760 \text{ mm-Hg}$$

$$V_1 = 1.0 \text{ L}$$

$$T_1 = (27 + 273) = 300 \text{ K}$$

$$P_2 = 420 \text{ mm-Hg}$$

$$T_2 = (-18 + 273) = 255 \text{ K}$$

$$\text{so } V_2 = \frac{760 \cancel{\text{mm-Hg}} \times 1.0 \text{ L} \times 255 \cancel{\text{K}}}{300 \cancel{\text{K}} \times 420 \cancel{\text{mm-Hg}}} = 1.5 \text{ L}$$

We find, then, that the decrease in pressure (causing expansion) produces a greater effect than the decrease in temperature (causing contraction).

KNOW BEFORE YOU GO

For each of these questions, assume ideal gas behavior.

5. Suppose we start again with a balloon filled to a volume of 1.0 L at a pressure of 760 mm-Hg and a temperature of 27°C—the same initial conditions as in the prior example. Now suppose we again reduce the external pressure to 420 mm-Hg. To what Celsius temperature would we have to cool the balloon to keep its volume at 1.0 L?
6. A shaving cream can bears the warning that it should not be exposed to temperatures above 50°C because of risk of the can rupturing from build-up of pressure. Assume the shaving cream contents of the can have been used up, but residual propellant gas remains inside. If the pressure within the can is 1.2 atm at 25°C, what would be the pressure (in atm) at 50°C? [Assume the volume of the can does not change (i.e., $V_1 = V_2$).]

Gas Compression and Expansion

When you manually inflate a bicycle tire, you may notice that the pump and tire often warm up in the process. The reason is that as you compress air, you reduce its volume. This increases the kinetic energy of the gas molecules and hence the temperature of the gas. Compressed into the limited volume of the tire, the molecules collide with one another and the wall of the tire more frequently, resulting in higher pressure. If you now vent the inflated tire, the released gas molecules expand into a larger volume. This lowers their kinetic energy and associated temperature.

The warming due to gas compression and cooling due to gas expansion have important practical applications. Most air conditioners and refrigerators, for example, operate according to these principles (**Figure 6.19**).

Until recently, the chemical compounds that constituted refrigerant gases were largely **chlorofluorocarbons (CFCs)**, which are compounds resembling hydrocarbons, but that have chlorine and fluorine atoms in place of many of the hydrogen atoms. These CFCs are normally chemically stable gases, but when they enter Earth's atmosphere, through leakage from the refrigerant machinery or by other means, they can be transformed by the action of sunlight into substances that are able to catalyze the depletion of the ozone layer, an atmospheric region that provides essential protection against the Sun's ultraviolet radiation. In the chapter on sustainability, we'll see how this occurs and examine the newer-generation refrigerant gases that have replaced CFCs.

We began this chapter by examining the effects of temperature on the physical state of a substance: on whether it exists as a solid, a liquid, or a gas. But pressure, too, plays an important part in influencing the physical state of a material. For example, in this section we saw that increasing the pressure on a refrigerant gas can cause the gas to condense to a liquid. This makes sense at the molecular level: As the compressor forces the gas molecules into an increasingly smaller volume of space, the distance between the gas molecules becomes smaller, eventually causing the gas itself to condense to a liquid. Applying even higher pressures to many types of liquids can cause them to solidify. *A Deeper Look* on page 181 explores the effects of both temperature and pressure on substances.

How refrigerators work • Figure 6.19

Most refrigerators, freezers, and air conditioners use a refrigerant gas that circulates through a continuous cycle of compression (heating) and expansion (cooling). The compressed gas releases heat to the outside of the device, and the expanded gas absorbs heat from the inside of the device.

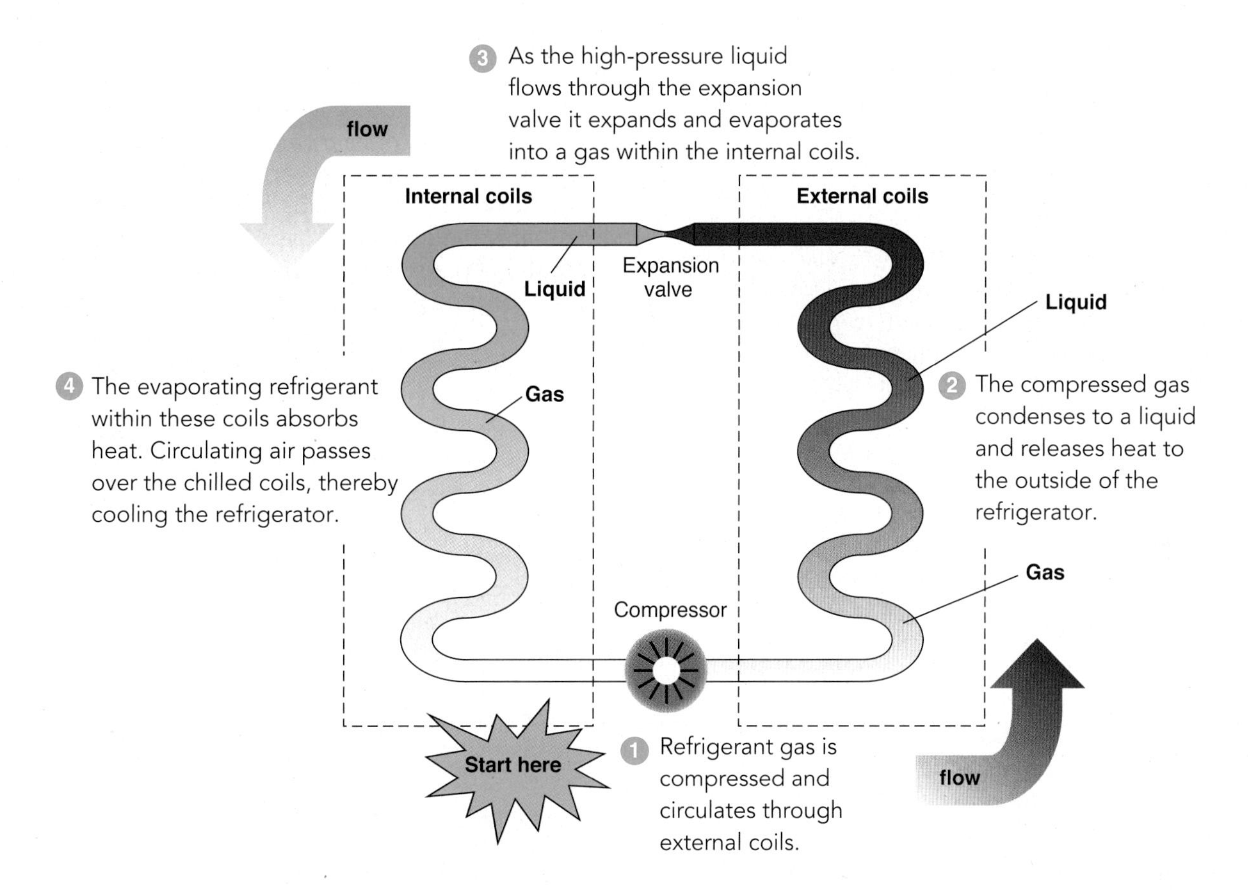

3 As the high-pressure liquid flows through the expansion valve it expands and evaporates into a gas within the internal coils.
flow
Internal coils
External coils
Liquid
Expansion valve
Liquid
Gas
4 The evaporating refrigerant within these coils absorbs heat. Circulating air passes over the chilled coils, thereby cooling the refrigerator.
2 The compressed gas condenses to a liquid and releases heat to the outside of the refrigerator.
Gas
Compressor
Start here
1 Refrigerant gas is compressed and circulates through external coils.
flow

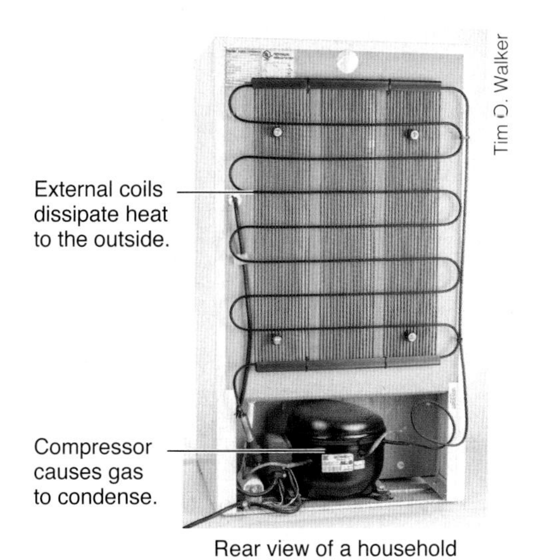

External coils dissipate heat to the outside.
Compressor causes gas to condense.
Tim D. Walker

Rear view of a household refrigerator.

PROCESS DIAGRAM

Ask Yourself

1. At what point in the cycle does a liquid evaporate? Is heat absorbed or released by the refrigerant at this point?
2. At what point in the cycle does a gas condense to a liquid? Is heat absorbed or released by the refrigerant at this point?

A DEEPER LOOK The Effects of Temperature and Pressure on Materials

A **phase diagram** shows graphically how both pressure and temperature influence the physical state of a substance. Insights into these diagrams provide an opportunity to create a wide range of interesting materials. We'll illustrate the value of phase diagrams with two examples of such materials: supercritical carbon dioxide and synthetic diamonds.

Supercritical carbon dioxide

In our everyday experience, dry ice (solid carbon dioxide) sublimes, turning directly from a solid into a gas. We can help explain this behavior by examining the phase diagram for CO_2 (**Figure a**). In a phase diagram, the regions (here, coded in shades of blue) indicate which phase or physical state the substance exhibits under a given set of temperature and pressure conditions.

In the upper right region of the figure—at pressures above 73 atm and at temperatures above 31°C— CO_2 exists as a **supercritical fluid**, exhibiting properties of both liquids and gases. Supercritical CO_2 ($scCO_2$) is particularly useful as a solvent, offering both environmental and practical benefits. For instance, $scCO_2$ is used to extract caffeine from coffee beans while allowing the beans to retain their flavorful compounds. $scCO_2$ is also used in dry cleaning, as an environmentally friendly alternative to traditional solvents. (We'll have more to say about solvents in Chapter 7.)

Synthetic diamonds

Diamonds are the hardest known substances, valued for both industrial cutting and grinding applications and for their brilliance as gemstones. Whether naturally formed or manufactured, a diamond represents a pure form, or **allotrope**, of carbon. Naturally occurring diamonds were formed perhaps a billion or more years ago, some 100 miles below the surface of Earth, where temperatures reach over 1000°C and pressures exceed 50,000 atm. Under these extreme conditions, carbon atoms coalesce into the tetrahedral bonding arrangement characteristic of all diamonds (**Figure b**). To reach Earth's surface, diamonds and other materials migrate upwards by way of subterranean magma flows.

Synthetic or manufactured diamonds are nonetheless real diamonds, having the same atomic ordering as found in those formed naturally. Synthetic diamonds have been commercially produced for some time, using *graphite*, another allotrope of carbon, as a starting material. But these synthetic diamonds have generally only been suitable for industrial uses, due to their small sizes. Recently, however, a few companies have identified the right temperature, pressure, and processing conditions to manufacture larger, gemstone-quality diamonds. These conditions involve high temperatures (1500°C) and high pressures (58,000 atm), similar to those under which natural diamonds form. Synthetic gemstone diamonds offer an option for those who might have concerns about the origins of traditionally mined diamonds.

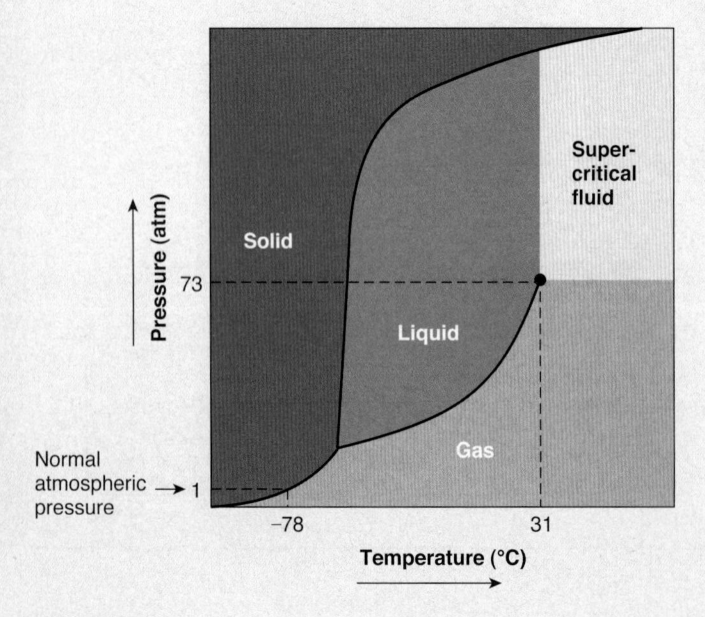

Figure a. The phase diagram of CO_2. Notice the regions in the diagram where CO_2 exists as a solid, a liquid, and a gas. At normal atmospheric pressure (1 atm), CO_2 is a solid below −78°C (shown in the bottom left corner of the diagram). At 1 atm and at temperatures above −78°C, CO_2 exists as a gas.

Interpret the Data

Examine the figure to predict what would happen to a sample of solid CO_2 held at a pressure of 73 atm if it is warmed from −78°C to room temperature (about 20°C). Under these conditions, would it be appropriate to call it *dry ice*?

Figure b. The structure of diamond. In diamond, each carbon atom covalently bonds to four others in a tetrahedral orientation. The arrangement shown here extends outward in space to form the crystal lattice of diamond.

CONCEPT CHECK STOP

1. **What** two gases account for about 99% of dry air?
2. **How** does heating a gas held under constant pressure affect its volume?
3. **Why** does expansion of a gas cause cooling?

6.3 Chemical Changes

LEARNING OBJECTIVES

1. **Describe** how chemical changes differ from physical changes.
2. **Explain** how to balance a chemical equation.

To this point we've examined how substances can undergo a variety of physical changes, such as melting, boiling, and evaporation. Although the substances change their physical form in each of these cases, they retain their chemical compositions. Now we turn our attention to chemical changes: processes that alter the chemical composition of substances.

chemical change A process that produces substances with new chemical compositions.

The Nature of Chemical Reactions

We saw in Chapter 5 that the carbon dioxide we exhale is a by-product of **metabolism**—the myriad of chemical reactions within our bodies that convert the macromolecules of our foods into energy that sustains life. Along with water vapor, carbon dioxide is also a product of the combustion of hydrocarbons, such as gasoline, as discussed in Chapter 4. The metabolism of nutrients and the combustion of hydrocarbons represent **chemical changes**: processes in which reactants are converted into new chemical species or products. These changes can take place quickly, as in the burning of gasoline, or slowly,

Chemistry InSight Chemical change • Figure 6.20

In a chemical reaction, one or more reactants are converted into one or more products. This process is often accompanied by a change in energy content, such as the release of heat (in an **exothermic** reaction) or the absorption of heat (in an **endothermic** reaction).

Here, hydrogen peroxide in the flask vigorously decomposes into oxygen gas and water vapor, releasing heat in the process. This decomposition reaction is facilitated by the addition of a small amount of catalyst, such as potassium iodide.

Some chemical reactions are initiated by the absorption of light, such as the reactions that produce atmospheric smog. Other reactions are accompanied by the production of light. For example, when you bend a glow stick to activate it, hydrogen peroxide mixes and reacts with the compound *diphenyl oxalate.* Energy released in this chemical reaction excites fluorescent dye molecules present in the tube, causing them to emit light. This is an example of *chemiluminescence.*

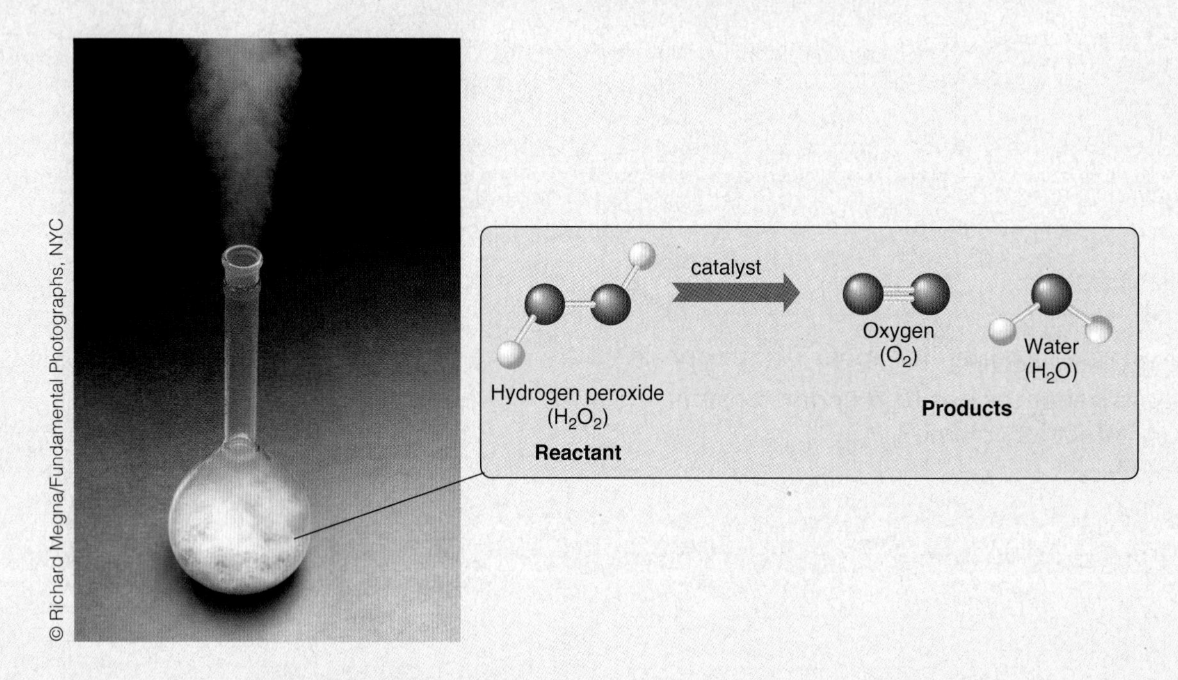

Think Critically In the chemical reaction shown, how do the products and the reactant differ in the bonding of their oxygen atoms?

as in the gradual rusting of a nail. The only true test of a chemical change or reaction is whether one or more new chemical entities are formed (**Figure 6.20**).

Chemical reactions provide enormous practical and economic value to society. Petrochemical companies, for example, convert simple hydrocarbons present in natural gas and oil into more complex compounds. These compounds, in turn, serve as starting materials for making such end products as plastics, textiles, and pharmaceuticals. In our own lives, we depend on a host of chemical reactions on an everyday basis, such as the combustion reactions that power our cars, trucks, and airplanes and the electrochemical reactions that power our electronic devices (**Figure 6.21**).

Balancing Chemical Equations

Conventional chemical reactions (as opposed to the nuclear reactions we'll discuss in Chapter 9) involve the sharing or transfer of valence electrons. Although new bonds may be formed, there is no change to the number of protons and neutrons in the nuclei of the atoms. As a

Chemical reactions in everyday life • Figure 6.21

Each of the reactions shown here has a variety of everyday applications and is further explored in a separate chapter of the book.

Electrochemical reactions provide the power for batteries, including the rechargeable lithium-ion cells widely used in consumer electronics.

Combustion reactions of hydrocarbons are widely used in transportation and heating.

An acid–base reaction produces carbon dioxide bubbles that help make these muffins rise as they bake.

Polymerization reactions produce plastics and other polymers, widely used for packaging and other materials.

Electrochemical reactions (Chapter 10) involve the transfer of electrons.
Combustion reactions (Chapter 4) combine fuels, such as hydrocarbons, with oxygen to release heat and gases.
Acid–base reactions (Chapter 8) involve the transfer of protons from acids to bases.
Polymerization reactions (Chapter 13) assemble large molecules called **polymers** from small repeating chemical units called **monomers**.
There are many other types of chemical reactions, including **photochemical** reactions—those initiated by light energy—an important topic with environmental implications that we'll address in Chapter 13.

result, the same number of atoms of a given element is found in the products of a chemical reaction as in the reactants. We observe this as a fundamental law of chemical reactions called **conservation of mass** (**Figure 6.22**).

The law of conservation of mass underlies the concept of **balanced chemical equations**, which is how we formally represent chemical reactions in writing. We can illustrate the process of balancing chemical equations with the decomposition of hydrogen peroxide shown in Figure 6.20.

conservation of mass Matter can neither be created nor destroyed as a result of a conventional (nonnuclear) chemical reaction.

balanced chemical equation An equation showing reactants, a reaction arrow, and products, with the same number of atoms of a given element on each side of the arrow.

Law of conservation of mass • Figure 6.22

This fundamental law of chemistry recognizes that mass can neither be created nor destroyed as a result of a chemical reaction. In other words, there must be exactly as much matter (and therefore exactly as many atoms of a given element) among the combined products of a chemical reaction as in its combined reactants.

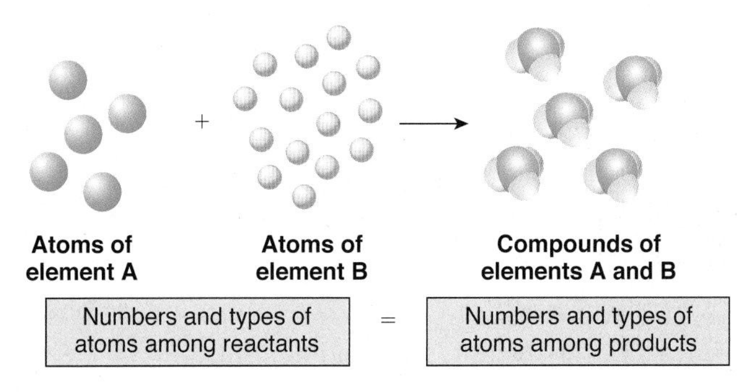

We begin by writing the reactant on the left and products on the right.

On the reactant side we find: 2 oxygen atoms (in red) and 2 hydrogen atoms (in white)

On the product side we find: 3 oxygen atoms (1 from H_2O and 2 from O_2) and 2 hydrogen atoms

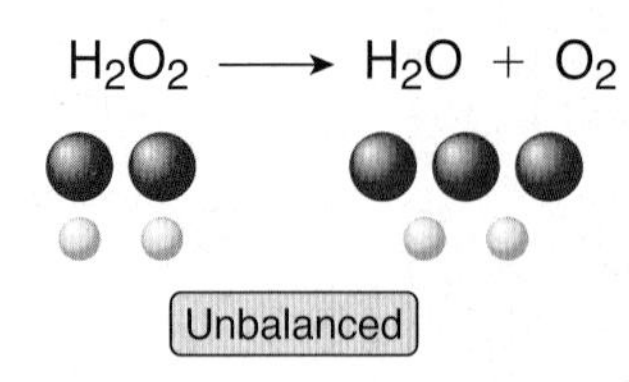

The reaction is unbalanced as written because more oxygen atoms are on the right side of the arrow than on the left side. We can't simply add an oxygen atom to the left, but we can add a **coefficient,** which is a whole number multiple greater than 1 in front of a reactant or product. We'll start by adding a 2 in front of H_2O_2 since we are short oxygen atoms on the left side. [A coefficient operates on all atoms in the formula that immediately follows it, so 2 H_2O_2 means (2 × H_2O_2), or 4 H atoms and 4 O atoms.]

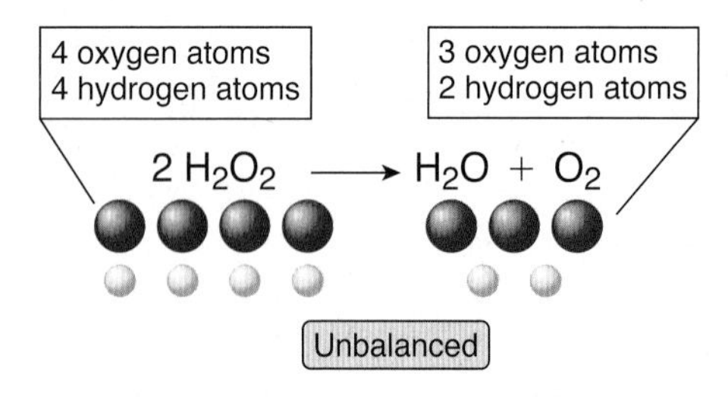

Now we have more oxygen and hydrogen atoms on the left side than on the right side. However, if we now add a coefficient of 2 in front of H_2O, we get:

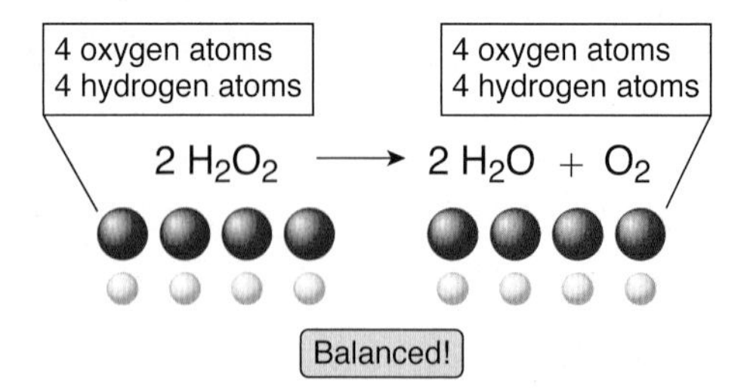

The equation is now balanced and tells us that:

2 H_2O_2 molecules react to produce 2 H_2O molecules and 1 O_2 molecule.

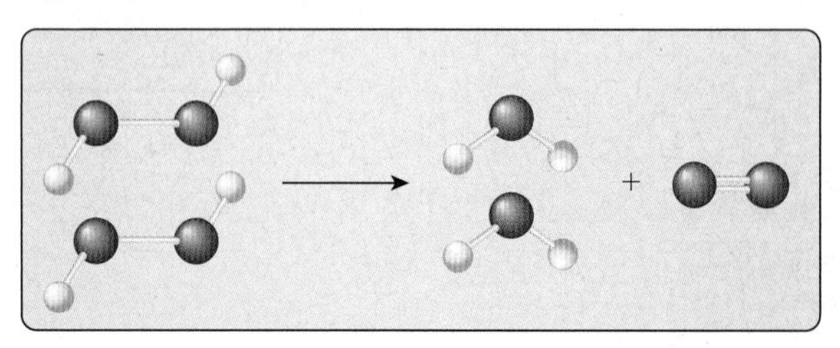

Bear in mind two important rules in balancing equations:

1. **Use the lowest whole number coefficients** that produce equal numbers of atoms of each element on each side of the arrow. Suppose we represented this chemical reaction as

$$4\,H_2O_2 \longrightarrow 4\,H_2O + 2\,O_2$$

The equation appears to balance, but it does not use the lowest coefficients possible, which we just established should be 2 for H_2O_2, 2 for H_2O and an implied 1 for O_2. [In a balanced equation, the absence of a coefficient in front of a chemical species implies the number 1.]

2. **Change only the coefficients** of the formulas involved; do not change any of the formulas themselves. For example, the changes shown in the following two equations are not permitted.

Here, the formula H_2O_2 was changed. ~~H_2O_3~~ $\longrightarrow H_2O + O_2$

$H_2O_2 \longrightarrow H_2O +$ ~~O~~ Here, the formula O_2 was changed.

In the next example, we balance the equation for the combustion of butane, C_4H_{10}, a common heating fuel. Hydrocarbons, such as butane, react with oxygen (O_2) to produce carbon dioxide (CO_2) and water vapor. This gives us:

$$\underset{\text{Butane}}{C_4H_{10}} + \underset{\text{Oxygen}}{O_2} \longrightarrow \underset{\text{Carbon dioxide}}{CO_2} + \underset{\text{Water}}{H_2O}$$

In balancing reactions of this type, **it's often helpful to start by balancing either the carbon or hydrogen atoms first**. Here, we'll start by adding a coefficient of 4 in front of CO_2 to balance the carbon atoms:

$$C_4H_{10} + O_2 \longrightarrow 4\,CO_2 + H_2O$$

	Reactants	**Products**
# of C atoms	4 ✓	4 ✓
# of H atoms	10	2
# of O atoms	2	(4 × 2) + 1 = **9**

We can now add a coefficient of 5 in front of H_2O to balance hydrogen atoms:

$$C_4H_{10} + O_2 \longrightarrow 4\ CO_2 + 5\ H_2O$$

	Reactants	Products
# of C atoms	4 ✓	4 ✓
# of H atoms	10 ✓	(5 × 2) = **10** ✓
# of O atoms	2	(4 × 2) + (5 × 1) = **13**

To balance the oxygen atoms, we start by adding a coefficient of 6.5 in front of O_2:

$$C_4H_{10} + 6.5\ O_2 \longrightarrow 4\ CO_2 + 5\ H_2O$$

	Reactants	Products
# of C atoms	4 ✓	4 ✓
# of H atoms	10 ✓	(5 × 2) = **10** ✓
# of O atoms	(6.5 × 2) = **13** ✓	(4 × 2) + (5 × 1) = **13** ✓

We now have the same number of oxygen atoms, 13, on both sides of the equation. However, since 6.5, the coefficient in front of O_2, is not a whole number, we multiply the entire equation by 2 to get the (lowest) whole number coefficients:

$$2 \times [C_4H_{10} + 6.5\ O_2 \longrightarrow 4\ CO_2 + 5\ H_2O]$$

$$= 2\ C_4H_{10} + 13\ O_2 \longrightarrow 8\ CO_2 + 10\ H_2O$$

	Reactants	Products
# of C atoms	(2 × 4) = **8** ✓	**8** ✓
# of H atoms	(2 × 10) = **20** ✓	(10 × 2) = **20** ✓
# of O atoms	(13 × 2) = **26** ✓	(8 × 2) + (10 × 1) = **26** ✓

Balanced!

KNOW BEFORE YOU GO

7. The hydrocarbon acetylene (C_2H_2) is used in oxy-acetylene torches for welding. Balance the following equation for the combustion of acetylene:

$$C_2H_2 + O_2 \longrightarrow CO_2 + H_2O$$

8. Automobile air bags inflate rapidly as sodium azide (NaN_3) decomposes into nitrogen gas (N_2). (This reaction also produces sodium metal [Na].) Balance the following equation for the decomposition of sodium azide:

$$NaN_3 \longrightarrow N_2 + Na$$

CONCEPT CHECK

STOP

1. **What** characteristic is shared by all chemical changes?
2. **What** physical law underlies the notion of a balanced chemical equation?

6.4 Counting Atoms and Molecules

LEARNING OBJECTIVES

1. **Explain** how the mass of a pure substance relates to the number of chemical particles it contains.
2. **Using** a balanced equation, determine the amounts of reactants consumed or products formed in a chemical reaction.
3. **Calculate** the amount of heat released in a combustion reaction.

In this section we explore a few quantitative aspects of chemistry to help shed light on some basic properties of matter and its changes. In this process we will work with balanced chemical equations to see how the quantities of reactants relate to the quantities of products in a chemical reaction.

In Words, Math, and Pictures

Counting by weighing • Figure 6.23

Suppose you happen to pass a display in a storefront window that reads, *"Come closest to guessing the number of jellybeans in the jar and you could be a winner!"*
One way to approach this type of problem (although it's unlikely you would be permitted to do this in an actual contest) is by a method called *counting by weighing*. Here's how it works:

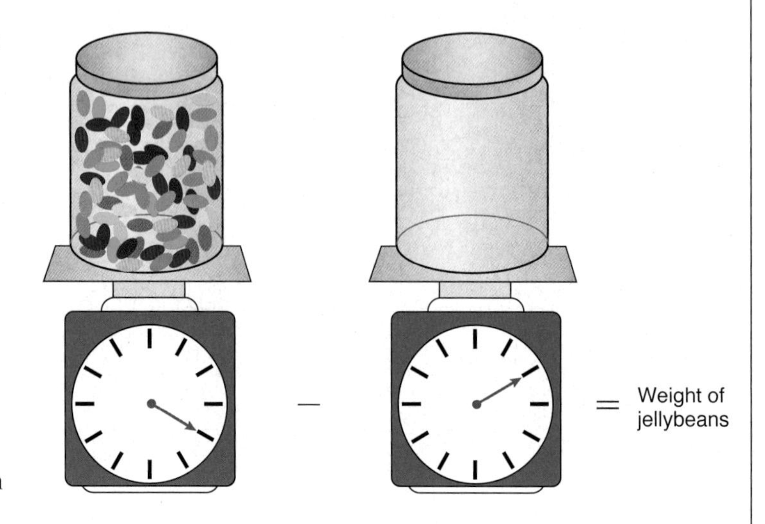

1. Find the weight of all the jellybeans.
If you weigh the filled jar, then dump the candy out and weigh the empty jar, by subtraction you would get the weight of the jellybeans themselves, as shown here.

2. Find the average weight of a single jellybean.
If you weigh a dozen randomly chosen jellybeans and divide this weight by 12, you would get the weight of one average, representative jellybean. Suppose the weight of 12 jellybeans is 16 g. Then the average weight of a typical jellybean is 16 g/12 = 1.3 g.

3. Use unit cancellation to find the number of jellybeans.
If you know the weight of a pile of similar objects, and you know the average weight of one of the objects, you can calculate the total number. In our example, if the jellybeans weigh 5.2 kg and the average weight of a single jellybean is 1.3 g, then the number of jellybeans is

$$\underbrace{5.2\ \cancel{\text{kg}}}_{\text{Total weight}} \times \underbrace{\frac{1000\ \cancel{\text{g}}}{1\ \cancel{\text{kg}}}}_{\text{Conversion factor from kg to g}} \times \underbrace{\frac{1\ \text{jellybean}}{1.3\ \cancel{\text{g}}}}_{\text{Average weight of a single jellybean}} = \frac{5.2 \times 1000 \times 1}{1 \times 1.3}\ \text{jellybeans} = \mathbf{4000\ jellybeans}$$

Again, we've estimated the number of objects not by counting, but by weighing.

You can do the same sort of thing with atoms and molecules.

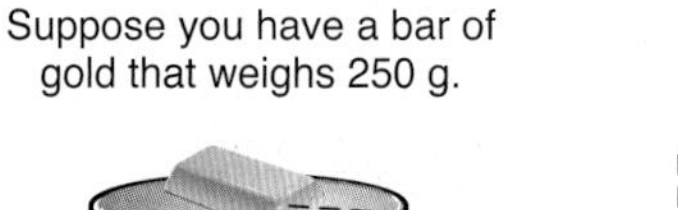
Suppose you have a bar of gold that weighs 250 g.

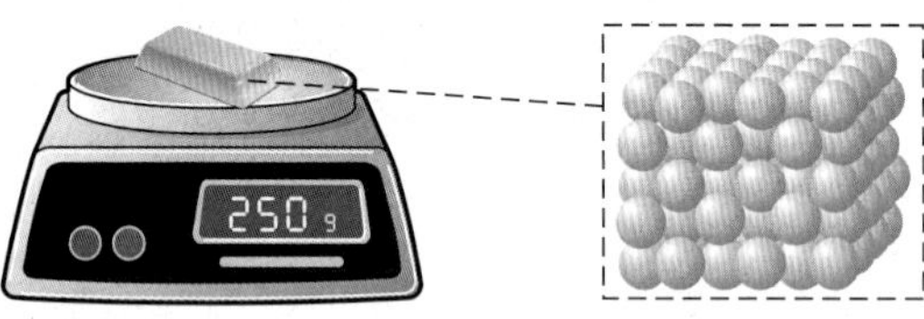

The gold is made up of a very large number of atoms.

A periodic table shows that the average weight of a gold atom is 197.0 atomic mass units (u).

We can estimate the number of gold atoms in the bar of gold using the same approach as before. Here, we use the fact that 1 atomic mass unit (u) is equivalent to 1.66×10^{-24} gram (g):

$$\underbrace{250\ \cancel{\text{g}}}_{\text{Total weight}} \times \underbrace{\frac{1\ \cancel{\text{u}}}{1.66 \times 10^{-24}\ \cancel{\text{g}}}}_{\text{Conversion factor from g to u}} \times \underbrace{\frac{1\ \text{gold atom}}{197\ \cancel{\text{u}}}}_{\text{Average weight of a single gold atom}} = \frac{250 \times 1 \times 1}{(1.66 \times 10^{-24}) \times 197} = \mathbf{7.64 \times 10^{23}\ gold\ atoms}$$

Chemical Accounting

We buy, sell, and count things in units. They can be units of weight, as in a pound of potatoes; units of volume, as in 15 gallons of gas; units of length, as in 50 feet of rope; or units of area, as in 100 acres of farmland. Sometimes the units reflect a natural or a common way of using things. We buy socks in pairs, for example. Since we have two feet, the pair is a good unit for socks. Lose one sock and you'll have to buy two of them to replace it. Soft drinks are sold in six-packs and eggs by the dozen. Sometimes we buy and sell very small, inexpensive things in packages of 50 or 100 simply for convenience. We're not likely to find run-of-the-mill paper clips for sale in ones and twos. More likely we'd buy a box of 100 or so.

Chemical particles, such as atoms, molecules, and ions, present a problem with units on a scale unlike any other. In even the smallest quantity of material we might handle, the enormous numbers of chemical particles present simply overwhelm our ability to count them. Terms like pair and dozen are useless; counting into the thousands or millions or billions is hopeless. There are more atoms of carbon in a single lump of charcoal than there are stars in the sky. Clearly we need some other approach to capture the sheer number of chemical particles in a sample of matter. The approach that we use is to *count by weighing* (**Figure 6.23**).

Answers to these types of problems can give us astonishingly large numbers because atoms are so minute. Fortunately, we have a simpler way to approach this, through the use of the **mole** concept. Just as two socks constitute a *pair* of socks, 12 donuts form a *dozen* donuts, and 500 sheets of paper represent a *ream* of paper, we define 6.02×10^{23} chemical particles as a mole of chemical particles, no matter whether they are atoms, ions, or molecules. This number, 6.02×10^{23}, is known as **Avogadro's number**[2] and is defined as the number of atoms in exactly 12 g of C-12, an isotope of carbon.

mole An amount of substance that contains 6.02×10^{23} chemical particles.

Avogadro's number is staggeringly large (**Figure 6.24**). In terms of money, just one mole of pennies, distributed equally to everyone in the world's population of about 7 billion people, would give each one of us nearly a trillion (1×10^{12}) dollars. If you could stack a mole of pennies in a single pile, it would reach to a height of 9.3×10^{17} km. It would take a beam of light 100,000 years to travel this distance.

[2] This number is named in honor of Amadeo Avogadro, an early-19th-century Italian physicist. Avogadro did not discover the number that bears his name, but instead recognized that equal volumes of gases (at the same temperature and pressure) contain equal numbers of molecules, a finding that laid the groundwork for the mole concept.

Avogadro's number • Figure 6.24

6.02×10^{23}—Avogadro's number—of individual units of any macroscopic object would have a colossal mass and volume. That many marshmallows, for instance, would have about the same mass as all of the water on Earth. Alternatively, Avogadro's number of atoms—a **mole** of atoms—is something you can often hold in the palm of your hand. For instance, 12 g of carbon, which is a mole of carbon atoms, is about what you would find in a couple of charcoal briquettes.

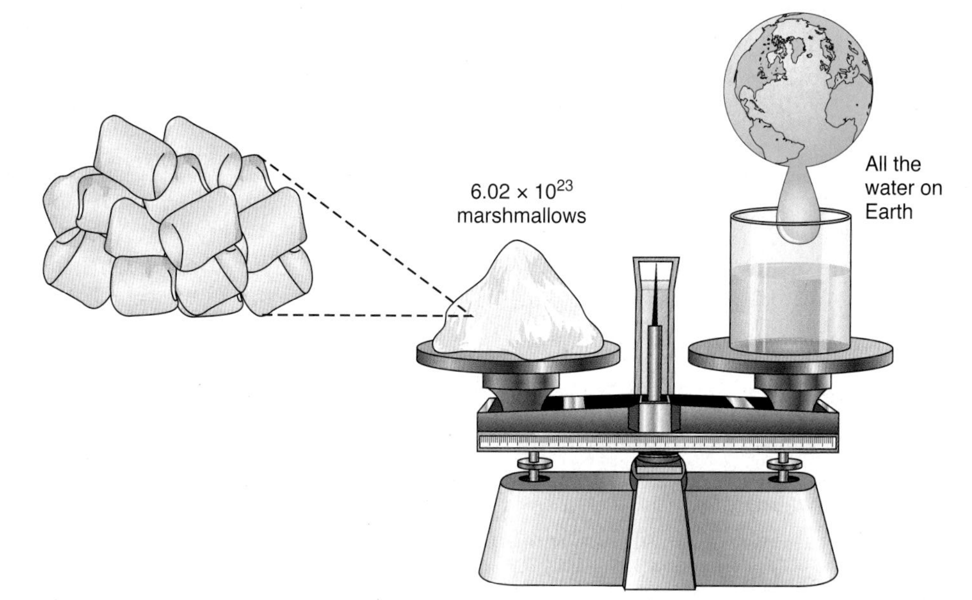

In Words, Math, and Pictures

Converting from mass to moles • Figure 6.25

In practice, if you weigh out a quantity of an element or a compound equal to its atomic or molecular mass expressed in grams, you have one mole of atoms or molecules of the substance.

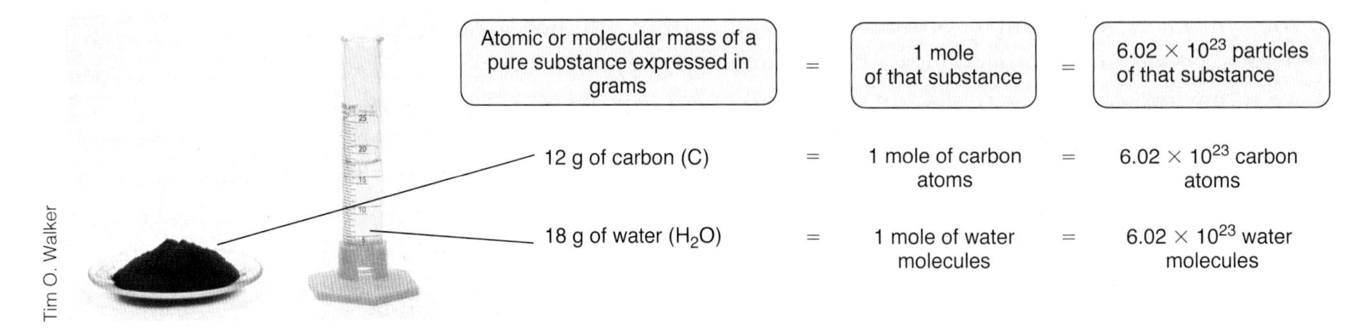

If you have a bar of gold with a mass of 250 g, the number of moles of gold atoms in the bar is

$$250 \text{ g} \times \underbrace{\frac{1 \text{ mole of gold}}{197 \text{ g}}}_{\text{Atomic mass of gold (Au)}} = \mathbf{1.27 \text{ moles of gold}}$$

Also, if you know the number of moles of a substance, you can calculate the number of chemical particles present:

$$1.27 \text{ moles of gold} \times \frac{6.02 \times 10^{23} \text{ gold atoms}}{1 \text{ mole of gold}} = \mathbf{7.64 \times 10^{23} \text{ gold atoms}}$$

Ask Yourself

1. How many moles of water molecules are in 100 g of water?
2. How many individual water molecules are in 100 g of water?

We can expand the concept of a mole by recognizing that if we have an amount of any pure chemical substance equal to its atomic or molecular mass *expressed in grams*, then we have one mole of that substance (again, 6.02×10^{23} of the chemical particles themselves) (**Figure 6.25**).

To use the mole successfully, it's helpful to recognize that one mole of any chemical substance contains the same number of chemical particles as one mole of any other chemical substance. For instance:

- One mole of carbon atoms (12 g of C) contains the same number of atoms as 1 mole of neon gas (20.2 g of Ne).
- One mole of water molecules (18 g of H_2O) contains the same number of molecules as one mole of sucrose (342 g of $C_{12}H_{22}O_{11}$).

KNOW BEFORE YOU GO

9. Consider the elements gold (Au) and lead (Pb).

   (a) Which has more atoms, one mole of gold or one mole of lead?
   (b) Which has a greater mass, one mole of gold (Au) or one mole of lead (Pb)?

10. Consider the molecules hydrogen (H_2) and oxygen (O_2). Which represents a greater number of moles, 50 g of H_2 or 50 g of O_2?

11. Consider an empty aluminum soft drink can with a mass of 16.5 g.

    (a) How many moles of aluminum atoms does it contain?
    (b) How many individual aluminum atoms does it contain?

Quantities and Chemical Reactions

For chemical reactions, we often want to know the quantity of chemical reactants needed to produce a certain amount of product. We also might want to know the amount of product that will form from a given amount of reactants. For this information, we rely on balanced chemical equations. The coefficient ratios in a balanced equation are in many ways similar to ratios of ingredients called for in a recipe (**Figure 6.26**).

The coefficients in a balanced equation not only tell us the ratio of individual chemical particles (such as molecules) involved in a reaction, but they can also refer to the ratio of *moles* of these particles. For instance, the chemical equation in Figure 6.24 indicates that for every 2 moles of H_2O_2 molecules that react, 2 moles of H_2O molecules and 1 mole of oxygen molecules are formed. In this way, we can use the coefficients in a balanced equation to predict how many moles of a chemical substance are consumed or produced in a chemical reaction.

For example, consider the balanced equation for the combustion of propane, C_3H_8, a common heating fuel:

$$C_3H_8 + 5\,O_2 \rightarrow 3\,CO_2 + 4\,H_2O$$

According to this equation, for every mole of C_3H_8 molecules that react:

5 moles of O_2 are consumed.
3 moles of CO_2 are produced.
4 moles of H_2O are produced.

Suppose we want to know how many moles of CO_2 are produced if 12 moles (abbreviated mol) of C_3H_8 burn in sufficient oxygen. Using the values required by the balanced equation, we can apply unit cancellation as follows:

$$12\,\cancel{\text{mol } C_3H_8} \times \underbrace{\frac{3\text{ mol } CO_2}{1\,\cancel{\text{mol } C_3H_8}}}_{\text{From the balanced equation}} = \frac{12 \times 3}{1}\text{ mol } CO_2 = \mathbf{36\text{ mol } CO_2}$$

We can use a similar approach to determine how much reactant is consumed for a given amount of product formed. For example, if we wanted to know how many moles of C_3H_8 are consumed when 50 moles of H_2O are produced by the same reaction, we would use the following unit cancellation:

$$50\,\cancel{\text{mol } H_2O} \times \underbrace{\frac{1\text{ mol } C_3H_8}{4\,\cancel{\text{mol } H_2O}}}_{\text{From the balanced equation}} = \frac{50 \times 1}{4}\text{ mol } C_3H_8 = \mathbf{12.5\text{ mol } C_3H_8}$$

An analogy for balanced chemical equations • Figure 6.26

Both a food recipe and a balanced chemical equation involve ratios of the quantities of their components.

a. A recipe tells us the ratio of ingredients needed to produce a certain amount of food.

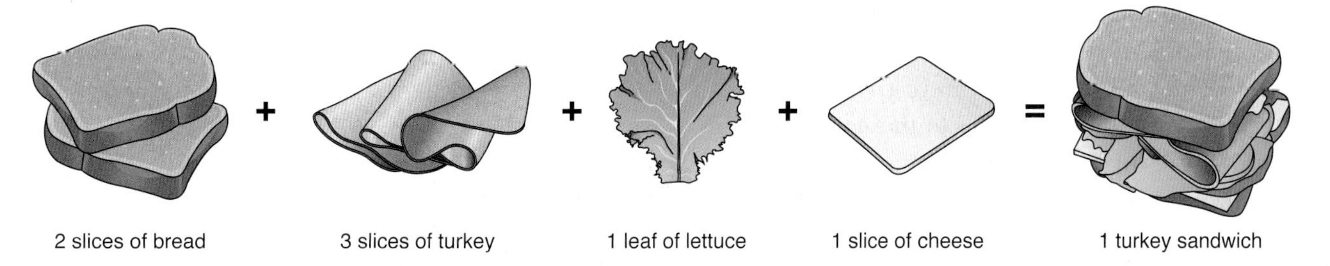

b. In a similar sense, the coefficients in a balanced equation tell us the ratios of chemical particles of reactants consumed and products formed in a chemical reaction.

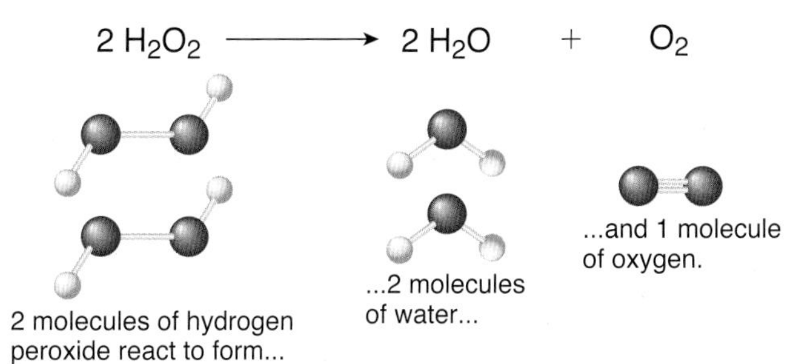

Ask Yourself

1. According to the recipe shown here, how many dozens of slices of bread would be required to make 1 dozen turkey sandwiches?
2. According to the equation shown here, how many dozens of H_2O_2 molecules would be required to produce 2 dozen H_2O molecules and 1 dozen O_2 molecules?

KNOW BEFORE YOU GO

12. How many moles of O_2 react when 6.2 moles of C_3H_8 burn according to the following balanced equation for the combustion of propane?

$$C_3H_8 + 5\,O_2 \longrightarrow 3\,CO_2 + 4\,H_2O$$

13. Fermentation of carbohydrates by yeast is used in manufacturing breads and other products.

(a) Balance the following equation for the fermentation of the carbohydrate, glucose.

$$\underset{\text{Glucose}}{C_6H_{12}O_6} \xrightarrow{\text{yeast}} \underset{\text{Ethanol}}{C_2H_6O} + \underset{\text{Carbon dioxide}}{CO_2}$$

(b) How many moles of CO_2 are produced when 8 moles of glucose undergo fermentation?

As a practical matter, we often want to know the masses of reactants needed to produce a given mass of product, or alternatively, the mass of product that can form, starting from certain known masses of reactants. For example, consider the following balanced equation for a reaction that converts silver sulfide (the black tarnish sometimes visible on silver) to silver metal.

$$\underset{\text{Silver sulfide}}{3\,Ag_2S} + \underset{\text{Aluminum}}{2\,Al} \longrightarrow \underset{\text{Silver}}{6\,Ag} + \underset{\text{Aluminum sulfide}}{Al_2S_3}$$

Suppose you want to know how many grams of silver are produced if 1.0 g of silver sulfide (Ag_2S) reacts completely according to the equation. In considering this kind of question, it's important to have a periodic table or similar resource on hand to determine the molecular (or atomic) masses of the substances of interest (expressed in units of g/mol). We sometimes refer to these values as **molar masses** because they represent the mass of 1 mole of a pure chemical substance. In this case:

$$\text{Molar mass of } Ag_2S = (2 \times 107.9) + (1 \times 32.0) = 247.8 \text{ g/mol}$$

$$\text{Molar mass of Ag} = 107.9 \text{ g/mol}$$

We can arrive at the answer using three successive unit conversions:

- We start by converting the mass of the first substance to moles.
- Then we convert from moles of the first substance to moles of the second, using coefficients in the balanced equation.
- Finally we convert moles of the second substance to mass.

$$\underbrace{1.0\text{ g } \cancel{Ag_2S}}_{\text{Provided in the question}} \times \underbrace{\overset{①}{\frac{1\text{ mol } \cancel{Ag_2S}}{247.8\text{ g } \cancel{Ag_2S}}}}_{\text{From the molar mass of } Ag_2S} \times \underbrace{\overset{②}{\frac{6\text{ } \cancel{\text{mol Ag}}}{3\text{ } \cancel{\text{mol } Ag_2S}}}}_{\text{From the coefficients in the balanced equation}} \times \underbrace{\overset{③}{\frac{107.9\text{ g Ag}}{1\text{ } \cancel{\text{mol Ag}}}}}_{\text{From the molar mass of Ag}}$$

$$= \frac{(1.0 \times 1 \times 6 \times 107.9)\text{g Ag}}{(247.8 \times 3 \times 1)} = \mathbf{0.87\ g\ Ag}$$

For another example, suppose you want to know how many grams of methane (CH_4) must react with oxygen to produce 5.0 g of carbon dioxide (CO_2). You start with the following balanced equation, which describes the reaction:

$$CH_4 + 2\,O_2 \longrightarrow CO_2 + 2\,H_2O$$

Then calculate the molar masses of the substances of interest:

$$\text{Molar mass of } CH_4 = (1 \times 12.0) + (4 \times 1.0) = 16.0 \text{ g/mol}$$

$$\text{Molar mass of } CO_2 = (1 \times 12.0) + (2 \times 16.0) = 44.0 \text{ g/mol}$$

Then set up a series of unit conversions as follows:

$$\underbrace{5.0\text{ g } \cancel{CO_2}}_{\text{Given in the question}} \times \underbrace{\frac{1\text{ mol } \cancel{CO_2}}{44.0\text{ g } \cancel{CO_2}}}_{\text{From the molar mass of } CO_2} \times \underbrace{\frac{1\text{ } \cancel{\text{mol } CH_4}}{1\text{ } \cancel{\text{mol } CO_2}}}_{\text{From the coefficients in the balanced equation}} \times \underbrace{\frac{16.0\text{ g } CH_4}{1\text{ } \cancel{\text{mol } CH_4}}}_{\text{From the molar mass of } CH_4}$$

$$= \frac{(5.0 \times 1 \times 1 \times 16.0)\text{g } CH_4}{(44.0 \times 1 \times 1)} = 1.8\text{ g } CH_4$$

The Energy of Chemical Reactions

One of the most important of all chemical reactions is the combustion of hydrocarbon fuels (discussed in Chapter 4). We look now at the quantitative connection between the amount of fuel that's used and the amount of energy generated.

Combustion, like all chemical reactions, follows the **law of conservation of energy**. This law holds that energy can neither be created nor destroyed as a result of chemical transformations. The sum of all the energy present in the products (including any that is liberated as the reaction progresses) must equal the sum of all the energy in the reactants (including any that is added to produce the reaction).

Why hydrocarbons release heat when they burn • Figure 6.27

The combustion of propane, C_3H_8, is represented here. Note the bonds in the structures of the reactants and products.

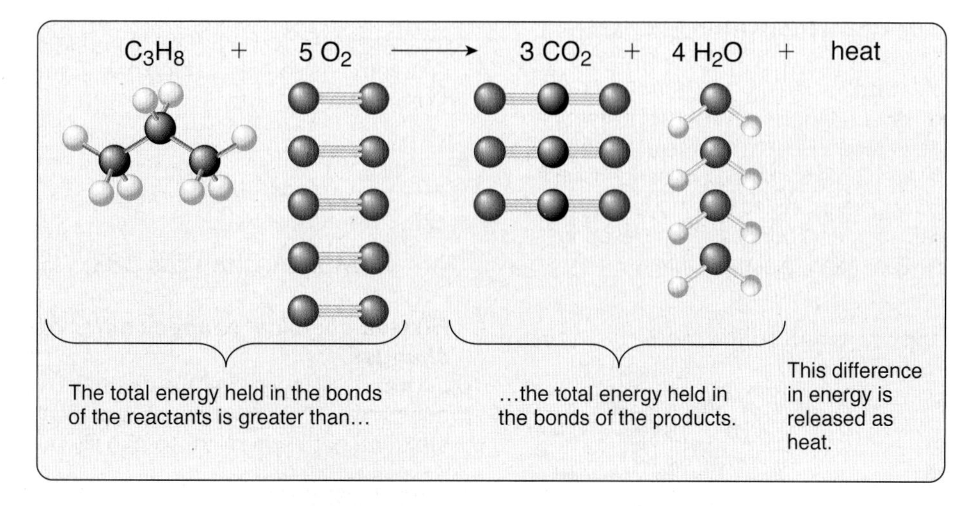

Ask Yourself

How many C—C bonds and C—H bonds are present in a molecule of propane?

KNOW BEFORE YOU GO

14. Natural gas, a widely used energy source, is made up mostly of methane, CH_4. How many grams of CO_2 are produced when 10.0 g of methane reacts according to the following equation?

$$CH_4 + 2O_2 \longrightarrow CO_2 + 2H_2O$$

As a practical matter, almost all the chemical energy contained in a molecule is stored in its chemical bonds. In a chemical reaction, if the total amount of energy contained in all the bonds of the products is less than the total amount of energy in all the bonds of the reactants, the difference in energy is released to the environment. For instance, when a hydrocarbon reacts with oxygen to produce CO_2 and H_2O, the total energy contained within all the C=O bonds and O—H bonds of the products is less than the total energy of all the C—H bonds, C—C bonds, and O=O bonds of the reactants. As a result, the reaction is exothermic, releasing the energy difference as heat (**Figure 6.27**).

For every mole of propane (44 g of C_3H_8) that burns, 531 kilocalories (kcal) of energy is released. (This is roughly equivalent to the caloric content of a Big Mac.) We can use this value, 531 kcal/mol, referred to as the **heat of combustion** of propane, to predict how much energy is released when a given amount of propane burns. For example, a typical 20-lb tank of propane used for outdoor grills and other heating purposes contains about 207 moles of propane. The energy released when the entire content of the tank burns is:

$$207\,\cancel{\text{mol } C_3H_8} \times \frac{531\text{ kcal}}{1\,\cancel{\text{mol } C_3H_8}} \approx 110{,}000\text{ kcal}$$

KNOW BEFORE YOU GO

15. The heat of combustion of methane (CH_4) is 213 kcal/mol. How many kcal of energy are released when 32 g of CH_4 burns in an excess of O_2?

CONCEPT CHECK STOP

1. How do you determine the number of moles in a sample of a pure chemical substance if you know the sample's mass?
2. What do the coefficients in a balanced chemical equation represent?
3. What does the heat of combustion of a hydrocarbon represent?

Summary

1 States of Matter and Their Changes 164

- ***What are changes of state?***
  Changes of state represent changes in the physical state of a substance. This can be in response to changes in temperature, pressure, or both. Melting and freezing describe transitions between the solid and liquid states (as shown here at the molecular level). Boiling and condensation describe transitions between the liquid and gaseous states.

Figure 6.2 • What happens when solids melt

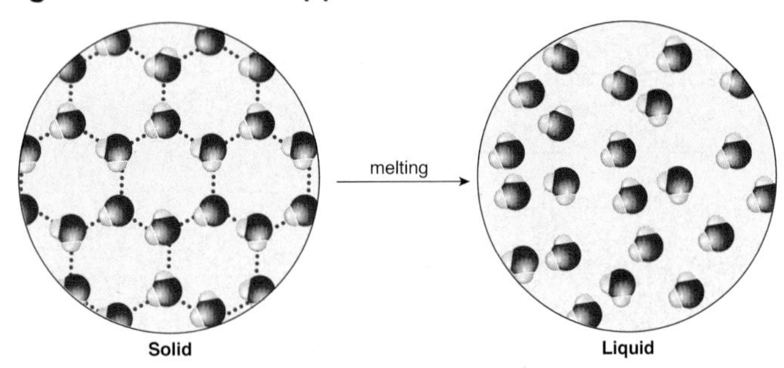

- ***What is density, and what factors influence this property?***
  Density is a measure of how much mass is in a given volume of a substance. The gaseous state of a substance has a much lower density than its liquid or solid states. Thermal expansion is an example of the effect of temperature on density.

2 Gas Behavior 172

How do gases behave in response to changes in pressure and temperature?
At constant temperature, the volume of a gas is inversely proportional to its pressure (Boyle's Law) , as shown here. At constant pressure, the volume is directly proportional to its Kelvin temperature (Charles's Law).

Figure 6.13 • Boyle's Law

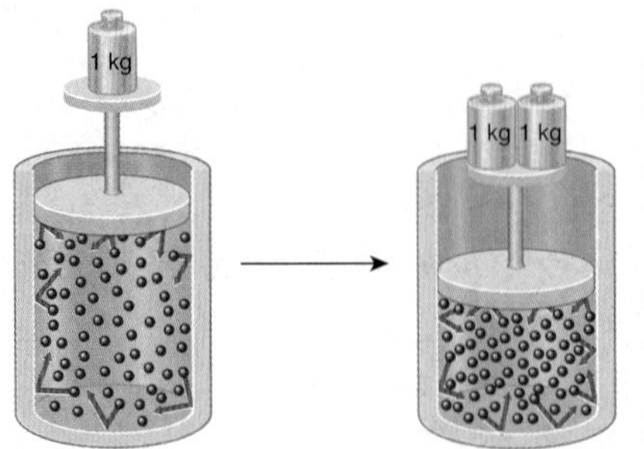

- ***How are gases used as refrigerants?***
  Gases release heat when they are compressed and absorb heat when they expand. Gases within the coils of refrigerators and other cooling devices undergo cycles of expansion and compression, absorbing heat from the inside and dissipating it to the outside.

3 Chemical Changes 183

- ***How do chemical changes differ from physical changes?***
  Chemical changes alter the chemical composition of a substance. Physical changes represent changes in physical state but not in chemical composition.
- ***What does it mean to balance a chemical equation?***
  In a chemical reaction, mass is conserved, as shown here. A balanced chemical equation shows coefficients preceding the formula or symbol for each chemical species. The coefficients allow for the same number of atoms of a given element on each side of the reaction arrow.

Figure 6.22 • Law of conservation of mass

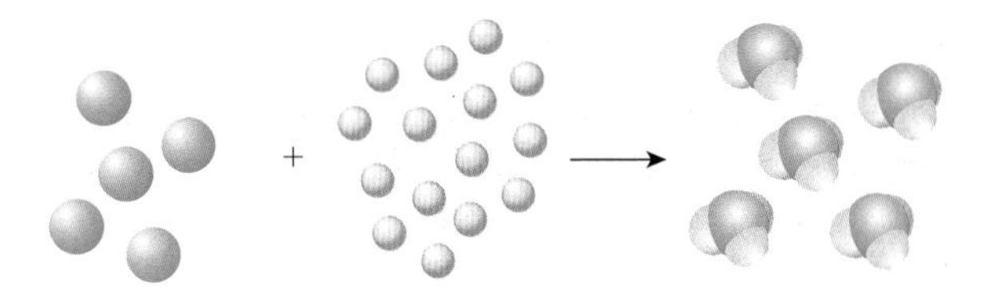

4 Counting Atoms and Molecules 187

- ***How can weighing a sample of a pure substance tell you the number of chemical particles it contains?***
  The atomic or molecular mass of a pure substance expressed in grams represents 1 mole, or 6.02×10^{23} chemical particles of that substance, as shown here.

Figure 6.25 • Converting from mass to moles

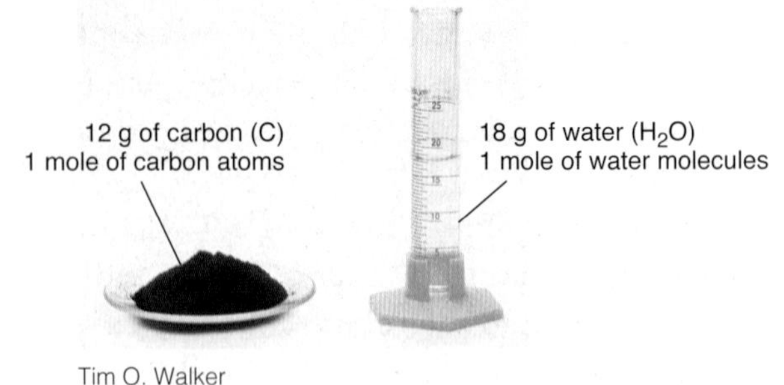

Tim O. Walker

- ***How can you predict the amount of reactant consumed or product formed in a chemical reaction?*** The coefficients in a balanced chemical equation tell us the molar ratios of the chemical species involved in a reaction. To determine the ratio of masses: (1) convert the mass of one substance into the number of moles it represents; (2) convert moles of that substance into moles of a second substance, using coefficients from the balanced equation; and (3) convert the moles of the second substance into the mass it represents.

Key Terms

What is happening in this picture?

Here, a worker injects **epoxy** adhesive into a seam of a massive stone sculpture to help prevent the stone from crumbling during transport. [This took place during the relocation of an entire stone temple to higher ground in advance of flooding expected from the building of the Aswan Dam in Egypt.] Epoxy adhesives use two chemical components: a polymer *resin* and a small molecule *hardener.* In hardware stores, you may find epoxy sold within a double-barrel syringe. This is to separate the liquid resin and hardener until they are ready to be applied. As the applied liquids mix and make contact with the surfaces to be bonded, the epoxy begins to *cure,* as the two components react with each other to form a solid polymeric product. The odor of epoxy is largely due to the evaporation of solvents.

Think Critically Does the use of the adhesive shown in this image involve chemical changes, physical changes, or both? Explain.

Exercises

Review

1. Describe how:
   a. the volume of a gas varies with changes in pressure under constant temperature;
   b. the volume of a gas varies with changes in temperature under constant pressure.
2. What determines the energy of the molecules of an ideal gas?
3. Rank the molecules of water, ice, and steam in order of the amount of kinetic energy of each. Start with the substance whose molecules have the highest kinetic energy.

4. What gas constitutes the largest percentage of the air we inhale?

5. What are the three kinds of kinetic energy a molecule can exhibit? Which one of these three resembles a ballet dancer twirling about in one spot on the stage? Which one resembles the energy of a sprinting runner? Which one is analogous to the energy you expend while your chest expands and contracts as you breathe, yet you remain standing in one place?

6. Describe, in terms of temperature, kinetic energy, and attractive forces between molecules, what happens as steam is cooled to water and the water is then cooled to ice.

7. a. Which state of matter of water is represented in the figure shown here?

   b. What do the dotted lines represent?

   c. Why do these lines appear only between a hydrogen atom in one molecule and an oxygen atom in an adjacent molecule?

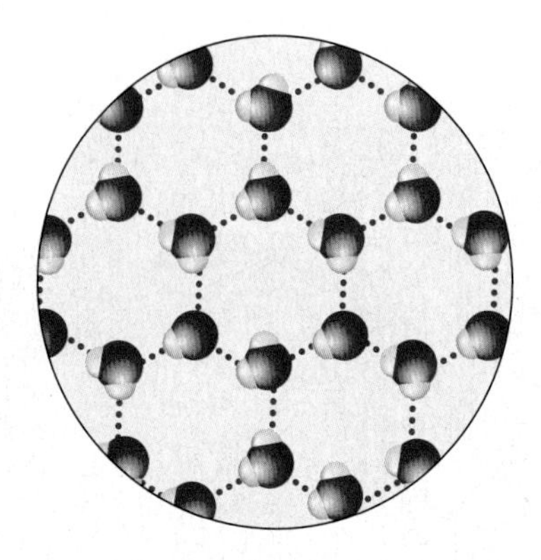

8. As shown in Table 6.1, acetic acid and water boil at higher temperatures than either ammonia or propane. What do these boiling points indicate about the relative strengths of intermolecular forces in acetic acid and water as compared to those in ammonia or propane?

9. Rank the following hydrocarbons from highest to lowest boiling points and explain your reasoning: methane (CH_4); butane (C_4H_{10}); heptane (C_7H_{16}); decane ($C_{10}H_{22}$)

10. What is the average barometric pressure at sea level:

    a. in mm-Hg;

    b. in atm;

    c. in psi (pounds per square inch; lb/in^2)?

11. What is the numerical value of absolute zero in degrees Celsius?

12. Why does ice float in water?

13. Does the density of a solid sample of a pure substance depend on its size?

14. What explains the hexagonal shape of snowflakes?

15. How do you determine the molar mass of a pure substance?

16. What physical process describes a change from:

    a. solid directly to gas

    b. gas to liquid

    c. solid to liquid

    d. liquid to solid

    e. liquid to gas, but only at the surface of the liquid

    f. liquid to gas throughout the bulk of the liquid

17. What is the function of the coefficients of a chemical equation?

Think

18. How are evaporation and boiling

    a. similar?

    b. different?

19. If a helium balloon breaks loose, it rises into the atmosphere and at some point it bursts. Explain why.

20. The solid line represents the temperature-volume relationship of a gas held at constant pressure.

    a. What occurs at the point where the solid line becomes dashed?

    b. What is one difference between the chemical particles of a real gas and the chemical particles of an ideal gas?

    c. What does Charles's Law predict about the volume of an ideal gas at absolute zero?

    d. What can you predict about the relative kinetic energies of the gas particles at 300 K as compared to at 100 K?

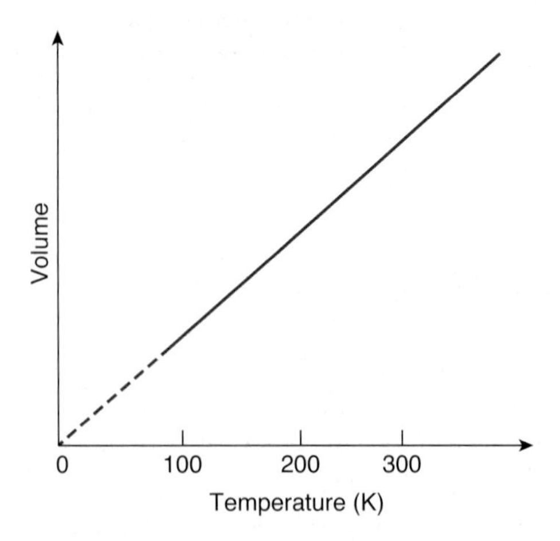

21. Every known element and compound can exist in a solid or a liquid form. Explain why this demonstrates that no real gas behaves exactly as an ideal gas would be expected to behave.

22. Why is it not possible, either in theory or in practice, to calibrate a scale in units of moles?

23. Suppose you have a scale that you know to be accurate, but that doesn't indicate what units of weight it measures. You have a quantity of table sugar (sucrose, $C_{12}H_{22}O_{11}$) and you want to measure a quantity of water that contains the same number of water molecules as there are sucrose molecules in your sample of sucrose. Could you use the scale to do that? Explain.

24. Why is the air pressurized inside an airplane flying at high altitudes?

25. Condensation trails (or contrails for short) may occur when water vapor from a high-altitude airplane's exhaust condenses in the frigid air, creating a suspension of fine water droplets that appear as white streaks in the sky. You may notice that the trails usually dissipate within a short period of time. What physical process causes the trails to disappear?

Calculate

26. Refer to Table 6.2 to help identify:
    a. the volume occupied by a 12-g solid piece of gold
    b. the mass (in kg) of a piece of oak lumber with dimensions 150 cm × 15 cm × 5 cm.

27. The Hope Diamond, a famous specimen located in the Smithsonian Museum of Natural History, weighs 45.5 carats (1 carat = 0.2 g). What is its volume? [Refer to Table 6.2].

28. Convert the following:
    a. 20°C to kelvins (K).
    b. 20 K to °C.
    c. 520 mm-Hg to atm
    d. 29.4 psi to atm

29. What would happen to the volume of the gas in the figure shown, if you:
    a. added a 2-kg weight
    b. removed a 1-kg weight
    c. added a 1-kg weight

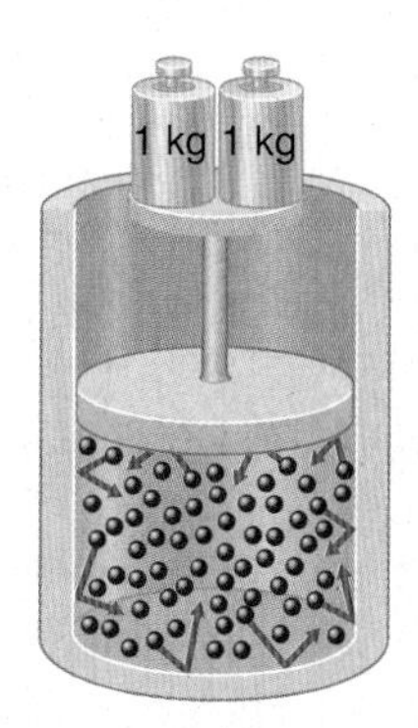

Adapted with permission of John Wiley & Sons, Inc. from Jespersen, N., Brady, J., and Hyslop, A., *Chemistry: The Molecular Nature of Matter*, Sixth Edition, p. 511. Copyright 2012.

30 In the figure below, the blue spheres represent element X, the white spheres represent element Y, and the product is XY_3.
    a. Write a balanced chemical equation for the reaction exactly as it is depicted here.
    b. Rewrite your equation using lowest whole number coefficients.

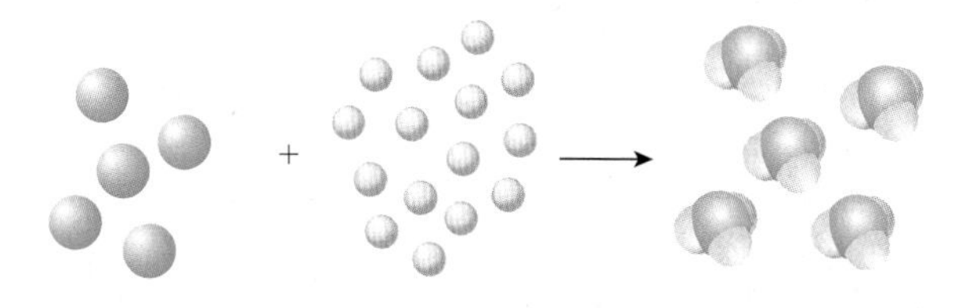

31. Suppose that the smallest decrease in the volume of a balloon that you could see easily is 20%. (That is, the balloon would have to shrink to 80% of its original volume for the decrease in volume to be visible.) To what Celsius temperature would you have to cool a balloon that was filled at 27°C, so that you could actually see it contract? Assume that atmospheric pressure remains constant.

32. You have a certain volume of gas kept at atmospheric pressure. You increase the temperature of the gas by 100°C and find that its volume doubles. What was the original temperature of the gas?

33. Suppose you fill a cold tire to 32 psi (lb/in^2) at 17°C and run it until it reaches a temperature of 37°C. What's the new pressure? (Assume that the volume of the tire doesn't change.)

34. A mountain climber fills three 1-L balloons with air at 760 mm-Hg and 27°C and carries one to the top of Mount Everest, the highest mountain in the world; one to Mount McKinley, the highest mountain in the United States; and one to Mount Kosciuszko, the highest mountain in Australia. Due to altitude and weather, the air pressure and temperature on the day she reaches each summit are:
Everest (245 mm-Hg, −36°C), McKinley (345 mm-Hg, −42°C), and Kosciuszko (590 mm-Hg, 7°C).

a. Calculate the volume of the balloon at each summit.

b. For each peak, is the drop in pressure or the drop in temperature the dominant factor in determining the final volume of the balloon?

35. Suppose you take a 1-L balloon filled with air at 760 mm-Hg and 27°C down to the bottom of Death Valley, California, the lowest point in the United States (86 m below sea level) on a day when the pressure reads 770 mm-Hg and the temperature is 57°C. (This temperature, the highest ever recorded in the Western Hemisphere, was reached in Death Valley in 1913). Will the balloon expand or contract, and what will the new volume be?

36. Balance any of the following equations that are not already balanced.

a. $Cl_2 + NaBr \longrightarrow NaCl + Br_2$

b. $2\ C_2H_6 + 7\ O_2 \longrightarrow 4\ CO_2 + 6\ H_2O$

c. $Cu + O_2 \longrightarrow CuO$

37. How many:

a. left shoes are there in seven pairs of shoes?

b. dozen yolks would you get from half a dozen eggs?

c. moles of sodium cations are there in half a mole of sodium chloride?

38. How many moles of oxygen atoms does it take to produce one mole of oxygen molecules (O_2)?

39. How many grams are represented by:

a. 1 mole of neon (Ne)?

b. 2 moles of silver (Ag)?

c. 1 mole of nitrogen gas (N_2)?

d. 3 moles of oxygen gas (O_2)?

e. 0.5 mole of chloride ions (Cl^-)?

f. 2 moles of chlorine gas (Cl_2)?

g. 1 mole of neutrons?

40. How many moles are contained in:

a. 1 g of Vitamin C (ascorbic acid, $C_6H_8O_6$)?

b. 0.325 g of aspirin (acetylsalicylic acid, $C_9H_8O_4$)?

c. 0.50 g of acetaminophen ($C_8H_9NO_2$), the pain reliever present in Tylenol?

d. How many individual molecules does each of your answers in parts a–c represent?

41. a. How many moles of CO_2 are produced when 1.5 moles of carbon atoms combine with 1.5 moles of oxygen gas (O_2)?

b. How many grams of CO_2 are produced by the reaction in part a?

42. a. Which of the substances in the figure—the marshmallows, the water, or both—represents a mole?

b. Estimate the mass of all the water on Earth (in kg) if the mass of an individual marshmallow is 7 g.

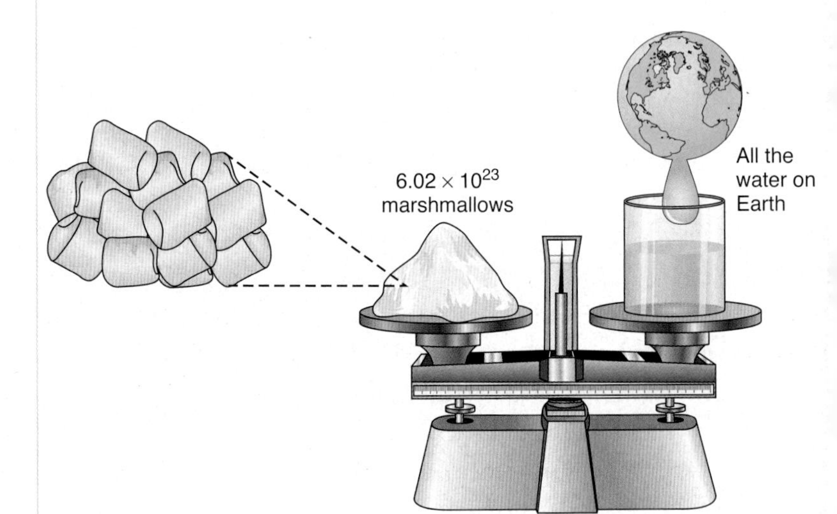

43. Chlorine gas consists of diatomic molecules, Cl_2. The balanced chemical equation for the reaction of sodium metal with chlorine gas to produce sodium chloride is $2\ Na + Cl_2 \longrightarrow 2\ NaCl$.

a. How many grams of chlorine gas does it take to react completely with 4.5 g of sodium?

b How many grams of sodium does it take to react completely with 1.42 g of chlorine gas?

c. If we started with exactly 10 g of sodium and 10 g of chlorine gas, would either of these elements be left over at the end of the reaction? If so, which one?

44. The use of fuels with small amounts of sulfur-containing impurities results in the formation of sulfur trioxide, SO_3, an atmospheric pollutant. The SO_3 dissolves in rain and other forms of atmospheric water to form sulfuric acid, H_2SO_4, according to the chemical equation $SO_3 + H_2O \longrightarrow H_2SO_4$. This reaction contributes to the formation of acid rain (Section 8.3)

   a. How many moles of SO_3 are there in 8 g of sulfur trioxide?

   b. How many moles of sulfuric acid form when 8 g of sulfur trioxide dissolve in water?

   c. How many grams of sulfuric acid does your answer in part b correspond to?

45. Protons and neutrons each have a molar mass of 1 g/mol.

   a. How many total *moles* of protons and neutrons combined are there in a person weighing 155 pounds (lb)? (2.2 lb = 1.0 kg) Assume that the person's electrons make a negligible contribution to her weight.

   b. How many total protons and neutrons combined does this represent?

46. Carbon monoxide (CO) is produced by the incomplete combustion of gasoline in a car's engine. To help reduce air pollution, catalytic converters (Section 4.2) are used in cars to promote the combination of carbon monoxide with oxygen, thus transforming the CO into CO_2.

   a. Write the balanced equation for the reaction of CO with O_2 to form CO_2.

   b. How many moles of CO react with each mole of O_2?

   c. How many grams of CO react with each gram of O_2?

47. The structures of a series of hydrocarbons are shown here, along with their heats of combustion, the amount of heat released when one mole of each burns.

   a. Examine the structures of methane and ethane. Confirm that methane has 4 covalent bonds and that ethane has a total of 7, comprising 1 C— C bond and 6 C— H bonds. Identify the total number of bonds in propane, butane, and pentane, respectively.

   b. For each compound in the chart, divide its heat of combustion by the total number of bonds established in part a. Approximately what value do you find for each? Note that this is not a rigorous analysis in that it does not take into account the covalent bonds of the O_2 consumed and the CO_2 and H_2O produced in these reactions. Nevertheless, do your results support the fact that there is energy stored within covalent bonds?

   c. How much heat is released in the combustion of (i) 100 g of methane; (ii) 100 g of pentane?

48. a. Write two balanced equations: one for the complete combustion of ethanol (C_2H_6O) and the other for the complete combustion of pentane (C_5H_{12}, a major hydrocarbon of gasoline).

   b. In each case, how many grams of carbon dioxide are produced by burning 100 g of the compound?

49. The average distance from the Sun to Earth is 150,000,000 km, and the diameter of a penny is 1.9 cm. If Avogadro's number of pennies were used to build a road from Earth to the Sun, and the road were just one layer of pennies deep, how many pennies wide would the road be?

Name	Formula	Structure	Heat of Combustion (kcal/mol)
Methane	CH_4		213
Ethane	C_2H_6		373
Propane	C_3H_8		531
Butane	C_4H_{10}		687
Pentane	C_5H_{12}		845

7

Water and Other Solutions

Earth's oceans consist of solutions of sodium chloride and other salts dissolved in water. They contain over 97% of all the water on our planet. Most of the remainder of Earth's water is frozen in glaciers and polar ice, leaving less than 1% as the readily accessible fresh water of our lakes, rivers, and underground reservoirs. We consume the largest proportion of this fresh water in cultivating food through ranching and farming.

Today over 1 billion people live in areas seriously deficient in fresh water, with the number expected to

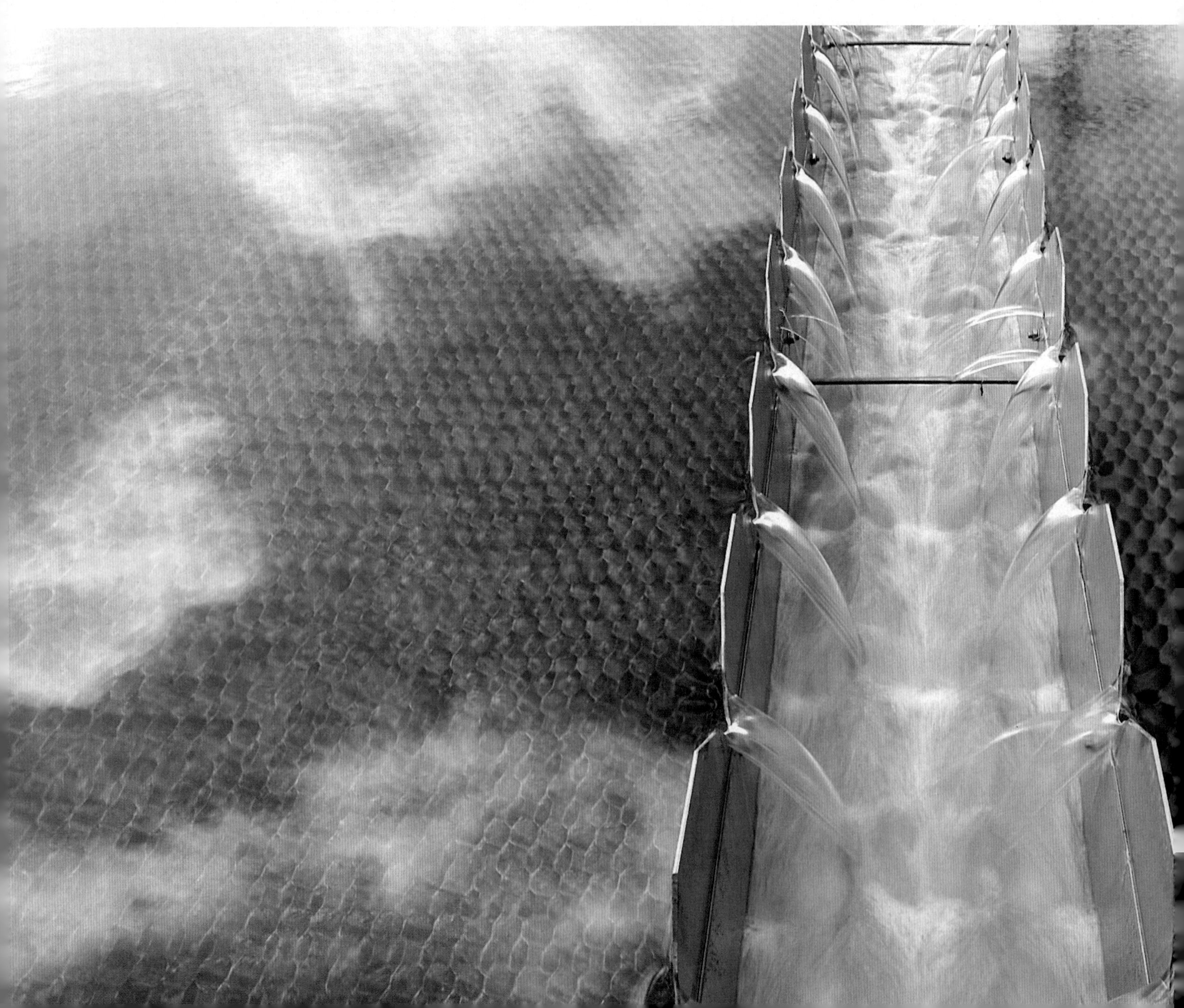

increase to 2 billion by 2025 if water usage rates continue to grow. With the abundance of salt water on the planet, one potential source of additional fresh water is desalination, which is the removal of salt from salt water. Over 14,000 desalination plants are currently operating worldwide, with most in the Middle East. In this image of a desalination plant, saline water in open-air tanks undergoes the first of a series of prefiltration steps prior to undergoing *reverse osmosis*, a process described shortly.

In this chapter we explore the broad topic of solutions—mixtures of solvents and dissolved substances—and their importance in our everyday lives.

Peter Macdiarmid/Getty Images

CHAPTER OUTLINE

7.1 Solutions and Other Mixtures

LEARNING OBJECTIVES

1. **Identify** the basic components of a solution.
2. **Describe** how water acts as a solvent.
3. **Explain** how colloids and dispersions differ from solutions.

Chemists often say there is a bit of something in just about anything else, meaning that the substances we find around us are typically mixtures. It is quite rare to find something that is truly chemically pure. In this section, we explore types of mixtures called solutions and discuss related topics.

solution A homogeneous mixture of two or more substances.

solvent A substance present in greater amount in a solution.

solute A substance present in lesser amount in a solution.

Types of Solutions

If you steep a tea bag in a cup of hot water, components of the tea leaves are **extracted** into the water to make a **solution**. Water is the **solvent**, and the various flavored compounds, pigments, and caffeine naturally present in the tea leaves are the **solutes**—the substances that dissolve within the solvent. Sweetening the tea with table sugar adds yet another solute, sucrose, to the solution. Swirling the tea with a spoon ensures that the mixture is **homogeneous**, having a uniform composition throughout, a property common to all true solutions.

The word solution derives from the Latin *solutio*, meaning a loosening or unfastening, which reflects what appears to happen to a substance as it dissolves in a solvent. But keep in mind that solution formation is a *physical* process—a change in form of the substances involved. Solution formation does not involve any chemical reactions.

We typically think of solutions as liquids, but solutions can also be gases or solids (**Figure 7.1**). Everyday consumer products, from personal care items to cleaning agents, consist

Solutions in different states • Figure 7.1

Solutions exist as a single state: either gaseous, liquid, or solid. The physical state of the solvent determines the physical state of the resulting solution.

Gaseous solutions
Air is a gaseous solution containing nitrogen, oxygen, water vapor, and trace gases.

Liquid solutions
Beverages are aqueous solutions. Champagne, for instance, is an aqueous solution containing various types of solutes, such as natural sugars, ethanol (a liquid), and carbon dioxide (a gas).

Solid solutions
Alloys are solid solutions of metals. For example, 18-karat gold is an alloy containing 75% gold and 25% other metals, such as copper.

Ask Yourself

1. Identify the solvent in each of the three solutions identified here. How does the physical state of each solvent compare to that of its respective solution?
2. (a) In air, what is the major solute gas? **(b)** In 18K gold, what is one type of solute?
3. Provide an example of: **(a)** a solid dissolved in a liquid, **(b)** a liquid dissolved in a liquid, **(c)** a gas dissolved in a liquid, and **(d)** a solid dissolved in a solid.

of a variety of solutions (**Figure 7.2**). Many if not most of these solutions are **aqueous**, and some contain **surfactants**, which are ingredients added to improve their cleaning abilities. (We'll explore surfactants, which include soaps, in Chapter 11.)

> **aqueous** Water-based or having water as solvent.
>
> **alloy** A solid metal solution containing two or more elements.

Solutions can exist as solids as well, including the many metals that are **alloyed** with other elements to improve their properties. For instance, pure gold (24 karat) is too **malleable** (soft) for use in jewelry. Alloying gold with other metals, such as silver, copper, or nickel, can improve its hardness as well as alter its color. (The purity of gold is indicated by its karats (K), with pure gold represented by 24K. Gold alloys include 18K, which is 75% gold, 14K which is 58% gold, and so on.)

Steel, which is used in construction and a host of other applications demanding strength and durability, represents another important alloy. In its most fundamental form, steel is an alloy of iron containing up to 2% carbon. Stainless steels, noted for their resistance to corrosion, are used in flatware, cookware, and other applications. These alloys of iron contain other metals as well, such as chromium and nickel.

Solubility

If you stir a teaspoon of table sugar or table salt into a glass of water at room temperature, it readily dissolves, yielding a clear solution. However, if you try dissolving cooking oil in water, you'll find that the oil does not dissolve but rather floats to the surface, just as an oil spill in a body of water forms an oil slick on the water's surface. We say that salt and sugar have a high **solubility** in water, but that oil does not.

> **solubility** A measure of the degree to which a solute dissolves in a solvent.

Common household solutions • Figure 7.2

Each of these solutions contains one or more solutes dissolved in water.

Windshield washer fluid can contain methanol, CH_3OH, and surfactants.
Shampoos, liquid soaps, and body washes contain cleaning agents called surfactants, as well as other ingredients.
Bleach is a solution of sodium hypochlorite, NaOCl.
Many cleaning solutions contain ammonia, NH_3.
Suave
NATURALS
hydrogen peroxide
CONCENTRATED
CLOROX
HEINZ
VINEGAR
PEAK
WINDSHIELD WASH
Ammonia
Janitorial Strength Formula
Net 32 oz
70% Isopropyl Alcohol
Tim O. Walker
Solutions of hydrogen peroxide, H_2O_2, are often used for cleaning and bleaching.
Vinegar is a solution of acetic acid, CH_3CO_2H.
Rubbing alcohol is a solution of isopropyl alcohol, $(CH_3)_2CHOH$.

Chemistry InSight Polarity of water and the basis for solubility in aqueous solutions • Figure 7.3

The angular shape of the water molecule and the difference in electronegativity between oxygen and hydrogen make water a polar compound.

a. A space-filling model of water.

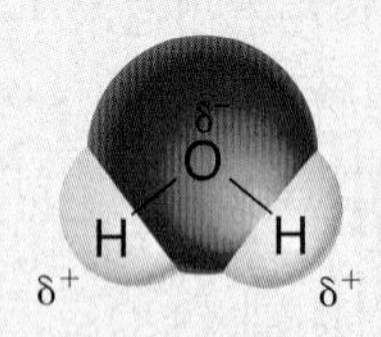

Water molecules associating with ions of sodium chloride (NaCl) in solution.

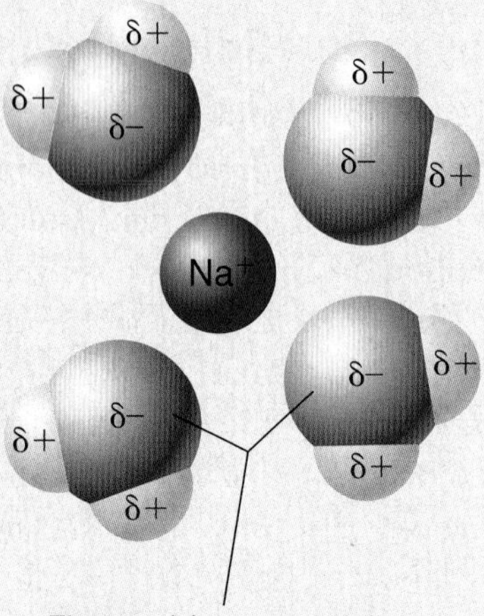

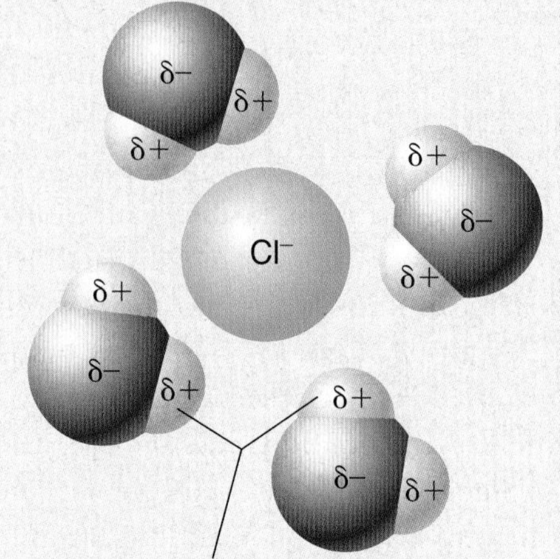

Electron-rich oxygen atoms of water surround positively charged ions, such as Na^+.

Electron-poor hydrogen atoms of water surround negatively charged ions, such as Cl^-.

Adapted with permission of John Wiley & Sons, Inc. from Jespersen, N., Brady, J., and Hyslop, A., Chemistry: *The Molecular Nature of Matter*, Sixth Edition, p. 535. Copyright 2012.

Water molecules associating with a molecule of glucose ($C_6H_{12}O_6$) in solution.

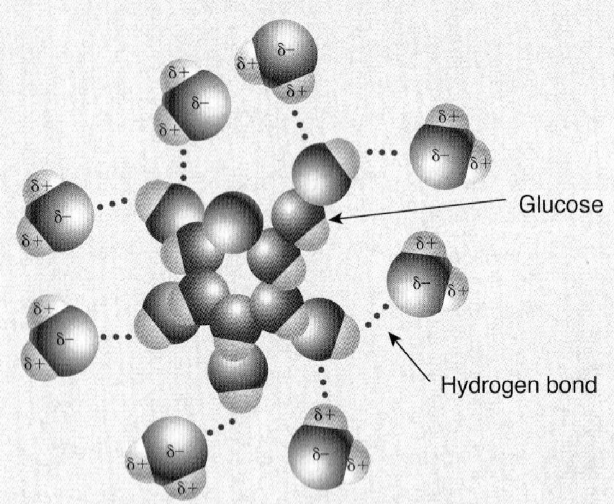

Intermolecular forces called *hydrogen bonds* can form between water molecules and the polar O–H groups of glucose.

b. The polarity of water enables it to form attractive interactions with ions in solution, such as the Na^+ and Cl^- ions of table salt (above), and with polar molecules, such as glucose (left). As a result, many ionic compounds and polar molecules are highly soluble in water.

Ask Yourself

1. Table sugar, or sucrose, discussed in Section 5.3, is highly soluble in water. Which type of intermolecular force discussed in this figure helps account for this solubility?
2. Water can form hydrogen bonds with glucose using either its own oxygen or hydrogen atoms. Explain why.

Sports drinks such as Gatorade are aqueous solutions containing:
- electrolytes such as Na^+ and Cl^- ions;
- sugars such as glucose;
- additional ingredients

Paul Johnson/E+/Getty Images

The intermolecular forces (discussed in Chapter 6) that determine whether a substance exists as a solid, liquid or gas at a given temperature, also play a key role in governing solubility. A useful rule of thumb for predicting solubility is that **like dissolves like**. This means that a solvent will generally dissolve a given solute if sufficiently strong attractive or intermolecular forces can form between solvent and solute particles. In our example, water (the solvent) and sugar and salt (the solutes) are all **polar** chemical species. Water is able to form strong attractions with the ions of salt and with the molecules of sugar (**Figure 7.3**). The molecules that make up oil, however, are **nonpolar** and are not able to form attractive forces with water, so oil does not dissolve in water (**Figure 7.4**).

polar Having a separation of charge due to differences of electronegativities of component atoms or ions.

nonpolar Lacking an overall separation of charge within a molecule.

The simple rule of "like dissolves like" has important practical applications. If you've ever tried to clean grease off your hands, dishes, or other surfaces, you know that water alone doesn't cut the grease, meaning that water doesn't dissolve grease. Soapy water, on the other hand, does wash away the grease, in part because soap molecules contain nonpolar hydrocarbon chains that associate with similar nonpolar chains of grease and oil molecules. (We'll have more to say about how soaps work in a later chapter.) For tougher stains rich in nonpolar compounds, such as the road grime and tar found on cars and bicycles, we sometimes use cleaners that contain **petroleum distillates**—hydrocarbons derived from petroleum. These nonpolar compounds belong to a broader class of chemicals called **volatile organic compounds** (VOCs), carbon-based compounds that are prone to evaporate easily. VOCs are used as solvents in a variety of everyday products, including paints and cosmetics.

Other factors also influence solubility, such as temperature. For instance, table sugar (sucrose) is far more soluble in hot water than in cold. As an example, a given volume of water at 95°C can dissolve more than twice as much sucrose as the same volume of water at 20°C.

We say a solution is **saturated** when it contains the maximum possible amount of dissolved solute at a given temperature. If this saturated solution is then cooled or solvent is allowed to evaporate, the solution can become

Why oil and water don't mix • Figure 7.4

The hydrocarbon-rich, nonpolar molecules of oils and the polar molecules of water do not attract each other. When oil and water (or aqueous solutions) are mixed, they separate into two layers, or **phases**, with the more dense aqueous phase beneath the less dense oil phase.

Vegetable oils are **triglycerides**, hydrocarbon-rich molecules (Section 5.1).
Vinegar is an aqueous solution.
Two separate layers form, with the denser layer (vinegar) on the bottom.

QUAYSIDE/iStockphoto

Supersaturated solutions and crystal formation • Figure 7.5

In supersaturated solutions—those in which the solvent contains an excess of solute—the excess solute crystallizes out of the solution until a (merely) saturated solution remains.

a. Here, rock candy (crystalline sucrose) forms on a submerged stick in a supersaturated sucrose solution. For more on how this is prepared, see *Wiley Plus*.

Ted Kinsman/Science Source

b. The Dead Sea has extremely high levels of **salinity** or dissolved salts. As the water evaporates, it leaves behind white, crystalline salt deposits along the shoreline as shown here. (The Dead Sea is also rich in a dark mineral mud often used as a skin treatment).

Eitan Simanor/Photolibrary/GettyImages

Think Critically **1.** Why does a hot, saturated aqueous sugar solution become supersaturated as it cools?
2. How do these examples of crystallization support the notion that solution formation is a physical process rather than one involving chemical reactions?

a **supersaturated**, or overly saturated solution. In this case, crystals of the solute eventually begin to form (**Figure 7.5**), although it is sometimes necessary to introduce a few "seed" crystals to initiate the crystallization process.

Colloids and Dispersions

Many types of liquid mixtures are not solutions at all, but are rather **colloidal suspensions**, or **colloids**, consisting of fine particles of one substance suspended in another. The easiest way to see the difference between a solution and a colloid is to shine a light through each of them. Shining a flashlight beam through a solution of sodium chloride in water doesn't show much.The beam passes through the clear solution without producing any observable effect. But shine the beam through a mixture of soap and water—which forms a colloid—and you can see the beam's path, especially if you view the mixture against a dark background. This phenomenon, called the **Tyndall effect**, was discovered in the 19th century by John Tyndall, a British physicist (**Figure 7.6**).

colloid A stable dispersion of fine particles of one substance in another.

Tyndall effect The scattering or illumination of a light beam passing through a colloidal mixture.

The Tyndall effect isn't limited to soapy water. A few drops of milk in a glass of water also show the path of the light. Milk is essentially an aqueous mixture of colloidal fats and proteins, along with dissolved lactose (milk sugar) and minerals. You can even see the Tyndall effect outdoors on a foggy night as the fog—a colloidal dispersion of water in air—scatters the headlight beams of automobiles.

The Tyndall effect • Figure 7.6

The Tyndall effect can be observed by shining a light beam through a colloidal suspension.

The light beam is not visible as it passes through the solution, because there are no suspended particles to scatter the light...
Light source
... but the beam is observable as it passes through the colloidal suspension, because the fine, dispersed particles scatter the light.
Solution
Colloidal suspension

Charles D.Winters/Science Photo Library

An **emulsion** is a type of colloid in which microscopic particles of one liquid, such as oil, are dispersed in another liquid, such as water (**Figure 7.7**). Examples of emulsions include homogenized milk, emulsified salad dressings, mayonnaise, and hand creams and lotions.

Emulsifiers, agents that help form and stabilize emulsions, contain compounds with two different kinds of functionality at the molecular level:

- polar, **hydrophilic** groups (water-loving, from the Greek *hydro* (water) and *philos* (love)) and
- nonpolar, **hydrophobic** groups (water-fearing, from the Greek *hydro* (water) and *phobos* (fear)).

The dual functionality of emulsifiers enables these molecules to form attractions to both oil and water molecules. The result is the formation of an emulsion. Two types of emulsions are possible:

Emulsions in everyday life • Figure 7.7

Oil and water do not form solutions, but they can form emulsions of a uniform consistency. Emulsions can be produced by homogenization, as in the case of milk, or by the addition of emulsifiers, as in the case of salad dressings and mayonnaise.

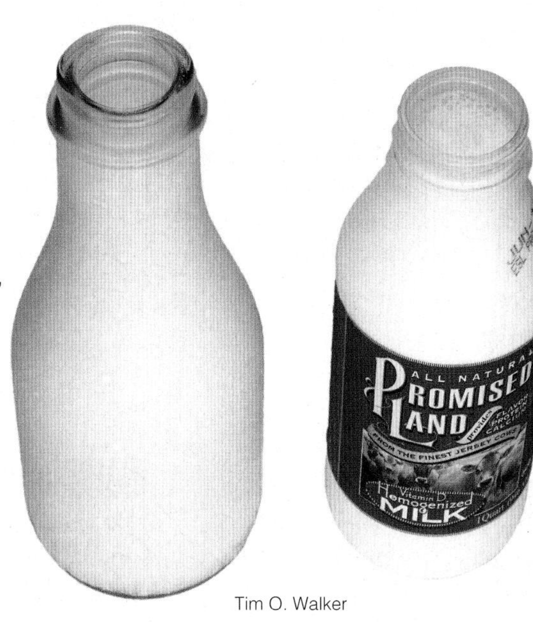

Tim O. Walker

Chemistry InSight How emulsions form •

Figure 7.8

Whisking together water, oil and a small amount of an emulsifier, such as mustard, forms an emulsion.

a. In an oil-in-water emulsion, shown here, the emulsifier surrounds each droplet of oil, enabling these droplets to remain dispersed in the water phase.

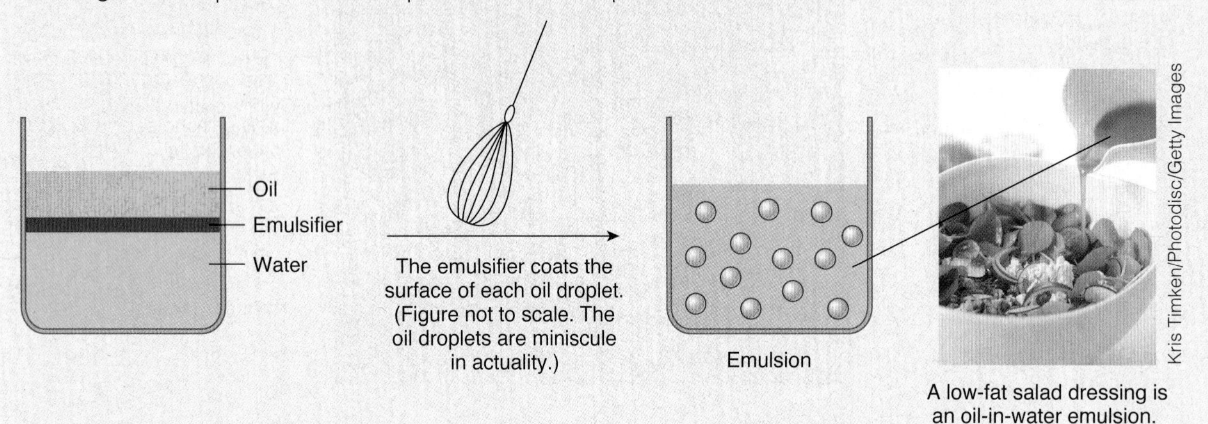

A low-fat salad dressing is an oil-in-water emulsion.

b. Emulsifiers, such as *lecithin* (present in egg yolks), contain both polar and nonpolar groups within the same molecule. A lecithin molecule has the following structure (hydrogen atoms have been omitted for clarity):

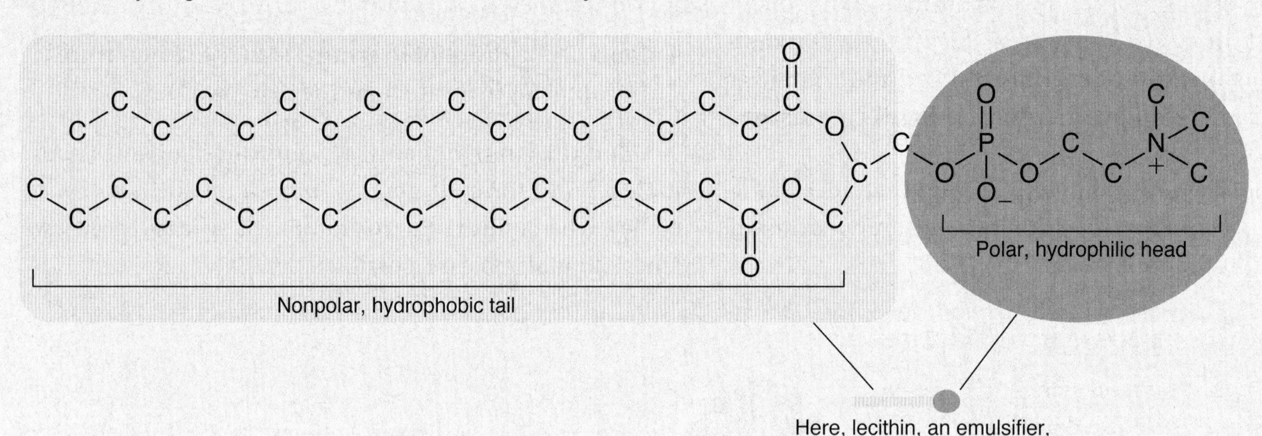

Here, lecithin, an emulsifier, is represented schematically with a nonpolar tail and a polar head.

c. The nonpolar tails of emulsifiers such as lecithin associate with the interior of oil droplets, leaving the polar heads of the emulsifier on the surface of the oil droplets. There they are available to form associations with the surrounding water.

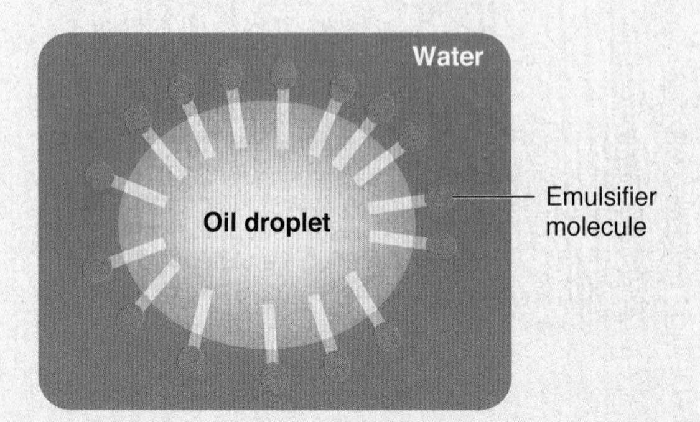

Ask Yourself

Mayonnaise is an emulsion containing oil, egg yolk, and either lemon juice or vinegar. What role does the egg yolk serve in stabilizing this emulsion?

Blood: A suspension and a solution • Figure 7.9

Blood is an aqueous suspension of cells, platelets, and other components. When blood is centrifuged, red blood cells settle to the bottom, leaving a straw-colored serum solution above, containing a variety of dissolved components.

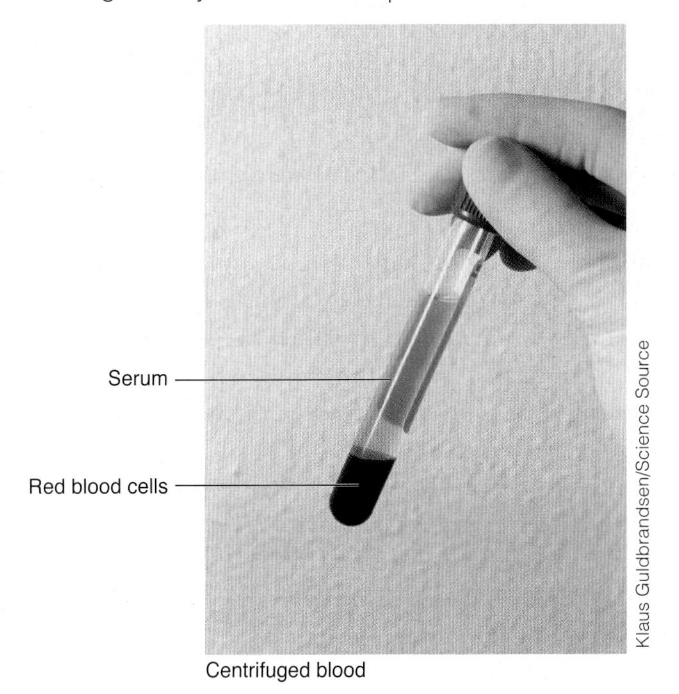

Centrifuged blood

- An oil-in-water emulsion, in which miniscule droplets of oil are suspended in water. Mayonnaise, for instance, is an oil-in-water emulsion (**Figure 7.8**).
- A water-in-oil emulsion, in which microscopic droplets of water remain suspended in oil. Butter and margarine, for example, are water-in-oil emulsions.

Suspensions are another type of dispersion. They typically contain fine particles of a solid dispersed in a liquid. These particles are generally slightly larger than those found in colloids, so the particles may settle out over time if the mixture is not periodically shaken or agitated. Common examples of suspensions include **milk of magnesia**, a milky white dispersion of magnesium hydroxide particles in water; **calamine lotion**, a dispersion of zinc oxide and iron oxide particles in water; and **Pepto-Bismol**, a dispersion of bismuth subsalicylate in water. Our blood is a complex fluid with characteristics of both suspensions and solutions (**Figure 7.9**).

> **suspension** A dispersion of solid particles in a fluid that tends to settle out over time.

CONCEPT CHECK STOP

1. **What** factor determines whether a solution exists as a solid, liquid, or gas?
2. **How** do solvent water molecules form attractive forces to ions and polar compounds in solution?
3. **What** can you predict about the relative sizes of the dispersed chemical particles in suspensions as compared to those dissolved in solutions?

7.2 Dissolved Gases

LEARNING OBJECTIVES

1. **Describe** Henry's Law.
2. **Provide** common examples of gases dissolved in liquids.

The familiar effervescence of carbonated beverages is due to bubbles of carbon dioxide gas coming out of solution. Have you ever wondered how soft drinks are carbonated to begin with and why they go flat over time? In this section we'll find answers to these questions and explore other common examples of gases dissolved in liquids.

Henry's Law of Dissolved Gases

William Henry, born in England in 1775, was an important early investigator of the nature of gases. Though he trained as a physician, he maintained a lifelong interest in chemistry, following in the steps of his father, a physician-apothecary who manufactured antacids

Henry's Law • Figure 7.10

At constant temperature, the amount of gas dissolved in a liquid is directly proportional to the pressure of the gas above the liquid.

Interpret the Data

1. In each of the figures above, what is the ratio of the number of gas particles (represented by circles) in the gas phase versus in the liquid phase?
2. How do your answers to Question 1 provide evidence for Henry's Law?

Less gas pressure above the surface of a liquid means that...

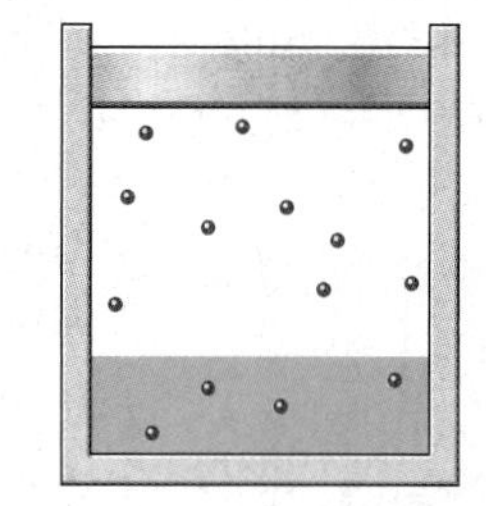

...*less* gas dissolves in the liquid.

More gas pressure above the surface of a liquid means that...

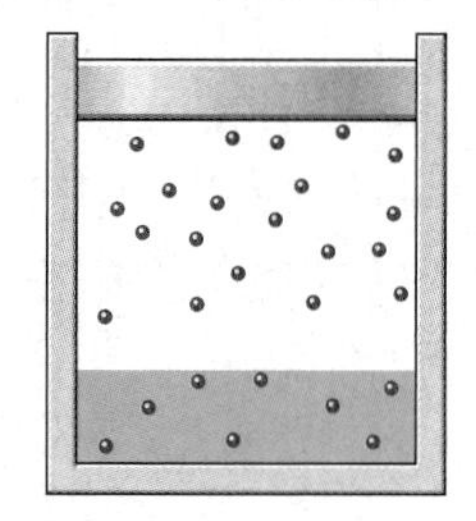

...*more* gas dissolves in the liquid.

Adapted with permission of John Wiley & Sons, Inc. from Jespersen, N., Brady, J., and Hyslop, A., *Chemistry: The Molecular Nature of Matter*, Sixth Edition, p. 595. Copyright 2012.

concentration A measure of the amount of solute dissolved within a given quantity of solvent or solution.

and soda water. Henry discovered that at a fixed temperature, as the pressure of a gas above a liquid increased, so did the amount or **concentration** of gas dissolved in the liquid. This relationship is known as **Henry's Law** (**Figure 7.10**). This Law forms the basis for how water is carbonated and why carbonated beverages eventually go flat if allowed to stand open to the atmosphere (**Figure 7.11**).

As the pressure of a gas above a liquid changes, Henry's Law helps predict how much of the gas dissolves in the liquid; but it says nothing about how fast the gas dissolves or how fast it comes out of solution. We know from experience that an open carbonated beverage loses its CO_2 over time. We can watch the bubbles form on the sides and bottom of the container and rise to the top. Shaking a bottle or a can of soda, especially a warm one in which the solubility of the CO_2 is low, often causes the drink to foam up and spill. This is due to a process called **nucleation**. Shaking the soda causes microscopic bubbles of the gas that occupies the space above the drink to enter the liquid and become trapped within it. These small, trapped bubbles serve as nuclei around which dissolved CO_2 can leave the solution. Opening the container results in a sudden drop in external pressure, causing the rapid formation of a large number of CO_2 bubbles around the gaseous nuclei.

nucleation A process that initiates the formation of a distinct state of matter within a solution, such as bubbles (gas) or crystals (solid).

Carbonated beverages • Figure 7.11

Soft drinks and other carbonated beverages are produced under a high pressure of carbon dioxide (CO_2) gas. According to Henry's Law, the amount of CO_2 dissolved in the liquid is proportional to the pressure of CO_2 above the liquid.

a. The high pressure of CO_2 in the unopened bottle (about 2.5 atm CO_2) helps maintain a high level of dissolved CO_2, or carbonation within the liquid.

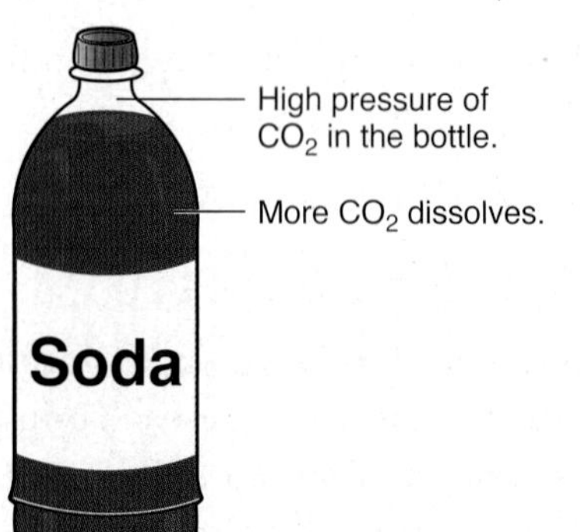

b. The very low pressure of CO_2 in the atmosphere (0.0004 atm CO_2) means that once the bottle is opened, relatively little CO_2 is soluble in the fluid. As a result, over time most of the CO_2 escapes, causing the soda to go flat.

Put It Together *Review Section 6.1 and answer the question.*

The familiar rush of escaping gas at the moment a bottled or canned soft drink is first opened is due in part to the expansion, or increase in volume, of the contained gas as the external pressure drops. What gas law describes this relationship between the pressure and volume of a gas?

Nucleation and release of dissolved gas • Figure 7.12

Dropping Mentos candies into an open bottle of a diet soda produces a geyser from the rapid release of dissolved CO_2 gas.

This electron microscope image of the surface of Mentos candy shows detail on the micrometer (μm) scale. The image reveals a rough, irregular surface that provides numerous nucleation sites for dissolved CO_2 to bubble out of solution.

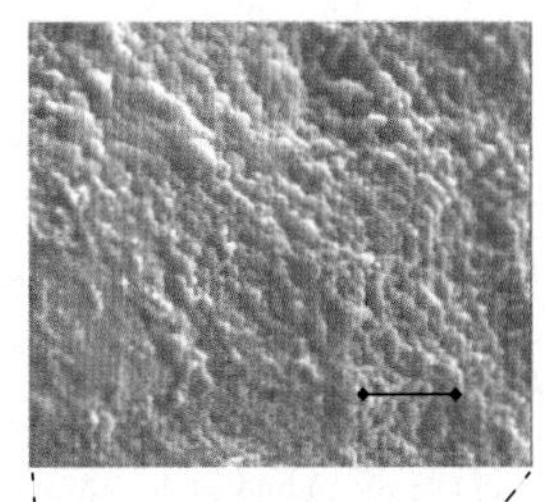

Coffey, T. S, *American Journal of Physics*, Vol. 76, Issue 6, pp. 551-557, 2008. Copyright 2008, American Association of Physics Teachers

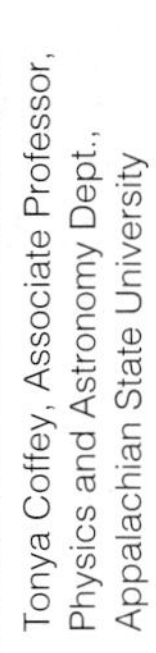

Tonya Coffey, Associate Professor, Physics and Astronomy Dept., Appalachian State University

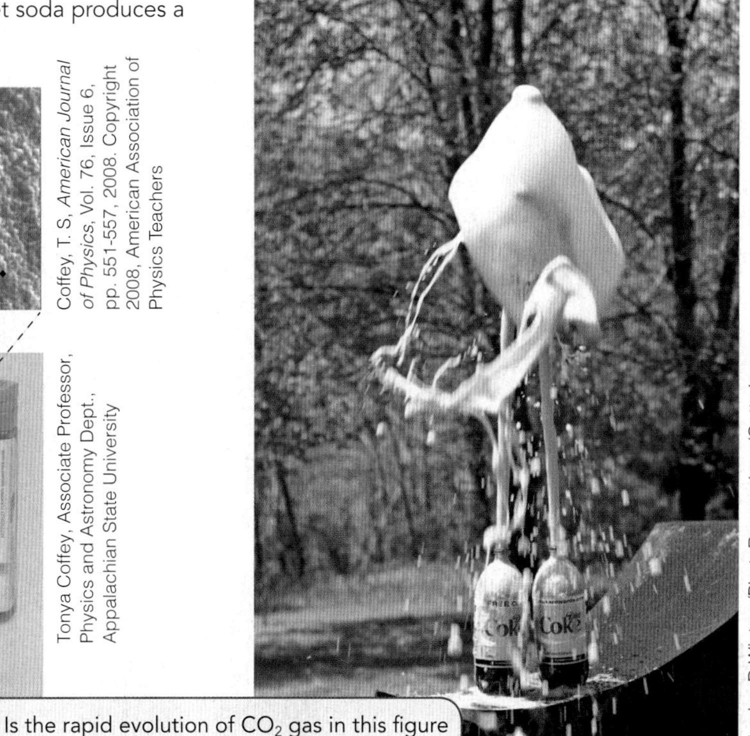

Charles D Winters/PhotoResearchers/Getty Images

Think Critically Is the rapid evolution of CO_2 gas in this figure due to a physical process or due to a chemical reaction?

Nucleation can also occur on the surface of a solid. Try adding a few crystals of sugar to a freshly poured soft drink. As soon as the granules enter the liquid, they provide a surface for nucleation and result in a rush of gas bubbles. **Figure 7.12** shows an even more dramatic effect involving nucleation.

The rate at which a gas dissolves in a liquid depends partly on the surface area of contact between liquid and gas. If all other conditions are equal, the larger the surface area, the more rapidly the gas diffuses into and dissolves in the liquid. The actual amount of gas that can dissolve in a liquid—its solubility—depends on various factors. For instance, Henry's Law predicts that as we increase the pressure of the gas over a liquid, its solubility increases. Also, lower temperatures help in this case, since gases, unlike most solids, are more soluble in colder liquids than warmer ones. This is especially important in aquatic environments. Fish, like other animals, need oxygen for life; they depend on the atmospheric oxygen that dissolves through the surface of bodies of water. The lower the temperature of the water, the greater the amount of oxygen that dissolves. In summer, ponds and other bodies of water warm up, which can limit the amount of dissolved oxygen. Fish in oxygen-deficient waters often show adaptive behaviors, such as pooling near the surface or gulping for air to obtain oxygen. If oxygen levels get too low, fish can die.

Other Common Examples of Dissolved Gases

Other important examples of dissolved gases include one that is vital to our own lives: the exchange of gases into and out of the bloodstream. We breathe simply to exchange the oxygen of the air for the carbon dioxide produced by the body's cells as they metabolize macronutrients. As an illustration, the cellular oxidation of glucose produces water, carbon dioxide, and energy.

$$\underset{\text{Glucose}}{C_6H_{12}O_6} + \underset{\text{Oxygen}}{6\,O_2} \longrightarrow \underset{\text{Water}}{6\,H_2O} + \underset{\text{Carbon dioxide}}{6\,CO_2} + \text{energy}$$

With each breath, oxygen enters our lungs and is transported by our blood to cells throughout the body. Our cells consume the oxygen through metabolic processes and use or store the energy generated by the metabolism. The water by-product of these processes becomes part of our general physical inventory. However, the carbon dioxide, a waste product, has to be transported by the blood from our cells back to the lungs so that it can be eliminated in exhaled breath.

We can use Henry's Law to describe concentrations of oxygen and carbon dioxide at various points in our bloodstream. Recall that at a given temperature, the concentration of a gas in a liquid solution is directly related to its pressure above the fluid. In a normal, healthy adult, blood just leaving the lungs carries its cargo of oxygen at a concentration equivalent to what we would find in a solution at normal body temperature that's under a pressure of 100 mm-Hg of O_2. Knowing this, we can use "100 mm-Hg" to define the concentration of O_2 in the blood leaving our lungs (**Figure 7.13**). [Note that oxygen is transported within the blood mainly by hemoglobin, a protein molecule present in red blood cells. In a later chapter we'll have more to say about how this process works.]

blood alcohol content The concentration of ethanol in the blood, typically measured in grams of ethanol per 100 mL blood.

Another application of Henry's Law is in determining **blood alcohol content** (BAC) using

PROCESS DIAGRAM

Gas transport within our bodies • Figure 7.13

As we inhale, air passes through the bronchial network of our lungs to the interior of **alveoli**, tiny cluster-shaped sacs. Here, oxygen (O_2) transfers to capillaries of the bloodstream and carbon dioxide (CO_2) transfers from the capillaries into the lungs (Step 1).

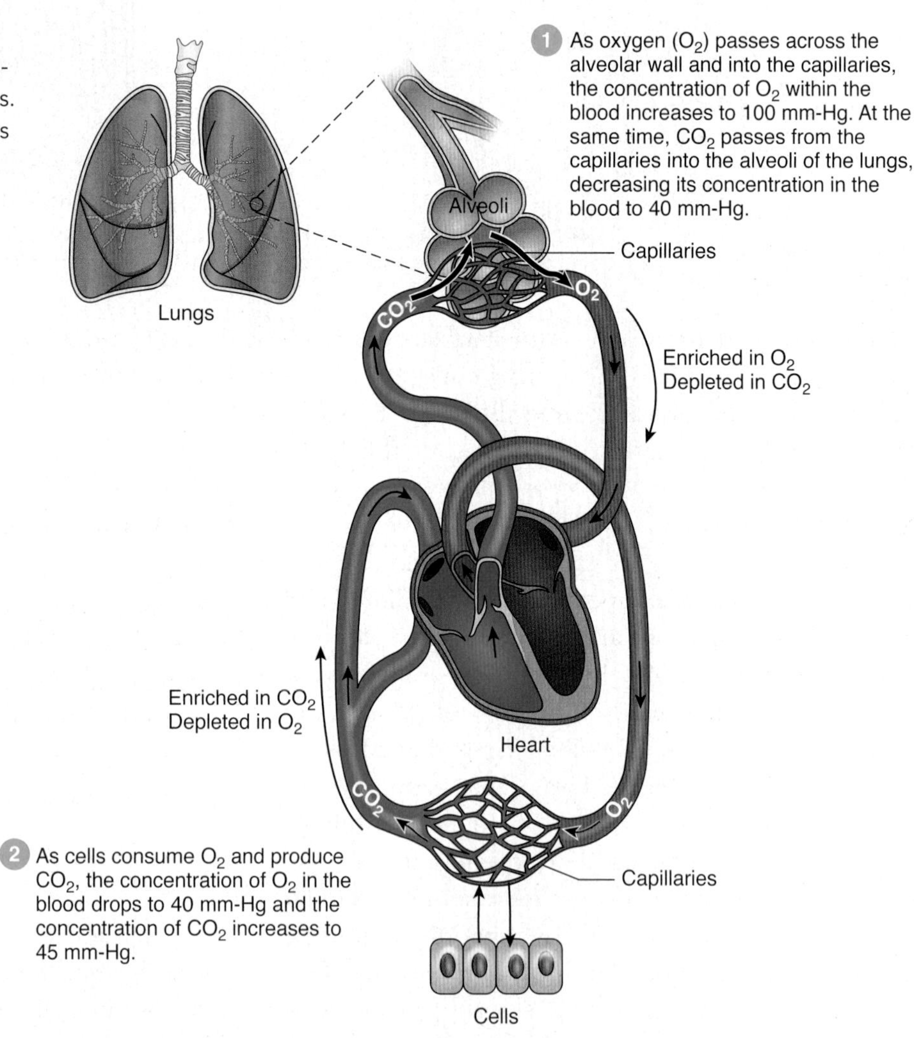

1 As oxygen (O_2) passes across the alveolar wall and into the capillaries, the concentration of O_2 within the blood increases to 100 mm-Hg. At the same time, CO_2 passes from the capillaries into the alveoli of the lungs, decreasing its concentration in the blood to 40 mm-Hg.

2 As cells consume O_2 and produce CO_2, the concentration of O_2 in the blood drops to 40 mm-Hg and the concentration of CO_2 increases to 45 mm-Hg.

Ask Yourself

1. What are the normal ranges in the concentrations of oxygen and carbon dioxide in the blood, expressed in mm-Hg?

2. Two gases constitute 99% of the (dry) air we inhale: nitrogen (78.1%) and oxygen (20.9%). Four gases make up virtually all of the air we exhale: nitrogen (74.9%), oxygen (15.3%), carbon dioxide (3.7%), and another gas that accounts for just over 6% of our exhaled breath. What is this fourth gas that makes up an even larger fraction of our breath than does carbon dioxide, and where does it come from?

WHAT A CHEMIST SEES

Blood Alcohol Testing Using a Breath Analyzer

When alcohol is ingested, it enters the bloodstream and eventually reaches the lungs, where small amounts of alcohol pass from the pulmonary capillaries into the alveoli of the lungs. In this process, a fixed ratio develops between the concentration of alcohol in the blood and the concentration of alcohol present in the **alveolar breath**, the exhaled breath that originates deep within the lungs.

a. A subject is required to exhale deeply into a breath tester so that alveolar breath can be analyzed. Any alcohol vapor present in the exhaled breath generates an electrical current within the breath tester, as a result of an electrochemical reaction. The magnitude of this current corresponds to the level of alcohol in the breath and therefore to the level of alcohol in the subject's blood as well.

DarrenMower/iStockphoto

b. Henry's Law describes the *partitioning* or distribution of a gas between the air above a fluid and within the fluid. Here, the concentration of alcohol in alveolar breath is proportional to the concentration of alcohol in the blood by a factor of about 1 to 2100.

A given volume of alveolar breath contains a little less than 0.05% of the ethanol in...

...the same volume of blood.

Interpret the Data

Suppose 0.01 milligrams (mg) of ethanol is present in 100 milliliters (mL) of a subject's alveolar breath. How many milligrams of ethanol would you expect to be present in 100 mL of the subject's blood? How many grams (g) of ethanol does this represent?

breath analyzer devices. Used by law enforcement officers as part of field sobriety tests, these instruments measure the concentration of ethanol vapor in a subject's exhaled breath. The concentration of ethanol in the breath can be correlated with the concentration of ethanol in the blood based on principles of Henry's Law (see *What a Chemist Sees*).

We've seen that environmental water contains dissolved oxygen, vital to the life of marine animals. Other gases present in the atmosphere also dissolve in water, such as nitrogen and carbon dioxide. Near room temperature, a maximum of about 0.01 g of nitrogen, 3.4 g of carbon dioxide, and 0.05 g of oxygen can dissolve in 1 L of water. (The small bubbles that rise from heated tap water, just before it starts to boil, are bubbles of these very same dissolved gases, which escape from the solution at higher temperatures.)

As rain forms in clouds and falls to Earth, it absorbs atmospheric gases. Atmospheric carbon dioxide that enters rainwater reacts with the water itself to form carbonic acid (H_2CO_3), contributing a slight acidity to all rainwater. The absorption of still other atmospheric gases, especially industrial pollutants, can increase the acidity of rainwater far beyond normal levels to produce what is known as **acid rain**. The world's oceans also absorb atmospheric carbon dioxide, leading to changes in the oceans themselves. We'll explore these important environmental topics in detail in the next chapter.

CONCEPT CHECK STOP

1. **How** does Henry's Law account for why an opened carbonated beverage eventually goes flat?
2. **What** similarities occur in the transport of carbon dioxide and of ethanol from blood into exhaled breath?

7.3 Solution Concentrations

LEARNING OBJECTIVES

1. **Define** solution molarity.
2. **Explain** how percentage concentrations are used to describe common solutions.
3. **Explain** what units are used to express exceedingly small concentrations.

The terms and units available for expressing solution concentrations are almost as varied as the kinds of solutions that can exist. These descriptions can range from the not very precise "heaping teaspoon per glass" for something you might prepare at home to more precise terms such as the percentages that appear on the labels of consumer products and units based on the concept of the mole. In this section, we'll explore various ways of describing solution concentrations and see their importance in everyday life.

Molarity

Chemists commonly express solution concentrations in **molarity** (*M*). The molarity of a solution refers simply to the number of moles of solute per liter of solution. A 1 *M* solution (pronounced "one molar") contains 1 mole of solute in each liter of solution; a 2 *M* solution contains 2 moles of solute per liter of solution; and so on.

> **molarity** A measure of solution concentration defined as the number of moles of solute per liter of solution.

To determine molarity, simply divide the number of moles of solute by the volume of the solution (in liters). For example, suppose 1.5 moles of solute is dissolved in enough solvent to give a solution volume of 0.4 L. The molarity of this solution is

$$\frac{1.5 \text{ mol solute}}{0.4 \text{ L solution}} = 3.75 \frac{\text{mol solute}}{\text{L solution}} = 3.75\ M$$

Knowing the molarity of a solution gives us a straightforward way to determine the number of moles of solute in a given volume of the solution. Simply multiply the volume (in liters) by its concentration (in moles per liter) to get the number of moles of solute. For instance, 0.4 L of a solution with a concentration of 3.75 moles per liter contains the following number of moles of solute:

$$\underbrace{0.4 \text{ L }\cancel{\text{solution}}}_{\text{Volume}} \times \underbrace{\frac{3.75 \text{ mol solute}}{1 \text{ L }\cancel{\text{solution}}}}_{\text{Concentration in molarity}} = \underbrace{1.5 \text{ mol solute}}_{\text{Moles of solute}}$$

Notice how we canceled units to arrive at the final answer, in units of moles.

For a given solution, if you know any two of the three values—molarity, volume, and number of moles of solute—you can determine the remaining value using unit cancellation. For example, to find the volume (in liters) of a 3.75 *M* solution that contains 1.5 moles of solute, you can use unit cancellation as follows:

$$1.5\ \cancel{\text{mol solute}} \times \frac{1 \text{ L solution}}{3.75\ \cancel{\text{mol solute}}} = 0.4 \text{ L solution}$$

You can use this same relationship if you want to prepare a specific volume of a solution of known molarity. For instance, suppose you wanted to prepare 5.0 liters of a 0.154 *M* NaCl solution. [This concentration of NaCl is the same as that used for intravenous saline solutions.] To prepare this solution, you need to weigh out the proper mass of NaCl, then dissolve it in enough water to make 5.0 liters of solution. The question is, how much NaCl should you weigh out? Using unit cancellation, you first determine the number of moles of NaCl required:

$$5.0 \text{ L }\cancel{\text{solution}} \times \frac{0.154 \text{ mol NaCl}}{1 \text{ L }\cancel{\text{solution}}} = 0.77 \text{ mol NaCl}$$

To convert from moles of NaCl to grams of NaCl, you need the molar mass of NaCl. You can calculate the molar mass of NaCl from the appropriate atomic masses:

$$\begin{aligned} \text{Na: } 1 \times 23.0 &= 23.0 \\ \text{Cl: } 1 \times 35.5 &= \underline{35.5} \\ &\ 58.5 \text{ g/mol} \end{aligned}$$

Knowing this molar mass, you now convert from moles of NaCl to grams of NaCl:

$$0.77\ \cancel{\text{mol NaCl}} \times \underbrace{\frac{58.5 \text{ g NaCl}}{1\ \cancel{\text{mol NaCl}}}}_{\text{Molar mass of NaCl}} = 45.0 \text{ g NaCl}$$

As a result of these calculations, you would dissolve 45 g of NaCl in somewhat less than 5.0 L of water, and then dilute the resulting solution with water until the volume of the NaCl solution reaches exactly 5.0 L.

KNOW BEFORE YOU GO

1. Household ammonia, a common cleaning agent, is a solution of ammonia, NH_3, in water. How many moles of NH_3 are in 2.0 L of a 0.50 *M* ammonia solution?
2. What volume (in L) of a 0.50 *M* ammonia solution contains 1.8 mol NH_3?
3. Sports drinks, such as Gatorade, contain sugars and electrolytes. You can prepare a simple homemade version of such a drink by mixing ¾ tsp salt (4 g of NaCl), ½ cup table sugar (95 g of sucrose), and 2 Tbsp lemon juice in enough water to make 2 quarts (1.9 L) of solution. In this solution, what is the molarity of a) NaCl? b) sucrose? (Use 342 g/mol for the molar mass of sucrose.)

Percentage Concentrations

Although molarity is a common measure of concentration in scientific studies, the concentrations of solutes listed on the labels of commercial products are often expressed as a percentage (%) of weight or volume. For example, the label of a typical bottle of vinegar, a dilute solution of acetic acid in water, indicates a concentration of "5% acidity." This shorthand notation means that vinegar typically contains 5 g of acetic acid in every 100 g of vinegar solution. Here we've expressed the weight of solute for every 100 units of weight of solution, which is a weight/weight percentage, or w/w %. Hydrogen peroxide solutions, commonly available in pharmacies, are typically sold as a 3% (w/w) solution of hydrogen peroxide, H_2O_2, in water. This means that every 100 g of solution contains 3 g of H_2O_2.

Another common way of expressing solution concentration is volume/volume percentage, or v/v %. For example, rubbing alcohol, a solution of isopropanol (often listed as "isopropyl alcohol") and water, is typically sold as a 70% (v/v) solution. This means that it contains 70 mL isopropanol in every 100 mL of solution.

The concentrations of alcoholic beverages (aqueous solutions of ethanol) are also described in terms of v/v %. For instance, most commercial beers have a concentration of ethanol between 4 and 6% (v/v). Twelve fluid ounces (355 mL) of such a beer contains 4–6% ethanol by volume, or about 15–20 mL ethanol, slightly more than a tablespoon. Most wine contains about 12% ethanol by volume, or 12% (v/v). The concentration of ethanol in hard liquor is often described in terms of "proof," which in the United States is defined as twice the volume-percentage value. For instance, 80-proof liquor contains ethanol at a concentration of 40% (v/v). The term *proof* was established early in the 16th century in England for purposes of taxing liquors according to alcohol content. The term originated in a test to determine whether someone may have tampered with the liquor by diluting it with water. The liquor in question was added to a small amount of gunpowder. If the wet gunpowder ignited, it was considered *proof* that the liquor met or exceeded a minimum expected alcohol content (usually about 50% alcohol by volume).

Figure 7.14 summarizes several ways of using percentages to define solution concentrations.

Percentage concentrations • Figure 7.14

Many types of commercial solutions are described in terms of percentage concentration.

Vinegar is a 5% w/w solution of acetic acid in water. Each 100 g of solution contains 5 g of acetic acid.

Tim O. Walker

Rubbing alcohol is commonly sold as a 70% v/v solution in water. Each 100 mL of solution contains 70 mL of isopropyl alcohol (isopropanol).

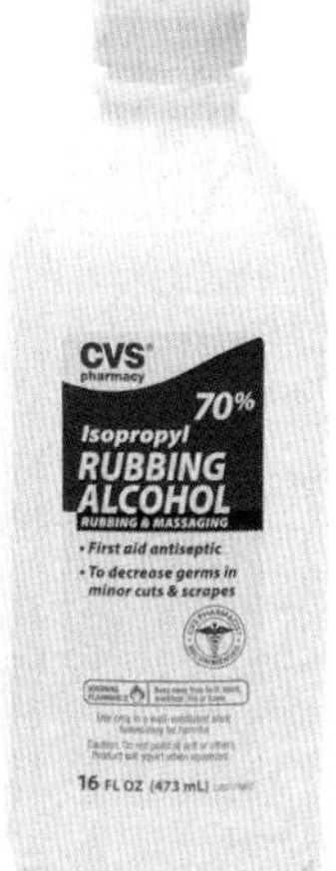

Tim O. Walker

Intravenous saline is a 0.9% w/v sterile solution of sodium chloride, NaCl, in water. Each 100 mL of solution contains 0.9 g of NaCl.

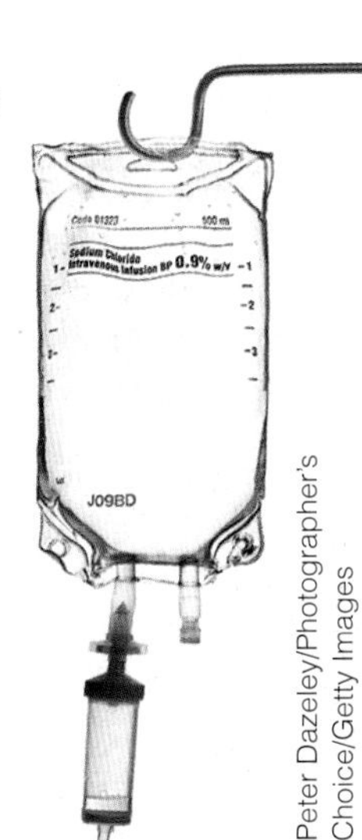

Peter Dazeley/Photographer's Choice/Getty Images

Ask Yourself

1. In the rubbing alcohol shown, which component is the **(a)** solvent, **(b)** solute?
2. Which of the solutions shown here are described in terms of volume of solution?

KNOW BEFORE YOU GO

4. A typical 8-fluid-ounce cup of coffee weighs about 240 g, and a teaspoon of sugar weighs about 6 g. What is the w/w % concentration of sugar in the coffee when you sweeten a typical cup of coffee with two teaspoons of sugar? (Don't forget to add the weight of the added sugar to the total weight of the solution.)

Concentrations of Solutes within Blood

To maintain proper health, our bodies operate according to the principle of **homeostasis**, which means maintaining our body temperature and blood composition within certain limits. The composition of blood includes cellular components as well as a host of dissolved substances, such as:

- blood sugar (glucose),
- electrolytes, such as sodium, potassium, and calcium ions,
- enzymes and other proteins,
- triglycerides and cholesterol,
- hormones, and various other components.

Normal, healthy body functioning maintains the concentration of each of these substances within a certain range. For instance, glucose levels are normally between 80 and 130 mg per 100 mL of blood, varying somewhat throughout the day as we eat food and engage in physical activity. For those with diabetes, though, blood glucose levels tend to run higher than normal and can lead to a variety of health problems if the condition is not treated.

Whenever you ingest food or medicine, compounds within those substances are largely absorbed by the blood. For instance, when you take aspirin, this medication's active ingredient, *acetylsalicylic acid,* enters the blood and circulates throughout the body. Concentrations of acetylsalicylic acid rise initially within the body as the medication enters the bloodstream, and then fall gradually as the body metabolizes and ultimately excretes the substance and its by-products.

The consumption of alcohol (ethanol) follows a similar pattern, producing an initial rise in blood alcohol content (BAC) as ethanol is absorbed by the body. In the United States, BAC is measured as grams of ethanol per 100 mL of blood, which is a weight-per-volume percentage (w/v %). Because alcohol can impair judgment and motor function, drivers in most countries are subject to legal limits for blood alcohol content. In the United States the limit is 0.08%, unless the driver is operating a commercial vehicle, in which case the limit is 0.04% or even less in some cases. The rate at which alcohol is absorbed and processed by the body varies, but it generally

Blood alcohol content • Figure 7.15

Levels of blood alcohol rise within a few minutes of consumption, eventually reaching a peak and then falling over time. Blood alcohol levels depend on several factors, including the amount of alcohol consumed, the length of time since the alcohol was consumed, your body weight, and gender.

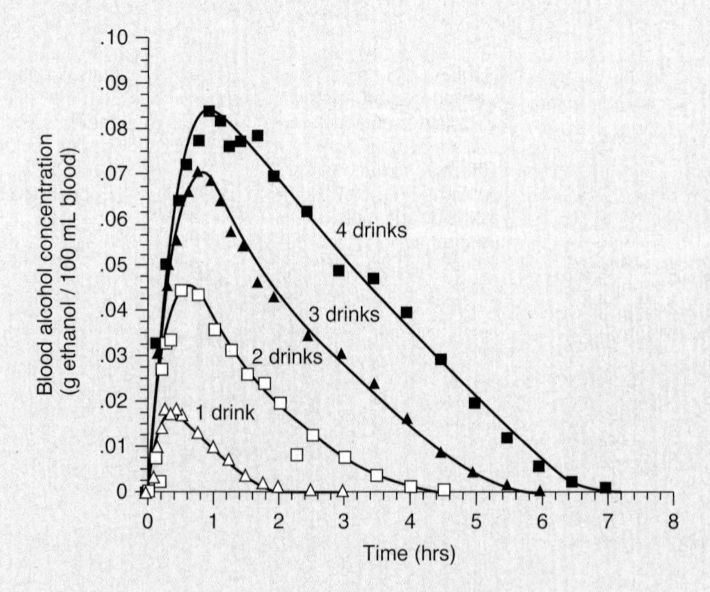

a. *Blood alcohol concentration as a function of time after consumption.* Each curve represents a different quantity of alcohol consumed. (The data is based on average responses from eight fasting, male subjects.) Source: National Institute on Alcohol Abuse and Alcoholism. *Note:* Since alcohol is absorbed faster from an empty stomach, consuming alcohol from a full stomach results in comparatively lower *peak* BAC levels, but the alcohol remains in the bloodstream for longer periods.

takes several hours after alcohol is consumed for blood alcohol content to dissipate (**Figure 7.15**).

Expressing Exceedingly Small Concentrations

Molarity is a useful concentration term for chemists, and percentages are informative for many consumer products. However, the very small concentrations of the pollutants that affect our environment and drinking water require a different approach.

It's useful to express exceedingly small concentrations in terms of parts per thousand, parts per million, parts per billion, and so on. These terms are closely related to the percentages we've already examined. The term *percent* literally means "per hundred." A one-percent solution contains one unit of solute in each 100 units of solution, whatever the units happen to be. We can also speak of concentrations in terms of parts per thousand, which are the same as tenths of a percent. Ocean water, for example, has a salinity, or concentration of dissolved salts, of about 0.3% ("three tenths of a percent"). This value, 0.3/100, is the same as 3/1000, which is the same as 3 parts per thousand.

part per million (ppm) A solution concentration equal to 1 mg of solute in 1 kg of solvent.

One part per million (one **ppm**) represents a particularly convenient concentration unit because it's the concentration of 1 milligram (1/1000 g) of one substance distributed throughout 1 kilogram (1000 g) of another:

$$\frac{1\ \text{mg}}{1\ \text{kg}} = \frac{\frac{1}{1000}\ \text{g}}{1000\ \text{g}} = \frac{1}{1000 \times 1000} = \frac{1}{1{,}000{,}000}$$

$$= 1\ \text{part per million (ppm)}$$

As we'll see in the next section, concentrations in parts per million (ppm) are commonly used to measure levels of dissolved solutes in drinking water. Concentrations in parts per billion (ppb) are even smaller by a factor of 1000, with 1 ppb representing one thousandth of 1 ppm (**Figure 7.16** on the next page).

KNOW BEFORE YOU GO

5. Zinc is important to human health. A deficiency of zinc can lead to stunted growth, incomplete development of sexual organs, and poor healing of wounds. Among foods richest in this element are liver, eggs, and shellfish, which contain zinc at levels ranging from about 2 to 6 mg per 100 g. Express this range of zinc concentrations in parts per million. (Note that an excess of zinc can be just as harmful as a deficiency, but a balanced diet provides most of us with all the zinc we need.)

b. *What is one drink?* One standard drink contains 0.6 fluid ounces (fl oz) of ethanol, equivalent to 14 g. Because of differences in alcohol content between beer, wine, and hard liquor, the different volumes of each drink shown here contain the same amount of ethanol, 0.6 fl oz.

Think Critically **Figure a** does not indicate the average body weight of the subjects under study. How would the data compare for a group of larger individuals (with a greater average total blood volume) as compared to a group of smaller individuals, other things being equal?

In Words, Math, and Pictures

Working with minute concentrations • Figure 7.16

smartstock/iStockphoto

This label reads: ***This salt supplies iodide, a necessary nutrient.***

a. Converting a concentration into parts per million (ppm)
Traces of iodide ion (I^-) in the diet help prevent the enlargement of the thyroid gland, a condition known as goiter. To provide this dietary iodide, potassium iodide, KI, is added to commercial table salt, NaCl, to the extent of about 7.6×10^{-5} g of KI per gram of NaCl. The following example shows how we can convert this concentration into ppm.

We know that the concentration of KI in table salt is

$$\frac{\text{mass KI}}{\text{mass NaCl}} = \frac{7.6 \times 10^{-5}\ \text{g KI}}{1\ \text{g NaCl}}$$

In this case, the quickest way to convert this to ppm is to multiply both the numerator and the denominator by one million, or 10^6. This then tells us how many grams of KI there are in one million grams of NaCl:

$$\frac{7.6 \times 10^{-5}\ \text{g KI}}{1\ \text{g NaCl}} \times \frac{10^6}{10^6} = \frac{76\ \text{g KI}}{10^6\ \text{g NaCl}} = \mathbf{76\ ppm\ KI}$$

Alternatively, we can use unit cancellation to convert the concentration into units of mg/kg, which also represents ppm:

$$\frac{7.6 \times 10^{-5}\ \cancel{\text{g KI}}}{1\ \cancel{\text{g NaCl}}} \times \frac{10^3\ \text{mg KI}}{1\ \cancel{\text{g KI}}} \times \frac{10^3\ \cancel{\text{g NaCl}}}{1\ \text{kg NaCl}}$$

$$= \frac{76\ \text{mg KI}}{1\ \text{kg NaCl}} = \mathbf{76\ ppm\ KI}$$

b. Converting a concentration into parts per billion (ppb)
Botulinum toxin, the virulent poison of spoiled food, provides another illustration of dealing with extremely low concentrations of substances. This substance is the most powerful biologically produced toxin known. A dose of as little as 1×10^{-9} g of Botulinum toxin can kill a mouse. This is one nanogram (ng), or a billionth of a gram.

In the 1970s, researchers discovered that injecting minute quantities of this toxin into muscle tissue interferes with muscle contraction, essentially paralyzing the muscle for a period of up to a few months. Now marketed under the trade name Botox, this substance is best known for reducing facial wrinkles, although it has other medically approved uses.

A vial of Botox contains 5×10^{-9} g (5 ng) of botulinum toxin. Prior to use, the contents of the vial are typically dissolved in 1 or 2 mL of sterile saline fluid, depending on the application. Assuming 2 mL of fluid is used, the concentration of toxin is

$$\frac{5\ \text{ng botulinum toxin}}{2\ \text{mL fluid}}$$

Since the density of the saline fluid used is 1 g/mL, the 2 mL of fluid used has a mass of 2 g. The concentration of the botulinum toxin in this mass of fluid is

$$\frac{5\ \text{ng botulinum toxin}}{2\ \text{g fluid}} = \frac{2.5\ \text{ng botulinum toxin}}{1\ \text{g fluid}}$$

$$= \mathbf{2.5\ ppb\ botulinum\ toxin}$$

The units of ng/g represent parts per billion (ppb), since a gram (g) is one billion times as large as a nanogram (ng).

CONCEPT CHECK

1. **What** two values do you need to know to calculate the molarity of a solution?
2. **What** is the major solute in household vinegar, and what is its approximate concentration?
3. **What** unit of concentration is 1 milligram of solute per kilogram of solution equivalent to?

7.4 Water in Our World

LEARNING OBJECTIVES

1. **Appreciate** that fresh water is a relatively scarce resource.
2. **Understand** that fresh water naturally contains dissolved substances.
3. **Define** polluted water.
4. **Describe** how fresh water can be produced from salt water.

Life on Earth would be impossible without water. All living things contain water and rely on water for their survival. Our own human bodies consist of 60% water, by weight. We rely on water for cooking, cleaning, recreation, and for industrial and agricultural processes. Virtually all the water we encounter in our daily lives exists as the solvent of a variety of solutions. In this section we'll explore the nature of these common, aqueous solutions, their chemical compositions, and the means we use to derive pure, uncontaminated water from them.

The Relative Scarcity of Fresh Water

Earth's crust holds about 2×10^{21} kg (2×10^{18} tons) of water dispersed over more than 70% of its surface; 97.5% of this water is salt water, unfit for drinking. Although 2.5% is fresh water, most of this is inaccessible, locked in glaciers or otherwise too remote. Only about 0.3% of Earth's water is readily accessible and available for use as fresh, drinkable water (**Figure 7.17**).

Currently, the population of Earth removes—and largely returns—about 4300 km^3 (or 1000 cubic miles) of

Water on Earth • Figure 7.17

The vast majority of Earth's water exists in the oceans.

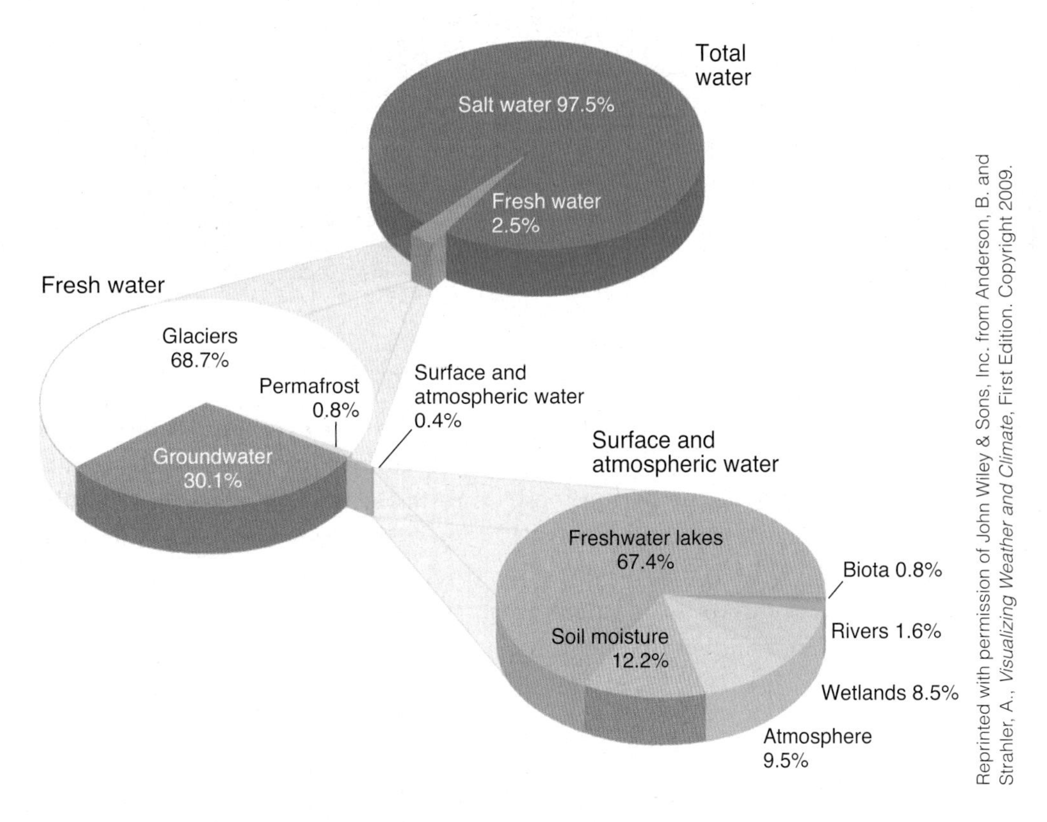

Interpret the Data

Of the total water present on Earth, what percentage exists as **(a)** groundwater? **(b)** freshwater lakes?

PROCESS DIAGRAM

Hydrologic cycle • Figure 7.18

Energy from the Sun helps drive the natural flow of water among the oceans (hydrosphere), the air (atmosphere), the land (lithosphere), and all living things (biosphere).

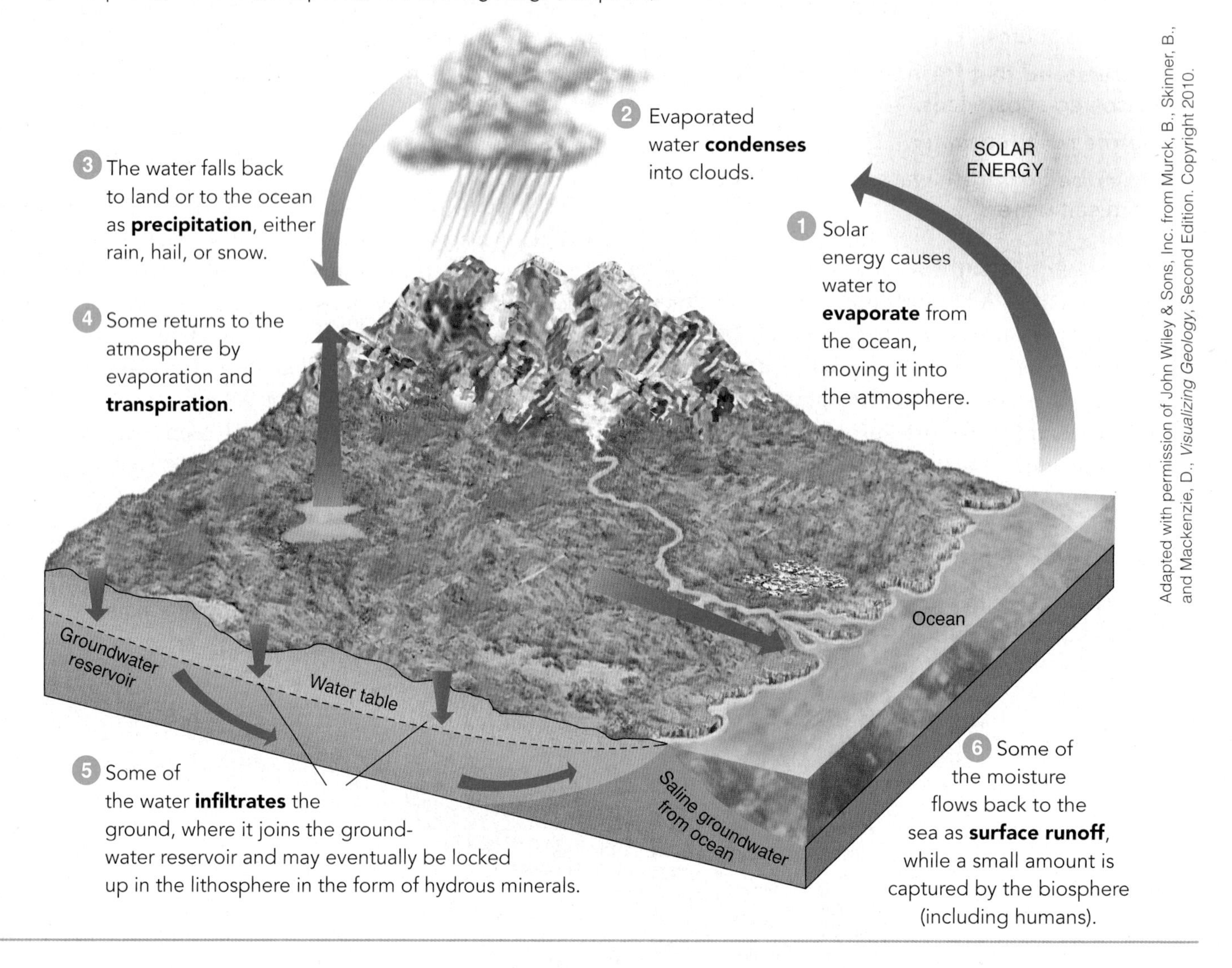

1 Solar energy causes water to **evaporate** from the ocean, moving it into the atmosphere.

2 Evaporated water **condenses** into clouds.

3 The water falls back to land or to the ocean as **precipitation**, either rain, hail, or snow.

4 Some returns to the atmosphere by evaporation and **transpiration**.

5 Some of the water **infiltrates** the ground, where it joins the groundwater reservoir and may eventually be locked up in the lithosphere in the form of hydrous minerals.

6 Some of the moisture flows back to the sea as **surface runoff**, while a small amount is captured by the biosphere (including humans).

freshwater each year from our lakes, rivers, and streams and from the groundwater (water held underground). With all the water we use for drinking, preparing food, and washing, we might think that fresh water is mostly used for these purposes. However, this is not so. Globally, about 67% of all the fresh water drawn is for irrigation and other agricultural needs. Domestic activities account for only about 10% of freshwater use.

Does Pure Water Exist?

When you think of pure water, you may imagine fresh rainwater falling from a crisp sky or clear water bubbling up from a mountain spring. The very purest water on Earth doesn't come from either of these sources, but rather from a chemist's laboratory. This is water that has passed through columns of specially prepared cleansing resins or water that has been distilled repeatedly under carefully controlled laboratory conditions to remove all traces of contaminants. All other water, including fresh rainwater and the very purest drinking water, carries a variety of impurities in a range of concentrations.

Rain that falls onto land soaks into the earth, runs off to rivers, lakes, and other bodies of water, or simply evaporates and reenters the atmosphere. Much of the rain that soaks into the earth enters layers of porous rock lying just below the surface, where it accumulates as large reservoirs of groundwater. This flow of water forms part of a larger process known as the **hydrologic cycle** (**Figure 7.18**).

The fresh water coming from your kitchen faucet or from a water fountain was probably drawn up from the reservoir of groundwater. Reaching the surface through wells or springs, this source provides the drinking water for about half the people in the United States and about three-quarters of those of us who live in large cities. More water is present in this subsurface layer than in all of our rivers, streams, and lakes combined.

In chemical terms, groundwater and the water we draw from our rivers and other bodies of water are solutions. Water is the solvent, and the substances that water picks up in its travels from the clouds, through the earth, and into our faucets are the solutes.

Partly because of rainwater's normal acidity (resulting from the carbonic acid it contains) and partly because water itself is a very good solvent for many substances, the rainfall that passes through the soil picks up a variety of minerals from the earth itself. As a result, all the waters of Earth, including those that feed our public and private water supplies and those that furnish commercially bottled waters, contain a variety of minerals and other dissolved solids in a range of concentrations (**Figure 7.19**).

Water naturally varies in the amounts of dissolved solids, depending on its source and the particular geography of the region. **Hard water** refers to water with especially high concentrations of dissolved minerals, such as calcium and magnesium ions. **Soft water** has low levels of these minerals.

Drinking water can also contain additives. For instance, most of the municipal drinking water in the United States contains added fluoride (F^-), as a preventative measure against tooth decay. Fluoridated water typically contains between 0.7 and 1.2 ppm fluoride, but recently the Department of Health and Human Services has recommended that a level of 0.7 ppm offers sufficient benefits while reducing potential risks. Excess fluoride consumption can produce *fluorosis*, a condition affecting tooth enamel.

Drinking water • Figure 7.19

All purified drinking water, whether from the tap or bottled, contains small amounts of dissolved minerals. These give drinking water a subtle, but agreeable body and taste compared to distilled water, which lacks dissolved minerals. These minerals include cations, such as calcium (Ca^{2+}), magnesium (Mg^{2+}), potassium (K^+), and sodium (Na^+); and anions such as chloride (Cl^-), bicarbonate (HCO_3^-), and sulfate (SO_4^{2-}). Ions such as these are usually harmless or even beneficial to most of us at their typical levels in natural and in commercially bottled drinking water.

The values shown are for **total dissolved solids** (TDS), the total concentration of dissolved minerals in each sample.

Concentrations of dissolved solids in drinking water are typically reported in parts per million (ppm). 1 ppm = 1 mg/kg = 1 mg/L.

Ask Yourself

For very dilute aqueous solutions, such as those shown here, a concentration of 1 mg/kg is equivalent to 1 mg/L. Why is this the case? (*Hint:* Consider the density of water.)

Defining Polluted Water

If we could detect and measure the most exceedingly small quantities of contaminants in some substance, we'd find a bit of whatever we might look for in anything we choose to examine. With regard to drinking water, the question is: What potential contaminants should we be concerned about, and at what levels do they pose an unacceptable risk to the public?

We know that a few of the chemicals that can enter our water supplies are particularly toxic, even at what may seem to be very low concentrations. Contamination by these substances occurs as rainwater accumulates residues of agricultural fertilizers and pesticides (known as agricultural runoff), and as industrial, urban, and household wastes – dumped onto the surface of the ground or injected just below it—seep into our water supplies. [In a later chapter, we explore how various forms of pollution enter our water supplies.]

To ensure the safety and high quality of public drinking water, the U.S. Congress passed the Safe Drinking Water Act of 1974, which establishes, among other things, **maximum contaminant levels** (MCLs) for specific, potentially hazardous chemical contaminants. The Act, enforced by the Environmental Protection Agency (EPA), has been amended twice since it was originally enacted, as newer information has become available. Table 7.1 lists a few of the substances controlled by the Act. Curiously, this law also applies to some kinds of commercially bottled water, such as the large jugs that are dispensed through water coolers, although most bottled water is regulated by the Food and Drug Administration.

The current version of the Act does not specify maximum allowable levels for sodium ions (Na^+), since the small amounts of sodium we typically ingest through drinking water do not generally pose health concerns. Nevertheless, the EPA recommends that sodium levels in drinking water not exceed 20 ppm for those on sodium-restricted diets, and more generally, should not exceed 250 ppm, the level at which many begin to notice the taste of salt in water.

You can test your own sensitivity to the taste of salt in water by preparing a series of salt-water solutions, with each one more dilute than the previous one by the same factor. If you sample each solution, starting from the most dilute and moving progressively to more concentrated samples, you can determine when you first notice the taste of salt. **Figure 7.20** shows how to prepare these solutions and how to perform this taste test.

Maximum contaminant levels (MCLs) allowed in public drinking water Table 7.1

Safety standards established by the Safe Drinking Water Act limit the amounts of various contaminants permitted in public water supplies. Maximum allowable levels for some of the mineral contaminants are as follows.

Contaminant	MCL mg/L (ppm)
Arsenic	0.01
Barium	2.
Cadmium	0.005
Chromium	0.1
Lead	0.015
Mercury	0.002
Selenium	0.05

The Act also establishes maximum allowable levels for microbial contaminants, organic compounds (including many agricultural chemicals and solvents), and radiological contaminants, such as uranium.

Interpret the Data

Which of the contaminants shown in the table, if any, exceed federally permitted levels when they are present in public water supplies at a level of 0.05 mg/L?

Meeting Water Needs

The world faces increasing challenges in meeting water needs. As the global population grows and standards of living improve, so does the demand for water. Food production, through farming and ranching, accounts for the majority of water use. Also, as nations become wealthier, their diets generally include more animal protein, which puts additional strain on water resources. For example, the production of grain-fed beef is particularly water-intensive, requiring an estimated 4000 gallons of water for each pound of beef generated.

Many countries are already experiencing water scarcity or shortages. By 2030 nearly half the global population could be facing water scarcity according to projections by the United Nations. Protecting freshwater resources and conserving water are likely the best strategies for addressing these concerns.

Technology plays an important role in making water **potable**, or safe for consumption. For example, a series of purification steps are required to produce municipal or tap water, which typically originates in groundwater,

Detecting the taste of sodium in water • Figure 7.20

Here we carry out a **serial dilution**, the preparation of a series of solutions, one more dilute than the next by a constant factor. We start by dissolving ½ tsp. of salt (about 3 g of NaCl) in a glass containing 1 cup (237 mL) of distilled water. We call this solution #1.

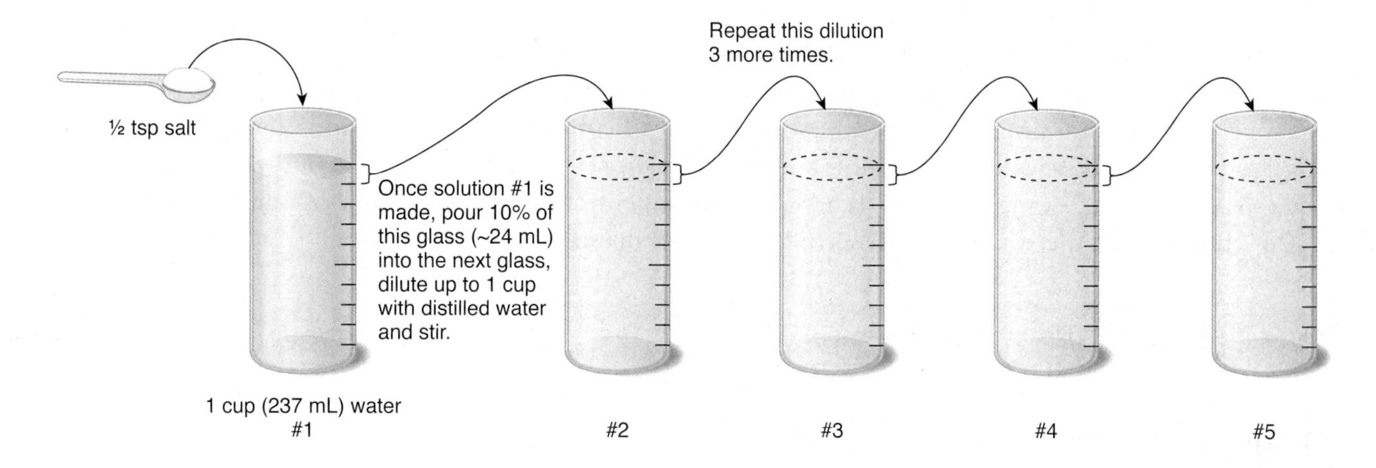

½ tsp salt
Repeat this dilution 3 more times.
Once solution #1 is made, pour 10% of this glass (~24 mL) into the next glass, dilute up to 1 cup with distilled water and stir.
1 cup (237 mL) water
#1
#2
#3
#4
#5

Once all five solutions are prepared, taste the water in each glass, starting with the most dilute solution, #5. Repeat the taste test with each successively lower-numbered glass (with higher concentrations of salt) until you're sure you can taste the salt. [It may be helpful to have a **control** sample of distilled water available as a comparison.]

The concentrations of sodium in each glass can be determined as follows:

Since the mass of a sodium ion is about 23.0 amu and the mass of a chloride ion is about 35.5 amu, the sodium ion makes up roughly 39% of the mass of any quantity of NaCl:

$$\frac{23.0 \text{ u}}{(23.0 + 35.5) \text{ u}} \times 100\% = 39\%$$

(Numerator: Mass Na; denominator: Mass NaCl)

A half-teaspoon of table salt has a mass of about 3 g. This means that the amount of sodium ions (Na^+) in the table salt we added to the first glass has a mass of about 39% of 3 g, or roughly 1.2 g, which is equivalent to 1200 mg.

One cup of water (237 mL) is equivalent to a volume of 0.237 L, which has a mass of 0.237 kg. The concentration of Na^+ in solution #1 is then:

$$\frac{1200 \text{ mg } Na^+}{0.237 \text{ kg water}} \approx 5{,}000 \text{ ppm } Na^+$$

Each subsequent dilution we made reduces the sodium ion concentration by a factor of 10.

Interpret the Data

1. What are the sodium ion concentrations in each of the solutions?
2. Starting from solution #5, which is the first glass that would appear to be "polluted" according to the EPA's recommendations for **(a)** those on sodium-restricted diets; **(b)** the general public?

lakes, rivers, or reservoirs (**Figure 7.21**). Over 85% of the U.S. population relies on tap water.

Municipal wastewater undergoes a series of processing steps at sewage treatment plants. Among other purposes, this is to ensure the treated water is safe to return to the environment.

In many locations throughout the world, groundwater and other freshwater resources are insufficient to meet water demand, in terms of either quality or quantity. Because seawater is plentiful as well as available to much of the world's population, **desalination**—the removal of salt from salt water—represents another approach to addressing growing water needs. However, because of the high affinity water and dissolved salts have for each other, desalination methods generally require large amounts of energy.

Reverse osmosis is currently the method of choice for converting salt water to fresh water. It uses less energy and is therefore less expensive than other desalination methods. To appreciate how reverse osmosis works, it's helpful to first understand **osmosis** (**Figure 7.22**), a process that underlies a variety of common phenomena, such as the absorption of water through the roots of plants.

osmosis The tendency for a solvent, such as water, to migrate through a semipermeable membrane from a dilute solution to a more concentrated solution.

PROCESS DIAGRAM

Municipal water purification • Figure 7.21

Fresh water is filtered and treated to remove suspended particles, organic contaminants, and microorganisms to make it safe for consumption.

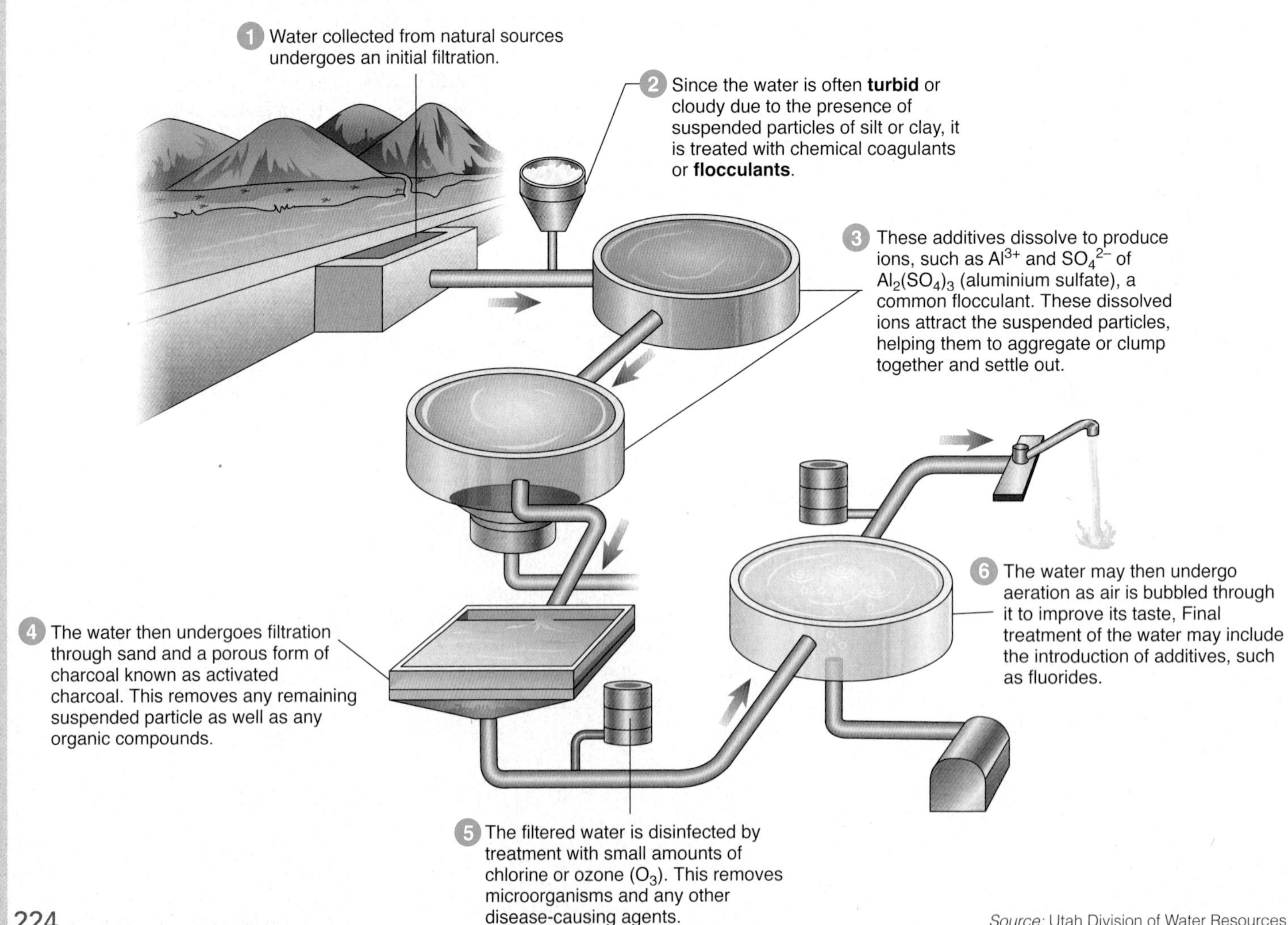

1 Water collected from natural sources undergoes an initial filtration.
2 Since the water is often **turbid** or cloudy due to the presence of suspended particles of silt or clay, it is treated with chemical coagulants or **flocculants**.
3 These additives dissolve to produce ions, such as Al^{3+} and SO_4^{2-} of $Al_2(SO_4)_3$ (aluminium sulfate), a common flocculant. These dissolved ions attract the suspended particles, helping them to aggregate or clump together and settle out.
4 The water then undergoes filtration through sand and a porous form of charcoal known as activated charcoal. This removes any remaining suspended particle as well as any organic compounds.
5 The filtered water is disinfected by treatment with small amounts of chlorine or ozone (O_3). This removes microorganisms and any other disease-causing agents.
6 The water may then undergo aeration as air is bubbled through it to improve its taste, Final treatment of the water may include the introduction of additives, such as fluorides.

Source: Utah Division of Water Resources

Osmosis • Figure 7.22

Osmosis causes a net flow of solvent through a semipermeable membrane from a less concentrated solution to a more concentrated solution.

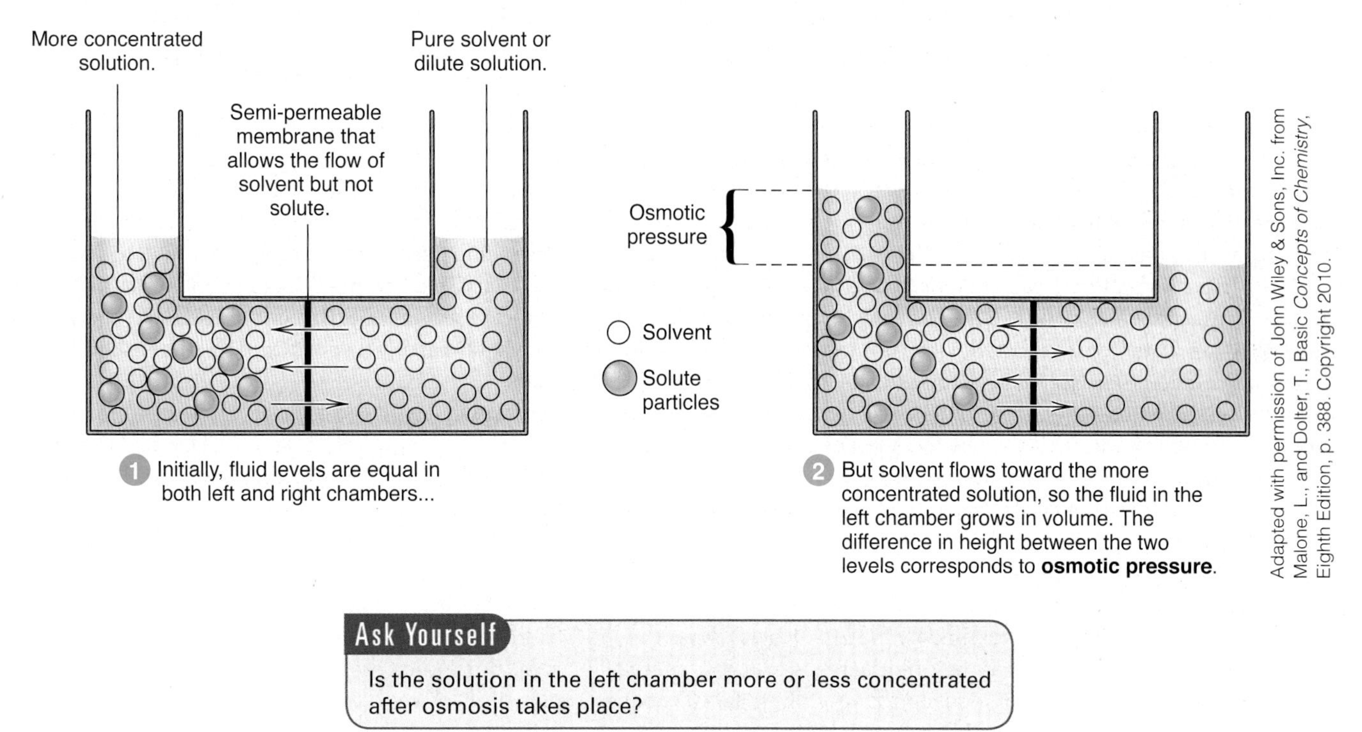

More concentrated solution.
Pure solvent or dilute solution.
Semi-permeable membrane that allows the flow of solvent but not solute.
Osmotic pressure
Solvent
Solute particles
1 Initially, fluid levels are equal in both left and right chambers...
2 But solvent flows toward the more concentrated solution, so the fluid in the left chamber grows in volume. The difference in height between the two levels corresponds to osmotic pressure.

Ask Yourself

Is the solution in the left chamber more or less concentrated after osmosis takes place?

To purify salt water through **reverse osmosis**, high pressure is applied to salt water in contact with a semipermeable membrane. This applied pressure counteracts and overcomes the force of osmosis, causing nearly pure water to pass through the membrane where it can be collected (**Figure 7.23**).

In 2011, desalination plants throughout the world collectively produced about 25 km^3 (or 6.6 trillion U.S. gallons) of fresh water, approximately two-thirds of which was generated through reverse osmosis. Global desalination capacity continues to increase each year but for various reasons is not a panacea for meeting water needs.

Reverse osmosis • Figure 7.23

In reverse osmosis, applied pressure forces water through a semipermeable membrane that contains extremely fine pores. This allows the passage of water but blocks the passage of most dissolved solids. Commercial reverse osmosis desalination plants pump seawater under high pressure (of about 1000 psi) through tubes containing semipermeable membranes. These membranes allow the passage of water but restrict the flow of salt.

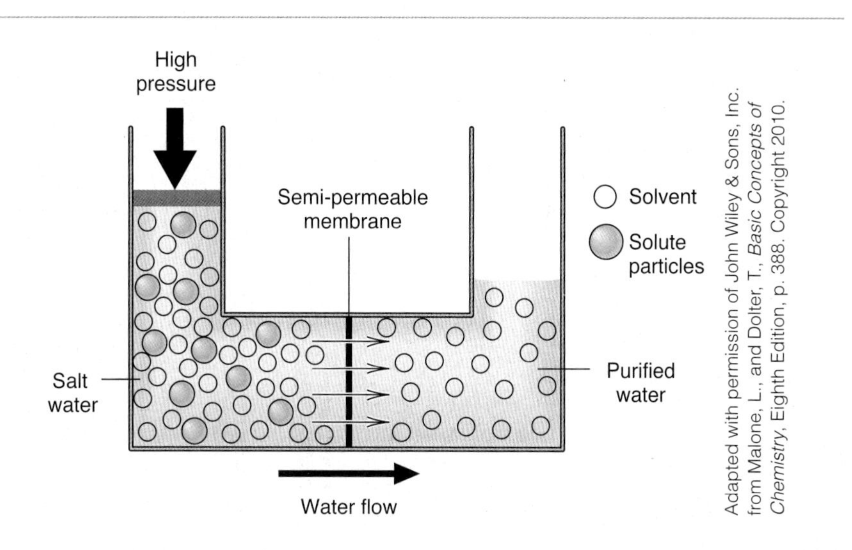

High pressure
Semi-permeable membrane
Solvent
Solute particles
Salt water
Purified water
Water flow

DID YOU **KNOW?**

Bottled water can cost over 1000 times as much as tap water, yet is not necessarily any safer or healthier

The demand for bottled water continues to grow each year, with Americans consuming on average more than 30 gallons of bottled water per person annually. A common perception is that bottled water is safer or healthier than tap water. Yet, all public drinking water (that is, tap or municipal water) is required to comply with the stringent quality standards of the Safe Drinking Water Act. These standards specify maximum contamination levels allowed in tap water for 88 substances known to pose potential health risks. The municipal water authorities responsible for meeting these standards are subject by law to rigorous quality control and reporting requirements, not the least of which is to provide annual water-quality reports to the public.

Although tap water is regulated by the Environmental Protection Agency, most bottled water falls under the auspices of the Food and Drug Administration (FDA). The FDA requires bottled water manufacturers to comply with essentially the same maximum contamination levels established by the EPA. However, if water bottlers find at any time that their product fails to meet these quality standards, they are not required to report these failures either to regulators or to the public (although they would not be permitted by law to sell such water). Public water authorities, however, must report any quality violations to authorities as well as make this information available to the public. As a result, although bottled water is essentially no less safe than tap water, it can be considered to be subject to less stringent reporting requirements.

Generally, bottled water is perceived to be *spring* or *mineral* water, obtained from a natural, protected source. Although many brands truly fit this description, a large segment of the bottled water market is classified by the FDA as "purified water," which is simply tap water that has undergone further purification, including the removal of dissolved solids by distillation (see **Figure**) or reverse osmosis.

Consumers may notice differences in taste between bottled and tap waters, originating from a variety of factors. For one, tap water is typically disinfected by chlorination, which can sometimes leave a residual taste. Also, the local plumbing within a house, especially if it is old, can contribute tastes or odors to tap water. On the other hand, consumers sometimes report noticing a residual taste of plastic in certain bottled waters.

To determine for yourself whether bottled water is truly worth its substantial price premium over tap water, you can carry out a blind tasting of a few bottled water brands and your local tap water. Under these unbiased tasting conditions, you may find yourself surprised by the results!

In **distillation** (Figure a.), water is boiled and the water vapor is removed and condensed, leaving contaminants behind with the residual, undistilled liquid.

The production of bottled "purified drinking water" involves distillation or reverse osmosis of the water to remove dissolved solids. Trace amounts of minerals are then typically introduced to adjust the water's taste. The water is then disinfected by ozonolysis (treatment with ozone gas, O_3) or by irradiation with ultraviolet light.

a. Water distillation

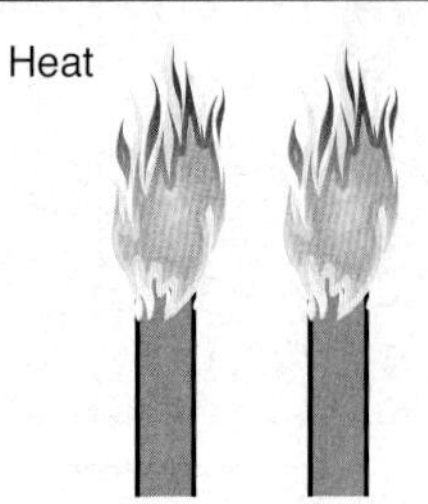

Ask Yourself

1. Both distillation and reverse osmosis require significant input of energy. What type of energy is required for each process, respectively?
2. Use the Internet to find the most recent copy of the annual water-quality report from the water system in your area and then answer these questions. **(a)** From what location does the water utility obtain its water? **(b)** Under which main categories do you find the contaminants listed? **(c)** What do the abbreviations ppm, ppb, MCL, and ND mean?

One is an issue of scale—desalination supplies less than 1% of the fresh water consumed globally. Another is a matter of cost—in many locations, desalination is simply far too expensive. In addition, much of the world's population lives inland or at higher altitudes, making the distribution of desalinated water through extensive pipelines impractical.

In the industrialized world and elsewhere, bottled water has been gaining in popularity. Many enjoy the convenience and taste of bottled water, but its use comes at a cost. This includes the enormous quantities of plastic water bottles produced and discarded each year. In addition, and contrary to many popular beliefs, bottled water is not necessarily any healthier to drink than tap water (see *Did You Know?*).

CONCEPT CHECK

1. **What** percentage of Earth's water is readily accessible and available as fresh water?
2. **What** are three common (i) anions, (ii) cations present in drinking water?
3. **Which** federal agency establishes limits for contaminants in municipal water?
4. **Why** does pressure need to be applied to salt water to purify it through reverse osmosis?

Summary

1 Solutions and Other Mixtures 202

- ***What are solutions?***
  Solutions are homogeneous mixtures of two or more substances, meaning that solutions are uniform throughout and exist as a single phase, either solid, liquid, or gas. All solutions contain a **solvent**, the substance present in greater amount, and one or more **solutes**, the substances dissolved within the solvent. The state of the solvent determines the state of the solution.

Figure 7.3 • Polarity of water and the basis for solubility in aqueous solutions

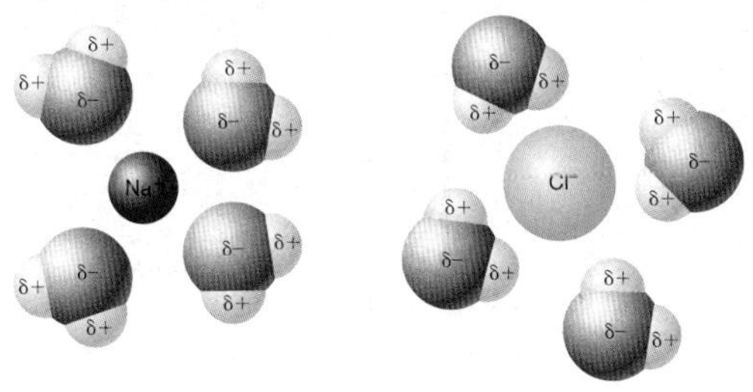

Adapted with permission of John Wiley & Sons, Inc. from Jespersen, N., Brady, J., and Hyslop, A., *Chemistry: The Molecular Nature of Matter*, Sixth Edition, p. 535. Copyright 2012.

- ***What factors influence solubility?***
  We often say that **like dissolves like**, meaning that polar solvents dissolve polar solutes more effectively and nonpolar solvents are more efficient at dissolving nonpolar solutes. Temperature also affects solubility. Solids are generally more soluble in warmer liquids, and gases are more soluble in colder liquids.

- ***What properties of water make it such an important solvent?***
  Water is a highly polar solvent able to dissolve ionic compounds, such as salts (as shown here), and polar compounds, such as sugars.

2 Dissolved Gases 209

- ***What is Henry's Law?***
  Henry's Law refers to the solubility of gases in liquids. At a given temperature, the higher the pressure of a gas above a liquid, the greater the amount of gas that dissolves in the liquid, as shown here.

Figure 7.10 • Henry's Law

Lower gas pressure means that...

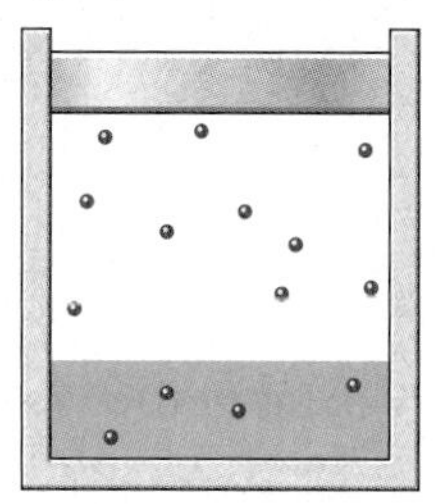

...*less* gas dissolves in the liquid.

Higher gas pressure means that...

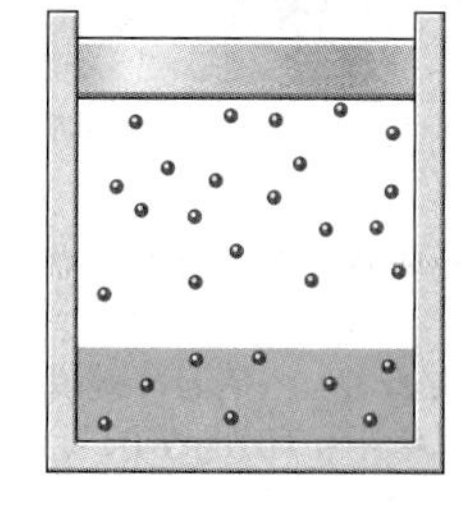

...*more* gas dissolves in the liquid.

Adapted with permission of John Wiley & Sons, Inc. from Jespersen, N., Brady, J., and Hyslop, A., *Chemistry: The Molecular Nature of Matter*, Sixth Edition, p. 595. Copyright 2012.

- ***What are some common examples of gases dissolved in liquids?***
  Water absorbs a variety of gases naturally present in the atmosphere, including oxygen and carbon dioxide. Dissolved oxygen in water supports fish and other aquatic animals. More generally, the lives of all animals rely on the oxygen carried through their bodies by their blood. Rain is naturally acidic due to the aborption of CO_2. Carbonated beverages are produced under a high pressure of CO_2. Metabolically generated CO_2 is transported by the blood to the lungs, where it is expelled from the body.

3 Solution Concentrations 214

- ***What units are used to express the concentrations of solutions?***
  Solution concentrations can be expressed in a variety of units, including **molarity** (mol/L) and percentage concentration by weight or volume. Exceedingly small concentrations can be represented as **parts per million** (ppm) or **parts per billion** (ppb). Concentrations of dissolved minerals in drinking water are typically reported in mg/L, equivalent to ppm, as shown here.

Figure 7.19 • Drinking water

1 ppm = 1 mg/kg = 1 mg/L.

4 Water in Our World 219

- ***Which dissolved solids commonly exist in the water solutions found in nature?***
  Over 97% of Earth's water can be found in the oceans as solutions containing sodium chloride and other solutes.

  The tiny fraction of Earth's water that is accessible as fresh water—either underground or in lakes, rivers and streams—naturally contains a variety of minerals from the earth itself. Hard water refers to water with especially high concentrations of dissolved minerals, such as calcium and magnesium ions. Soft water has low levels of these minerals.

- ***What safety standards apply to public drinking water?***
  The Safe Drinking Water Act, enforced by the Environmental Protection Agency, establishes maximum contaminant levels (MCLs) for specific, potentially hazardous contaminants. These substances include minerals, microbial contaminants, organic compounds (including many agricultural chemicals and solvents), and radiological contaminants.

- ***How is water purified?***
  Purification of municipal water involves: (i) coagulation to remove suspended particles, (ii) filtration to remove any remaining particles, and (iii) disinfection to remove microbiological contaminants. Desalination of water can be accomplished by **reverse osmosis**, as shown here. In this process, high pressure is applied to a water solution in contact with a semipermeable membrane. This membrane allows the flow of water molecules but restricts the flow of most dissolved substances.

Figure 7.23 • Reverse osmosis

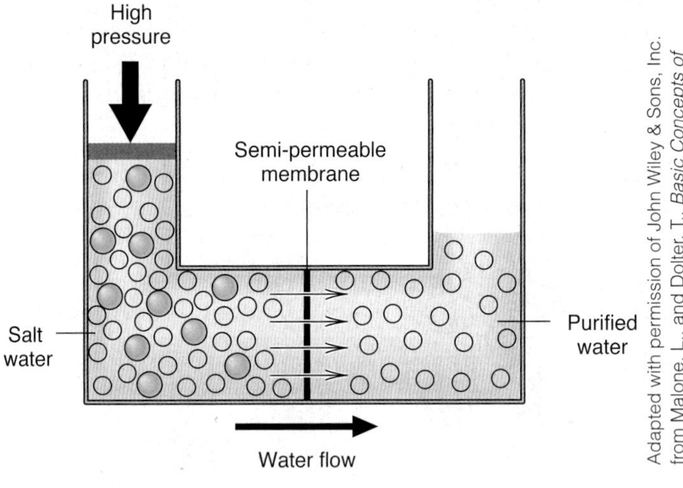

Key Terms

- alloy 203
- aqueous 203
- blood alcohol content 212
- colloid 206
- concentration 210
- molarity 214
- nonpolar 205
- nucleation 210
- osmosis 224
- parts per million (ppm) 217
- polar 205
- solubility 203
- solute 202
- solution 202
- solvent 202
- suspension 209
- Tyndall effect 206

What is happening in this picture?

Lechugilla cave is the fifth-longest cave in the world. It is noted for its unusual mineral deposits, such as these **stalactites** composed of the mineral gypsum, a form of calcium sulfate, $CaSO_4$. These chandelier-like formations result from a series of very slow processes occurring over a long period of time. In this case, hydrogen sulfide gas, H_2S, is emitted from oil deposits below these caverns and dissolves in underground water to form solutions of sulfuric acid, H_2SO_4. This sulfuric acid reacts with limestone deposits (calcium carbonate, $CaCO_3$) naturally present within the cave to form solutions of calcium sulfate, $CaSO_4$. As the water table drops, these minerals crystallize out of solution.

Think Critically Identify two solutes in the aqueous solutions mentioned in this passage. Which one of these solutes forms the white mineral deposits depicted here?

Exercises

Review

1. What do the terms a. *homogeneous* and b. *aqueous* mean as they pertain to solutions?
2. What dictates whether a solution is a gas, a liquid, or a solid?
3. Why is air considered to be a solution?
4. What term describes a solid solution consisting of two or more metals?
5. What is the primary element in steel? What other element is present in all steel?
6. The figure here depicts sodium chloride dissolving in water. With which ion does the
   a. oxygen atom of water;
   b. hydrogen atoms of water, associate? Why?

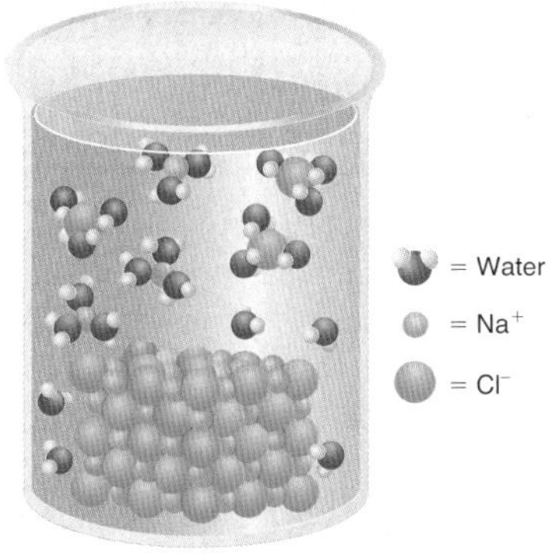

Adapted with permission of John Wiley & Sons, Inc. from Hein, M., and Arena, S., *Foundations of College Chemistry*, Thirteenth Edition, p. 320. Copyright 2011.

7. Why are colloids and suspensions considered to be mixtures but not true solutions? What is the Tyndall effect?
8. In what aspects is blood considered
   a. a suspension?
   b. a solution?
9. Are nonpolar compounds hydrophobic or hydrophilic? Explain your answer.
10. Why is water an effective solvent for
    a. ionic compounds, such as sodium chloride,
    b. polar compounds, such as sugars and alcohols?
11. State Henry's Law and provide a common example of this law.
12. In addition to temperature, what is a factor that determines how much N_2 can dissolve in a given quantity of water? What is a factor that determines how fast the N_2 dissolves?
13. Why don't oil and water mix?
14. What characteristic of emulsifiers allows them to form attractions to both water and oils?
15. What unit of concentration is given by moles of solute per liter of solution?
16. What type of concentration unit is used most frequently in consumer products?

17. Name a unit of concentration useful in discussions of extremely small concentrations, as in pollution studies.

18. What unit of concentration is equivalent to 1 milligram of solute per kilogram of solution?

19. What act of the U.S. Congress set forth quality standards for public drinking water?

20. Describe what is meant by a serial dilution.

21. What do the processes of distillation and reverse osmosis of water both accomplish? How are these processes different?

22. What is the name of the process that initiates the formation of a new state within a solution, such as bubbles of gas developing within a liquid or crystals forming within a liquid solution?

23. What would you expect to observe if a glass of seawater is allowed to evaporate completely?

24. Identify two ways a solution can become supersaturated.

Think

25. Consider the diagram shown here, representing the distribution of gas particles within a liquid held at a fixed temperature and in the gas above the liquid.
    a. What is the ratio of particles within the gas phase as compared to those within the liquid phase?
    b. Would this ratio increase, decrease, or remain constant if the liquid were warmed? (*Hint*: Consider whether gases are generally more soluble in warmer or cooler fluids.)
    c. Now consider breath test analyzers for blood alcohol concentration. These instruments rely on an assumption of how ethanol is distributed between blood and alveolar breath at normal body temperatures. If a subject being tested is running a fever, could this affect the reading provided by the instrument? If so, what might the impact be?

Adapted with permission of John Wiley & Sons, Inc. from Jespersen, N., Brady, J., and Hyslop, A., *Chemistry: The Molecular Nature of Matter*, Sixth Edition, p. 595. Copyright 2012.

26. As scuba divers descend to greater depths, the pressure of the air they breathe increases. Under higher pressure, more of the inhaled air, including nitrogen, the main component of air, dissolves in the blood. As scuba divers ascend, they must do so slowly or take periodic breaks on their ascent to allow the dissolved nitrogen to diffuse out of their blood. If they ascend rapidly, dissolved nitrogen can form bubbles within their blood, causing painful and sometimes fatal symptoms, known as the bends or decompression sickness.
    a. How is Henry's Law related to the phenomenon of decompression sickness?
    b. *Hyperbaric oxygen therapy*, breathing oxygen at higher than normal pressure, is used to treat decompression sickness. What effect would this treatment have on dissolved oxygen levels in the blood?

27. Ethanol (CH_3CH_2OH) and water are *miscible*, meaning they form a solution when mixed together in any proportions. For each figure, indicate whether the indicated interaction between the molecules shows a plausible positive interaction. Explain your answers.

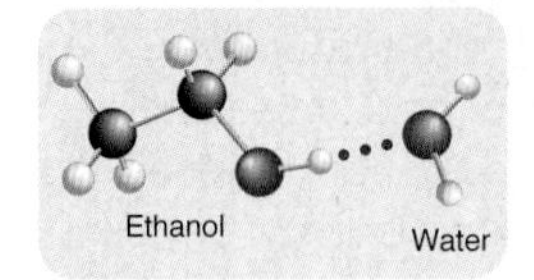

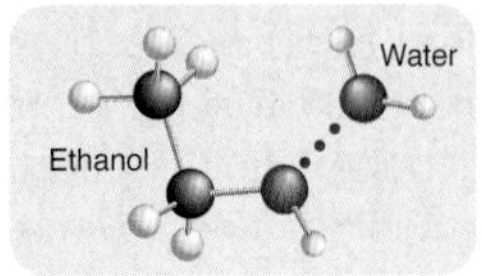

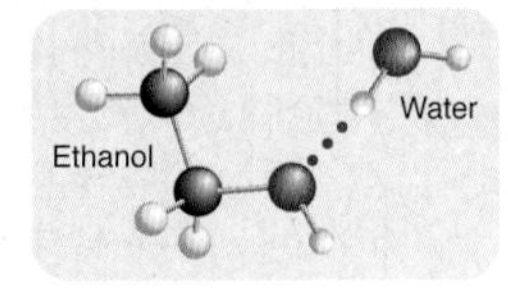

28. Some brands of peanut butters and salad dressing appear uniform in consistency throughout, whereas others need to be stirred or shaken before use because they have separated into two layers.
    a. What type of agent has likely been added to peanut butter or salad dressing so that they don't require mixing before use?
    b. Use the Internet to find the molecular structure of a monoglyceride, an additive sometimes used in processed foods. Identify the polar and nonpolar regions of the molecule and describe its purpose as a food additive.

29. Sometimes a sticky adhesive residue remains when you remove a stick-on label from a surface. Various "home remedies" have been suggested to dissolve and remove this residue from nonporous surfaces, including the use of small amounts of peanut butter, mineral oil, or Vaseline, each of which is rich in nonpolar compounds. If these substances work as suggested, what can you infer about the polarity of the compounds in the adhesives of stick-on labels?

30. Assume you have an analytical instrument that could tell you exactly how many molecules of an agricultural insecticide there are in a glass of water. What standard(s)—the number of insecticide molecules, the odor, color, taste of the water, or any other criterion—would you use to determine whether the water is fit to drink? (No governmental drinking-water standards have yet been set for this insecticide.)

31. Regardless of the recommendations suggested by the Environmental Protection Agency for sodium in drinking water, which glasses in Figure 7.16 would you consider to be polluted with salt? Which, if any, would you consider safe to drink?

32. In preparing the solutions shown in Figure 7.16, why was it recommended that distilled water should be used?

33. a. Under what conditions would you consider water to be "polluted"?
    b. Would you say that all "polluted" water is unfit to drink?
    c. Can water that is unfit to drink not be "polluted"?
    d. Can water that is "polluted" be safe to drink?

34. Describe two specific actions you would recommend to improve the purity of your drinking water.

35. We normally think of community water purification processes as designed to remove impurities from water. Yet they can also add chemicals to water to benefit the public health and welfare. Many communities, for example, add chlorine to water to kill disease-causing microorganisms. Some add fluoride salts to retard tooth decay.
    a. Do you favor the addition of chemicals to public drinking water to promote the public health and welfare?
    b. Would you agree that small quantities of essential nutrients or vitamins ought to be added to drinking water?

c. Do you believe that nothing ought to be added to the water since everyone in the community must drink the same water and some may not wish to have anything added to their drinking water?

d. Do you think that the public welfare ought to be more important than individual wishes in making these decisions? Explain your answers.

Calculate

36. Labels on fruit drinks and on a variety of other bottled and canned foods often indicate the presence of 0.1% sodium benzoate, the sodium salt of benzoic acid, as an added preservative. Express this concentration in terms of parts per thousand.

37. You are one individual person among all the people on Earth. With the world's total population of roughly 7 billion humans, you yourself constitute roughly one person in 7 billion. Represent your own "concentration" among all those now living in terms of a. parts per billion, b. parts per trillion.

38. About 0.01 g of nitrogen (N_2) and 0.05 g of oxygen (O_2) can dissolve in 1 L of water. Assuming that drinking water has a density of 1000 g per liter, what is the maximum concentration, in ppm, of N_2 and of O_2 in the water we drink?

39. Use the total dissolved solid concentrations shown in the figure to determine which of the following samples contains the greatest mass of total dissolved solids: one 0.33-L bottle of Evian water, one 1.5-L bottle of Icelandic water, or three 0.85 L bottles of Voss water.

40. Assuming that rubbing alcohol is 70% (w/w) isopropyl alcohol and that its density is 1 kg/L,

a. How many grams of water and how many grams of isopropyl alcohol are there in a liter of rubbing alcohol?

b. How many moles of water and how many moles of isopropyl alcohol (molecular weight 60) are there in a liter of rubbing alcohol?

c. If we decide which is the solute on the basis of weight, is isopropyl alcohol the solute or the solvent?

d. If we decide on the basis of the number of moles, is isopropyl alcohol the solute or the solvent?

e. Considering water as the solute and the isopropyl alcohol as the solvent, what is the molarity of the water in the solution?

f. Considering isopropyl alcohol as the solute and water as the solvent, what is the molarity of the isopropyl alcohol in the solution?

41. A typical aspirin tablet contains 0.325 g of the active ingredient, acetylsalicylic acid. What is the average concentration, expressed as a percentage by weight (w/w %), of the acetylsalicylic acid in the body of a 165-lb (75-kg) person who has just taken two aspirin tablets?

42. a. How many moles of solute are in 2.5 L of solution that has a concentration of 0.4 *M*?

b. How many grams of glucose (molar mass = 180.0 g/mol) would you need to prepare 2.5 L of a glucose solution with a concentration of 0.4*M*?

c. How many liters of a 0.8 *M* solution contain 6.0 moles of solute?

43. Suppose you dissolve three teaspoons of sugar (about 15 g of sucrose) into a large cup of coffee (20-fl oz ; with a mass of about 600 g).

a. What is the total mass of the solution?

b. What is the volume of this solution (in liters) if the density is 1050 g per liter?

c. How many moles of sucrose were dissolved (molar mass of sucrose = 342 g/mol)?

d. Use your answers to parts b and c to determine the molarity of sucrose in this solution.

44. You have 1 L of a solution that contains selenium at a level of 90 parts per billion and 1 L of another solution that contains barium at a level of 3 parts per million. You now mix the two solutions together to produce 2 L of a single solution. Does the new solution produced by mixing the two original liters together meet federal drinking water standards for selenium? For barium?

45. The Environmental Protection Agency recommends a limit of no more than 250 mg of sodium per liter of drinking water based on taste considerations. What is the molarity of sodium chloride in water containing this limit of 250 mg sodium per liter?

46. If you continued the serial dilutions shown in Figure 7.20 up to 7 glasses, what would be the concentration of sodium ions in glass #7 expressed in

a. parts per million (ppm), b. parts per billion (ppb)?

47. Prepare a table showing the molarity of the sodium chloride in each of the five glasses shown in Figure 7.20.

48. Suppose a person's blood alcohol content (BAC) is 0.04%.

a. How many grams of ethanol are present in 100 mL of this person's blood?

b. Assuming the person has a total blood volume of 5.0 L, how many grams of ethanol are present in the blood?

49. To support a healthy environment for fish, the water chemistry in fish tanks must be maintained within certain limits. Home testing kits can be used to monitor levels of important dissolved chemicals within this water, such as ammonia (NH_3), nitrite ions (NO_2^-), and nitrate ions (NO_3^-). (Fish waste products contain ammonia, which is converted to nitrite and then to nitrate by bacteria present within the tank.) What is the concentration (in ppm) of nitrite in a 10-gallon (37.9-L) fish tank if the water contains 9.5 mg of nitrite?

8 Acids and Bases

The properties of acids and bases are often closely tied to a substance that covers most of Earth's surface: water. The underwater world of our planet's oceans, with its varieties of marine plants and animals, including coral structures, can be quite sensitive to the interplay of acids and bases.

The oceans may seem remote from our surface environment, but this watery world is closely affected by our own activities above its surface and the pollutants we produce and release into our environment. Much of the waste material our industries generate finds its way into our rivers and streams and is eventually carried by the moving water into the oceans.

In this chapter, we will learn how some of the pollutants we release into our surface environment—including the carbon dioxide we generate through the combustion of fossil fuels—can change into acids, affecting the marine life and coral structures of the oceans. We will also see that a knowledge of the chemistry of acids and bases is important not just for understanding the effects of pollution but also for combating these effects and reducing pollution of our air and water.

This reef, exposed naturally to a constant stream of carbon dioxide bubbles from vents in the ocean floor, has far less marine life as a result of the way acids affect coral.

Katharina Fabricius, Australian Institute of Marine Science

Water chemistry is critical to the health of coral reefs. Here, a diver notes the diversity of marine species, a measure of the health of coral reef ecosystems.

Brian J. Skerry/NG Image Collection

CHAPTER OUTLINE

8.1 Acids, Bases, and Neutralization

LEARNING OBJECTIVES

1. **Describe** three simple ways to distinguish between acids and bases.
2. **Explain** acid-base neutralization.
3. **Define** acids and bases at the molecular level.

When you bite into a lemon or other citrus fruit, you experience the familiar sour taste of acids. On the other hand, if you've ever tasted a pinch of baking soda or had the misfortune of getting soap in your mouth, you have experienced the bitter flavor of bases. We'll elaborate on simple observations such as these that help us differentiate between acids and bases. Then we will explore the molecular basis for why these two classes of compounds behave as they do.

Acids and Bases—Preliminary Observations

A sour taste, in fact, may be the earliest known test for acidity. Our word **acid** comes from the Latin *acidus*, an adjective meaning "sour" or "having a sharp taste." In addition, we know that water solutions of acids:

- turn **litmus paper** red and
- react with certain metals, such as magnesium, or zinc, to liberate hydrogen gas (**Figure 8.1**).

acid A chemical compound that turns litmus paper red and reacts with bases to form salts.

litmus paper An absorbent paper strip containing a natural dye that turns red in acid solution and blue in basic solution.

base A chemical compound that turns litmus paper blue and reacts with acids to form salts.

In addition to tasting bitter, water solutions of all **bases**:

- turn litmus paper blue and
- feel slippery.

Of all these diagnostic tests, only the litmus test is safe enough for general use with a substance of uncertain identity. It is far too dangerous to taste a substance you're not sure of or to rub it between your fingers to learn whether it has the properties of an acid or a base.

Although many common foods—lemons and vinegar, for example—contain acids and taste distinctly acidic, other substances may be composed of acids (or bases) powerful enough to destroy skin and mucous membranes. Some substances we may come into contact with are highly poisonous in ways that have nothing to do with acidity or basicity. So a good rule of thumb is simply ***never taste or touch any material unless you are sure it's safe.***

We begin our examination of acids and bases by looking at some we are likely to find in our everyday lives (**Figure 8.2**). If we focus on the chemical formulas of acids and bases, we can make the following observations:

- Common acids (of the type we'll see in this book) have one or more hydrogen atoms, although not all compounds with hydrogen atoms are necessarily acidic. For example, *hydrochloric acid* (HCl) and *acetic acid* (CH_3CO_2H) are acids, whereas *ammonia* (NH_3) is not.
- Ionic compounds containing the *hydroxide* ion (OH^-) or *carbonate* ion (CO_3^{2-}) are basic; examples are *sodium hydroxide* ($NaOH$) and *calcium carbonate* ($CaCO_3$). Other kinds of compounds, without OH^- or CO_3^{2-} ions, can also be basic. The covalent compound ammonia (NH_3) for example, is a common base.

We will explore the reasons behind these observations shortly.

A test for acids • Figure 8.1

A strip of magnesium immersed in an acid solution liberates hydrogen gas.

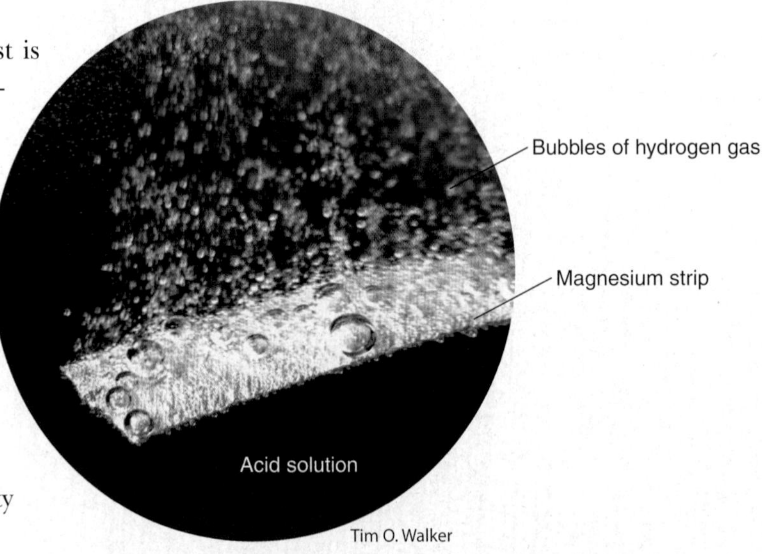

Tim O. Walker

Acids and bases in everyday products • Figure 8.2

Acids and bases are common in products we use every day, such as our cleaning agents, medicines, foods, and a variety of other materials.

ACIDS

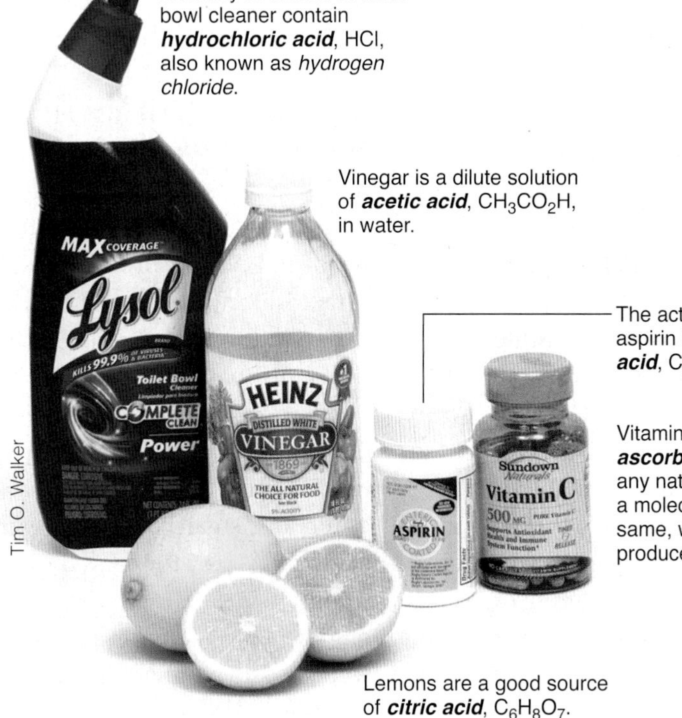

A variety of brands of toilet bowl cleaner contain ***hydrochloric acid***, HCl, also known as *hydrogen chloride.*

Vinegar is a dilute solution of ***acetic acid***, CH_3CO_2H, in water.

The active ingredient in aspirin is ***acetylsalicylic acid***, $C_9H_8O_4$.

Vitamin C is also known as ***ascorbic acid***, $C_6H_8O_6$. As with any naturally occurring compound, a molecule of ascorbic acid is the same, whether found in nature or produced synthetically.

Lemons are a good source of ***citric acid***, $C_6H_8O_7$.

BASES

Drain and oven cleaners can contain ***sodium hydroxide***, NaOH, also known as lye.

Household ammonia is a dilute solution of the covalent compound, ***ammonia***, NH_3, in water.

Baking soda is also known as ***sodium bicarbonate***, $NaHCO_3$.

Calcium carbonate, $CaCO_3$, is but one of a variety of bases used to combat excess stomach acid and relieve heartburn. $CaCO_3$ is also the main component of seashells, eggshells, and chalk.

Alkaline batteries contain the basic or *alkaline* component, ***potassium hydroxide***, KOH.

Ask Yourself

Identify two acids and one base that are found in nature.

An acid-base neutralization reaction • Figure 8.3

An acid and a base can react with each other to form a salt and water.

$$HCl + NaOH \longrightarrow NaCl + H_2O$$

Hydrochloric acid + Sodium hydroxide → Sodium chloride, a **salt** + Water

Neutralization Reactions

In addition to the characteristics we've just described, another useful property that distinguishes acids and bases from all other kinds of chemical substances is their ability to neutralize each other. That is, acids can neutralize bases and bases can neutralize acids.

neutralization A reaction between an acid and a base to produce a salt and water.

A particularly simple acid-base **neutralization** occurs, for example, when HCl reacts with NaOH (**Figure 8.3**). Hydrochloric acid, usually available in hardware stores under its commercial name, muriatic acid, is a strong acid used for cleaning metals, masonry, cement, and stucco. Sodium hydroxide, better known as household lye, is a powerful base used to clear clogged drains and clean ovens. Both NaOH and HCl are dangerous, corrosive chemicals that should be handled with care.

Hydrochloric acid and sodium hydroxide react with each other to form sodium chloride, a **salt**, and water. Notice that the H of the HCl and the OH of the NaOH combine to form water, whereas a combination of the Cl of the HCl and the Na of the NaOH produces NaCl. This resulting sodium chloride is a perfectly neutral salt—neither acidic nor basic.

salt An ionic compound produced by the reaction of an acid and a base. In common usage, **salt** or **table salt** refers specifically to sodium chloride (NaCl).

A variety of neutralization reactions take place within our bodies. For example, our pancreas produces a digestive fluid rich in sodium bicarbonate, a base. This fluid enters our small intestines, where it neutralizes stomach acids carried by foods as they exit the stomach and enter the intestines. We can also find a variety of practical everyday uses for neutralization reactions, one example of which is shown in **Figure 8.4**.

Neutralization reactions can produce a variety of other salts, as shown in **Figure 8.5**, including *potassium iodide* (KI), which provides the iodide in iodized salt, and *magnesium sulfate* ($MgSO_4$), better known as Epsom salts.

Using acetic acid (vinegar) to clean a showerhead • Figure 8.4

Mineral deposits, such as calcium and magnesium carbonates, in pipes, faucets, and showerheads can build up over time and impede the flow of water. We can use common vinegar in an acid-base neutralization reaction to remove these deposits, as shown in the following sample equation:

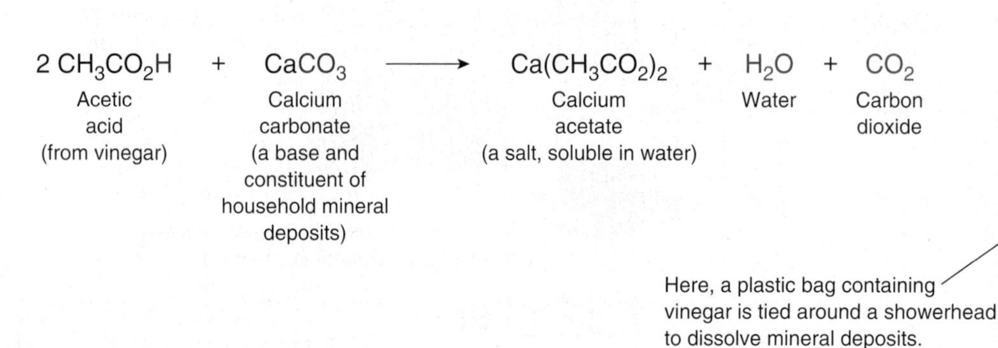

Here, a plastic bag containing vinegar is tied around a showerhead to dissolve mineral deposits.

Tim O. Walker

Acid-base neutralization reactions produce salts • Figure 8.5

The base taking part in a neutralization reaction provides the cation (positive ion) of the generated salt; the acid provides the salt's anion (negative ion).

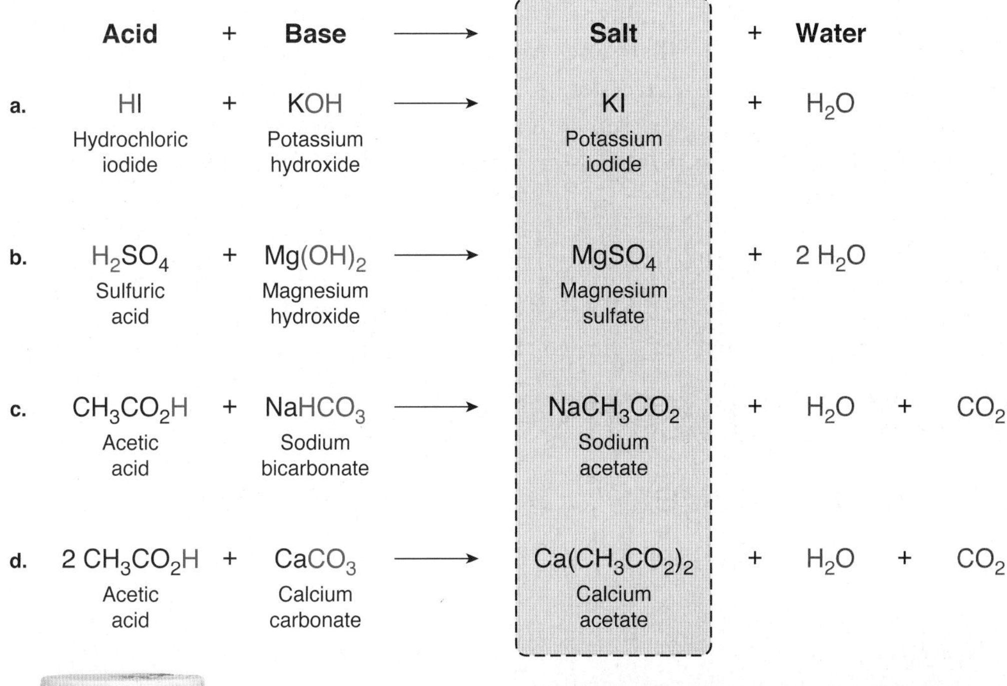

	Acid	+	Base	→	Salt	+	Water		
a.	HI Hydrochloric iodide	+	KOH Potassium hydroxide	→	KI Potassium iodide	+	H_2O		
b.	H_2SO_4 Sulfuric acid	+	$Mg(OH)_2$ Magnesium hydroxide	→	$MgSO_4$ Magnesium sulfate	+	$2\ H_2O$		
c.	CH_3CO_2H Acetic acid	+	$NaHCO_3$ Sodium bicarbonate	→	$NaCH_3CO_2$ Sodium acetate	+	H_2O	+	CO_2
d.	$2\ CH_3CO_2H$ Acetic acid	+	$CaCO_3$ Calcium carbonate	→	$Ca(CH_3CO_2)_2$ Calcium acetate	+	H_2O	+	CO_2

Tim O. Walker

a. Iodized salt contains added potassium iodide.

Tim O. Walker

b. Epsom salts (magnesium sulfate) are used as bath salts and in several other products.

Tim O. Walker

c. The fizz of this homemade geyser comes from rapidly generated carbon dioxide gas (CO_2), a by-product of the reaction of vinegar (acetic acid solution) and baking soda (sodium bicarbonate).

d. Soaking an egg in vinegar slowly dissolves the shell. Here we see CO_2 bubbles, generated by the reaction of acetic acid (from vinegar) and calcium carbonate of the eggshell.

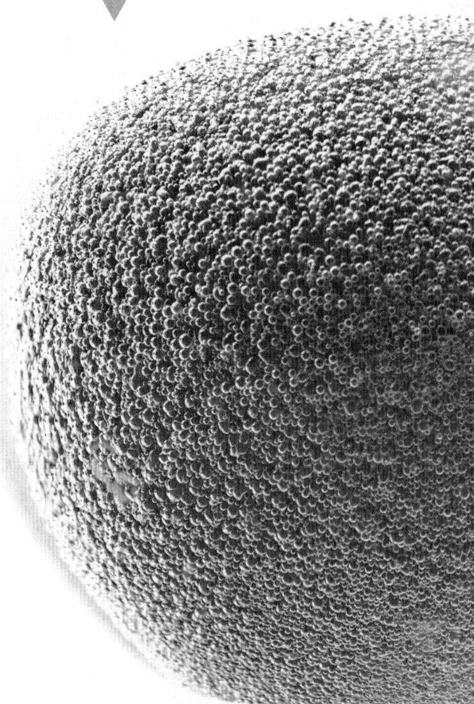

Tim O. Walker

Acids and Bases at the Molecular Level

We don't need to know anything at all about the molecular structure of a substance to decide whether it's an acid. All we need to do is dissolve a bit of it in water and test it with litmus paper, observe whether it reacts with zinc or similar metals to produce hydrogen gas, or determine whether it neutralizes bases. If it does any of these, it's an acid. Similarly, we can determine whether something is a base by observing a few of its properties.

However, chemists are interested in more fundamental phenomena than color changes and the like. In chemistry, the question: "What are acids and bases?" has a deeper meaning, more like: "What is the single characteristic of a molecule that gives it all the properties of an acid or of a base?"

The answer isn't simple. In fact, there is no single, unequivocal answer. Attempts at defining an acid at a fundamental level started long ago, early in the history of chemistry. In 1778, Antoine Lavoisier (pronounced lav-WAH-zee-ay), a French chemist, proposed that all acids are formed by the combination of oxygen—an element that had only recently been discovered—with other elements. The name given to this new element, *oxygen,* is a word formed by a combination of the Greek prefix *oxy-* ("sharp or pungent") and the Greek suffix *-gen* ("forming or producing"). It reflected Lavoisier's belief that this new element is necessary for the generation of acids. The connection between oxygen and acidity was reasonable, given the state of knowledge at that time: All acids whose compositions were known to Lavoisier did, indeed, contain oxygen. A few examples of oxygen-containing acids (some of which were not known to Lavoisier) appear in *What a Chemist Sees.*

WHAT A CHEMIST SEES

Common Uses of Oxygen-Containing Acids

Although many acids contain oxygen atoms, not all acids do, as we'll see shortly.

Sulfuric acid (H_2SO_4)

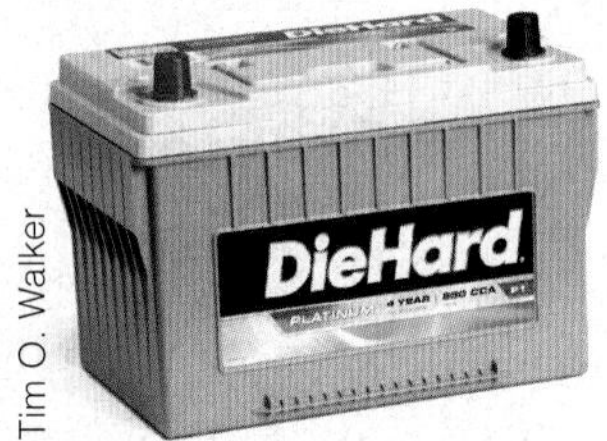

Tim O. Walker

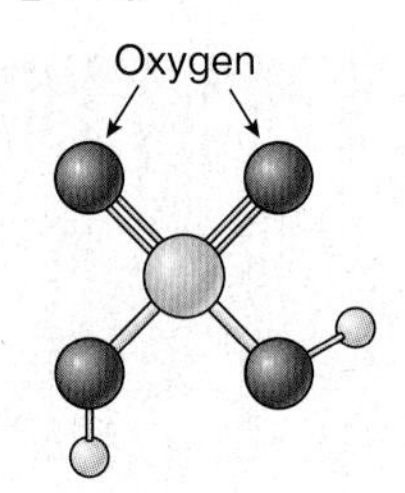

Standard lead-acid car batteries use sulfuric acid.

Nitric acid (HNO_3)

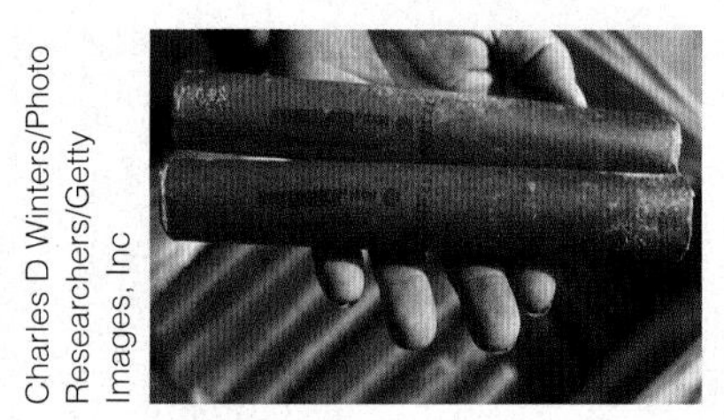

Charles D Winters/Photo Researchers/Getty Images, Inc

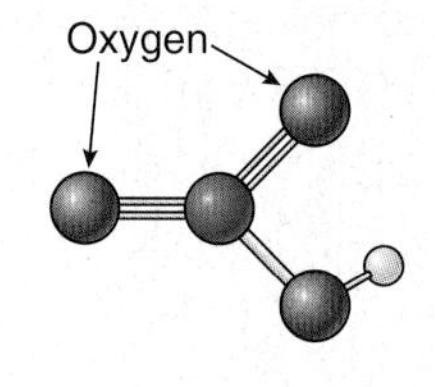

Nitric acid is used in the production of explosives, like TNT, and in making fertilizers.

Boric acid (H_3BO_3)

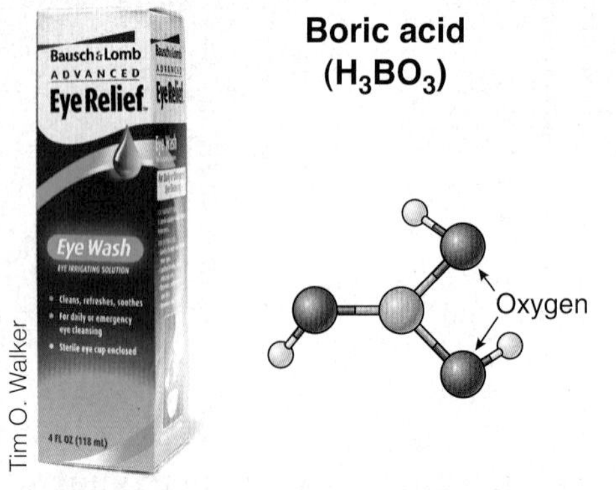

Tim O. Walker

Various types of eye washes and drops have boric acid as an ingredient.

Carbonic acid (H_2CO_3)

Tim O. Walker

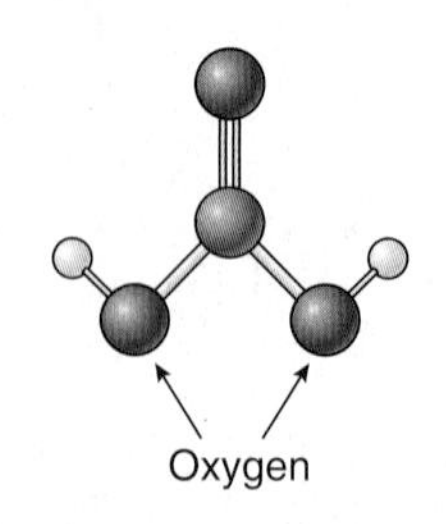

Small amounts of carbonic acid are present in all carbonated beverages.

Ask Yourself

Each of these four acids has a central atom of a different element. Identify the element in each case.

Ionization of HCl in water • Figure 8.6

The atoms of an HCl molecule are held together by a bond. When HCl dissolves in water, this bond breaks.

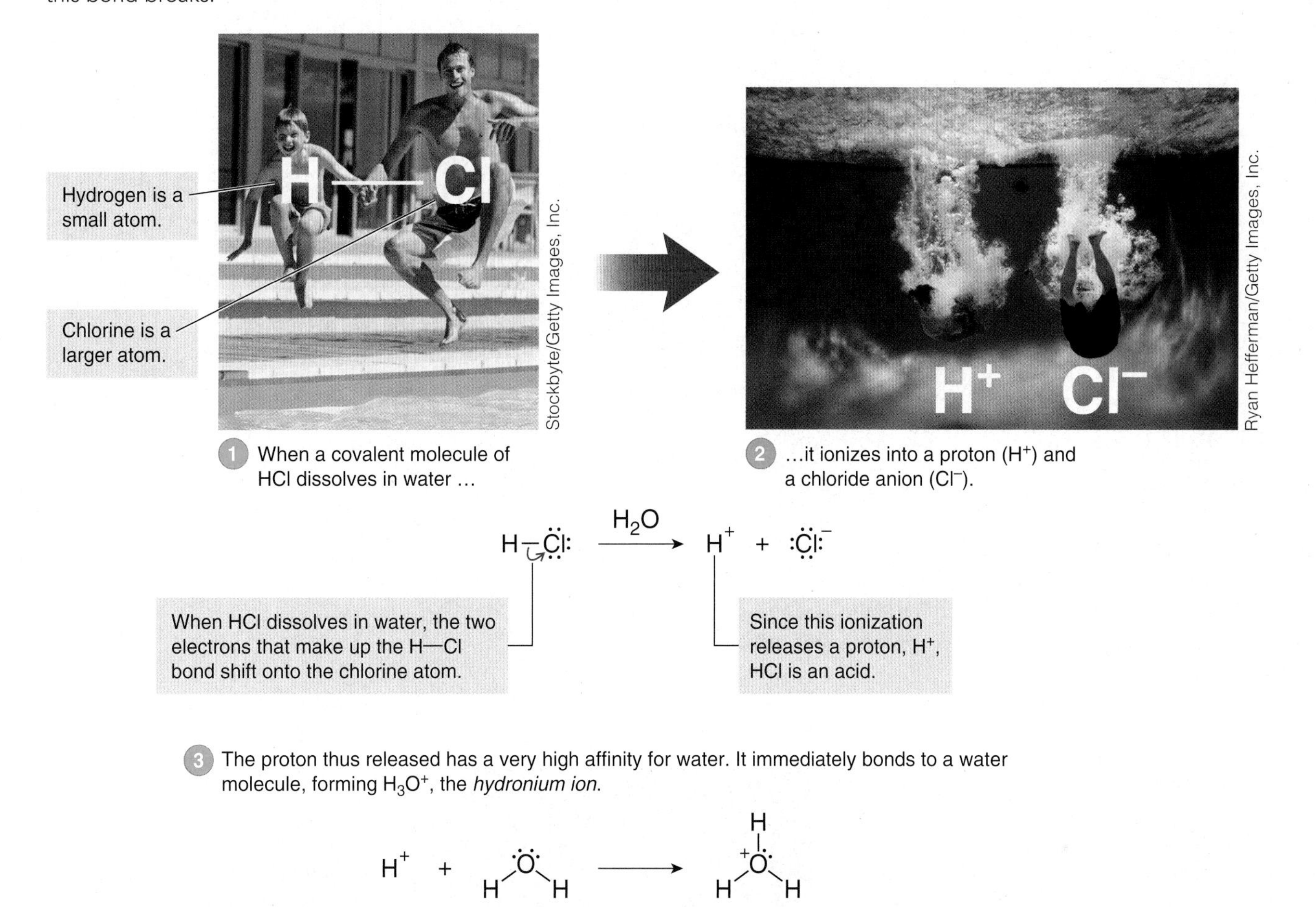

Hydrogen is a small atom.
Chlorine is a larger atom.
H—Cl
Stockbyte/Getty Images, Inc.
H⁺ Cl⁻
Ryan Hefferman/Getty Images, Inc.
1 When a covalent molecule of HCl dissolves in water ...
2 ...it ionizes into a proton (H⁺) and a chloride anion (Cl⁻).
H—Cl: H₂O → H⁺ + :Cl:⁻
When HCl dissolves in water, the two electrons that make up the H—Cl bond shift onto the chlorine atom.
Since this ionization releases a proton, H⁺, HCl is an acid.
3 The proton thus released has a very high affinity for water. It immediately bonds to a water molecule, forming H₃O⁺, the hydronium ion.
H⁺ + H–O–H → H₃O⁺
Hydronium ion

Think Critically How many hydronium ions are produced each time an HCl molecule ionizes in water?

By the early 1800s, hydrochloric acid was found to be composed of the elements hydrogen and chlorine alone (HCl), with no oxygen present. This discovery shifted attention from oxygen to hydrogen as the acid-forming element. Partly to accommodate the role of hydrogen, the Swedish chemist and physicist Svante Arrhenius proposed in 1887 that an acid is anything that produces hydrogen ions (H^+) in water. (The hydrogen ion is simply a proton, the very same particle that forms the nucleus of the hydrogen atom). A hydrogen ion does not remain isolated in water but immediately bonds with a water molecule to form the **hydronium ion** (H_3O^+) (**Figure 8.6**). Arrhenius also proposed that a base is anything that produces **hydroxide ions** (OH^-) in water.

Unlike Lavoisier's proposal, Arrhenius's view is perfectly valid. When dissolved in water, an acid does generate protons (and a base does generate hydroxide ions). But since much important chemistry takes place

hydronium ion The cation H_3O^+, which is produced when acids release hydrogen ions (H^+) in water.

hydroxide ion The anion OH^-, which is produced when bases dissolve in water.

Chemistry InSight Definitions of acids and bases • Figure 8.7

Arrhenius provided the first modern or molecular description of acids and bases. Later, a broader definition was developed by Brønsted and Lowry.

a. Arrhenius definition of an acid

In 1887, Arrhenius first defined an **acid** as **any substance that produces hydrogen ions, H^+, in water**. For example, when hydrochloric acid (HCl) dissolves in water, it dissociates into hydrogen ion (H^+) and chloride ion (Cl^-). The H^+ bonds to H_2O, forming H_3O^+, hydronium ion.

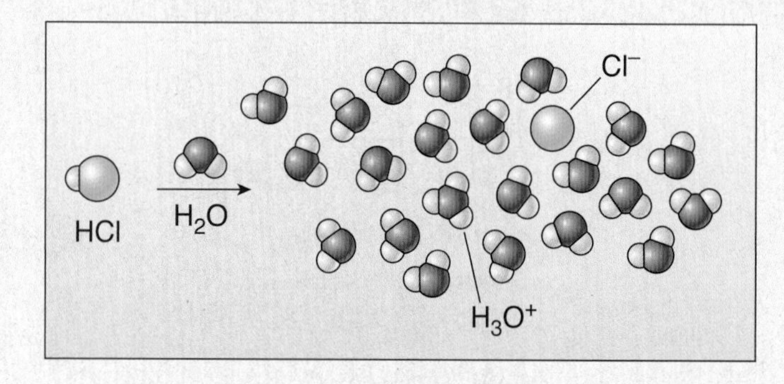

HCl
H2O
Cl−
H3O+

b. Brønsted-Lowry definition of an acid

The Arrhenius definition of acids and bases is restricted to **aqueous**, or water-based, solutions. However, since aqueous solutions are commonly encountered in a wide variety of circumstances, including biological, environmental, and many industrial systems, the Arrhenius framework still has broad relevance.

In 1923, Brønsted and Lowry independently proposed that **acids *donate* hydrogen ions (protons or H^+) whereas bases *accept* them**.

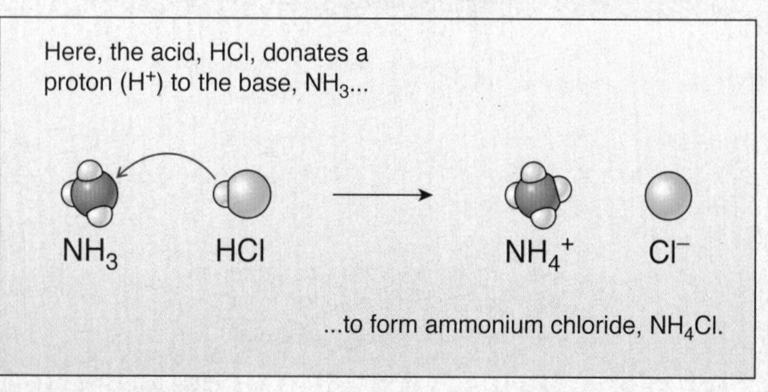

Here, the acid, HCl, donates a proton (H+) to the base, NH3...
NH3
HCl
NH4+
Cl−
...to form ammonium chloride, NH4Cl.

c. Generality of the Brønsted-Lowry definition

The Brønsted-Lowry definition of acids and bases is more general than that proposed by Arrhenius because it can apply to solutions of any kind, both aqueous and nonaqueous, as well as to gas-phase reactions, as shown here.

Andy Washnik

What's that smoke? As ammonia vapors (NH_3) rise from one bottle, they meet hydrochloric acid vapors (HCl) ascending from the other. Together they form ammonium chloride (NH_4Cl), which appears as a white smoke. This acid-base reaction takes place in the gas phase; it doesn't require a solvent.

in the complete absence of water, a broader definition was ultimately developed. In 1923, two chemists, the Dane Johannes Brønsted and the Englishman Thomas M. Lowry, independently defined an acid as anything that can transfer a proton to another chemical species and a base as anything that can accept a proton from another chemical species. Water may be the solvent, but it needn't be (**Figure 8.7**).

Notice that acids and bases are intimately connected by the Brønsted-Lowry definition. For anything to act as an acid, a base must be present to accept a proton; for anything to act as a base, an acid must be present to provide the proton. In essence, the Brønsted-Lowry definition views an acid-base reaction as no more than the simple transfer of a proton (H^+) from one substance (the acid) to another (the base).

The definitions of acids and bases can be explored even further, but for us the Brønsted-Lowry framework will serve to explain the rest of the acid-base reactions we describe in this book.

KNOW BEFORE YOU GO

1. For each of the following reactions, identify the acid and base and then indicate the products.
   a. $LiOH + HI \rightarrow$?
   b. $HNO_3 + NH_3 \rightarrow$?

CONCEPT CHECK STOP

1. **What** color would you expect litmus paper to turn when exposed to the following?
   a. vinegar; b. a slurry of baking soda in water; c. household ammonia
2. **What** chemical species is always produced in a neutralization reaction?
3. **Why** is the Brønsted-Lowry definition of acids and bases considered to be more general than that proposed by Arrhenius?

8.2 The pH Scale

LEARNING OBJECTIVES

1. **Describe** the acid-base properties of pure water.
2. **Define** pH, and describe its relationship to acidity and basicity.
3. **Differentiate** between strong and weak acids.

We have examined some of the properties that identify acids and bases experimentally and some of the definitions that focus on molecular structures. Now let's turn to the unusual acid-base properties of one of our most ordinary substances—water.

We ingest more water than any other substance, by far. Depending on your age and the amount of fat in your body, water makes up half to three-quarters of your weight. About 2.5 liters of water enter your body directly each day through the food and drink you consume. Another quarter of a liter comes from the chemical oxidation of food inside your body. Through several routes, some 75,000 liters of water pass through your body in 75 years. We start by asking: Is all this water we consume an acid, a base, neither, or both?

Amphoteric Water

In Figure 3.10, we saw that water ionizes reversibly to provide both protons and hydroxide ions. This makes water both an acid and a base by the Arrhenius definition (**Figure 8.8**). Water and other substances that can behave in this way are called **amphoteric**, from the Greek amphoteros (meaning "either of two").

amphoteric Able to act as either an acid or a base.

Water can act as an acid or a base • Figure 8.8

Water ionizes to a very small extent, producing both hydronium and hydroxide ions.

a. Water ionizes and then re-forms itself. At any given moment, the amount of ionized products is exceedingly small. Nevertheless, water is both an acid and a base by the Arrhenius definition because it provides both protons and hydroxide ions (in water, of course):

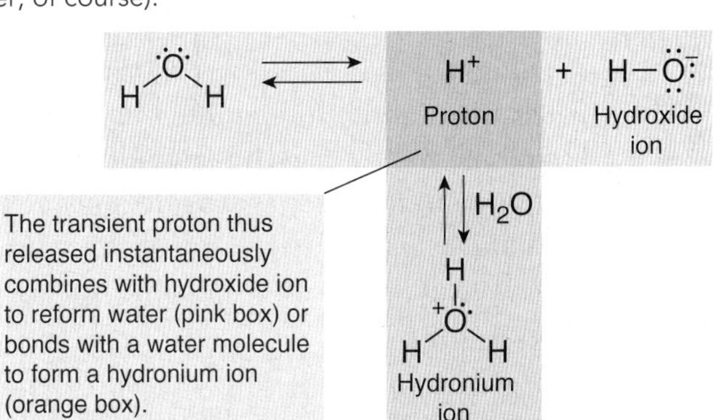

H O H
H+
Proton
+ H—O
Hydroxide ion
H2O
H
H O H
Hydronium ion
The transient proton thus released instantaneously combines with hydroxide ion to reform water (pink box) or bonds with a water molecule to form a hydronium ion (orange box).

b. Water is both an acid and a base by the Brønsted-Lowry definition as well since the proton released from one water molecule transfers to another water molecule.

The proton from one water molecule (acting as an acid) transfers to another water molecule (acting as a base)... to form hydronium ion... and hydroxide ion.

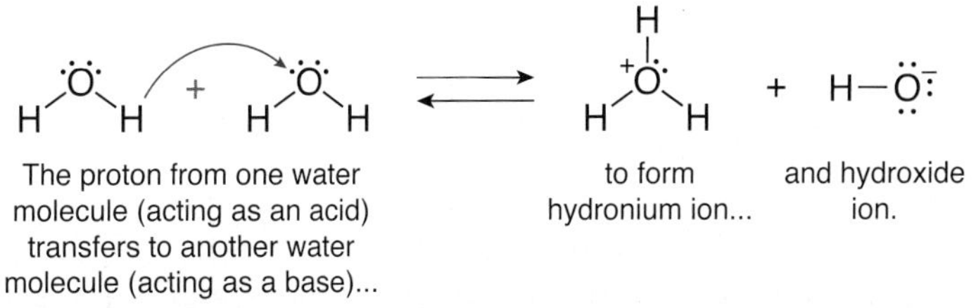

Think Critically From Figure 8.8b, determine the ratio of hydronium ions to hydroxide ions in a sample of pure water at any given moment.

An analogy for dynamic equilibrium • Figure 8.9

Imagine two islands, A and B. Suppose that every time a swimmer lands on an island, a person on that island swims to the other one, so that there is a continuous flow of people back and forth. If all swimmers travel the same distance and at the same speed between islands, the population of each island should remain the same.

For a sample of pure water, the number of water molecules is like the population of Island A and the number of ions is like the population of Island B. At equilibrium, the ratio of water molecules to ions remains the same despite constant transformations between the two.

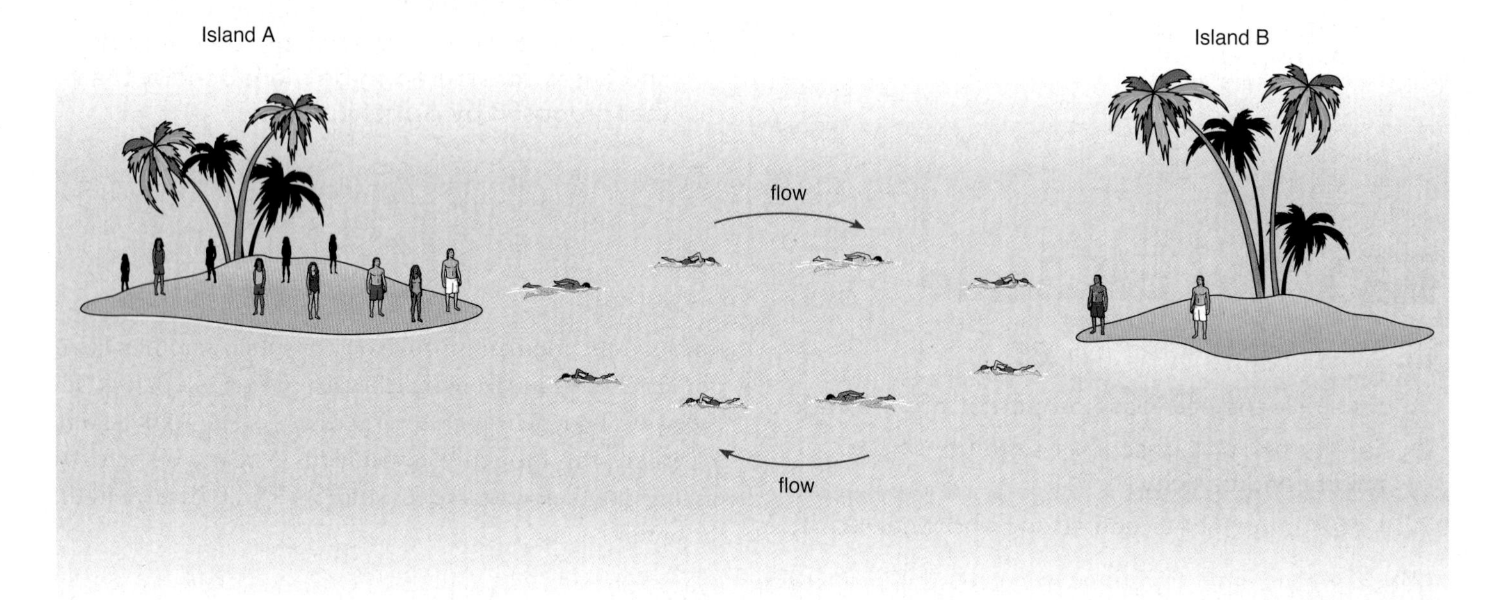

The actual ratio of water molecules to ions in a sample of pure water at room temperature is on the order of 500 million water molecules for each pair of ions! In other words, the relative number of water molecules is far greater than can be represented here.

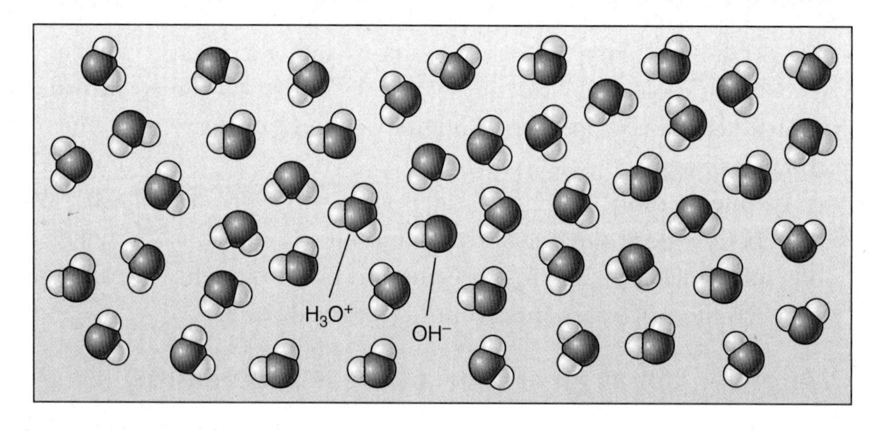

The simultaneous ionization and recombination of water is a dynamic process. At any instant, countless covalent water molecules are ionizing to hydronium and hydroxide ions throughout any sample of water. At the same instant, equally countless hydronium and hydroxide ions are recombining to regenerate covalent water molecules. Yet despite all the commotion of random ionizations and recombinations, the actual concentrations of both the hydronium and hydroxide ions remain constant in pure water at any given temperature. We call this state a **dynamic equilibrium** (**Figure 8.9**).

dynamic equilibrium A state in which the rate of a forward reaction (such as ionization) equals the rate of the reverse reaction (such as recombination) so that the concentrations of all chemical species remain constant.

In chemically pure water, the concentrations of all the transient hydronium (H_3O^+) and hydroxide (OH^-) ions that exist at equilibrium are not only fixed at any given temperature but are also equal to each other. They must always equal each other in pure water because the ionization of each water molecule produces an equal number of ions—one hydronium ion and one hydroxide ion. Experimental measurements show that each of these ions is present in pure, neutral water at a concentration of 0.0000001 moles per liter at 25°C. For brevity, we write $[H_3O^+]$ for "the molar concentration of H_3O^+" and we express the value of the molar concentration in exponential notation, using the italicized capital *M* as the symbol for moles/liter.

pH A measure of acidity, defined as the negative logarithm of the hydronium ion concentration: $pH = -\log [H_3O^+]$.

$$\text{Molar concentration of hydronium ion} = 0.0000001\ M$$

$$[H_3O^+] = 1 \times 10^{-7}\ M$$

$$\text{Equally, } [OH^-] = 1 \times 10^{-7}\ M$$

Thus at 25°C, pure water—which is perfectly neutral and neither the slightest bit acidic nor basic—contains both OH^- anions and H_3O^+ cations, each at a concentration of 1×10^{-7} *M*. We should note that this value of 1×10^{-7} holds only at a temperature of 25°C. At higher temperatures, the values of both $[H_3O^+]$ and $[OH^-]$ are somewhat higher; at lower temperatures they are both a bit lower.

KNOW BEFORE YOU GO

2. Suppose you have a solution of HCl in water that has a hydronium ion concentration of $[H_3O^+] = 0.001$ *M*. Express this value in exponential notation.
3. Now suppose you have a solution in which $[H_3O^+] = 1 \times 10^{-5}$ *M*. Express this value in decimal notation. How many moles of hydronium ions are in 2 liters of this solution?

pH: The Measure of Acidity

As we just saw, we can make things simpler for ourselves by writing the value of $[H_3O^+]$ as an exponent of 10. For example, a hydronium ion concentration of 0.0000001 *M* is the same as 1×10^{-7} *M*, which is simply 10^{-7} *M*.

We can carry this a step further by dispensing with both the 10 and the negative sign. This is exactly what the Danish biochemist Søren Sørensen did in 1909 when he proposed that concentrations of H^+ (or, as we now know, H_3O^+) be treated as exponential values. (Mathematically, this process of using the exponent or power of a number as a value is called "taking the logarithm" of that number.) Following Sørensen's recommendation, we now consider the concentration of hydronium ion, $[H_3O^+]$, in terms of **pH** (**Figure 8.10**).

The letters *pH* represent the "power of the Hydrogen (or Hydronium) ion." As a symbol for acidity, pH reflects nicely the international character of chemistry. The letter *p* begins the English word *power* as well as its French and German equivalents, *puissance* and *Potenz*. At the time of Sørensen's suggestion, English, French, and German were the world's dominant scientific languages.

KNOW BEFORE YOU GO

4. What is the pH of pure water at a temperature of 25°C?

   Answer: In a sample of pure water at 25°C, $[H_3O^+] = 1 \times 10^{-7}$ *M*, which equals 10^{-7} *M*.

   To get the pH, we take the exponent, −7, and reverse the sign; so pH = 7.

5. Fresh eggs typically have a hydronium ion concentration of about 0.00000001 *M*. What is the typical pH of a fresh egg?
6. The pH of a typical soft drink is about 3. What is the value of $[H_3O^+]$ for a soft drink?

Determining pH from hydronium ion concentration • Figure 8.10

We can determine pH as follows:

If hydronium ion concentration is expressed as a power of ten...

$[H_3O^+] = 10^{-x}$

...simply take the exponent and reverse the sign to get pH.

$pH = x$

For example, if:

$[H_3O^+] = 10^{-9}\ M$

$pH = 9$

In Words, Math, and Pictures

The pH scale • Figure 8.11

In water solutions at 25°C, the product of the hydronium ion and hydroxide ion molar concentrations,

$$[H_3O^+] \times [OH^-], \text{ is always } 10^{-14}.$$

When both $[H_3O^+]$ and $[OH^-]$ equal 10^{-7} *M*, water is neutral, with a pH of 7.

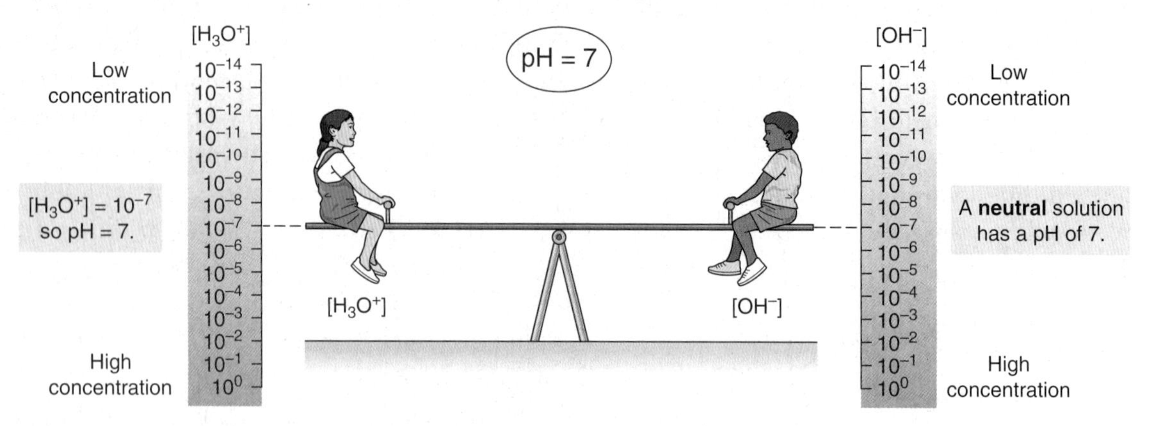

Any increase in the hydronium ion concentration above 10^{-7} (with an associated decrease in the hydroxide ion concentration) produces an **acidic** solution:

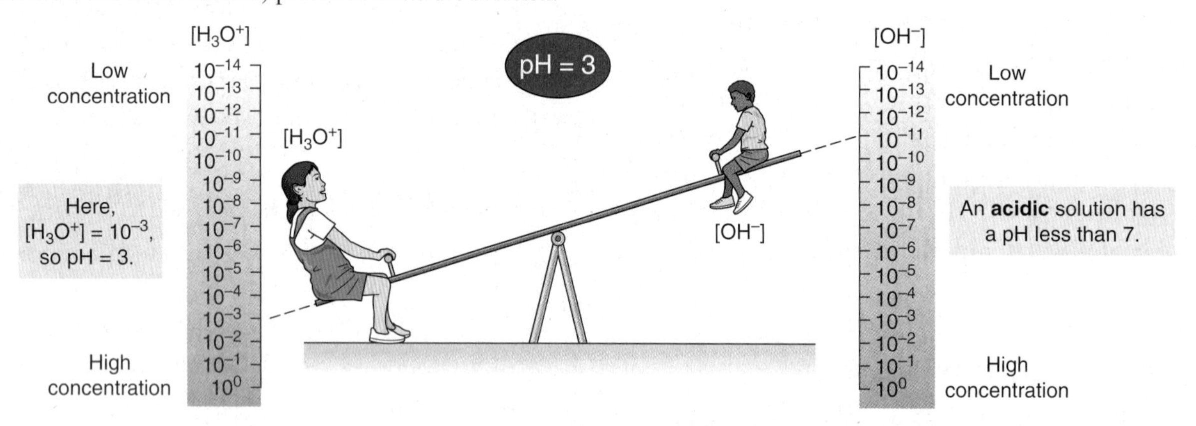

Any decrease in the hydronium ion concentration below 10^{-7} (with an associated increase in the hydroxide ion concentration) produces a **basic** solution:

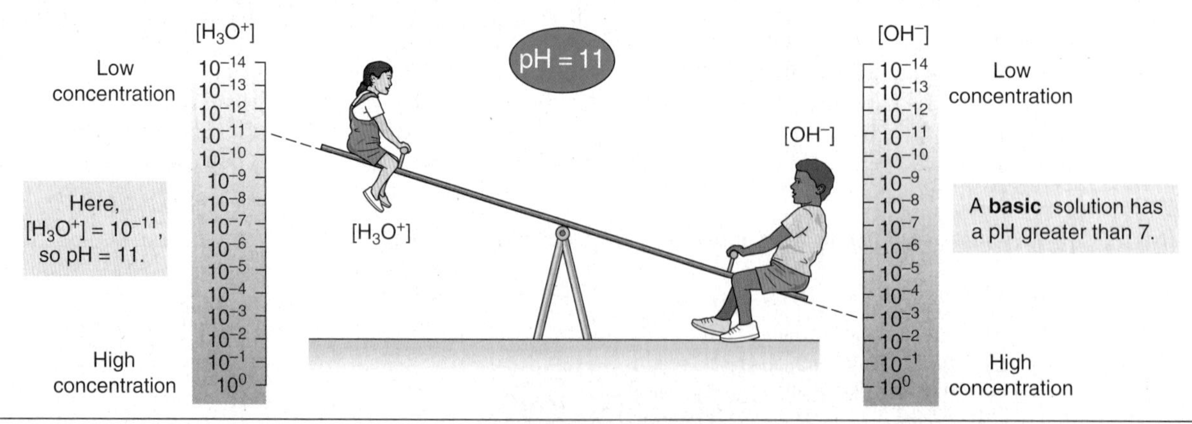

Interpret the Data

1. For each of these three examples, show that $[H_3O^+] \times [OH^-] = 10^{-14}$.
2. If a water solution at 25°C has $[OH^-] = 10^{-5}$ *M*, what is $[H_3O^+]$ and what is the pH? Is this solution acidic or basic?

We can increase the H_3O^+ concentration of a solution by adding acid, such as HCl. Conversely, we can increase the OH^- concentration by adding base, such as NaOH. Furthermore, the value obtained by multiplying the hydronium ion concentration of a solution, $[H_3O^+]$, by its hydroxide ion concentration, $[OH^-]$, is *always* constant, regardless of the addition of acid or base to the water. Since the value of this product remains fixed at any specific temperature, adding acid not only increases $[H_3O^+]$ but lowers $[OH^-]$ as well. Similarly, adding base to water increases $[OH^-]$ and lowers $[H_3O^+]$. We can see this seesaw effect and its relation to pH in **Figure 8.11** *In Words, Math, and Pictures.*

To summarize, at 25°C:

- The pH of a neutral solution equals 7.
- The pH of an acidic solution is less than 7.
- The pH of a basic solution is greater than 7.

We can see the relationship between hydronium ion concentration and pH in **Figure 8.12**.

The pH scale • Figure 8.12

pH values run from 0 (highly acidic) to 14 (highly basic). A neutral solution has a pH of 7.

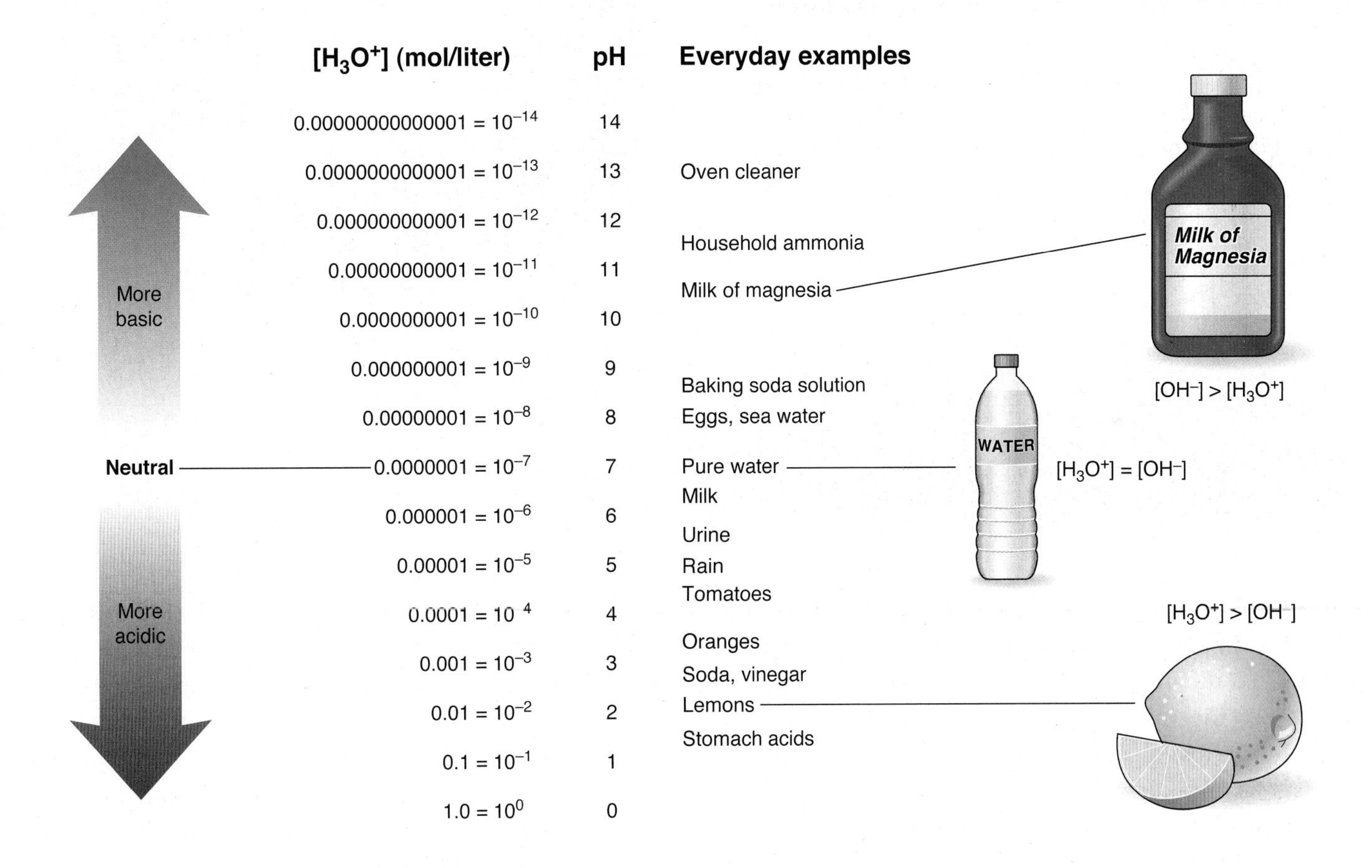

Interpret the Data

1. Are each of the following substances neutral, acidic, or basic? **(a)** eggs, **(b)** rain, **(c)** Diet Coke
2. As pH increases, does the hydronium ion concentration increase or decrease?
3. By what factor does the hydronium ion concentration change if we decrease pH by **(a)** one unit? **(b)** two units? **(c)** three units?

Measuring pH and acid-base indicators • Figure 8.13

pH test strips are quick and convenient, but less accurate than pH meters.

a. A pH meter very accurately measures hydrogen ion concentration. Here a meter records the pH of a sample of yogurt.

Tim O. Walker

b. "Universal" test strips are saturated with combinations of acid-base indicators that turn various colors as the pH changes. Although universal strips are more sensitive than litmus paper (which simply indicates acidity or basicity), they are less accurate than pH meters.

© Sabine Kappel/iStockphoto

c. Certain plants, such as beets, blueberries, cherries, and red cabbage, contain pigments that act as acid-base indicators. The figure shows the color of red cabbage extract at various pH levels.

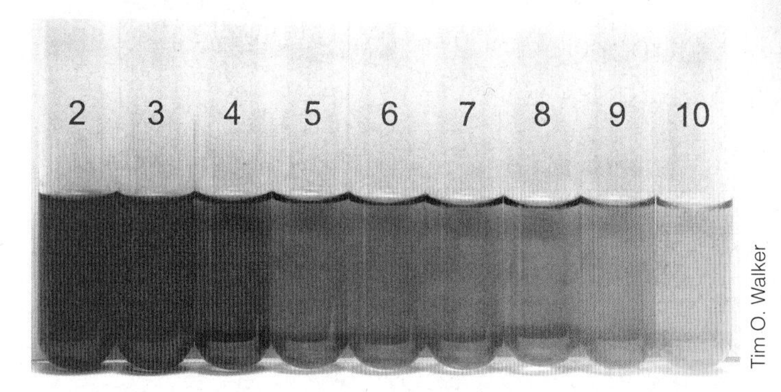

Tim O. Walker

Usually pH is measured with either a pH meter or a strip of test paper. For more precise measurements, the pH meter is the instrument of choice. Most common pH meters provide values accurate to about 0.01 pH unit. Color test strips, on the other hand, give only approximate values, although they are fast, convenient, and inexpensive. We take a further look into how we measure pH in **Figure 8.13**.

Strong Versus Weak Acids

We have seen that the very slight ionization of water generates a very small concentration of hydronium ions in pure water. In contrast, the ionization of HCl in water produces a relatively large number of hydronium ions since every HCl molecule that enters water produces one Cl^- ion and one H_3O^+ ion (see Figure 8.7). For example, dissolving 0.01 mole of pure HCl in 1 liter of water gives a solution with a hydronium ion concentration of 0.01 *M*.

Between these two extremes—pure water, which ionizes to form hydronium atoms only in the most marginal sense, and HCl, which completely ionizes in water to form hydronium atoms—we can find many examples of acids that occupy a middle ground. For example, acetic acid (the acid in vinegar) ionizes reversibly and therefore only partially in water (**Figure 8.14**). Acids such as acetic acid, which ionize only partially, are called **weak acids**. Acids such as hydrochloric acid, which ionize completely, are called **strong acids**.

weak acid An acid that ionizes partially in water.

strong acid An acid that ionizes completely in water to form an equivalent amount of hydronium ions.

Weak versus strong acids • Figure 8.14

A weak acid ionizes only partially in water, whereas a strong acid ionizes completely.

a. Acetic acid ionizes reversibly and partially in water, making it a **weak acid**.

$$H_3C-\overset{\overset{\displaystyle O}{\|}}{C}-O-H + H_2O \rightleftharpoons H_3C-\overset{\overset{\displaystyle O}{\|}}{C}-O^- + H_3O^+$$

Acetic acid, non-ionized; Water; Acetate ion; Hydronium ion

The vertical bars in the following diagram show relative amounts of acetic acid and its ionized forms before and after the acid dissolves in water.

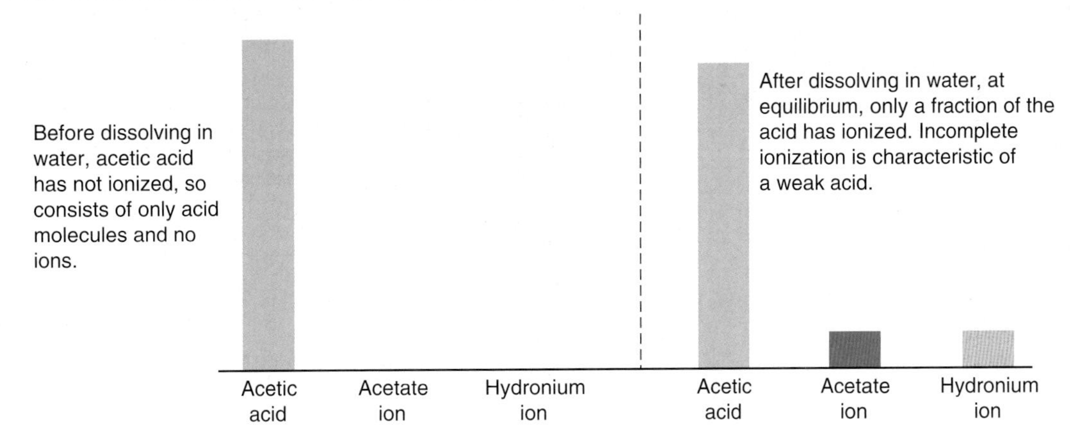
Before dissolving in water, acetic acid has not ionized, so consists of only acid molecules and no ions.
After dissolving in water, at equilibrium, only a fraction of the acid has ionized. Incomplete ionization is characteristic of a weak acid.
Acetic acid
Acetate ion
Hydronium ion
Acetic acid
Acetate ion
Hydronium ion

b. Hydrochloric acid ionizes irreversibly and completely in water, making it a **strong acid**.

$$H-Cl \xrightarrow{H_2O} Cl^- + H_3O^+$$

Hydrochloric acid, non-ionized; Chloride ion; Hydronium ion

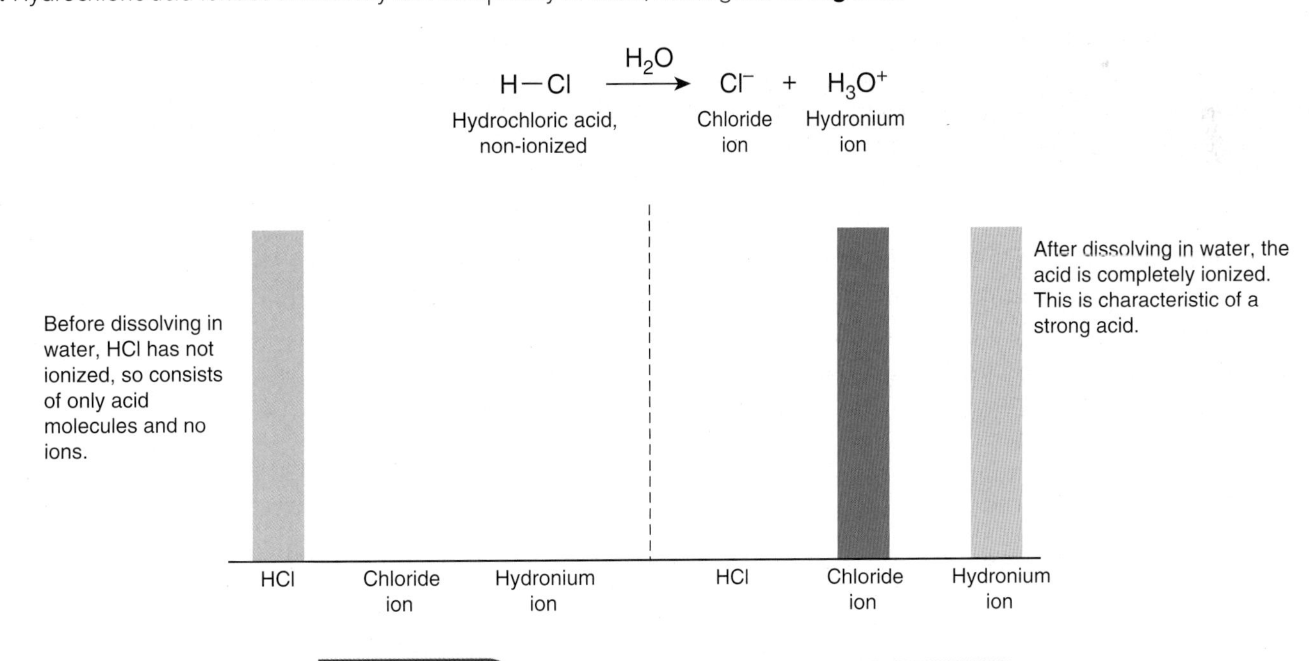
Before dissolving in water, HCl has not ionized, so consists of only acid molecules and no ions.
After dissolving in water, the acid is completely ionized. This is characteristic of a strong acid.
HCl
Chloride ion
Hydronium ion
HCl
Chloride ion
Hydronium ion

Ask Yourself

Which of the following procedures would you expect to produce the larger hydronium ion concentration? **(a)** dissolving 0.01 mole of HCl in 1 liter of water, **(b)** dissolving 0.01 mole of acetic acid in 1 liter of water

Ammonia, a weak base • Figure 8.15

Like other bases, ammonia accepts a proton from water, forming the hydroxide ion—in this case, ammonium hydroxide. Ammonia ionizes reversibly and partially in water, making it a weak base. At equilibrium, both ammonia and ammonium hydroxide are present.

$$NH_3 + H_2O \rightleftharpoons NH_4^+ \; OH^-$$

Ammonia, a weak base   Water   Ammonium hydroxide

Household ammonia is a dilute solution of ammonia in water.

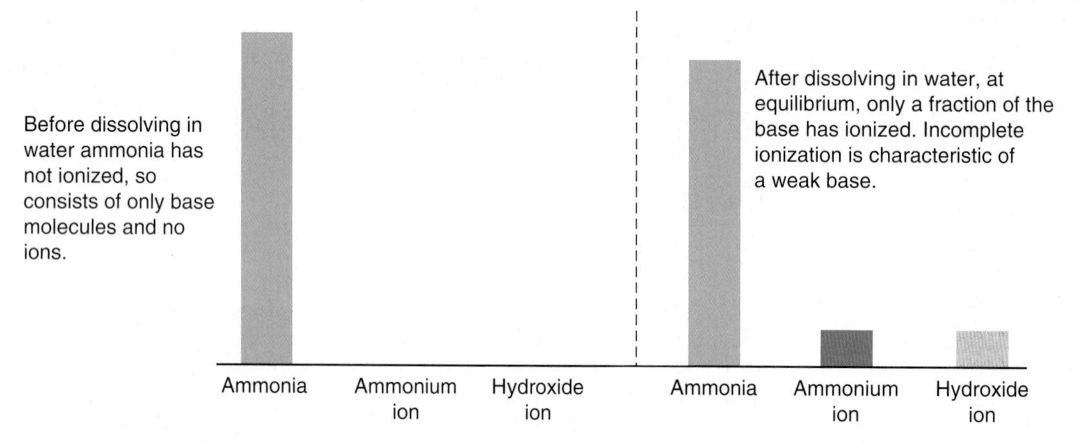

Before dissolving in water ammonia has not ionized, so consists of only base molecules and no ions.
After dissolving in water, at equilibrium, only a fraction of the base has ionized. Incomplete ionization is characteristic of a weak base.
Ammonia
Ammonium ion
Hydroxide ion
Ammonia
Ammonium ion
Hydroxide ion

Since hydrochloric acid ionizes completely in water and acetic acid ionizes partially in water, a solution of hydrochloric acid is more acidic than an equally concentrated solution of acetic acid. Strong acids include *hydrochloric* (HCl), *sulfuric* (H_2SO_4), and *nitric* (HNO_3) acids. Weak acids include *acetic* (CH_3CO_2H), *boric* (H_3BO_3), and *carbonic* (H_2CO_3) acids.

The acidity of a solution—the concentration of the hydronium ion—depends not only on the extent to which the acid ionizes but also on the concentration of the acid. A more concentrated solution of a given acid is more acidic than a less concentrated solution of the same acid.

KNOW BEFORE YOU GO

7. Which member (if either) of each of the following pairs of solutions would you expect to show a higher hydronium ion concentration?
   a. 0.01 *M* HCl or 0.0001 *M* HCl
   b. 0.01 *M* acetic acid or 0.0001 *M* acetic acid
   c. 0.01 *M* HNO_3 or 0.01 *M* acetic acid

Weak bases are similar to weak acids in that they too ionize reversibly and partially in water. The difference, however, is that bases accept protons from water as opposed to donating them (**Figure 8.15**).

In addition, **strong bases**, such as sodium hydroxide (NaOH), are similar to strong acids such as HCl in that they dissociate completely in water.

weak base A base that ionizes partially in water.

strong base A base that dissociates completely in water to form an equivalent amount of hydroxide ions.

CONCEPT CHECK

1. **Why** are the concentrations of hydronium and hydroxide ions in pure water constant (at a given temperature) and equal to each other?
2. **What** is the pH of a solution at 25°C when the *hydroxide* ion concentration is 10^{-8} *M*?
3. **Why** is the hydronium ion concentration different in the case of a 0.04 *M* solution of acetic acid than it is in a 0.04 *M* solution of sulfuric acid?

8.3 Acids and Bases in Everyday Life

LEARNING OBJECTIVES

1. **Identify** the pH range of gastric juices and that of typical foods.
2. **List** examples of carboxylic acids and the foods in which they occur.
3. **Describe** how antacids and buffers work.
4. **Explain** how acid rain forms and the basis for ocean acidification.

The acids and bases of our everyday lives fall largely into two broad categories: (1) our bodies, foods, and consumer products; and (2) our environment.

Bodies, Foods, and Consumer Products

The strongest acids we encounter every day—the acids with the lowest pH—are those of our stomach fluids, especially the HCl of the gastric juices secreted by the stomach's lining. The digestion of proteins in our foods requires an acidic environment. The stomach enzyme pepsin, which cleaves te large protein molecules of foods into smaller, more easily handled fragments, does its best work at a pH of about 1.5–2.5. At higher pH values, around 4 or 5, pepsin is no longer effective. For good health and good digestion, our gastric juices must be highly acidic, with a pH of about 1.0–2.0.

To help us maintain an acidic digestive environment, our foods are generally acidic to neutral, with a few in the weakly basic category. Citrus fruit, vinegar, pickles, soft drinks, wine, and tomatoes generally run in a pH range from about 2.0 to 4.5. Cow's milk is slightly acidic, with a pH in the range of 6.4–6.8, whereas fresh eggs are slightly basic, with a pH of about 7.5–8.0 (**Figure 8.16**).

The pH of foods • Figure 8.16

Most foods are mildly to moderately acidic, with some exceptions such as eggs, which are mildly basic.

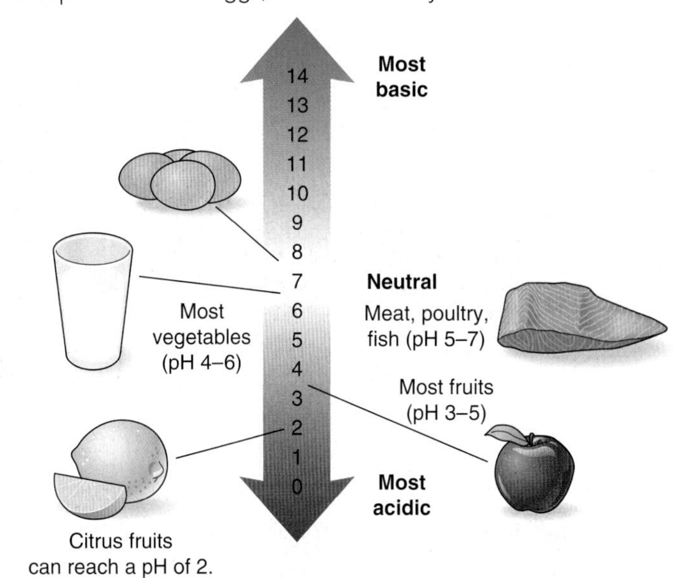

Many of the acids present in our foods are **carboxylic acids**. These acids are characterized by molecular structures containing a carbon doubly bonded to an oxygen and singly bonded to a **hydroxyl group** (—OH) (**Figure 8.17**).

> **carboxylic acid** A carbon-bearing acid characterized by the presence of the carboxyl group, $-CO_2H$.

Carboxylic acids • Figure 8.17

Carboxylic acids are characterized by the presence of a **carboxyl** group. The carboxyl group can also be shown as —COOH or $—CO_2H$.

$$-\overset{\overset{\displaystyle O}{\|}}{C}-OH$$

Carboxyl group

a. One of the simplest carboxylic acids is acetic acid, the principal acid of vinegar. Acetic acid consists of a carboxyl group bonded to a **methyl** group, $—CH_3$. With only a single carboxyl group per molecule, acetic acid is a **monocarboxylic** acid.

$$CH_3-\overset{\overset{\displaystyle O}{\|}}{C}-OH \quad \text{or} \quad CH_3-CO_2H$$

Acetic acid

b. Oxalic acid, which occurs in green leafy vegetables such as spinach, represents a **dicarboxylic** acid, with two carboxyl groups per molecule. In oxalic acid, the two carboxyl groups are bonded to each other.

$$HO-\overset{\overset{\displaystyle O}{\|}}{C}-\overset{\overset{\displaystyle O}{\|}}{C}-OH \quad \text{or} \quad HO_2C-CO_2H$$

Oxalic acid

c. Citric acid is an example of a **tricarboxylic** acid, with three carboxyl groups in each molecule. Citric acid is the principal acid of citrus fruits such as oranges, limes, and lemons.

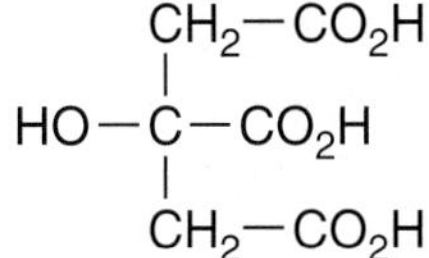

Citric acid

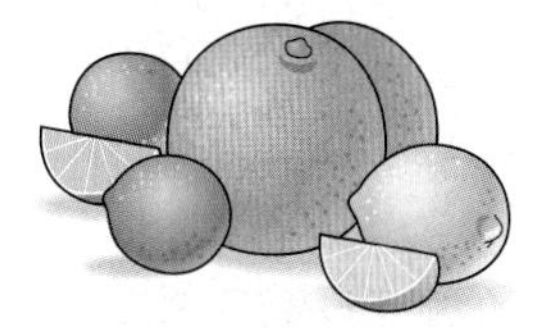

Alpha-hydroxy acids in cosmetics • Figure 8.18

Cosmetics containing alpha-hydroxy acids (AHAs) exfoliate the skin, meaning they shed surface skin cells and can help improve overall complexion.

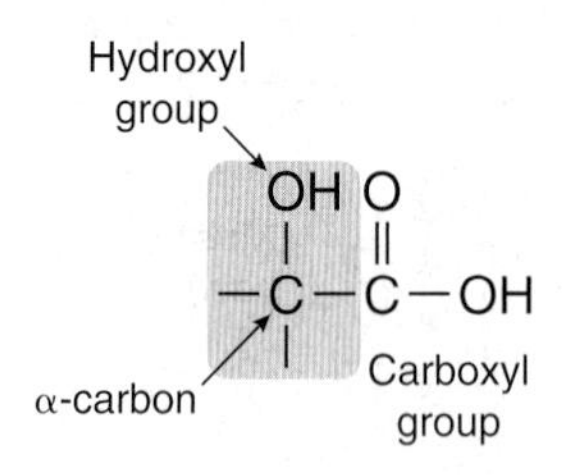

Alpha-hydroxy acids have a hydroxyl group (–OH) attached to the alpha (α) position, the carbon adjacent to a carboxyl group. AHAs can be found in a variety of foods such as yogurt, grapes, and citrus.

This label states an AHA concentration of 4%, which is within the typical range for over-the-counter cosmetics of this type. The greater the AHA content, the stronger the exfoliating effect.

Tim O. Walker

Put It Together Review Figure 8.17c and answer this question. In citric acid, to which alpha-carbon is the hydroxyl group bound?

Citric acid is an example of an **alpha-hydroxy acid**. Compounds of this type are used in a variety of skin and facial creams to improve skin texture and reduce wrinkles (**Figure 8.18**).

Antacids Although an acidic stomach is essential to good health, excess stomach acidity can range from uncomfortable to dangerous. To help bring an excessively low stomach pH back to a more normal range, we can use **antacids**. Each year Americans buy a quarter of a billion dollars worth of antacids, including Alka-Seltzer, Rolaids, and Tums, for the relief of heartburn. Antacids contain weak bases that neutralize excess stomach acid (**Figure 8.19**). These include *sodium bicarbonate* ($NaHCO_3$; pH 8.4 as a saturated solution in water), *calcium carbonate* ($CaCO_3$; pH 9.4 as a saturated solution), and *magnesium hydroxide* ($Mg(OH)_2$; pH 10.5 as a milky-white dispersion in water, known as milk of magnesia).

Antacids • Figure 8.19

The active ingredients of all antacids are weakly basic compounds.

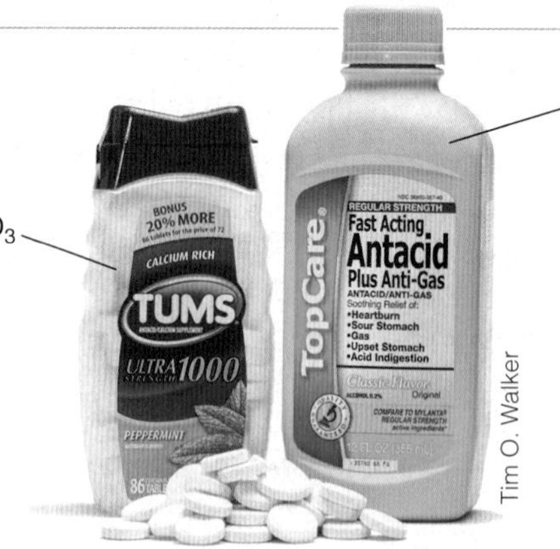

A representative neutralization reaction of antacids appears below. When calcium carbonate neutralizes excess stomach acid, it produces calcium chloride, a salt that is readily soluble in water.

$$CaCO_3 + 2\,HCl \longrightarrow CaCl_2 + H_2O + CO_2$$

$CaCO_3$	$2\,HCl$	$CaCl_2$
Calcium carbonate, from an antacid	Hydrochloric acid, from stomach fluids	Calcium chloride, a salt

Ask Yourself

What salt is produced when magnesium hydroxide neutralizes excess stomach acid?

One of the most basic of our everyday substances is household ammonia, a solution of NH_3 in water, pH 10.5–12, that is a useful cleaning agent. (Household ammonia has no substantial medical use.) Solutions of sodium hydroxide (NaOH), pH 12–14, are even more basic and are commonly used as drain cleaners.

Buffers Living organisms are generally sensitive to changes in pH and have chemical mechanisms in place, called **buffers**, to help maintain a stable pH. For example, our blood is buffered to maintain a pH of about 7.4, which is slightly basic. Buffers maintain a nearly constant pH through a combination of a weak acid and its complementary, or **conjugate**, base. The weak acid neutralizes exposure to any additional base and the conjugate base neutralizes exposure to any additional acid. Blood, for instance, is buffered by the presence of carbonic acid (H_2CO_3) and its conjugate base, bicarbonate ion (HCO_3^-) (**Figure 8.20**).

> **buffer** A combination of a weak acid and its conjugate base that resists changes in pH.

Buffers can exist in consumer products as well as in biological systems. For instance, many shampoos and cosmetics contain buffers to help maintain a pH desirable for the product. We'll revisit this topic in Chapter 11.

Conjugate acid-base pairs and buffers • Figure 8.20

A pair of chemical species that differ by the presence of a proton (H^+) is an **acid-base conjugate pair**. Conjugate pairs involving weak acids and bases play an important role in buffers.

a. Acid-base conjugate pairs

H_2CO_3	HCO_3^-	CH_3CO_2H	$CH_3CO_2^-$	NH_3	NH_4^+
Carbonic acid	Carbonate ion	Acetic acid	Acetate ion	Ammonia, a base	Ammonium ion

b. The carbonic acid/bicarbonate ion buffer system helps maintain the pH of blood close to 7.4.

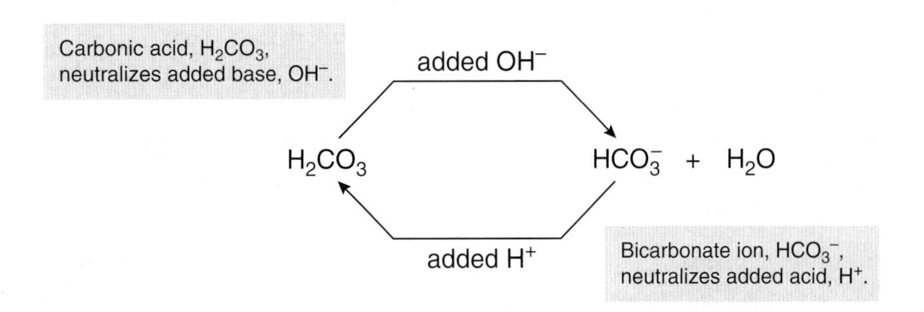

c. The body helps regulate the amounts of H_2CO_3 and HCO_3^- in the blood.

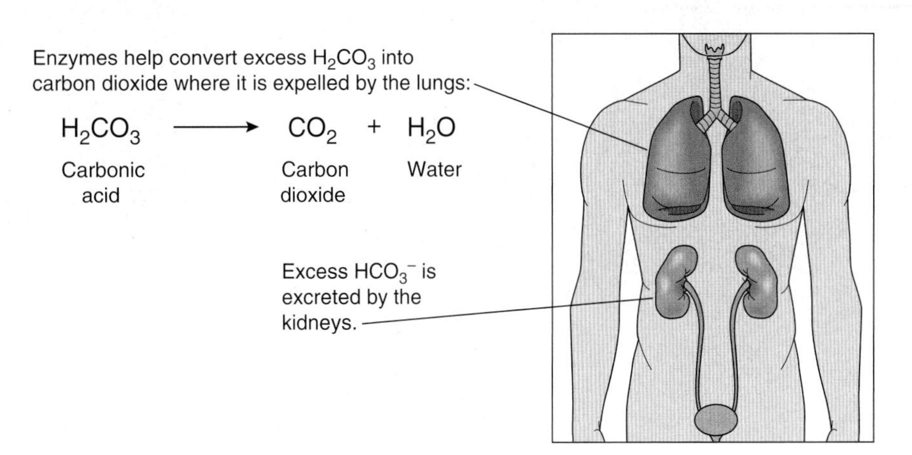

Think Critically Which of the following pairs of compounds represent acid-base conjugate pairs?
a. H_2O/H_3O^+; **b.** H_3O^+/OH^-; **c.** citric acid ($C_6H_8O_7$)/citrate ion ($C_6H_7O_7^-$)

Acids in Our Environment

Although many of our common foods are acidic, as are the gastric fluids that help us digest them, acids are often destructive to our environment.

acid rain Rain with a pH lower than 5.6, typically due to sulfuric and nitric acid contamination.

Acid rain is a typical example. In reality, all rainfall is naturally acidic simply because, as the raindrops fall through Earth's atmosphere, the water absorbs atmospheric carbon dioxide

The pH of rain • Figure 8.21

The pH of rain is always less than 7, but the extent of the decrease below 7 depends on absorption of certain atmospheric gases.

a. Rain is naturally acidic due to absorption of carbon dioxide.

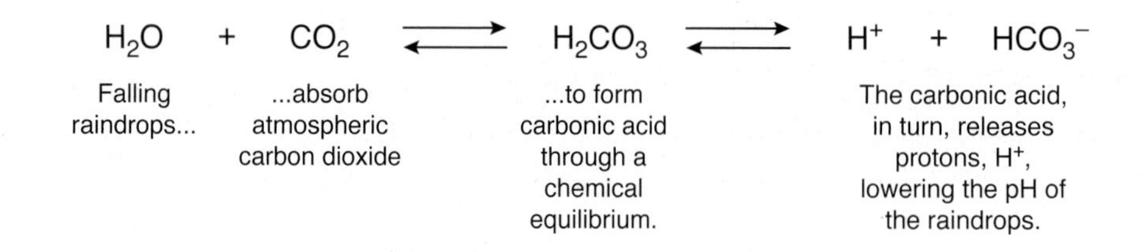

H_2O + CO_2 ⇌ H_2CO_3 ⇌ H^+ + HCO_3^-
Falling raindrops...
...absorb atmospheric carbon dioxide
...to form carbonic acid through a chemical equilibrium.
The carbonic acid, in turn, releases protons, H^+, lowering the pH of the raindrops.

b. The acidity of rain falling in the U.S. has decreased during past two decades largely as a result of pollutant restrictions included in the Clean Air Act. The largest decreases have occurred in the industrial Midwest and Northeast as a result of fewer pollutants released by coal-burning power plants common in those regions. (Source: National Atmospheric Deposition Program.)

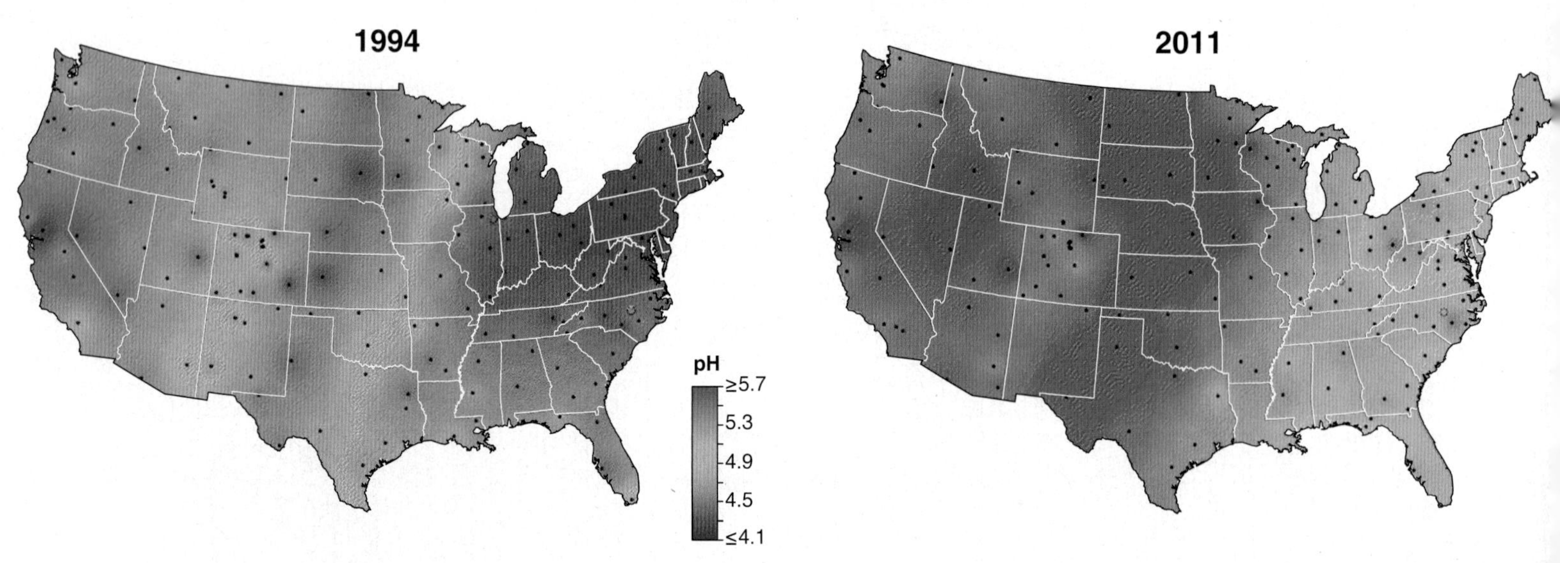

Average pH of rainfall
1994
2011
pH
≥5.7
5.3
4.9
4.5
≤4.1

Formation of atmospheric sulfuric acid, a component of acid rain • Figure 8.22

Gases released from the burning of fossil fuels in our factories and cars contribute to the formation of acid rain.

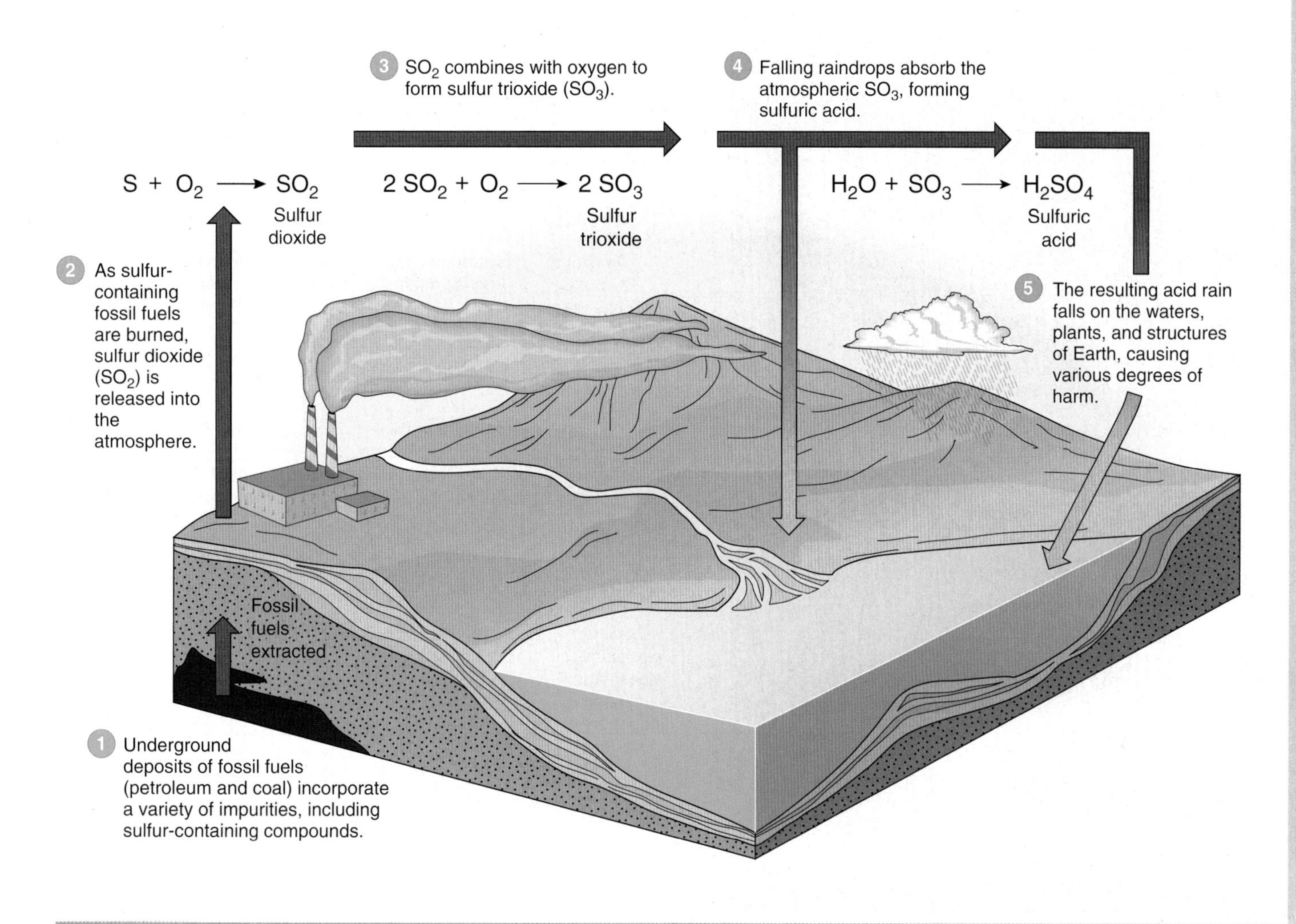

3 SO_2 combines with oxygen to form sulfur trioxide (SO_3).
4 Falling raindrops absorb the atmospheric SO_3, forming sulfuric acid.
$S + O_2 \longrightarrow SO_2$
Sulfur dioxide
$2\,SO_2 + O_2 \longrightarrow 2\,SO_3$
Sulfur trioxide
$H_2O + SO_3 \longrightarrow H_2SO_4$
Sulfuric acid
2 As sulfur-containing fossil fuels are burned, sulfur dioxide (SO_2) is released into the atmosphere.
5 The resulting acid rain falls on the waters, plants, and structures of Earth, causing various degrees of harm.
Fossil fuels extracted
1 Underground deposits of fossil fuels (petroleum and coal) incorporate a variety of impurities, including sulfur-containing compounds.

(CO_2), which reacts with the water to form carbonic acid (H_2CO_3) (**Figure 8.21**). With a high enough concentration of CO_2, the pH of rainwater can fall as low as 5.6. Rainwater with a pH lower than 5.6—because of the presence of other acidic components—is considered to be acid rain.

Acid rain can be generated by the absorption of a variety of atmospheric pollutants—generated by both human and natural activities—into falling rain. For example, both volcanic eruptions and the burning of fossil fuels by humans can result in the release of *sulfur dioxide* (SO_2) into the atmosphere. Reactions of this atmospheric sulfur dioxide can produce *sulfuric acid* (H_2SO_4), a strong acid (**Figure 8.22**).

Another important contributor to acid rain is nitric acid (HNO_3), formed when *nitrogen oxide* (NO_x) gases combine with oxygen (O_2) and water in the atmosphere:

$$\underset{\text{Nitrogen oxides}}{NO_x} \xrightarrow[H_2O]{O_2} \underset{\text{Nitric acid}}{HNO_3}$$

(The notation NO_x represents several different oxides of nitrogen, such as NO and NO_2.)

Effects of acid rain • Figure 8.23

Due to wind and weather patterns, the impacts of acid rain may be felt hundreds of miles from the sources of SO_2 and NO_x emissions. Lakes, streams, forests, and various construction materials are at particular risk from the effects of acid rain.

James H. Robinson/Photo Researchers, Inc.

a. This mountainside forest in North Carolina shows damage from acid rain. This acidity not only directly damages leaves and bark but can also alter soil chemistry, adding further stress to plants. For example, acid rain can leach beneficial minerals, such as calcium ions (Ca^{2+}), out of soil and release harmful minerals, such as aluminum ions (Al^{3+}), into the soil. The latter is a neutralization reaction:

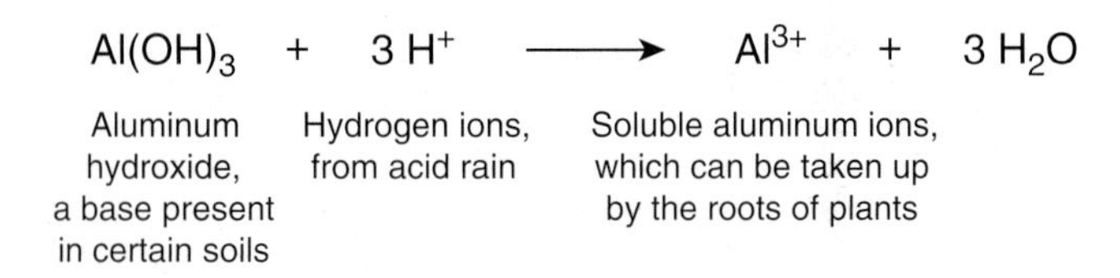

Adam Hart-Davis/Photo Researchers, Inc.

b. This stone statue in England shows severe erosion by acid rain. Calcium carbonate in the limestone combines with acid in a neutralization reaction:

$$CaCO_3 + 2H^+ \longrightarrow Ca^{2+} + H_2O + CO_2$$

Calcium carbonate, a base and primary component of limestone — Hydrogen ions, from acid rain — Soluble calcium ions, which wash away

Although natural processes, including lightning and bacterial action, can produce NO_x, human activities, such as the combustion of fossil fuels, are more important sources of these gases. Human-generated NO_x emissions result from the combination of nitrogen (N_2) and oxygen (O_2) in air under the high temperatures of combustion. Cars and trucks are the primary sources of these NO_x emissions, but fossil-fuel power plants and other industrial sources contribute as well.

The presence of nitric and sulfuric acids in the atmosphere can produce rainfall with a pH as low as that of common vinegar—and occasionally even lower. Acid rain poses risks to the environment, to buildings, and to other structures (**Figure 8.23**).

The world's oceans present a somewhat different picture of the environmental acid-base balance. The oceans absorb a significant amount—more than a quarter—of the CO_2 generated by the burning of fossil fuels. As these CO_2 emissions dissolve in the oceans, they convert water (H_2O) into carbonic acid (H_2CO_3), a process called **ocean acidification**. (CO_2 absorption is also the basis for the natural acidity of rain [Figure 8.21a].) Although ocean water is not expected to turn acidic (defined by a pH below 7.0), its pH has been predicted to drop from its current 8.1 to somewhere near 7.8 as we approach the next century. Although still slightly basic, a pH this low can harm ocean life in a variety of ways, including the slowing of growth or even causing the decay of the shells

ocean acidification A decrease in the pH of the oceans due to the absorption of atmospheric CO_2.

of many marine organisms, leading to a decrease in their populations (**Figure 8.24**).

Just as our knowledge of acids and bases has allowed us to understand the mechanisms underlying acid rain and ocean acidification, this knowledge can also be used to address these concerns. For example, since the early 1980s, acid rain has decreased significantly in the United States due to a combination of regulatory and technical measures designed to limit the release of sulfur and nitrogen oxide gases. (This is encouraging in that annual consumption of fossil fuels has actually increased over this time period.)

With regard to the oceans, note that ocean uptake of CO_2 reduces atmospheric levels of CO_2, and thereby helps to lessen the climate impact of this greenhouse gas. Here, the cost—ocean acidification—although of real concern, is associated with benefits in terms of moderating the pace of climate change. As the world grapples with the more fundamental issue of increasing CO_2 emissions from fossil fuel consumption, expertise from a variety of scientific disciplines, including chemistry, will help provide solutions to the problems created by these emissions (a topic we address in Chapter 4).

The concept of acids and bases can help us understand the chemistry of our bodies, foods, consumer products, and our environment. Acid-base concepts have even influenced the very language we use (see *Did You Know?* on the next page).

Ocean acidification • Figure 8.24

As the oceans continue to absorb carbon dioxide, the pH of the water decreases, putting stress on coral and other marine life.

a. Dissolved carbon dioxide lowers the pH of ocean water, which decreases levels of carbonate ions (CO_3^{2-}) in the water.

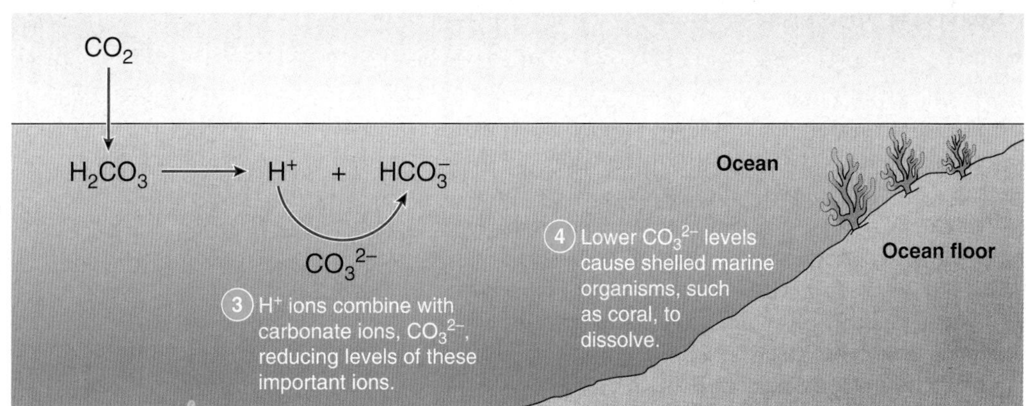

b. Pteropods, tiny sea snails, are said to be especially vulnerable to the lowering of the ocean's pH because their shells are made of a very soluble type of calcium carbonate ($CaCO_3$), which dissolves as the pH drops. Here we see the decay of a pteropod shell as it is kept for several weeks in seawater with a pH maintained near 7.8. Computer models predict ocean waters may turn inhospitable for many shell-building organisms by the end of the century, with impacts higher up the food chain. Pteropods, for example, are a food source for fish such as salmon.

David Liittschwager/NG Image Collection

David Liittschwager/NG Image Collection

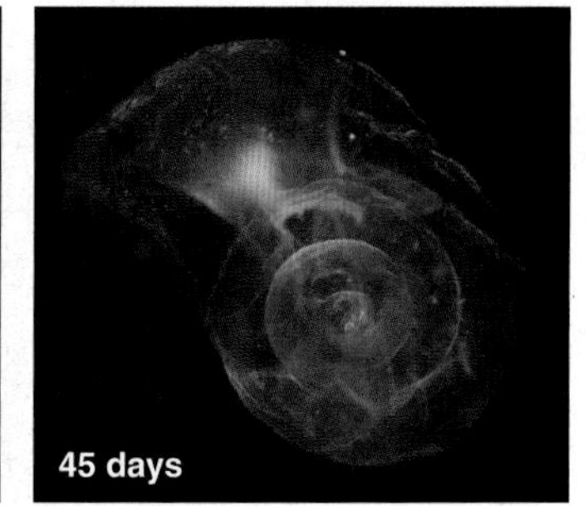

David Liittschwager/NG Image Collection

Ask Yourself

As solid calcium carbonate ($CaCO_3$) dissolves, what ions from this compound are released into the water?

DID YOU **KNOW?**

A variety of common expressions get their meanings from chemistry

The term **litmus test** (**Figure a**) originated in the test that tells us simply and quickly whether a substance is an acid or a base. But the term has since expanded to describe simple and definitive tests of social issues, political candidates, and economic policies. The following quotation captures this broader use of the term: "The extent and sensitivity to which we attend the interests of people in later life is a good **litmus test** for any plural and caring society."[1] The litmus test of social issues is often a simple question whose answer determines how we (or society, in this case) stand on an entire issue, just as the chemical litmus test quickly addresses whether we have an acid or a base.

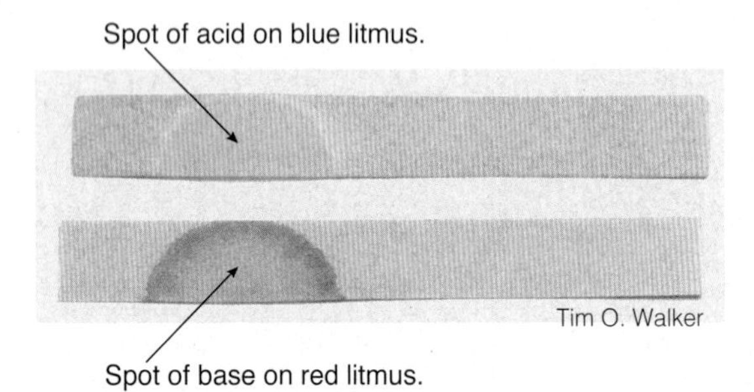

Tim O. Walker

a. Litmus test
Small strips of paper saturated with litmus serve as a quick, convenient test for acidity and basicity. To accentuate the color changes, commercial litmus paper is color pretreated: red litmus paper tests for bases, giving a blue spot against a red background, and blue litmus paper tests for acids, giving a red spot that shows up clearly against the blue background.

Like **litmus test**, a variety of expressions from the world of acids and bases have gained broader meaning in the English language, although the origins of these terms are not always apparent. For example, when a celebrity is in the **limelight**, it means he or she is in the public eye. The term originates in a type of stage lighting widely used in theaters before the advent of electric lights. The basic compound calcium oxide (CaO), also known as lime or quicklime, gives off a brilliant white light when heated and for a time was incorporated into highly effective spotlights **(Figure b)**.

b. Limelight
Here, a chunk of the basic compound, calcium oxide, also known as lime, gives off a brilliant glow when heated.

Adam Hart-Davis/
Photo Researchers, Inc.

[1]Dr. Mike McCarthy, *The Sunday Telegraph* (London), October 17, 2010.

Some terms used in chemistry may well have originated in everyday use, but nevertheless continue to share meanings in both senses. For example, when someone makes a **caustic** remark, he or she is conveying insult or scorn. The term **caustic** means corrosive and describes chemicals that burn the skin. The basic (and corrosive) compounds sodium hydroxide and potassium hydroxide are sometimes simply referred to as **caustic**. Another term is **buffer**, which means "to shield or cushion." A **buffer zone**, for example, is a neutral zone at the border between rival nations that cushions against hostilities. As discussed in Section 8.3, a buffer describes a solution that stays within a limited pH range even when exposed to moderate amounts of acid or base. Some might say it is *cushioned* against pH changes.

It shouldn't surprise you that a variety of common expressions use the word *acid*, no doubt reflecting the sharp taste of acids. Thus, *acid-tongued* means language that is critical or sharp, whereas **acid test** is used in an everyday sense like *litmus test*. The expression *acid test* originates in a simple assay used to determine the authenticity or purity of gold and other precious metals (**Figure c**).

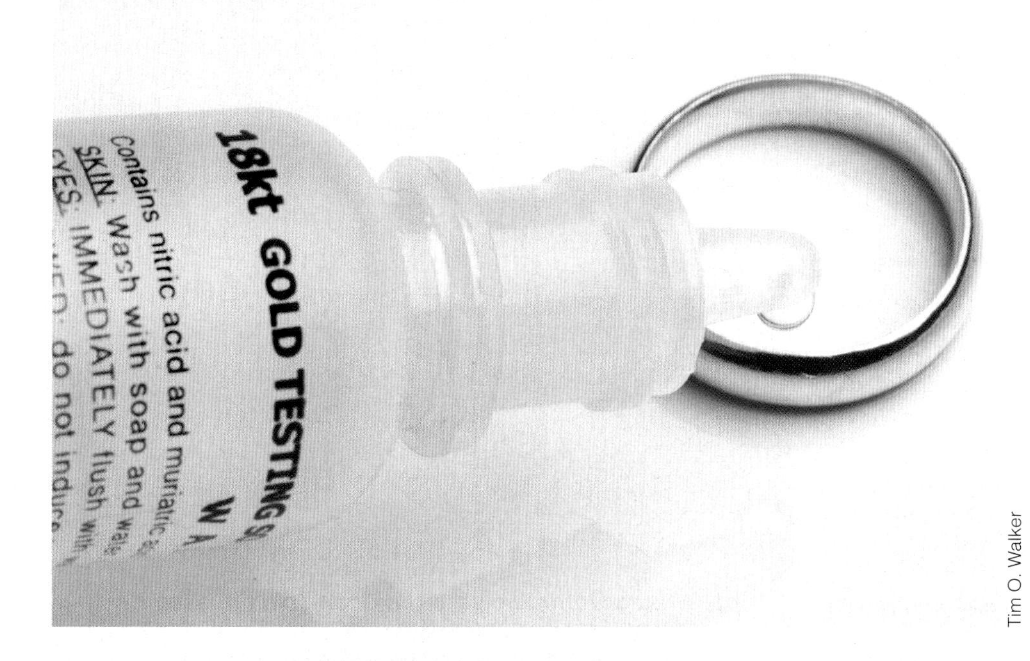

Tim O. Walker

c. Acid test
Solutions of nitric acid are used to test for the purity of gold. Here, a particular concentration of nitric acid is used to test for 18-carat (18K) gold. If the sample reacts (as evidenced by small bubbles), it is less pure than 18K. The basis for the test is that gold itself does not react with the acid but lesser metals do.

Think Critically

1. When lime (CaO) is added to water, it reacts rapidly to produce a suspension of calcium hydroxide ($Ca(OH)_2$). What color would a drop of this liquid turn litmus paper?
2. Eye drops and contact lens solution are examples of buffered solutions. What can we predict about the stability of the pH of these products? Why might this be important?
3. Identify at least one metal present in gold alloy that could react with an acid test solution (i.e., nitric acid). Use the Internet to find your answer.

CONCEPT CHECK 

1. **Which** category describes most foods: acidic, neutral, or basic?
2. **What** structural feature is common to all carboxylic acids?
3. **Why** does an antacid neutralize some, but not all, stomach acid?
4. **What** two gaseous emissions are primarily responsible for acid rain?

Summary

1 Acids, Bases, and Neutralization 234

- ***What simple tests can we use to distinguish an acid from a base?***
  Acids turn **litmus paper** red; react with certain metals, such as magnesium or zinc, to liberate hydrogen gas (as shown here); and can taste sour. **Bases** turn litmus paper blue, feel slippery on the skin, and can taste bitter. **(Never touch or taste any material unless you are sure it is safe.)**

Figure 8.1 • A test for acids

Tim O. Walker

- ***How do acids and bases react with one another?***
  Acids are proton donors and bases are proton acceptors. When acids and bases react, they **neutralize** one another; the acid donates a proton to the base to produce a **salt** (an ionic compound) and water.

2 The pH Scale 241

- ***What does pH tell us?***
  pH is a measure of the **hydronium ion** concentration, $[H_3O^+]$, of a substance. Since pH equals the ***negative*** of the logarithm of $[H_3O^+]$, the larger the $[H_3O^+]$, the smaller the pH. As shown here, substances with a pH less than 7 are acidic, with more hydronium than **hydroxide ions**. Those with a pH greater than 7 are basic, with more hydroxide than hydronium ions. A pH equal to 7 is neutral, neither acidic nor basic, since $[H_3O^+] = [OH^-]$.

Figure 8.12 • The pH scale

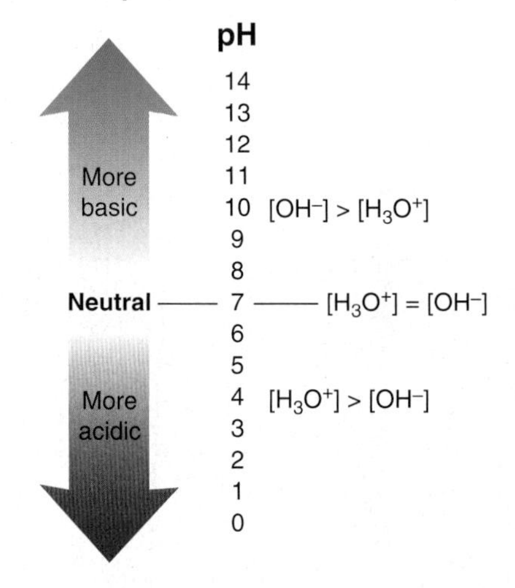

- ***What governs the strength of an acid or base?***
  Strong acids and **strong bases** ionize completely in water, whereas **weak acids** and **weak bases** ionize partially and reversibly in water. The hydronium ion concentration of a strongly acidic solution is the same as the concentration of the acid itself since virtually every acid molecule produces a hydronium ion. The hydronium ion concentration of a weakly acidic solution is much less than the concentration of the acid itself since only a fraction of the acid molecules produces hydronium ions at any given moment.

3 Acids and Bases in Everyday Life 249

- ***What roles do acids and bases play in our everyday lives and in the environment?***

  Acids and bases exist in our bodies, our foods, our consumer products, and the environment. Stomach fluids are highly acidic, with a pH of 1–2. The pH of foods varies by type, but most foods are mildly acidic. **Carboxylic acids**, characterized by the $-CO_2H$ group (as shown here) are present in many of our foods. Everyday bases include antacids, which neutralize excess stomach acids. **Acid rain** is caused by the formation of sulfuric and nitric acids. Ocean acidification is caused by the absorption of CO_2, which forms carbonic acid.

Figure 8.17 • Carboxylic acids

CH_2-CO_2H
$HO-C-CO_2H$
CH_2-CO_2H

Citric acid

Key Terms

- acid 234
- acid rain 252
- amphoteric 241
- base 234
- buffer 251
- carboxylic acid 249
- dynamic equilibrium 242
- hydronium ion 239
- hydroxide ion 239
- litmus paper 234
- neutralization 236
- ocean acidification 254
- pH 243
- salt 236
- strong acid 246
- strong base 248
- weak acid 246
- weak base 248

What is happening in this picture?

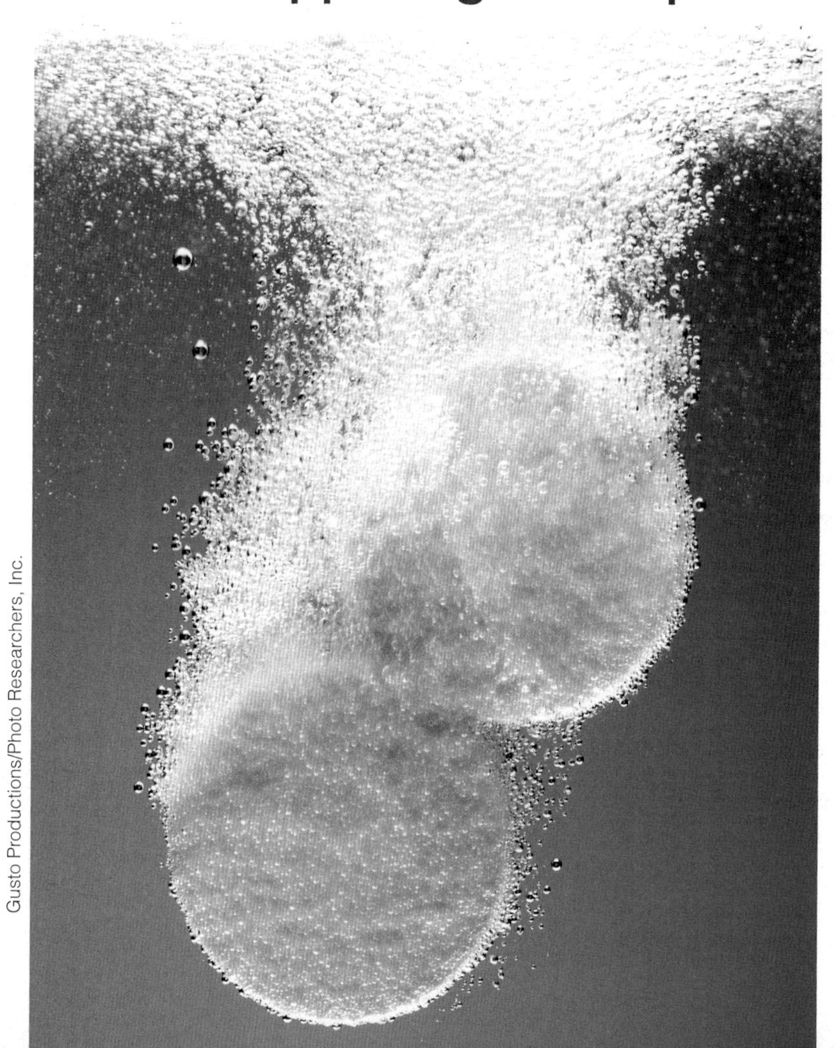

Gusto Productions/Photo Researchers, Inc.

Alka-Seltzer consists of a combination of aspirin, the base sodium bicarbonate ($NaHCO_3$), and a small amount of citric acid. When you drop the tablet into water, the sodium bicarbonate and citric acid dissolve. They react with each other in solution to produce carbonic acid (H_2CO_3), which decomposes into a gas and water (H_2O). The fizz of the Alka-Seltzer tablet comes from the chemical release of this gas. (You can produce the very same effervescence by squeezing a little lemon juice onto some household baking soda, which is sodium bicarbonate. The citric acid of the lemon juice reacts with the sodium bicarbonate to produce the same gas bubbles as when you drop an Alka-Seltzer tablet into water.)

Think Critically

1. What gas is produced in the picture?
2. When citric acid ($C_6H_8O_7$) reacts with a base, it forms the citrate ion ($C_6H_7O_7^-$). What salt is formed when citric acid reacts with sodium bicarbonate?

Exercises

Review

1. Define or identify each of the following terms.

a. amphoteric	f. hydroxide ion
b. antacid	g. neutralization
c. weak acid	h. a salt
d. carboxylic acid	i. dynamic equilibrium
e. hydronium ion	

2. Indicate the acid or base associated with each of the following everyday items:

a. car batteries	e. seltzer water
b. alkaline batteries	f. vinegar
c. vitamin C	g. oranges
d. baking soda	h. typical oven cleaners

3. What color—red or blue—would you expect each of the following liquids to produce when added to litmus: (a) milk of magnesia, (b) wine, (c) seawater, (d) a soft drink, (e) tomato juice?

4. Identify each of the following compounds as an acid, a base, or a salt; give its chemical name; and name a consumer product in which it occurs:

a. $CaCO_3$	g. $Mg(OH)_2$
b. CH_3CO_2H	h. $MgSO_4$
c. H_2SO_4	i. KI
d. HCl	j. NaOH
e. $NaHCO_3$	k. NH_3
f. H_3BO_3	

5. Define a salt in terms of acids and bases.

6. Suppose the nucleus of a hydrogen atom enters a quantity of water. What is produced by the interaction of the hydrogen nucleus with a water molecule?

7. What do we mean when we say that water is amphoteric?

8. What is the difference between a hydrogen ion and a hydronium ion?

9. What gas is formed when you drop an Alka-Seltzer tablet into water? Write the chemical reaction that produces this gas.

10. Hydrogen fluoride (HF), an acid, reacts with sodium hydroxide to produce a salt that is used in toothpaste to help prevent tooth decay. Write the chemical equation for the neutralization of hydrogen fluoride with sodium hydroxide. What is the name of the salt that forms in this reaction?

11. What characteristic of molecular structure is common to all carboxylic acids?

12. Give the names of two carboxylic acids that can be found in foods.

13. What is the difference between a weak acid and a strong acid? Give an example of each.

14. What type of reaction is the following equation? Identify each compound (other than water) as an acid, a base, or a salt.

$$CH_3CO_2H + NaOH \rightarrow Na^+CH_3CO_2^- + H_2O$$

15. a. What acid makes rainwater normally acidic?
    b. What acids give rise to acid rain?

16. Write the chemical reaction for the generation of fizz or foam when you squeeze lemon juice into water containing dissolved baking soda.

17. The Arrhenius definitions of acids and bases assume the presence of what substance?

18. This image depicts a common chemical test. Identify the test and indicate whether the pH of the solution is above, below, or equal to 7.

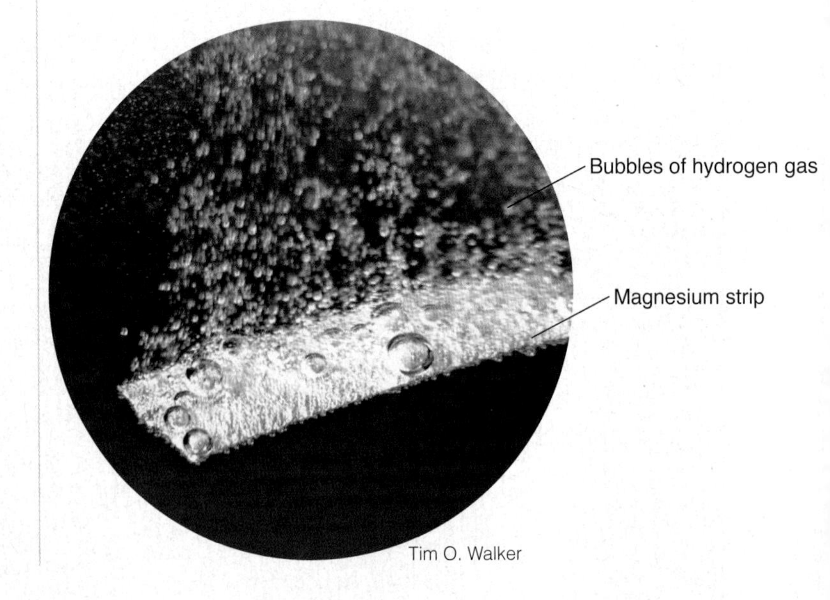

Tim O. Walker

19. Indicate which diagram depicts a solution of a strong acid in water and which depicts a solution of a weak acid in water.

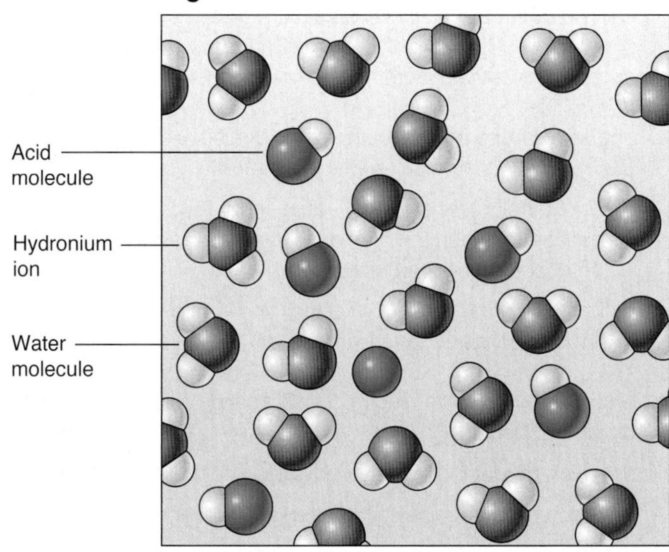

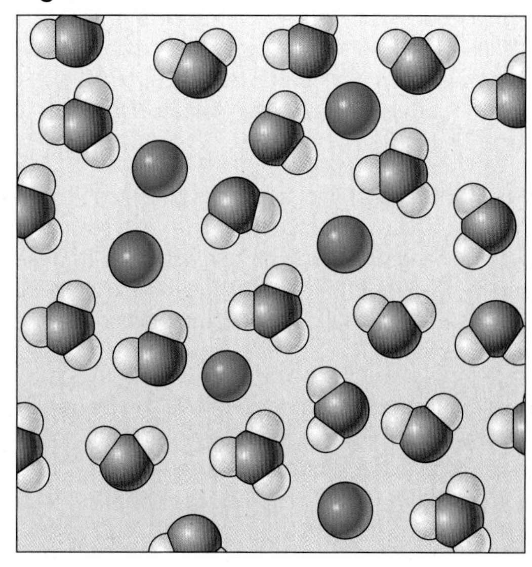

Think

20. In Section 8.1 we reviewed three tests that can be used to determine whether a substance is an acid. A fourth test comes from a reaction of the substance in question with ordinary chalk, which is mostly calcium carbonate, $CaCO_3$. Complete and balance the following equation for the reaction of HCl with $CaCO_3$, and explain how the test could be used to determine whether the substance is an acid.

$$HCl + CaCO_3 \rightarrow CaCl_2 + ?$$

21. Acids and bases react with each other to produce neutral solutions. Could you obtain a neutral solution by mixing, in the proper proportions: (a) vinegar and sodium bicarbonate; (b) the juice of sour pickles and carbonated water; (c) seawater and milk of magnesia?

22. Why is it so much more convenient to express acidities in terms of pH rather than in terms of the molar concentration of the hydrogen ion?

23. For each of the following sets, pick the single compound that does not belong with the others. Explain your choice in each case.

1. a. HCl
   b. CH_3CO_2H
   c. H_2CO_3
   d. NH_3
   e. H_2SO_4
2. a. acetic acid
   b. citric acid
   c. nitric acid
   d. oxalic acid
   e. lactic acid (the acid in sour milk; find its structure on the Internet)
3. a. boric acid
   b. sodium bicarbonate
   c. carbonic acid
   d. ammonia
   e. sodium hydroxide

24. Galvanized nails are coated with zinc to protect them against corrosion and rust. If you cover several galvanized nails with vinegar, you will soon see small bubbles of gas forming on the nails and rising to the surface of the vinegar. Why? Describe the chemical reactions taking place. Why could this be a hazardous thing to do? Would you expect the same result if you used iron nails that are not galvanized? Would you expect the same result if you used lemon juice instead of vinegar? Explain your answers.

25. Many acid-base indicators change color as a result of changes in their molecular structure as the pH of a solution changes. Some change color with an increase in pH as a result of structural changes brought on as the molecule releases a proton, H^+. To what general class of compounds do such indicators belong?

26. Acid deposits around a car's battery terminals can be cleaned with a solution of sodium bicarbonate. Assuming that the acid is sulfuric acid, complete and balance the equation for the neutralization reaction:

$$NaHCO_3 + H_2SO_4 \rightarrow ?$$

27. Malonic acid, with a molecular formula $C_3H_4O_4$, is a dicarboxylic acid containing three carbons. Draw its molecular structure.

28. Blood is buffered by the presence of the acid H_2CO_3 and its conjugate base HCO_3^-. Another combination of an acid and a base—$H_2PO_4^-$ and HPO_4^{2-}—helps maintain the fluid within our cells (**intracellular fluid**) within a narrow pH range. Recognizing that $H_2PO_4^-$ is the more acidic of these two components, and that HPO_4^{2-} is the more basic component, write a chemical reaction that shows how this system changes as the intracellular fluid becomes (a) more acidic and (b) more basic.

29. Can a substance act as an acid in the absence of a base? Explain your answer.

30. Baking powder, one of several types of leavening agents used in preparing bread, cakes and similar baked goods, causes dough to rise during the baking process by releasing tiny bubbles of gas. This gas is generated through an acid-base neutralization between two components of the baking powder: sodium bicarbonate ($NaHCO_3$), and an acid, such as calcium acid phosphate ($CaHPO_4$). The gas is released as these two chemicals react with each other within the wet dough. (a) What is the name and the chemical formula of the gas released during this reaction? (b) What role does the sodium bicarbonate serve in this process?

31. Describe the amphoterism of water in terms of the Brønsted-Lowry definitions of acids and bases.

32. The term **litmus test** is sometimes used in describing a political issue that, by itself, can determine a voter's support for a candidate. Give several examples of litmus test issues that can, by themselves, determine a voter's choice of candidates.

33. The active ingredient in the household mineral stain remover LIME-A-WAY is sulfamic acid (H_3NSO_3).

    a. Write a balanced chemical equation for the neutralization reaction between 2 moles of sulfamic acid and 1 mole of the base calcium carbonate.

    b. Identify the salt formed. You may find it helpful to refer to a similar neutralization reaction shown in Figure 8.5.

34. Shown are two swimming pool chemical products, sold as pH Up and pH Down.

    a. Which product lowers hydronium ion concentrations?

    b. Which product raises hydronium ion concentrations?

    c. Which product contains acid?

    d. Which product contains base?

Tim O. Walker

35. a. **Figure a** represents an acid common in certain foods. Identify the compound.

b. **Figure b** represents the active ingredient in aspirin. Identify the compound.

c. Which represents a carboxylic acid: Figure a, Figure b, both, or neither?

Figure a.

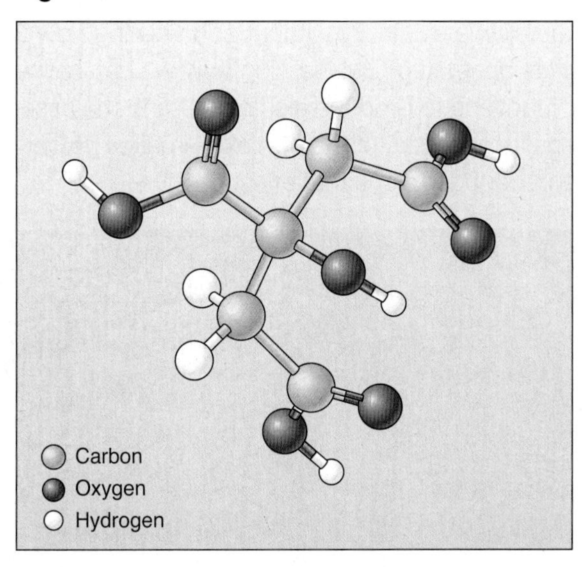

Figure b.

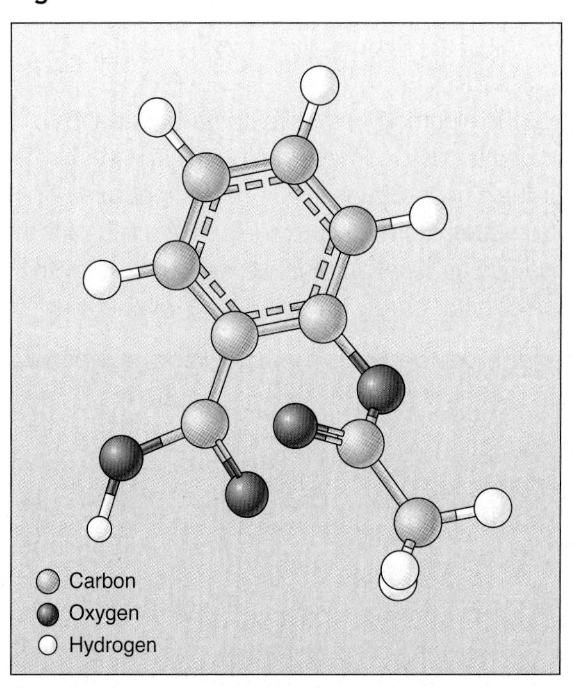

Calculate

36. For a solution at 25°C, what is the pH when (a) $[H_3O^+]$ is 0.0001 *M*? (b) $[H_3O^+]$ is 0.0000000001 *M*? (c) equal amounts of hydronium and hydroxide ions are present?

37. At 25°C, what is $[H_3O^+]$ in a solution whose pH is (a) 3? (b) 10?

38. Suppose you prepare a solution by dissolving 1 mole of pure HCl gas in 100 liters of water. (Recall that HCl ionizes completely in water.) What is (a) the $[H_3O^+]$ in this solution? (b) the pH of this solution?

39. Suppose you prepare a solution by dissolving 1 mole of pure NaOH in 100 liters of water. (NaOH ionizes completely in water.) What is (a) the $[OH^-]$ in this solution? (b) the $[H_3O^+]$ in this solution? (c) the pH of this solution?

40. What is the concentration of Cl^- in a solution of HCl that has a pH of 2?

41. HF is a weak acid; it does not ionize completely. What can you say about the pH of a solution prepared by dissolving 1 mole of HF in 100 liters of water? What can you say about the molar concentration of the fluoride anion in this solution?

42. How many liters of 0.1 M NaOH solution does it take to neutralize (a) 1 liter of 0.1 *M* HCl? (b) 0.5 liter of 0.2 *M* HCl? (c) 3 liters of 0.01 *M* HCl?

43. How many liters of 0.5 M HCl solution does it take to neutralize (a) 0.5 liter of 0.1 *M* NaOH? (b) 1 liter of 0.5 *M* NaOH? (c) 0.1 liter of 2 *M* NaOH?

44. By analogy to pH, how would you define pOH?

45. Based on your definition of pOH from Exercise 44, what will always be true of the sum pH + pOH in any acidic or basic solution in water at 25°C?

9 Nuclear Chemistry

Much of the electricity we have come to depend on, day after day, comes from nuclear power. Throughout the United States, 100 nuclear reactors, located in 30 states, currently provide about a fifth of the electricity we use as a nation. Just as we might enjoy the convenience of a car or rely on computers without fully understanding how they work, we can enjoy the convenience of reliable electricity without appreciating how this power was generated. But learning how we can harness nuclear reactions to produce electricity—and the associated benefits and risks of the process—makes us better informed about decisions that affect us all.

In this chapter we'll explore some of the chemistry and applications of nuclear processes, including those that take place within a nuclear power plant. We will also examine some of the deep concerns associated with these technologies, such as the risk of a nuclear accident and nuclear proliferation.

One successful approach to reducing nuclear proliferation involved a 20-year agreement with the United States. In this agreement, Russia dismantled thousands of their nuclear weapons, converted the uranium within them to material suitable for generating electricity, and sold this fuel to the United States. Shown here, U.S. officials commemorate the final shipment of uranium under this arrangement.

Dmitry Lovetsky/Associated Press

CHAPTER OUTLINE

9.1 Radioactivity

LEARNING OBJECTIVES

1. **Summarize** how radioactivity was discovered.
2. **Explain** what makes atomic nuclei prone to radioactive decay.
3. **List** the three common types of radioactive decay.
4. **Describe** a radioactive process with the use of nuclear notation.

Becquerel's experiments leading to the discovery of radioactivity • Figure 9.1

Serendipity played an important role in Becquerel's discovery of radioactivity.

a. Becquerel first demonstrated that he could protect a photographic plate from exposure to light by wrapping it in black paper.

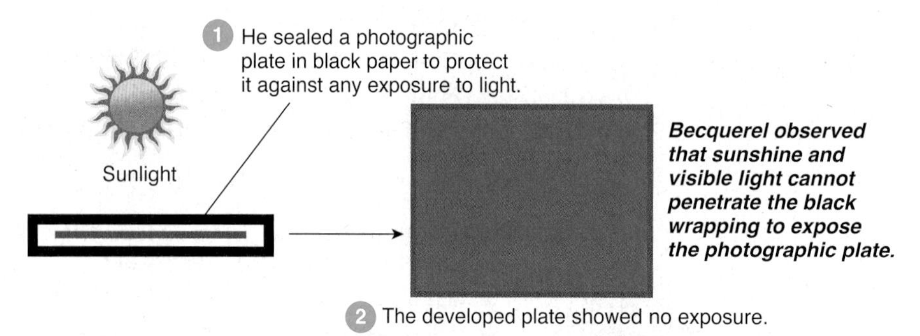

1 He sealed a photographic plate in black paper to protect it against any exposure to light.
Sunlight
Becquerel observed that sunshine and visible light cannot penetrate the black wrapping to expose the photographic plate.
2 The developed plate showed no exposure.

b. Having demonstrated that he could protect a photographic plate effectively against light, he investigated the result of light-induced phosphorescence.

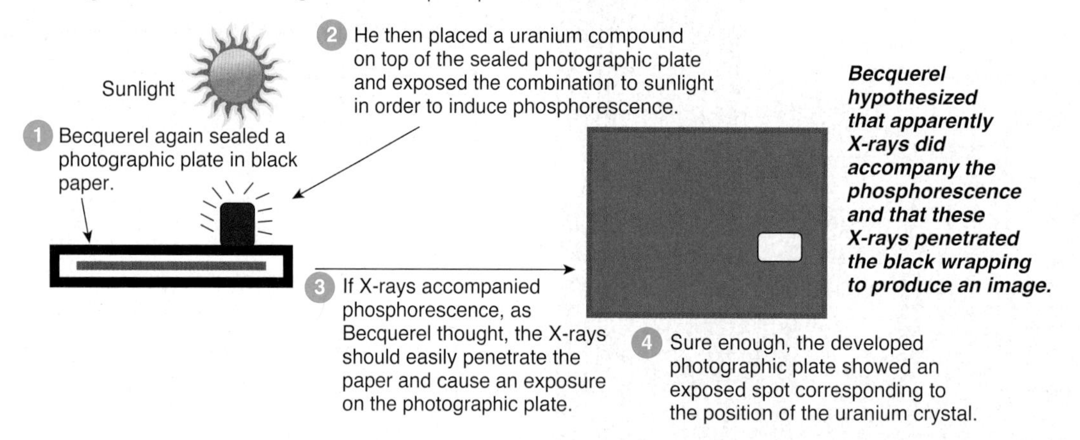

Sunlight
1 Becquerel again sealed a photographic plate in black paper.
2 He then placed a uranium compound on top of the sealed photographic plate and exposed the combination to sunlight in order to induce phosphorescence.
Becquerel hypothesized that apparently X-rays did accompany the phosphorescence and that these X-rays penetrated the black wrapping to produce an image.
3 If X-rays accompanied phosphorescence, as Becquerel thought, the X-rays should easily penetrate the paper and cause an exposure on the photographic plate.
4 Sure enough, the developed photographic plate showed an exposed spot corresponding to the position of the uranium crystal.

c. The truth, however, was revealed in a case of serendipity. Because of a lack of sunlight during a cloudy period, Becquerel suspended his work and stored his materials in a dark desk drawer, with the uranium compound standing next to the wrapped photographic plate. He developed the plate several days later.

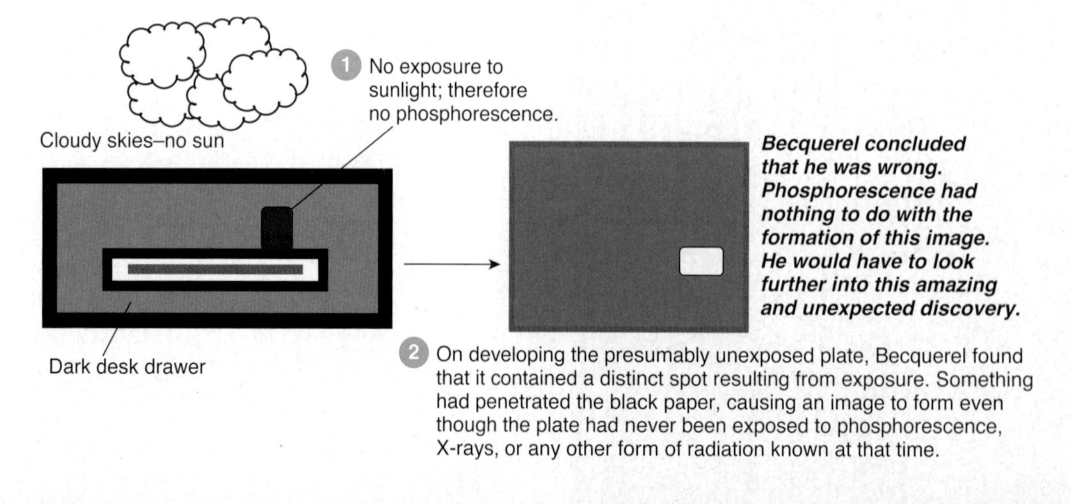

Cloudy skies–no sun
1 No exposure to sunlight; therefore no phosphorescence.
Becquerel concluded that he was wrong. Phosphorescence had nothing to do with the formation of this image. He would have to look further into this amazing and unexpected discovery.
Dark desk drawer
2 On developing the presumably unexposed plate, Becquerel found that it contained a distinct spot resulting from exposure. Something had penetrated the black paper, causing an image to form even though the plate had never been exposed to phosphorescence, X-rays, or any other form of radiation known at that time.

As we've seen in previous chapters, atoms constitute the fundamental units of the matter that forms our everyday world, and atoms can combine with one another to form a variety of chemical compounds. These combinations of atoms occur through the transfer or the sharing of electrons in the valence shells—the outermost quantum shells—of the combining atoms.

In this section, we shift our focus from these outlying electrons to the nucleus, which (as we've also seen in earlier chapters) occupies a remarkably small volume at the center of the atom. We'll begin our examination of the nucleus with the discovery of **radioactivity**, which emanates from within the nucleus itself. This remarkable discovery, which occurred near the dawn of the 20th century, set in motion a series of events that would usher in the atomic age.

Discovery of Radioactive Decay

In 1896, Antoine Henri Becquerel, a French physicist working in Paris, made an unexpected and startling discovery. Becquerel was investigating a phenomenon called phosphorescence, which causes some substances to glow visibly for a short time after they have been exposed to some forms of radiation, including the ultraviolet radiation of sunlight. He thought, *incorrectly* as it turns out, that X-rays might accompany this phosphorescence. (X-rays had been discovered only recently, in 1895.) To test his hypothesis, Becquerel carried out the experiments described in **Figure 9.1**.

Once Becquerel had confirmed that the element uranium itself emitted this strange new form of radiation, a graduate student named Marie Curie took up the investigation and coined the term **radioactivity** to describe this phenomenon. Furthermore, she proposed that atoms themselves fundamentally change in the process of exhibiting radioactivity. Working with her husband, Pierre, Marie soon discovered two new radioactive elements, radium and polonium, that are present in crude uranium ore (**Figure 9.2**). The Curies arduously processed tons of uranium ore in order to purify minute amounts of these elements. Today we recognize that the radiation emitted by these elements originates within their unstable nuclei as the nuclei undergo spontaneous radioactive decay.

radioactivity The emission of radiation by the spontaneous decay of an unstable atomic nucleus.

nuclide The nucleus of an isotope, characterized by the number of protons and neutrons it contains.

What is it about certain atomic nuclei, also known as **nuclides**, that makes them unstable and prone to radioactive decay? We generally find either of two conditions underlying radioactivity:

1. Radioactive nuclei have high atomic numbers. All nuclides with 84 or more protons are radioactive. This means that all elements in the periodic table

The Curies • Figure 9.2

In 1903, Marie and Pierre Curie shared the Nobel Prize in physics with Becquerel for their investigations into radioactivity. Marie, or *Madame Curie* as she came to be known, received a second Nobel Prize in 1911, this time in chemistry, for her discovery of the elements radium and polonium, the latter named for Poland, her native country.

Physics Today Collection/American Institute of Physics/Science Source

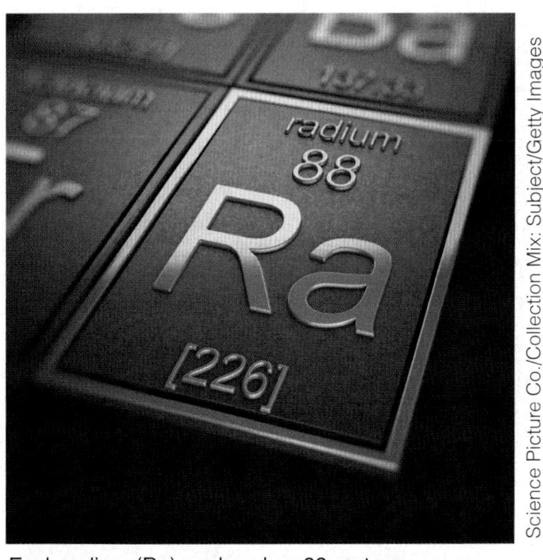

Science Picture Co./Collection Mix: Subject/Getty Images

Each radium (Ra) nucleus has 88 protons and 138 neutrons.

Alpha decay • Figure 9.3

In alpha decay, a new element is formed with two fewer protons (and two fewer neutrons). Here, Americium-241 decays by discharging an alpha particle from its nucleus. The product is an atom with two fewer protons, making it an atom of a different element: neptunium, Np.

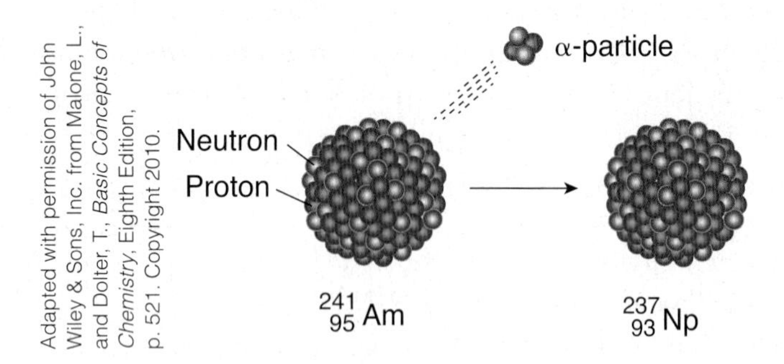

Adapted with permission of John Wiley & Sons, Inc. from Malone, L., and Dolter, T., *Basic Concepts of Chemistry*, Eighth Edition, p. 521. Copyright 2010.

Ask Yourself

1. What is the difference in the mass numbers of the americium and neptunium nuclei shown? What accounts for this difference?
2. Does atomic number increase or decrease when an element undergoes α decay and why?

from polonium through all those of higher atomic numbers are radioactive.

2. The ratio of neutrons to protons within a radioactive nucleus is either above or below a stable range. For the element carbon, for instance, this stable range is 6 or 7 neutrons for carbon's 6 protons. C-14, an isotope with 8 neutrons and 6 protons—a high ratio for a carbon atom—is radioactive. At the other end of the range, C-11, with 5 neutrons and 6 protons, a relatively low ratio, is also radioactive. However, C-12, with 6 neutrons and 6 protons is a stable, nonradioactive isotope of carbon, as is C-13, with 7 neutrons and 6 protons.

Today we know of slightly more than 250 stable nuclides. All others, whether naturally occurring or artificially produced, are unstable and decay through a variety of radioactive processes.

Types of Radioactive Decay

In 1899, shortly after Becquerel and the Curies made their seminal discoveries, Ernest Rutherford showed that radioactivity can consist of more than one form of radiation. He identified two kinds: α (alpha) radiation and β (beta) radiation. A third kind, γ (gamma) radiation, was discovered a short time later by the French scientist Paul Villard.

As discussed in Section 2.1, Rutherford gave us our modern view of the atom: a chemical particle consisting of a dense, positively charged central nucleus surrounded by clouds of negatively charged electrons. As a result of the work done by Rutherford and his colleagues, we also now know that the radiation discovered by Becquerel comes directly from the nuclei of the atoms of radioactive isotopes, also known as **radioisotopes**. We also know that as the radioactive atoms emit these rays, their atomic numbers and mass numbers often change. Since each atomic number corresponds to a specific element, radioactivity often results in the transformation of atoms of one element into atoms of (one or more) other elements. To understand how one element can convert into another through radioactive decay, we'll examine some common types of radiation.

Alpha Radiation Alpha (α) radiation results from the discharge of **α particles**, which are clusters of two protons and two neutrons. Thus, α particles are equivalent to the nuclei of He-4 atoms and are sometimes represented as $^{4}_{2}$He. Alpha radiation is typical for elements of very high atomic number. Because the α particle ejected from the radioactive nucleus carries two protons with it, the radiation always results in the conversion of the radioactive element into a new element, two atomic numbers lower than the original. Americium (atomic number 95), for example, undergoes α decay to yield neptunium (atomic number 93), as shown in **Figure 9.3**.

α particle A positively charged particle consisting of two protons and two neutrons.

β particle A high-speed electron emitted during (β) radioactive decay.

Beta Radiation Beta (β) radiation, or β decay, results from the spontaneous discharge of **β particles**, which are nothing other than fast-moving electrons, traveling at 90% of the speed of light. But since β radiation, like α radiation and γ radiation, originates in atomic nuclei, which contain only protons and neutrons, you might reasonably ask, "How can electrons leave an atomic nucleus as β particles if there aren't any electrons in the nucleus in the first place?"

Beta decay • Figure 9.4

In beta decay, an atom loses an electron to form a new atom with the same atomic mass but one unit higher in atomic number.

a. As a source of beta emission, a neutron can be thought of as a combination of a proton and an electron.

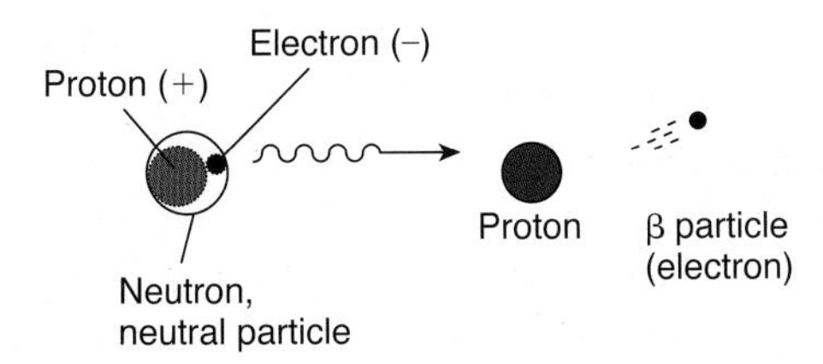

b. Cs-137 undergoes β decay to produce Ba-137. Notice the mass number (137 in this case) does not change because a neutron is converted into a proton. However, the atomic number increases by one (55 → 56).

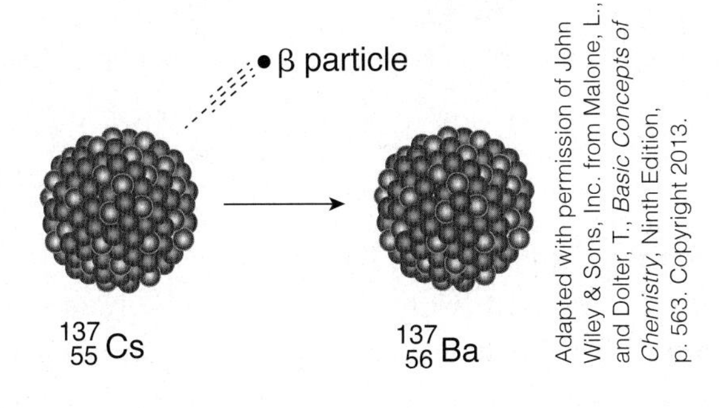

Put It Together Refer to a periodic table and answer the question: What element is produced when an isotope of iodine (atomic number 53) undergoes β decay?

To understand β decay, you can think of a neutron as an electrically neutral combination of a positively charged proton and a negatively charged electron (**Figure 9.4a**). As the newly released electron leaves the atom in the form of a β particle, the newly produced proton remains in the nucleus, generating a new element that's one atomic number higher than the original, but with the same mass number (**Figure 9.4b**).

The beta particle is represented as $^{0}_{-1}\beta$. The −1 of the $^{0}_{-1}\beta$ symbol indicates the −1, or single negative charge of its solitary electron. Similarly, the β particle's lack of protons and neutrons translates into a mass number of zero, shown at the upper left of the symbol.

To account for the particles of interest in any given nuclear process, such as radioactive decay, it's often convenient to represent the process as a **nuclear equation** (**Figure 9.5**). Like all chemical equations, nuclear equations must be balanced to ensure conservation of mass. For a nuclear equation, that means that the sums of the atomic numbers must be the same on both sides of the equation. The same is true for the sums of the mass numbers as well.

KNOW BEFORE YOU GO

1. Write the nuclear equation for (a) α decay of radium-226; (b) β decay of carbon-14.

Nuclear equations • Figure 9.5

Radioactive decay can be represented through nuclear equations, as shown here. As in all nuclear equations, the sum of the mass numbers to the left of the arrow equals the sum of the mass numbers to the right. This is true for the sums of the atomic numbers as well.

a. Alpha decay of americium-241, the same reaction shown graphically in Figure 9.3.

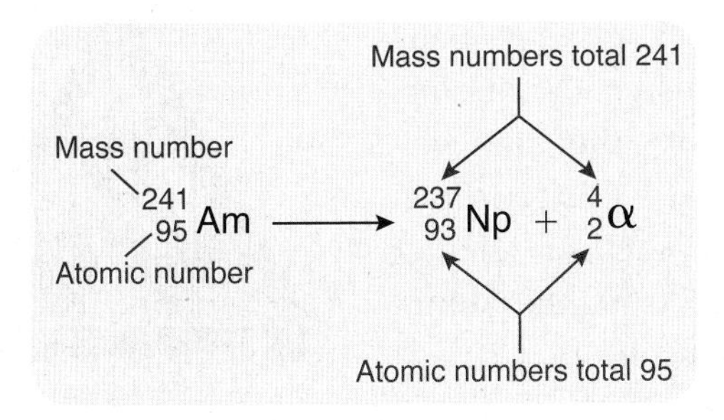

b. Beta decay of cesium-137, the same reaction shown graphically in Figure 9.4b.

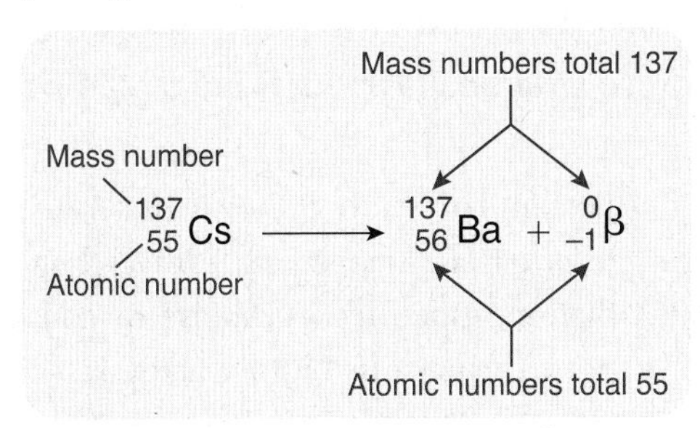

Gamma decay • Figure 9.6

An excited atomic nucleus can emit gamma (γ) radiation.

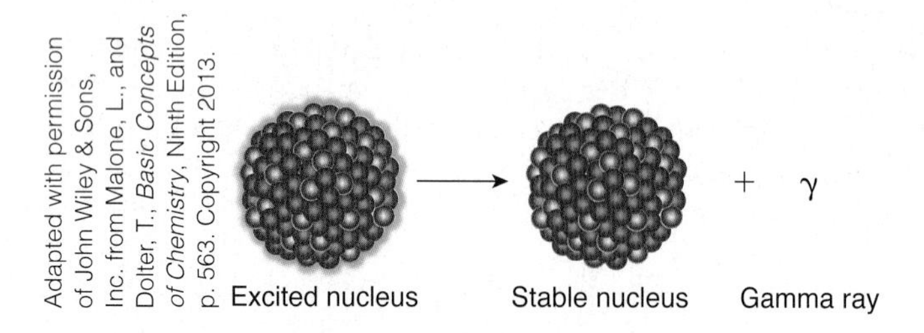

Adapted with permission of John Wiley & Sons, Inc. from Malone, L., and Dolter, T., *Basic Concepts of Chemistry*, Ninth Edition, p. 563. Copyright 2013.

Gamma Radiation Like visible light, radio and television waves, and X-rays, the **gamma (γ) rays** that constitute the third type of radioactive emissions are a form of electromagnetic radiation. Because of their high energies (even greater than the energies of X-rays) and their ability to penetrate deeply into matter, they can cause considerable biological harm. Without either mass or charge, gamma rays are denoted by ${}^{0}_{0}\gamma$ or simply γ. Gamma rays are emitted during the relaxation of nuclei from high-energy, or **excited** states, into lower energy, more stable states (**Figure 9.6**). The process resembles the one by which excited electrons in higher-lying quantum shells emit visible or UV-light as these excited electrons relax to lower-lying quantum shells (as shown in Figure 2.7).

gamma (γ) ray Very high-energy (short wavelength) radiation emitted by excited-state nuclei.

Emission of an α or β particle often leaves the nucleus in a transitory excited state. This excited nucleus can relax by emitting γ rays, so γ emission often accompanies α and β decay. For example, cobalt-60, a radioactive isotope with practical uses we'll examine shortly, undergoes β decay with emission of γ rays.

CONCEPT CHECK

1. **What** incorrect assumption did Becquerel make in his investigation of uranium, sunlight, and photographic paper? What fortuitous event led him to the insight for which he is remembered?
2. **In** what part of the atom does radioactivity originate, and what two general conditions tend to make atoms prone to radioactive decay?
3. **Which** type of decay, α or β produces (i) an element with a lower atomic number, (ii) an element with a higher atomic number?
4. **What** particle is emitted when I-131 (a radioactive isotope used to treat certain thyroid conditions) decays to Xe-131?

9.2 Ionizing Radiation: Effects and Applications

LEARNING OBJECTIVES

1. **Compare** the penetrating power of α, β, and γ rays.
2. **Describe** the biological effects of exposure to ionizing radiation.
3. **Explain** how radiocarbon dating works, using the concept of half-life.
4. **Describe** practical applications of radioactivity.

The familiar symbol shown here warns of the hazards of radioactivity. Although nuclear radiation, especially in intense or continued exposure, can cause biological damage, including death, the radiation itself has many useful, practical applications, especially in the field of medicine.

Ionizing Radiation and Its Effects on Health

Pi-Lens/Shutterstock

When α, β, or γ rays penetrate matter, including living tissue, they can generate ions by expelling the outer-shell or valence electrons from the atoms and molecules they strike.

Because of this property, we classify these rays—and other forms of high-energy radiation such as cosmic rays and X-rays—as **ionizing radiation.** When ionizing radiation penetrates living matter, the ions it generates can become converted into **free radicals**, which are highly reactive molecules containing unpaired valence electrons. Free radicals can react indiscriminately with biological molecules in their path, potentially causing damage to cells and tissue.

ionizing radiation Radiation (such as α, β, γ, x-, and cosmic rays) capable of ionizing atoms or molecules in its path.

free radical A chemical species that has an odd number of electrons.

penetrating power The ability of ionizing radiation to penetrate matter.

The biological effect of ionizing radiation depends in part on its **penetrating power**; that is, its ability to penetrate matter. Alpha particles are easily stopped by paper or even a few centimeters of air (as well as by skin and lightweight clothing). As a result, α radiation from an external source poses little, if any, risk. However, if ingested, α emitters can be deadly, as we'll soon see.

Beta particles can penetrate matter more readily than alpha particles. It would take a thick sheet of aluminum foil, a block of dense wood, or heavy clothing to stop β rays. Gamma rays—unlike α and β rays—have no mass and no electrical charge, and hence pass through most substances with ease. Blocking them requires a thick shield of lead or concrete (**Figure 9.7**).

The damage done to living things by ionizing radiation depends on the radiation's **ionizing power**—the ease with which it produces ionizations—as well as its penetrating power. Alpha particles, with their relatively large masses (four atomic mass units) and two positive charges, have enough mass and charge to do considerable damage to whatever atoms and molecules they strike. They have considerable ionizing power, but, as we've seen, they have little penetrating power. However, they do a considerable amount of damage in the short distances they travel. Beta and gamma rays have progressively weaker ionizing power than alpha rays but they penetrate matter further. For example, the relatively weakly ionizing γ rays do less damage than either α or β rays for a comparable distance traveled. But since they penetrate matter so readily, they nonetheless can generate considerable ionization and can do significant damage.

ionizing power A measure of how efficiently ionizing radiation causes ionization as it penetrates matter.

Radiation in Our Environment We began this chapter with a description of Becquerel's discovery of radioactivity through his investigation of the phosphorescence of uranium compounds. By far, the most common uranium

Penetrating power of alpha, beta, and gamma radiation • Figure 9.7

Among α, β and γ radiation, the α particles penetrate matter least effectively, the β particles penetrate somewhat more effectively, and the γ rays penetrate the most effectively of all.

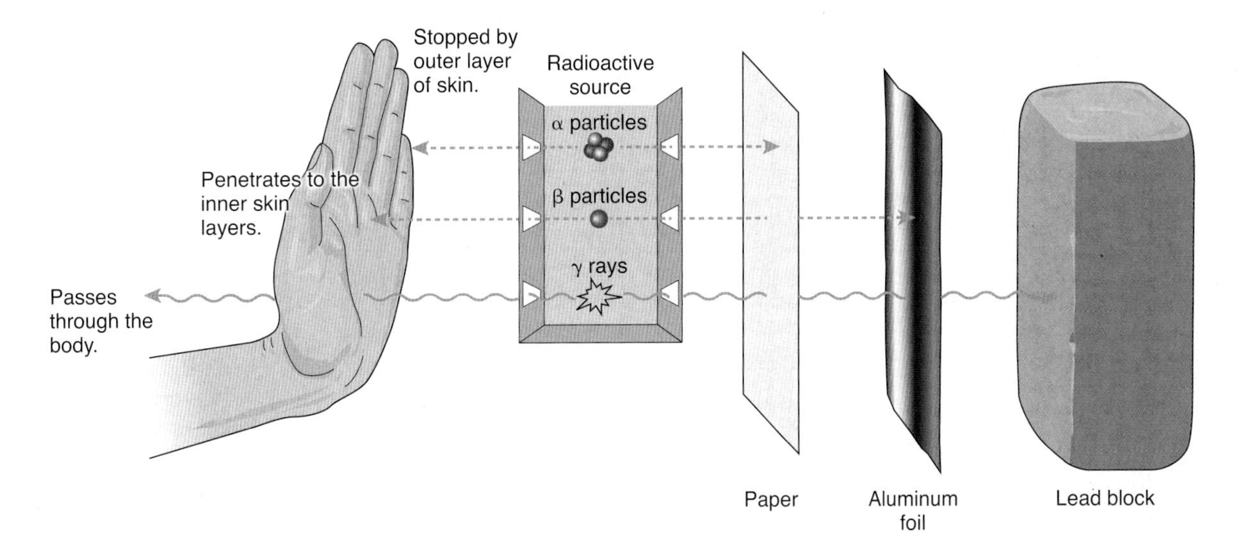

Alpha decay of uranium-238 • Figure 9.8

Uranium-238 undergoes alpha decay to produce thorium-234.

$$^{238}_{92}\text{U} \longrightarrow {}^{234}_{90}\text{Th} + {}^{4}_{2}\alpha$$

Note again that the sum of both atomic numbers and mass numbers is equal on both sides of the arrow.

isotope found in nature is U-238, with an isotopic abundance of 99.3%. The remaining fraction of a percent is mostly U-235, an important isotope we'll examine further later in this chapter. Alpha decay of uranium-238, accompanied by emission of γ radiation, produces thorium-234. Since this accompanying γ emission has neither mass nor charge and therefore doesn't affect the mass number or atomic number of any of the nuclei produced, it's often omitted from the equation (**Figure 9.8**). The Th-234 thus produced is itself unstable and undergoes β decay to Pa-234 with an increase of one in atomic number. This process of successive radioactive decay, from one isotope to the next, ends with the formation of lead-206, a stable (nonradioactive) isotope. The entire sequence, known as a **decay series**, appears in **Figure 9.9**.

Figure 9.9 shows that a radioactive isotope of radon, Rn-222, is a transient part of this decay chain, formed from the radioactive decay of radium (Ra). The Rn-222 decays, in turn, to polonium (Po-218). Radon itself is an invisible and odorless noble gas. There's concern that radon, formed by the decay of very small amounts of naturally occurring radioisotopes in some kinds of rocks and soils, may seep upward through the ground, enter homes and other buildings, and present a risk of producing lung cancer (**Figure 9.10**). Ordinarily, we might expect that any radon we inhale would leave our lungs when we exhale. But any radioactive decay of radon during the brief moment it resides inside our lungs—between our inhaling and exhaling—would produce minuscule amounts of polonium and other radioisotopes. Any traces of these radioactive elements that might remain within our bodies could do considerable biological damage through ionizing radiation. The U.S. Environmental Protection Agency (EPA) estimates that nearly 1 in 15 homes have elevated radon levels, defined as exceeding 4 picoCuries per liter of air. (A Curie is a unit of radioactivity, named in honor of Marie Curie; a picoCurie, pCi, is one trillionth, or 10^{-12}, of a Curie). The National Academy

Radioactive decay series of uranium-238 • Figure 9.9

This series of radioactive decays starts from U-238, a naturally occurring radioisotope of uranium, and ends with Pb-206, a stable isotope of lead.

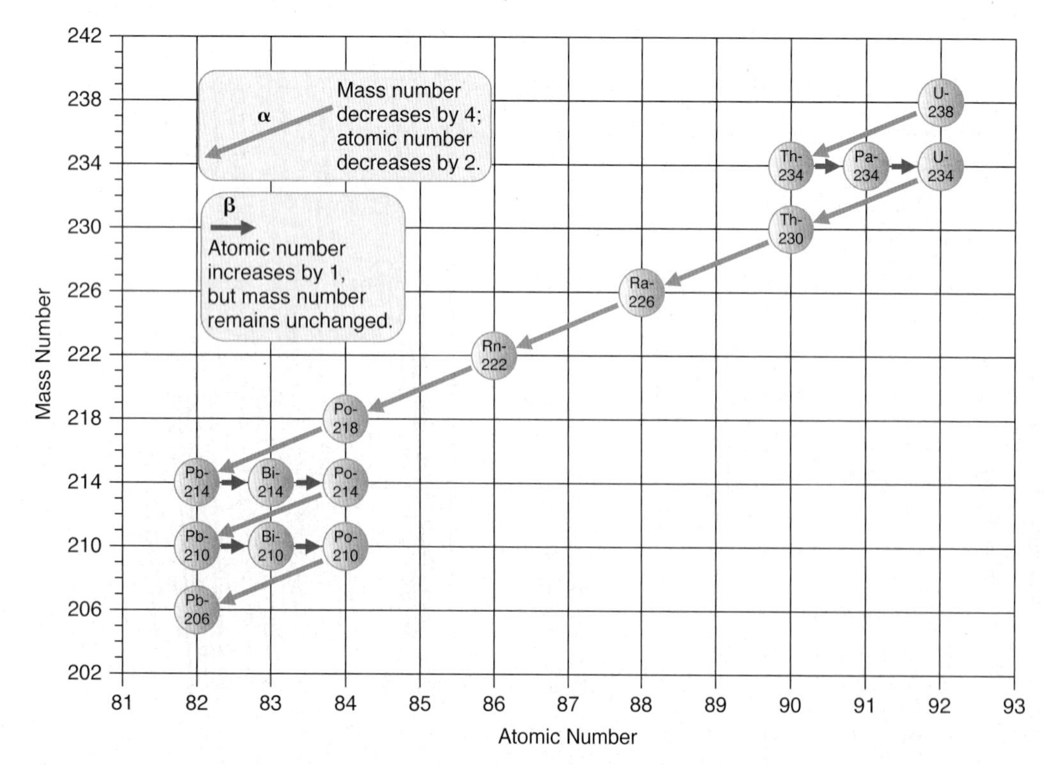

Interpret the Data

Notice that some elements, such as thorium (Th), polonium (Po), and lead (Pb), appear more than once in this figure. Explain how this happens.

Radon risk • Figure 9.10

Exposure to elevated levels of radon gas presents a risk of lung cancer.

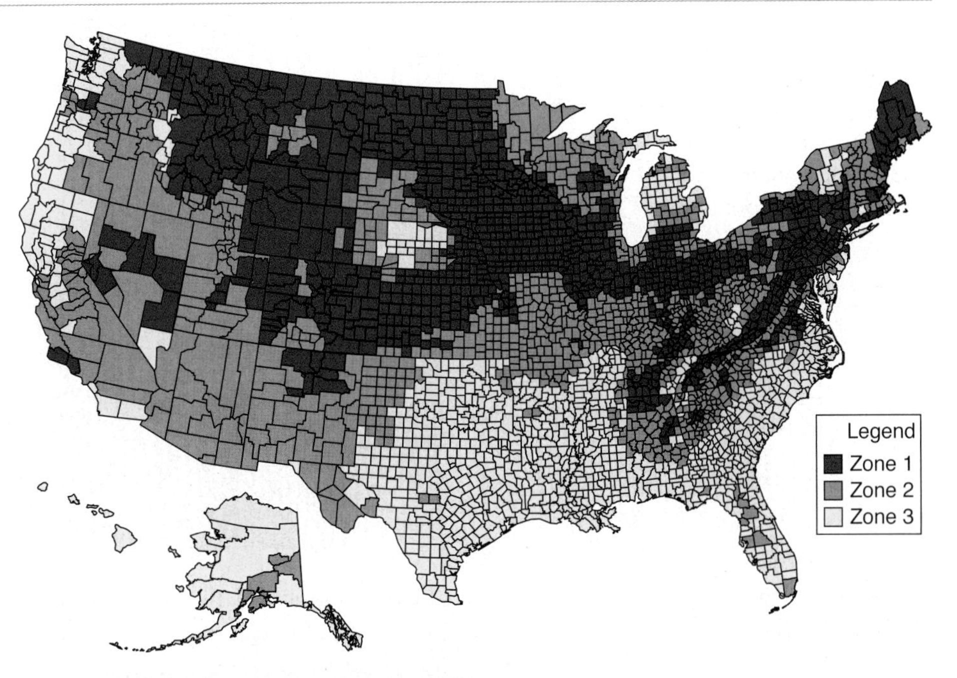

a. EPA map of radon zones In the United States, Zone 1 represents the highest potential, and Zone 3 represents the lowest potential for elevated radon levels. The EPA recommends testing for radon in all homes, as elevated levels of radon have been found in all three zones.

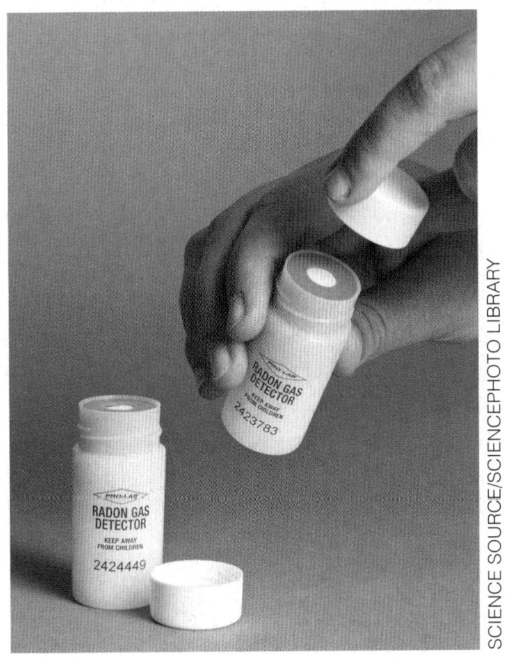

b. A home radon test kit This type of kit is left open and undisturbed within the home for a designated period of time to allow for any radon that may be present within the air to interact with the materials within the kit. The canister is then sealed and mailed to a laboratory for analysis.

of Sciences reports that indoor radon contributes to 10–15% of the estimated 157,000 lung cancer deaths each year in the United States.

background radiation lonizing radiation emanating from within us as well as from our environment and outer space.

Radiation released by the radon in our environment is an example of **background radiation**, very low levels of radiation from natural sources. Radiation from the radioisotope potassium-40 represents still another example of background radiation. Potassium, the most abundant cation within our cells (as K^+), is an element essential to life. A 60-kg (132-lb) person carries roughly 200 g of potassium at all times. We could not live without it. Yet just over 0.01% of all the potassium in the universe, including the potassium in our bodies, consists quite naturally of the radioisotope K-40. A simple calculation shows that each of us carries about 20 mg of radioactive potassium-40 within our bodies at all times. What's more, high-energy cosmic rays, carrying greater energy than γ rays, constantly bombard the surface

Sources of background radiation • Figure 9.11

Background radiation is ever present, both from outside and within our bodies. This radiation can come from the sky, the air, the ground, and our diet.

Cosmic radiation
Both high-energy ionizing radiation and protons traveling near the speed of light continuously enter Earth's atmosphere from outer space.
Internal radiation
Potassium-40 within both our foods and our bodies continuously emits radiation, which consists largely of beta particles accompanied by smaller amounts of gamma radiation.
Terrestrial radiation
Radioactive isotopes of uranium, polonium, radon, bismuth, and lead, naturally present within Earth's soil, emit alpha and beta particles, accompanied by gamma rays.

of Earth and everything on it, including our own bodies (**Figure 9.11**).

The biological damage caused by ionizing radiation depends on several factors, including the energy of the radiation, how long we're exposed to it, its penetrating and ionizing power, and whether it's coming from outside or from within our bodies. A widely accepted unit of measurement to describe the biological effectiveness of ionizing radiation on humans is the **rem**, short for Roentgen-equivalent-man. (Wilhelm Roentgen was the physicist whose discovery of X-rays in 1895 set the stage for Becquerel's discovery of radioactivity the following year.)

> **rem** A unit of ionizing radiation representing an equivalent dose to humans, regardless of the type of radiation.

Regardless of the type of ionizing radiation being considered (whether α, β, γ, cosmic rays, X-rays, etc.), a given rem of exposure represents the *same amount* of biological effect in human beings. In practice, because our exposures are typically so small, a more common unit is the millirem (mrem), one-thousandth of a rem. As a rough estimate, a typical resident of the United States receives an average of about 600 mrem per year, about half of which comes naturally from background radiation. Just 30 years ago, this value was estimated at 360 mrem but has been since revised upward, largely due to the more widespread use of certain medical imaging procedures, some of which we'll discuss shortly.

Acute Exposure to Radiation Large doses of ionizing radiation (> 50 rem) have the capacity to transform the finely detailed molecular structures of life—including enzymes, cell membranes, and DNA—into new and utterly useless or even harmful molecular forms. When this damage is extensive enough to cause illness or death, it is called **somatic damage** (**Figure 9.12**). The more immediate

Biological damage from ionizing radiation • Figure 9.12

Acute exposure to ionizing radiation can produce genetic defects and can result in death. Alexander Litvinenko, a former Russian intelligence officer living in exile in London, was poisoned with the radioisotope polonium-210 in 2006, dying just three weeks later. Some call this the first known case of nuclear terrorism.

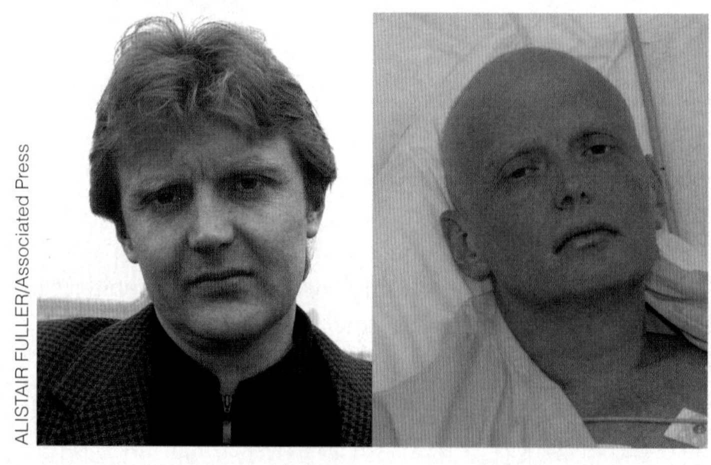

ALISTAIR FULLER/Associated Press

Natasja Weitsz/Getty Images

Put It Together Refer to Figure 9.9 and answer the question. How does polonium-210 decay, and what is its *daughter isotope,* that is, the isotope it directly produces?

results of somatic damage range from reduced white blood cell counts to symptoms including fatigue and nausea. Delayed effects appear as damage to bodily organs (the spleen, for example), glands (such as the thyroid), and bone marrow, and the development of leukemia and other cancers. Extreme somatic damage can lead to painful death.

Although **genetic damage** from radiation has been observed in laboratory animals for more than 75 years, indications that similar damage might occur in humans have appeared only recently. Analysis of DNA in families living near the Chernobyl nuclear reactor disaster (to be discussed later) suggests that radiation-induced genetic damage in parents may have been passed on to children born years after the explosion. Although the link between radiation and genetic damage is well established for laboratory animals, it's still tenuous for humans.

Half-life and Radiocarbon Dating

In addition to knowing what kinds of radiation are emitted during radioactive decay, and what daughter isotopes are generated as a radioisotope decays, it's also useful to know the *rate* at which this decay occurs. It's convenient to represent this rate as the radioisotope's **half-life**, which is the length of time it takes for exactly half of any quantity of the isotope to decay. Since each isotope decays at its own particular rate, each has its own, specific half-life. Polonium-210, for example, has a half-life of 138 days. If we start with 5.0 g of Po-210, for example, after 138 days, 2.5 g will remain; after another 138 days (a total of 276 days), 1.25 g will remain; after another 138 days (a total of 414 days), 0.625 g will be left, and so on. Radioactive half-life values range from fractions of a second for certain fleetingly short-lived isotopes to billions of years for especially long-lived ones. The radioisotope U-238, for example, has a half-life of 4.5 billion years.

half-life The length of time it takes for one-half of a given quantity of a radioactive isotope to decay. Every radioisotope has its own characteristic half-life.

Because each radioisotope decays at its own specific, characteristic rate, we can think of radioisotopes as chemical clocks, with the capacity to mark time. Carbon-14, for example, with its characteristic half-life of 5730 years, is useful for determining the age of residues of living or once-living things. **Figure 9.13** shows how an initial quantity of 10 g of

Radioactive decay of carbon-14 • Figure 9.13

The half-life of carbon-14 is 5730 years. As a result, after each 5730-year period has passed, half of the sample that was present at the beginning of the period remains and half has undergone radioactive decay.

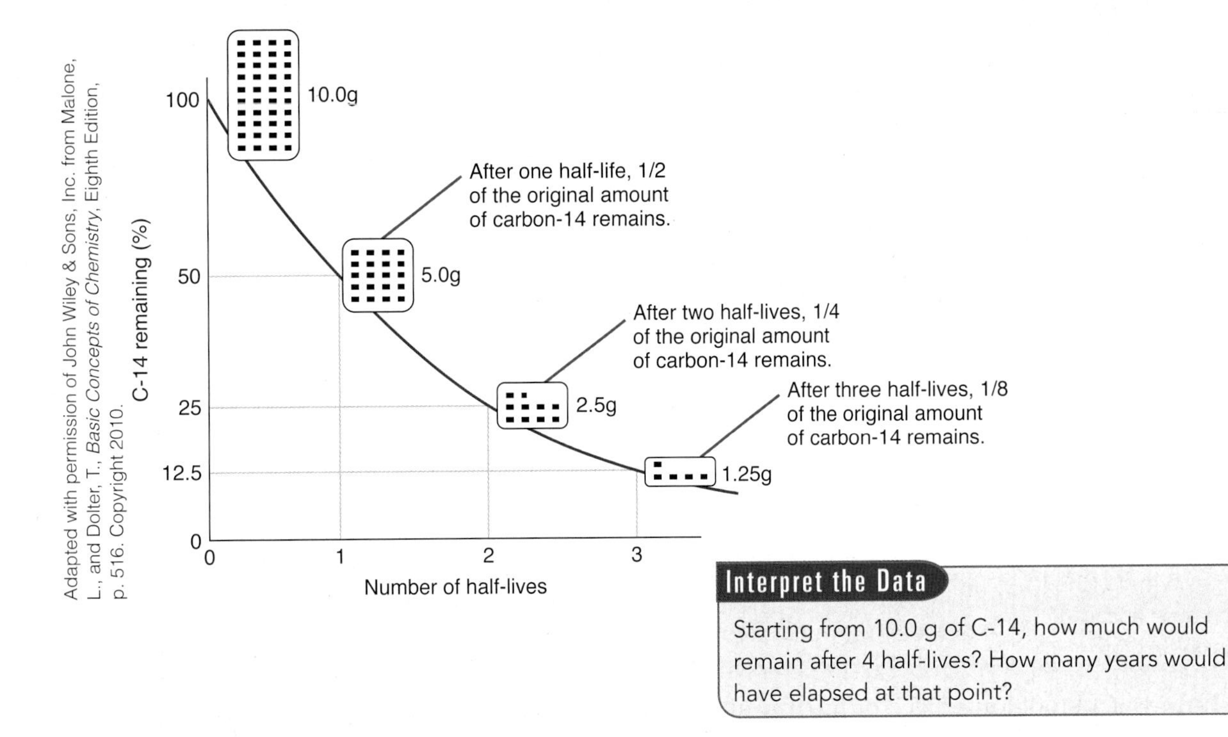
C-14 remaining (%)
100
50
25
12.5
0
10.0g
After one half-life, 1/2 of the original amount of carbon-14 remains.
5.0g
After two half-lives, 1/4 of the original amount of carbon-14 remains.
2.5g
After three half-lives, 1/8 of the original amount of carbon-14 remains.
1.25g
0
1
2
3
Number of half-lives

Interpret the Data

Starting from 10.0 g of C-14, how much would remain after 4 half-lives? How many years would have elapsed at that point?

PROCESS DIAGRAM

Radiocarbon dating • Figure 9.14

The use of radiocarbon dating underlies much of our present-day understanding of early human history and prehistory.

1 Cosmic rays from outer space react with the upper atmosphere to continually regenerate C-14. The C-14 gets incorporated into atmospheric carbon dioxide, CO_2.
2 As plants live and grow, they incorporate CO_2 into their structures. The carbon of the CO_2 consists of three isotopes: C-12, C-13, and radioactive C-14 (i.e., $^{12}CO_2$, $^{14}CO_2$ and $^{14}CO_2$). The ratio of $^{12}CO_2$ to $^{14}CO_2$ in the atmosphere is constant at about $10^{12}/1$ (a trillion to one).
3 As a result, the carbon-rich material that makes up most of plants' structure has this same C-12/C-14 ratio of about $10^{12}/1$.
4 Since animals eat plants (or eat other animals who have eaten plants), they, too, maintain this same C-12/C-14 ratio.

5 When the plant or animal dies, the capture of carbon ends. The C-12/C-14 ratio in any resulting artifacts of plant or animal origin begins to climb as the stable C-12 retains its orginal concentration, but the C-14 continues its steady decay into its daughter isotopes. After 5730 years, the ratio of C-12 to C-14 becomes $10^{12}/0.5$; after 2 half-lives (11,460 years), the ratio increases to $10^{12}/0.25$, and so on.

6 Thus if we can measure the C-12 to C-14 ratio in an ancient object of interest (up to about 50,000 years old), we can determine its approxmiate age. Beyond this time, such a small proportion of C-14 remains that the technique is no longer reliable.

Ask Yourself

If one half-life of C-14 represents 5730 years, two half-lives represents twice as long or 11,460 years, and so on, approximately how many half-lives of C-14 will have elapsed in 50,000 years? (Round your answer to the nearest whole number.)

C-14 decays over the course of a few half-lives. In a technique known as **radiocarbon dating**, we can use C-14's known half-life to establish the ages of ancient objects of plant or animal origin (**Figure 9.14**).

radiocarbon dating A technique, based on the characteristic decay rate of carbon-14, for determining the age of an artifact of biological origin.

Radiocarbon dating has served as useful tool in dating artifacts in such diverse fields as archeology and art history (**Figure 9.15**). For instance, the Shroud of Turin is a 14-foot-long piece of linen in the shape of a burial cloth purported to be the actual burial cloth of Jesus. Independent radiocarbon dating analysis on several portions of the cloth by researchers at Oxford University, the University of Arizona, and the Swiss Federal Institute of Technology concluded that the linen originated sometime between AD 1260–1390. This evidence strongly suggests that this specimen is not an authentic relic. Nevertheless, some people, unconvinced by these results, maintain that biological material of relatively recent origin may have contaminated the Shroud, thereby invalidating the analytical data.

Radiocarbon dating • Figure 9.15

Levels of carbon-14 naturally present in artifacts of plant or animal origin can be used to estimate the age of these specimens.

This whole-body human corpse, dubbed Ötzi the iceman, was discovered in 1991 in the Austrian Alps. Naturally mummified in permafrost, it was remarkably well preserved given its age. Radiocarbon dating indicates that it lived about 5300 years ago.

Paul HANNY/Gamma-Rapho/Getty Images

Medical and Safety Applications of Radioisotopes

The most valuable use of radioisotopes may well lie in their medical applications, such as in imaging and diagnostic procedures and in medical therapies. The radioisotope most widely used in medical imaging and diagnosis is technetium-99m (^{99m}Tc). The *m* of the mass-number superscript refers to the metastable or excited state of this isotope's nucleus. It decays with loss of γ rays alone, without any change in its atomic or mass number:

$$^{99m}_{43}\mathrm{Tc} \longrightarrow {}^{99}_{43}\mathrm{Tc} + {}^{0}_{0}\gamma$$

Incorporated into appropriate chemical compounds, Tc-99m can be introduced into various organs, including the heart, kidneys, liver, and lungs, and into glands such as the thyroid. Because Tc-99m emits only γ rays, and since these rays penetrate tissue and exit the body as readily as they enter it, this radioisotope is particularly useful in generating diagnostic images. With a half-life of just six hours, Tc-99m is one example of a class of agents called **radiotracers**, substances with half-lives so short that their radiation disappears from the body rapidly, thereby diminishing the risk of biological damage.

Another important class of radiotracers are the **positron emitters**, which are used in **positron emission tomography**—known more commonly as PET—a powerful medical imaging technique. A **positron** is a particle indistinguishable from an electron except that it carries a positive charge. Certain radioisotopes, such as carbon-11, emit positrons, which disappear in a burst of γ radiation when they collide with nearby electrons. (The electrons vanish as well.) A positron emitter inserted into a bodily organ produces a cascade of γ rays from the organ. As these rays leave the body, they can be converted by nearby computerized radiation detectors into powerful diagnostic images representing slices or planes through the organ (**Figure 9.16** on the next page). (The *tomography* of positron emission tomography comes from the Greek word *tomos*, meaning "slice" or "section.")

In one important PET application, an atom of the positron emitter fluorine-18 is incorporated into a molecule of glucose, the sugar used by the brain as its exclusive nutritional fuel. This tagged glucose enters the brain along with ordinary glucose and emits its positrons, which collide with electrons to produce bursts of γ rays. Analysis of images obtained by this process allows physicians to follow the path of the glucose within the brain and to diagnose and treat brain abnormalities.

positron emitters Radioisotopes that decay with positron emission, some of which are used in medical imaging.

positron A positively charged analog of an electron.

Cancer Therapy Because cancer cells divide more rapidly than normal cells and are more active metabolically, they are also more susceptible to damage by ionizing radiation. Cancers located deep within the body are sometimes treated with sharply focused beams

Chemistry InSight Positron emission •

Figure 9.16

Carbon-11, a short-lived radioisotope, emits positrons (a) and is often used in PET studies (b).

Adapted with permission of John Wiley & Sons, Inc. from Jespersen, N., Brady, J., and Hyslop, A., *Chemistry: The Molecular Nature of Matter*, Sixth Edition, p. 985. Copyright 2012.

a. Radioactive decay of Carbon-11.

Step 1: Positron emission

$$^{11}_{6}C \longrightarrow\ ^{11}_{5}B +\ ^{0}_{1}e$$

(Positron)

When a positron is emitted, the atomic number decreases by one, and a new element is formed — in this case, boron.

Step 2: Positron/electron annihilation

$$^{0}_{1}e +\ ^{0}_{-1}e \longrightarrow 2\ ^{0}_{0}\gamma$$

(Positron, Electron)

The positron emitted in Step 1 collides immediately with a nearby electron, annihilating both particles and producing two gamma rays in the process.

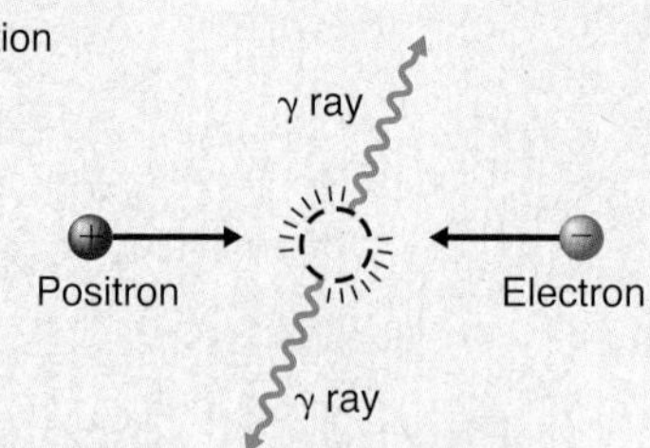

b. Positron emission tomography (PET) scans of a nonsmoker and smoker.

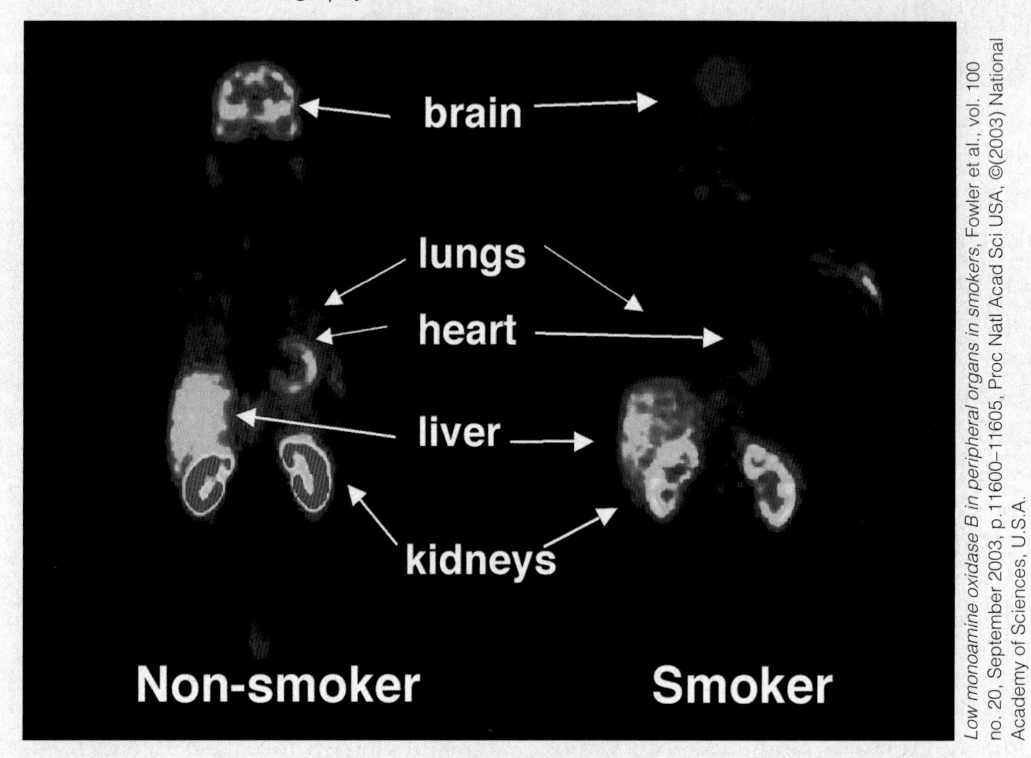

Low monoamine oxidase B in peripheral organs in smokers, Fowler et al., vol. 100 no. 20, September 2003, p.11600–11605, Proc Natl Acad Sci USA, ©(2003) National Academy of Sciences, U.S.A.

These whole-body PET scans show significantly less uptake of a carbon-11 radiotracer within various organs of a smoker as compared to a nonsmoker. This study reveals that smoking suppresses enzyme activity not only within the lungs, but also within other organs such as the brain, heart, liver and kidneys.

of radiation emitted by an external source, such as γ rays from cobalt-60.

In one technique known as gamma knife therapy, a large number of γ ray beams, each of very low intensity, are focused onto a brain tumor from different angles. Individually, the beams are too weak to damage the surrounding, healthy tissues. But together, the beams converge to deliver a large dose of γ radiation directly to the tumor. In brachytherapy (from the Greek *brachys* for short distance), minute metal pellets, called seeds, containing γ–emitting radioisotopes are inserted at the site of a tumor, thereby focusing the ionizing radiation where it is most effective in destroying the growth. Used for a variety of tumors, including prostate cancers, the seeds can often be left in the patient for an extended period as their radioactivity gradually diminishes.

Safety Applications of Radioisotopes One of the most widely used applications of ionizing radiation is in household smoke detectors. Alpha radiation from an americium-241 source inside these devices ionizes air molecules. The ionized air thus formed can carry a current, completing an electrical circuit to the battery. The presence of smoke interferes with the ionized air, disrupting the current and thus triggering an alarm.

Although not in widespread practice, irradiation of raw foods, such as poultry, meat, fruits and vegetables, with low levels of ionizing radiation is a proven way to reduce levels of foodborne pathogens, such as bacteria, thereby improving food safety and increasing shelf-life. This irradiation—typically γ rays from cobalt-60—does not compromise nutritional content, taste, or texture (nor leave the food in any way radioactive). Various medical organizations, including the Centers for Disease Control and Prevention and the World Health Organization, have endorsed the practice.

CONCEPT CHECK

1. **Rank** the penetrating power of α, β and γ rays, respectively. How does this compare to their relative ionizing power?
2. **What** is the source of environmental radon, and why is environmental radon a cause for concern?
3. **Which** naturally occurring isotope of carbon undergoes radioactive decay?
4. **A** commonly used radiotracer in PET studies is oxygen-15. Into what element is an atom of O-15 converted when it emits a positron?

9.3 Mass Defect and Binding Energy

LEARNING OBJECTIVES

1. **Explain** the relationship between mass and energy.
2. **Describe** how mass defect relates to binding energy.

The nuclear processes we've examined to this point involve relatively small changes to atomic nuclei, such as the emission of small, subatomic particles and electromagnetic radiation. But *nuclear fission* is another matter entirely. In nuclear fission, a nucleus of a heavy element splits into two or more sizable pieces. The process releases the energy that sustains nuclear power plants and provides the explosive force of atomic bombs and other nuclear weapons. It's the same energy that holds a nucleus intact despite the immense repulsive forces of large numbers of adjacent protons within the nucleus, all pushing against each other because they all bear the same (positive) electrical charge.

We discuss nuclear fission in the next section. First, in this current section, we examine how nuclear fission takes place.

Mass and Energy

In conventional chemical reactions—changes that involve the sharing or transfer of valence electrons—we invariably observe the operation of two fundamental laws of chemistry:

1. **Conservation of Mass** recognizes that mass can neither be created nor destroyed as a result of a chemical reaction. In other words, there must be exactly as much matter (and therefore exactly as many atoms) among the combined products of a chemical reaction as in its combined reactants.

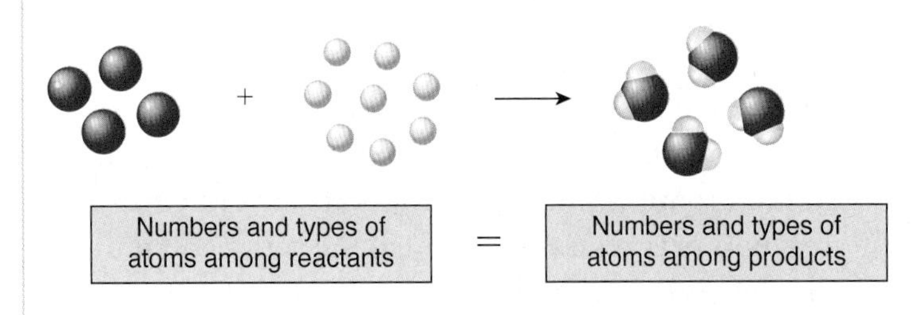

In Words, Math, and Pictures

Mass defect and binding energy • Figure 9.17

The mass defect of an atomic nucleus is equivalent to the energy that binds its component particles to one another.

An atomic nucleus weighs less...
...than the sum of the individual particles that compose it.

We can illustrate the concepts of mass defect and binding energy with uranium-235, an enormously important isotope. To determine the mass defect of this isotope, we first calculate the mass of its constituent particles:

1. Determine how many protons, neutrons, and electrons constitute a single atom. The number of protons is simply the atomic number, 92. The number of neutrons, 143, is the mass number minus the atomic number:

Mass number → $^{235}_{92}U$ ← Atomic number

$$\begin{array}{l} 235 \text{ (protons + neutrons)} \\ -92 \text{ protons} \\ \hline = 143 \text{ neutrons} \end{array}$$

The number of electrons equals the number of protons, 92.

2. Add the masses of all the particles: 92 protons, 143 neutrons, and 92 electrons:

$$\left.\begin{array}{r} 92 \text{ protons} \times 1.007 \text{ u (mass of an individual proton)} = 92.64 \text{ u} \\ 143 \text{ neutrons} \times 1.009 \text{ u (mass of an individual neutron)} = 144.29 \text{ u} \\ 92 \text{ electrons} \times 0.0005 \text{ u (mass of an individual electron)} = 0.0460 \text{ u} \end{array}\right\} \text{The sum of all these masses is 236.98 u.}$$

To summarize, the *calculated* mass of the U-235 atom—arrived at by adding up the masses of all the subatomic particles that compose it— is 236.98 u. We can round this off to **237.0 u**.

The *actual* mass of an atom of U-235, determined experimentally (and extremely accurately), is only 235.043924 u. For our purposes, we can round this to **235.0 u**.

The mass defect, then, is the difference between the calculated mass and the actual, measured mass: about **2.0 u**.

Recall that this mass defect, or missing mass, is equivalent to the binding energy, the energy required to bind the U-235 nucleus together. We could now calculate the binding energy by using Einstein's equation, $E = mc^2$, where E is the binding energy, m is the mass defect, and c is the speed of light. Instead we'll simply point out that an atom's binding energy is the energy equivalent of its mass defect. Importantly, a portion of this binding energy is released in the course of certain nuclear reactions, a topic we address in the next section.

2. **Conservation of Energy** states that energy can neither be created nor destroyed as a result of chemical transformations. The sum of all the energy present in the products (including any that is liberated as the reaction progresses) must equal the sum of all the energy in the reactants (including any that is added to produce the reaction).

Although matter can be converted from one substance (a reactant) to another (a product) in a chemical reaction, matter can neither be created nor destroyed in the process. Similarly, although energy can be transferred or changed from one form to another in a chemical reaction, energy can neither be created nor destroyed in the process. These laws hold as long as the nuclei of atoms remain intact. However, in reactions involving atomic nuclei—radioactivity and nuclear fission, for example—we can observe the creation of energy and the disappearance of mass as matter and energy are interconverted.

Albert Einstein, a theoretical physicist, first described these interchanges in mathematical terms. He found that matter and energy are themselves equivalent through the equation $E = mc^2$, where E represents energy, m represents mass, and c represents the speed of light. Einstein's work was entirely theoretical. Later experiments demonstrated that energy could indeed be converted to mass, and mass into energy.

Converting Mass to Energy within the Nucleus

We now know that matter and energy are equivalent and that a specific quantity of mass is equivalent to a specific quantity of energy. This enables us to find the source of the energy that binds the protons and neutrons into an atom's compact, dense nucleus, and to examine the source of the energy released by nuclear fission.

We start with the key observation that if we were to compare the mass of an atomic nucleus with the total mass of the individual protons and neutrons that make up the nucleus, we would find that the mass of the nucleus is less than the sum of the masses of all the particles that compose it. This missing mass—the difference between the mass of the nucleus and the sum of the masses of its component parts—is called the **mass defect**. The question is, what happened to the missing mass? Einstein has the answer for us: The missing mass has been converted into energy, the very energy that binds all of the protons and neutrons together to form a compact, coherent nucleus. This **binding energy**—the cement that holds the nucleus together comes from the conversion into energy of a very small fraction of the masses of the protons and neutrons that compose the atom. We calculate a typical example in **Figure 9.17**.

mass defect The difference between the mass of an atomic nucleus and the (larger) sum of the masses of its constituent particles.

binding energy The energy required to hold the protons and neutrons together in the nucleus of an atom.

CONCEPT CHECK STOP

1. **The** burning of wood to produce heat energy results from conventional chemical changes involving valence electrons. Would this be an example of how matter and energy can be interconverted?
2. **The** atomic number of plutonium (Pu) is 94. Calculate the mass defect of an atom of Pu-239, which has a measured mass of 239.05 u.

9.4 Unleashing the Power of the Nucleus

LEARNING OBJECTIVES

1. **Describe** nuclear fission.
2. **Explain** why isotopic enrichment of uranium is a necessary part of nuclear energy production and how the enrichment process operates.
3. **Show** how a nuclear chain reaction propagates and releases energy as it proceeds.
4. **Summarize** the advantages and disadvantages of nuclear power.
5. **Describe** the difference between nuclear fusion and nuclear fission.

Some 40 years after Becquerel's discovery of radioactivity, another groundbreaking event occurred that would forever change the world: the discovery of nuclear fission. This process derives its enormous energy from the conversion of the matter (mass) of the nucleus into energy. In this section we examine the discovery of fission, the building of the first nuclear weapons, and the development of nuclear power.

Nuclear Fission

Beginning in Berlin in 1908, physicist Lise Meitner and chemist Otto Hahn enjoyed what would become a 30-year working relationship as they investigated radioactive isotopes and nuclear processes. By 1938, Meitner, who was of Jewish descent, had fled to Denmark as the Nazis took power. But in that very year, Hahn and his co-worker, Fritz Strassmann, made a remarkable discovery. Among the products produced when they bombarded uranium with neutrons was barium, a much lighter element than uranium.

nuclear fission A reaction that splits a relatively massive nucleus into two or more sizable fragments, releasing energy in the process.

Uncertain how to interpret the generation of barium by the addition of neutrons to the uranium nucleus, but convinced of its significance, Hahn and Strassmann sent word of their discovery to Lise Meitner. Working with her nephew, Otto Frisch, Meitner concluded that the neutrons had actually caused the uranium nucleus to cleave into two or more large fragments, a transformation without precedent in all previous atomic studies. (A few weeks later Hahn and Strassmann came to the same conclusion as well.) Meitner and Frisch coined the term **nuclear fission** for the splitting of an atomic nucleus (**Figure 9.18**).

The fission of U-235 also releases several neutrons. Any one (or more) of these could penetrate into another U-235 nucleus and continue the chain of energy-releasing fission reactions. In this way, the fission of a single U-235 atom could begin a cascading **chain reaction** (**Figure 9.19**). Such a reaction would consume the U-235 present and release energy in amounts never before achieved by humans.

Discovery of nuclear fission • Figure 9.18

The physicist Lise Meitner (a) correctly interpreted Otto Hahn and Fritz Strassmann's discovery as nuclear fission (b), the process that would come to underlie the atomic bomb and nuclear power.

a. Otto Hahn and Lise Meitner. Hahn was awarded the Nobel Prize in chemistry in 1944 for discovering nuclear fission, although many felt Meitner should have shared the prize.

bpk, Berlin/Art Resource

b. In one type of nuclear fission, a U-235 nucleus absorbs a neutron, then splits into two fragments (Kr-92 and Ba-141) and releases three neutrons.

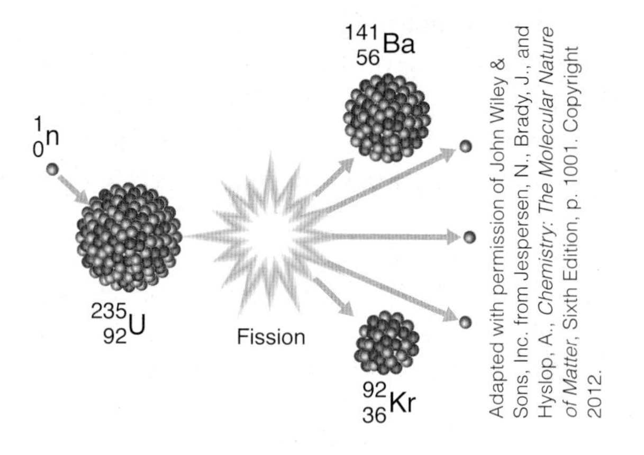

Adapted with permission of John Wiley & Sons, Inc. from Jespersen, N., Brady, J., and Hyslop, A., *Chemistry: The Molecular Nature of Matter*, Sixth Edition, p. 1001. Copyright 2012.

Interpret the Data

Complete the following nuclear equation for the reaction depicted in part b of this figure.

$$^{1}_{0}n + ^{235}_{92}U \longrightarrow$$

Nuclear fission chain reaction • Figure 9.19

Each time a U-235 nucleus absorbs a neutron, it splits into smaller nuclei (shown in orange) and releases more free neutrons.

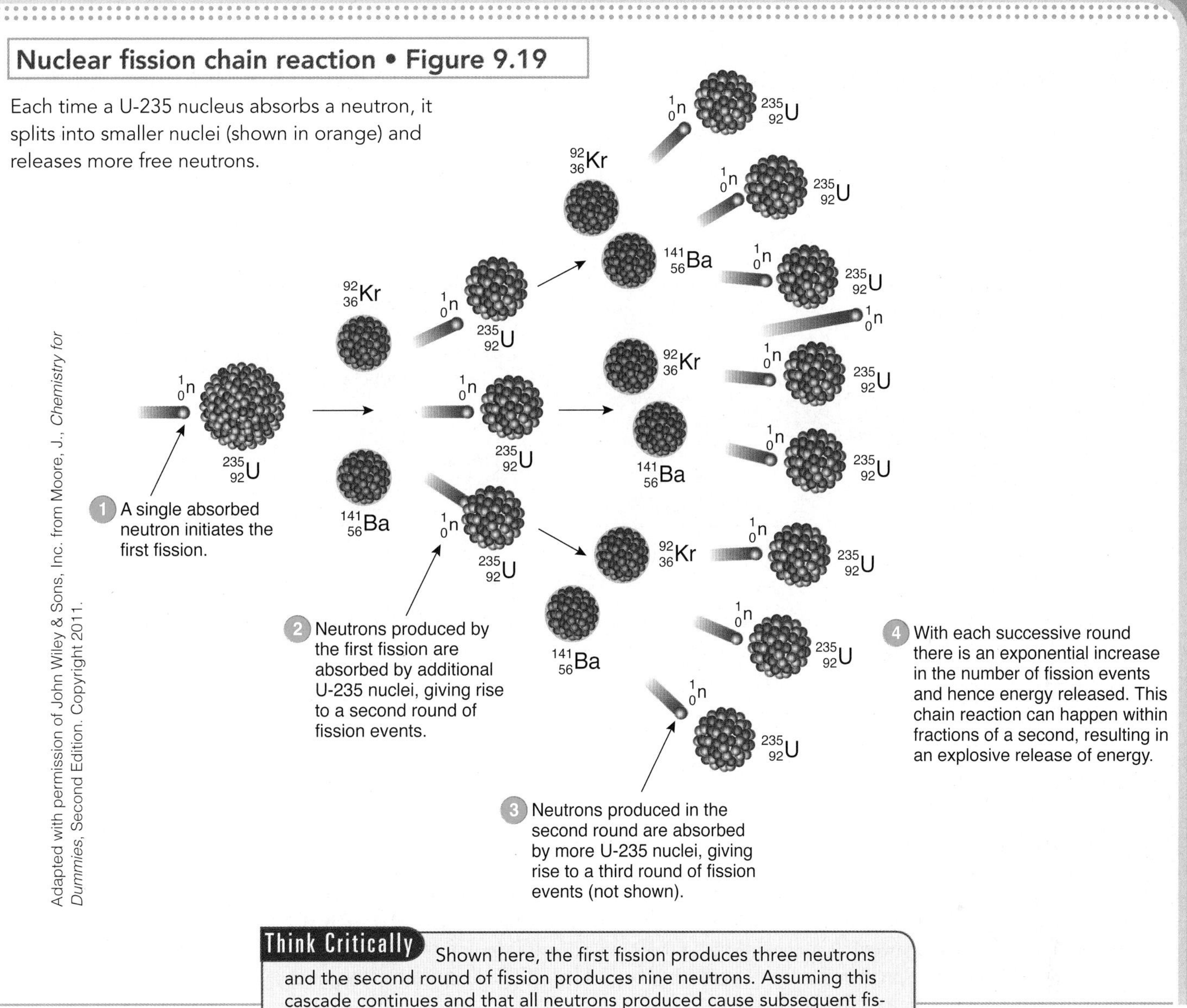

$^{1}_{0}n$
$^{235}_{92}U$
$^{92}_{36}Kr$
$^{141}_{56}Ba$
1 A single absorbed neutron initiates the first fission.
2 Neutrons produced by the first fission are absorbed by additional U-235 nuclei, giving rise to a second round of fission events.
3 Neutrons produced in the second round are absorbed by more U-235 nuclei, giving rise to a third round of fission events (not shown).
4 With each successive round there is an exponential increase in the number of fission events and hence energy released. This chain reaction can happen within fractions of a second, resulting in an explosive release of energy.

Think Critically Shown here, the first fission produces three neutrons and the second round of fission produces nine neutrons. Assuming this cascade continues and that all neutrons produced cause subsequent fission, how many neutrons would be produced in the fifth round of fission?

With the recognition of the power inherent in nuclear fission (**Figure 9.20** on the next page), and with the coming of World War II, research on the atomic nucleus accelerated. If a sustained sequence of fission reactions could be maintained, consuming an entire package of fissionable material in an instant and releasing the accompanying energy instantaneously, the result might be enormously explosive. Thoughts quickly turned toward the new war and to the building of an atomic bomb that would convert the sudden release of energy from a rapid chain reaction into the immensely explosive force of a devastating weapon. What was needed was a **critical mass** of U-235.

critical mass A mass of fissionable material large enough to ensure that released neutrons are absorbed by other fissionable nuclei, thereby sustaining a nuclear chain reaction.

The First Nuclear Bomb

By 1942, spurred on by the recognition that a weapon as powerful as an atomic bomb could determine the outcome of World War II, and with the fear that Germany might be making rapid progress toward the same goal, the United States set out to build an atomic weapon as quickly as possible. Control of the project was given to the U.S. Army Corps of Engineers. Initially located in a New York office, the entire operation quickly became known as the Manhattan Project. Various secret production facilities for fissile material were soon established across the country. The center that would actually design and assemble the bomb was located at Los Alamos, New Mexico.

In Words, Math, and Pictures

The energy of uranium fission • Figure 9.20

We can calculate the amount of energy released in nuclear fission by examining the neutron-induced fission of U-235 to barium, krypton, and three neutrons.

$${}^{1}_{0}n + {}^{235}_{92}U \rightarrow {}^{141}_{56}Ba + {}^{92}_{36}Kr + 3\,{}^{1}_{0}n$$

The total mass of the combined products of the fission of a single U-235 atom is 0.187 u less than the total mass of the combined reactants:

Reactants
Particle Mass (u)
neutron 1.009
U-235 235.044
Total 236.053
Products
Particle(s) Mass (u)
3 neutrons 3.026
Kr-92 91.926
Ba-141 140.914
Total 235.866

Difference in mass = (236.053 − 235.866) u = 0.187 u

This represents a loss of a little less than one-tenth of 1% of the mass of the combined reactants. This lost mass is converted into the energy released by the fission.

Mass lost in fission

$$\frac{0.187 \text{ u}}{236.053 \text{ u}} \times 100 = 0.079\%, \text{ the portion of the mass converted into energy}$$

Combined mass of the reactants

Einstein's equation, $E = mc^2$, reveals that if a single gram of U-235—about the mass of one paper clip—were to undergo fission by this reaction, even the conversion of a bit less than one-tenth of 1% of reactant mass into energy would generate enough power to keep a 100-Watt light bulb glowing for about 23 years! In comparison, burning 1 g of gasoline efficiently would generate only enough energy to keep the same bulb lit for a mere 8 minutes.

Photo Quest/Archive Photos/Getty Images

Release instantaneously through the detonation of an atomic bomb, the energy of uranium fission generates enormous destructive power.

RAYMOND GEHMAN/National Geographic Creative

Released slowly, under the controlled conditions of a nuclear power plant, the energy of uranium fission produces a convenient source of commercial electricity.

The world's first self-sustaining nuclear chain reaction • Figure 9.21

This artist's depiction shows the events of December 2, 1942. The Defense Department was informed of the success by the coded message, "The Italian navigator has just landed in the new world."

Enrico Fermi surrounded by scientists of the Manhattan Project.
Assistants stood ready to douse the pile with cadmium solution, if all else failed, to smother any runaway reaction.
Withdrawing the control rods.

Chicago History Museum, ICHi-33305; Gary Sheahan, artist

In December of 1942, a major milestone in the project was achieved—the first self-sustaining, controlled nuclear chain reaction (**Figure 9.21**). This event took place at the University of Chicago, one of many sites of the Manhattan Project, under the direction of Italian physicist Enrico Fermi. The 432-ton **atomic pile** that Fermi and his team built consisted of a matrix of uranium-based fuel rods and bricks of graphite (a form of carbon). Rapidly moving neutrons slow down as they pass through graphite, which increases the chance that some would be absorbed by U-235 nuclei to produce fission. Also located throughout the pile were ten **control rods** of cadmium, an element that absorbs large quantities of neutrons. These cadmium rods, which prevented any chain reaction from starting, were removed in stages to start the neutrons flowing and begin the chain reaction. As the last remaining cadmium rod was withdrawn slowly, detectors stationed around the pile began responding with audible clicks, which soon became a continuous static, indicating neutrons were cascading freely. After running successfully for almost 5 minutes, the central cadmium rod was driven back into the pile and the experiment was shut down. It had worked, and the atomic age had begun.

By this time, investigators had known that uranium fission occurs most efficiently through the addition of a neutron to the U-235 isotope rather than the more common U-238. Since more than 99% of all naturally occurring uranium is U-238 and less than 1% is the more fissionable U-235, the challenge was (and still is) to effectively increase the relative abundance of U-235 in a given sample of both isotopes The process is known as **isotopic enrichment**, often referred to as "uranium enrichment" (**Figure 9.22**).

isotopic enrichment A process of increasing the abundance of a desired isotope in a mixture by removing the undesired isotope.

A model for isotopic enrichment • Figure 9.22

In this example, we start with a sample made up of two isotopes: 10% of a minor isotope, represented by red dots, and 90% of the major isotope, represented by blue dots. As we remove the blue dots in stages, the relative abundance of the red dots increases; in this case, from 10% to 15% to 30%.
In the case of uranium, the initial mixture would be 0.7% U-235 and 99.3% U-238, reflecting the natural abundance of each isotope. As enrichment continues, U-235 can eventually constitute a majority of the mixture. In fact, the very first atomic bomb used a mixture enriched to 80% U-235.

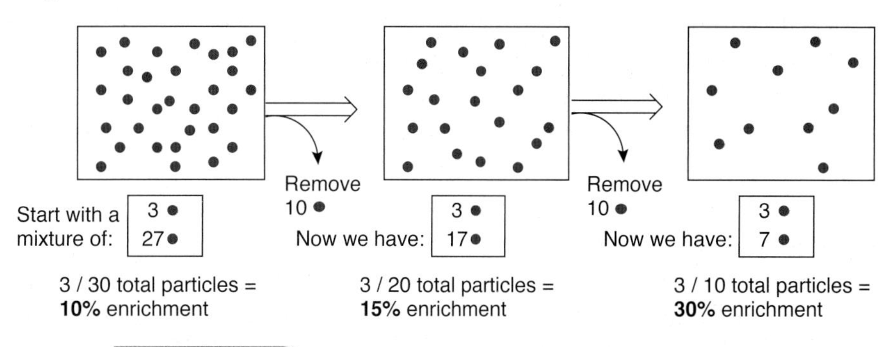

Think Critically How many more particles of the isotope represented by blue dots would have to be removed from the 30% enriched mixture to get to a mixture that is 60% enriched?

Uranium enrichment • Figure 9.23

Uranium is converted into-gaseous uranium hexafluoride, UF_6, which is then isotopically enriched in rapidly spinning cylinders, a process called centrifugation.

a. Gas centrifugation

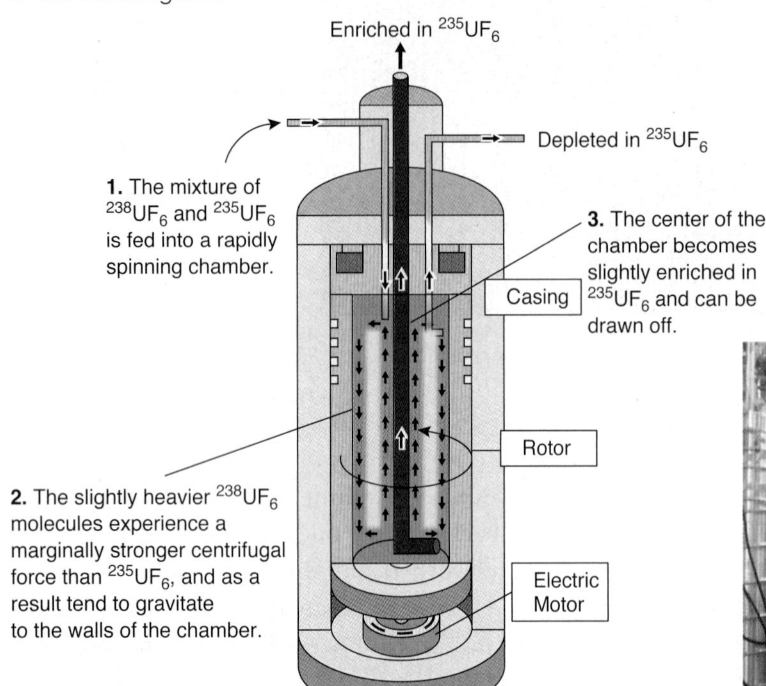

Enriched in $^{235}UF_6$
Depleted in $^{235}UF_6$
1. The mixture of $^{238}UF_6$ and $^{235}UF_6$ is fed into a rapidly spinning chamber.
3. The center of the chamber becomes slightly enriched in $^{235}UF_6$ and can be drawn off.
Casing
Rotor
2. The slightly heavier $^{238}UF_6$ molecules experience a marginally stronger centrifugal force than $^{235}UF_6$, and as a result tend to gravitate to the walls of the chamber.
Electric Motor

b. Gas centrifugation requires many successive stages to enrich uranium to adequate levels. Shown here is an array of centrifuges at an enrichment demonstration facility in Tehran. Uranium from one centrifuge enters the next, where it is further enriched—a process that continues on throughout the array.

Reuters/Caren Firouz

In actual practice, common uranium—a mixture of isotopes—is first converted into UF_6, uranium hexafluoride gas, consisting of $^{238}UF_6$ and $^{235}UF_6$, which is then isotopically enriched. At the time of the Manhattan Project, a slow process known as **gaseous diffusion** was used to isolate the desired $^{235}UF_6$. Today, **gas centrifugation** is more commonly used. Since molecules of $^{238}UF_6$ have a slightly larger mass than molecules of $^{235}UF_6$, they tend to migrate gradually to the walls of the spinning centrifuge. As a result, the gas at the center of the centrifuge becomes enriched slowly in the lighter $^{235}UF_6$ and can be drawn off (**Figure 9.23**). As this process is repeated in successive stages, it progressively enriches the mixture in $^{235}UF_6$ to desired levels. Chemical conversion of the enriched gas back to uranium itself yields material enriched in the desired fissionable isotope, U-235. The intended use of the enriched uranium determines the level of enrichment needed. Fuel for nuclear power plants requires enrichment to only about 3% U-235. Uranium to be used in nuclear weapons—**weapons grade** uranium—requires enrichment to about 90% U-235.

Because no atomic bomb had ever been designed or built before World War II, and because of wartime urgency, the Manhattan Project pursued a second path as well: a bomb based on an isotope of plutonium, Pu-239. Both designs—U-235 and Pu-239—required that fissionable material exist as **subcritical** masses (that is, masses incapable of chain reactions) until the moment of detonation. At that point, two or more subcritical masses would be combined rapidly to form a critical mass, produce a chain reaction, and thus detonate a nuclear explosion. In the absence of a single critical mass, no chain reaction could occur, so a bomb could not explode.

On August 6 and 9, 1945, atomic bombs were dropped on the cities of Hiroshima and Nagasaki, Japan, respectively, destroying both cities and killing or wounding an estimated 200,000 people. The bombs were history's first atomic weapons. They produced unimaginable destruction and deaths, both immediately from the blast and fireball and for decades beyond, from lingering effects of **radioactive fallout**. The uranium bomb dropped on Hiroshima packed an explosive power equivalent to an estimated 15,000 tons of TNT. The plutonium bomb dropped over Nagasaki was even more explosive. Destruction of Hiroshima was virtually total for a distance of over a mile from the center of the blast; temperatures of more than 3000°C incinerated buildings and people within 2 miles of the center. Seeing the blast and the mushroom cloud from the cockpit of the plane that dropped the bomb over Hiroshima, the co-pilot of the flight wrote in his journal, "My God, what have we done?" Within a few days, Japan surrendered and World War II ended.

radioactive fallout Airborne radioactive particles, from a nuclear explosion or accident, which settle to the ground. Depending on the severity of the incident and weather patterns, fallout can potentially distribute across the globe.

pressurized water reactor A reactor in which circulating water is kept at high pressure to prevent it from boiling as it's heated by an atomic pile. Steam is then generated in a secondary loop.

Nuclear Power

The very same fission chain reaction that releases energy instantaneously in the case of nuclear weapons can be designed to release its energy in a slow, controlled fashion, much as Fermi's atomic pile achieved. The heat generated by nuclear fission within the pile can be used to convert water into steam, which turns turbine blades attached to an electrical generator, thereby producing electricity. Two-thirds of the nuclear power plants currently operating in the United States are **pressurized water reactors** (PWR) (**Figure 9.24**) that operate in a similar manner.

PROCESS DIAGRAM

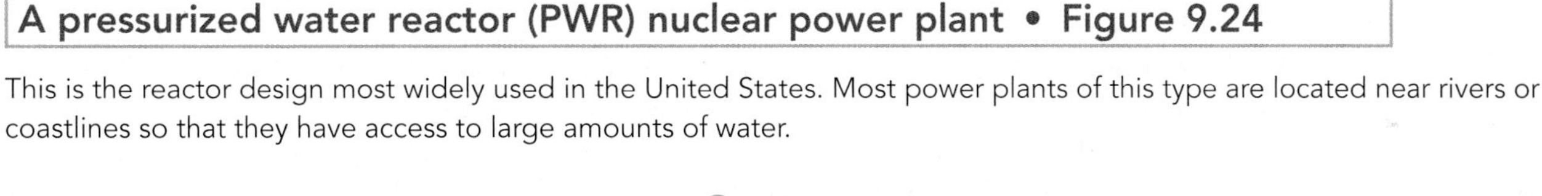

A pressurized water reactor (PWR) nuclear power plant • Figure 9.24

This is the reactor design most widely used in the United States. Most power plants of this type are located near rivers or coastlines so that they have access to large amounts of water.

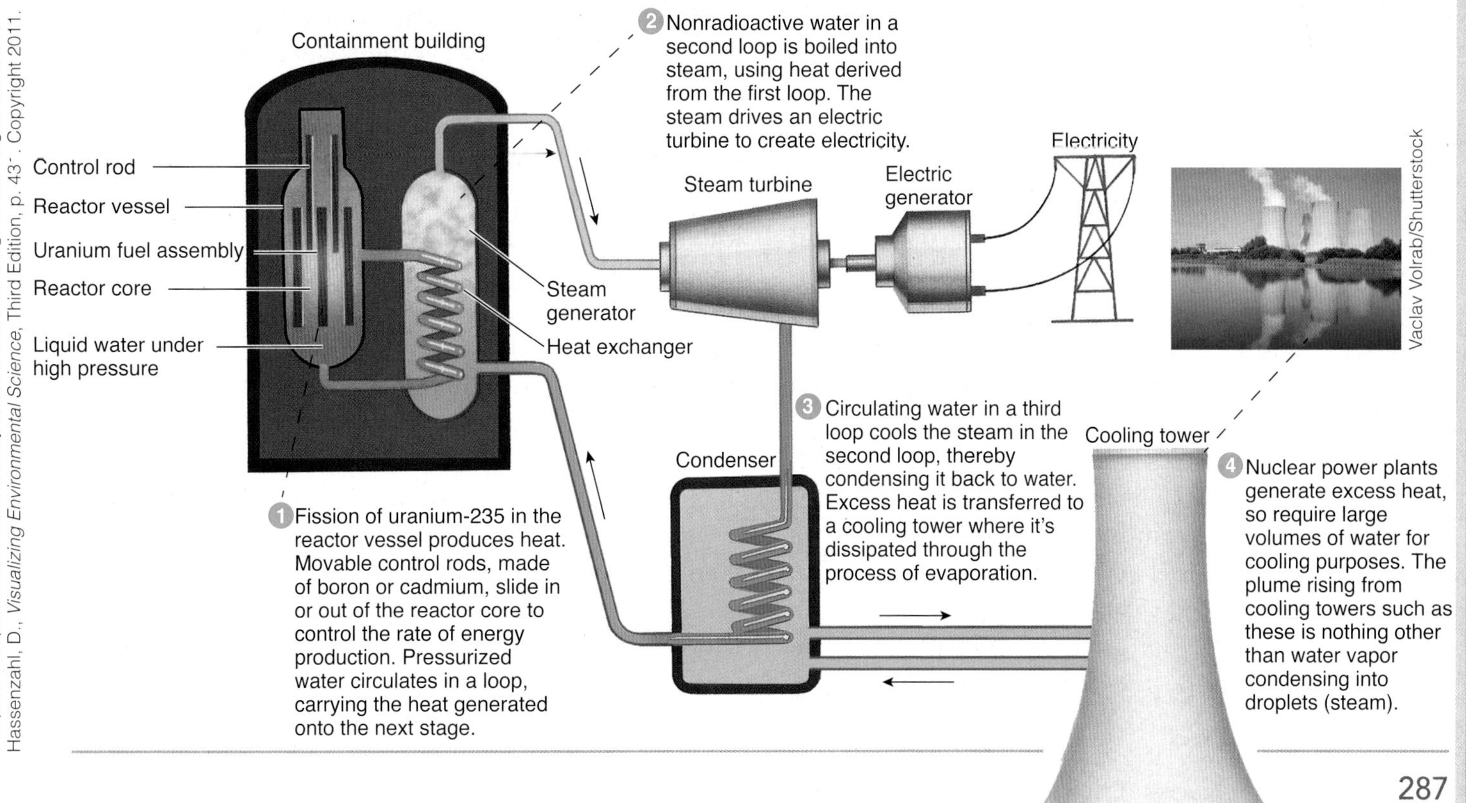

Containment building
Control rod
Reactor vessel
Uranium fuel assembly
Reactor core
Liquid water under high pressure
Steam generator
Heat exchanger
2 Nonradioactive water in a second loop is boiled into steam, using heat derived from the first loop. The steam drives an electric turbine to create electricity.
Steam turbine
Electric generator
Electricity
Condenser
Cooling tower
1 Fission of uranium-235 in the reactor vessel produces heat. Movable control rods, made of boron or cadmium, slide in or out of the reactor core to control the rate of energy production. Pressurized water circulates in a loop, carrying the heat generated onto the next stage.
3 Circulating water in a third loop cools the steam in the second loop, thereby condensing it back to water. Excess heat is transferred to a cooling tower where it's dissipated through the process of evaporation.
4 Nuclear power plants generate excess heat, so require large volumes of water for cooling purposes. The plume rising from cooling towers such as these is nothing other than water vapor condensing into droplets (steam).
Vaclav Volrab/Shutterstock

Nuclear power generating capacity by country • Figure 9.25

Thirty countries throughout the world generate nuclear power, with the United States, France, and Japan the leading producers. Values shown are in gigawatts (GW), with 1 GW equal to one billion (10^9) watts. 1 GW can power a city the size of San Francisco.

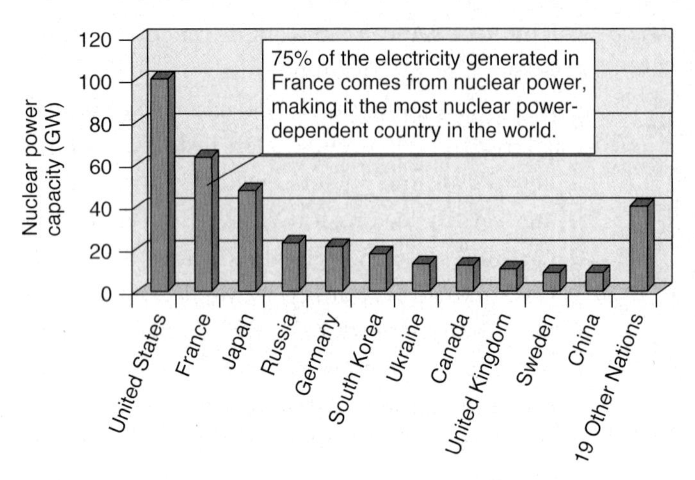

When nuclear power was introduced in the United States in the mid-20th century, it was welcomed as a source of cheap, plentiful electricity produced by a fuel that would free us from the smoke, smog, and other pollutants of fossil fuels. But the original promise of cheap, plentiful, pollution-free power soon gave way to problems and fears that have stunted the growth of commercial nuclear power. Nevertheless, the United States still has the largest number of operating nuclear reactors in the world (with 100 in 2014) and is the largest producer of nuclear power (**Figure 9.25**). Nuclear power generates about 20% of the electricity in the United States, a value that has remained reasonably constant since 1990.

Nuclear power engenders a variety of concerns and fears, but at least one fear is baseless—that somehow a nuclear power plant might explode into a nuclear fireball. A nuclear explosion requires the almost instantaneous release of nuclear energy as the fissionable material is compacted into an explosive critical mass. This requires highly enriched uranium—with a purity on the order of 90%. A fission pile within a nuclear reactor, on the other hand, is designed to produce a slow and continuous release of energy. The fuel rods of nuclear piles contain only about 3% enriched U-235.

Concerns about other kinds of accidents are more realistic. These include the accidental overheating of the core, resulting in **meltdown**, and the occurrence of nonnuclear fires or explosions that might release radioactive material (**Figure 9.26**).

meltdown Accidental overheating of the fuel in a nuclear reactor, leading to a melting of the reactor core.

One of the chief concerns about radioactive fallout is its potential ability to enter the food supply. Iodine-131, for example, poses a particular concern due to its effects on children. The thyroid, a gland located in the neck, is especially sensitive to I-131 because most of the iodine contained in our food and drink concentrates in this gland. Children who drink milk containing I-131—produced by cows that graze on grass contaminated by fallout—are particularly at risk of thyroid cancer. According to the International Atomic Energy Agency, among Chernobyl victims, at least 1800 cases of thyroid cancer in children who were under 14 when the accident occurred have been documented, a number of cases far higher than normal.

Nuclear Waste Perhaps the most troubling issue concerning nuclear power is the safe disposal of radioactive wastes. These consist of a large number of radioactive isotopes produced as by-products from the operation of nuclear reactors, along with residual nuclear fuel whose radioactivity has dropped below useful levels but is nonetheless still hazardous. Some of these radioisotopes remain dangerously radioactive for a very long time due to their long half lives, which can span up to hundreds of thousands of years. Since we can't increase the rate of their decay (so as to convert them quickly to stable or nonradioactive isotopes), large quantities of radioactive nuclear wastes will be with us, and with those who come after us, for generations almost beyond measure. To protect ourselves and future generations from this radioactivity, we need a safe way to store these wastes so that they remain out of contact with our environment until they no longer pose a threat or until we devise other ways to dispose of them.

In plans currently under study, these high-level long-lived wastes would be **vitrified**—fused with molten glass—then packed into secure, corrosion-resistant containers, and stored deep underground in a geologically stable, national waste repository. Until then, high-level wastes must be stored on the grounds of nuclear power plants

Nuclear power plant accidents • Figure 9.26

These serious accidents at nuclear power plants illustrate the dangers associated with meltdown and radioactive contamination.

a. Three Mile Island, Pennsylvania (1979). Equipment failures and human errors led to a loss of coolant in one of the two reactors. The reactor core overheated to about 2000°C, producing a partial meltdown of the uranium fuel. The accident released small amounts of radioactive gas into the atmosphere, but no deaths were reported. Decommissioning the reactor site took nearly 15 years at a cost of nearly one billion dollars. The event catalyzed opposition to nuclear power in the United States and led to calls for more stringent safety measures in the industry.

These two cooling towers form part of the decommissioned nuclear reactor at Three Mile Island. The other nuclear reactor at this site continues to operate.

Peter Essick

b. Chernobyl, Ukraine (1986). Radioactive material spewed into the atmosphere, prompting evacuation of over 100,000 citizens and producing radioactive fallout that dispersed across national borders. Thirty people died in the immediate aftermath, and thousands of extra cancer deaths are expected over time. An 18-mile "exclusion zone" around the site is still largely uninhabited due to residual radioactivity.

Flash overheating of a nuclear reactor vaporized coolant water, causing an explosion that blew off its 1000-ton containment cover.

MCT via Getty Images

c. Fukushima, Japan (2011). A tsunami disabled cooling systems, leading to the meltdowns of three nuclear reactors. Radioactive material was released into the air, ground, and seawater. Although total levels are believed to be well below those released in Chernobyl, the long-term environmental and health effects of this disaster may not be known for years.

Accumulation of hydrogen gas in the aftermath of the meltdown caused an explosion of this reactor containment building.

Bloomberg via Getty Images

Waste nuclear fuel storage • Figure 9.27

In the United States, most wastes derived from nuclear fuel are stored in pools located onsite at reactor facilities (a). Potential long-term storage options include dry casks (b) and underground depositories (c).

a. Spent fuel pools. The spent nuclear fuel from nuclear power plants can be temporarily stored onsite for a number of years under cooled pools of water treated with neutron-absorbing agents. The pools dissipate excess heat and absorb radiation.

U-238 represents just over 95% of the mass of spent nuclear fuel. Most of the remainder consists of a host of radioisotopes from the fission of U-235.

b. Dry cask storage. After storage in cooling pools, spent nuclear fuel can be temporarily housed onsite in above-ground dry casks. Each cask can hold upwards of 20 tons of spent fuel.

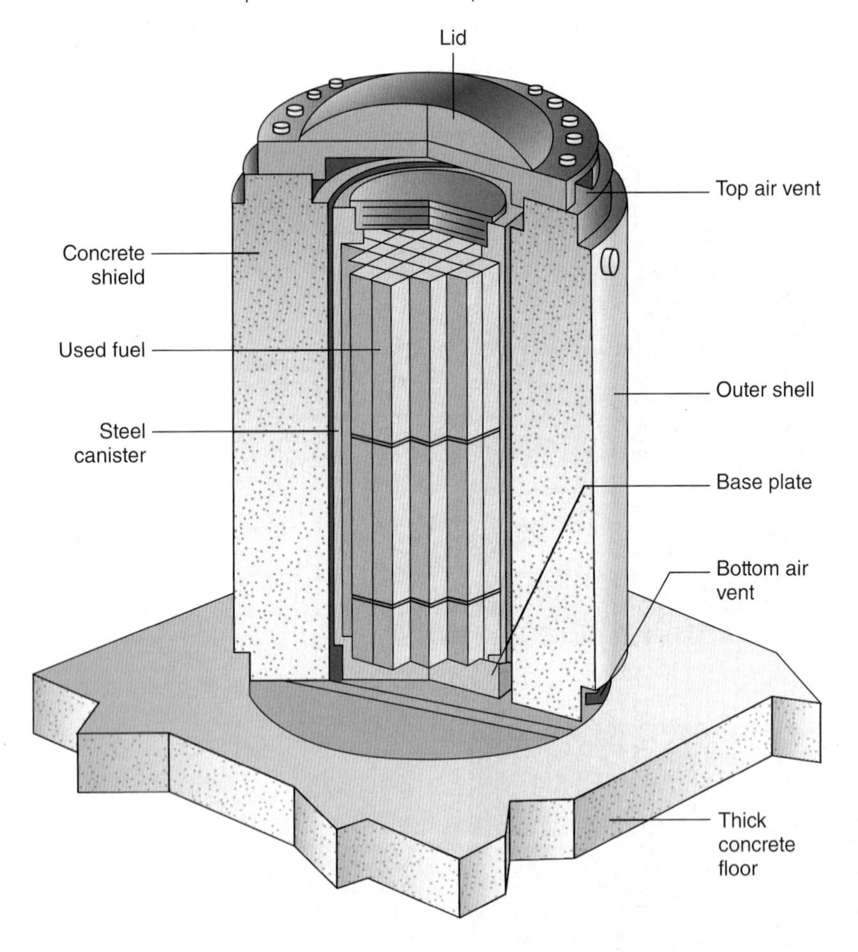

c. Deep, geological storage represents perhaps the best option for long-term storage of spent nuclear fuel Shown here is an entrance tunnel (leading some 300 meters underground) at a proposed federal high-level radioactive waste repository at Yucca Mountain, Nevada. In spite of nearly 30 years of planning, this site has been mired in controversy.

Ask Yourself

Why does U-238 make up the vast majority of spent nuclear fuel?

and at sites that had been used for the production of nuclear weapons (**Figure 9.27**).

Pros and Cons of Nuclear Power The United States depends on nuclear power for about 20% of its electricity needs, but growth of the nuclear power industry has largely stagnated for a variety of reasons, including concerns about long-term waste storage. Worldwide, the events at Fukushima, Japan, may have chilled developments in the industry even further, with Germany vowing to shut down its nuclear power facilities by 2022. Nevertheless, rising concerns about fossil fuel emissions and their effects on climate change have not only stimulated investment in a variety of renewable energies but have also shed a fresh light on nuclear power. At issue is whether the risks associated with climate change outweigh the risks posed by nuclear power.

Proponents of nuclear power argue that it is an established technology with even safer, more efficient designs (known as Generation III reactors) on the horizon. Perhaps the most promising of these reactors is the European Pressurized Reactor (EPR), which contains four independent cooling systems and a built-in life expectancy of 60 years (as compared to 40 years for prior designs). Improved operating efficiencies require less uranium fuel to generate the same level of power.

Nuclear power also has some surprising advocates among environmentalists. If further expanded, nuclear power could reduce carbon emissions significantly. Power from new nuclear reactors, for example, could partially offset the need for coal-fired power plants, which emit large amounts of the greenhouse gas carbon dioxide. Proponents also argue that nuclear power generation makes a far smaller demand on land area than would the building of renewable energy facilities (such as wind or solar) of equivalent capacity.

Opponents to nuclear power express a variety of concerns, including:

- **Waste storage**: In the United States, the problem of long-term storage of nuclear waste is still unresolved. Onsite waste storage at nuclear power facilities continues to grow, well beyond initial estimates.
- **Questions of safety**: In spite of increased safety measures and improved designs, accidents can still occur. Also, nuclear reactor and waste storage facilities could be targets for terrorist attacks.
- **Long lead time to build new plants**: Some energy experts argue that nuclear power does not necessarily offer a quick solution to our energy needs. Due to regulatory and other issues, it can take 20 years or more from the planning to the final commissioning of a new nuclear reactor.
- **Nuclear proliferation**: Globally, there is deep concern about the nuclear initiatives of some countries. The same uranium enrichment techniques that create fuel for a national nuclear power program could potentially be used to make weapons-grade material.

What is certain is that there is renewed discussion about nuclear power. Whether climate concerns or other factors are powerful enough to offset objections to further growth of the industry remains to be seen.

Nuclear Fusion

We've seen that the nuclei of certain isotopes of relatively large mass split apart into smaller fragments during fission, releasing energy through the conversion of matter into energy. In **nuclear fusion**, a transformation resembling the reverse of fission, several atoms of small mass fuse together at extremely high temperatures (100 million degrees Celsius or higher) to form larger nuclei. As in fission, nuclear fusion also results in the conversion of matter into energy. For example, the fusion of hydrogen atoms to form a helium atom results in a fractional loss of mass, which is released as energy. This fusion reaction—the conversion of hydrogen into helium—produces the energy of the sun and is part of the life cycle for all other stars. Gram for gram, the fusion of hydrogen atoms converts mass into energy more effectively than does the fission of uranium. Because no materials or designs are yet available for structures that can withstand the extraordinarily high temperatures required for commercial nuclear fusion, no fusion power plant currently exists. If and when one can be created, it may well be based on the fusion of deuterium and tritium, both of which are isotopes of hydrogen (**Figure 9.28**).

nuclear fusion A nuclear reaction in which light atomic nuclei fuse together with the release of energy.

Deuterium-tritium fusion • Figure 9.28

In addition to energy that would be captured as heat, each fusion event involving deuterium and tritium nuclei produces a He-4 nucleus and a free neutron.

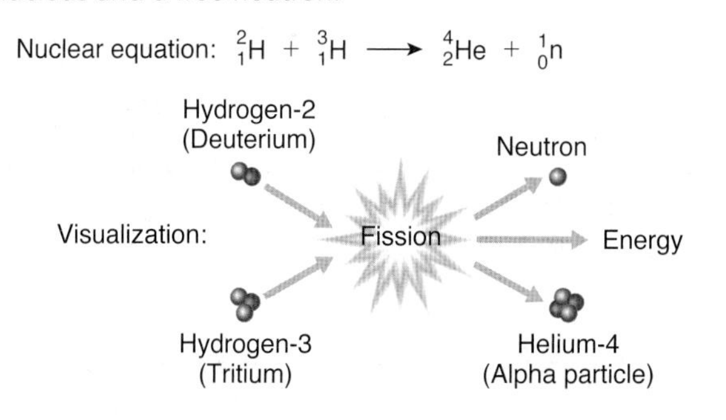

Results obtained in experimental test facilities indicate that more energy may be needed to create and sustain the nuclear fusion reaction than can be obtained from the reaction itself. Because of this and a variety of other technical hurdles, some experts maintain that fusion power will not become a practical reality in the foreseeable future.

CONCEPT CHECK

1. **What** is needed to initiate fission of a critical mass of U-235?
2. **Why** is isotopic enrichment of uranium necessary, and how can it be accomplished?
3. **How** does a nuclear reactor produce electricity?
4. **What** is the general recommendation for long-term storage of high-level radioactive waste?
5. **How** is fusion different from fission?

Summary

1 Radioactivity 266

- ***What is radioactivity?***
  Radioactivity results from the spontaneous decay of unstable atomic nuclei.

- ***What types of radioactive decay do atomic nuclei undergo?***
  Common types of radioactivity include emission of α **particles** (fast-moving He-4 nuclei, as shown here), β **particles** (fast-moving electrons), and γ **rays** (high-energy electromagnetic radiation).

Figure 9.3 • Alpha decay

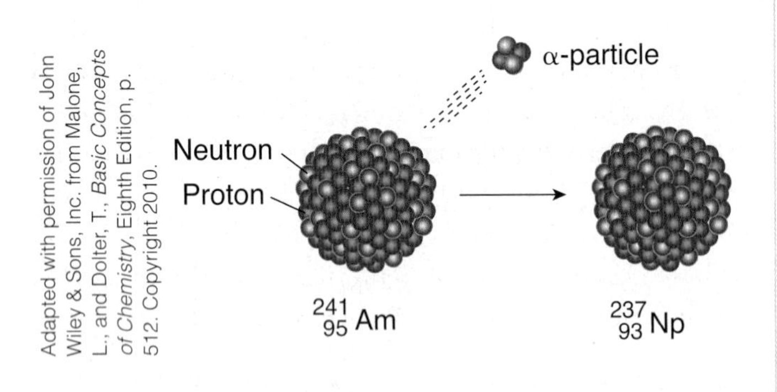

2 Radiation Effects and Applications 270

- ***How does ionizing radiation affect living things?***
  Ionizing radiation can cause biological damage, with consequences that depend on the radiation's **penetrating power**, **ionizing power**, and other factors. Acute exposure can cause sickness, genetic damage, and/or death. Low levels of ionizing radiation naturally present in the environment constitute **background radiation**. Radon, a natural by-product of the decay of uranium, is one of many sources of background radiation.

- ***What is half-life?***
  Half-life is the length of time it takes for half of a given quantity of a radioactive isotope to decay, as shown here. Each radioactive isotope has a characteristic half-life.

Figure 9.13 • Radioactive decay of carbon-14

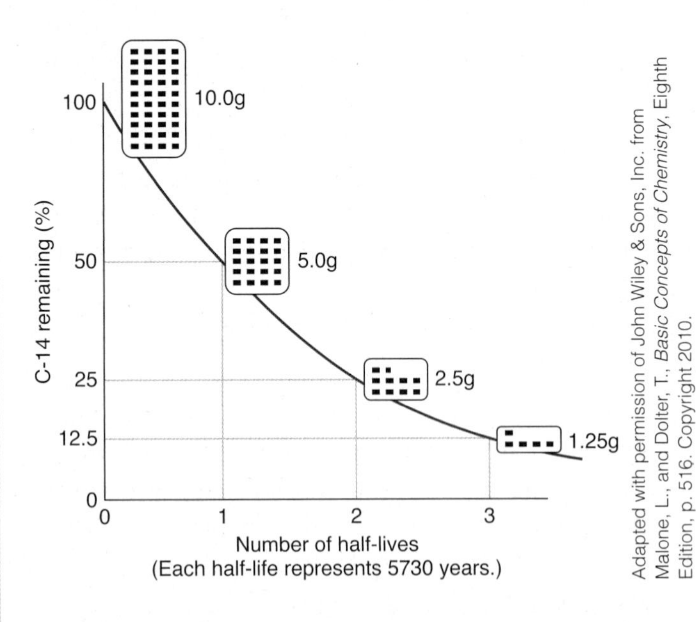

- ***What are some practical uses of ionizing radiation?***
  Radiocarbon dating, based on the half-life of carbon-14, can be used to estimate the age of artifacts of biological origin. Radioisotopes are used in a variety of medical applications, including cancer therapy, medical imaging, and diagnostics.

3 Mass Defect and Binding Energy 279

- ***Why is the mass of a nucleus less than the sum of the masses of its individual particles?***
  This missing mass, or **mass defect** (represented here), has turned into **binding energy**, which holds the particles of the nucleus together. The equivalence of mass and energy is the basis for the energy released in nuclear reactions.

Figure 9.17 • Mass defect

4 Unleashing the Power of the Nucleus 281

- ***How can we harness the energy of nuclear reactions?***
  When a uranium-235 nucleus absorbs a neutron, it undergoes **nuclear fission**, releasing energy. A **critical mass** of fissionable material is needed for a self-sustaining nuclear chain reaction (shown here). The degree of **isotopic enrichment** of uranium determines how it's to be used. A low level of enrichment is used for nuclear power; a much higher level is used for nuclear weapons. Nuclear power produces by-products, including many fission products (radioisotopes) with long half-lives, requiring long-term storage. In **nuclear fusion**, the nuclei of very light isotopes fuse together, releasing energy.

Figure 9.19 • Nuclear fission chain reaction

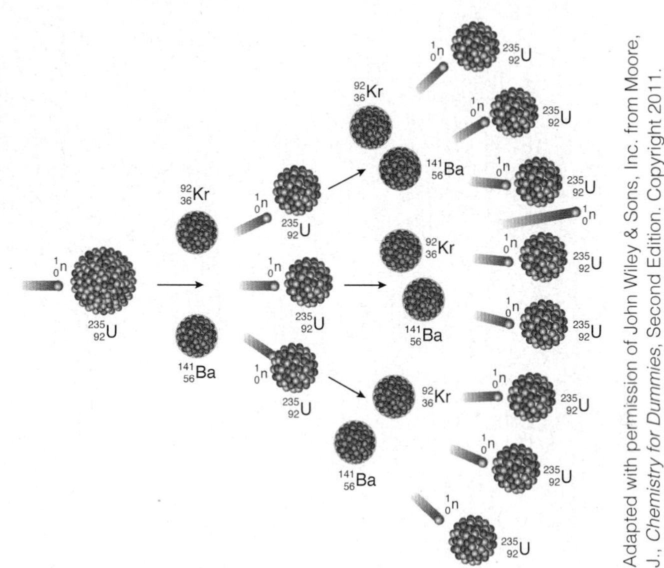

Key Terms

- α particle 268
- β particle 268
- background radiation 273
- binding energy 281
- critical mass 283
- free radical 271
- gamma (γ) ray 270
- half-life 275
- ionizing power 271
- ionizing radiation 271
- isotopic enrichment 285
- mass defect 281
- meltdown 288
- nuclear fission 282
- nuclear fusion 291
- nuclide 267
- penetrating power 271
- positron 277
- positron emitter 277
- pressurized water reactor 287
- radioactive fallout 287
- radioactivity 267
- radiocarbon dating 276
- rem 274

What is happening in this picture?

Here, a *Geiger counter* records radioactivity emitted by *trinitite*, glassy-rock material produced by the fireball of the first nuclear bomb test, code-named Trinity. This photograph was taken in 2006, 61 years after the explosion, which took place in the New Mexico desert in 1945, yet the rocks still show activity. [When ionizing radiation passes through the test window of a *Geiger counter,* it ionizes atoms of a gas inside, momentarily completing an electrical circuit and recording a count and/or producing an audible click. The number of counts received per minute (CPM) is one measure of radioactivity.]

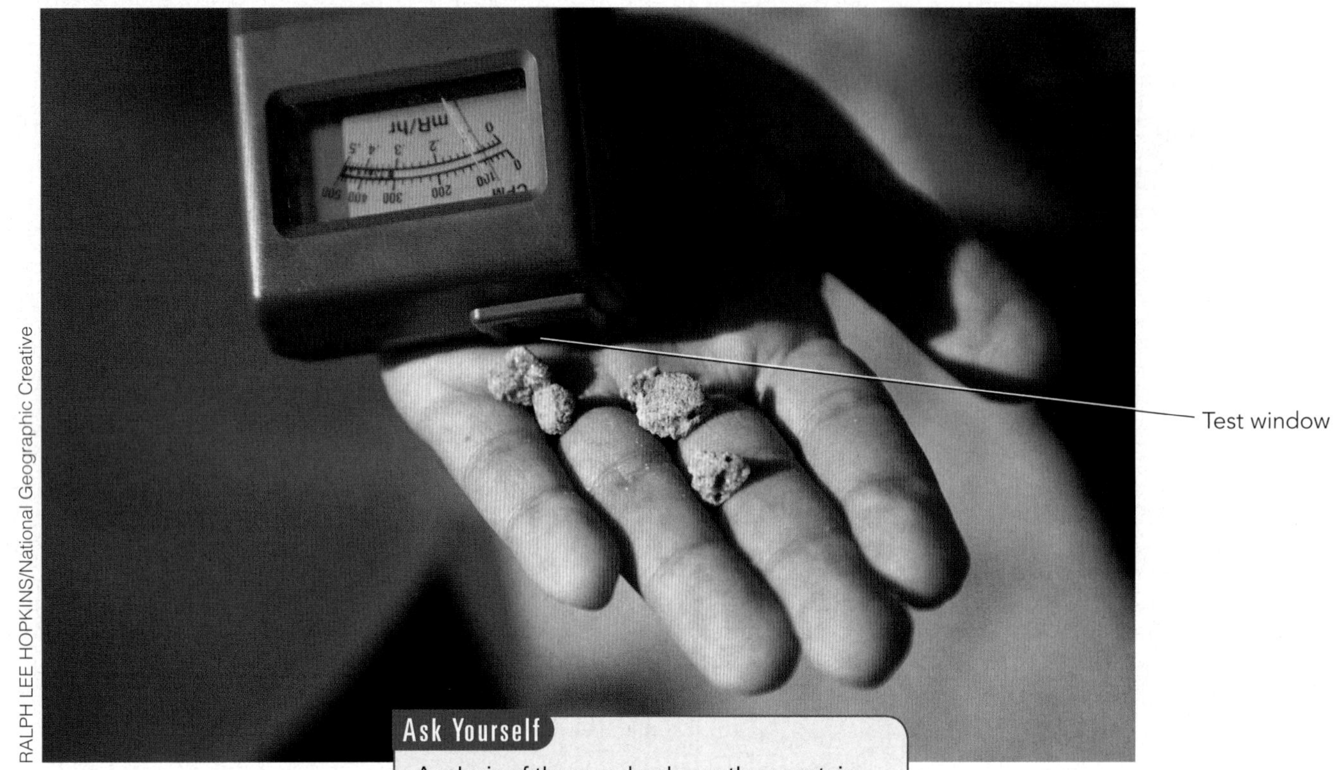

RALPH LEE HOPKINS/National Geographic Creative

Ask Yourself

Analysis of these rocks shows they contain the plutonium (bomb) fission products, strontium-90 (Sr-90) and cesium-137 (Cs-137). The half-lives of these radioisotopes are coincidentally about 30 years each. At the time this image was taken, roughly what percentage of the Sr-90 and Cs-137 generated by the explosion remained in these rocks?

Exercises

Review

1. What major discovery was made by Antoine Henri Becquerel and further investigated by Marie and Pierre Curie?
2. What major discovery was made by the combined work of Otto Hahn, Fritz Strassmann, Lise Meitner, and Otto Frisch?
3. Describe two characteristics of an atomic nucleus that are likely to result in radioactive decay.
4. What kind of nuclear radiation produces each of the following changes in a nucleus:

   a. a decrease in both atomic number and mass number

   b. an increase in atomic number but no change in mass number

   c. a decrease in atomic number but no change in mass number

   d. no change in either atomic number or mass number

5. Complete the following nuclear equations.

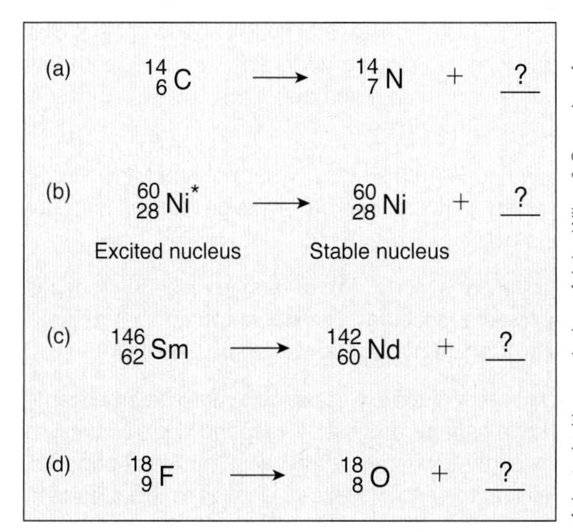

Adapted with permission of John Wiley & Sons, Inc. from Jespersen, N., Brady, J., and Hyslop, A., *Chemistry: The Molecular Nature of Matter*, Sixth Edition, p. 981. Copyright 2012.

6. Why are radioactive emissions classified as forms of ionizing radiation?

7. How is each of the following used in medical imaging, diagnosis, and/or therapy?

a. C-11; b. Tc-99m; c. Co-60

8. Of the set of α, β, and γ rays, which would be the most hazardous to a person standing a few feet away from its source? Which would be the least hazardous?

9. Identify four factors that affect the amount of harm that ionizing radiation can do to our bodies.

10. Identify four sources of background radiation that might affect someone standing outdoors in a rocky field.

11. Describe the kind of harm that results from *somatic* damage.

12. Identify a radioisotope with a half-life

a. less than one day

b. between one day and one year

c. between 1000 and 10,000 years

d. greater than one billion years

(You can find half-lives of various isotopes on the Internet.)

13. In principle, which of the following can be dated by radiocarbon techniques?

a. rock; b. a leather slipper; c. a wooden boat; d. a mummified body; e. a silver spoon. Describe your reasoning.

14. Identify a radioisotope that emits only γ rays, describe its major application, and explain why this radioisotope is useful for this application.

15. What is PET, and what two radioisotopes are useful in PET studies?

16. Why is ionizing radiation especially useful in treating cancerous growths?

17. What is *gamma knife therapy*? What radioisotope is especially useful in this kind of therapy?

18. The experimentally measured mass of an atom is always smaller than the mass calculated as the sum of the masses of all the protons, neutrons, and electrons that compose it. Why?

19. What is a radioactive decay series?

20. a. What process involving U-235 is depicted here?

b. Explain the importance of the capture and release of neutrons in this process.

c. Would a subcritical mass of U-235 be expected to behave as shown here? Explain.

21. What was the objective of the Manhattan Project?

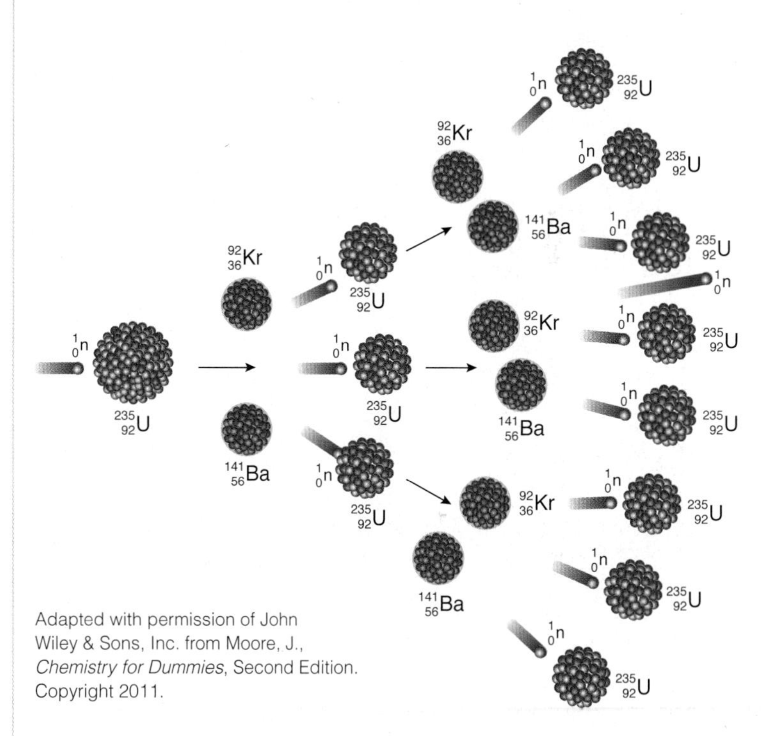

Adapted with permission of John Wiley & Sons, Inc. from Moore, J., *Chemistry for Dummies*, Second Edition. Copyright 2011.

22. In the history of the development of nuclear power and nuclear weapons, what events took place at:

a. the University of Chicago

b. Los Alamos, New Mexico

c. Hiroshima, Japan

23. What method is currently most used to isotopically enrich uranium hexafluoride gas?

24. What is the advantage of using U-235 rather than U-238 as the fissionable material in building a fission bomb? What is the disadvantage to using U-235 rather than U-238 as the fissionable material?

25. Explain why a nuclear explosion cannot occur in a commercial fission reactor. What kinds of accidents can occur?

26. Label each of the following as nuclear fission or nuclear fusion. How are these processes (i) similar and (ii) different?

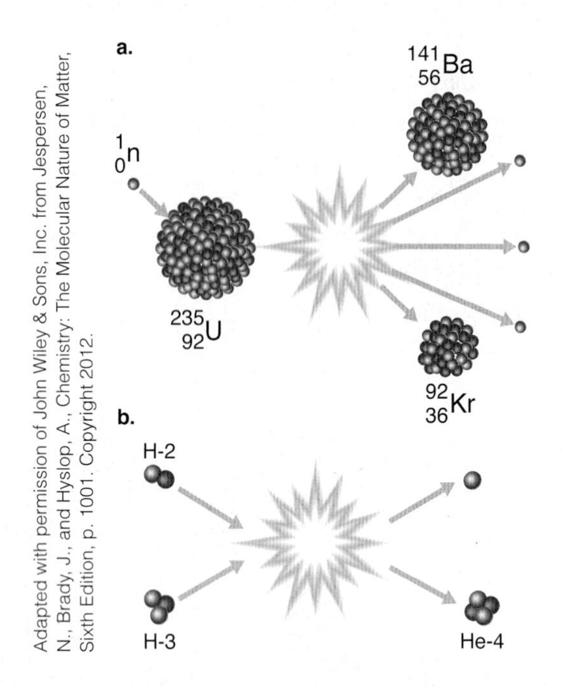

Adapted with permission of John Wiley & Sons, Inc. from Jespersen, N., Brady, J., and Hyslop, A., Chemistry: The Molecular Nature of Matter, Sixth Edition, p. 1001. Copyright 2012.

27. a. How is rate of energy production within a nuclear reactor controlled? b. Why are nuclear reactors typically situated near rivers or large bodies of water?

28. Write a reaction for the nuclear process that provides the energy of the sun.

29. What results from the collision of an electron with a positron? Explain why this is an example of the conversion of mass into energy.

30. Why are radioisotopes of iodine particularly useful in the diagnosis and therapy of disorders of the thyroid gland?

31. Why does the EPA recommend testing for radon levels in all homes?

Think

32. Of α, β, and γ radiation, which is the most dangerous kind of *external* radiation for a human to be exposed to? Which is the most dangerous kind of *internal* radiation?

33. Why is lead found in all deposits of uranium ore?

34. What is the ultimate fate of every radioactive atom now in existence?

35. It has been suggested that in areas with certain kinds of soil, underground rock, and mineral deposits, it may be more hazardous to live in a house that is well sealed against drafts than in one with loose-fitting doors and windows that allow a continual flow of air into and out of the house. Suggest a reason why this may be so.

36. How did Enrico Fermi's studies of nuclear reactions contribute to the use of atomic energy for the production of commercial power?

37. What is the most serious form of damage that could occur if a natural disaster such as a hurricane, tornado, earthquake, or tsunami struck a nuclear power plant? Explain.

38. Do you think that nuclear weapons ought to be an acceptable part of a nation's arsenal of weapons? Do you think that all nations ought to renounce the use of nuclear weapons, destroy all that they now possess, and refrain from building or helping others to build more? If you think that nations ought to maintain nuclear weapons, do you think these weapons ought to be considered as conventional weapons, like bullets and nonnuclear bombs, or that they ought to be considered as unconventional weapons, like chemical and biological warfare agents? Explain your answers.

39. What are the advantages and disadvantages of building new nuclear reactors to help diversify a nation's energy resources? Would you support such a policy? Explain your answer. If you would support building new reactors, would your support remain unchanged if a nuclear reactor were to be installed close to where you live or work, say within 15 miles? If you do not support building new reactors, would you change your views if a new reactor were to be built in a desert or a similar location far from any inhabitants? Describe your reasoning.

40. The half-life of C-14 is 5730 years. This suggests that there is a 50% chance that any particular atom of C-14 will undergo radioactive decay after 5730 years have passed. It also suggests that if you have a sample of exactly four C-14 atoms, *two* of the four will decay after a period of 5730 years have passed, then only *one* of the remaining two will decay after another period of 5730 years have passed. Show that these two conclusions are compatible with each other.

Calculate

41. If a radioisotope has a half-life of 8 hours, does this mean that a sample of all of it will be gone in 16 hrs? Explain.

42. If one radioisotope has a half-life of 50 years and another has a half-life of 70 years, which one decays more quickly?

43. When a radioactive nucleus ejects an α particle, both the atomic number and the mass number decrease. By what quantity does each decrease?

44. When a radioactive nucleus ejects a β particle, only the atomic number or mass number changes. Which one? Does it increase or decrease? By what quantity?

45. What subatomic particle is lost by Po-210 as it is converted into Pb-206 in the final step of the decay sequence of U-238? Write the reaction for this step.

46. Into what element is an atom of O-15 converted when it emits a positron?

47. You have just examined a wooden utensil recovered from an ancient archeological site and have found that the ratio of C-12 to C-14 is 16 times the ratio of C-12 to C-14 measured in a branch just cut from a nearby tree. What, if anything, can you conclude about the age of the wooden utensil?

48. Fe-56, the most common isotope of iron, has one of the greatest mass defects and one of the highest binding energies of all atoms. Given that the experimentally measured atomic mass of this isotope of iron is 55.9349, calculate its mass defect. The atomic number of iron is 26.

49. The cleavage of a U-235 nucleus can occur in a variety of ways. For example when a U-235 nucleus absorbs one neutron, it can produce a Te-137 nucleus and a Zr-97 nucleus and release neutrons and energy in the process (but no α or β particles). Write the balanced nuclear equation for this reaction and determine the number of neutrons released. (Using a periodic table may help.)

50. U-235 can undergo the following fission reaction:

$$^{1}_{0}n + ^{235}_{92}U \rightarrow ^{139}_{54}Xe + ^{94}_{38}Sr + 3\,^{1}_{0}n$$

Answer the following using the masses provided below:

a. In this reaction, the reactants have a greater mass than the products. Calculate the difference in mass. This value represents the *lost mass* of the reaction, which is the source of the energy released by fission.

b. The lost mass of this reaction represents what percentage of the mass of the reactants?

Particle or nuclide	Mass (atomic mass units, u)
$^{1}_{0}n$	1.01
$^{235}_{92}U$	235.04
$^{139}_{54}Xe$	138.92
$^{94}_{38}Sr$	93.92

10

Energy from Electron Transfer

The landscape of salt flats shown here, located in Bolivia, contains the world's largest known reserves of lithium. This important element is an essential component of lithium ion batteries, which are widely used in portable electronic devices, including smartphones and tablets, and in vehicles, such as hybrid and electric cars. These batteries are useful in these applications because of their high-energy density, which means they generally

store more energy and are lighter than comparable batteries of equivalent size.

In this chapter we'll examine how the energy provided by all batteries, including those made from lithium, is created through oxidation and reduction reactions, two chemical processes that always occur in tandem. We'll see that a combination of oxidation and reduction, or redox for short, also underlie other important energy technologies, such as fuel cells and photovoltaic cells. In addition, a variety of everyday phenomena involve redox processes, including the whitening of clothes by bleach and the rusting of iron.

Noah Friedman-Rudovsky/The New York Times/Redux

CHAPTER OUTLINE

10.1 Oxidation and Reduction

LEARNING OBJECTIVES

1. **Distinguish** between oxidation and reduction and explain why these processes always occur together.
2. **Provide** examples of oxidation and reduction in everyday life.

We live and breathe in a world abundant in atmospheric oxygen, the diatomic molecule O_2. This molecule is absorbed through our lungs and transported throughout our bodies for the oxidation of the macronutrients in foods, in a process called **cellular respiration**. Oxygen also combines with fossil fuels in the **combustion** reactions that supply most of the energy needs of society. In this section, we'll explore oxidation reactions such as these. We will see that although the process of oxidation does not necessarily always involve oxygen, it is always accompanied by a related process, known as reduction.

Electron Transfer in Oxidation and Reduction

The terms **oxidation** and **reduction** were first proposed in the late 18th century by the French chemist Antoine Lavoisier. Through careful observations of the combustion of charcoal, the reactions of metals with air to form oxides, and human respiration and metabolism

> **oxidation** Loss of electrons.
>
> **reduction** Gain of electrons.

Chemistry InSight Defining oxidation and reduction • Figure 10.1

Oxidation is the loss of electrons and reduction is the gain of electrons. However, these processes may also be observed in other ways, such as in the transfer of oxygen and hydrogen atoms.

a. If you imagine a baseball as representing an electron, pitching can be likened to electron loss (oxidation), and catching can be seen as electron gain (reduction).

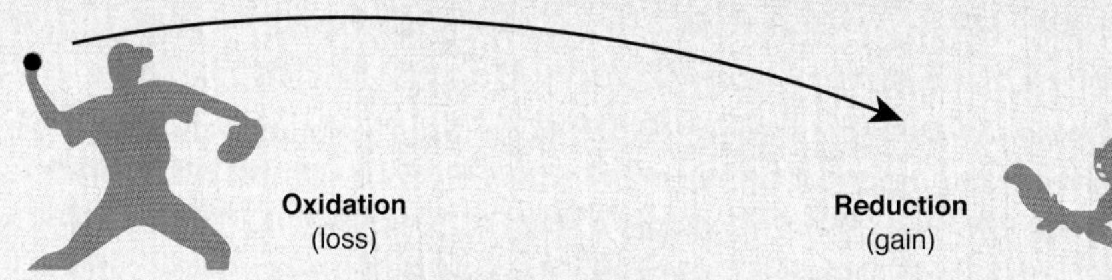

b. The formation of ionic compounds (discussed in Section 3.3) involves oxidation and reduction. For example, in the formation of lithium fluoride, lithium (Li) is oxidized and fluorine (F) is reduced.

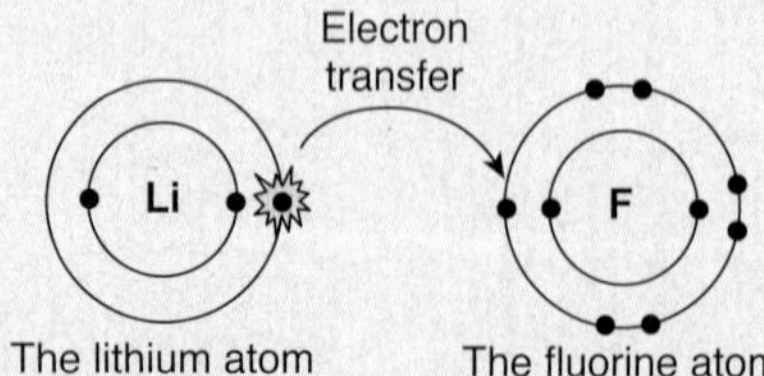

The lithium atom loses an electron and so is oxidized.

The fluorine atom gains an electron and so is reduced.

c. We can also observe oxidation in terms of loss of hydrogen or gain in oxygen. Reduction is the complementary process.

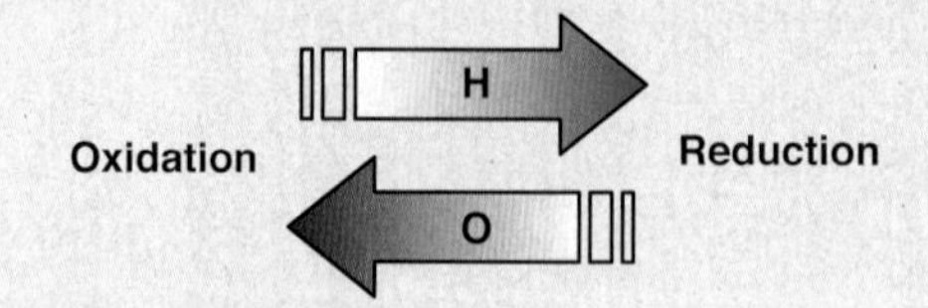

d. The combustion of hydrocarbons, such as the methane of natural gas, involves oxidation and reduction.

Chepko Danil Vitalevich/ Shutterstock

A natural gas stove.

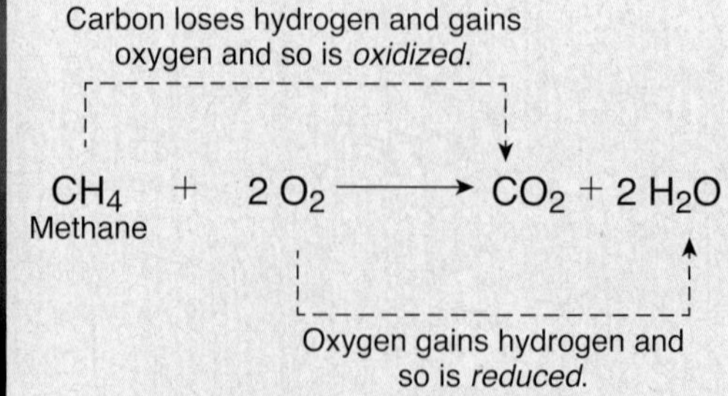

> **Ask Yourself**
>
> What ions are produced as a result of the oxidation and reduction shown in Figure **b**?

A simple redox demonstration • Figure 10.2

In this demonstration, the color of iodine disappears due to reduction by vitamin C and then reappears due to oxidation by bleach.

a. Vitamin C (ascorbic acid) reduces iodine (I_2), to form the colorless iodide ion (I^-).

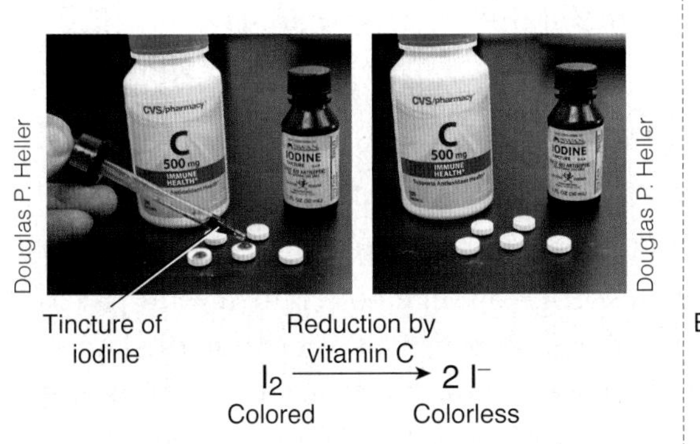

b. The liquid bleach added contains the hypochlorite anion (ClO^-), which oxidizes the iodide ion (I^-) back to iodine (I_2), restoring the color.

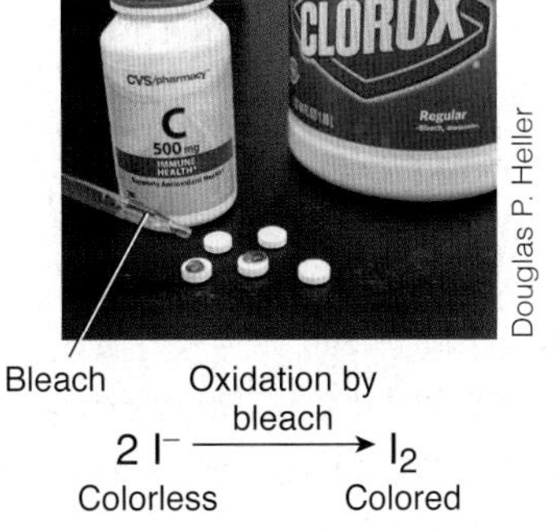

Adapted with permission of John Wiley & Sons, Inc. from Snyder, C., *The Extraordinary Chemistry of Ordinary Things*, Fourth Edition. Copyright 2003.

Think Critically
1. In **a**, does the vitamin C gain or lose electrons? Is the vitamin C oxidized or reduced?
2. In **b**, does the hypochlorite ion of the bleach become oxidized or reduced?

of foods, Lavoisier discovered that all of these processes require the oxygen that's present in the air. Furthermore, each of these oxidations is accompanied by a complementary process called reduction. Over time, other scientists came to recognize that these and other oxidation and reduction reactions all involve the transfer of electrons. **Figure 10.1** shows several different ways of describing oxidation and reduction.

For an easy way to remember what happens to electrons in oxidations and reductions, think of an OIL RIG. You'll know instantly that Oxidation Is Loss (of electrons), and Reduction Is Gain (again, of electrons). Furthermore, **oxidation and reduction always occur in tandem.** For a substance to be oxidized, another must be reduced.

A substance that causes another to be oxidized is an **oxidizing agent** or oxidant, and a substance that causes another to be reduced is a **reducing agent** or reductant. For example, atoms of most metals readily donate electrons, causing other chemical particles to accept them. Because of this tendency to lose electrons, most metals are good reducing agents. On the other hand, nonmetals, such as oxygen and chlorine, readily accept electrons, causing other substances to lose them. Because of this tendency to gain electrons, oxygen and chlorine are good oxidizing agents.

oxidizing agent
A substance that causes another to be oxidized.

reducing agent
A substance that causes another to be reduced.

Oxidation and Reduction in Everyday Life

Oxidation and reduction reactions, or **redox reactions**, are quite common within our everyday experience. You can familiarize yourself with redox reactions by carrying out the following demonstration involving common items. To do this you'll need:

- a vitamin C tablet;
- tincture of iodine (an antiseptic commonly sold in drugstores);
- household bleach.

Place a drop or two of iodine on the vitamin C tablet. Notice that the dark color of the iodine disappears as the iodine reacts with the vitamin C. Now add a few drops of bleach to the same spot. You'll see that the color of the iodine reappears. **Figure 10.2** helps explain these observations.

Antioxidants As shown in Figure 10.2, vitamin C is easily oxidized. In fact, vitamin C is so easily oxidized that it serves an important role in the body as an **antioxidant**. This class of compounds, which includes not only vitamin C, but also vitamin E, beta carotene, and other substances, helps protect cells within the body from oxidative damage.

antioxidant
A substance that readily undergoes oxidation, thereby preventing other compounds from being oxidized.

free radical A chemical species that has an odd number of electrons.

Within the body, antioxidants serve as protective agents against **free radicals**, which are highly reactive chemical species with odd numbers of electrons in their valence shells. Free radicals can damage molecules within the body, including DNA, by oxidizing them. As shown in **Figure 10.3**, antioxidants serve a protective role against free radicals by donating electrons to them. This reduces them, gives them even numbers of electrons in their valence shells, and thereby quenches their reactivity.

Some foods are known to be rich in antioxidants, including berries and other highly colored fruits and vegetables. Other notable dietary sources of antioxidants include green tea, red wine, and dark chocolate. Research suggests that consuming antioxidants may reduce the risk of certain diseases, such as cancer and heart disease. However, to reap these potential benefits, you need to consume antioxidant-rich foods rather than take these compounds in pill or supplement form. It is still not fully understood why antioxidant supplements do not appear to offer substantial health benefits. Clearly, more research is needed to better understand the role of these compounds in our bodies and in disease prevention.

In the food industry, antioxidants are also widely used as preservatives, helping to protect foods from the oxidation that naturally occurs upon exposure to the air. For example, a class of fats, present in certain types of foods—the unsaturated fats—can slowly react over time with oxygen in the air to impart a rancid taste to food. This oxidation can be suppressed by the addition of antioxidants, such as *alpha-tocopherols* (vitamin E derivatives). Other types of antioxidants used as food preservatives include sulfur-based compounds, such as *sulfites* (substances containing the SO_3^{2-} ion) and *sulfur dioxide* (SO_2). These compounds are commonly added to wine and dried fruits to help prevent discoloration.

You can observe the oxidative discoloring of foods and see how to inhibit this process by using:

- an apple, pear, banana, potato, or avocado as your test fruit or vegetable, and
- a lemon or lemon juice. (Other types of citrus juice work as well.)

Slice your test fruit or vegetable into several pieces and divide the slices into two groups. Squeeze the lemon or other citrus juice generously on one set of slices, but not on the other. Allow both sets to sit open in the air for a couple of hours. You'll notice that the surface of the untreated fruit turns brown, but the citrus-treated fruit does not discolor. The reason is that the slicing exposes cells on the surface of the fruits to atmospheric oxygen. Aided by enzymes within the fruit, the oxygen oxidizes compounds within the fruit to brown-colored by-products. This reaction is inhibited by the presence of citrus juice, which contains the antioxidants vitamin C (ascorbic acid) and citric acid. These two compounds readily undergo oxidation themselves, thereby inhibiting the oxidation of compounds on the exposed slices.

Oxidants Antioxidants are reducing agents, because they cause other chemical species to be reduced. Now we'll examine oxidizing agents, or oxidants for short. *Chlorine* (Cl_2), for example, is an oxidant that is commonly added

How antioxidants work • Figure 10.3

Free radicals initiate chain reactions of oxidations, damaging molecules and tissue. Antioxidants block the harmful effects of free radicals by reducing them.

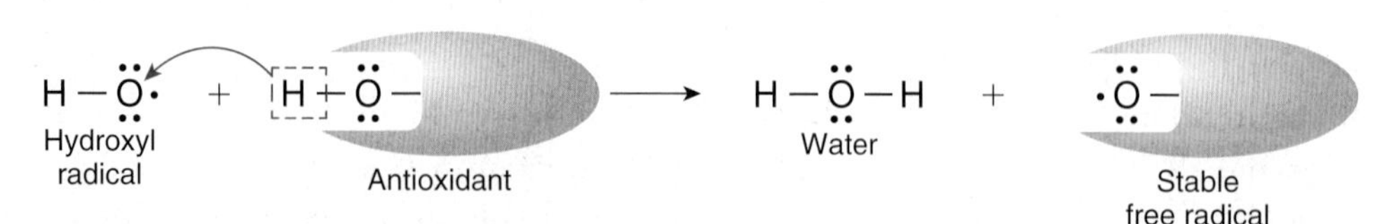

Free radicals, such as the hydroxyl radical, have unpaired electrons, making them highly reactive.

Antioxidants, reduce free radicals. A common method by which this occurs is shown here, in which a *hydrogen atom* (one proton and one electron) of the antioxidant is donated to the free radical.

The free radical has been quenched. In this case, a water molecule is formed.

The antioxidant molecule itself has been transformed into a free radical. However, this type of free radical is unusually stable due to the particular molecular structure of the antioxidant.

Ask Yourself

Vitamin E is one of a variety of antioxidants that reduce free radicals by the mechanism shown here. Use the Internet to find the structure of vitamin E and identify the hydrogen atom this molecule uses in this reduction. (*Hint:* The hydrogen atom used for this purpose is not bonded to carbon.)

Household oxidizing agents • Figure 10.4

Oxidizing agents, such as bleach and peroxide compounds, serve as disinfectants and whitening agents.

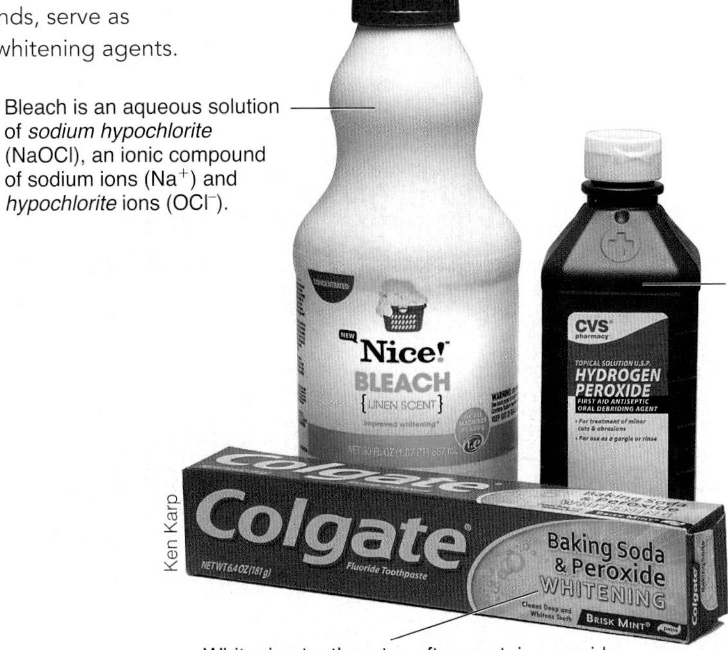

Bleach is an aqueous solution of *sodium hypochlorite* (NaOCl), an ionic compound of sodium ions (Na^+) and *hypochlorite* ions (OCl^-).
Hydrogen peroxide is an aqueous solution of the covalent compound, H_2O_2.
Whitening toothpastes often contain peroxide compounds, such as calcium peroxide (CaO_2), which help whiten teeth by oxidizing colored compounds in tooth enamel.
Nice!
BLEACH
CVS
HYDROGEN PEROXIDE
Colgate
Baking Soda & Peroxide
WHITENING
Ken Karp

Ask Yourself

A certain brand of toothpaste claims to whiten teeth with oxygen bubbles. What is the likely source of these oxygen bubbles?

to municipal water supplies and to swimming pools to disinfect the water. Chlorine kills a variety of bacteria through oxidative damage. When chlorine is added to water, it produces the *hypochlorite ion* (OCl^-), which is the same ion found in household bleach, an aqueous solution of *sodium hypochlorite* (NaOCl). Bleach and other common oxidizing agents, such as *hydrogen peroxide* (H_2O_2), have a variety of household uses (**Figure 10.4**).

Some resistant food stains on clothing, such as those produced by spaghetti sauce, can be removed if carefully treated with oxidizing agents, such as those contained in bleach or peroxide. The stains disappear when they are treated with oxidizing agents, because the rich colors of certain foods, such as the red color of tomatoes, are due to the presence of hydrocarbon molecules with alternating single and double bonds (**Figure 10.5**). Compounds such as these absorb some of the energy of visible light, giving rise to a colored appearance. Oxidizing agents can react with the carbon–carbon double bonds in these compounds, rendering the compounds colorless. [Note that you should always be careful in using bleach, even diluted, on colored clothes, as the bleach can also oxidize dyes in the fabrics, causing permanent white stains.]

Lycopene • Figure 10.5

Bleaching agents can react with compounds responsible for food stains, such as lycopene, which is the red pigment in tomatoes. Bleach and peroxide disrupt the alternating pattern of double bonds, which changes the structure and optical properties of this molecule, causing it to lose its ability to produce color.

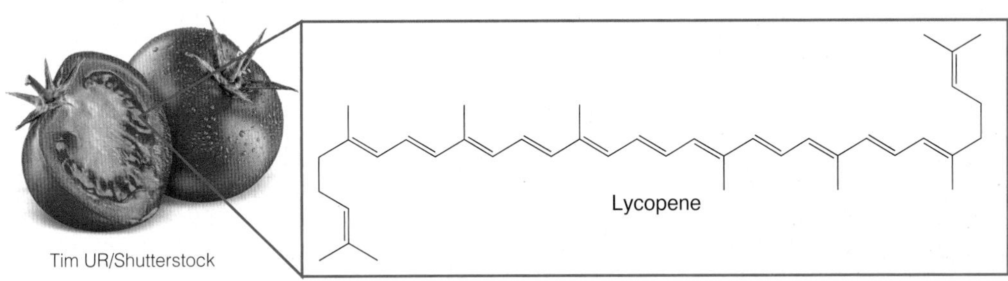

Glass photochromic lenses • Figure 10.6

Ultraviolet (UV) radiation, present in sunlight, initiates a redox reaction that causes glass photochromic lenses to darken. The absense of UV light causes these lenses to clear.

a. Lens darkening. In the presence of UV light, chloride ions reduce silver ions to neutral silver atoms, which impart a dark color.

Electron transfer initiated by ultraviolet light

$$Ag^{+} + :\ddot{\underset{\cdot\cdot}{Cl}}:^{-} \longrightarrow \dot{A}g + \cdot\ddot{\underset{\cdot\cdot}{Cl}}:$$

Silver ion + Chloride ion → Silver atom + Chloride atom

Neustockimages/iStockphoto

b. Lens clearing. The oxidation of silver atoms back to silver ions is mediated by Cu^{2+} ions present within the glass.

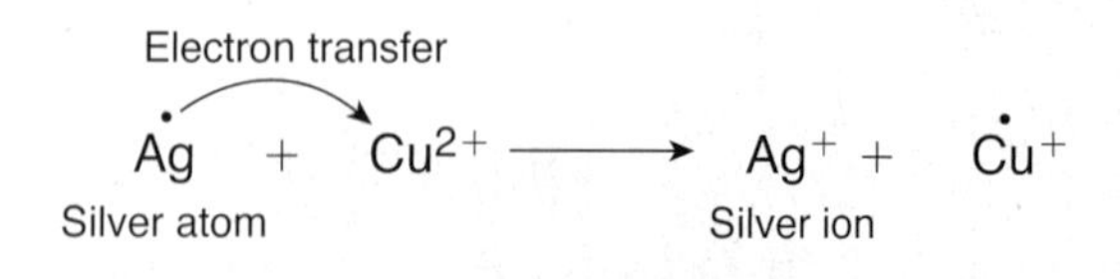

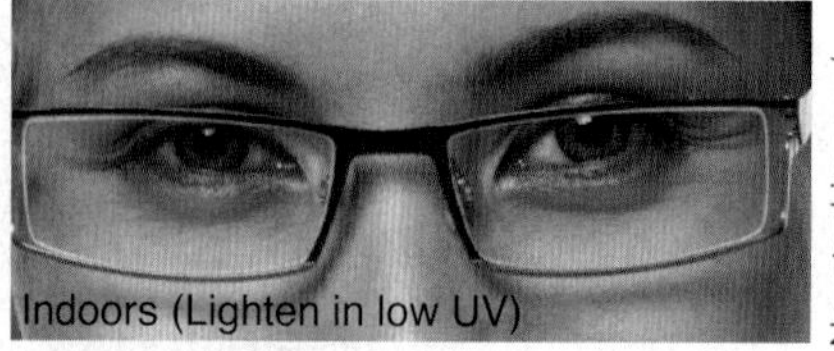

Neustockimages/iStockphoto

Think Critically 1. Identify the oxidizing and reducing agents in each of the reactions shown here.
2. Why does the silver ion shown have a single positive charge whereas the silver atom bears no charge?

Other Everyday Redox Reactions Redox reactions play an important role in other consumer products, such as batteries (a topic we'll discuss in Section 10.2). They are also involved in **photochromic** or **transition** sunglasses, which have lenses that automatically darken in sunlight and lighten in indoor settings. Glass lenses of this type contain very fine, embedded particles of silver chloride (AgCl), a compound composed of silver ions (Ag^{+}) and chloride ions (Cl^{-}). Ultraviolet (UV) radiation present in sunlight initiates a redox reaction between the silver and chloride ions, producing silver atoms. These atoms give a dark hue to the lenses. **Figure 10.6** illustrates this process.

Just as the oxygen of our atmosphere can oxidize some substances, other gases in the air can do so as well. For example, gaseous sulfur compounds can oxidize silver, causing tarnishing or discoloration of silver jewelry and other items made of silver (**Figure 10.7**).

The tarnishing of silver • Figure 10.7

Silver reacts with *hydrogen sulfide* (H_2S), a compound sometimes found in polluted air, to form *silver sulfide* (Ag_2S), the black compound responsible for tarnish.

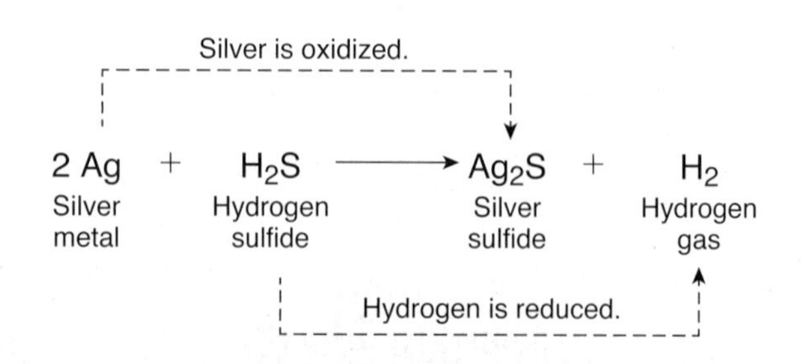

This silver kettle has developed tarnish, a thin coating of silver sulfide, but the spoon has been polished to restore the natural luster of silver.

James L Amos/Photo Researchers/Getty Images

Think Critically Redox reactions are sometimes characterized in terms of gain and loss of oxygen or other elements. **(a)** Based on the equation shown here, can you describe oxidation and reduction in terms of gain and loss, respectively, of an element other than oxygen? **(b)** If so, where does this element appear in the periodic table with respect to oxygen?

The **combustion** of fuels, such as wood and fossil fuels, involves redox reactions between the carbon compounds of these fuels and oxygen. **Cellular respiration** involves redox reactions between the macronutrients of foods (fats, carbohydrates, and proteins) and oxygen within living organisms. In both cases, carbon is oxidized to carbon dioxide (CO_2) and oxygen is reduced to water (H_2O). However, one key difference between these two processes is that combustion occurs at high temperatures, while cellular respiration occurs under much lower body temperatures, facilitated by biological catalysts called **enzymes**.

Like other macronutrients, ethanol (also known as grain alcohol) is also oxidized within the body with the aid of enzymes. After consumption of ethanol, the blood alcohol concentration initially rises as this substance is absorbed into the bloodstream. It then falls as the alcohol is metabolized by the body. Blood alcohol concentrations within the body can be determined with a breathalyzer (discussed in the chapter on solutions), a device that oxidizes the alcohol vapor present in exhaled breath to measure its concentration. This concentration of alcohol vapor in exhaled breath correlates with the blood alcohol concentration within the body.

Redox in the Environment and in Commercial Processes Although our scientific understanding of oxidation and reduction is a product of modern times, redox reactions with naturally occurring substances have been carried out since antiquity. In early times, tin, lead, copper, iron, and bronze (a copper alloy) were produced by chemically reducing the cations (positively charged ions) of metallic elements found in naturally occurring ores. This reduction converted the cations to the metals themselves.

A common reducing agent used for this conversion of cations to metals (and still used in the same process today) is carbon, the principal element of charcoal. The process of reducing metal ores, called **smelting**, may have been discovered serendipitously by prehistoric humans as a result of heating metal ores in ancient hearths. Within the hearth, the combination of heat from the fire and the carbon from the wood or charcoal fuel could have reduced ores to their corresponding metals. **Figure 10.8** illustrates a redox reaction involved in smelting.

> **smelting** A process for producing metals from their ores.

The production of metals, including copper, nickel, tin, iron, and lead, through the mining of their ores and subsequent smelting is enormously important to society, yet raises environmental concerns. The mining of metal ores, for example, can degrade landscapes and release pollutants to both land and water. In addition, smelting facilities are highly energy intensive. Yet it is difficult to imagine our modern world functioning without the metals produced through smelting.

Smelting produces metals through the reduction of their ores, while corrosion results from the oxidation of metals, especially through atmospheric effects. Corrosion of various kinds does billions of dollars of damage each year in the United States alone. Because of the variety and complexity of the chemical reactions that lead to corrosion, we consider only the single kind that affects us most directly—the rusting of metals. This process requires exposure to both oxygen and water. Iron, for example, rusts more rapidly in humid air than in dry air. Rusting produces metal oxides, which are chemical combinations of the metal and oxygen. Perhaps surprisingly,

Smelting iron • Figure 10.8

A worker assists in producing iron—a redox reaction.

$$2\,Fe_2O_3(s) + 3\,C(s) \xrightarrow{\text{heat}} 4\,Fe(s) + 3\,CO_2(g)$$

Iron oxide, a component of iron ore, containing Fe^{3+} ions — Carbon — Iron metal — Carbon dioxide

Think Critically In the reaction shown, identify which substance is oxidized and which is reduced.

To produce iron, iron ores are treated at high temperatures with reducing agents, such as coke, a carbon-rich material derived from charcoal.

James L. Stanfield/National Geographic Creative

WHAT A CHEMIST SEES

Rusting

Rusting is generally undesirable, enacting a significant economic toll to society. This type of corrosion gradually weakens structures made of iron or steel, such as buildings, bridges, and cars.

These rust spots not only look bad, but weaken the strength of the car as well. Cars sold today are often coated with thin films to protect the metal from exposure and hence offer improved corrosion resistance.

E. Pals/Shutterstock

The rusting of iron requires both oxygen and water and proceeds by a three-step process:

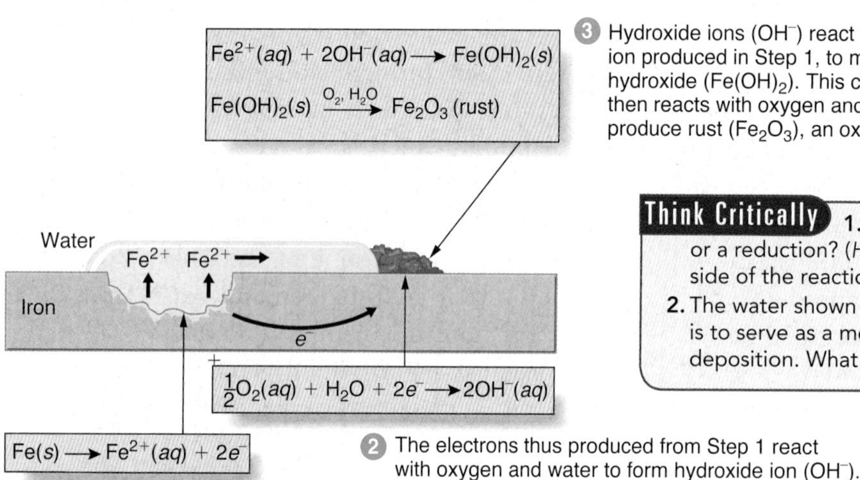

3 Hydroxide ions (OH^-) react with the Fe^{2+} ion produced in Step 1, to make iron hydroxide ($Fe(OH)_2$). This compound then reacts with oxygen and water to produce rust (Fe_2O_3), an oxide of iron.

2 The electrons thus produced from Step 1 react with oxygen and water to form hydroxide ion (OH^-).

1 Iron (Fe) is oxidized to an ionic form (Fe^{2+}), releasing electrons.

Think Critically

1. Is the reaction depicted in Step 2 an oxidation or a reduction? (*Hint*: Consider that electrons are shown on the left side of the reaction arrow.)
2. The water shown in the figure serves two purposes, one of which is to serve as a medium to carry iron ions (Fe^{2+}) to the site of rust deposition. What is water's other role in rusting?

the formation of these oxides can either accelerate continued corrosion or retard it, depending on the nature of the metal and the oxide it forms. Iron rusts rapidly because its oxides are granular and flaky. Once formed, the oxides separate easily from the remaining material, exposing fresh surfaces to additional corrosion. In *What a Chemist Sees* we explore the redox reactions involved in the rusting of iron.

CONCEPT CHECK 

1. **What** is the oxidizing agent, and what is the reducing agent when sodium metal (Na) and chlorine gas (Cl_2) combine to form sodium chloride (NaCl)?
2. **What** do combustion, cellular respiration, and rusting all have in common?

10.2 Redox Reactions and Electrical Current

LEARNING OBJECTIVES

1. **Define** oxidation states.
2. **Define** reduction and oxidation potentials.
3. **Explain** how redox reactions produce the energy of batteries.
4. **Describe** how electrolysis works and what its uses are.

In this section, we'll examine how redox reactions occuring within a battery produce electrical current. We'll then explore practical uses of the opposite process, known as electrolysis, in which an electrical current is used to generate redox reactions.

Oxidation States

In studying redox reactions, we often find it useful to assign an **oxidation state** (or number) to each type of atom involved, to help recognize which atoms are being reduced and which are being oxidized, and to what extent. Oxidation states of the type we discuss here can be either zero or a positive integer (such as +1, +2, +3, etc.). All elements in their free, neutral states have an oxidation number of zero. For example, atoms within the elemental forms of hydrogen (H_2), nitrogen (N_2), oxygen (O_2), and chlorine (Cl_2) are assigned an oxidation number of zero. Similarly, all metals in their neutral, elemental states are also assigned an oxidation number of zero. The zero is sometimes shown as a superscript to the right of the chemical symbol, such as in the metals, copper (Cu^0), iron (Fe^0), zinc (Zn^0), and lead (Pb^0). When a metal becomes ionized, its oxidation number is simply the same number as its ionic charge. For example, the atoms of Cu^{2+} exist at an oxidation state of 2+, and those of Fe^{3+} are at an oxidation state of 3+.

oxidation state A number describing the degree of oxidation of an atom.

Putting Oxidation and Reduction to Work

When you switch on a flashlight or turn on any portable electronic device, the battery within the device becomes part of an **electrical circuit**, which is simply the path by which electrons flow through the device. This flow or movement of electrons through a circuit constitutes an **electric current**, which powers the device. What actually causes the electrons to flow, and in which direction they flow through the circuit, are questions we address here.

Recall that oxidation is electron loss and reduction is electron gain. When a battery is connected to a circuit, an oxidation reaction within one part of the battery releases electrons to the circuit, while at the same time, a reduction reaction occurs within another part of the battery as it accepts electrons from the circuit. Electrons leave the battery and enter the circuit at the battery's **anode** or negative terminal. These electrons return from the circuit and enter the battery at its **cathode** or positive terminal.

anode A battery's negative terminal.

cathode A battery's positive terminal.

One of the earliest commercial batteries, invented in 1836 by the British chemist John Frederic Daniell, is known as a **Daniell cell**. It provided power to the earliest telegraph machines, precursors to telephones, and is an example of an **electrochemical or galvanic cell**. You can create a version of a Daniell cell by using two beakers: one containing an aqueous solution of *zinc sulfate* and the other containng an aqueous solution of *copper sulfate*. A strip of zinc metal is partly submerged in the zinc sulfate solution, and a strip of copper metal is partly submerged in the copper sulfate solution.

galvanic cell A device that creates electrical energy from a chemical reaction.

Zinc sulfate ($ZnSO_4$) and copper sulfate ($CuSO_4$) are **electrolytes**, which means they produce ions when they dissolve in water. $ZnSO_4$ produces Zn^{2+} cations and SO_4^{2-} anions; $CuSO_4$ generates Cu^{2+} cations and SO_4^{2-} anions. [The sulfate anion, SO_4^{2-}, is an example of a polyatomic ion, discussed in Section 3.4.] The two solutions are joined by a **salt bridge** that allows ions to diffuse from each beaker into the other. This bridge need be no more than a strip of cloth soaked in a solution of sodium chloride, with its ends dipping into the two solutions. (A semipermeable glass frit connecting the two beakers also serves this function.) The battery circuit is completed by

PROCESS DIAGRAM

A Daniell cell • Figure 10.9

This apparatus is modeled after the Daniell cell, one of the earliest batteries invented.

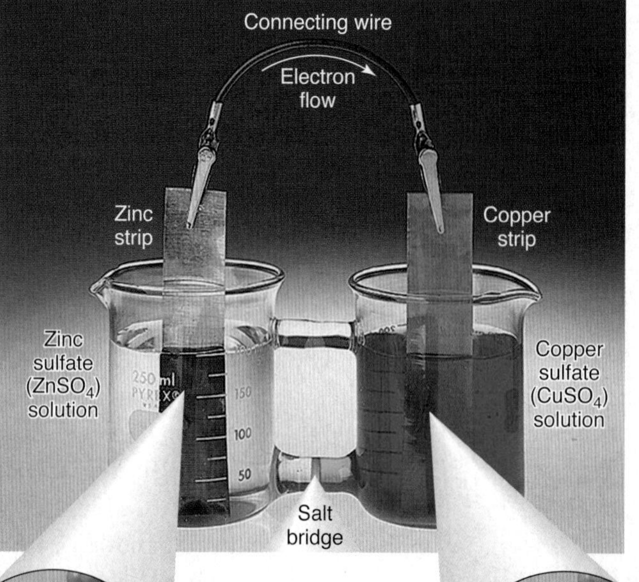

1 Zn atoms in the zinc strip become oxidized, each losing two electrons and releasing a Zn^{2+} ion into the solution. This oxidation is represented as:

$$Zn^0 \longrightarrow Zn^{2+} + 2e^-$$

2 The electrons thus released travel upwards through the zinc strip and across the connecting wire to the copper strip.

3 Cu^{2+} ions in the copper sulfate solution become *reduced*, each combining with two electrons which were released from the zinc oxidation and traveled to the copper strip. Their combination with a Cu^{2+} cation produces a Cu atom. This reduction is represented as:

$$Cu^{2+} + 2e^- \longrightarrow Cu^0$$

4 The newly formed copper atoms deposit on the copper strip.

2+ 2e⁻ 2+ 2+

2+ 2e⁻ 2+ 2+

Ask Yourself

1. Which metal is the anode and which metal is the cathode in the Daniell cell?
2. Refer to a periodic table to identify the number of protons and electrons in each of the following: **(a)** Zn^0; **(b)** Zn^{2+}; **(c)** Cu^0; **(d)** Cu^{2+}

connecting the two metal strips to each other with a copper wire, as shown in **Figure 10.9**.

Initially, the zinc and the copper strips appear clean and bright and the copper sulfate solution shows the intense blue color of its Cu^{2+} ions. Slowly, however, as the Daniell cell produces current, the deep blue color of the copper sulfate solution fades and the copper plate darkens and grows thicker. In the other beaker of the cell, the zinc strip erodes and eventually decays. Why do these changes occur?

As the Daniell cell produces an electric current, each of the zinc atoms of the zinc strip loses two electrons and becomes a zinc ion (Zn^{2+}), which dissolves in the solution. This explains why the zinc strip slowly erodes. In addition, each of the blue copper ions (Cu^{2+}), which are also soluble in water, gains the two electrons lost by each zinc atom. In this way the copper ions turn into particles of copper metal, which apprear as rust-colored or black granules on the copper strip. The blue color of the solution disappears because the copper ions (the source of the color) are converted into copper atoms.

As with all batteries, the redox reactions of the Daniell cell produce a flow of electrons. Electrons are released at the **anode**, the site of **oxidation**, and travel to the **cathode**, the site of **reduction**.

In the case of the Daniell cell, the sulfate anions (SO_4^{2-}) of both solutions remain unchanged; they don't take part in the redox reacton in any way. However, as the atoms of the zinc strip lose their electrons and enter the zinc sulfate soution as zinc cations (Zn^{2+}), a **surplus** of cations develops in this solution. At the same time, a **deficit** of cations develops in the copper sulfate solution, as copper cations (Cu^{2+}) gain electrons and are reduced to copper atoms, which leave this solution to become part of the metallic copper strip.

Since a surplus of cations (positively charged ions) develops in the zinc sulfate solution and a deficit of cations (again, positively charged ions) develops in the copper sulfate solution, there has to be a way to reestablish an electrical balance. That's where the salt bridge comes in. It allows cations to travel away from the surplus of positive charge in the zinc sulfate solution and toward the deficit of positive charge in the copper sulfate solution. This charge imbalance is also corrected by the passage of anions through the salt bridge in the opposite direction. As these migrations occur, the movement of ions through the salt bridge maintains electrical neutrality in both solutions. The cell continues to produce an electrical current as long as there are any zinc atoms left to be oxidized and copper ions left to be reduced.

KNOW BEFORE YOU GO

1. If the salt bridge in our Daniel cell were a strip of cloth that had been soaked in salt water (sodium chloride solution), would you expect the sodium and chloride ions of the bridge to migrate across the cloth strip as the cell generates an electrical current? If your answer is yes, in which direction would the sodium ions move and in which direction would the chloride ions move? If your answer is no, explain why not.

Electrical Voltage: Putting Pressure on an Electron If you were to remove the connecting wire of the Daniell cell and insert a **voltmeter** (an instrument that measures voltage) in its place, you would find that the meter records a little under 1.1 volts. Understanding just what this voltage represents leads to an understanding of how batteries operate.

The **volt** is a unit of electrical potential energy, which represents the tendency of electrons to move from one point on the circuit to another. The greater the voltage, the greater the pressure that moves the electron through the circuit. It's a bit like water pressure. The greater the water pressure in a pipe, the greater the force that moves the water along.

volt A measure of electrical potential.

The word "volt" honors Alessandro Volta, an Italian physicist who published a description of the world's first electrical battery, called the Voltaic pile, in 1800. Volta's pile consisted of a series of disks made of (1) silver, (2) paper moistened with a salt solution, and (3) zinc. This trio was repeated over and over to form a tall pile. (In later versions, copper successfully replaced the silver.)

The volt is a unit of electrical potential, while the **ampere** (which is often shortened simply to amp) measures the rate of flow of an electrical current, in much the same way that a unit like gallons/minute measures the rate of flow of water. As amperage increases, the number of electrons traveling through a circuit during any particular period of time also increases. André-Marie Ampère, the French physicist for whom the unit is named, was a contemporary of Volta and, like the Italian physicist, is remembered for his pioneering work in electricity and magnetism. **Figure 10.10** uses waterfalls as visual analogies to describe volts and amps.

ampere A measure of the rate of flow of an electrical current.

Under normal conditions, it takes a combination of a high voltage and a high amperage, such as we might find in a lightning bolt, to pose a hazard to humans. The voltage of ordinary consumer batteries, though, is too low

Volts and amps • Figure 10.10

Voltage, a measure of potential energy, can be likened to the height of a waterfall. By analogy, amperage, or electrical current, is similar to the rate of water flow.

a. Niagara Falls is a relatively low waterfall, but has the highest flow rate of any known waterfall, in part due to its width. This is analogous to low voltage and high amperage.

Glen Allison/Photodisc/Getty Images

b. Angel Falls, in Venezuela, is the world's highest waterfall but has a very low water flow rate due to its narrow width. We can liken this scenario to high voltage and low amperage.

Kevin Schafer/The Image Bank/Getty Images

to do us any harm. An automobile battery, for example, can deliver a current measuring in hundreds of amperes, but at such a low potential, 12 volts, that it simply can't cause us serious injury under ordinary conditions. A high voltage that delivers an insignificant number of amperes is equally harmless to people. The spark you produce as you walk across a carpet on a dry day and touch another person or a light switch carries thousands of volts, but its infinitesimal current can't hurt us. (Note that household electricity, as in a wall socket, has enough voltage and current to be extemely dangerous. You should always be very careful with household appliances and outlets.)

Battery Voltage Returning to our example of the Daniell cell, we conclude that the electrons that pass through the voltmeter with an electrical potential of 1.1 volts must be moving from the zinc strip to the copper strip, rather than in the reverse direction. To take our conclusion one step further, there must be a greater potential for electrons to leave zinc atoms than there is for electrons to leave copper atoms. Similarly, the Cu^{2+} cations must have a greater potential for acquiring two electrons than the Zn^{2+} cations have.

We've observed indirectly, then, that copper cations, which acquire electrons spontaneously in the cell, have a greater electrical potential for reduction, or a greater **reduction potential**, than do the zinc cations. Similarly, the zinc atoms, which lose electrons spontaneously to the copper cations, have a greater electrical potential for oxidation, or a greater **oxidation potential**, than the copper atoms. We can measure these potentials for the gain and loss of electrons as electrical voltages, much as we measure the potential for the movement of water from one point to another in terms of water pressures.

We can't isolate an individual oxidation or reduction reaction from the redox combination, but we are able to measure the electrochemical potential of any one of these individual **half-cell reactions** relative to a universally accepted standard reduction reaction: the reduction of two hydrogen ions to a hydrogen molecule, shown here.

half-cell reaction Either the reduction or oxidation reaction within an electrochemical cell.

Reduction of hydrogen ions: $2\ H^+ + 2\ e^- \rightarrow H_2$
Standard reduction reaction

In this reduction of hydrogen ions, the addition of two electrons, one to each of two protons, produces two hydrogen atoms that combine with each other through the sharing of these two added electrons to form a diatomic hydrogen molecule, H_2. If we now define the electrical potential of this particular reduction (arbitrarily) as exactly zero volts, we can measure all other half-cell potentials with respect to it. These measurements, standardized at 25°C and at specific concentrations of the ions, produce a series of **standard reduction potentials** (**Table 10.1**) that are universally applicable.

The numerical values of Table 10.1 express the electrical potential for the gain or loss of electrons from the chemical particles shown in the table, all in units of volts. The more positive the standard reduction potential, the greater the driving force for the ion, atom, or molecule to acquire electrons and for the reduction to occur. Similarly, the less positive (or more negative) the standard reduction potential, the smaller the driving force for the reactant to acquire an electron and to be reduced.

standard reduction potential A standard measure of how readily a chemical particle accepts electrons.

In other words, any chemical with a large, positive standard reduction potential has a strong tendency to acquire electrons from some other substance (that's capable of releasing electrons) and to oxidize it. As a result, a chemical with a large, positive standard reduction potential is itself a strong oxidizing agent, readily capable of oxidizing (removing electrons from) many other substances. To summarize, **the more positive the standard reduction potential of a chemical, the greater its tendency to behave as an oxidizing agent.**

KNOW BEFORE YOU GO

2. Why is a chemical species that is more easily reduced considered a stronger oxidizing agent?
3. (a) Which element, atom, molecule, or ion in Table 10.1 is the strongest oxidizing agent?

   (b) Which chemical species in Table 10.1 is the weakest oxidizing agent?

The reverse of each of the reductions in Table 10.1 is an oxidation. Reversing any of the reduction half-cells, then, gives us an oxidation half-cell. In reversing the direction of the chemical equation, we must also reverse the sign of the voltage associated with it. For example, the +2.87 volts for the reduction of molecular fluorine (F_2) becomes −2.87 volts for the oxidation potential of the fluoride anion (F^-).

Oxidation of fluoride ion: $2\ F^- \rightarrow F_2 + 2e^-\ (-2.87\ V)$

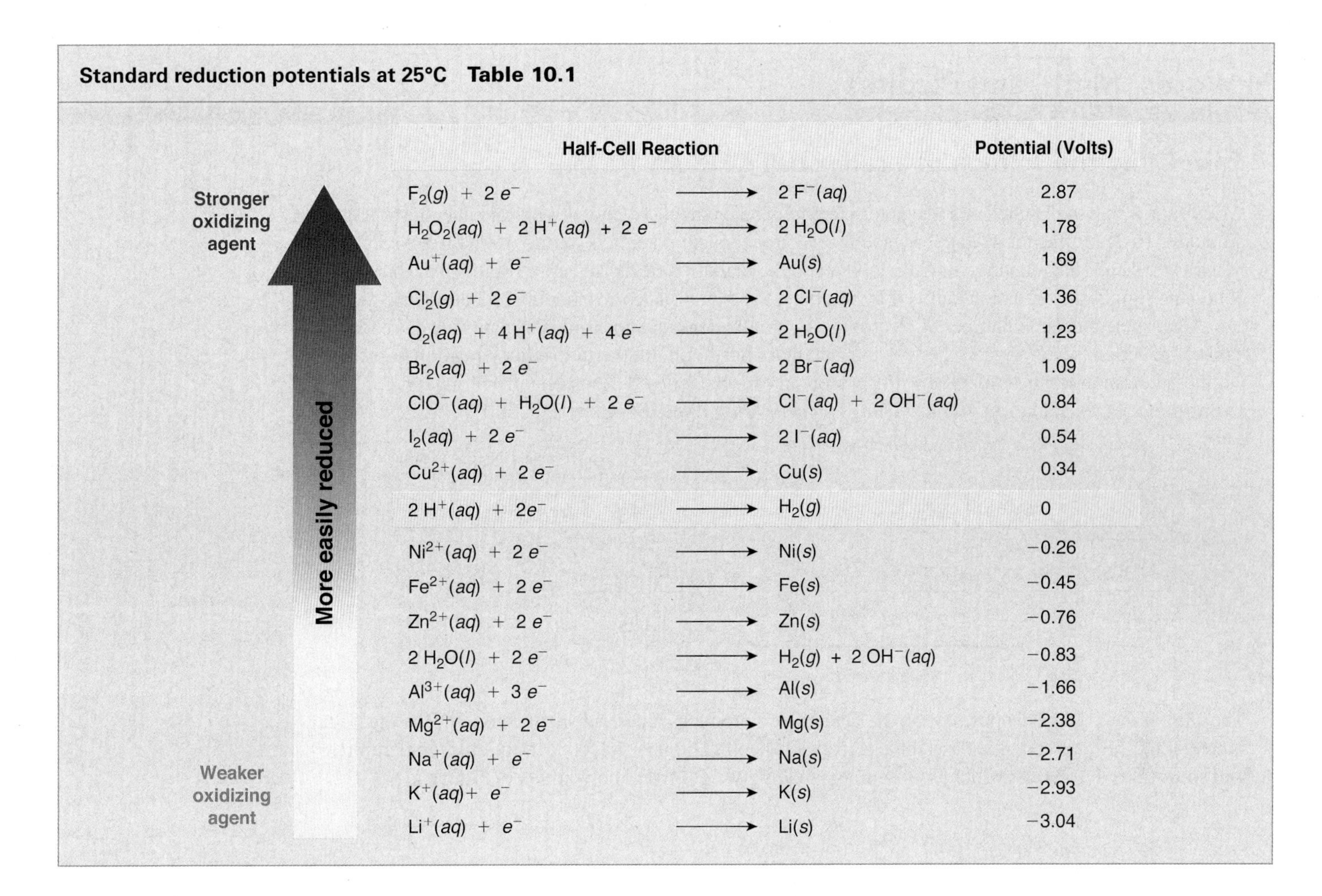

Standard reduction potentials at 25°C Table 10.1

Half-Cell Reaction			Potential (Volts)
$F_2(g) + 2e^-$	$\longrightarrow$	$2F^-(aq)$	2.87
$H_2O_2(aq) + 2H^+(aq) + 2e^-$	$\longrightarrow$	$2H_2O(l)$	1.78
$Au^+(aq) + e^-$	$\longrightarrow$	$Au(s)$	1.69
$Cl_2(g) + 2e^-$	$\longrightarrow$	$2Cl^-(aq)$	1.36
$O_2(aq) + 4H^+(aq) + 4e^-$	$\longrightarrow$	$2H_2O(l)$	1.23
$Br_2(aq) + 2e^-$	$\longrightarrow$	$2Br^-(aq)$	1.09
$ClO^-(aq) + H_2O(l) + 2e^-$	$\longrightarrow$	$Cl^-(aq) + 2OH^-(aq)$	0.84
$I_2(aq) + 2e^-$	$\longrightarrow$	$2I^-(aq)$	0.54
$Cu^{2+}(aq) + 2e^-$	$\longrightarrow$	$Cu(s)$	0.34
$2H^+(aq) + 2e^-$	$\longrightarrow$	$H_2(g)$	0
$Ni^{2+}(aq) + 2e^-$	$\longrightarrow$	$Ni(s)$	−0.26
$Fe^{2+}(aq) + 2e^-$	$\longrightarrow$	$Fe(s)$	−0.45
$Zn^{2+}(aq) + 2e^-$	$\longrightarrow$	$Zn(s)$	−0.76
$2H_2O(l) + 2e^-$	$\longrightarrow$	$H_2(g) + 2OH^-(aq)$	−0.83
$Al^{3+}(aq) + 3e^-$	$\longrightarrow$	$Al(s)$	−1.66
$Mg^{2+}(aq) + 2e^-$	$\longrightarrow$	$Mg(s)$	−2.38
$Na^+(aq) + e^-$	$\longrightarrow$	$Na(s)$	−2.71
$K^+(aq) + e^-$	$\longrightarrow$	$K(s)$	−2.93
$Li^+(aq) + e^-$	$\longrightarrow$	$Li(s)$	−3.04

In parallel to standard reduction potentials, this relatively large, negative number indicates that there's very little tendency either for a fluoride anion to release its acquired electron or for two fluoride anions to be oxidized to F_2. The fluoride anion is such an extremely weak reducing agent that it's not normally considered to be a reducing agent at all. After all, once a powerful oxidizing agent acquires an electron from another substance, it's hardly likely to give up that acquired electron easily. Because of this, it can't be much of a reducing agent once it has secured the electron.

KNOW BEFORE YOU GO

4. What is the most powerful reducing agent in Table 10.1? Explain.

We can now use our understanding of redox reactions to see just how and why batteries work. Batteries work because electrons flow spontaneously from reducing agents to oxidizing agents. When we turn on a battery-powered electrical device, we provide a circuit for the electrons to follow as they move from the battery's reducing agent to its oxidizing agent. In flowing through that circuit, the electrons provide the electrical current that runs the device. What's required is **a redox reaction that takes place spontaneously** within the battery and causes the battery to send its electrons out from the oxidation half-cell, through the electrical circuit, and into the reduction half-cell.

To determine whether any particular redox reaction occurs spontaneously, simply add the voltages of its two component half-cells: the reduction half-cell and the oxidation half-cell. If the resulting sum of voltages is positive, the two half-cell reactions can occur spontaneously with the release of energy (see **Figure 10.11** on the next page.) If the sum is negative, the redox reaction cannot occur spontaneously and no current flows. Nevertheless, we can cause a reaction with a negative redox potential to take place by adding electricity to the cell from an external source. We'll explore this process, called **electrolysis**, shortly.

KNOW BEFORE YOU GO

5. Use Table 10.1 to calculate the voltage expected from the redox reaction involving the reduction of iodine (I_2) and the oxidation of zinc (Zn). Write the net redox equation for this reaction.

Calculating the voltage of a redox cell • Figure 10.11

We can show that the Daniell cell generates electricity spontaneously, and we can calculate the voltage it produces. To determine the voltage, we add the electrochemical potentials for the two half-cell reactions of the Danell cell: the reduction of Cu^{2+} to Cu^0 and the oxidation of Zn^0 to Zn^{2+}. (Remember that we obtain the sign of the oxidation half-cell by reversing the sign of its corresponding reduction half-cell.)

Also, since the same number of electrons—two in this case—appear on alternate sides of each half-cell reaction, when we add the two half-cell reactions to obtain the **net redox equation**, we cancel out these common terms. (Adapted with permission of John Wiley & Sons, Inc. from Snyder, C., *The Extraordinary Chemistry of Ordinary Things, Fourth Edition*, p. 264. Copyright 2003.)

We use the value from Table 10.1 because this involves a reduction.

Reduction: $Cu^{2+} + \cancel{2e^-} \longrightarrow Cu^0$ $\quad + 0.34V$

We use the opposite sign of the value from Table 10.1 because this involves an oxidation.

Oxidation: $Zn^0 \longrightarrow Zn^{2+} + \cancel{2e^-}$ $\quad + 0.76V$

Redox sum: $Cu^{2+} + Zn^0 \longrightarrow Cu^0 + Zn^{2+}$ $\quad + 1.10V$

Net redox equation

The sum of the voltages is positive, so the Daniell cell produces a current spontaneously. The electrical potential the cell delivers if we assemble it to exact specifications of ion concentrations, temperature, and so forth, is 1.10 volts, which is close to the reading we get from the voltmeter (**Figure a**).

a. Measuring the voltage of a Daniell cell

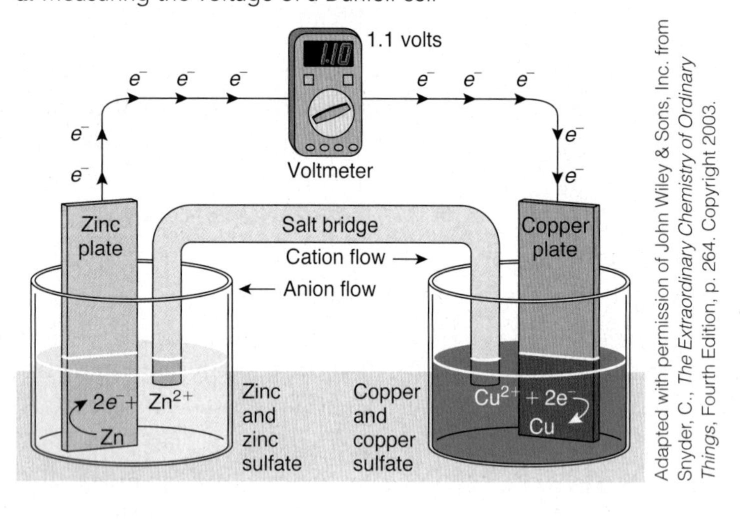

Battery Design

Almost anywhere you go, you can find batteries that power flashlights, watches, cameras, smartphones, tablets, and laptops, as well as start our cars. These compact, transportable storehouses of electricity are available in a variety of forms and sizes, and include nonrechargeable and rechargeable types. A common nonrechargeable cell is the alkaline battery, used in flashlights, toys, remote controls, and the like. Rechargeable batteries include the lead–acid batteries of cars, lithium–ion batteries in portable electronic devices, and the nickel–cadmium batteries in rechargeable power tools. In this section we'll examine how alkaline, lithium–ion, and lead–acid car batteries work.

Alkaline Batteries The common alkaline battery gets its name from the use of the basic or *alkaline* substance, potassium hydroxide (KOH). A solution of this compound serves as the battery's **electrolyte**, allowing the flow of ions within the battery. Although more expensive than standard *zinc–carbon* flashlight batteries, alkaline batteries are generally favored due to their significantly longer life. Alkaline cells now account for over 80% of the consumer batteries sold in the United States.

The two half-reactions and the net redox equation of the alkaline battery are

When this battery is connected to a circuit, the zinc (Zn^0) of the anode becomes oxidized to zinc oxide (ZnO), an ionic substance containing Zn^{2+} ions. Meanwhile, the *managanese dioxide* (MnO_2) of the cathode—an ionic substance containing Mn^{4+} ions—is reduced to Mn_2O_3, an ionic substance containing Mn^{3+} ions. The interior of an alkaline battery looks much like what is shown in **Figure 10.12**.

Lead–Acid Batteries Lead–acid batteries are the kind of battery found most often in cars and similar vehicles. This battery is relatively inexpensive, can be discharged and recharged repeatedly, lasts from about three to five

Reduction (cathode): $2\ MnO_2 + \cancel{H_2O} + \cancel{2e^-} \longrightarrow Mn_2O_3 + \cancel{2\ OH^-}$

Oxidation (anode): $Zn^0 + \cancel{2\ OH^-} \longrightarrow ZnO + \cancel{H_2O} + \cancel{2e^-}$

Net redox equation: $Zn^0 + 2\ MnO_2 \longrightarrow ZnO + Mn_2O_3$

In adding these two half-reactions to show the net redox equation, we cancel common terms, in this case, the waters (H_2O), hydroxide ions (OH^-) and electrons (e^-).

An alkaline battery • Figure 10.12

Alkaline batteries use manganese compounds and zinc to generate power. This design generates about 1.5 volts.

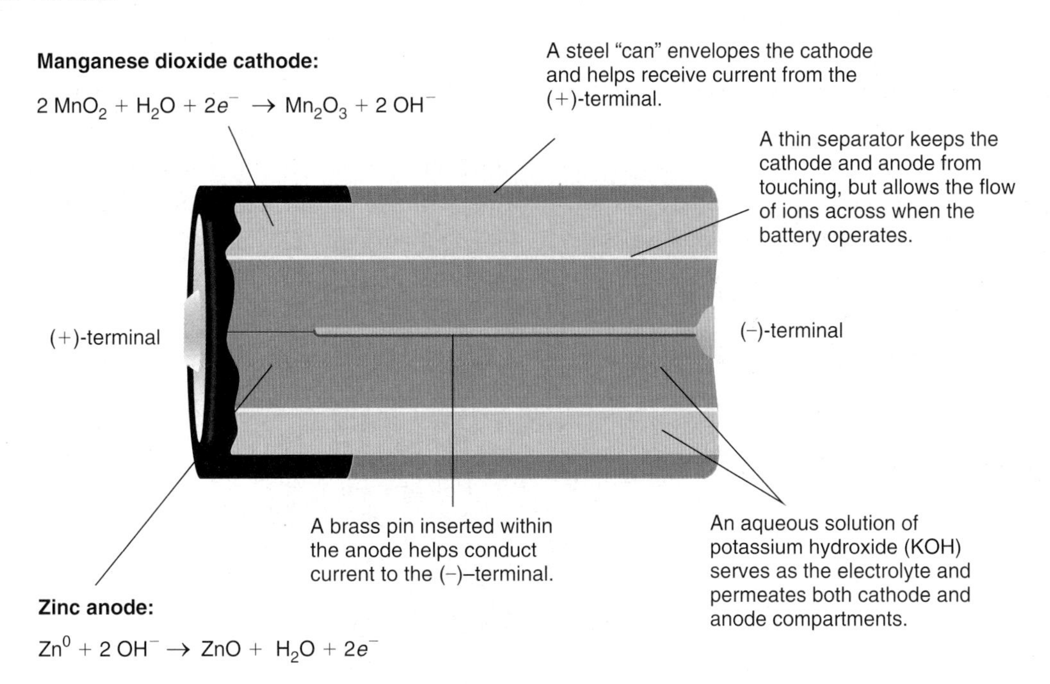

Manganese dioxide cathode:
$2\ MnO_2 + H_2O + 2e^- \rightarrow Mn_2O_3 + 2\ OH^-$
A steel "can" envelopes the cathode and helps receive current from the (+)-terminal.
A thin separator keeps the cathode and anode from touching, but allows the flow of ions across when the battery operates.
(+)-terminal
(–)-terminal
A brass pin inserted within the anode helps conduct current to the (–)–terminal.
An aqueous solution of potassium hydroxide (KOH) serves as the electrolyte and permeates both cathode and anode compartments.
Zinc anode:
$Zn^0 + 2\ OH^- \rightarrow ZnO + H_2O + 2e^-$

Think Critically When the battery is connected to a circuit:
(a) Are electrons released from the zinc anode or the manganese dioxide cathode?
(b) From which terminal do the electrons enter the circuit?

years (depending on its construction and use), and is small enough to be installed in cars and trucks. If you've ever tried to lift or move one of these batteries, you know that they are heavy for their size. This is in part because they contain lead, a dense metal.

Inside this battery, two sets of plates stand immersed in an aqueous solution of sulfuric acid. One set, made of a spongy form of metallic lead, serves as the anode; the cathode consists of plates of lead dioxide, PbO_2. The two half-reactions and the net redox equation of the lead–acid car battery are

contains six of these cells, each of which produces 2.0 volts. These six 2-volt cells are connected within the battery compartment in *series*, head-to-tail in a sense, as in a flashlight. This combination produces a total of 12 volts (**Figure 10.13**).

The lead–acid battery offers one great advantage over the common flashlight battery: It can be recharged. As long as the engine is running, whether you're driving the car or simply letting it idle at a stoplight or in neutral, the electric current produced by the generator or the alternator restores the electrochemical energy of the battery by

$$\text{Reduction (cathode):} \quad PbO_2 + H_2SO_4 + \cancel{2H^+} + \cancel{2e^-} \longrightarrow PbSO_4 + 2\ H_2O \quad (+1.68V)$$

$$\text{Oxidation (anode):} \quad Pb^0 + H_2SO_4 \longrightarrow PbSO_4 + \cancel{2H^+} + \cancel{2e^-} \ (+0.36V)$$

$$\text{Net redox equation (discharge):} \quad Pb^0 + PbO_2 + 2\ H_2SO_4 \longrightarrow 2\ PbSO_4 + 2\ H_2O \quad (+2.04V)$$

In the reduction half-cell, or cathode, the lead dioxide (PbO_2) reacts with sulfuric acid (H_2SO_4) and two protons (H^+) to form lead sulfate ($PbSO_4$) and water. In this case, the Pb^{4+} ions of the lead dioxide pick up two electrons and are reduced from an oxidation state of 4+ (in the PbO_2) to 2+ (in the $PbSO_4$).

In the oxidation half-cell, or anode, the spongy lead (Pb^0) reacts with sulfuric acid (H_2SO_4), producing lead sulfate ($PbSO_4$) and releasing two protons (H^+) and two electrons in the process. The lead is oxidized, from an oxidation state of zero (Pb^0) to an oxidation state of 2+ (in the $PbSO_4$).

The overall standard voltage for the redox equation of the lead–acid cell is +2.04 volts, which we can round off to +2.0 volts for convenience. A typical car battery

reversing the redox reaction. The current enters the battery and reconverts the lead sulfate and water into sulfuric acid, spongy lead metal, and lead dioxide, according to the following equation, which represents the reverse of the lead–acid battery discharge equation we just saw.

Net redox equation (recharge):

$$2\ PbSO_4 + 2\ H_2O \rightarrow Pb^0 + PbO_2 + 2\ H_2SO_4 \ (-2.04V)$$

Again, we can round this voltage to −2.0 volts. Notice that this negative potential tells us that a battery won't recharge itself spontaneously. (Common sense tells us the same thing.) To carry out this redox reaction, electrical energy must be put back into the battery in the form of electical current. This current is supplied by the car's alternator or generator.

A cell of the lead-acid battery • Figure 10.13

When you start a car, the battery jolts the engine into action by a combination of oxidation of the lead metal (Pb) of the anode and reduction of the lead dioxide (PbO_2) of the cathode.

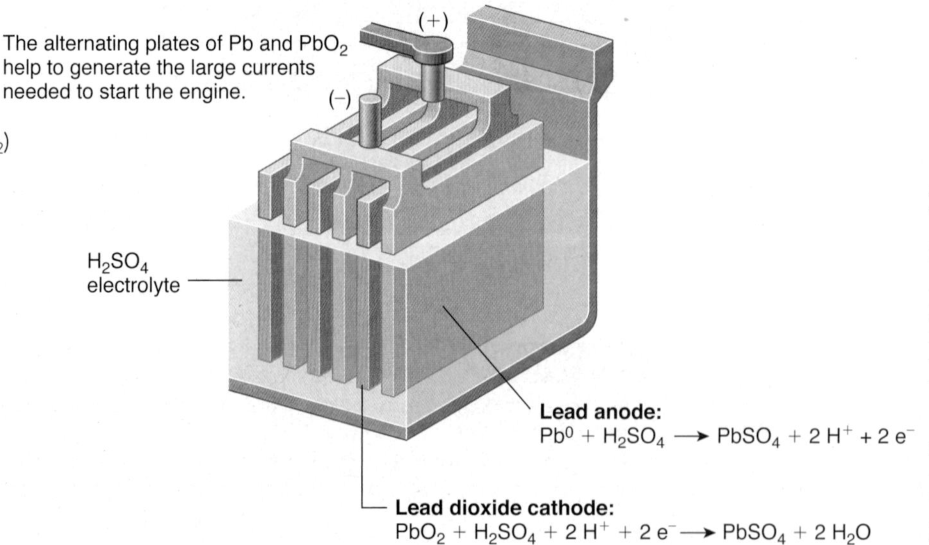

A lithium–ion battery • Figure 10.14

This type of battery contains an anode composed of a form of carbon called *graphite*. The material that forms the cathode can vary, depending on the specific use for which the battery is designed. Here we show a common cathode based on the cobalt-containing compound, $LiCoO_2$. Part **a** illustrates the battery as it discharges. Part **b** shows the battery as it is being recharged.

a. Discharge mode. When the battery is connected to a circuit, lithium ions spontaneously migrate from the anode, through the electrolyte, to the cathode. This internal flow of ions is accompanied by the external flow of electrons through the circuit.

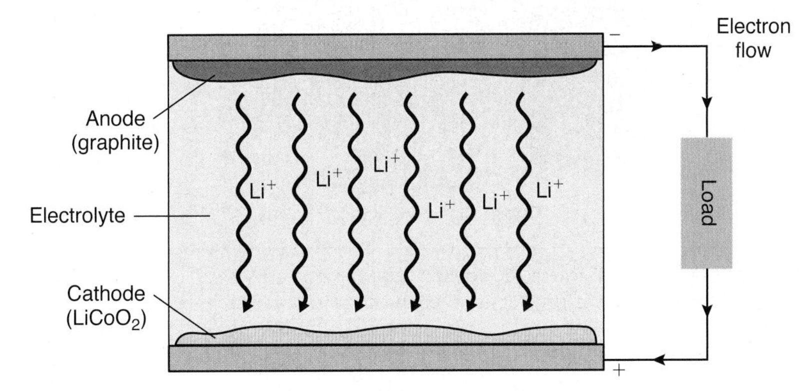

b. Recharge mode. When the battery is being recharged, electrical current flows in the opposite direction, causing lithium ions to migrate back to the graphite electrode. The electrical potential or voltage of this battery results in part from lithium ions being stored within this graphite electrode.

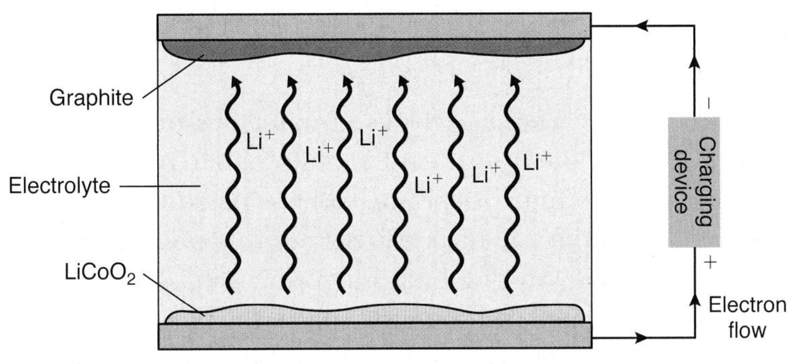

Ask Yourself

Why do lithium ions flow to the graphite electrode when the battery is being recharged?

Lithium–Ion Batteries Lithium is the least dense metal and also has the highest oxidation potential (+3.04 volts) of any metal. As a result, lithium batteries generally weigh less and store more energy than most other batteries of an equivalent size, making them highly valued, especially for portable applications.

The most common types of lithium batteries are single-use cells (designed to be discarded once their charge has been depleted), and rechargeable cells. Each of these relies on somewhat different chemical reactions. We focus here on the **lithium–ion** cell, a rechargeable battery widely used in portable electronic devices, as well as in hybrid and electric cars. The details of the chemistry involved in the operation of this battery are beyond the scope of our discussion, but we can describe its general design. The cell's operation relies on the shuttling or migration of lithium cations (Li^+) from one of its electrodes to the other. When the battery is connected to a circuit, lithium cations migrate from the anode, through the electrolyte, to the cathode. This internal flow of lithium ions is accompanied by an external flow of electrons from the anode, through an external circuit, to the cathode. This flow of (negatively charged) electrons in the circuit outside the battery balances the flow of (positively charged) lithium ions within the battery (**Figure 10.14a**). When the battery is recharged, the power supplied causes electrons to flow in the opposite direction, which also reverses the flow of lithium ions (**Figure 10.14b**).

Electrolysis of water • Figure 10.15

An electric current can decompose water molecules into hydrogen (H_2) and oxygen (O_2) gases.

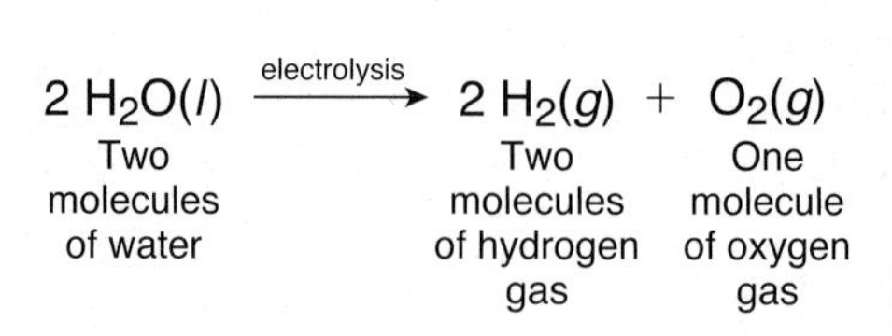

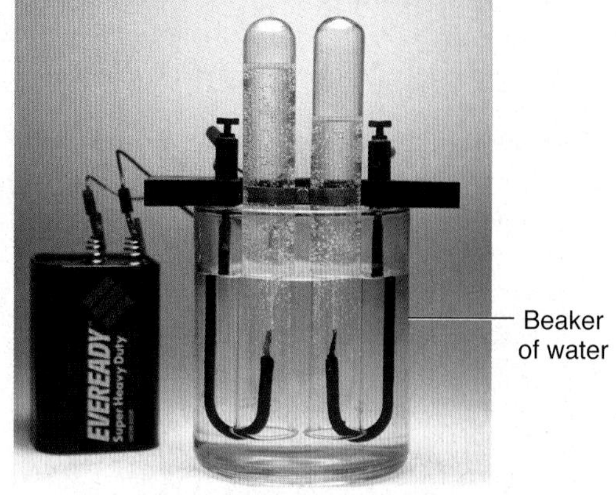

Charles D. Winters/Science Source

Think Critically In the apparatus shown here, the inverted test tubes were initially completely filled with water. Identify which gas is produced in each of the test tubes.

In general, lithium batteries have proven to be highly practical and reliable, although there have been instances of malfunction, some quite serious. For instance, under unusual circumstances, lithium–ion batteries have dangerously overheated or short-circuited, reportedly due to manufacturing defects, causing them to smoke or catch fire. In 2013, the Federal Aviation Administration temporarily grounded all U.S. flights of a newly introduced aircraft, the Boeing Dreamliner 787, as a result of two separate instances of electrical fires involving the plane's lithium–ion batteries.

Electrolysis

We have seen that all batteries use chemical reactions to produce electrical energy. The opposite process—using electrical energy to cause chemical change—is called **electrolysis**. For example, when we recharge a battery, we use electrical energy (in the form of current supplied to the battery) to produce a chemical change within the battery, thereby allowing it to continue (or resume) producing energy. When we recharge a lead–acid battery, for instance, the electrical energy we apply to the battery causes lead sulfate and water within the battery to be transformed into lead, lead dioxide, and sulfuric acid. These chemical changes would not occur in the absence of electrical energy supplied to the battery.

electrolysis Chemical changes caused by passing an electric current through a substance.

Electrical energy can also be used to generate a variety of other (nonspontaneous) chemical changes. In the electrolysis of water, for example, the water is converted into hydrogen and oxygen. The suffix *-lysis* implies a cleavage or rupture, so electrolysis produces the decomposition of a substance into its component parts by means of electricity.

The electrolysis of water was first reported in 1800 by the British natural philosopher William Nicholson. At the time, Nicholson was fascinated by a newly published report of the earliest known battery, called the Voltaic pile, and he built one of his own. Nicholson then connected platinum wires from the pile to a container of water and observed bubbles of what proved to be hydrogen gas at one wire and oxygen at the other. These gases were produced in a constant ratio of two volumes of hydrogen to one of oxygen, providing early evidence that a water molecule contains twice as many hydrogen atoms as oxygen atoms. **Figure 10.15** shows a modern version of this experiment.

Today, electrolysis has several practical industrial uses, including the production of commercially important chemicals, the purification of metals, and the electroplating of metals, which we'll discuss shortly. Electrolysis is also sometimes used in personal care, as a hair removal technique. We'll explore this application as well.

Chemical Production Perhaps the most commercially important application of electrolysis occurs in the conversion of concentrated solutions of sodium chloride in water (also known as *brine*) into three valuable chemicals: chlorine gas (Cl_2), hydrogen gas (H_2), and sodium hydroxde (NaOH). **Figure 10.16** illustrates this process. Carried out on a commercial scale, this reaction consumes vast amounts of electricity and is economically viable only where electricity is cheap and plentiful, such as near sources of hydroelectric power.

Chlorine, a strong oxidizing agent, is used to disinfect public water supplies, swimming pools, and sewage. In this capacity, chlorine has played a major role worldwide in reducing the incidence of some waterborne diseases. Chlorine is also a valuable raw material for manufacturing several consumer products, ranging from the simple, inexpensive plastic, *polyvinyl chloride* (PVC), which is used as plumbing tubing, to pharmaceuticals with complex molecular structures.

Hydrogen has several industrial uses, including the conversion of vegetable oils to shortening (discussed in Chapter 5). Hydrogen is also a clean-burning fuel, releasing only water vapor when it burns. The term **hydrogen economy** refers to a possible future in which hydrogen is widely used as a replacement for fossil fuels in transportation, heating, and the generation of electricity. However, since hydrogen is a highly flammable gas, safety concerns regarding its storage and handling currently limit its use as a fuel of choice in everyday life.

Sodium hydroxide (also known as *lye*) is an important raw material useful in the production of a variety of commercial products, including soaps (discussed in Chapter 11). It is also an ingredient in household drain cleaners and oven cleaners. Since sodium hydroxide is extremely corrosive, it must be used with great care.

Metal Production and Electroplating Many of our most important raw metals are produced through electrolysis. For example, aluminum is obtained through the electrolysis of *aluminum oxide*, Al_2O_3. This method of generating aluminum metal was discovered in 1886 by a 23-year-old American inventor named Charles Martin Hall. (The French scientist Paul Héroult independently developed this method at nearly the same time as Hall.) Before these discoveries, aluminum had been a rare,

Electrolysis of brine • Figure 10.16

Electricity converts brine, a concentrated solution of sodium chloride, into three products: chlorine (Cl_2), hydrogen (H_2), and sodium hydroxide (NaOH).

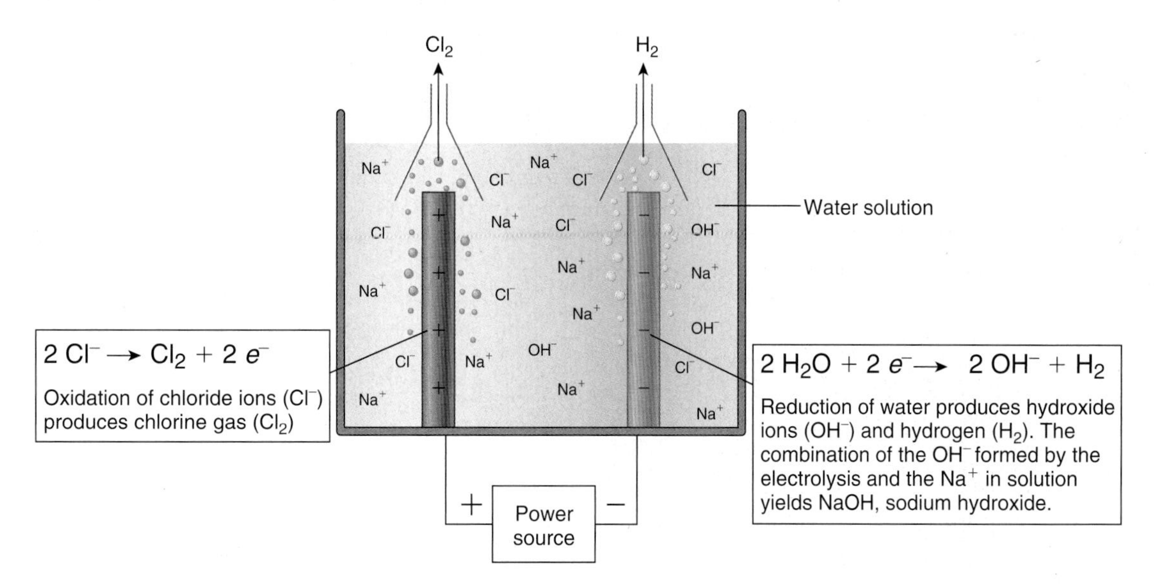

Think Critically **1.** Excluding the trace amounts of hydronium and hydroxide ions naturally present due to the autoionization of water, what ions are in solution before the electrolysis begins?
2. What ion is produced in solution as a result of the electrolysis?
3. What three chemicals would you expect to be produced by the electrolysis of an aqueous solution of sodium bromide?
4. Buffalo, New York, located near Niagara Falls, has historically been an important location for the electrolytic production of chemicals such as chlorine. Can you suggest a reason why?

expensive material. The process Hall and Héroult discovered is still in use today and relies on an important mineral called **bauxite** as the source of the aluminum oxide (**Figure 10.17**). Aluminum metal is valued for its high strength-to-weight ratio and its resistance to corrosion. This metal is used in a variety of consumer applications, including aluminum cans and foil, in construction and transportation materials, and in the fabrication of airplanes.

Electroplating uses electrolysis to deposit a very thin coating of one metal over another. This technique is typically employed to prevent corrosion of the underlying metal or to improve its appearance. Various metals can be electroplated onto others, including gold, chromium, copper, tin, and zinc.

To understand how electroplating operates, it may be helpful to review what happens in a standard galvanic cell, such as the Daniell cell (Figure 10.9). Here, we'll focus on what occurs at the zinc electrode. As this cell operates, electrons flow away from the zinc electrode (to the copper electrode), and the zinc metal is oxidized, releasing zinc ions (Zn^{2+}) into solution. Over time, we observe that the zinc strip erodes (and the copper electrode grows in size). Now, imagine what would happen if we reversed the flow of electrons in this cell by connecting a power source to the wire between the two half-cells. In this case, zinc ions in solution would combine with the electrons flowing into the zinc electrode to produce zinc atoms. As these newly formed zinc atoms deposit on the zinc electrode, this strip gradually grows larger.

Now suppose we turn off the external power source and replace the zinc strip with a strip of iron. As we turn the power source back on, electrons flow into the iron electrode. However, since the solution still contains zinc ions (Zn^{2+}), these ions will continue to combine with the incoming electrons to produce zinc metal, which adheres to the iron electrode, forming a thin layer of zinc.

What we've just described is an example of electroplating—using a power source to reduce metal ions in solution, depositing the resulting metal atoms onto an electrode. An electroplating apparatus can be as simple as a combination of a single beaker, an electrolyte solution, a battery, and two electrodes, as shown in **Figure 10.18**. Various metals are used in electroplating. Deposits of gold and silver improve the appearance of jewelry and trophies; electroplating an exterior layer of chromium onto the steel surfaces of motorcycles and car parts imparts a shiny appearance and prevents rusting.

Coating metals, such as iron or steel, with a thin layer of zinc to protect them from corrosion is called **galvanization,** a process that represents the largest commercial use for metallic zinc. The zinc coating can be applied through electroplating or by other means, such as "hot-dipping"—immersing the

> **galvanization** Applying a thin coating of zinc to a base metal, such as iron or steel, in order to inhibit corrosion.

Electrolytic production of aluminum • Figure 10.17

Electrolysis of a molten solution of aluminum oxide, Al_2O_3, produces aluminum. Like other commercial electrolytic processes, production of aluminum by this method is highly energy intensive.

Robert Nickelsberg/Getty Images

Bauxite, seen here, is refined into aluminum oxide.

$$2\,Al_2O_3 \xrightarrow[1000°C]{\text{electrolysis}} 4\,Al + 3\,O_2$$

Aluminum oxide → Aluminum + Oxygen

Ask Yourself

1. Al_2O_3 is an ionic substance containing Al^{3+} ions. Does this ion become reduced or oxidized in the process shown here?
2. Why is it more economical to produce aluminum materials through the recycling of waste aluminum than through the electrolysis of aluminum oxide?

Electroplating • Figure 10.18

Chrome-plated steel and silver-plated jewelry are just two of the many products obtained through electroplating. In this technique, the metal object to be coated serves as the cathode, or site of reduction

a. A chrome-plated motorcycle.
Chrome plating involves the reduction of chromium ions (Cr^{6+} or Cr^{3+}) to chromium atoms (Cr^0). The typical thickness of the deposited chrome layer is a few millionths of an inch, enough to be decorative and provide corrosion protection.

Sami Sarkis/Photographer's Choice RF/Getty Images

Think Critically What changes, if any, would you expect to observe in the silver electrode as the cell shown in Figure b continues to operate for some length of time?

b. A silver-plating apparatus.
Here, silver ions (Ag^+) in solution are reduced to silver atoms (Ag^0) by electrons entering the cathode, which in this case happens to be a spoon. The newly formed silver atoms deposit on the spoon.

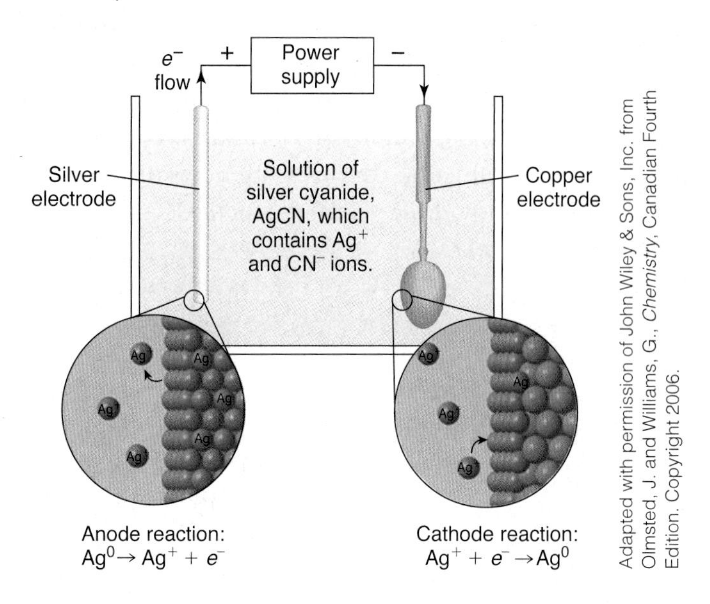

Adapted with permission of John Wiley & Sons, Inc. from Olmsted, J. and Williams, G., *Chemistry*, Canadian Fourth Edition. Copyright 2006.

object to be coated into molten zinc. Galvanized products include nails, screws, and sheet metal. The zinc coating protects the underlying material by reacting with components of the atmosphere to form a protective film. If the coating of zinc should crack in spots so that the iron's surface becomes exposed to the atmosphere, the zinc metal corrodes before the iron does, since the zinc oxidizes more readily. In essence, the zinc is used as a sacrificial metal to protect the more valuable underlying material.

Because galvanizing a large metal object presents practical difficulties, metal tanks used for underground storage of gasoline and other hazardous liquids are often protected from oxidation by attaching a sacrificial block of aluminum, magnesium, or zinc to the outside of the tank. These three metals are more readily oxidized than the iron of the tanks and thereby protect the iron from corrosion.

Electrolysis for hair removal Another example of electrolysis occurs in a process developed to permanently remove unwanted body hair. In this technique, a tiny probe inserted into individual hair follicles, one at a time, delivers a small electrical current. This current reacts with trace amounts of sodium chloride and moisture naturally present within the follicles. Recall from Figure 10.15 that the electrolysis of brine, or sodium chloride solution, produces sodium hydroxide (NaOH) and other chemicals. Electrolysis produces these same chemical changes within the hair follicle but on a microscale. The sodium hydroxide thus formed destroys the hair follicle, preventing future hair growth.

CONCEPT CHECK

1. **How** do the oxidation states of the reactants in a Daniell cell change as the cell operates?
2. **Which** ion has a greater tendency to be reduced: Au^+ or Na^+?
3. **Why** do batteries spontaneouly produce current when connected to a circuit?
4. **Why** is recharging a battery an example of electrolysis?

10.3 Fuel Cells and Solar Cells

LEARNING OBJECTIVES

1. **Describe** how fuel cells and batteries resemble each other and yet are different from each other.
2. **Explain** how photovoltaic cells work.

Growth in energy demand, coupled with our dependence on nonsustainable energy sources, such as fossil fuels, has led to increased demand for more sustainable methods of energy production. In this section, we discuss two devices that employ these methods: fuel cells and solar, or **photovoltaic,** cells.

Fuel Cells

A nonchargeable battery runs out of power when the chemical reactants inside have been essentially depleted and there is no way to restore them. On the other hand, it's possible to restore power to rechargeable batteries by regenerating the chemicals that give them their power. However, even rechargeable batteries have limitations. Unless they are continuously recharged, they can only discharge for a limited length of time before they run out of power, and the recharging itself takes time. Also, with extended use, the repeated cycles of discharging and recharging take their toll on battery life. As a result, even rechargeable batteries need to be replaced eventually.

Fuel cells address some of these limitations. Like batteries, fuel cells are galvanic cells that use chemical reactions to generate electrical energy. But in the case of fuel cells, the chemical reactants are continuously supplied (from an external source), so in theory, as long this supply of reactants is maintained, the fuel cell can continue to operate indefinitely.

fuel cell An electrochemical cell that produces current as long as it is supplied with fuel.

Fuel cells are often used in applications where a long-term, uninterrupted supply of energy is required, as well as in remote locations, where access to electricity is typically unavailable. Fuel cells are also used as commercial power generators, often where large-scale continuous power is required, as in data centers. Many large, well-known corporations use fuel cells for this purpose.

Fuel cells • Figure 10.19

Hydrogen fuel cells produce energy and water vapor. The diagram shows the chemical reactions that produce this energy.

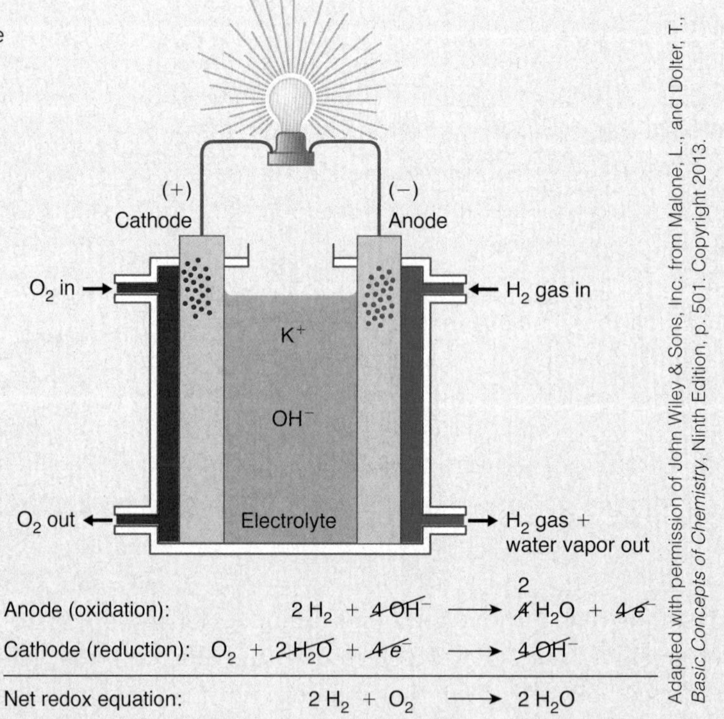

a. In this fuel cell, hydrogen (H_2) reacts at the anode and oxygen (O_2) reacts at the cathode. When adding these half-cell reactions, we cancel common terms as shown. The net reaction yields water vapor.

From an environmental standpoint, perhaps the most promising use of fuel cells is in transportation. In a typical fuel cell used for this purpose, hydrogen (H_2) and oxygen (O_2) react on the surface of catalysts within the fuel cell (**Figure 10.19**). Since the only chemical product of this redox reaction is water, the cell is pollution-free. Cars and buses powered by hydrogen fuels are commercially available, but currently on a limited basis, largely because of technical, cost, and safety issues. As we noted in our discussion of electrolysis in the preceding section, hydrogen is a highly flammable gas and must be handled cautiously. Nonetheless, these various challenges continue to be addressed, and many car makers have fuel cell models in development.

Fuel-cell-powered cars, like battery-powered, or electric cars, use electric motors, so they don't require gasoline (or engine oil, for that matter). However, fuel-cell-powered cars need to be refueled with hydrogen periodically, just as electric cars need to be regularly recharged.

Hybrid cars have battery-powered electric motors as well as gasoline engines. These cars use **regenerative braking**, which means that when you apply the brakes, the cars are able to convert the kinetic energy of their motion into electrical energy, which they use to recharge their batteries. Hybrid cars are currently far more popular than fuel-cell and electric vehicles, but electric cars are starting to be seen more commonly on the roads (see *Did You Know?* on the next page) and fuel cell cars may not be far behind.

Solar Cells

The sun's radiant energy supports all life on Earth. By some estimates, the amount of solar energy hitting Earth's surface in just one hour is equivalent to the total amount of energy consumed by the entire world in one year. Sunlight is free, essentially limitless, and abundant, although some places on Earth inherently receive more sunlight than others. How can we harness this solar energy in a practical and economical manner?

Various technologies are available for capturing solar energy, including **photothermal** methods, which convert light into heat energy, and **photovoltaic** methods, which convert light into electrical energy. Here we discuss photovoltaic methods.

> **photovoltaic cell** A cell that converts solar energy into electrical energy.

Photovoltaic cells, like galvanic cells, produce electrical current, but instead of doing so through chemical reactions, they do so by absorbing light. To understand how a photovoltaic cell—also known as a PV or solar cell—works, it's helpful to first discuss a phenomenon called the **photoelectric effect**. This effect describes the observation that light of a certain energy can eject electrons from the surface of a metal

b. A network of hydrogen filling stations will be needed for fuel cell vehicles. Since hydrogen is stored as a compressed gas, the nozzle used for its fueling, shown here, forms a tight seal with the vehicle.

Ask Yourself

In the fuel cell shown, do the electrons flow from the cathode to the anode, or vice versa?

DID YOU **KNOW?**

Electric cars were at one time more popular than gasoline-powered vehicles

Given concerns about the widespread use of petroleum-based fuels for transportation and the fact that batteries are a proven alternative technology, we might naturally wonder why we don't see more battery-powered cars on the road. Years ago, these cars were actually relatively commonplace. At the beginning of the 20th century, cars that ran on electric batteries and steam-powered cars dominated the marketplace, sharing the road almost equally. Combined, they outsold cars with gasoline engines by better than three to one. Yet by 1917, only 17 years later, more than 98% of all cars were running on the same sort of gasoline engines we use today. Steam-powered cars had all but disappeared, and the electric car was dying rapidly. The causes of the death of the electric car were varied and complex. They included limited driving ranges and low speeds, and the great weight and long recharge times of the batteries. Continuous improvements to the gasoline-powered car, as well as other factors, contributed to the displacement of the electric car.

Yet the internal combustion engine brought its own problems, principally air pollution. In 1990, the state of California was the first state to enact more stringent vehicle emissions standards than required by federal law, over concerns about air quality in its metropolitan areas. These standards, which continue to be updated periodically, encourage the introduction of very low-emission cars (such as hybrids) and zero-emission cars (such as electrics) into the new-car market. Various other states, including several in the Northeast, have adopted these standards. We can expect cleaner air as these newer cars displace older, polluting vehicles from our roads and highways.

Figure a. The dashboard of the Nissan Leaf, an electric car. This car is powered by a 600-pound lithium–ion battery pack, which gives it an estimated driving range of 70–100 miles, depending on driving conditions.

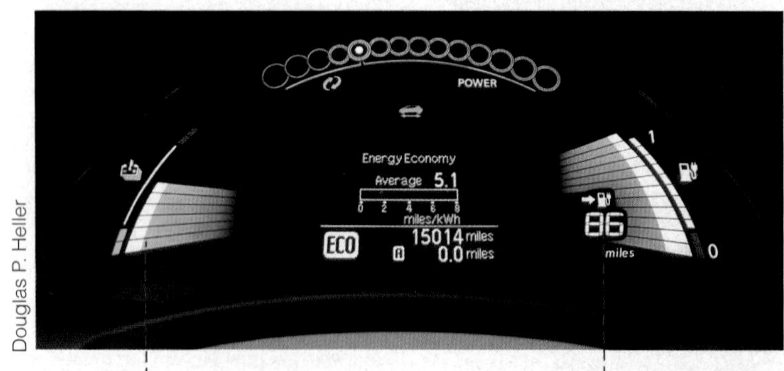

This gauge reflects battery temperature. Here, it shows the battery is within an acceptable temperature range, neither too hot nor too cold. Batteries deliver lower voltage and power at colder temperatures and can produce unwanted chemical reactions if too hot.

Instead of a gas gauge, this car has a range gauge, which provides an estimate of the driving range remaining until the battery's charge is depleted. The number of bars displayed reflects the state of charge. Here, the battery is close to full charge.

Almost all automobile manufacturers now produce hybrid cars, which run on both battery power and gasoline. But even hybrids emit some exhaust. Electric cars, by contrast, have no emissions, and recent technological improvements in these cars suggest they may be poised to make a comeback. For example, the Nissan Leaf (**Figure a**) is a standard four-door hatchback that happens to be fully electric. Tesla Motors sells fully electric luxury sedans and sports cars. The Chevrolet Volt is an electric car with a back-up gasoline-powered engine for extended range.

Whether consumers adopt electric cars in large numbers remains to be seen. In terms of acceleration and handling, these cars can perform as well as their gasoline-powered counterparts. However, one of their chief limitations has been driving range. (The term *range anxiety* describes how some drivers of these cars reportedly feel as a result of concern about the car's limited energy supply.) Yet driving ranges on a single charge of the battery continue to improve, with some models reporting well over 100 miles of estimated range, which is perfectly adequate for most daily commutes. Networks of charging stations in public places are growing as well.

Electric cars use substantially less energy and resources per mile driven than even hybrid cars, yet they are not a panacea for the environmental challenges posed by transportation. Ultimately, the electricity used by electric cars has to be generated somewhere, and in the United States, this electricity is still produced mostly by fossil fuels, which emit greenhouse gases and other pollutants. In the future, other green technologies, such as hydrogen-powered fuel cell cars, may prove to be commercially viable. Even these cars are not free of potential environmental impact (at least, not currently), because the hydrogen fuel they require is produced commercially from methane and other hydrocarbons in a process that yields carbon dioxide as a by-product. Nevertheless, considering the history of the automobile, if we learn to use electrochemical reactions of one kind or another (whether batteries or fuel cells) to power our vehicles more efficiently, the car of the future could well turn out to be the car of the distant past.

Ask Yourself

Batteries and fuel cells both use *electrochemical reactions.* Explain this statement.

The photoelectric effect • Figure 10.20

Under the right conditions, light waves hitting the surface of a metal or other materials can cause electrons to be ejected from the surface. Only photons with sufficient energy can cause this to occur.

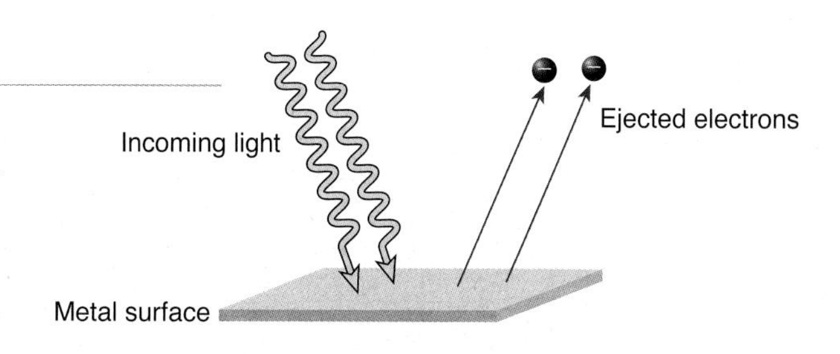

(**Figure 10.20**). In 1905, Albert Einstein recognized that transmitted light (and more broadly, **electromagnetic radiation**, which is wave-like energy that includes radio waves, visible light, and X-rays) can be thought of as a stream of **photons**, or tiny discrete packets of electromagnetic or light energy. This insight helped to explain the photoelectric effect, and ultimately earned Einstein a Nobel Prize in physics in 1921.

Solar cells, first invented in the 1950s, are based on a phenomenon called the photovoltaic effect, which resembles the photoelectric effect in that in both cases incident light causes a flow of electrons. In the PV cell, however, the electrons are not ejected, but rather flow through the cell and out through a connected circuit. The strength of the resulting current—its amperage—depends on several factors, including the frequency of the light hitting the cell, the surface area of the cell, and the **efficiency** of the cell, which is a measurement of how efficiently the cell converts absorbed light into electrical current. Since the current produced by the cell depends on the intensity of sunlight hitting it, this current naturally varies with the time of day and weather conditions, such as cloudiness, seasonal factors, and geography.

The efficiency of a solar cell depends on the cell's particular design and the materials that compose it. Silicon, the most common element used in PV cells, is a semiconductor, which means that in its pure state it does not conduct electricity very well. We can improve silicon's conductivity substantially by adding small amounts of impurities called **dopants**, such as boron or phosphorus. At this stage it's helpful to review where boron, silicon, and phosphorus reside in the periodic table. Boron is a Group 13 element, containing 3 valence electrons; silicon is in Group 14, with 4 valence electrons, and phophorus is in Group 15, with 5 valence electrons.

An atom can use its valence electrons to form bonds with other atoms. In the case of pure silicon, each silicon atom uses its four valence electrons to bond to four other silicon atoms, as shown in **Figure 10.21a**. When silicon is doped with either boron or phosphorus, the addition of

Doping silicon • Figure 10.21

Pure silicon (**a**) is free of defects. Adding dopants to silicon introduces defects, which alter the silicon's electrical conductivity. Boron-doped silicon is called **p-type** (**b**), and phosphorus-doped silicon is called **n-type** (**c**).

a. In pure silicon (Si), each silicon atom uses its four valence electrons to bond to four other silicon atoms.

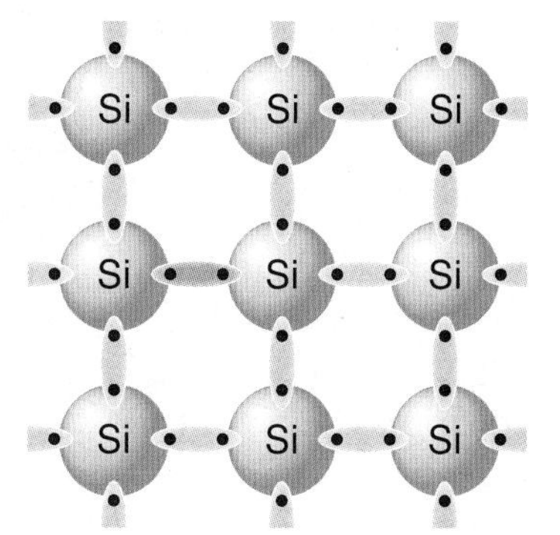

b. Introducing an impurity such as boron (B) into silicon creates a defect known as a **hole**, because elements such as boron have only three valence electrons, one less than silicon. Semiconductor materials such as this are called **positive-** or **p-type.**

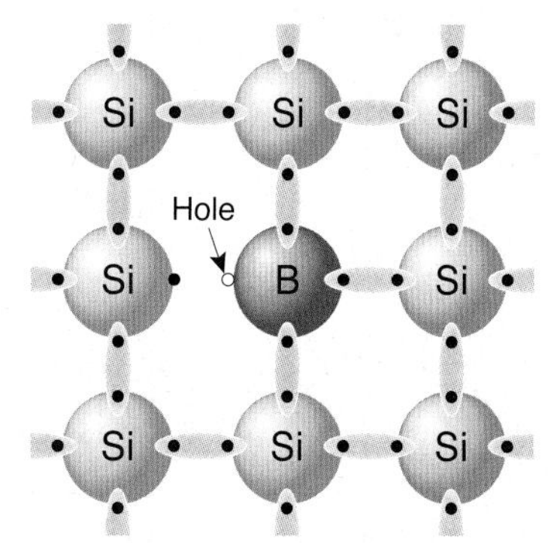

c. Introducing an impurity such as phosphorus (P) into silicon creates a defect known as a **free** or **extra electron**, because elements such as phosphorus have five valence electrons, one more than silicon. Semiconductor materials such as this are called **negative**- or **n-type.**

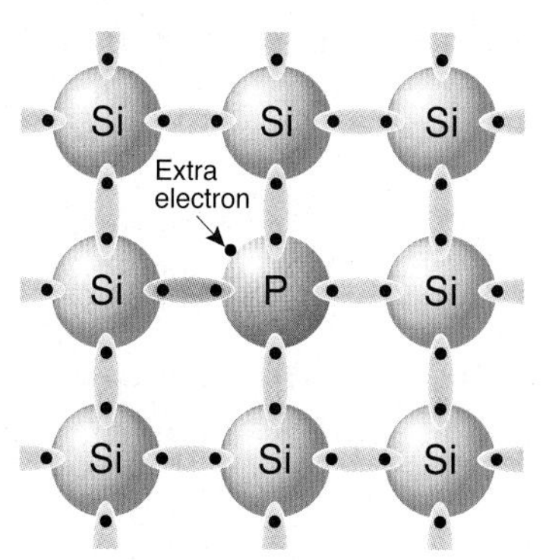

Solar cells • Figure 10.22

(a) Solar cells are assembled into panels, which then may be installed in large arrays to increase overall power output. **(b)** Photons of light hitting the cell cause electrons to flow through a connected circuit.

a. A large-scale solar installation for generating electricity. Desert locations, such as this, are often chosen for this purpose, due to the favorable sunlight conditions and the relatively low cost of the land.

John Burcham/National Geographic/Getty Images

b. A photovoltaic cell.

n-type semiconductor

Current →

Antireflective coating

Cover glass

p-n junction

p-type semiconductor

← Current

The mobilized electrons then pass through a circuit (**4**) and reenter the photovoltaic cell, where they recombine with the electron holes.

Incident light (**1**) absorbed by a photovoltaic cell...

...causes electrons to migrate from the p-layer (**2**) to the n-layer (**3**), leaving sites of positive charge —known as "electron holes"—in the p-layer.

Ask Yourself Recent improvements in antireflective coatings used in PV cells have helped improve the efficiency of these cells. Explain.

these impurities creates defects or imperfections in the silicon's bonding structure, as shown in **Figure 10.21b, c.** The resulting material can conduct positive charge (p-type) or negative charge (n-type) more easily than pure silicon.

Solar cells are constructed of adjacent layers of p- and n-type semiconductors. The boundary formed between these layers is called a **p-n junction**. This junction acts essentially as a one-way gate, allowing the flow of electrons from the p-side to the n-side, but not in the reverse direction. When light strikes the solar cell, it causes electrons to migrate from the p-side to the n-side. Since electrons are negatively charged, this migration builds negative charge on the n-side and leaves sites of positive charge remaining in the p-side. What we've described is nothing more than the creation of an electrical potential, which results in a flow of elections from the n-side, out into a connected circuit, then back into the cell on the p-side (**Figure 10.22**).

CONCEPT CHECK

STOP

1. **What** product is released during the operation of a hydrogen fuel cell?
2. **Why** is it necessary to dope silicon in the manufacturing of silicon-based photovoltaic cells?

Summary

1 Oxidation and Reduction 300

- ***What are oxidation and reduction?***
  Oxidation is defined as the loss of electrons, and **reduction** as the gain of electrons, as shown here. Oxidation can also be observed as a loss of hydrogen or gain in oxygen. Reduction is the complementary process. Oxidation and reduction reactions always occur in tandem. A substance that causes another to be oxidized is an **oxidizing agent**, and a substance that causes another to be reduced is a **reducing agent**.

Figure 10.1 • Defining oxidation and reduction

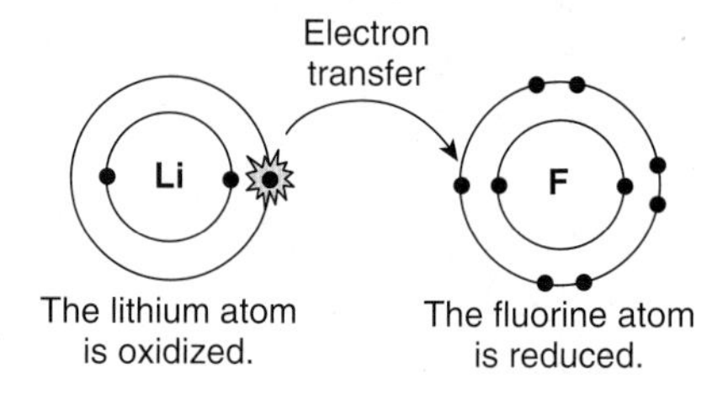

- ***Where do oxidation and reduction occur in everyday life?***
  Reactions that involve both oxidation and reduction, which are also known as redox reactions, are everywhere. Examples include the combustion of fuels; the metabolism of foods; the action of **antioxidants** (substances found in foods and our bodies that inhibit the oxidation of other substances); and the whitening and disinfecting action of oxidizing agents such as bleach and peroxide compounds. In addition, batteries use redox reactions, **smelting** involves the reduction of ores to produce metals, and rusting is due to the oxidation of iron.

2 Redox Reactions and Electrical Current 307

- ***What are standard reduction potentials?***
  Standard reduction potentials are measured in **volts**, a unit of electrical potential, and reflect how readily a chemical particle accepts electrons. The reduction of hydrogen ions (H^+) to produce hydrogen gas (H_2) is assigned a reference value of zero volts. A substance that is more readily reduced than a hydrogen ion has a positive standard reduction potential, and a substance that is less readily reduced has a negative standard reduction potential.

- ***How do batteries work?***
  Batteries, also known as **galvanic cells**, produce electrical energy through the operation of chemical reactions, such as shown here. All galvanic cells contain an **anode**, the site of oxidation; a **cathode**, the site of reduction; and an electrolyte, in which the anode and the cathode are immersed. We can calculate the standard voltage of a galvanic cell by adding the voltages of the reduction and oxidation **half-cell reactions.** If the standard voltage has a positive value, the cell spontaneously produces current when connected to a circuit. In **electrolysis**, electrical energy is used to produce a chemical change, as when a battery is recharged. **Electrolysis** is an important industrial process used to produce chemicals and metals.

Figure 10.9 • A Daniell cell

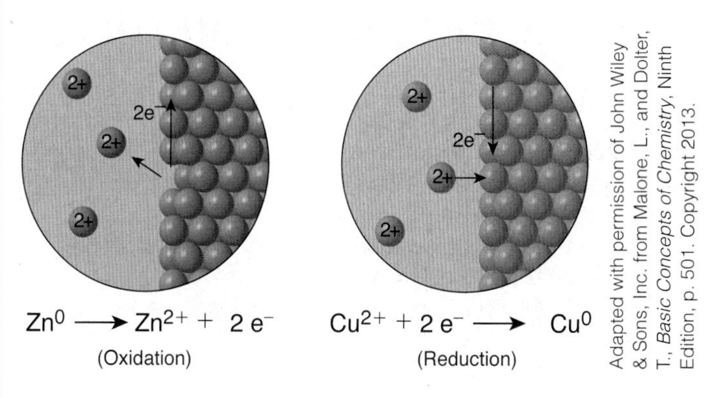

3 Fuel Cells and Solar Cells 320

- ***How are fuel cells different from batteries?***
  Both **fuel cells** and batteries are galvanic cells, using chemical reactions to produce electrical energy. However, the chemical reactants of fuel cells are supplied externally, so fuel cells can operate continuously. Since the chemical reactants of batteries cannot be replenished externally, batteries cease producing electrical power when their chemical reactants are exhausted.

- ***How do photovoltaic cells work?***
  Photovoltaic cells convert the radiant energy of sunlight into electrical energy. The cells typically consist of silicon containing **dopants**, which alter the silicon's electrical properties. Within each PV cell two adjacent layers of doped material form a **p-n junction,** which restricts the flow of electrons through this boundary to a single direction. As shown here, incident light: (**1**) causes electrons to pass from the p-layer (**2**) to the n-layer (**3**), leaving "electron holes" or sites of positive charge in the p-layer. The mobilized electrons then pass through a circuit (**4**), thereby providing current, before reentering the PV cell and recombining with electron holes.

Figure 10.22 • Solar cells

n-type semiconductor

Current →

p-n junction

p-type semiconductor

← Current

Key Terms

- ampere (or amp) 309
- anode 307
- antioxidant 301
- cathode 307
- electrolysis 316
- free radical 302
- fuel cell 320
- galvanic or electrochemical cell 307
- galvanization 318
- half-cell reaction 310
- oxidation 300
- oxidation state 307
- oxidizing agent 301
- photovoltaic cell 321
- reducing agent 301
- reduction 300
- smelting 305
- standard reduction potential 310
- volt 309

What is happening in this picture?

This long-exposure image captures the swirling patterns created by glow sticks. The light from these devices is due to a phenomenon called **chemiluminescence**, the production of light (but not heat) due to a chemical reaction. In this case, hydrogen peroxide (H_2O_2) undergoes a redox reaction with an organic compound called *diphenyl oxalate*. One of the resulting products of this reaction decomposes, releasing energy to a nearby dye molecule. This energized dye molecule then emits light (through a process called fluorescence), the color of which depends on the dye's chemical structure. A dye molecule used to emit blue light, for example, is shown here.

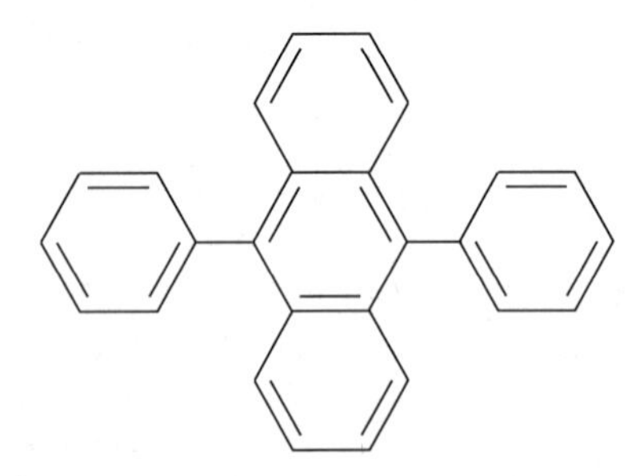

Think Critically **1.** Is hydrogen peroxide an oxidizing agent or a reducing agent?

2. Note the pattern of alternating single and double carbon–carbon bonds present in the dye molecule. What other compound discussed in this chapter has a similar pattern of bonds? Does this compound exhibit any optical properties?

Roberto Schmidt/Staff/AFP/Getty Images

Exercises

Review

1. Define, describe, or explain each of the following terms:

a. anode
b. cathode
c. electrical circuit
d. electric current
e. net redox equation
f. electrolysis
g. fuel cell
h. oxidizing agent
i. reducing agent
j. salt bridge

2. What happens when the anode and cathode of an electrochemical cell are connected to each other?

3. What element is used to coat "galvanized" metal products?

4. What's the common term used for the commercial electrochemical cells that provide power to our flashlights and other portable electronic devices?

5. What chemical substance gives an "alkaline" battery its name?

6. Identify the oxidizing agent and the reducing agent in the following figure:

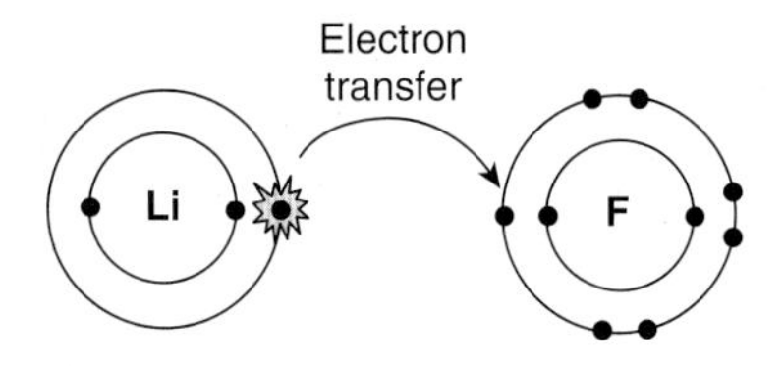

7. Do electrons move from oxidizing agents to reducing agents or from reducing agents to oxidizing agents?

8. a. When a battery provides power to a circuit, do electrons move from the anode to the cathode or from the cathode to the anode? b. When recharging a battery, do electrons move from the anode to the cathode or from the cathode to the anode?

9. What's the significance of a positive potential for a redox reaction? A negative potential?

10. How can you cause a reaction with a negative redox potential to occur?

11. You are given a table of standard reduction potentials. a. How do you determine the reduction potential of a reduction half-cell? b. How do you determine the oxidation potential of an oxidation half-cell? c. How do you determine the redox potential of a redox reaction?

12. As discussed in Chapter 3, elemental sodium and elemental chlorine react with each other to form sodium chloride. Although chlorine normally exists as diatomic molecules (Cl_2), we can write the reaction of sodium atoms with chlorine atoms as $Na + Cl \rightarrow Na^+ + Cl^-$. In this reaction as written, what is being oxidized and what is being reduced? What is the oxidizing agent and what is the reducing agent? Refer to Table 10.1 to write the individual half-cell reactions that combine to form the redox reaction.

13. Refer to Table 10.1 to answer the following questions. a.) Which one of the following chemicals has the greatest tendency to acquire an electron? b. Which has the greatest tendency to lose an electron?

$$Zn, Cl^-, Br_2, K^+, H^+, H_2O, Cl_2.$$

14. Name and write the chemical formulas of all the products produced on electrolysis of a. pure water, b. a solution of sodium chloride in water, c. a solution of potassium chloride in water.

15. a. Which type of semiconductor, p-type or n-type, is represented in Figures a. and b., respectively?

b. Which of the following elements would you expect to serve a similar function as the dopant used in Figure a: calcium (Ca); carbon (C); arsenic (As); aluminum (Al)?

c. Which of the following elements would you expect to serve a similar function as the dopant used in Figure b: gallium (Ga); selenium (Se); lead (Pb); sulfur (S)?

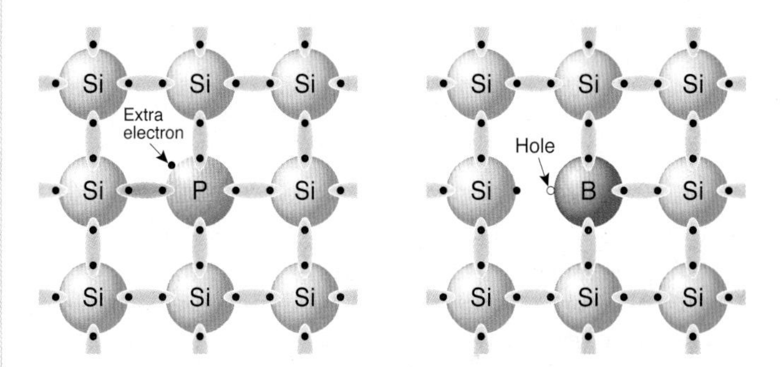

16. Some brands of toothpaste claim to be particularly effective in brightening your teeth, eliminating mouth odors, killing oral bacteria, and so on, because of a special oxidizing ingredient they contain. What is this oxidizing ingredient?

17. a. Give an example of oxidation of a metal that causes corrosion.

b. Give an example of oxidation of a metal that protects against corrosion.

18. Large metal storage tanks used for holding gasoline underground are too large to be coated with zinc by any practical galvanizing process. How are they protected from corrosion?

19. What's the principal advantage of an electric car or fuel cell car over conventional cars? What were some of the factors that led to the displacement of electric cars, early in the 20th century, by cars with internal combustion engines?

20. Blueberries, dark chocolate, and green tea are noted for being good sources of what types of compounds?

Think

21. By increasing the areas of the zinc and copper plates in the Daniell cell, we can increase the amperage of the current flowing through the circuit without affecting the cell's voltage.

a. Why does increasing the size of the zinc and copper plates have no effect on the voltage?

b. How does increasing the size of the plates affect the amperage of the current flowing through the circuit?

22. What determines whether a substance that can act as either an oxidizing agent or a reducing agent does, in fact, behaves as an oxidizing agent or as a reducing agent?

23. If you remove an iodine stain from a piece of cotton by rubbing the stain with a moist vitamin C tablet and neglect to rinse the area afterwards, a dark color can reappear soon. Suggest a reason for this behavior.

24. Household bleach is normally used to remove colored stains from fabrics. Yet when we decolorize a solution of tincture of iodine with vitamin C, as we did in Figure 10.2, adding household bleach produced a color in the nearly colorless solution. Explain why this happened.

25. Liquid household bleaches are good oxidizing agents. If you add a few drops of a liquid bleach to a clear, colorless solution of KBr in water, the solution turns orange-brown. What do you conclude from this observation? (*Note*: The products of the reduction of the liquid household bleach are colorless.)

26. If we carry out the electrolysis of pure water, the reaction proceeds very slowly and produces oxygen at the positive terminal. If we add table salt to the water, the electrolysis proceeds much faster and produces chlorine at the positive terminal. Explain both of these phenomena and show how they are related to each other.

27. A freshly constructed Daniell cell made of copper metal, copper sulfate, zinc metal, and zinc sulfate generates a voltage of about 1.1 volts. What voltage is produced by the same cell after it has been operating long enough to discharge completely the blue color of the copper sulfate solution and turn the solution completely colorless? Explain your reasoning.

28. Would a Daniell cell work equally well if the salt bridge were replaced by an ordinary copper wire of the type used with household electrical appliances? Explain.

29. Food that stands in contact with air for a long time sometimes deteriorates, as some of the components of the food react with the oxygen of the air. Vitamin C is added to some foods not only as a supplemental nutrient, but also to protect against this kind of deterioration. How does vitamin C provide this protection?

30. How can a rechargeable battery act as both a galvanic cell and an electrolytic cell?

31. *Butylated hydroxyanisole* (BHA) and *butylated hydroxytoluene* (BHT) are two antioxidants used as food preservatives. Use the Internet to find the molecular structure of each.

a. What structural features do these two compounds have in common?

b. For each of these compounds, identify the hydrogen atom that quenches free radicals.

32. The use of natural gas as a fuel for an internal combustion engine produces very little of the kinds of pollution generated by gasoline. Natural gas, which consists largely of methane, burns according to the balanced equation:

$$CH_4 + 2\,O_2 \longrightarrow CO_2 + 2\,H_2O$$

With this in mind, would an electric car powered by a hydrogen fuel cell have any advantage over a similar car that uses natural gas as a fuel for its internal combustion engine?

33. A fuel cell using the chemical combination of hydrogen and oxygen to generate electricity might use the oxygen of the air. How might the hydrogen be obtained?

34. What do you believe is the greatest advantage of an electric car as compared with a gasoline-powered car? What do you believe is the greatest advantage of a gasoline-powered car compared with an electric car?

35. Would you be in favor of a federal regulation requiring that all cars now on the road be replaced eventually by electric cars? Explain.

36. Conventional submarines, as opposed to those that run on nuclear power, run on battery-powered engines when they are submerged and on diesel engines (a variation of an internal combustion engine) when they are on the surface. Why don't they run on battery-powered engines all the time?

37. Some chemists claim that all chemical reactions are redox reactions. In the simple ionization of water, what is being oxidized according to this view? What is being reduced? Refer to Section 3.3.

38. What is the current status of the use of fuel cells in automobiles and other vehicles? Use the Internet to find your answer.

Calculate

39. What standard voltage would you obtain from a galvanic cell made of copper metal, copper sulfate, magnesium metal, and magnesium sulfate?

40. When a zinc strip is submerged in a solution of copper sulfate, over time the zinc strip erodes, dark granules of copper appear, and the blue color of the solution fades. By contrast, when a copper strip is submerged in a zinc sulfate solution, no change occurs. Use information in Table 10.1 to explain these results.

41. The lead–acid automobile battery consists of plates of spongy lead (Pb) and plates of lead dioxide (PbO_2). As the battery discharges, both the lead and the lead dioxide are converted to lead sulfate ($PbSO_4$). Would a similar battery consisting of plates of spongy lead and plates of lead sulfate provide any current? If your answer is yes, what voltage would you expect? If your answer is no, explain why it would not work. Repeat this question for a battery consisting of plates of lead dioxide and lead sulfate.

42. Potassium is a hazardous metal that reacts rapidly and explosively with water. In contact with water, potassium produces potassium hydroxide (KOH) and hydrogen gas, which ignites as a result of the heat generated in the reaction.

a. In view of this (and referring to Table 10.1), can water act as an oxidizing agent?

b. Can water also act as a reducing agent?

c. Name two halogens that water is capable of reducing.

43. How do we manage to get 12 volts from a car battery when the redox voltage of the electrochemical reaction taking place within it is only 2 volts?

44. Suppose we modify Table 10.1 by defining the reduction of F_2 to fluoride ions as our standard, replacing the reduction of hydrogen ions to H_2, which is the current standard. If we now define the reduction potential for fluorine as zero (arbitrarily, just as the reduction potential of hydrogen ions to H_2 was originally defined arbitrarily as zero):

a. How would this change all the other values in the table?

b. What would be the new reduction potential for hydrogen ions?

c. How would this change affect the measured value for the voltage of the Daniell cell?

d. How would this change affect the calculated value for the voltage of the Daniell cell?

e. Would we still be able to use the new table to determine whether a redox reaction can occur spontaneously?

f. If your answer to part e is Yes, explain how we would now use our modified table.

g. If your answer to part e is No, explain why we would not be able to use our modified table for this purpose.

h. Does it matter which reduction reaction we choose as our standard? Describe your reasoning.

45. a. In the first two photos shown below, molecular iodine (I_2) is reduced to iodide ions (I^-) by vitamin C. The vitamin C (*ascorbic acid*) is itself oxidized to a compound called *dehydroascorbic acid* according to the following half-cell reaction:

	Standard oxidation potential
ascorbic acid $\rightarrow$ dehydroascorbic acid $+ 2e^- + 2\,H^+$	-0.06 V

Identify the iodine (I_2) reduction half-cell reaction from Table 10.1. Write the net redox equation between iodine and ascorbic acid and calculate the standard voltage. Does it make sense that the reduction of iodine by vitamin C is spontaneous? Explain your reasoning.

b. In the third panel, iodide ion (I^-) is oxidized to molecular iodine (I_2), and the perchlorate ion (OCl^-) of bleach is reduced. Using Table 10.1, identify the half-cell reaction for the reduction of the perchlroate ion (OCl^-). Write the net redox equation between the iodide ion and perhchlorate ion and calculate the standard voltage. Does it make sense that the oxidation of iodide by bleach is spontaneous? Explain your answer.

c. Use Table 10.1 identify the half-cell reaction for the reduction of hydrogen peroxide (H_2O_2). Would hydrogen peroxide be expected to oxidize iodide ion (I^-) as well?

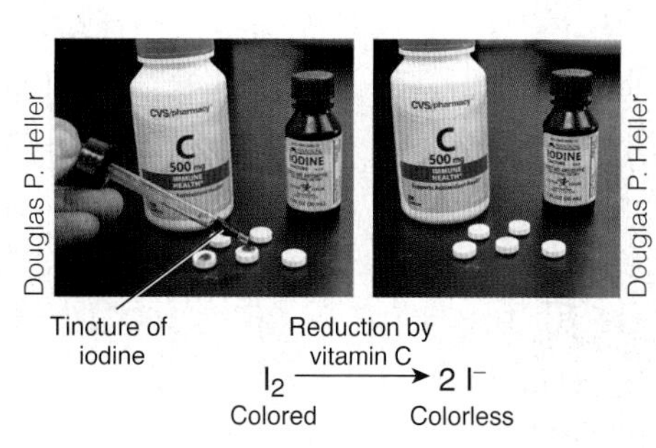

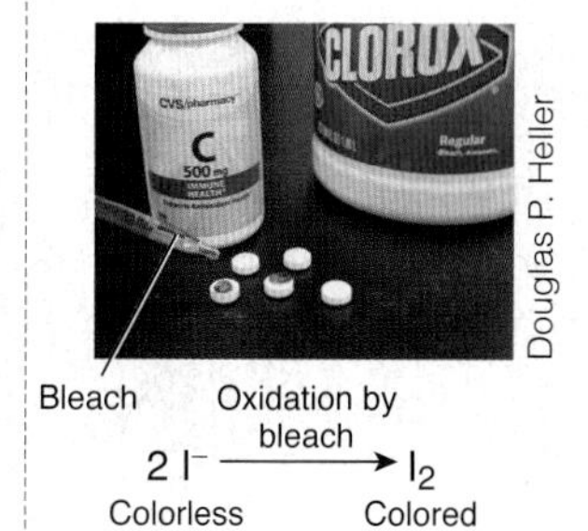

11 Cleaning Agents, Personal Care, and Cosmetics

Most of us clean and prepare ourselves to face the world virtually every day. Just consider all the soaps, shampoos, toothpastes, mouthwashes, cosmetics, moisturizers, and other skin and hair products we use, not to mention laundry detergents and household cleaners. Consumption of these products is essential to how we take care of ourselves, our clothes, and our living spaces, as well as how we present ourselves to the world. As a result, the worldwide market for household and personal care products is enormous—nearly

450 billion dollars annually and growing. Every day new personal care products and cleaners seem to become available, each claiming new and/or improved benefits, such as moisturizing our skin and making it more radiant, or delivering cleaner, brighter clothes. Here, a worker in a perfume and cosmetics research center investigates a new formulation of a hand cream.

In this chapter we'll explore the chemistry behind many of these personal care products, including soaps, shampoos, laundry detergents, fabric softeners, hand sanitizers, shaving creams, mouthwashes, and skin lotions. All of these products incorporate chemical *surfactants*. We'll see that the various applications of these surfactants derive from differences in their chemical structures.

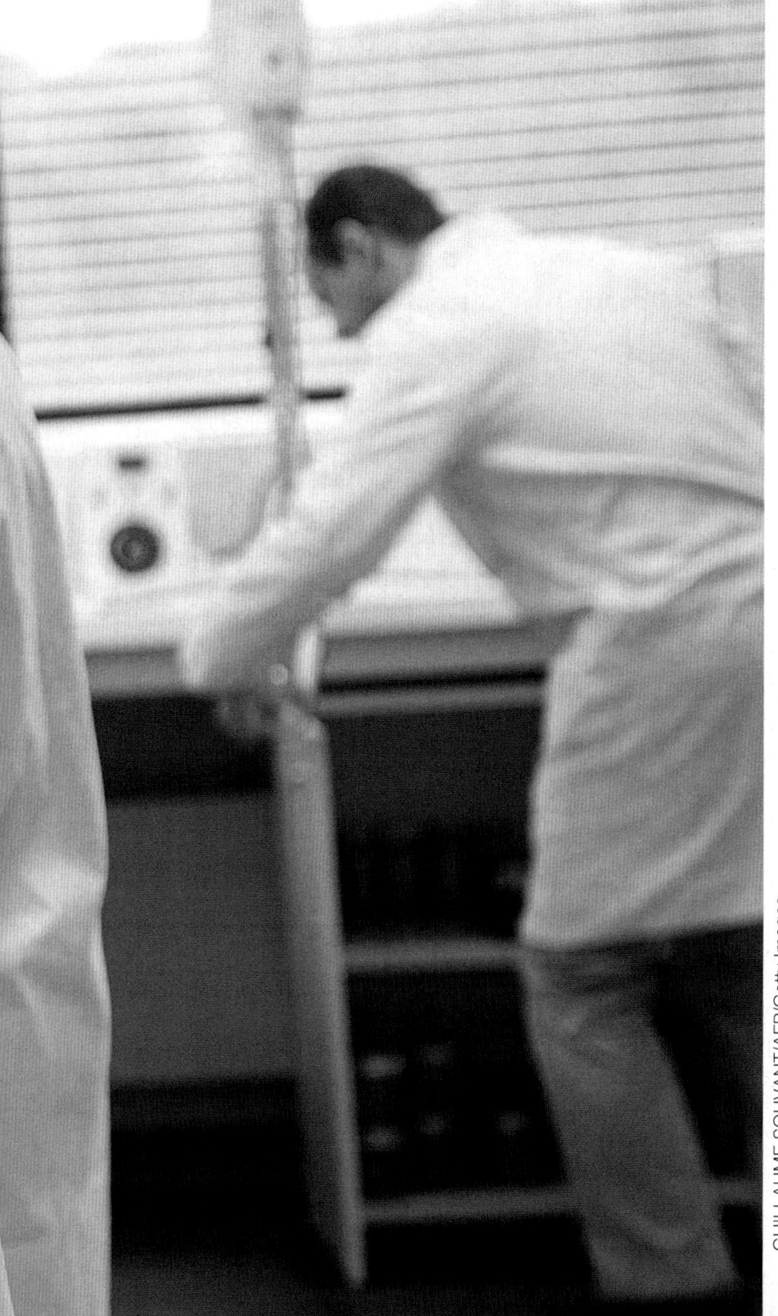

GUILLAUME SOUVANT/AFP/Getty Images

CHAPTER OUTLINE

11.1 Soaps and Surfactants

LEARNING OBJECTIVES

1. **Define** surface tension.
2. **Describe** how soap works.
3. **Differentiate** between anionic, cationic, and nonionic surfactants.
4. **Explain** what the term *hard water* means and how water is softened.

Every day we commonly use a variety of personal care and cleaning products. These compounds, which include soaps and synthetic detergents, can serve a variety of functions, but all share the common property of lowering the *surface tension* of water. In this section we'll first explore the meaning of surface tension and then see how soaps, detergents, and similar products perform their functions.

Surface Tension

If you've ever seen an insect called a water strider skimming across the surface of a puddle or pond, or noticed the beading of water on a freshly waxed car or on a waxy leaf, you've witnessed a phenomenon called **surface tension**. You can explore this property of liquids by very gently "floating" a conventional metal thumbtack, pin side up, on the surface of a glass of cold water. Because the metal tack is more dense than water, you might expect it to sink to the bottom of the glass. But with a little practice you should be able to get it to rest gently on the water's surface, supported by the water's surface tension.

surface tension A property of liquids that causes them to behave as if the surface is covered by a thin membrane.

Chemistry InSight Surface tension • Figure 11.1

The molecules that make up liquids exert cohesive, attractive forces between neighboring molecules. This is especially pronounced in water and gives rise to surface tension.

Water molecules exert large forces of attraction on neighboring molecules. Within the liquid, the total force of attraction exerted by any one water molecule on its neighbors is dispersed in all directions.

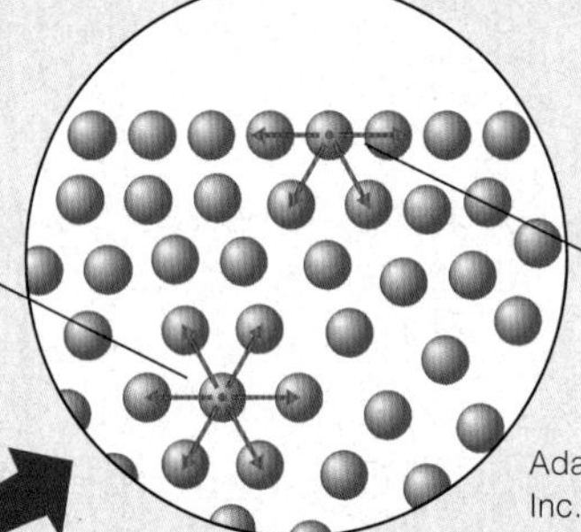

At the water's surface, though, nearby water molecules are below and to the sides. This produces particularly strong attractions of the surface molecules immediately below them, and results in a cohesion of the surface or surface tension.

Adapted with permission of John Wiley & Sons, Inc. from Jespersen, N., Brady, J., and Hyslop, A., *Chemistry: The Molecular Nature of Matter*, Sixth Edition, p. 538. Copyright 2012.

John Churchman/Photolibrary/ Getty Images

As a result of surface tension, drops of water tend to form spherical droplets to minimize their surface area. This effect is more pronounced when the water is on a waxy surface, such as on this leaf or on a freshly waxed car.

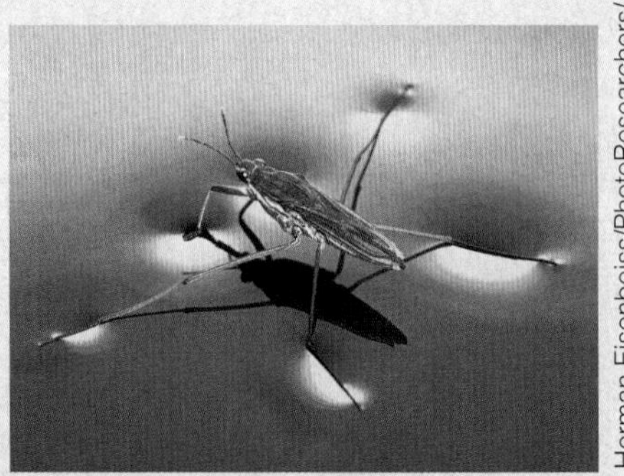

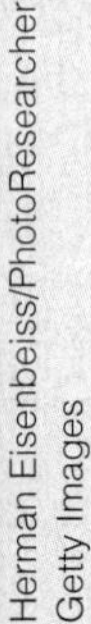

Surface tension is what prevents insects, such as this water strider, from sinking into the water.

Molecular structure of soap • Figure 11.2

Sodium stearate, the sodium salt of stearic acid, represents a typical soap molecule. The compound contains a hydrophilic (water-seeking) carboxylate ion head and a hydrophobic (water-avoiding) hydrocarbon tail.

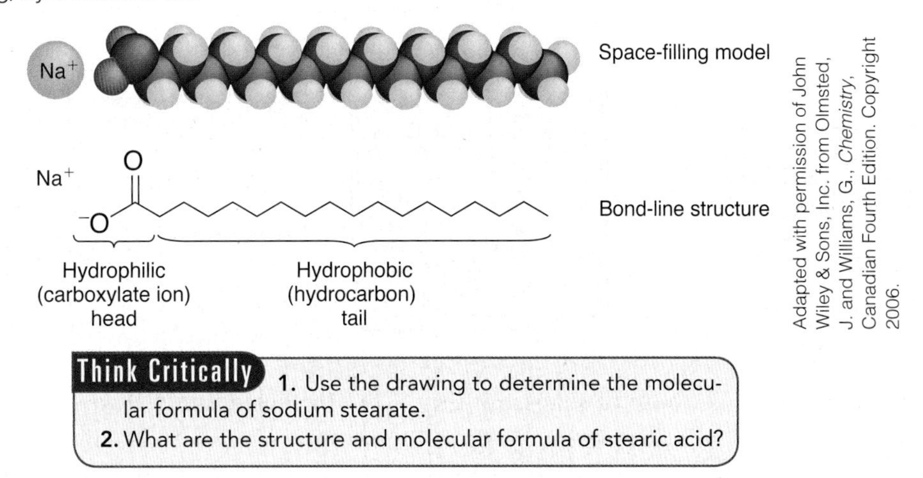

Think Critically **1.** Use the drawing to determine the molecular formula of sodium stearate.
2. What are the structure and molecular formula of stearic acid?

You might want to challenge a friend to duplicate what you're about to do next. Push a wooden toothpick ever so gently through the surface about midway between the tack and the edge of the glass. With care, you should be able to do this without sinking the tack. Now hand the toothpick to your friend. As soon as the tip of the toothpick enters the water, no matter how it was inserted, the tack drops immediately to the bottom of the glass. The secret of the trick lies in a twist of chemistry. Here's how it works.

Before beginning this demonstration, secretly coat one end of the toothpick with a drop of liquid soap or detergent. You push the *dry* end of the toothpick into the water, but as you hand it to your friend, you invert the toothpick (discreetly, unobserved) so that your friend sticks the detergent-coated end into the water. As the detergent molecules come into contact with the water, they sharply lower the water's surface tension and the tack drops.

To understand what gives rise to the surface tension that supported the tack in the first place, consider that the water molecules at the surface of water (and the surface of any liquid, for that matter) behave a little differently from those that constitute the bulk of the water within the glass. **Figure 11.1** illustrates this difference.

Soaps and Detergents

Each time you use a bar of soap, you go back a long way in chemical history. There's reason to believe the Babylonians knew how to make soap almost 5000 years ago. The Phoenicians and ancient Egyptians may have manufactured it as well, but most historians give the Romans credit for discovering soap, or at least for recording the details of its preparation. The Romans knew that heating goat fat with extracts of wood ashes, which contain strong bases, produces soap. They also used *sodium hydroxide* (NaOH), also known as *lye*, a stronger base than the ash extracts and more effective in converting fat into soap.

Soaps are detergents in the sense that they help clean oily and greasy dirt from various substances, including fabrics, dishes, skin, and hair. But soaps make up a very narrow class of detergents. We restrict the term **soap** to the sodium salts of long chain carboxylic acids. As we saw in Section 8.3, a **carboxylic acid** is marked by the presence of a carboxyl group ($-CO_2H$). Loss of a proton (H^+) from this carboxyl group gives the carboxylate anion ($-CO_2^-$). In the case of soap molecules, the negative charge of this carboxylate anion is balanced by the positive charge of a sodium cation (Na^+) as shown in **Figure 11.2**.

soap The sodium salt of a long-chain carboxylic acid.

Note the **bond-line** structure in Figure 11.2. In this shorthand notation:

- bonds are represented as lines between atoms,
- carbon atoms are not explicitly shown; rather, each point where lines meet or the endpoint of a line represents a carbon atom.

For more on bond-line structures and practice in interpreting them, see WileyPLUS.

The effect of detergents on surface tension • Figure 11.3

When detergents are added to water, they form a monolayer—a layer one molecule thick—at the surface.

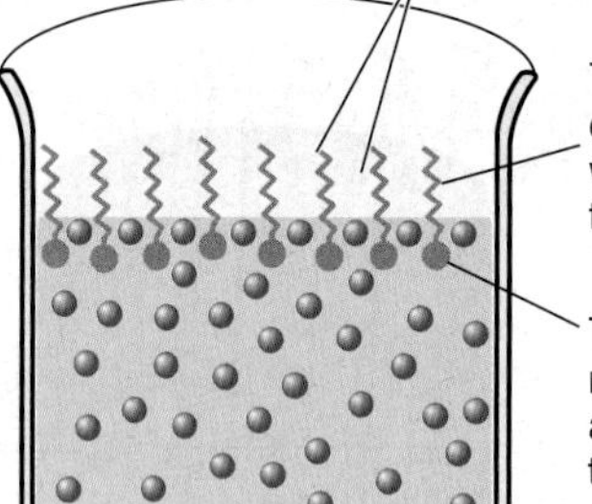

Detergent molecules disrupt attractive forces among water molecules at the surface and thereby lower surface tension.
The nonpolar, hydrophobic tails of the detergent molecules protrude out of the water's surface, thereby disassociating themselves from the water molecules.
The polar, hydrophilic heads of the detergent molecules are attracted to water molecules, and therefore they remain embedded among the water molecules of the water's surface.

Think Critically Why do detergent molecules experience both attractive and repulsive interactions with water molecules?

All **detergent** molecules, like those of soaps, consist of a hydrophilic portion and a hydrophobic portion. When they enter water, detergent molecules tend to accumulate at the surface of the water, where their **hydrophilic** (water-seeking) carboxylate groups can become embedded among the water molecules of the surface, and their **hydrophobic** (water-avoiding) hydrocarbon chains can protrude through and away from the water molecules at the surface. As shown in **Figure 11.3**, the detergent molecules become interspersed among the molecules at the water's surface. In disrupting this tightly knit layer of water molecules, the detergent interferes with the strong attractive forces that the surface water molecules normally exert on each other, thereby lowering the surface tension. This is why the tack drops when the toothpick introduces detergent into the water.

detergent A cleansing agent consisting of molecules that contain hydrophilic heads and hydrophobic tails.

Soaps, detergents, and other substances that accumulate at surfaces of bodies of water and change their surface properties sharply, especially by lowering the surface tension, are called surface-active agents or, more briefly, **surfactants**. Even a small amount of a surface-active agent can produce dramatic effects. At a concentration of 0.1%, for example, soap lowers water's surface tension by almost 70%.

surfactant Shortened form of surface-active agent; a chemical that accumulates at a liquid's surface and changes the properties of that surface.

Micelles As you add detergent to a body of water, the surface becomes saturated with surfactant molecules. Introducing more detergent causes the newly added detergent molecules to accumulate within the interior of the body of water. These molecules shield their hydrophilic tails from contact with water molecules by a process different from the one that takes place at the surface. As detergent molecules accumulate within the body of water, they coalesce into **micelles**—microscopic spheres of one substance distributed throughout another, usually a liquid—with their hydrophobic tails pointed toward the interior of the micelles. Thus the tails are shielded from contact with water molecules. The hydrophilic heads form the surface of the micelle and are in close contact with the surrounding water molecules (**Figure 11.4**).

micelles Spherical aggregations of detergent molecules.

Micelles • Figure 11.4

When detergent is added to water, the detergent molecules can shield their hydrophobic tails from water by aggregating into microscopic, spherical structures called micelles. A cutaway view of a micelle is shown here.

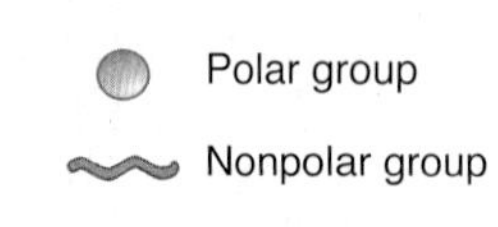

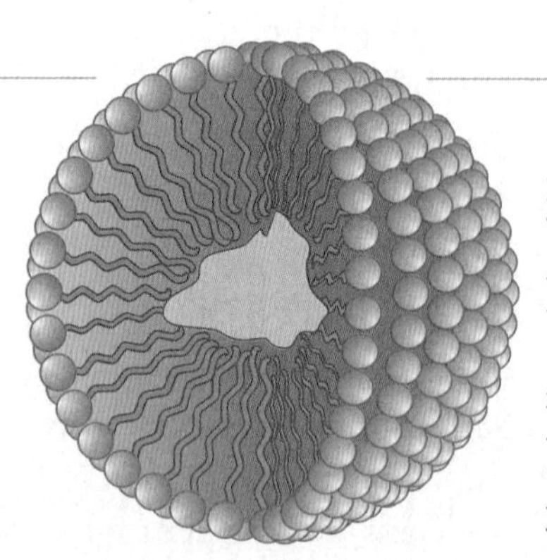

Although detergent micelles are well dispersed in water, they don't actually dissolve. They're present as a **colloid** rather than as the solute of a solution. [As discussed in an earlier chapter, a colloid is a stable dispersion of fine particles of one substance in another.]

How soaps and detergents clean Common sense tells us that water wets whatever it touches. However, if you look closely, you find that water isn't a particularly effective wetting agent after all. Examine the surface of a freshly waxed car just after a rain and you'll see the water forming small beads instead of spreading out in a thin layer covering the car's surface. Look at your clothing or at the top of an umbrella after you've been out in a light drizzle and you'll see the rain forming small beads on the fabric before it penetrates into the cloth.

In washing, though, water must penetrate well and deeply into the substance that we want to clean. A detergent, such as soap, lowers water's surface tension, allowing the water—which carries the detergent micelles—to get into the substance (such as a fabric) we want to clean and reach the dirt in its interior. When the micelles reach the embedded dirt, they trap these greasy or oily dirt particles within the micelles' hydrophobic interiors and then become dispersed or suspended again in the soapy water. This process prevents the greasy or oily particles from coalescing back to globules of grease that could redeposit on a clean surface and allows them to be rinsed away with fresh water. **Figure 11.5** illustrates this process.

Micelles formed by soaps have surfaces covered by the negatively charged carboxylate groups (the hydrophilic $-CO_2^-$ groups) of the embedded soap molecules. With a coating of negative electrical charges enveloping the entire surface of each micelle, the grease droplets repel each other and remain suspended in the wash water instead of coalescing and redepositing on the material being cleaned. Since the anionic carboxylate groups of the soap molecules place a cover of negative electrical charge on the micelles' surfaces, soap falls into the class of **anionic detergents**.

PROCESS DIAGRAM

How detergents clean • Figure 11.5

Detergents help water penetrate fabrics, and the detergent micelles trap grease particles within themselves.

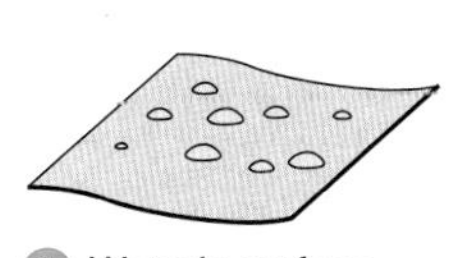

1 Water's surface tension causes it to bead on the surface of fabrics.

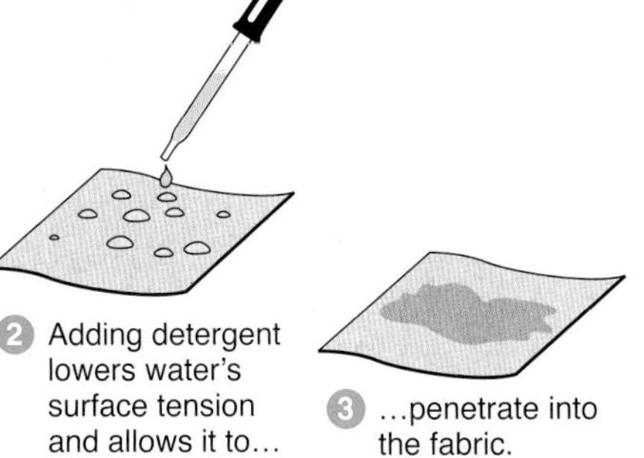

2 Adding detergent lowers water's surface tension and allows it to...

3 ...penetrate into the fabric.

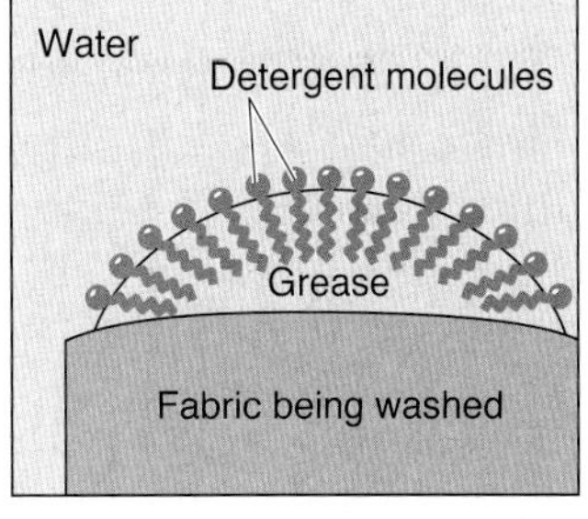

4 The detergent molecules form micelles, trapping dirt, grease, and oil particles within the hydrophobic interiors of these micelles.

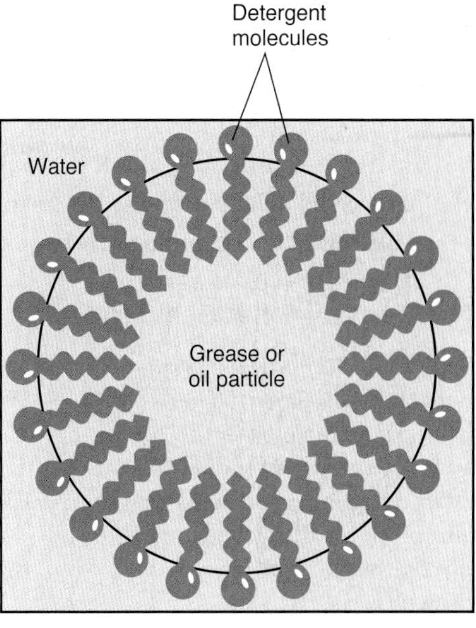

5 The grease- or dirt-embedded micelles disperse throughout the bulk of the water and are carried away by rinsing with fresh water.

Ask Yourself

Which portion of the detergent molecule associates with the **(a)** greasy dirt, **(b)** water?

Ester hydrolysis • Figure 11.6

Heating an ester with water in the presence of a base, such as sodium hydroxide, hydrolyzes the ester into an alcohol and the salt of a carboxylic acid.

$$CH_3CH_2O(C{=}O)CH_3 + NaOH \xrightarrow[\text{heat}]{H_2O} CH_3CH_2OH + Na^{+-}O(C{=}O)CH_3$$

Ethyl acetate, an ester · Sodium hydroxide, a base · Ethanol, an alcohol · Sodium acetate, the salt of a carboxylic acid

Ask Yourself

What alcohol and carboxylic acid salt would be produced on hydrolysis of the ester *methyl acetate*, $CH_3O(C{=}O)CH_3$, in the presence of sodium hydroxide?

How soap is made However soap is made—whether by boiling goat fat and extracts of wood ashes in ancient kettles or through more modern methods—it's all the same chemically. Soap forms from the hydrolysis of naturally occurring fats and oils, which are themselves triglycerides, or triesters of glycerol and fatty acids. (We discussed the structure of triglycerides in Chapter 5.)

To understand the process of soap-making, it may be helpful to first review the structure of esters. We can form an ester by the reaction of an alcohol and a carboxylic acid, as shown here.

$$CH_3CH_2OH + HO(C{=}O)CH_3 \rightleftharpoons CH_3CH_2O(C{=}O)CH_3 + H_2O$$

Ethanol, an alcohol · Acetic acid, a carboxylic acid · Ethyl acetate, an ester (Ester bond)

In this example, the alcohol, *ethanol*, combines with the carboxylic acid, *acetic acid*, to form the ester, *ethyl acetate*. This reaction is facilitated by heat and the presence of a catalyst, such as an acid. A molecule of water is released in the process. Note that this reaction can also proceed in the reverse direction, in which a molecule of water reacts with an ester to form an alcohol and a carboxylic acid. This represents a **hydrolysis**, which is the decomposition of a substance—in this case, an ester—through the action of water. When the hydrolysis of an ester is performed in the presence of a base, such as sodium hydroxide, the products are an alcohol and the salt of the carboxylic acid, rather than the acid itself (**Figure 11.6**).

We can hydrolyze a variety of esters by heating them with water in the presence of sodium hydroxide (lye). When we perform this hydrolysis on triglycerides—the ester molecules that make up all fats and oils—we produce glycerol and soap molecules. This process, called **saponification**, brings us to the chemistry of soap-making (**Figure 11.7**).

saponification Heating fats or oils in the presence of lye to produce soap and glycerol.

How soap is made • Figure 11.7

Soap is produced by heating a triglyceride—an animal fat or a vegetable oil—in an aqueous solution of sodium hydroxide (lye). The reaction also produces the alcohol, *glycerol*, also known as *glycerin*.

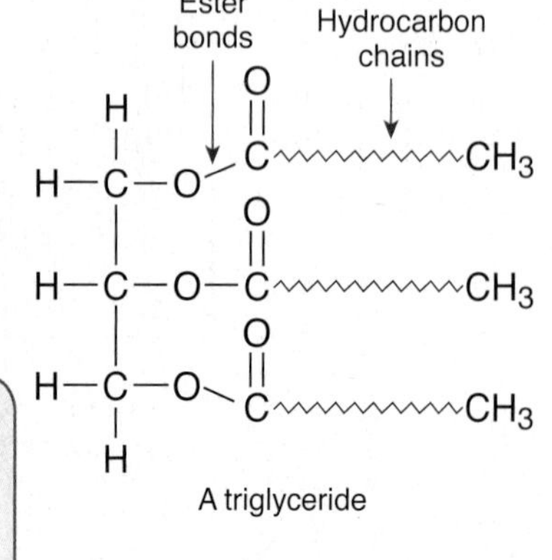

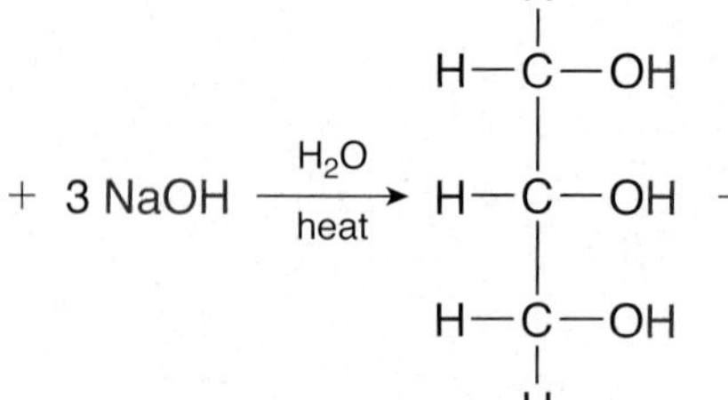

$$+ \; 3\,Na^{+-}O{-}(C{=}O)\sim\sim\sim CH_3$$

Soap molecules, the sodium salts of fatty acids

Ask Yourself

What parts of the chemical structure of any specific triglyceride are common to all triglycerides?

The contents of a bar of soap • Figure 11.8

A typical bar of soap can also contain skin conditioners, fragrances, stabilizers, and colorants, in addition to the primary ingredient: sodium salts of fatty acids containing even-numbered, unbranched chains of 10–18 carbons.

This soap comes from the saponification of tallow (animal fat), coconut oil, and/or palm kernel oil.

INGREDIENTS:
SOAP (SODIUM TALLOWATE, SODIUM COCOATE, AND/OR SODIUM PALM KERNELATE), WATER, GLYCERIN (SKIN CONDITIONER), HYDROGENATED TALLOW ACID (SKIN CONDITIONER), PETROLATUM, COCONUT ACID, FRAGRANCE, SODIUM CHLORIDE, POLYQUATERNIUM-6, ALOE BARBADENSIS LEAF EXTRACT, PENTASODIUM PENTETATE, PENTAERYTHRITYL TETRA-DI-T-BUTYL HYDROXYHYDROCINNAMATE, TITANIUM DIOXIDE, CHROMIUM OXIDE GREENS

These ingredients provide color.

Ask Yourself

Some soaps, including this one, include glycerin, also known as glycerol. How is this ingredient made and what role does it serve?

To sum up, then, heating an animal fat or a vegetable oil in an aqueous solution of sodium hydroxide generates both glycerol, which remains dissolved in the water, and the sodium salts of the various long-chain carboxylic acids (or **fatty acids**) that make up the triglyceride. These salts of the fatty acids congeal at the surface as the mixture cools and are removed as soap. Virtually all of these salts of naturally occurring fatty acids consist of straight, unbranched chains containing even numbers of carbons. Sodium salts of the fatty acids containing 10–18 carbons make the best soaps. A typical bar of soap or bottle of liquid soap contains molecules such as these, as well as other ingredients (**Figure 11.8**).

Other Surfactants

Figure 11.9 summarizes the relationships among the soaps, detergents, and surfactants we have been examining. Surfactants, or surface-active agents, constitute the broadest category of the three. These chemicals produce

Classifying soaps, detergents, and surfactants • Figure 11.9

All surfactants act at surfaces. Some surfactants are detergents, and some detergents are soaps. The structures of a soap, *sodium laurate*, and a common detergent, *sodium lauryl sulfate*, are shown here.

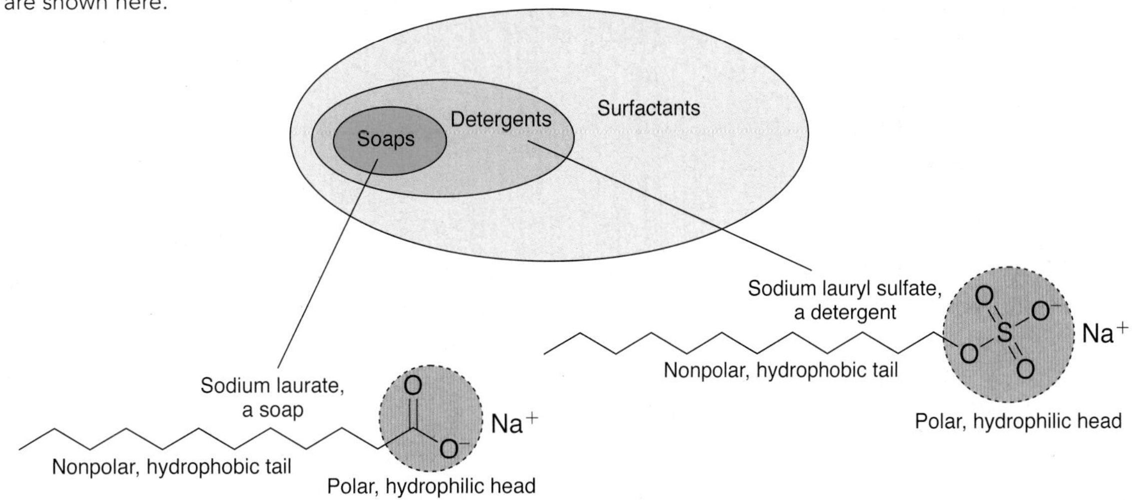

Think Critically

1. Are all soaps detergents?
2. Are all detergents soaps?
3. Are all surfactants detergents?
4. How many carbon atoms are present in each of the molecules shown?
5. What similarities exist within the structures of each of the molecules shown? What differences exist between these molecules?

Anionic, cationic, and nonionic surfactants • Figure 11.10

Surfactant molecules are classified into one of three categories, each of which provides distinctive benefits. Each of the products shown contains multiple ingredients.

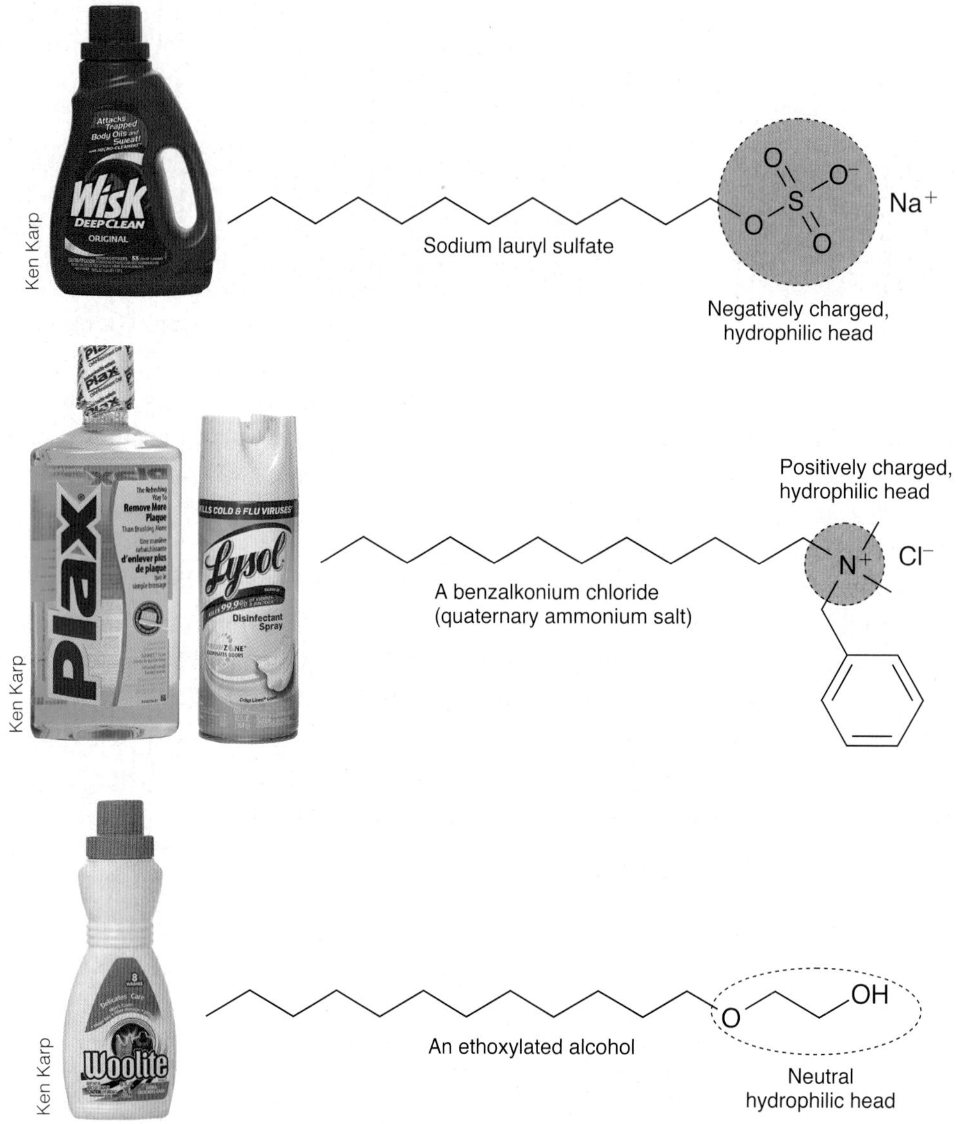

Ken Karp

Sodium lauryl sulfate

Na^+

Negatively charged, hydrophilic head

Ken Karp

Positively charged, hydrophilic head

Cl^-

A benzalkonium chloride (quaternary ammonium salt)

Ken Karp

An ethoxylated alcohol

Neutral hydrophilic head

their greatest effect at surfaces and, more specifically, have the ability to lower the surface tension of water. Detergents are a subclass of surfactants that includes the active ingredients of dishwashing liquids and laundry detergents. They serve nicely as cleansing agents. Soaps, obtained by the hydrolysis of triglycerides (a form of fat), constitute a subclass of detergents.

Note that all detergents (including soap molecules) have a typical molecular structure in common: a long, hydrophobic carbon chain that resembles a molecular "tail," which is connected to a hydrophilic "head." If this hydrophilic head bears a negative charge, these molecules fall within the class of **anionic surfactants.** If the hydrophilic head bears a positive charge, these molecules are categorized as **cationic surfactants**. A third class of surfactant molecules, which do not bear electrical charge, are called **nonionic surfactants**.

The great bulk of detergents used today consist of anionic surfactants. These anionic compounds—each with a negatively charged hydrophilic head—are particularly effective at cleaning fabrics that absorb water readily, such as those made of natural fibers of cotton, silk, and wool.

Most cationic surfactants are *quaternary ammonium salts* (or *quats* for short), which contain a nitrogen atom bonded to four *alkyl* or hydrocarbon groups. Cationic surfactants are effective germicides used in products such as disinfectant sprays, wet naps, and certain mouthwashes. Because the positive charges of the cationic surfactants adhere to many fabrics that normally carry negative electrical charges, these compounds also form the active ingredients in fabric softeners. The nonpolar, hydrocarbon chains of these cationic detergents tend to remain detached from the surface of the fabric, which helps to impart a soft, silky feel.

Nonionic surfactants typically have large numbers of covalently bonded oxygens in their hydrophilic structures. Detergents that fall within this class are especially useful in cleaning synthetic fabrics, such as *polyesters.* These nonionic detergents typically produce little foam and are often used in combination with anionic detergents in the formulations of dishwashing liquids and liquid laundry detergents. **Figure 11.10** shows the structures of some anionic, cationic, and nonionic surfactants and products in which they are used.

Hard Water

As discussed in the chapter on solutions, the drinking water of many regions contains various minerals, dissolved in slightly acidic rainwater as it filters through the soil. Water that's rich in the salts of calcium, magnesium, and/or iron is called **hard water**. (**Soft water**, on the other hand, is virtually free of these minerals.) When we use soap in hard water, the mineral cations of this water, such as Ca^{2+} and Mg^{2+} ions, combine with the fatty acid anions of soap molecules to form insoluble salts that precipitate out of solution as a waxy, insoluble material. In regions where the water is particularly hard, you can actually see these precipitates deposited as gray rings, known as **soap scum** or **curd**, around bathtubs and sinks after washing with soap. This curd is made up of the calcium, magnesium, and/or iron salts of the fatty acids of soap (**Figure 11.11**).

Using soap as a cleaning agent with hard water wastes soap because of this formation of soap scum or curd. In effect, much of the soap added to the water acts as a removal agent for the water's Ca^{2+}, Mg^{2+}, or Fe^{2+} ions rather than as a detergent. Not until these ions have been removed, by precipitating along with the soap molecules, can additional soap exert its cleaning effect. What's more, as this curd deposits on the surface of laundered clothes, it dulls the surface of the cloth, giving washed goods a slightly gray appearance.

You can see the effect of hard water on soap with the following exercise. Obtain an empty, plastic 1- or 2-liter soft drink bottle and fill it halfway with distilled water, which you can get in a supermarket or convenience store. Now add a small chunk of bar soap to the bottle, cap it, and shake. The distilled water is free of the mineral ions of hard water and should generate a good deal of suds with the soap. Now uncap the bottle and add about one tablespoon of milk, which contributes calcium ions and increases the hardness of the water. The suds should vanish as you swirl the milk into the soapy water because the fatty acid anions are converted into their insoluble calcium salts. You'll probably be able to see the curd swirling in the water.

Soap scum • Figure 11.11

Mineral ions present in hard water lower the solubility of soap, causing insoluble deposits, sometimes observed as *ring around the bathtub* or *sink*. The process is simply an ion exchange in which the Na^+ ions of soap are exchanged for the Ca^{2+}, Mg^{2+}, or Fe^{2+} ions of hard water. Here we illustrate the process with calcium ions.

Water softening • Figure 11.12

A typical home water softener contains polymeric ion exchange resins. During use (Step 1), ions in hard water get trapped by the resin, displacing sodium ions already on the resin. During recharge (Step 2), the resin is saturated with sodium ions, thus displacing the hard water ions.

From water supply: Mg^{2+}, Ca^{2+}, Fe^{2+} ions

Input valve

Ion exchange resin

Output valve

To household fixtures: Low levels of Na^+ ions

1. During use, water flows through an ion exchange resin. This resin has anionic (negatively charged) groups, which bind the incoming Mg^{2+}, Ca^{2+}, Fe^{2+} ions of hard water and release Na^+ ions, thus producing soft water.

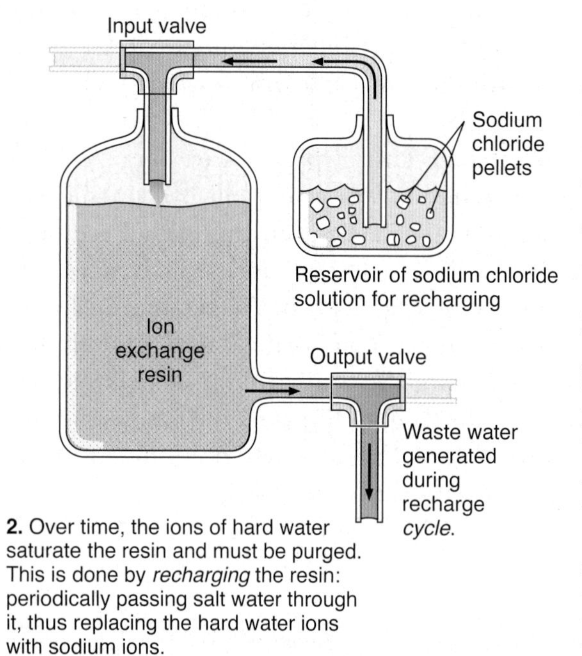

2. Over time, the ions of hard water saturate the resin and must be purged. This is done by *recharging* the resin: periodically passing salt water through it, thus replacing the hard water ions with sodium ions.

Ask Yourself

What happens to the calcium, magnesium, and iron ions of hard water that have accumulated on the ion-exchange resin as they are replaced by sodium ions during the recharging process? Where do these cations of hard water go?

Water softeners One solution to the problem caused by hard water is to remove the cations that produce the hardness. Commercial water softeners do this by replacing the water-hardening cations—the calcium, magnesium, iron, and similar ions—with sodium ions, thereby converting hard water to soft water.

A typical home water softener consists of a tank containing a specially prepared polymer that has a large number of anionic functional groups, which allow it to hold a large number of cations. You activate or charge the water softener by running concentrated salt water through it. The sodium ions of the salt water displace whatever cations might have been on the polymer and remain with the polymer in the tank. Then, when hard water passes through the tank, an exchange of cations takes place between the polymer and the water. The sodium cations leave the polymer and enter the water just as the water-hardening calcium and/or magnesium cations leave the water and take the places on the polymer that have been vacated by the sodium ions, thereby softening the water (**Figure 11.12**).

Synthetic detergents An alternative to softening hard water is simply to use synthetic detergents. These compounds maintain their solubility and cleansing abilities in hard water (in contrast to soaps, which do not). A common type of synthetic detergent contains a sulfonate functional group, $-SO_3^-$, rather than a carboxylate group, $-CO_2^-$, as the hydrophilic head of the molecule.

The advantage of the sulfonate anion is the great water solubility of virtually every one of its mineral salts. Unlike soaps, synthetic detergents, such as *alkylbenzenesulfonates* (ABS), remain dispersed and effective in hard water. As a result, these detergents are widely used anionic surfactants found in laundry detergents and other cleaning products. However, they tend to accumulate in water released by commercial cleaning and sewage treatment plants, causing suds to appear in streams and rivers. As a result, ABS detergents are being replaced with linear alkylbenzenesulfonate (LAS) detergents, which are biodegradable. Microorganisms in water can decompose these molecules into SO_4^{2-} ions and common substances such as H_2O and CO_2 (**Figure 11.13**).

Synthetic detergents • Figure 11.13

Alkylbenzenesulfonates, a type of synthetic detergent, are valued for their excellent detergent properties and their ability to remain soluble in hard water. In addition, those with linear alkyl chains, such as shown here, tend to be more biodegradable than those with branched chains.

A linear alkylbenzenesulfonate (LAS) detergent easily decomposed by microorganisms.

Linear alkyl (hydrocarbon) tail — Benzene ring — Sulfonate group ($SO_3^- Na^+$)

Think Critically **1.** Identify the hydrophilic and hydrophobic portions of the alkylbenzenesulfonate shown.
2. Draw the magnesium salt of this compound. Would the structure you have drawn be expected to be more soluble or less soluble in water than the magnesium salt of a soap?

Now, with understanding of what makes for an effective detergent molecule, we can ask: What's actually in a typical laundry detergent? Naturally, the most important ingredient of a commercial detergent is the surfactant itself. Historically, anionic alkylbenzenesulfonates have been the principal surfactants in detergent formulations. Today, though, most detergents contain a mixture of anionic and nonionic surfactants, as the result of a growing concern for the environment. Among the environmental advantages of nonionic surfactants is their smaller need for **builders**, which are compounds that effectively soften water by sequestering the mineral ions associated with hard water. (The term *builders* reflects the ability of these added agents to "build up" the detergent power of the surfactant.) Although all classes of synthetic detergents are superior to soaps in hard water, the nonionic detergents are even more effective than the anionic detergents. With less builder needed for the nonionic detergents, their formulations can be more concentrated and their packages smaller. Compact packaging decreases the amount of discarded packing material in trash and landfills, while higher concentrations of their formulations result in smaller amounts of detergents passed into the environment with the waste wash water.

For many years, **phosphates**—compounds structurally related to phosphoric acid, H_3PO_4—seemed to be the perfect builders. Their low cost, very low toxicity, and general absence of hazard, coupled with their ability to bind firmly to the ions of hard water and thereby reduce their chemical activity, made them the leading builders. Yet phosphates suffer from one overwhelming defect: They are superb nutrients for algae and other small plants that grow on the surfaces of lakes and streams. Algae, nourished by a steady supply of phosphates, can cover the surface of a body of water and prevent atmospheric oxygen from reaching the marine life below the surface, a process called **eutrophication**. The resulting death of fish and other aquatic animals, sometimes occurring on a large scale in lakes and rivers covered by algae, led many states to ban the use of phosphates as detergent builders (**Figure 11.14**).

Phosphate use and its alternatives • Figure 11.14

The use of phosphates in detergents has led to eutrophication, harming aquatic life in lakes, ponds, and rivers. Zeolites serve the same water softening functions as phosphates, yet do not appear to pose environmental concerns.

Surface covering of algae nourished by phosphates from agricultural runoff.
Chris R. Sharp/Science Source

DID YOU **KNOW?**

"Green" cleaners may offer environmental or safety benefits, but are not "chemical free"

The labels on "green" cleaners commonly tout the use of *natural* or *organic* ingredients, or the absence of *toxic or harsh chemicals, parabens, phthalates,* or *petrochemicals.* Market observers have labeled this trend "the greening of consumer products," although uniform criteria for what makes a product "green" have yet to be established. (The term *organic*, on the other hand, as it applies to food, involves very specific standards as set forth by the U.S. Department of Agriculture.).

Companies that manufacture and market environmentally friendly products are aware that it's not enough simply to be "green" in order to gain market share. Their products must successfully compete on performance, convenience, and other factors important to consumers. With our understanding of the typical ingredients in mass-market cleaning products, we can examine their counterparts found in "green" cleaners.

Ken Karp

Green cleaners typically contain plant-derived surfactants as well as essential oils.

The table provides one such set of comparisons. It shows some of the components—both naturally derived and synthetic—that are found in conventional and in "green" laundry detergents. Although the two types of detergents contain many of the same ingredients, the "green" products do not incorporate optical whiteners, and they use fragrances of biological origin rather than the synthetic compounds found in conventional detergents.

The surfactants used in green laundry detergents are *semisynthetic*, meaning that plant-derived feedstocks, such as *lauric acid* (common in coconut and palm oils), are converted chemically to the desired detergent ingredients. In addition, some green laundry detergents use very small amounts of synthetic preservatives to prevent the growth of mold and bacteria. This underscores the importance of product performance and stability. Occasionally, allowances are made for the use of synthetic ingredients when natural alternatives are not available.

For common household cleaning tasks, you can create your own "green" cleaners using various recipes available on the Internet. For example, baking soda (sodium bicarbonate, a base) is an effective mild abrasive and deodorizer. Vinegar (a dilute solution of acetic acid) can be mixed with equal amounts of water and a few drops of essential oils to create an all-purpose glass and surface cleaner.

Component	Conventional laundry detergent ingredients	"Green" laundry detergent ingredients
Surfactants	Alkylbenzenesulfonates (anionic) Laureth-9 (nonionic)	Sodium lauryl sulfate (anionic) Laureth-6 (nonionic)
Builders (water softeners)	Borax, DTPA, citric acid	Boric acid, sodium citrate (a derivative of citric acid)
Enzymes	Protease, amylase, and mannanase	Protease, amylase, and mannanase
Suspension agents/ Enzyme stabilizers	Polyethylene glycol, propylene glycol, ethanolamine	Glycerin, calcium chloride
Optical whiteners	Disodium diaminostilbene disulfonate	None
Fragrance	Synthetic compounds	Essential oils and botanical extracts
Preservatives	Methylisothiazolinone	Methylisothiazolinone

Think Critically Would you characterize soap molecules as natural, semisynthetic, or synthetic? Explain your reasoning.

The most promising substitute for phosphates is a class of compounds known as **zeolites**. These replacements for phosphates are cage-like compounds composed of aluminum, silicon, and oxygen. When added to hard water as part of a commercial detergent, the sodium salts of zeolites exchange their sodium ions for the ions that produce the water's hardness, just as ion-exchange resins do, thereby softening the water.

Other ingredients widely used in laundry detergents include stain-removing substances, such as oxygen-containing bleaches and enzymes that can decompose the stain-generating proteins of blood, grass, and other

Components of a typical laundry detergent Table 11.1		
Component	**Example**	**Function**
Surfactants	Alkylbenzenesulfonates	Provide detergency.
Builders	Zeolites	Soften water and increases surfactant's efficiency.
Enzymes	Protease and amylase	Help remove stains, such as from blood and grass.
Suspension agents	Carboxymethylcellulose (CMC), a carbohydrate polymer	Help keep dirt from redepositing on fabric.
Color-safe bleaches	Perborates	Help remove stains.
Optical whiteners	Fluorescent dyes	Make white clothes appear brighter.
Fragrance	—	Adds pleasing odor to both detergent and fabrics.

biological materials; and suspension (or antideposition) agents to help prevent suspended grease, dirt, and grime particles from redepositing on fabrics. All of these contribute to the surfactant's ability to produce a clean product.

Some of the remaining ingredients may add more to the appearance of a clean wash than to its reality. These include optical whiteners, which are organic compounds that coat the fabrics and convert the invisible ultraviolet component of sunlight into an almost imperceptible blue tint. The effect of this blue tint (which can also be produced by blue coloring agents added to the detergent) is to add a bit of brilliance to white fabrics and give them the appearance of extra cleanliness. Finally, added fragrances produce what many regard as a pleasant odor in the finished wash. **Table 11.1** summarizes common ingredients in laundry detergent.

The greening of consumer products With increasing concerns about the environmental impacts of our consumer-driven society, many consumer products, including those used for cleaning and personal care, have been reformulated and repackaged to make these products more environmentally friendly. This involves a life cycle or cradle-to-grave assessment (discussed in Section 1.2) of the raw materials used, how the product is formulated and packaged, and how much energy is consumed in the manufacturing, transportation, and use of the product. In the case of "green" cleaners, for instance, sustainably sourced ingredients, including plant-derived surfactants, solvents, and fragrances, are often incorporated into the product, while chemicals believed to pose health or environmental risks are avoided. (See *Did You Know?*)

CONCEPT CHECK

1. **How** does soap lower the surface tension of water?
2. **What** interactions between molecules exist in the interior of micelles?
3. **Which** part of a surfactant molecule determines whether the surfactant is anionic or cationic?
4. **Which** ions are typically exchanged when water is softened?

11.2 Cosmetics and Skin Care

LEARNING OBJECTIVES

1. **Define** the term *cosmetic*.
2. **Describe** the structure of skin and the chemistry involved in common skin-care products.

The urge to improve our appearance—to look good and to smell nice, especially by applying chemicals to our bodies—is common to the entire human race, just as it has been throughout history. In this section, we'll explore the kinds of skin-care products we use for these purposes and their underlying chemistry.

Cosmetics: Then and Now

Peoples of all regions and of all times have painted themselves—sometimes with plant extracts, sometimes with powdered metals—to appear more vivid and to appeal to the eyes of others. Henna, a hair dye extracted from the henna plants of Africa, India, and the Middle East, and

Henna • Figure 11.15

The pigments naturally present in henna extract have been used for skin and hair dyeing since ancient times and are still in use today.

Jodi Cobb/National Geographic Creative

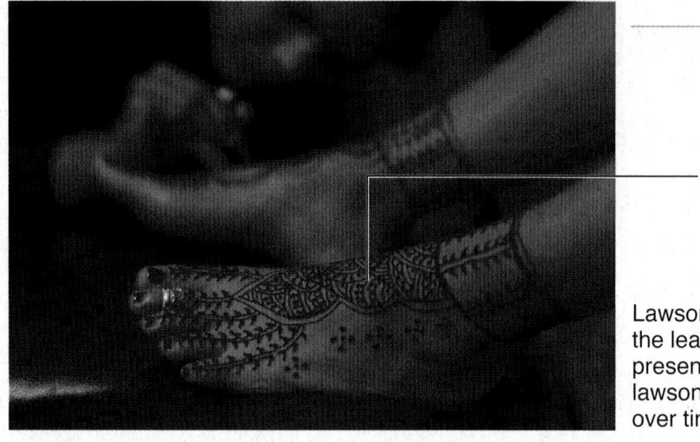

Lawsone, $C_{10}H_6O_2$, is a reddish dye found in the leaves of the henna plant. Keratin protein present in skin and hair chemically bonds to lawsone in the henna extract. The dye fades over time as surface skin layers are shed.

still in current use (**Figure 11.15**), once colored the hair of Egyptians now mummified. Vases of cosmetics, oils, and ointments were sealed into the tombs of the kings of ancient Egypt to make them more presentable in the afterlife.

The word *cosmetic* derives from the ancient Greek word *kosmos*, meaning order, adornment, and the universe itself. The connection among the three meanings comes from the belief of the Greek philosophers that order has beauty and that the universe is an ordered place and therefore beautiful. Today, according to the Federal Food, Drug and Cosmetic Act, a cosmetic is anything intended to be applied directly to the human body for "cleansing, beautifying, promoting attractiveness, or altering the appearance" (presumably for the better). According to the Act, soap isn't legally considered to be a cosmetic. But for our purposes, we can consider that anything we apply directly to our bodies to make us more attractive fits the definition of a cosmetic.

The growing appreciation that a healthy body is an attractive body can lead to confusion between a cosmetic and a medicine. We'll examine medicines more closely in a later chapter, but for now we can say that cosmetics make people more attractive, whereas medicines are used specifically for treating diseases, illnesses, and injuries. Cosmetic companies that claim their products improve health must be able to verify their claims to the Food and Drug Administration. Manufacturers of skin-care products, for example, sometimes claim that their creams and ointments lead to healthier or more youthful skin. In fact, unless the substance can actually cure a disease of the skin, it's a cosmetic rather than a medicine.

In the United States, four categories of products account for about half of the total sales of all personal care products: hair-care products; skin-care products; hand, face, and body soaps; and perfumes and colognes. The manufacturing and marketing of these products can be quite lucrative. Overall, only about a tenth of the retail cost of the average cosmetic or toiletry goes toward its raw materials: detergents, fragrances, moisturizers, solvents, and the like. Manufacturers bear other costs, including packaging, distribution, and marketing, as well as research and development expenses.

Of all the chemicals used in personal care product formulations, two classes, aside from water and common solvents, stand out:

- Surfactants constitute the largest single category of chemicals used in the manufacture of personal care products and are the critical components of shampoos, soaps, toothpastes, and related products.
- Fragrances, though contributing less than 1% to the total weight of all personal care products combined, make up about 25% of the total costs of all the raw materials used in their manufacture. Gram for gram, fragrances are by far the most expensive of all the ingredients.

To meet our broader expectations, each cosmetic and toiletry is designed to perform a set of at least three different kinds of tasks, each important to the commercial success of the product. First comes the principal function. For example, we expect shampoos to produce suds and to clean hair. Second is a set of expectations relating to specific applications on hair, teeth, skin, or nails. We expect shampoos, for example, to provide body and luster to hair, as well as to preserve at least some of its oils as it cleans; we expect toothpastes to give us a fresh-tasting mouth and clean-smelling breath along with clean, cavity-resistant teeth. Finally, there are the more subtle qualities expected by consumers. Personal care products must provide perceived value, must be convenient to use, and must convince people that they do their jobs well and effectively. Along with all this—whether we're consciously aware of it or not—the advertising campaigns and brand identities of the products we select for personal care should reflect and reinforce our self-image, values, and aspirations.

Skin Care

Skin is the body's largest organ. On average, about a quarter of every dollar we spend for cosmetics goes toward skin care through the use of moisturizers and emollients (softeners); hand, face, and body soaps; and deodorants and antiperspirants.

Skin's most obvious functions are covering and containing the body and protecting it from damage by foreign matter and microorganisms. But it also has the more subtle tasks of regulating body temperature, sensing stimuli provided by the outside world, and even synthesizing compounds such as vitamin D (discussed in Chapter 14). The skin itself consists of two major layers. The underlying layer, known as the **dermis**, contains nerves, blood vessels, sweat glands, and the active portion of hair follicles. It also supports the upper layer, called the **epidermis** (**Figure 11.16**). The epidermis consists of several tiers of cells. At its bottom, resting on top of the dermis, is a single sheet of cells that divide continuously, always pushing upward. As they move toward the outside of the skin, driven along by new cells coming up from beneath, they lose their ability to divide and they eventually die. By the time they become the lifeless, outermost layer of the skin, called the **stratum corneum**, they have been transformed into keratin, the same protein that forms hair.

Skin lotions Commercial skin creams and lotions help keep the skin soft and moist, in much the same was as sebum does. These personal care products are **emulsions**, or colloidal dispersions of two or more liquids that are insoluble in each other (usually an oil and water, as discussed in Chapter 7). If the first ingredient listed on the cream or lotion is an oil, it's likely a **water-in-oil emulsion**, with an oil content greater than 50%. If the first ingredient listed is water, it's probably an **oil-in-water emulsion**, with a water content greater than 50% (**Figure 11.17**). A cold cream, one of the oldest and most common skin moisturizers, consists of an emulsion with mineral oil as the major ingredient and water and waxes as minor components. In the more freely flowing hand and body lotions, water and oils replace some of the waxes of the creams. Since the function of these lotions and creams is to keep the outer layer of skin moist, they are most effective when applied after a bath or shower, while the skin is still damp.

> **emulsion** A colloidal dispersion of two or more liquids that are insoluble in each other.

Antiperspirants and deodorants The body controls its temperature by perspiring or sweating. The evaporation of the water contained in perspiration is what cools the body (see Figure 6.3). When you're warm or tense, your sweat

Skin structure • Figure 11.16

The soaps and cosmetics we apply to our skin coat the outermost layer, called the *stratum corneum*.

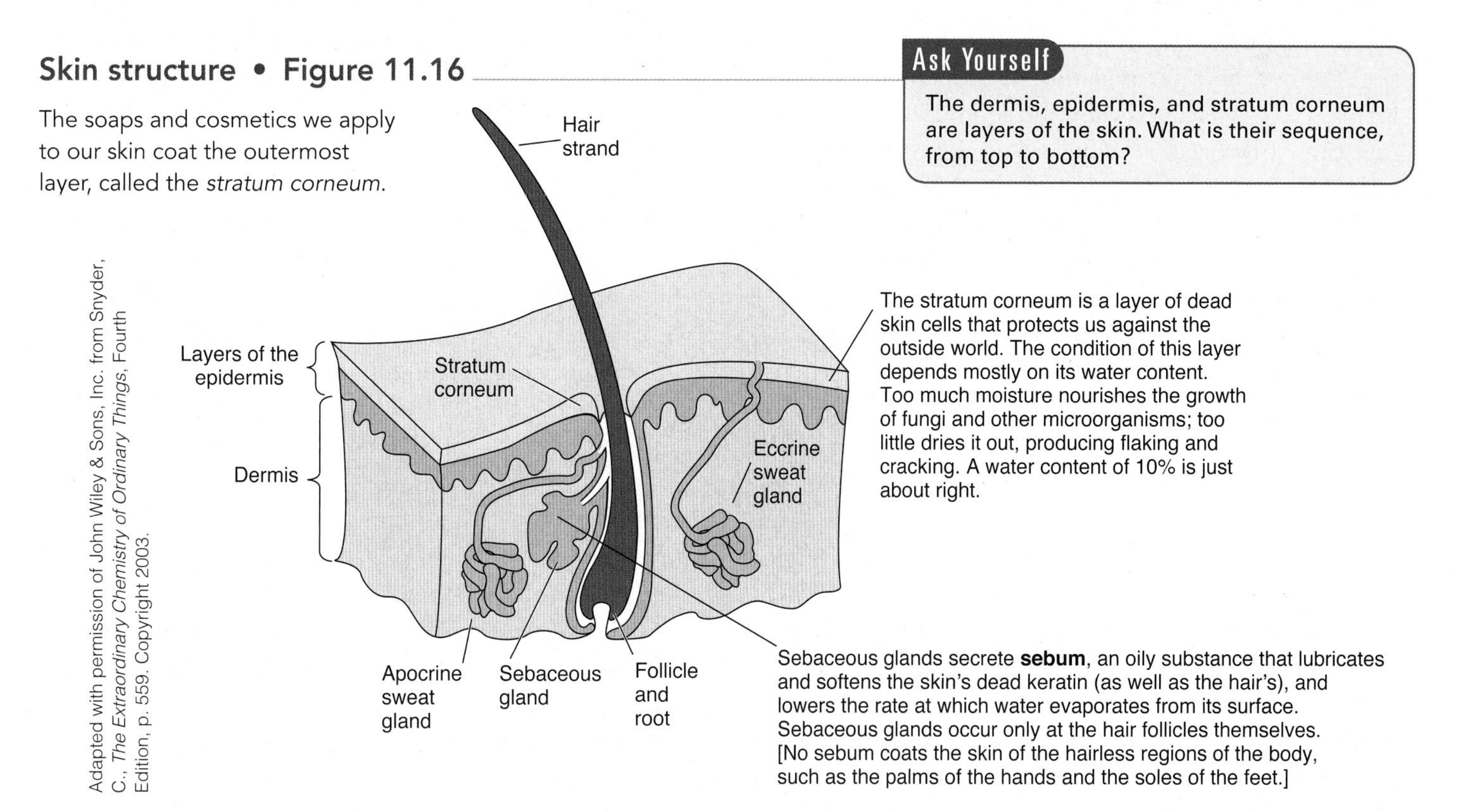

Hair strand
Layers of the epidermis
Dermis
Stratum corneum
Eccrine sweat gland
Apocrine sweat gland
Sebaceous gland
Follicle and root
The stratum corneum is a layer of dead skin cells that protects us against the outside world. The condition of this layer depends mostly on its water content. Too much moisture nourishes the growth of fungi and other microorganisms; too little dries it out, producing flaking and cracking. A water content of 10% is just about right.
Sebaceous glands secrete **sebum**, an oily substance that lubricates and softens the skin's dead keratin (as well as the hair's), and lowers the rate at which water evaporates from its surface. Sebaceous glands occur only at the hair follicles themselves. [No sebum coats the skin of the hairless regions of the body, such as the palms of the hands and the soles of the feet.]

Adapted with permission of John Wiley & Sons, Inc. from Snyder, C., *The Extraordinary Chemistry of Ordinary Things*, Fourth Edition, p. 559. Copyright 2003.

Ask Yourself

The dermis, epidermis, and stratum corneum are layers of the skin. What is their sequence, from top to bottom?

Skin lotions and creams • Figure 11.17

Ingredients in cosmetics are listed in order of their abundance. For skin lotions and creams, the first ingredient is the solvent or substance present in the greatest amount, which determines the type of emulsion present.

Think Critically Triethanolamine, $(HOCH_2CH_2)_3N$, and glycerin, $HOCH_2CH(OH)CH_2OH$, are often used in skin lotions (as well as in many other cosmetics). These compounds act as humectants, attracting water and helping to moisturize skin. Through what type of interaction do these molecules attract water?

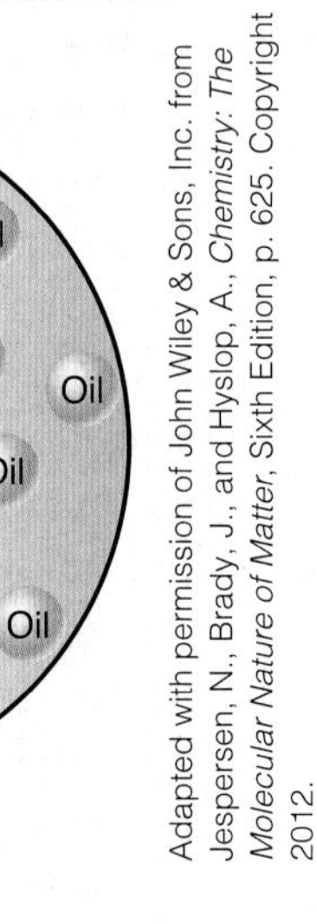

	A typical cold cream	A typical moisturizing lotion
Ingredients listed earliest	Mineral oil, water, ceresin (a wax), beeswax, triethanolamine	Water, glycerin, petrolatum, stearic acid, glycol stearate

glands begin to excrete perspiration. One set of these glands, the **eccrine glands** (sometimes called the "true" sweat glands; see Figure 11.16), covers most of the skin. They're especially dense on the forehead, face, palms, soles, and armpits, and they secrete a slightly acidic, very dilute solution of inorganic ions (largely sodium, potassium, and chloride), lactic acid ($CH_3—CHOH—CO_2H$), some urea ($H_2N—CO—NH_2$), and a little glucose.

Another set of sweat glands, the **apocrine glands** (see Figure 11.16), release a different kind of substance, one that can easily become disagreeable. Like the sebaceous glands, these apocrine glands secrete their fluids into the hair follicles. But unlike the sebaceous glands, which occur wherever hair grows, apocrine glands lie almost exclusively under the arms, in the groin, and in a few other smaller regions of the body. Their secretions produce little or no odors in themselves, but bacteria that accumulate in the nearby strands of hair can degrade the contents of the apocrine fluids into unpleasant-smelling products. To control perspiration and any of its associated odors, we can use antiperspirants and deodorants. Antiperspirants inhibit sweating and keep the body relatively dry. Deodorants, on the other hand, directly attack odors themselves.

The most widely used antiperspirants are ionic compounds containing aluminum cations (Al^{3+}), such as aluminum chlorohydrate. As these ions enter skin cells, water is drawn in and the cells swell, causing the ducts of the eccrine glands to squeeze shut, thereby preventing the release of sweat. These aluminum compounds also reduce the odor associated with the apocrine glands, probably by killing the bacteria that decompose the organic portion of the fluid.

Deodorants mask or eliminate odors with pleasant fragrances and also with antibacterial agents that remove the bacteria. In addition to the aluminum salts, deodorants include antibacterial compounds such as *triclosan,* also widely used in antibacterial soaps. Since the odor itself comes from the action of bacteria on the accumulated perspiration, daily washing alone often solves many of these problems.

Cosmetics that add color Coloring cosmetics include lipsticks, eye colorings, nail polish, and face powder. About half the weight of a typical lipstick consists of highly purified castor oil (a mixture of triglycerides obtained from plant seeds). This oil serves to dissolve the dyes and, more importantly, to give the substance the ability to remain a waxy solid in its container yet flow smoothly as it touches the warmer lips. Most of the remainder of the stick is a mixture of oils, waxes, and polymers, all formulated to produce a desirable texture, allow the film to flow onto the lips, and help keep the lips moist and soft. The remaining ingredients—dyes, perfumes, and preservatives—make up a very small percentage of the weight. *What a Chemist Sees* explores the nature of some of these dyes, as well as the composition of other cosmetics, including nail polish and eye shadow.

WHAT A CHEMIST SEES

Cosmetics That Add Color

Most dyes and pigments are classified as organic (carbon-based) or inorganic (metal-based). Organic dyes tend to be highly **conjugated**, having an extended system of alternating single and double bonds. These structures absorb components of visible light, resulting in the generation of color. Inorganic pigments contain ions of **transition metals**, such as iron, chromium, and titanium (elements found in the middle region of the periodic table). These metal ions can absorb or scatter light, thereby producing color.

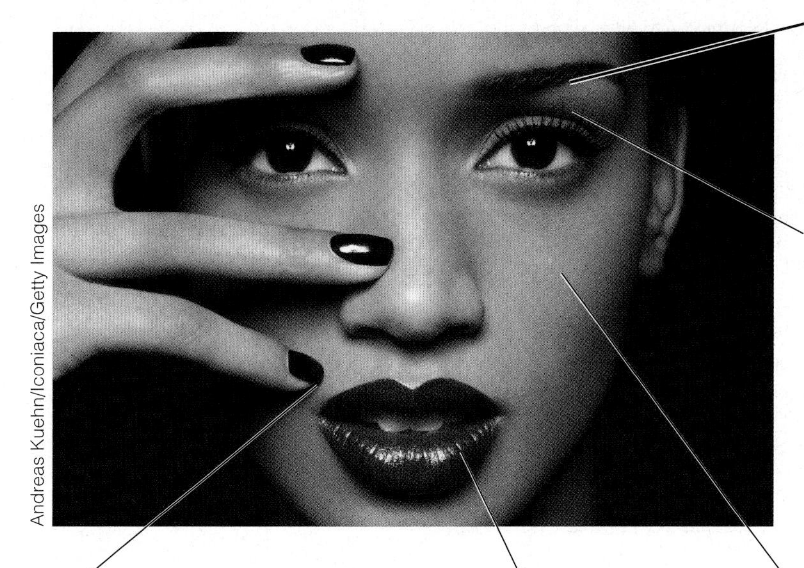

Andreas Kuehn/Iconiaca/Getty Images

Eye shadows use largely inorganic dyes including ultramarine blue (an inorganic polymer containing aluminum, oxygen, silicon, sodium, and sulfur), iron oxides of various shades of red, carbon black (a form of carbon resembling powdered charcoal), and titanium dioxide, TiO_2, a white powder.

Eyelashes turn pale near their ends, so they tend to become invisible against the background of the skin and eyes. Mascaras make the lashes appear longer and thicker by darkening both the ends and shaft, for more contrast and volume. To protect the eyes against potential allergic reactions to synthetic dyes, the FDA requires that only natural dyes, inorganic pigments, and carbon black be used as mascara colorants.

Nail polish is a flexible lacquer that can bend with the nail rather than crack and flake. Its pigments include ultramarine blue, carbon black, organic dyes, and various metal oxides. These, together with nitrocellulose (to help form a smooth film), resins (to help the lacquer harden and stick to the nails), and plasticizers (to help give flexibility to the dried polish), are mixed in organic solvents. These solvents include the esters, butyl acetate and ethyl acetate. As they evaporate they leave the dyes embedded in a polymeric film that grips the nail.

Ethyl group
Butyl group
Ethyl acetate
Butyl acetate

Volatile organic solvents in nail polish.

Face powders are designed to dull the glossy shine of sebum and perspiration, while at the same time adding a pleasant tint, texture, and odor of their own. A typical face powder includes the minerals talc, kaolin, and zinc oxide, the chemical formulas and functions of which are shown here.

Ingredient	Formula	Function
Talc	$Mg_3(Si_2O_5)_2(OH)_2$	Provides bulk and improves spreadability.
Kaolin	Al_2SiO_5	Absorbs water.
Zinc oxide	ZnO	Helps mask blemishes.

Lipstick dyes include D&C Orange No. 5 (dibromofluorescein), and D&C Red No. 22 (tetrabromofluorescein). The D&C indicates that the Food and Drug Administration has approved these dyes for use in **d**rugs and **c**osmetics.

Dibromofluorescein (D&C Orange #5)

Tetrabromofluorescein (D&C Red #22)

Organic dyes in lipstick and other cosmetics. The interaction of light with the systems of alternating single and double bonds within these molecule produces color.

Ask Yourself

1. Mascara doesn't actually lengthen eyelashes, but it does make them appear to be longer. How does mascara produce this effect?
2. Identify the conjugated system within tetrabromofluorescein. What observable property of the compound does this conjugated system produce?

Fragrance notes • Figure 11.18

The musical metaphor of notes is used to describe the three successive phases over which the aromas of a perfume or cologne reach our senses.

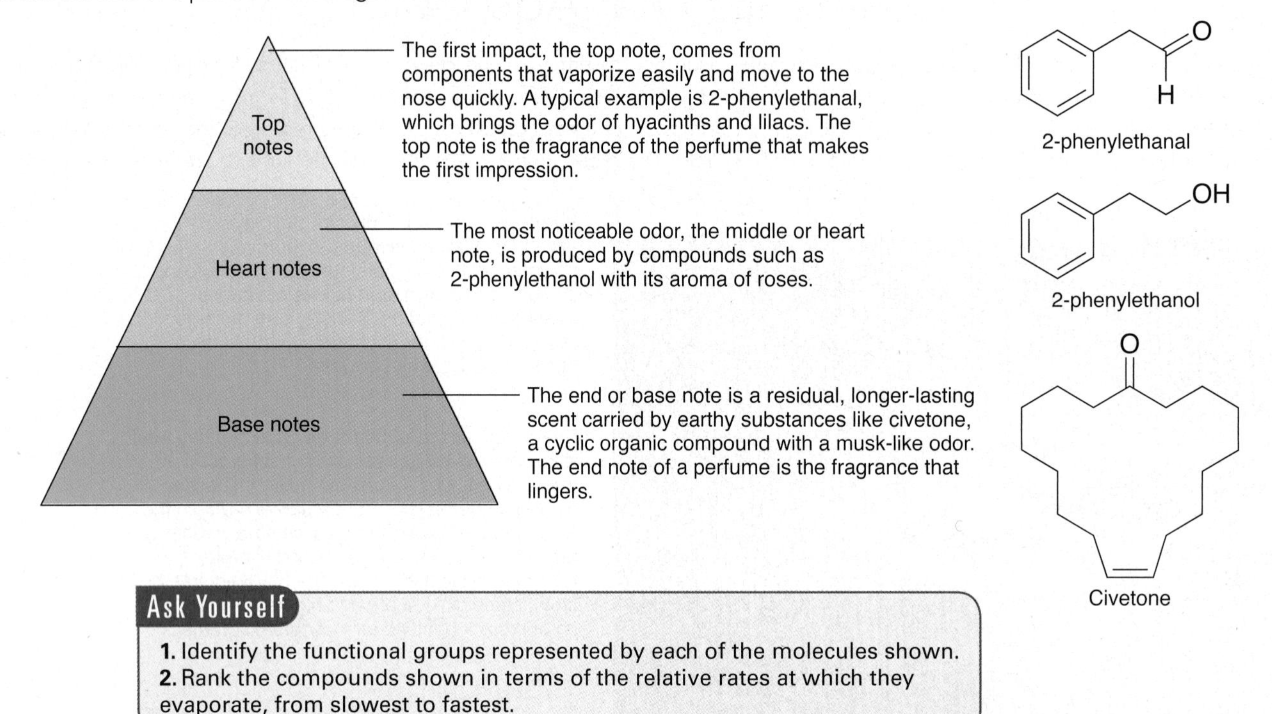

Top notes
Heart notes
Base notes
The first impact, the top note, comes from components that vaporize easily and move to the nose quickly. A typical example is 2-phenylethanal, which brings the odor of hyacinths and lilacs. The top note is the fragrance of the perfume that makes the first impression.
The most noticeable odor, the middle or heart note, is produced by compounds such as 2-phenylethanol with its aroma of roses.
The end or base note is a residual, longer-lasting scent carried by earthy substances like civetone, a cyclic organic compound with a musk-like odor. The end note of a perfume is the fragrance that lingers.
O
H
2-phenylethanal
OH
2-phenylethanol
O
Civetone

Ask Yourself

1. Identify the functional groups represented by each of the molecules shown.
2. Rank the compounds shown in terms of the relative rates at which they evaporate, from slowest to fastest.

Fragrances Like other body adornments, perfumes and colognes have their origins in antiquity. The word "perfume" itself comes from the Latin *per* (through) and *fumus* (smoke), and may have applied originally to scents carried by the smoke of incense and odorous plants used in sacred ceremonies. Cologne is a much more dilute and generally less expensive version of a perfume, with concentrations of the fragrant oils running about a tenth of those used in perfumes. The term itself refers to the city of Cologne, Germany, where in the early 1700s, a lotion based on citrus fruit was manufactured.

Today's perfumes are the results of a long history of changes in the popularity of different sorts of odors and of the methods of blending them. Modern perfumes are mixtures of various synthetic chemicals and plant extracts, all dissolved as 10–25% solutions in alcohol. These substances are **volatile**, meaning they evaporate readily, thus mixing with the air and allowing us to smell them. The rates at which these compounds evaporate can vary, depending on their structures, but overall, their volatility is increased by body warmth. The mixed fragrances are designed to reach the nose in three *notes* or phases (**Figure 11.18**).

Some of the more common organic functional groups represented by the compounds in perfumes and colognes include

$R{-}OH$	$R{-}\overset{O}{\overset{\|\|}{C}}{-}H$	$R{-}\overset{O}{\overset{\|\|}{C}}{-}R'$	$R{-}\overset{O}{\overset{\|\|}{C}}{-}OR'$
Alcohols	Aldehydes	Ketones	Esters

Here, the R and R′ stand for any carbon-based groups. Notice that aldehydes, ketones, and esters contain the carbonyl (car-bo-NEEL) group (C═O), with a carbon atom double bonded to an oxygen atom.

The aromas represented by these fragrance molecules can vary widely, but some general characteristics exist. For example, many of the esters are known for their fruity aromas, such as *propyl acetate*, $CH_3CH_2CH_2O(C{=}O)CH_3$, which smells like pears.

The choice of specific compounds used in a fragrance depends on the nature of the application, the chemical behavior of the available materials, and safety considerations. A nicely scented substance that oxidizes easily to a foul-smelling product could hardly be used in a face powder or laundry detergent, for instance. Consumers' perceptions and expectations come into play as well as technical factors. A deodorant soap, for example, might carry a strong smell suggesting the clean air of a forest, yet the scent of a lipstick must be more understated so that it does not interfere with the taste of food and drink.

Sunscreens To understand how our bodies respond to the sun's radiation—why we can get a sunburn or tan by exposure to the sun, and how sunscreens work—let's first take a closer look at the sun's rays. The sun's energy and light reach Earth as electromagnetic radiation, the same phenomenon that includes radio signals, microwaves, and X-rays. This radiation is wavelike, similar to waves of water. Waves of both electromagnetic energy and water travel at a measurable speed, with alternating crests and troughs. We can characterize the **wavelength** of these waves simply as the distance from one crest to another. Furthermore, since all forms of electromagnetic radiation travel at the speed of light, which is a physical constant, we can describe these waves in terms of their **frequency**, which is the number of waves that pass a fixed point in one second (**Figure 11.19**). These two characteristics of any wave motion, wavelength and frequency, are mutually reciprocal. That is, the larger the wavelength, the smaller the frequency (or the smaller the wavelength, the larger the frequency).

Wavelength and frequency • Figure 11.19

Electromagnetic radiation is characterized by both wavelength (a) and frequency (b).

a. Wavelength is the distance between two successive crests of a wave.

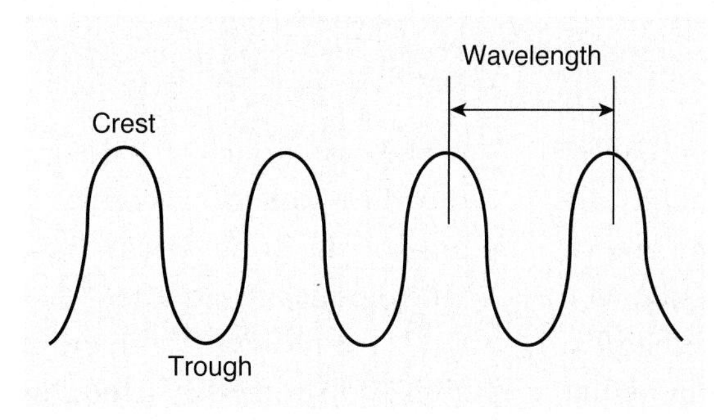

b. Frequency is the number of wave cycles per second. Here, four complete wave cycles pass a point in one second, so the frequency is four cycles per second.

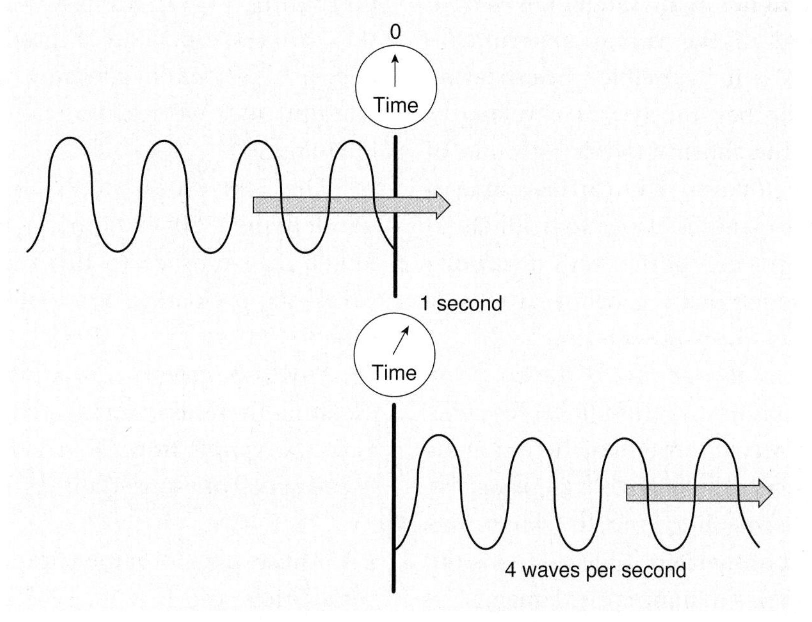

The electromagnetic spectrum • Figure 11.20

Electromagnetic radiation exists in a range of wavelengths (and associated frequencies), from highest-energy gamma rays to lowest-energy radio waves. Visible light spans the wavelength range of 400–700 nanometers (nm) or billionths of a meter.

Interpret the Data

Which is higher in energy: X-rays or microwaves?

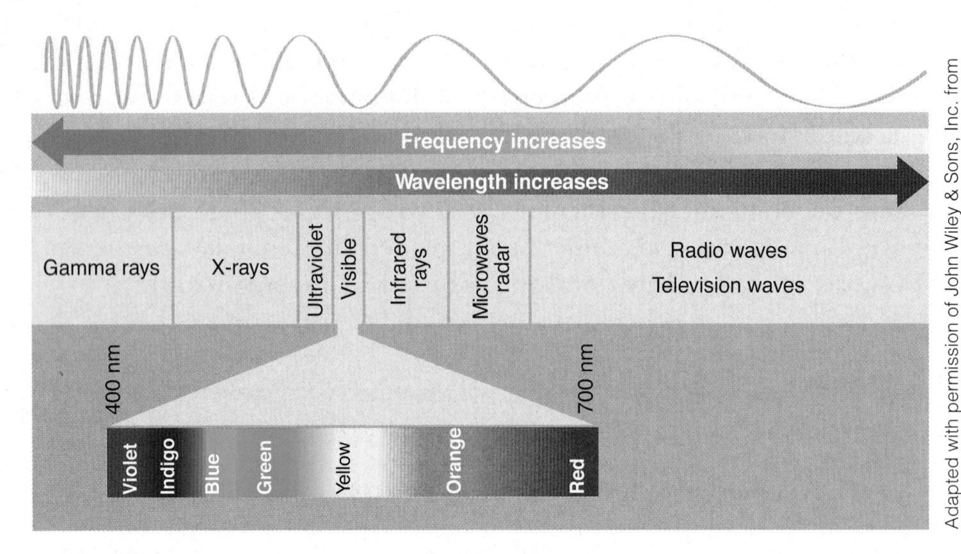

Electromagnetic radiation exists in a wide range of frequencies (and associated wavelengths) called the **electromagnetic spectrum** (**Figure 11.20**). The energy of a given radiation depends only on its frequency. The greater its frequency (or the shorter its wavelength), the higher its energy. X-rays, for example, have a higher frequency (and a shorter wavelength) and therefore a higher energy than radio waves.

electromagnetic spectrum The full range of frequencies that characterizes electromagnetic radiation.

Within the visible region, red light lies at the longer-wavelength end of the visible spectrum, at 700 nm. Invisible, heat-bearing infrared radiation lies even beyond that, at wavelengths longer than 700 nm. At the shortest-wavelength end of the visible spectrum, near 400 nm, light carries a blue-violet color. Still shorter wavelengths, running from about 290 to 400 nm, form the portion of the sun's **ultraviolet (UV) radiation,** which penetrates the ozone layer and reaches Earth's surface and our exposed skin.

A suntan is simply the visible evidence of the body's attempt to protect itself from the harm this ultraviolet radiation can cause. In fact, the very absorption of this energetic ultraviolet radiation stimulates the body to take protective measures. As one means of shielding itself, the skin begins producing the dark pigment **melanin**, which screens out part of the radiation and helps minimize the damage.

The potential for damage from the absorption of ultraviolet radiation comes from two adjoining segments of the sun's UV radiation, called the UV-A and UV-B regions, which affect the skin in slightly different ways. UV-B radiation, the more powerful of the two, occupies the shorter wavelength region, from 290 to 320 nm. It does part of its damage quickly, with the remainder coming more slowly, over a longer term. The acute injury lies in the tissue destruction of sunburns, and the associated redness, blistering, and peeling. Repeated exposure to UV-B radiation over a longer period produces skin cancers, especially among fair-haired, light-skinned people. One beneficial effect of the absorption of small amounts of UV-B radiation, however, is the generation of vitamin D.

The less energetic UV-A radiation, with longer wavelengths (320–400 nm), doesn't cause sunburns. But long-term exposure to this radiation is associated with wrinkling, premature aging of the skin, and certain skin cancers.

Most sunscreens now available are **broad-spectrum**, meaning that they provide protection against both UV-B and UV-A radiation. The active ingredients of these broad-spectrum sunscreens fall into two major categories:

- **UV absorbers**: organic compounds that absorb UV radiation, and

- **Scattering agents**: inorganic oxides, such as zinc oxide (ZnO) and titanium dioxide (TiO_2). These scatter or block all the radiation, letting none through to the skin. They are usually applied to the nose and the tops of the ears, areas that receive the most direct exposure.

The UV absorbers *para-aminobenzoic acid* (or PABA for short) and its derivatives, were originally incorporated into sunscreens because of their excellent protection against UV-B radiation. However, these compounds have largely fallen out of favor because of allergenic and other risks. Among the more widely used UV absorbers are compounds shown in **Figure 11.21**.

Zinc oxide or titanium dioxide of sunblocks, the white pastes often used by lifeguards, provide protection from the sun by scattering and reflecting sunlight. More advanced, **nanotechnology** versions of these materials are now used in sunscreens and other applications. When zinc oxide and titanium dioxide particles of nanoscale proportions—on the order of tens of nanometers, or billionths of a meter—are produced, these materials become colorless, yet still offer protection against UV radiation.

> **nanotechnology** The science of developing materials with dimensions on the nanoscale (less than 100 nm in size) and often exhibiting novel properties.

Whatever specific ingredients the sunscreen may contain, the lotion's **sun protection factor (SPF)**, which appears on its container, serves as a measure of its ability to protect against UV-B radiation. A product with an SPF of 30, for example, reduces the amount of UV-B radiation reaching the skin by a factor of 30. This means that after 30 minutes in the sun, protected by a sunscreen lotion of SPF 30, you would have received the same amount of UV radiation as someone who spent only 1 minute in the same intensity of sunlight with no sunscreen at all. Although SPFs much higher than 30 are available, it's doubtful whether they provide significantly more protection to the average person. The American Academy of Dermatology recommends applying one ounce (about a shot-glass full) of water-resistant, SPF 30 (or greater) broad-spectrum lotion to cover the body of an average adult and reapplying it every two hours while in direct sunlight.

> **sun protection factor (SPF)** A measure of protection against UV-B radiation.

UV absorbers in sunscreens • Figure 11.21

Broad-spectrum sunscreens include UV-B and UV-A absorbers, some of which are shown here. These compounds often contain benzene rings bonded to carbonyl groups. The electrons that occupy the double bonds of these conjugated systems are responsible for the UV absorption.

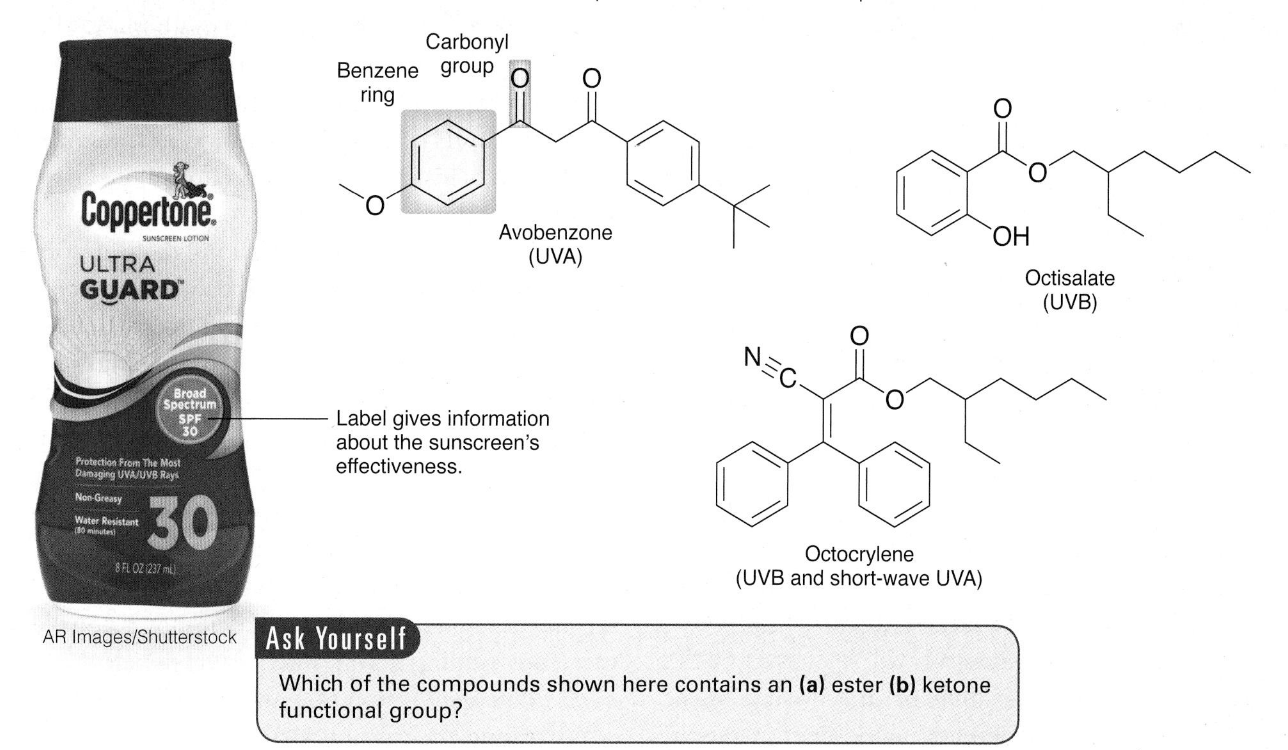

AR Images/Shutterstock

Ask Yourself

Which of the compounds shown here contains an **(a)** ester **(b)** ketone functional group?

Sunless tanners • Figure 11.22

Dihydroxyacetone, present in many sunless tanners, reacts with amino acids in the outer layer of skin to produce brown pigments.

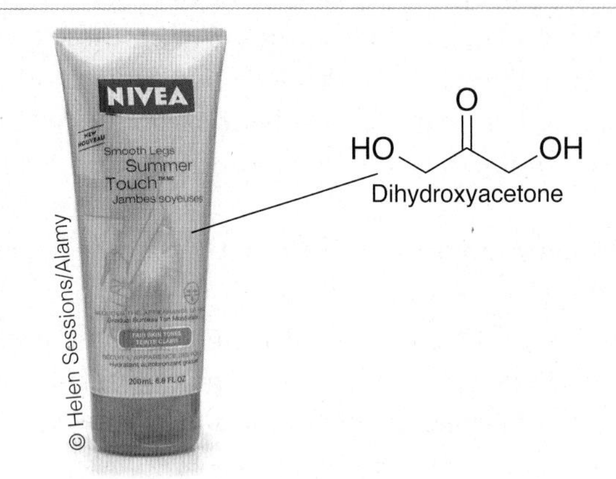

Ask Yourself

Which two kinds of organic functional groups are present in dihydroxyacetone?

It's also possible to develop what appears to be a tan without any exposure to the sun at all. Some chemicals, such as *dihydroxyacetone*, react with the amino acids of the skin to produce a brown pigment. This and other compounds that produce the same effect are components of "sunless tanning" products (**Figure 11.22**).

CONCEPT CHECK STOP

1. **Which** component of cosmetics is typically the most costly on the basis of weight?
2. **Which** type of emulsion is likely to be found in a moisturizing lotion?

11.3 Oral Care and Hair Care

LEARNING OBJECTIVES

1. **Describe** the chemical composition of teeth and the chemistry involved in oral-care products.
2. **Describe** the structure of hair and the chemistry involved in common hair-care products.

In this section we'll discuss our teeth and hair and describe the chemistry underlying the products we use to clean and maintain them.

Oral Care

Bright, clean teeth carry the image of healthy teeth. But although we see a rich foam as the clearest indication that a toothpaste is doing its job, the surfactants that produce the foam are among the least important cleansing agents of a good **dentifrice**. They aren't what keep teeth clean and healthy, and a few powdered dentifrices don't even contain them. In fact, the Food and Drug Administration doesn't consider a surfactant to be an active ingredient of a toothpaste. What's needed for a good dentifrice is an effective abrasive. To understand why takes a bit of dental chemistry.

dentifrice A substance, typically a paste, for cleaning teeth.

A healthy tooth is covered by a hard layer of material called enamel that serves both to grind food and to protect the interior regions of the tooth itself. As in bone, calcium makes up most of this extremely hard surface, which is composed largely of the mineral *hydroxyapatite*, $Ca_{10}(PO_4)_6(OH)_2$. Despite its toughness, enamel is susceptible to acids generated from a thin, clear, adhesive, polysaccharide film called **plaque**.

Bacteria in the mouth continuously convert some of the sugars of our food into this plaque, which attaches to the enamel and serves as a home to still other bacteria. These other bacteria, in turn, convert the plaque into an acid that erodes the calcium and the phosphate from the shield of hydroxyapatite that protects the tooth. When enough erosion has occurred, microorganisms can pass through the weakened barrier and attack the interior. The result is **dental caries**, better known as tooth decay or cavities.

The key to keeping teeth free of cavities lies in removing the accumulated plaque. Removing plaque depends more on grinding it away with a good dental abrasive than on the detergent action of a surfactant. With daily removal of the plaque, the calcium and phosphate normally present

Toothpaste formulations • Figure 11.23

Some typical components of toothpaste and their functions.

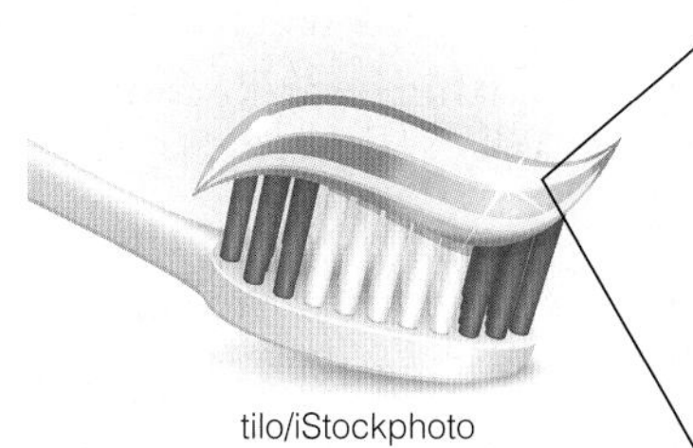
tilo/iStockphoto

Component	Example	Function
Fluorides	Stannous fluoride, SnF_2	Prevents cavities.
Mild abrasives	Hydrated silica, $SiO_2 \cdot H_2O$	Helps remove plaque from tooth enamel.
Surfactants	PEG-6 (nonionic) Sodium lauryl sulfate (anionic)	Helps provide detergency.
Flavoring agents	Sodium saccharine (artificial sweetener)	Provides pleasant taste, such as sweetness or mintiness.
Pigments	Titanium dioxide, TiO_2, a white pigment; Blue lake 1	Provides color.

in saliva replace any calcium and phosphate that might have been removed by mouth acids. As long as no bacteria have penetrated the body of the tooth itself, this reconstitution process, known as **remineralization**, returns the enamel to its original strength. To be useful as a dental abrasive, the grinding agent must be harsh enough to remove accumulated plaque, yet not grind away the enamel itself.

In addition, the active ingredient listed on most toothpastes is typically a compound containing fluoride, F^-, which helps to maintain the strength of the enamel. (We discussed water fluoridation, a related topic, in Chapter 7.) Fluorides are commonly present in toothpastes in the form of ionic compounds such as stannous fluoride (SnF_2) and sodium fluoride (NaF). The fluoride seems to act by:

- replacing some of the hydroxy groups (HO^-) in the enamel's hydroxyapatite with fluoride groups (F^-), which turns hydroxyapatite into a harder mineral, fluoroapatite, that is more resistant to erosion by acids; and
- suppressing the bacteria's ability to generate acids.

Although the surfactants of toothpaste formulations are not our principal protection against tooth decay, they do effectively remove loose debris from the mouth and also give us the sense of cleanliness. Almost all dentifrices contain a bit of saccharin and some flavoring or fragrance to leave us with a sense of sweetness and freshness after brushing. The components of a typical toothpaste are listed in **Figure 11.23**.

Hair Care

All mammals have hair, which serves to help protect the skin and regulate body temperature. Hair is a lifeless structure composed of a protein called keratin. A strand of hair contains no more living tissue than do our nails or the keratin-containing claws, horns, hoofs, or feathers of other animals. Like all human proteins, keratin is a polypeptide with monomeric units consisting of a combination of the 20 amino acids of Table 4.2. In keratin, the sulfur-containing amino acid, cysteine, dominates the polymer and accounts for about 15% of its amino acids. Along the polypeptide chain of keratin, the thiol groups (—SH) of cysteines readily form disulfide links (—S—S—) with the thiol groups of cysteines in neighboring polypeptide chains of the protein. These —S-S— links joining **cystine** groups in adjoining polypeptide chains give hair much of its strength by connecting the strands of protein and keeping them aligned with one another, as shown in **Figure 11.24**.

Cystine formation in keratin protein • Figure 11.24

Keratin is the major protein in hair, nails, and the outermost layer of skin. Disulfide links form between polypeptide chains of this molecule.

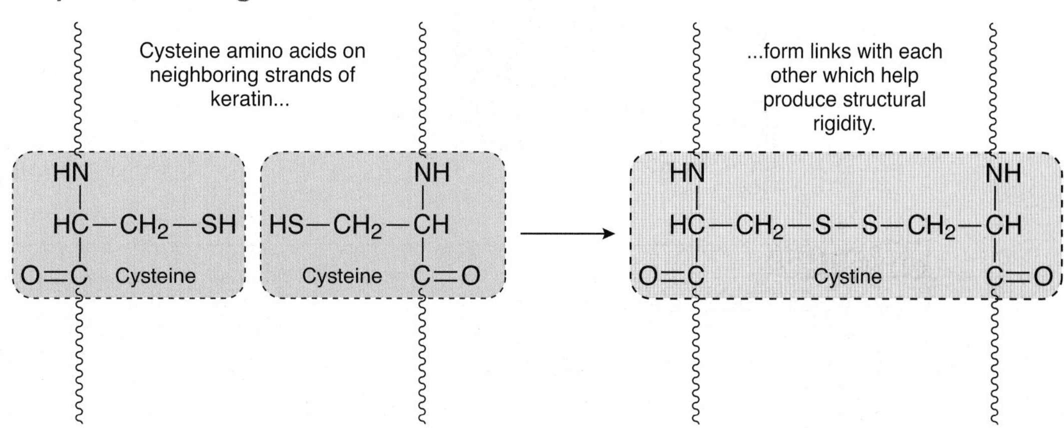

Ask Yourself

Covalent bonds between two atoms of which element are responsible for keeping strands of keratin aligned with each other?

The structure of hair • Figure 11.25

Each hair strand contains a central core called the cortex and shingle-like structures called the cuticle covering the outside.

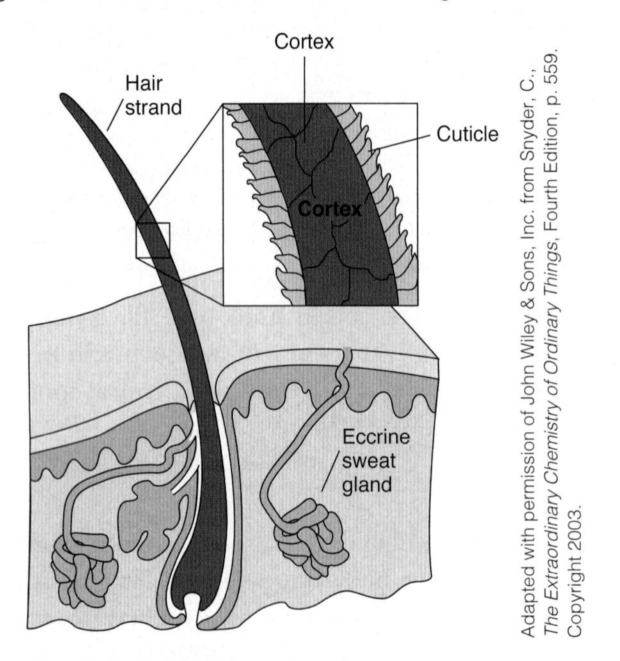

Adapted with permission of John Wiley & Sons, Inc. from Snyder, C., *The Extraordinary Chemistry of Ordinary Things*, Fourth Edition, p. 559. Copyright 2003.

Each strand of hair grows from a living root embedded within microscopic sacks or follicles, buried under the surface of the skin. The hair of the scalp grows outward at about one centimeter each month, or about a tenth of an inch per week. After several years of steady growth, each strand falls out, replaced by a new one as it begins its own growth.

Typically, each strand of hair consists of a central core, called the **cortex**, which forms the bulk of the fiber and contains its coloring pigments. Enveloping this cortex is a thin, translucent, scaly sheath, called the **cuticle** (**Figure 11.25**). Sebaceous glands lying in the skin near the follicle lubricate the emerging shaft with an oily sebum, which gives the hair a gloss, keeps the scales of the cuticle lying flat, and prevents the strand from drying out. Too much sebum makes the hair feel greasy and dirty; too little sebum and it's dry, dull, and wild. The detergent of a hair shampoo, then, must remove dirt from the hair and scalp, as well as enough accumulated sebum to keep the hair looking clean, but not so much as to remove all of the oil. The moderate detergent action of the lauryl sulfates and related surfactants accomplishes this tightrope act very nicely. Many shampoo formulations contain added conditioners to replace at least part of the lubricant that might be lost in the washing.

Shampoos and conditioners As you read the list of ingredients of a typical shampoo, you'll probably find among them ammonium lauryl sulfate or triethanolammonium (TEA) lauryl sulfate. These are anionic surfactants, with negatively charged hydrophilic heads. The negative charges of these anions are balanced by positively charged ammonium ions, NH_4^+, so we sometimes refer to these compounds as ammonium salts (**Figure 11.26**). Compared to the sodium salts of these same surfactants, the ammonium salts tend to be more soluble in

Common surfactants in shampoo • Figure 11.26

The cleansing properties that we expect from shampoo are provided by surfactants such as the compounds shown here.

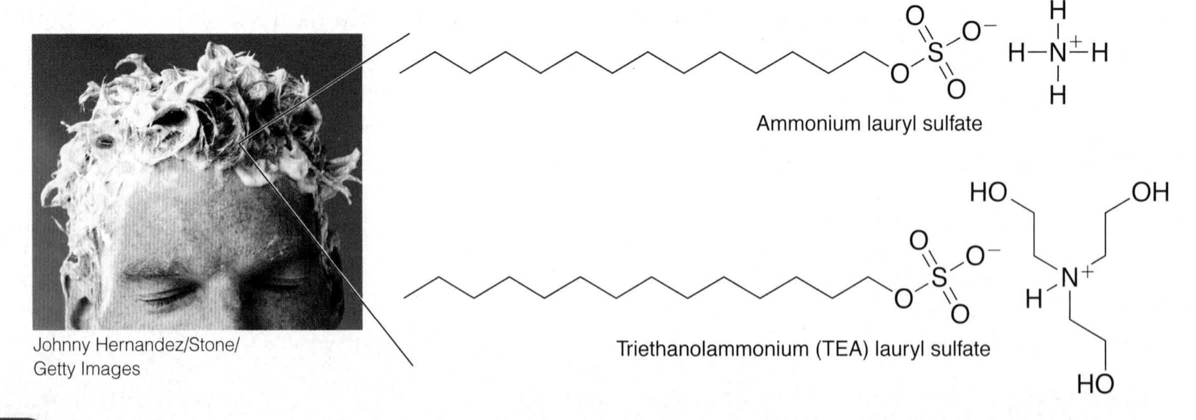

Johnny Hernandez/Stone/Getty Images

Ask Yourself

1. Why are these compounds classified as anionic surfactants?
2. What is structurally different about these two compounds?
3. What similarities exist among the cations of these compounds?
4. What are two properties of ammonium and triethanolammonium lauryl sulfates that make these anionic detergents superior to sodium salts for use in shampoos?

pH and the radiance of hair • Figure 11.27

Most shampoos and conditioners are formulated to be slightly acidic, with a pH of 4–6, so as to improve the luster of hair.

a. Lower pH.
At pH 4–6, the tighter, more organized cuticle reflects light coherently, giving hair a luster.

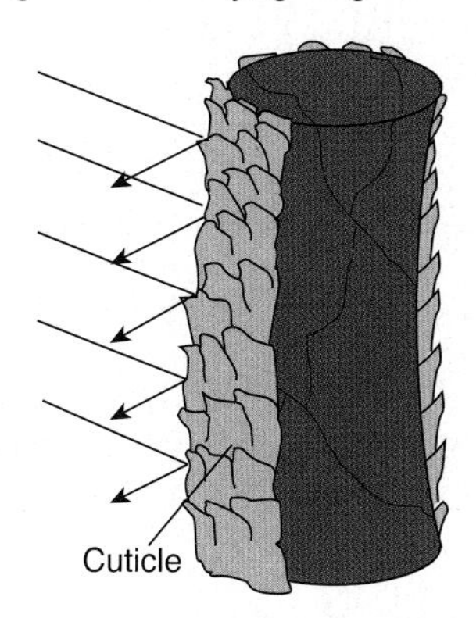

b. Higher pH.
At higher pH, the ruffled cuticle scatters light, making hair look flat and dull.

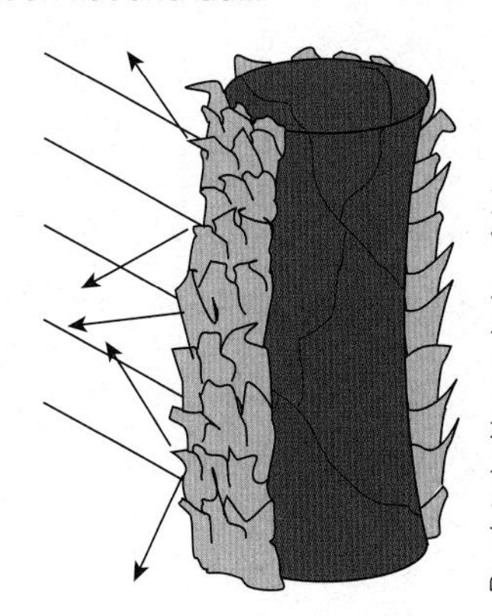

cold water and gentler on sensitive skin, and they don't dry out the hair quite as much.

Other shampoo ingredients help stabilize the foam, act as preservatives, give the shampoo itself a pleasing viscosity or body, adjust the pH, bind to the metal ions of hard water (which might degrade the surfactant's action), and add color and fragrance. The acidity of the shampoo has a lot to do with its ability to produce a good luster and resilient hair. Other things being equal, a strand of hair is strongest under slightly acidic conditions, at a pH of about 4–6. Moreover, the scales of the cuticle tend to swell up and fluff out under basic conditions. This condition causes reflected light to scatter, making the hair look dull. A tight, slightly acidic cuticle reflects light more coherently, giving the hair a pleasant luster. **Figure 11.27** shows this pH-dependent effect, and **Table 11.2** summarizes common components of shampoo.

As we saw earlier, the use of shampoo involves the removal of dirt as well as a balancing act between the removal of enough accumulated sebum to keep your hair clean, but not the removal of all of the oil. Nevertheless, if you shampoo without using a conditioner afterward, you may notice that your hair, though clean, might be somewhat dry, lacking body and sheen, and might tangle easily. Hair conditioners are designed to address some of these issues, including replenishing hair with oils that may have been stripped away by the shampoo. Conditioners also commonly contain **humectants**, which are compounds that help attract and retain moisture. Common

Components of a typical shampoo Table 11.2

Component	Example	Function
Solvent	Water	Helps dissolve or disperse other components and adds bulk.
Surfactants	Ammonium lauryl sulfates (anionic surfactants)	Provides detergency.
Acidifiers	Carboxylic acids, such as citric acid	Provides acidity.
Foaming agents	Fatty amides, such as cocamide DEA	Provides lather.
Moisturizing agents	Fatty alcohols, such as cetyl alcohol: $CH_3(CH_2)_{15}OH$	Helps retain moisture and replace lost oils.
Conditioners	Polyquaternium (cationic surfactants)	Provides softness and antistatic properties.
Fragrance	—	Adds pleasant aromas.
Preservatives	Methylisothiazolinone	Prevents the growth of microorganisms.

Components of a typical hair conditioner Table 11.3

Component	Example	Function
Solvent	Water	Helps dissolve or disperse other components and adds bulk.
Moisturizing agents, oils, and waxes	Humectants, such as glycerin: $HOCH_2CH(OH)CH_2OH$ Fatty alcohols, such as cetyl alcohol: $CH_3(CH_2)_{15}OH$ Fatty esters, such as cetyl esters: $CH_3(CH_2)_{15}O(C{=}O)(CH_2)_{14}CH_3$	Helps retain moisture and replace lost oils.
Conditioners	Polyquaternium (cationic surfactants)	Provides softness and antistatic properties.
Acidifiers	Carboxylic acids, such as citric acid	Provides acidity to improve hair luster.
Lubricants	Dimethicone, a silicon-based polymer	Helps prevent tangling and provides sheen.
Fragrance	—	Adds pleasant aromas.
Preservatives	Methylisothiazolinone	Prevents the growth of microorganisms.

ingredients in a typical hair conditioner are summarized in **Table 11.3**.

Hair Treatments Several issues come into play in determining whether your hair is naturally straight or curly, including genetics and hormonal factors, both of which can influence the shape of hair follicles. Straight hair tends to grow out of follicles that provide a circular opening for hair growth, whereas curly hair grows out of follicles that provide an oval outlet. In addition, several chemical characteristics of the keratin protein of the hair itself can influence the shape of the strand and determine whether the hair is straight, curly, or forms tight curls or loose waves. The most important of these chemical influences are:

- disulfide links—the sulfur-sulfur covalent bonds of the keratin's cystine;
- salt bridges—the ionic bonds that form between the acidic group of one amino acid and a basic (amino) group of another amino acid located somewhere else on the same or an adjacent protein molecule;
- hydrogen bonds—weak links that form between the H of an $-OH$ or a $-NH$ group and the O or N of another $-OH$ or $-NH$ group.

We discussed the disulfide links of cystine groups in keratin earlier (see Figure 11.23). The salt bridges form as the proton of a carboxyl group ($-CO_2H$) on an amino acid, such as aspartic acid or glutamic acid, transfers to a nearby amino acid containing a free amine group ($-NH_2$), such as the lysine or proline of another chain (see Table 5.1). This acid–base reaction produces the ionic bonding of a carboxylate anion ($-CO_2^-$) and an ammonium ion ($-NH_3^+$). This bond holds portions of adjacent keratin chains near each other, just as ionic bonds cement a crystal of sodium chloride.

The hydrogen bonds of keratin result from weak electrical attractions between the hydrogen of an $-NH$ or $-OH$ group and a nearby oxygen or nitrogen atom. **Figure 11.28** sums up these interactions as they exist in the strands of keratin (as well as other proteins).

Although individual hydrogen bonds are weaker than any ionic or covalent bonds, their enormous numbers give dry hair most of its strength. In wet hair, though, water molecules intrude between the keratin strands and disrupt the hydrogen bonds that keep the strands aligned with each other, allowing them to shift a bit. As hair dries, the water molecules leave and the hair retains its new shape, held intact by the combined forces of large numbers of hydrogen bonds in the newly realigned polymeric chains. Every time you wash your hair, set it, and then blow-dry it, you use a bit of chemistry to disrupt the keratin's hydrogen bonds, rearrange its strands, then remove the water molecules so that the keratin's own hydrogen bonds hold your hair in its new shape—at least until it gets wet again.

You can produce a longer-lasting rearrangement of the keratin strands by using either a commercial hair relaxer or a "perm" kit and, in this case, a bit of sulfur chemistry. The sulfur–sulfur bonds of the cystine units shared by adjacent keratin strands hold these polymers in fixed positions. Hair relaxers use alkaline compounds such as lithium hydroxide (LiOH) to disrupt these disulfide links, allowing curly hair to straighten. "Perms" simply break the sulfur–sulfur bonds, reorganize the strands, and then regenerate new sulfur–sulfur bonds (**Figure 11.29**). Unlike a wet wave, the

Attractive forces and bonds between protein strands • Figure 11.28

Different types of interactions between amino acids provide structural rigidity to proteins such as keratin. These interactions include disulfide linkages (a), salt links (b), and hydrogen bonds (c).

a. A disulfide linkage can form between two cysteine amino acids. This assembly is called cystine.

b. A salt link can form between the side chains of acidic amino acids, such as aspartic acid, and basic amino acids, such as lysine.

c. Hydrogen bonds within proteins commonly form by attractions between oxygen atoms on carbonyl groups (C=O) and hydrogen atoms bonded to nitrogen atoms within the chain.

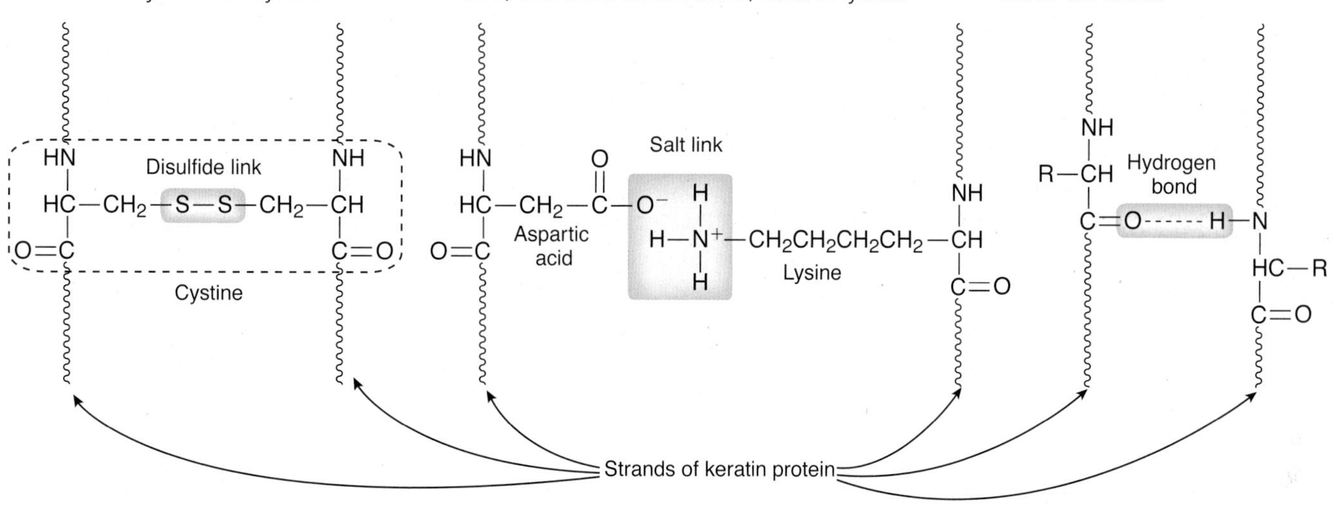

Chemistry of "perms" • Figure 11.29

Permanent wave hair treatments use a reducing agent to cleave the disulfide links, allowing hair to be curled. The hair is then set in this orientation with an oxidizing agent, which converts the cysteines back to cystines.

$HS-CH_2-C(=O)-OH$

1. Thioglycolic acid cleaves the disulfide links, reducing them to $-SH$ groups and separating the protein strands from each other.
2. Curling reorients the strands and the $-SH$ groups.
3. An oxidizing agent, such as a dilute solution of hydrogen peroxide, converts adjacent $-SH$ groups to disulfide links and fixes the new orientation of the strands of keratin.

H_2O_2

Adapted with permission of John Wiley & Sons, Inc. from Snyder, C., *The Extraordinary Chemistry of Ordinary Things*, Fourth Edition, p. 566. Copyright 2003.

PROCESS DIAGRAM

permanent wave keeps its shape through the newly formed covalent bonds. These remain firm whether the hair is wet or dry.

CONCEPT CHECK

1. **What** is the purpose of the mild abrasives commonly found in toothpaste?
2. **How** do shampoos and conditioners serve complementary functions?

Summary

1 Soaps and Surfactants 332

- ***What are surfactants?***
Surfactants lower **surface tension**, a property of water and other liquids that causes them to behave as though their surfaces are covered with a thin membrane. Surfactant molecules have a hydrophobic (*water-avoiding*) hydrocarbon tail and a hydrophilic (*water-seeking)* head. Surfactants are classified into one of three classes: anionic, cationic, and nonionic, based on the nature of the hydrophilic portion of the molecule.

- ***How does soap clean, and what are micelles?***
Soaps are sodium salts of long-chain carboxylic acids, also known as *fatty acids*. When mixed with water, soaps and detergents can aggregate into **micelles**. As shown here, the hydrophobic tails of soap or detergent molecules compose the micelle's interior, shielded from water, and the hydrophilic heads of these molecules form the surface, in contact with water. Soaps and detergents clean by (i) lowering surface tension, allowing water to better wet fabrics, (ii) trapping grease and dirt particles in the interiors of micelles, and (iii) keeping these particles in suspension until rinsed away.

Figure 11.4 • Micelles

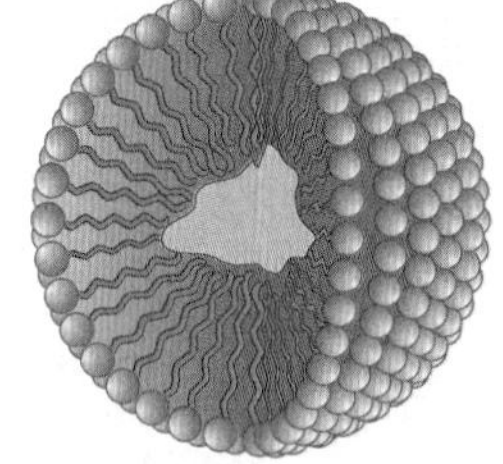

Adapted with permission of John Wiley & Sons, Inc. from Klein, D., Organic Chemistry, Second Edition, p. 41. Copyright 2015.

2 Cosmetics and Skin Care 343

- ***What is the structure of the skin?***
The skin itself consists of two major layers: the underlying **dermis**, which contains nerves, blood vessels, sweat glands, and the active portion of the hair follicles; and the upper layer, the **epidermis**. The outermost layer of the epidermis is the **stratum corneum**, made up of keratin, the same protein that forms hair.

- ***What are the major components of common skin-care products and cosmetics?***
Moisturizing lotions are commonly oil-in-water **emulsions**, with water as the main component. Antiperspirants inhibit sweating typically by the action of compounds that contain aluminum ions (Al^{3+}). Deodorants mask odors with fragrances as well as destroying odor-causing bacteria with agents such as *triclosan*. Colognes and perfumes contain volatile, readily evaporating organic compounds dissolved in alcohol. Cosmetics that provide color, such as lipstick, nail polish, and eye shadow, contain chemical dyes that may be dispersed in waxes, solvents, or powders.

- ***What is UV radiation, and how do sunscreens work?***
Ultraviolet radiation, or UV for short, is slightly higher in frequency (or energy) and shorter in wavelength than visible light (as seen in the electromagnetic spectrum shown here). Sunscreens block the sun's UV light by the action of two main types of ingredients: organic UV absorbers, such as avobenzone, and inorganic scattering agents, such as zinc oxide. Broad-spectrum sunscreens protect against UV-B radiation, which causes sunburns, and UV-A radiation, which is associated with premature skin aging. The SPF (Sun Protection Factor) rating of a sunscreen is a measure of the protection it provides against UV radiation.

Figure 11.20 • The electromagnetic spectrum

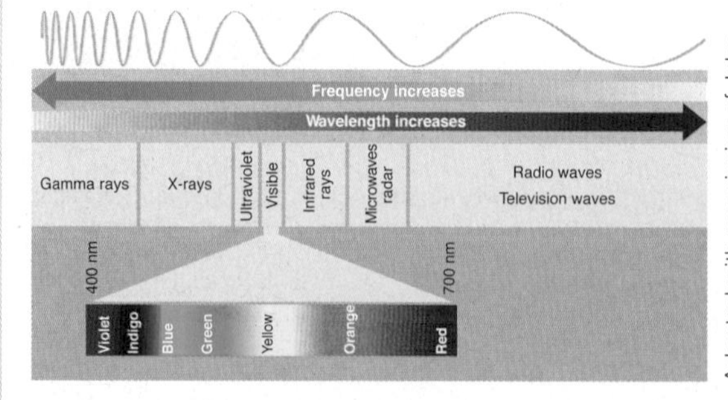

Adapted with permission of John Wiley & Sons, Inc. from Jespersen, N., Brady, J., and Hyslop, A., *Chemistry: The Molecular Nature of Matter*, Sixth Edition, p. 310. Copyright 2012.

3 Oral Care and and Hair Care 352

- ***What are the compositions of our teeth and hair?***
Teeth are largely composed of the calcium-based mineral

hydroxyapatite. A thin polysaccharide coating, called plaque, naturally forms on the tooth's enamel or exterior surface. Bacteria in the mouth convert plaque to acids, which can erode enamel and lead to tooth decay. Hair is composed mostly of the protein *keratin*. The most abundant amino acid within keratin is the sulfur-containing compound *cysteine*. Each hair shaft has a pigmented interior called a **cortex** and a transparent, shingle-like exterior called the **cuticle**. Hair (as well as skin) is naturally lubricated by **sebum**, an oily substance secreted by the sebaceous glands.

- ***What are the major components of toothpastes and shampoos?***
  Active cleansing agents in most toothpastes include mild abrasives (which help remove plaque) and surfactants (which provide sudsing and cleansing action). An important antidecay ingredient is fluoride ion, F^-, an agent that helps strengthen tooth enamel and prevent the formation of cavities. Toothpastes also contain flavoring agents to provide a pleasant taste.

  The primary components of shampoos, other than water, are surfactants, such as ammonium lauryl sulfate, an anionic surfactant (shown here). Shampoos also contain moisturizing agents (to counteract the drying effects of stripping the hair of its oils), and acidifying agents such as citric acid. The slightly acidic pH of shampoos flattens the cuticles of hair, providing luster.

Figure 11.26 • Common surfactants

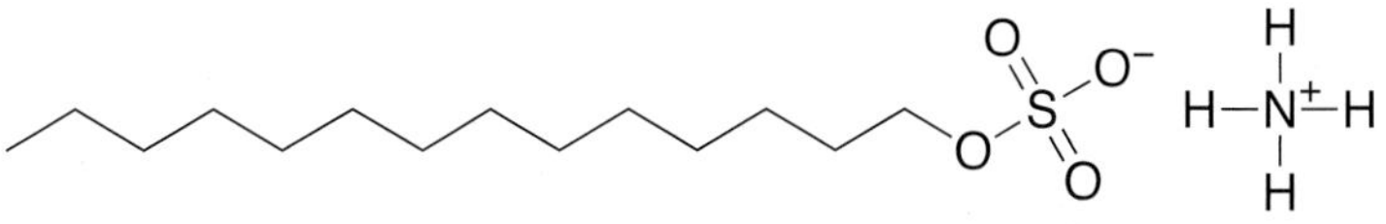

Ammonium lauryl sulfate

Key Terms

- dentifrice 352
- detergent 334
- electromagnetic spectrum 350
- emulsion 345
- micelle 334
- nanotechnology 351
- saponification 336
- soap 333
- sun protection factor (SPF) 351
- surface tension 332
- surfactant 334

What is happening in this picture?

In 2010, a catastophic malfunction and associated explosion of the *Deepwater Horizon* oil rig in the Gulf of Mexico led to the largest marine oil spill in history. It took three months to cap the gushing well, at which point approximately 200 million gallons of crude oil had been released into the environment.

Various measures were taken to lessen the impacts of the spill. In one approach, nearly 2 million gallons of **dispersants**, mixtures of detergents and petroleum-based solvents, were applied to the oil to help dissipate the slick. Some criticized this strategy on the grounds that the environmental risks posed by the release of the dispersants themselves could outweigh the potential benefits gained by the action of these dispersants on the oil.

Think Critically

1. By what means do detergents disperse or dissipate oil?
2. Corexit, the trade name for the dispersant used on this spill, contains an ingredient called Span 80. Use the web to identify the structure of this compound and identify its hydrophilic and hydrophobic regions. Based on its structure, what class of surfactant is it: anionic, cationic, or nonionic?

Exercises

Review

1. Define, identify, or explain each of the following terms:

a. micelles
b. hydrophobic
c. hydrophilic
d. soap
e. hard water
f. surfactant
g. colloidal disperson
h. ester
i. fatty acid
j. triglyceride
k. lye
l. eutrophication
m. saponification
n. soap scum
o. hydrolysis
p. glycerol
q. zeolite
r. alkylbenzenesulfonates

2. What is a simple way to lower the surface tension of water?

3. Explain what we mean when we say that a detergent molecule is a wetting agent.

4. a. Classify each of the following as an anionic, cationic, or nonionic surfactant.

i. O O OH

ii. O O$^-$ S O O Na$^+$

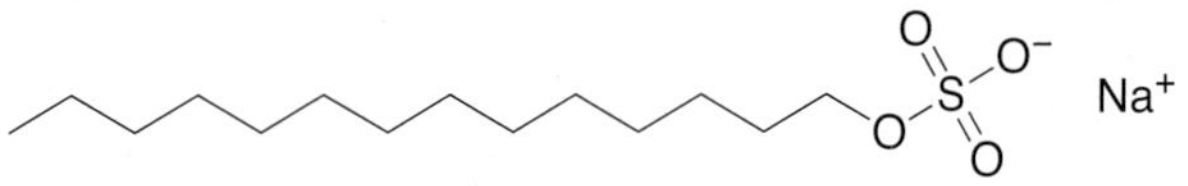

iii. N$^+$ Cl$^-$

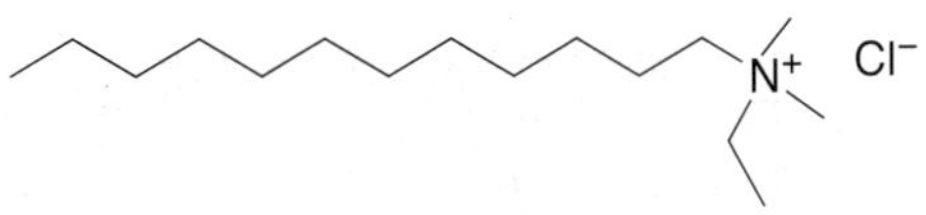

b. Identify the hydrophobic and hydrophilic regions of each molecule.

5. What two functional groups are generated when an ester is hydrolyzed in the presence of a. a strong acid? b. a strong base?

6. What alcohol is produced by the hydrolysis of a triglyceride?

7. The orange structures depicted here are soap molecules at the surface of water:

a. What is represented by: (i) the round heads and (ii) zig-zag tails of these structures?

b. Why do these structures appear at the surface of the water?

c. As more of these structures are added to the sample and the surface becomes too crowded with them to accommodate additional structures, how do they continue to aggregate?

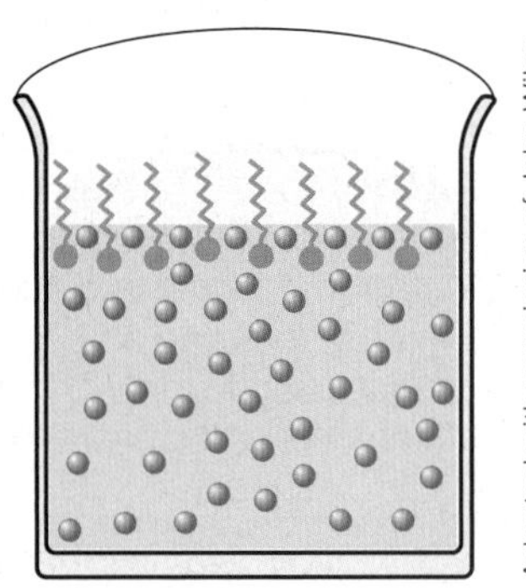

Adapted with permission of John Wiley & Sons, Inc. from Snyder, C., *The Extraordinary Chemistry of Ordinary Things*, Fourth Edition, p. 313. Copyright 2003.

8. When the ancient Romans made soap, what did they use as the source of

a. triglycerides?

b. bases?

9. What property of ammonium salts makes them useful as cationic detergents in mouthwashes?

10. Which three ions mentioned in the chapter can contribute to water hardness?

11. a. What causes the insoluble soapy deposits seen here?

b. What is a typical chemical composition of these deposits?

Martyn F. Chillmaid/Photo Researchers, Inc.

12. List six common components of a commercial laundry detergent, other than the water and surfactants, and describe the function of each.

13. Why were phosphates originally added to detergent formulations? Why were they subsequently removed?

14. a. What type of detergent molecule is shown here?

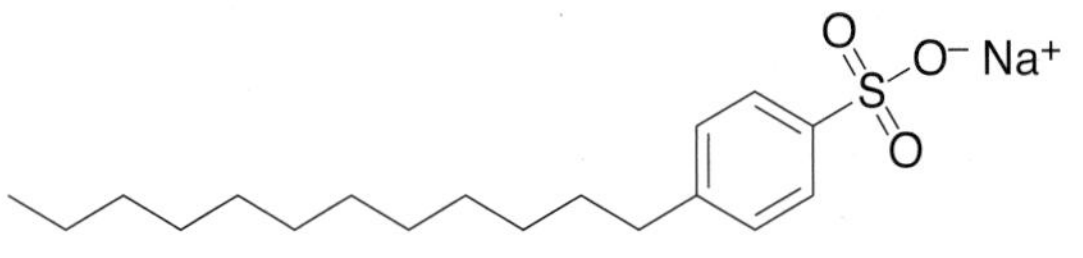

   b. What principle advantage does this molecule offer over soap?

15. How can zeolites help to improve the cleaning efficiency of detergents?

16. a. How does the Federal Food, Drug and Cosmetic Act define a cosmetic?

   b. How do we define it in this chapter?

17. What class of chemical makes up the largest category of compounds used in the manufacture of personal care products?

18. The curl of a permanent wave is produced by breaking covalent bonds between two strands of keratin, reorganizing the strands, then reforming the covalent bonds. Atoms of which element are connected by the covalent bonds that are broken and reformed?

19. What's the difference between the action of an antiperspirant and a deodorant?

20. How do perfumes and colognes compare in terms of the relative concentrations of fragrance present?

21. a. What are the functions or actions of the top note, middle note, and end note of a perfume?

   b. For each of these categories name a chemical that produces the desired effect.

22. Most commercial nail polish removers contain ordinary organic solvents, primarily ethyl acetate and acetone. Identify the functional group represented by each molecule.

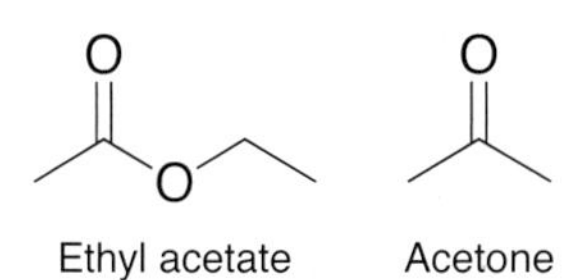

23. Define, identify, or explain each of the following terms:

   a. keratin
   b. cortex of hair
   c. follicle
   d. sebum
   e. apocrine gland
   f. stratum corneum
   g. dermis
   h. epidermis
   i. cuticle of hair
   j. humectant
   k. emollient
   l. emulsion
   m. dental abrasive

24. Define, identify, or explain each of the following terms:

   a. sodium lauryl sulfate
   b. cysteine
   c. hydroxyapatite
   d. sun protection factor
   e. plaque
   f. dental caries
   g. zinc oxide
   h. melanin
   i. broad-spectrum sunscreen
   j. dihydroxyacetone

25. a. How is the frequency of radiation related to its wavelength?

   b. How is the energy of radiation related to its frequency?

26. a. Label the indicated regions of the electromagnetic spectrum shown here.

   b. Which region of the entire spectrum has the lowest-energy radiation?

   c. Which region has the highest-energy radiation?

   d Which region is responsible for skin damage from solar radiation?

   e. What is the significance of the 400 nm and 700 nm labels?

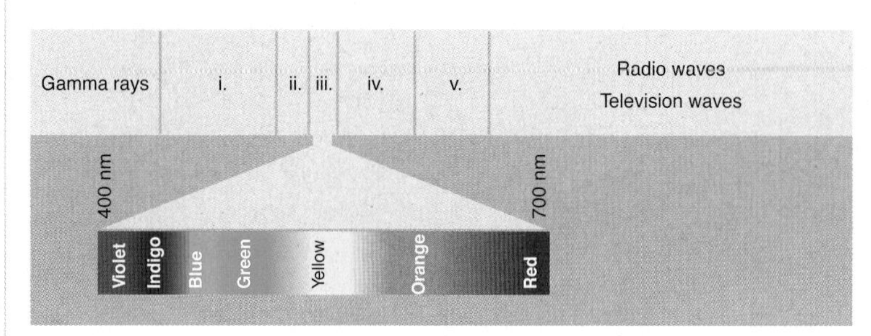

Adapted with permission of John Wiley & Sons, Inc. from Jespersen, N., Brady, J., and Hyslop, A., *Chemistry: The Molecular Nature of Matter*, Sixth Edition, p. 310. Copyright 2012.

27. Name each of the following compounds and describe its function in personal care products:

a. ZnO
b. TiO_2
c. NaF
d. HS-CH_2-CO_2H
e. LiOH

28. Name a hair dye that was used by the ancient Egyptians and is still in use throughout the world today.

29. On a weight basis, what are the most expensive ingredients of cosmetic and personal care products?

30. Why is either ammonium lauryl sulfate or triethanolammonium lauryl sulfate better suited than sodium lauryl sulfate for use as a detergent in a hair shampoo?

31. What is the difference between the cortex and the cuticle of a strand of hair?

32. Name three chemical forces that give the hair strength and explain how each works.

33. a. Provide the missing labels in the figure below.
b. Which gland produces sebum?
c. Which layer of the skin consists largely of dead cells?
d. Which gland is considered the "true sweat gland"?
e. Which gland secretes fluid that can be degraded by bacteria into foul-smelling products?

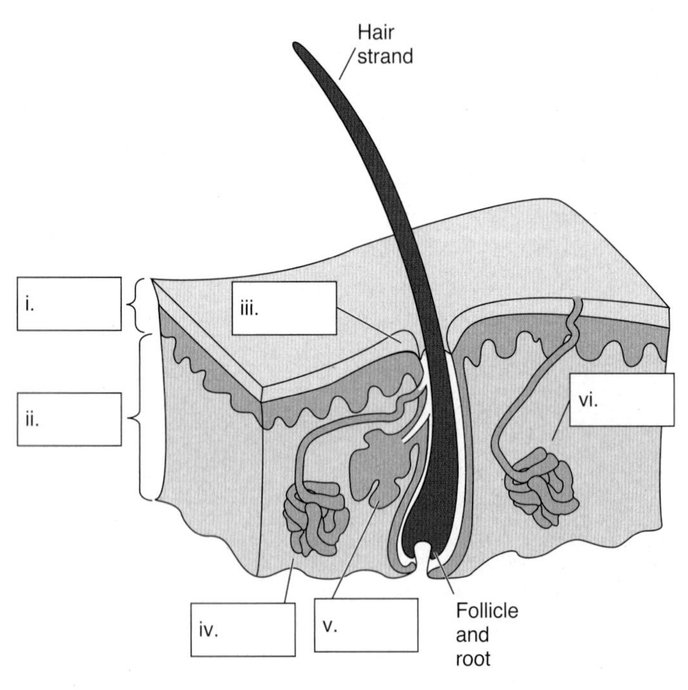

Adapted with permission of John Wiley & Sons, Inc. from Snyder, C., *The Extraordinary Chemistry of Ordinary Things*, Fourth Edition, p. 559. Copyright 2003.

34. a. What is the difference between UV-A and UV-B radiation?
b. What kind of skin damage results from overexposure to (i) UV-A radiation? (ii) UV-B radiation?

35. How do the aluminum compounds in antiperspirants act to reduce perspiration?

36. How does triclosan act to reduce body odors?

37. How do the following compounds differ in the means by which they protect against UV radiation?

a. zinc oxide,
b. octocrylene
c. avobenzone

38. A sunscreen that contains a combination of avobenzone and octisalate is likely to be labeled as offering what type of protection against UV radiation?

39. What advantage does nanoscale zinc oxide offer as a component of sunscreen, as compared to conventional-sized particles of this compound?

Think

40. Can a surfactant exist that is not a soap? Can a soap exist that is not a surfactant? Explain.

41. Would you expect the surface tension of cold water to be greater than, less than, or exactly the same as the surface tension of water that's near its boiling point? Explain.

42. What is the source of the mineral ions present in hard water?

43. Would you expect a soap to act as an effective detergent in seawater? Would you expect a synthetic detergent (such as an alkylbenzenesulfonate) to act as an effective detergent in seawater? Explain.

44. Why doesn't the identity or chemical behavior of *anions* present in water affect the water's hardness?

45. Assume you have two samples of water—one is very hard, with a high concentration of calcium ions, the other is not as hard, with a lower concentration of calcium ions. Show how you could use a solution of soap in distilled water and a dropper or other measuring device to determine which is the harder water.

46. The sodium chloride of a water softener's storage bin must be replaced periodically as the NaCl is consumed in the formation of salt water. What would result if calcium chloride rather than sodium chloride were mistakenly added to an empty storage bin?

47. Suppose that you live in an area with very hard water and have just found that you are allergic to all available synthetic detergents. You find that you must use soap for all your household cleaning. What would you do to use soap most effectively?

48. Brushing your teeth with ordinary baking soda ($NaHCO_3$) has been recommended as an alternative if toothpaste is unavailable. Although baking soda doesn't contain detergent (or a pleasant flavor or fragrance), it does serve as a mild abrasive. What additional benefit does baking soda offer as a dentrifice?

49. In what ways are the characteristics of a hair shampoo and a toothpaste similar? In what ways are they different?

50. Even though an SPF of 30 is generally regarded as providing sufficient protection from solar radiation at sea level, a person living at a high altitude or on a mountain-climbing expedition might want to use a much higher SPF. Why?

51. Describe the role of dental plaque in tooth decay.

52. Describe the chemical differences between a wet wave of the hair and a permanent wave.

53. What do you expect as the primary quality of an antiperspirant? What other, secondary qualities do you expect?

54. Explain why the pH of hair is important to its appearance.

55. The sulfur–sulfur bonds (-S—S-) of a single cystine molecule can be broken by the addition of a molecule of hydrogen to produce a new amino acid, containing an —SH group. What is the name of the amino acid produced by the addition of hydrogen to cystine?

Calculate

56. The label on a particular box of toothpaste indicates a fluoride concentration of "0.16% w/v fluoride," meaning that 0.16 g (160 mg) of fluoride is present in each 100 mL of toothpaste.

    a. What mass of fluoride is present in an entire tube of toothpaste if its volume is 125 mL?

    b. What mass of fluoride is present in the amount of toothpaste one might use in a single brushing, or about 0.4 mL?

57. Using a properly applied SPF 45 sunscreen, how long could you stay out in the sun and receive the same amount of UV radiation as someone who spent 5 minutes in the sun without any sunscreen on? (Assume that the intensity of the sunlight is the same in both scenarios and that the surface area of exposed skin is the same in both individuals.)

58. Light with a wavelength of 400 nm lies near the border between ultraviolet light and visible, violet light. If we increase the wavelength by 50%, to 600 nm

    a. do we increase or decrease the energy of the light?

    b. What color light do we obtain?

12

Genes, Medicines, and Drugs

For some athletes, the desire to win at all costs may lead them to engage in unsanctioned practices, such as the use of performance-enhancing drugs (PEDs). These substances can include both anabolic steroids, which help increase muscle mass; and the hormone erythropoetin, which increases red-cell production, thus boosting the oxygen-carrying capacity of the blood. This misuse of various drugs to enhance athletic abilities, a practice commonly known as doping, is familiar to us from professional sports. A well-known example is the

use of PEDs by Lance Armstrong (at far left in photo). After having won seven Tour de France titles, his victories were revoked when he admitted to drug doping.

In this chapter we'll examine the underlying chemistry and various uses of medicines and drugs. Since many diseases may be inherited, we'll also discuss the chemical principles of heredity and the roles that genetic factors play in health and disease. We'll see that although advances in the genetic, chemical, and medical sciences, such as genetic engineering and therapeutic cloning, offer innovative treatments for injury and disease, these same advances can also pose important ethical questions.

PASCAL PAVANI/AFP/Getty Images

CHAPTER OUTLINE

12.1 Nonprescription Medicines

LEARNING OBJECTIVES

1. **Describe** how aspirin was discovered and explain its mode of action.
2. **Distinguish** between decongestants and antihistamines.

In this section we'll explore some of the more common nonprescription medicines we use to treat aches and pains, as well as colds and allergies. These drugs are considered safe enough to be bought by consumers over the counter (OTC), but some still have potentially risky side effects.

Aspirin and Other Nonprescription Pain Relievers

For most of human existence, just about the only substances available to ease human suffering and illness were naturally derived substances, such as plant or animal extracts. Some were potent, some were not. A few were addictive or poisonous or both.

With the growth of the chemical industry in the latter part of the 19th century, **semisynthetic** drugs—those made by chemical modifications of **natural products**—started to become available. These drugs offered greater effectiveness and fewer side effects than those that came directly from plants or animals. Aspirin, one of the earliest semisynthetic drugs, was introduced commercially at the end of the nineteenth century.

semisynthetic A substance made by chemical modification of a natural product.

natural product A chemical compound produced by a living organism.

Aspirin was developed from a botanical remedy for aches and fevers that had been used since antiquity. Physicians as far back as Hippocrates—the ancient Greek healer known as the Father of Medicine—knew of the curative powers of the bark of the willow tree and closely related plants, and especially their **antipyretic** or fever-reducing properties. These plants contain *salicylates*, compounds structurally related to aspirin. (The Latin term for willow, *salix*, gives us the name of this family of compounds.) In 1827, chemists isolated *salicin*, the active ingredient in willow bark. Subsequent developments during the 19th century led to the first synthesis of the active ingredient in aspirin, *acetylsalicylic acid*, and the eventual marketing of aspirin (**Figure 12.1**).

Developments in the discovery and marketing of aspirin • Figure 12.1

Acetylsalicylic acid, also known as aspirin, was first prepared in 1853 but was not marketed until almost 50 years later.

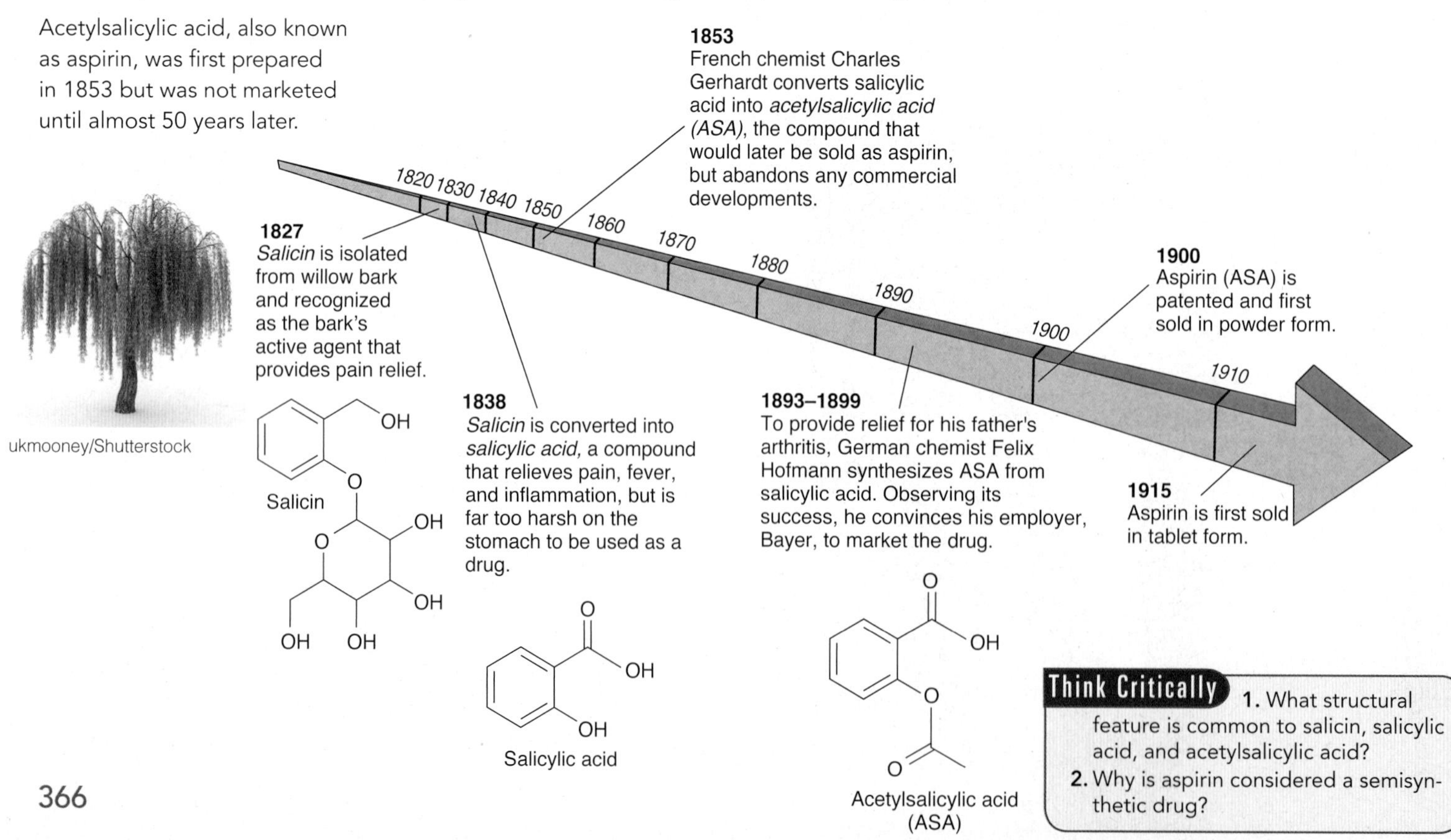

1853
French chemist Charles Gerhardt converts salicylic acid into *acetylsalicylic acid (ASA)*, the compound that would later be sold as aspirin, but abandons any commercial developments.
1820
1830
1840
1850
1860
1870
1880
1890
1900
1910
1827
Salicin is isolated from willow bark and recognized as the bark's active agent that provides pain relief.
1900
Aspirin (ASA) is patented and first sold in powder form.
ukmooney/Shutterstock
1838
Salicin is converted into *salicylic acid*, a compound that relieves pain, fever, and inflammation, but is far too harsh on the stomach to be used as a drug.
1893–1899
To provide relief for his father's arthritis, German chemist Felix Hofmann synthesizes ASA from salicylic acid. Observing its success, he convinces his employer, Bayer, to market the drug.
1915
Aspirin is first sold in tablet form.
OH
O
OH
OH
OH
OH
Salicin
O
OH
OH
Salicylic acid
O
OH
O
O
Acetylsalicylic acid (ASA)

Think Critically 1. What structural feature is common to salicin, salicylic acid, and acetylsalicylic acid?
2. Why is aspirin considered a semisynthetic drug?

Aspirin semisynthesis • Figure 12.2

Aspirin (acetylsalicylic acid) is prepared through the reaction of acetic anhydride with salicylic acid.

Acetic anhydride
Salicylic acid
Acetylsalicylic acid (aspirin)
Acetic acid
Acetic anhydride is a useful acetylating agent. Here, it transfers one of its two equivalent acetyl groups to salicylic acid. [Acetic anhydride itself is formed from the combination of two molecules of acetic acid with the loss of one molecule of water. The term *anyhydride* means without water.]
This chemical name tells us that an acetyl group has been added to salicylic acid.

Ask Yourself

1. Identify the carboxylic acid functional group in salicylic acid and in acetylsalicylic acid. (For a review of this functional group, see Section 8.3.)
2. What purpose does acetic anhydride serve in the chemical synthesis of aspirin?

Aspirin has become the most widely used of all drugs for the treatment of illness or injury. This compound acts as an **analgesic** (to relieve pain), as an **antipyretic** (to lower fever), and as an **anti-inflammatory agent** (to reduce inflammation). It's effective in treating rheumatoid arthritis and in preventing some types of strokes and heart attacks, such as those resulting from the accumulation of platelets in the blood vessels. The term *aspirin* was coined by adding an *a* (for acetylated) to a portion of an older name for salicylic acid, *spiraeic* acid, which derives from the name of a plant from which it was isolated. **Figure 12.2** shows that aspirin can be made by acetylating, or adding an *acetyl* group, to salicylic acid.

As you might expect, aspirin doesn't begin to do its work until it enters the bloodstream. The time it takes for the acetylsalicylic acid bound up in a solid aspirin tablet to enter the blood is governed largely by the rate at which the tablet disintegrates in the stomach. This rate, in turn, depends on pH (Section 8.2). The higher the pH, the faster the tablet breaks up and the faster the acetylsalicylic acid gets into the blood. To accelerate the disintegration of the tablet and the absorption of the aspirin itself, some companies produce a "buffered" aspirin. (A buffer is a combination of an acid or a base and one of its salts. By reacting with both acids and bases, this combination is able to maintain the pH of a solution within a narrow range.)

Despite the use of this term, buffered aspirin tablets aren't truly buffered. Rather than a combination of acetylsalicylic acid and one of its salts, which would constitute a true buffer, "buffered" aspirin consists of a combination of aspirin and one or more bases, including magnesium carbonate, $MgCO_3$, and magnesium hydroxide, $Mg(OH)_2$. Although the addition of these bases to the tablet does increase rates of disintegration and absorption, clinical studies have failed to produce evidence of faster or greater pain relief from "buffered" aspirin than from the "nonbuffered" variety. In clinical tests, ordinary aspirin works as well as buffered types. However, buffered aspirin does produce less stomach acidity than ordinary aspirin.

How aspirin works Aspirin and some other pain relievers, including ibuprofen (sold under the trade names Motrin and Advil), belong to a class of compounds called **NSAIDs**, short for non-steroidal anti-inflammatory drugs. As this name implies, these drugs are not classified as steroids. (We'll discuss the structures of steroids later in this chapter.) NSAIDs work by inhibiting the formation of **prostaglandins**, which are compounds named for the prostate tissue and seminal fluid in which they were discovered. Prostaglandins occur in virtually every tissue and fluid of the male and female body and participate, in one way or another, in almost every bodily function. They play particularly important roles in the sensation of pain and its transmission along the nervous system; in the generation of fevers; and in the swelling of inflammations. By interfering with the

NSAID A non-steroidal anti-inflammatory drug, typically used to treat pain, fever, or inflammation.

Chemistry InSight Enzyme inhibition • Figure 12.3

Enzymes normally help convert substrate molecules into other compounds. Enzyme inhibitors block this process.

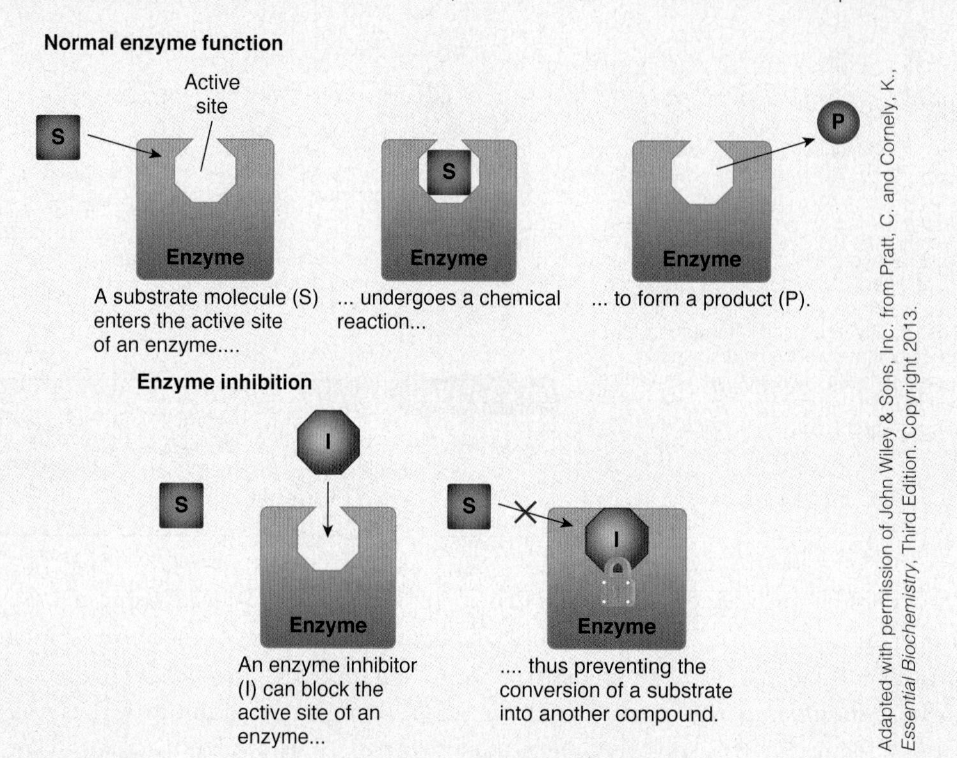

Normal enzyme function
Active site
S
Enzyme
S
Enzyme
P
Enzyme
A substrate molecule (S) enters the active site of an enzyme....
... undergoes a chemical reaction...
... to form a product (P).
Enzyme inhibition
I
S
Enzyme
S
I
Enzyme
An enzyme inhibitor (I) can block the active site of an enzyme...
.... thus preventing the conversion of a substrate into another compound.

Adapted with permission of John Wiley & Sons, Inc. from Pratt, C. and Cornely, K., *Essential Biochemistry*, Third Edition. Copyright 2013.

Drugs that inhibit enzymes

Non-steroidal anti-inflammatory drugs (NSAIDs) inhibit enzymes necessary for the formation of prostaglandins, compounds associated with pain, fever, and inflammation.

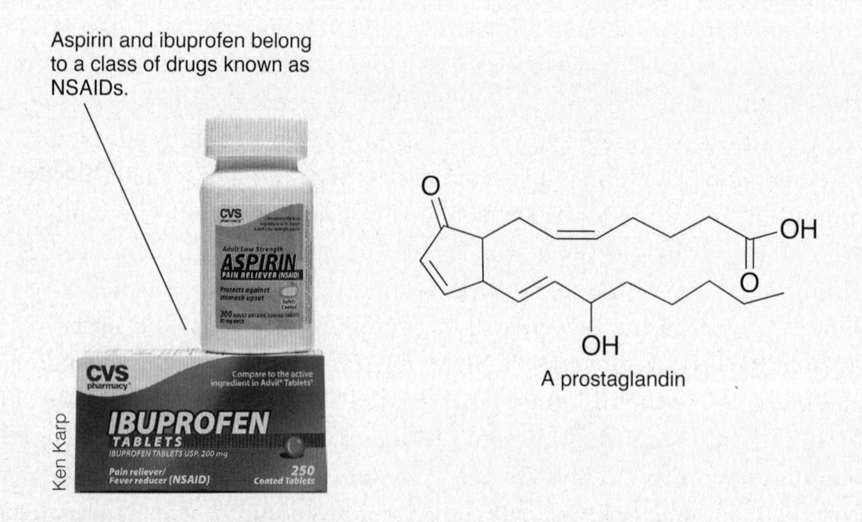

Aspirin and ibuprofen belong to a class of drugs known as NSAIDs.
CVS
ASPIRIN
PAIN RELIEVER (NSAID)
CVS pharmacy
IBUPROFEN
TABLETS
Pain reliever/ Fever reducer (NSAID)
250
Ken Karp
O
OH
O
OH
A prostaglandin

Ask Yourself

Identify each of the following as a substrate (S), inhibitor (I), or product (P): **(a)** aspirin **(b)** prostaglandin **(c)** ibuprofen

formation of the prostaglandins, NSAIDs relieve each of these conditions. The actual mechanism by which prostaglandin formation is suppressed is known as **enzyme inhibition** (**Figure 12.3**). Most drugs, including those used against HIV-AIDS, for example, work by inhibiting or deactivating enzymes.

enzyme inhibition Suppressing an enzyme's normal function.

Aspirin can produce side effects, including upset stomach and various allergic reactions. Some of the major risks in taking aspirin result from its effect on the blood's ability to clot. Acetylsalicylic acid inhibits the formation of blood platelets, which are small bodies within the blood serum that initiate clotting. Insufficient concentrations of platelets can lead to gastrointestinal bleeding, susceptibility to bruising, and an increased risk of hemorrhagic stroke, a condition brought about by the rupture of small blood vessels within the brain. To guard against excessive bleeding, surgery patients are often advised not to take aspirin for several days before the scheduled operation.

More serious than these side effects is Reye's syndrome, a rare and sometimes fatal condition experienced principally by children and adolescents recovering from chicken pox, the flu, and other viral infections. Aspirin and other salicylates appear to increase the risk of developing Reye's syndrome. Aspirin labels in the United States carry warnings to consult a physician before giving the drug to children or adolescents with symptoms of either chicken pox or the flu.

Other nonprescription analgesics For people who may be allergic to aspirin or who find that it produces stomach upset, alternative medications are available. These include the NSAID *ibuprofen*, which is the active ingredient in Advil and Motrin, and *acetaminophen*, the active ingredient in Tylenol. Acetaminophen is the least toxic member of a class of analgesic and antipyretic (but not anti-inflammatory) medications known as *p*-aminophenols (*para*-aminophenols). Compounds of this sort trace back to acetanilide, which was used to alleviate fever as early as 1886. Acetanilide proved too toxic for general use, yet it offered promise. In 1887, phenacetin, and, in 1893, acetaminophen—both of which are related structurally to acetanilide—were introduced. In 1983, almost a century after phenacetin was first used as an analgesic, the U.S. Food and Drug Administration banned it because of its tendency to damage the kidneys and to produce disorders of the blood when used excessively (**Figure 12.4**).

Despite acetaminophen's relatively low toxicity, at least as compared to other *p*-aminophenols, this comparatively safe medication carries its own risks. Acetaminophen can cause liver damage when taken in larger-than-recommended doses or in combination with alcoholic drinks. Also, although the occasional use of

Acetaminophen and related compounds • Figure 12.4

Acetaminophen traces its origins to structurally similar compounds used to reduce fever: acetanilide and phenacetin. Of these three compounds, acetaminophen showed the fewest side effects.

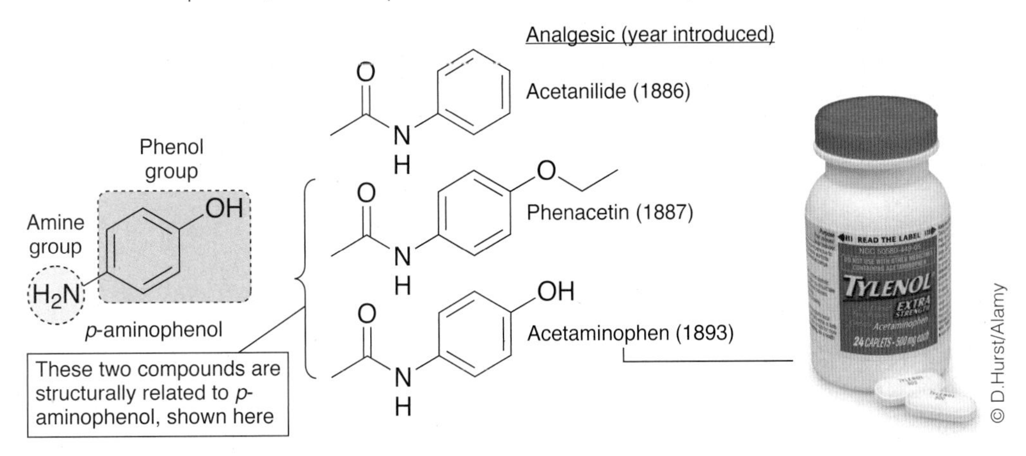

Ask Yourself

1. How are acetanilide, phenacetin, and acetaminophen structurally similar?
2. How are phenacetin and acetaminophen structurally similar to *p*-aminophenol?
3. The systematic chemical name for acetaminophen is *N*-acetyl-*p*-aminophenol. Suggest how this term gave rise to the trade name, Tylenol.

Decongestants and antihistamines • Figure 12.5

Symptoms of hay fever and seasonal allergies can be treated with (a) decongestants and (b) antihistamines.

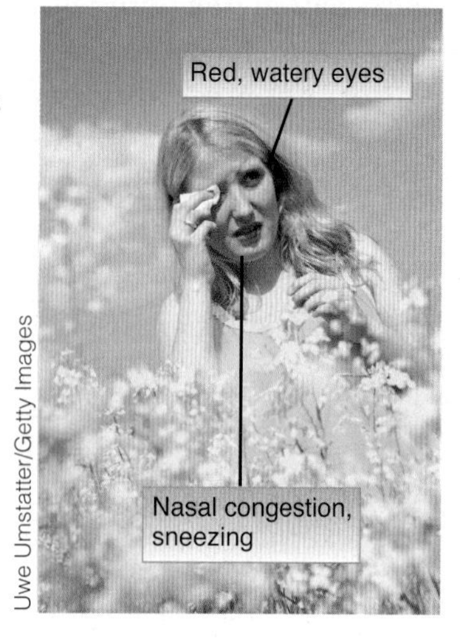

Phenylephrine, a decongestant

Phenyl group
Ethyl group
Amine group
NH_2
Phenylethylamine

a. Decongestants help reduce swelling of the sinuses. Phenylephrine, a common decongestant, falls within the class of psychoactive substances known as phenylethylamines, and as a result, can sometimes cause restlessness.

Loratidine, a non sedating antihistamine

b. Antihistamines block the action of histamine, a signaling compound involved in allergic response. Loratidine is a newer-generation antihistamine, lacking the drowsy side effects associated with first-generation antihistamines, such as diphenhydramine, the active ingredient in Benadryl.

acetaminophen may pose minimal risks, the daily consumption of acetaminophen over a long period can produce kidney damage.

Cold and Allergy Medicines

At some time, you've probably suffered from the discomforts of a cold or a seasonal allergy. Colds are caused by viruses. Seasonal allergies, on the other hand, may result from certain **allergens** (allergy-causing substances) present in the air, such as pollen, and often occur in the spring and summer. Although colds and allergies are distinctly different medical conditions, they can exhibit similar symptoms, including nasal congestion and sneezing.

There is no known cure for the common cold, but a variety of nonprescription medicines are available to relieve its discomforts. Many of these same products are often useful in treating symptoms of seasonal allergies. For example, **decongestants** help relieve swelling of the nasal passages by restricting blood flow to the sinuses. **Antihistamines** can help reduce the inflammation, watery eyes, and sneezing associated with allergies (**Figure 12.5**). Antihistamines work by blocking the action of a chemical, *histamine*, that triggers inflammation. It's a compound structurally related to the amino acid *histidine*.

CONCEPT CHECK STOP

1. **How** does aspirin suppress the formation of prostaglandins?
2. **What** is histamine?

12.2 Prescription Medicines

LEARNING OBJECTIVES

1. **Explain** how prescription medicines are developed.
2. **Describe** the chemical structures and functions of common prescription medicines.

When a doctor writes a prescription for you, it may link you to a story that began decades earlier with the discovery of the compound that provides the active agent of the medication. In this section we'll explore how prescription medicines are developed, and we'll also discuss the chemistry underlying common classes of medicines, especially antibiotics and steroids.

Drug Development

The notion of taking a pill to relieve aches, pains, and ailments is embedded in our culture. In the United States alone, more than four billion prescriptions are written each year, producing annual sales in excess of $320 billion. Even if you're not taking a prescription drug, you're probably aware of these products as a result of pharmaceutical ads. Since 2008, the U.S. Food and Drug Administration (FDA) has allowed these messages to appear on television and in other media.

The process of developing new medicines begins with identifying **drug candidates**, which are chemical compounds that show biological activity and have the potential for being developed into prescription drugs. Drug candidates may be discovered through a variety of means, including:

- **rational drug design**, a process that requires first identifying a biological molecule, such as a protein or an enzyme, involved in causing a disease, and then designing molecules that interact with this chemical target (many HIV-AIDs drugs, for instance, have been developed using this method);
- **high-throughput screening**, which involves applying robotics to screen thousands of compounds (a topic we discussed in Section 1.3);
- serendipity or chance (penicillin was discovered in this manner);
- chemical modifications of existing drugs in order to improve their efficacy and/or reduce side effects.

In drug development, literally thousands of compounds may be screened and tested for every one that ultimately receives marketing approval by the FDA. The process is also time consuming, involving many years of clinical studies. It can take 15 years or more from the time a compound is discovered to its approval for use. As a result, the process is also quite expensive. By some estimates, nearly $1 billion may be spent by a company on research and development for each of its drugs that reaches the pharmaceutical market. The greatest costs incurred during this process are for **clinical trials**, which test drug candidates in human subjects. **Figure 12.6** outlines the major steps in developing a new medicine.

The drug development process • Figure 12.6

Thousands of compounds may be screened and tested for every one that receives FDA approval.

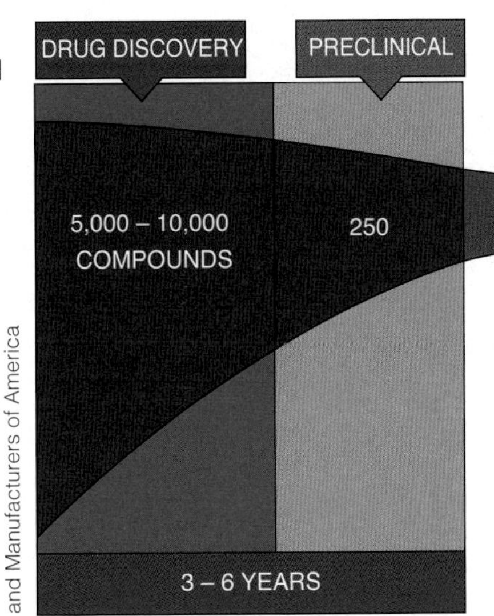

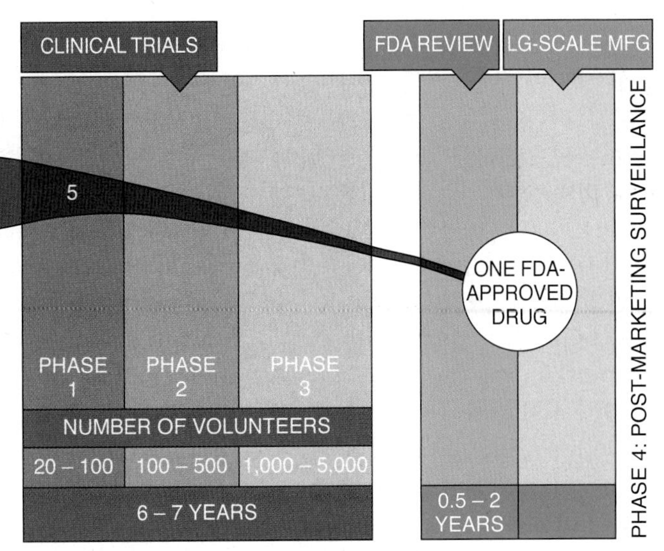

Adapted from Pharmaceutical Research and Manufacturers of America

During preclinical trials, drug candidates are tested in laboratory animals.

Data from preclinical studies are submitted to the FDA for review before permission may be granted to proceed with clinical trials.

Clinical trials proceed in three phases. A drug candidate may be pulled from testing at any point along this process if problems arise.
- Phase I focuses on the safety of the drug within a small set of individuals.
- Phase II evaluates dosing requirements, efficacy, side effects, and safety of the drug within a larger group.
- Phase III evaluates the drug in larger populations and at different locations over longer periods of time.

Data from clinical trials are submitted to the FDA for final review.

The results of clinical trials of a drug candidate largely determine whether the FDA approves the medicine. The overall objective of these clinical studies is to determine whether the drug candidate provides significantly better clinical outcomes than a **placebo**. It is well documented that administering a placebo (a word derived from the Latin meaning to please or to pacify) makes it appear to a patient that something important is being done and sets an expectation that symptoms will improve, which they sometimes do. This **placebo effect** has been shown to have stopped coughs, promoted the healing of wounds, relieved depression, lowered blood pressure and pulse rates, and even produced side effects, including dryness of the mouth, headaches, and drowsiness.

placebo An inert substance used as a scientific control in tests of an actual medication.

The individuals who participate in a clinical trial are divided into two groups by random assignment. One set, the control group, receives a placebo that may look identical to the drug being tested. The other, the test group, receives the actual drug. To avoid any potential for bias, the study is carried out **double-blind**, meaning that during the trial itself, no one, neither the investigators nor the participants, know whether any particular individual is in the control group or the test group. This requires an independent third party who codes each pill with a number that identifies its contents, and maintains a plan for their distribution. The investigators distributing the pills to test subjects remain "blind" as to the significance of the codes, thus preventing the investigators from revealing any hints, even subconsciously, to those receiving them. With this level of concealment, and with the pills appearing identical in all respects except for their coding, the placebo effect operates equally with everyone involved. Only once the test is over can it be revealed whether those receiving the authentic medication responded differently from those receiving the placebo. If a significant difference is observed, it must be due to the real potency of the medicine and not simply to a placebo effect.

Newly introduced drugs are sold under patent protection, which prevents other companies from making and selling the same medicine while the patent is in force. Once the drug's patent expires, however, other companies may sell a generic form of the drug. **Generic drugs** sold in the United States are regulated by the FDA and must show equivalence to their brand-name counterparts in all respects, including purity, safety, strength, and performance characteristics. Generic drugs are typically less expensive than their brand-name equivalents and are sold under their generic or chemical names. For instance, when the patent for the highly successful cholesterol-lowering drug, Lipitor, expired, the medication became available as a generic drug at lower cost, sold under its chemical name, atorvastatin.

generic drugs Drugs that are chemically equivalent to patented drugs.

Therapeutic Classes of Drugs

More than one thousand distinct medications are listed in *The Physician's Desk Reference*, a compendium of the most commonly prescribed drugs. These medicines are used to treat a wide variety of diseases and conditions, including cancers, infectious diseases, cardiovascular disorders (including high blood pressure), hormone-related disorders (such as diabetes), and mental disorders (including depression). Here we'll explore the chemistry underlying a few of these classes of medications.

Antibiotics Throughout history, humans and other living organisms have been beset by illness and disease as a result of infectious agents, including bacteria, viruses, and parasites. In the case of humans, two of the most serious epidemics of infectious disease were the bubonic plague, a bacterial-borne disease that took the lives of an estimated 75 million people during the Middle Ages; and smallpox, a viral-borne infection that killed over 300 million people in the 20th century alone. [Since 1979, however, no additional cases of smallpox have been reported, which is attributed to a worldwide vaccination program against the disease.] More recently, acquired immunodeficiency syndrome (AIDS), which is caused by the human immunodeficiency virus (HIV), has taken the lives of more than 25 million people since first being reported in the early 1980s.

Antibiotics (from the Greek *anti*, against, and *bios*, life) are substances produced naturally by microbes, such as molds and fungi, as chemical defenses against other microbes. Since the discovery, nearly 80 years ago, that antibiotics could be used to treat bacterial infections, these compounds have saved countless lives, including World War II soldiers suffering from wound infections and those suffering from illnesses, such as bacterial meningitis, bacterial pneumonia, and tuberculosis. Some of the more notable developments in the history of antibiotics are noted in **Figure 12.7**.

antibiotics Substances, originally derived from certain molds and fungi, that kill other microbes.

Chemistry InSight Selected developments in the history of antibiotics • Figure 12.7

Our bodies are host to and exposed to billions of bacteria, most of which are benign but some of which may be *pathogenic* or disease-causing. Antibiotics are often used to treat these bacterial infections.

a. Penicillin and its derivatives

In 1928, British bacteriologist Alexander Fleming noticed that a mold contaminating a petri dish had killed bacteria that had been growing there. He named this antibacterial agent **penicillin**, for the **Penicillium** fungus from which it derives.

A modern image of the same types of microorganisms Alexander Fleming observed in his discovery of penicillin.

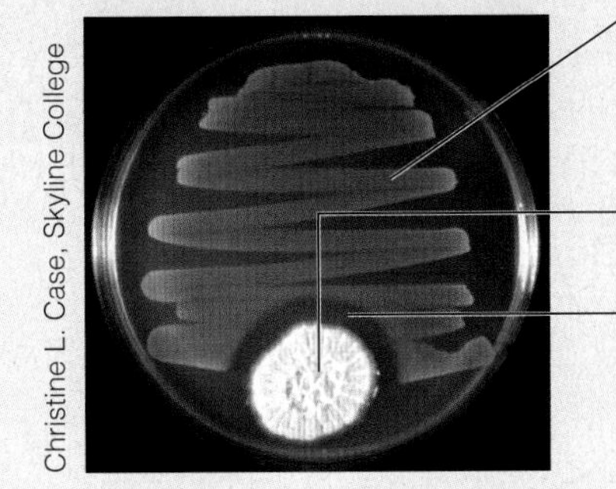

Christine L. Case, Skyline College

The wide, light-blue zig-zags are *Staphylococcus* bacteria growing on a petri dish.

The white circle is *Penicillium* fungus.

Notice that no bacteria grow in this banded region around the mold. They have been inhibited by the penicillin naturally secreted by the fungus.

It took over a decade after Fleming's initial discovery to determine the structure of penicillin and to produce sufficient quantities of the drug to conduct clinical trials and treat patients on a massive scale. Penicillin was used successfully to treat wounded Allied soldiers during World War II.

Many **analogs** of existing antibiotics, including penicillin, have been produced through semisynthesis. Analogs of a compound share the same basic structure, but have small modifications that may result in improved performance. Amoxycillin, one of the most prescribed antibiotics today, is an analog of penicillin.

Penicillin

Amoxycillin

b. Other classes of antibiotics

In the period from 1940 to 1960, several new classes of antibiotics were discovered, including *tetracyclines*, named for the four (*tetra*) ring (*cyclo*) structures shared by these compounds, and *cephalosporins*. In 1980, azithromycin, a member of a class of antibiotics called *macrolides*, was discovered. Marketed under the trade name Zithromax, it is now one of the world's best-selling antibiotics.

Tetracycline

Ask Yourself

Why is amoxycillin considered a drug analog? How is its structure different from that of penicillin?

Antiviral medication and HIV-AIDS • Figure 12.8

HIV drugs interfere with enzymes required for the human immunodeficiency virus to replicate within the body.

a. The very first drug to treat HIV infection, AZT, was introduced in 1987.

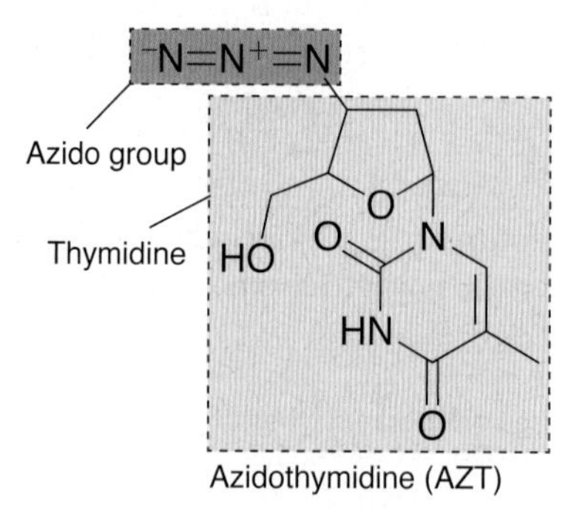

b. Since 1995, standard HIV treatment involves administering a *drug cocktail*, or mixture of compounds used to target various enzymes in the human immunodeficiency virus replication cycle. Since the introduction of this approach, deaths due to AIDS have declined in the United States, even though the number of those living with HIV infection has risen.

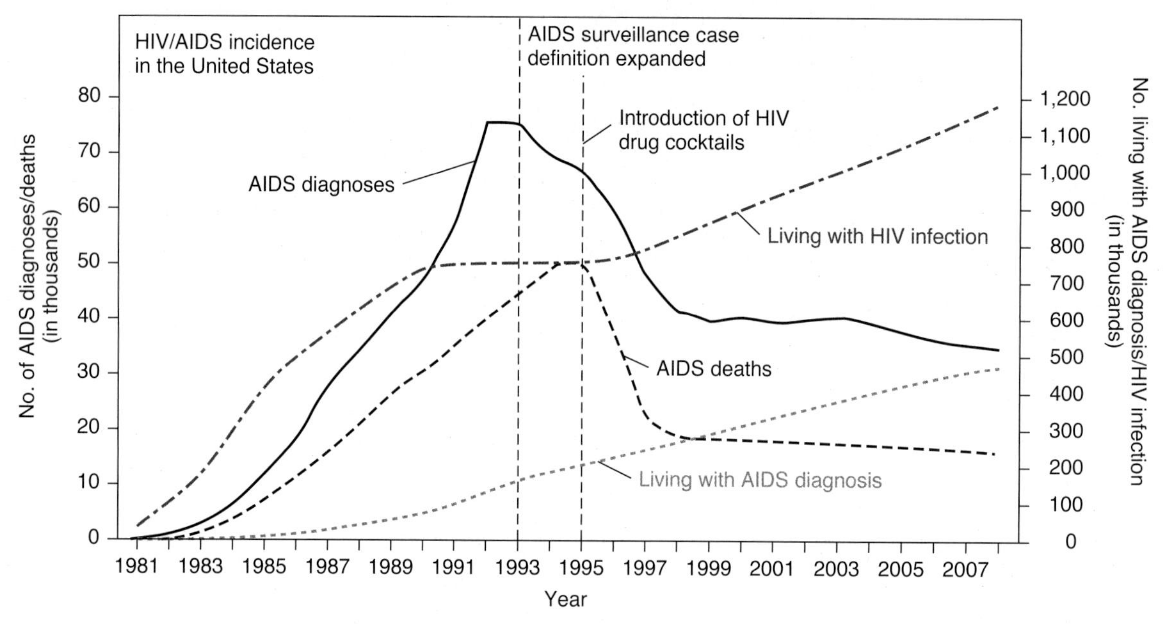

Source: Centers for Disease Control and Prevention. "HIV Surveillance—United States, 1981—2008." MMWR 2011;60: 689-693.

A century ago, infectious disease was the leading cause of death in the United States. This is no longer the case, due in large part to the efficacy of antibiotics. (Within the United States, deaths from heart disease and cancer now exceed those from infectious disease.) But overuse of antibiotics has exacerbated the problem of **antibiotic resistance**, a phenomenon in which bacteria develop a tolerance to the compounds designed to kill them. For instance, mortality rates due to tuberculosis have declined significantly since antibiotics were introduced to treat this disease, but these same drugs are now linked to a higher occurrence of antibiotic-resistant strains of the bacterium that causes tuberculosis. The prophylactic use of antibiotics—such as the small amounts of these drugs added to animal feed in the industrial production of meat—raises further concerns about the development of antibiotic-resistant bacteria.

The common use of antibiotics as antibacterial agents has led to an association of the term *antibiotic* with substances used to treat bacterial infections. However, antibiotics also include agents that target other microbes as well, including viruses. These antiviral medications have played an important role in combatting viral-borne illness and disease, such as AIDS (**Figure 12.8**), a disease that compromises the body's immune system. AIDS fatalities result from the body's inability to combat opportunistic infections and certain viral-based cancers.

Psychiatric medications Pain is often an indication of something wrong with the body, like a wound or a serious disease. But beyond the hurts of raw, physical pain, people are also vulnerable to mental and emotional anguish, ranging from anxiety and depression to psychoses and other psychiatric disorders. The kinds of drugs available to treat mental illness are as varied as the symptoms they relieve. Among the most commonly used are **tranquilizers** (which reduce anxiety) and **antidepressants** (which treat depression).

The most frequently prescribed tranquilizers fall within a class of compounds called *benzodiazepines*, with alprazolam (Xanax) among the most popular (**Figure 12.9a**). The earliest benzodiazepines were tested on laboratory animals in the late 1950s and were found to act as muscle relaxants and sedatives as well as tranquilizers. Compounds of this structure were soon used to treat disorders in humans. Though benzodiazepines, including Xanax and Valium, are highly effective tranquilizers, they have been shown to exhibit a high potential for abuse (a topic we'll explore in the next section).

Fluoxetine (sold under the trade name Prozac) was the first of a new class of antidepressants known as selective serotonin reuptake inhibitors (SSRIs). Since the introduction of Prozac in 1988, a variety of SSRIs, including citalopram (Celexa), have been developed. SSRIs received their long and complex name because they inhibit the reuptake (or reabsorption) of the neurotransmitter serotonin by neurons or nerve cells within the brain. The resulting higher levels of circulating serotonin within the brain appear to enhance the mood of those who are depressed and reduce anxiety for those suffering from anxiety disorders (**Figure 12.9b**). Although SSRIs can exhibit side effects, including nausea and loss of appetite, the benefits offered by these compounds generally far exceed these drawbacks.

Steroidal medications Steroids constitute a large class of organic compounds that share a common four-ring structure, shown in **Figure 12.10**. Hundreds of steroids are known, including those occurring naturally in both animals and plants, and those produced synthetically. Steroids exhibit a wide range of regulatory activities within the body, including the control of sexual development and reproduction; and the regulation of metabolism, inflammatory processes, and water retention.

Tranquilizers and antidepressants • Figure 12.9

(a) The most commonly prescribed tranquilizers fall within the class of benzodiazepines.
(b) The most prescribed antidepressents fall within the class of selective serotonin reuptake inhibitors (SSRIs).

a. Benzodiazepine tranquilizers, such as alprazolam (Xanax), share a common structure based on conjoined or fused benzene and diazepine ring systems.

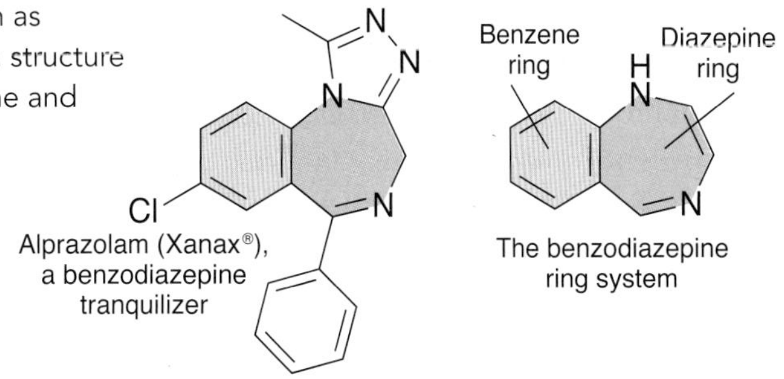

Alprazolam (Xanax®), a benzodiazepine tranquilizer

The benzodiazepine ring system

b. SSRIs, such as fluoxetine (Prozac), increase serotonin levels in the brain and are widely used to treat anxiety and depression.

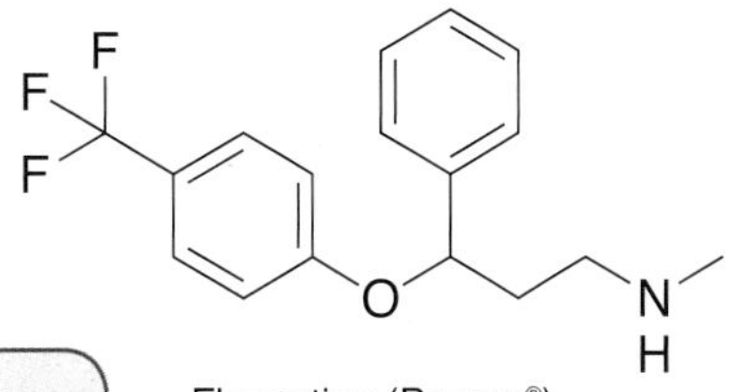

Fluoxetine (Prozac®)

Ask Yourself

What chemical element is present in fluoxetine that is not present in any of the other compounds we have examined so far in this chapter?

Steroidal structure • Figure 12.10

(a) All steroids share a common four-ring, hydrocarbon structure. (b) Testosterone is a steroidal compound.

a. The core steroid structure has three 6-member rings (A-C) and one 5-member ring (D) fused to one another in the arrangement shown here.

b. The steroid, testosterone, $C_{19}H_{28}O_2$, is the principal male sex hormone. Note the presence of the ketone functional group, which gives rise to the suffix of testosterone.

Ask Yourself

1. How many carbon atoms are present in the basic steroid structure shown in part a?
2. Where is the alcohol functional group in testosterone located?

The most common steroidal medications are called **corticosteroids**, which are synthetic compounds similar in structure to **cortisol**, a steroidal hormone produced by the adrenal gland. Corticosteroids reduce inflammation associated with a variety of conditions. For instance, injected into joints, corticosteroids can help relieve the pain and swelling of arthritis or tendinitis; inhaled as a mist, these drugs can treat asthma; and applied to the skin as a cream, they can help relieve the itching and inflammation of eczema. Prolonged use of corticosteroids, especially those taken internally, increases the likelihood of side effects, such as weight gain, insomnia, and lowered resistance to infection.

Sex hormones, including estrogen, progesterone, and testosterone, represent another major class of steroids. **Anabolic steroids** (discussed in the chapter opener) are synthetic compounds that act on the body similarly to testosterone. As a result, they help increase muscle mass and are sometimes used as performance-enhancing drugs. The use of anabolic steroids, however, can pose serious side effects, both mental (resulting in aggressive outbursts) and physical (resulting in unwanted changes to the body's appearance).

A great deal of research has focused on the creation of semisynthetic sex hormones, starting from steroids naturally found in plants and animals. As described next, these efforts have led to the introduction of oral contraceptives.

After inventing the octane rating system for gasolines in the 1920s, the American chemist Russell Marker began investigating steroidal hormones derived from plants. His field work took him to Texas and then to Mexico in search of plants yielding large quantities of a steroid known as **diosgenin**, a compound with a chemical structure similar to that of the female sex hormone, progesterone. Marker's plan was to modify the chemical structure of diosgenin to produce progesterone, which was at the time a highly coveted and expensive material, not easily available in a pure form. In 1942, Marker found his source of diosgenin in a species of native yam plants growing near Mexico City. Harvesting a few hundred pounds of the tubers of this plant enabled him to produce 3 kilograms of progesterone, an unheard-of quantity at the time.

In 1944, Marker co-founded a company called Syntex in Mexico City to produce progesterone and other sex hormones. He left the company a year later as a result of disputes with his business partners, but Syntex thrived in the ensuing years, recruiting notable organic chemists from Europe and the United States, including an Austrian-born, American-trained chemist named Carl Djerassi. In 1951, Djerassi co-led a team in the first synthesis of **norethindrone**, an orally active form of progesterone, meaning that it can be taken in pill form and acts on the body like progesterone (**Figure 12.11**). Norethindrone is one of a variety of **progestins**, or progesterone mimics.

Steroids as oral contraceptives • Figure 12.11

Norethindrone, used in oral contraceptives, is a progestin whose scientific origins trace back to Marker's original work with the Mexican yam.

a. Progesterone, a female sex hormone first prepared semisynthetically by Marker in 1943.

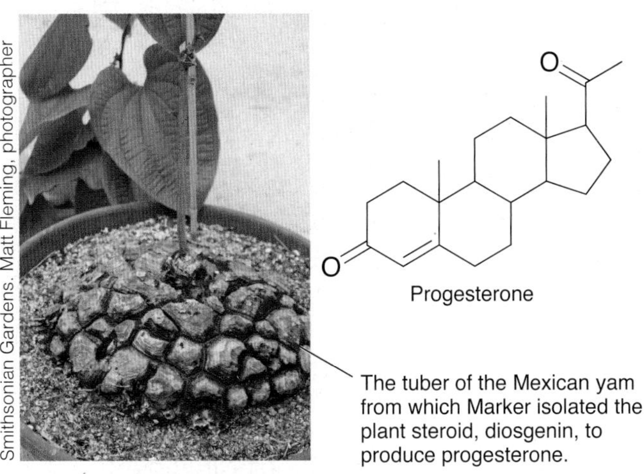

Smithsonian Gardens. Matt Fleming, photographer
O
O
Progesterone
The tuber of the Mexican yam from which Marker isolated the plant steroid, diosgenin, to produce progesterone.

b. Norethindrone, a progesterone mimic first prepared by Syntex in 1951.

OH

O

Norethindrone

Ask Yourself

1. How are the structures of progesterone and norethindrone **(a)** similar? **(b)** different?
2. Suggest why these two compounds share an *–one* suffix.

Around this same time, in 1952, a chemist named Frank Colton, working for the pharmaceutical company Searle, in Chicago, first synthesized **norethynodrel**, another progestin whose structure was almost identical to the norethindrone produced at Syntex. Executives at Searle were at the time also supporting the work of a Massachusetts-based endocrinologist, Gregory Pincus, who was investigating the effects of progesterone on animals. Pincus discovered that administering progesterone to female rabbits inhibited ovulation and thus could potentially be used to prevent conception. Subsequent clinical trials with norethynodrel ultimately led to the first approved oral contraceptive, Enovid, in 1960. Birth control pills commonly used today contain a combination of a progestin, such as norethindrone, and a synthetic estrogen.

CONCEPT CHECK STOP

1. **Why** are placebos used in clinical trials?
2. **How** was penicillin discovered?

12.3 Recreational, Illicit, and Abused Drugs

LEARNING OBJECTIVES

1. **Describe** how alcohol is metabolized by the body.
2. **Define** the term *alkaloid* and identify drugs that fall within this class of compounds.
3. **Identify** drugs that share the phenylethylamine structure.

So far we've used the terms *medicines* and *drugs* interchangeably. In some sense, they are virtually indistinguishable from each other and generally mean chemicals used medically for treating diseases and injuries. Yet the term *drugs*, unlike *medicines*, carries with it negative connotations, including addiction and illicit or criminal activities. In this section

Short-term effects of alcohol consumption • Figure 12.12

Alcohol's effects on the body depend on blood alcohol concentration levels (BACs). These levels depend on the rate and quantity of alcohol consumed and the time that has passed after consumption, among other factors. In most states, a BAC value of 0.08 is the legal limit for operating a motor vehicle.

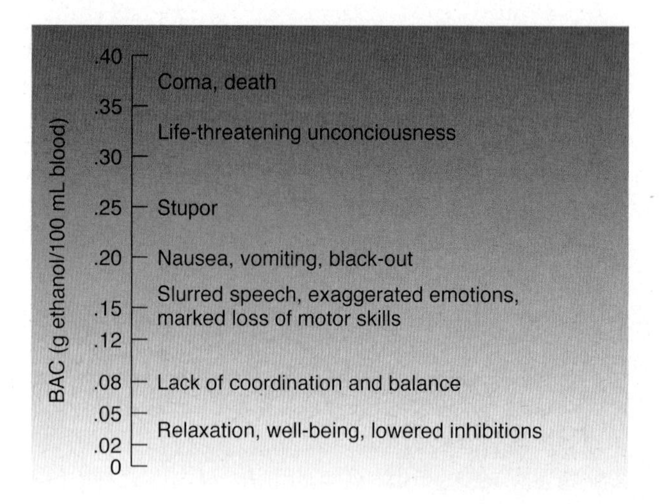

we'll examine several chemicals that seem more appropriately described as drugs than as medicines, though some of them do have legitimate medical uses.

Alcohol and Marijuana

In the United States, alcohol was declared illegal in 1920, but the demand for it gave rise to large-scale racketeering and crime. The prohibition against alcohol was repealed in 1933, and it has since remained legal. Marijuana is illegal in the United States under federal law, but recently, several states have legalized it for medical purposes (see below).

Effects of alcohol Alcohol, or more accurately ethanol, CH_3CH_2OH, is produced naturally by the fermentation of the carbohydrates of grains, fruits, and starchy vegetables. An example of the fermentation of glucose, a simple carbohydrate, is

$$\underset{\text{Glucose}}{C_6H_{12}O_6} \xrightarrow{\text{yeast}} \underset{\text{Ethanol}}{2\,CH_3CH_2OH} + \underset{\text{Carbon dioxide}}{2\,CO_2}$$

The type of alcoholic beverage produced depends on the nature of the plant material used in the fermentation process. For example, the fermentation of barley is used to make beer, whereas the fermentation of grapes generates wine. The typical level of alcohol in beer, measured as a percentage by volume, is 4–6%; the concentration of alcohol in wine is 12–14%. Hard liquor, such as vodka or rum, is produced through distillation of fermented plant materials such as potatoes or sugar cane, and has a typical alcohol concentration of 40%, also referred to as 80 proof. (The **proof** of an alcoholic beverage is generally twice the percentage of alcohol it contains.) A **standard drink** is defined as 12 fl oz (fluid ounces) of beer, 4 fl oz of wine, or 1 fl oz of 80 proof liquor. Each of these quantities contains the same amount of alcohol.

Alcohol acts as a central nervous system (CNS) depressant, slowing down signaling between the brain and the rest of the body. Its short-term mental and physical effects depend on the level of alcohol circulating in the blood, or **blood alcohol concentration**, BAC (**Figure 12.12**).

With the exception of caffeine, which is present in coffee and tea, alcohol is the most widely consumed drug. Nearly two-thirds of U.S. high school students and four-fifths of U.S. college students report having consumed alcohol within the prior year. Irresponsible alcohol consumption poses several societal risks, including a greater incidence of drunk driving and violent behaviors. **Binge drinking**, defined as consuming at least four standard drinks within a two-hour period, is of particular concern. In recent studies, over 40% of college students report having participated in binge drinking within the prior two weeks. Because of alcohol's effects on the entire central nervous system, habitual use of the substance can produce a powerful psychological addiction. Alcohol consumption while pregnant can also pose very serious risks to fetal development.

Despite the physical and psychological risks involved, the effects of alcohol consumption are not entirely negative. For example, for those at risk of cardiovascular disease, such as heart attack or stroke, recent studies suggest that light to moderate alcohol consumption—defined as no more than 1–2 drinks per day—may lower the risk of developing these conditions.

Within the body, alcohol is metabolized by the liver through a series of oxidation reactions facilitated by enzymes. The two initial steps of this process are

$$\underset{\text{Ethanol}}{CH_3CH_2OH} \xrightarrow{\text{oxidation}} \underset{\text{Acetaldehyde}}{CH_3CHO} \xrightarrow{\text{oxidation}} \underset{\text{Acetate ion}}{CH_3COO^-}$$

The second step—conversion of acetaldehyde to acetate ion—is generally slower than the first step. This can cause levels of acetaldehyde to increase in the body, especially when alcohol is consumed rapidly. The build-up of acetaldehyde within the blood is responsible for some of the unpleasant side effects associated with rapid or excessive drinking, including flushed face, nausea, headache, and vomiting.

Marijuana Marijuana refers to the dried leaves of the cannabis or hemp plant. Its active ingredient, tetrahydrocannabinol (THC), targets receptors in the brain associated with pleasure, thinking, and sensory perception. The resulting "high" experienced by users may also be associated with a distorted perception of time and interference with short-term memory, problem-solving skills, and judgment. Short-term physical responses to marijuana include reddened eyes, dry mouth, and increased appetite.

Marijuana is the most commonly used illicit drug in the United States. Approximately one-third of U.S. high school and college students report having used it within the prior year. Its legal status in the United States appears to be in transition: Some states have begun to allow its use for medical purposes or have decriminalized the possession of small amounts. Marijuana seems to have medical value in controlling glaucoma, a disease of the eyes in which increasing internal pressure leads to blindness, and in relieving the nausea of chemical treatment for cancers and other diseases.

Although marijuana does not appear to cause physical addiction, its habitual use can lead to psychological dependence, with attempts at withdrawal causing mood swings and irritability. Chronic use among adolescents has been associated with impaired brain development and increased risks of psychiatric disorders, especially for those with underlying predisposition toward these conditions.

The THC molecule, shown in **Figure 12.13**, is **lipophilic** or fat soluble. As a result, THC remaining within the body after marijuana use is stored within the body's fat cells. Though this residual amount does not produce the drug's effects, it can take a few days or weeks to clear the body, depending on the level of use.

Alkaloids

Alkaloids are a major category of plant compounds known for their alkalinity—that is, their behavior as bases in the presence of acids. At the molecular level, alkaloids all owe their basicity to one or more amine groups in their molecular structures. Most taste bitter and produce physiological reactions of various kinds and intensity. Amines contain a nitrogen atom with three bonds, as shown here.

> **alkaloid** A class of plant compounds, often bitter, that contain the basic, amine group.

$$-\ddot{N}-$$

An amine group

Marijuana • Figure 12.13

The principal active ingredient in marijuana is tetrahydrocannabinol (THC).

> **Ask Yourself**
> The fat solubility of THC is due to the nonpolar hydrocarbon-rich components of this molecule. Identify these molecular components of THC.

Doug Menuez/Photodisc/Getty Images

OH
O
Tetrahydrocannabinol (THC)

Caffeine and nicotine Probably the most widely used of all the alkaloids are **caffeine**, found in coffee beans and tea leaves, and **nicotine**, produced by tobacco plants (**Figure 12.14**). Caffeine stimulates the central nervous system and the heart, and heightens a sense of awareness. It's added to cola drinks at a concentration of about 60 mg per 16-oz serving, and it occurs naturally in tea and coffee. The amount of caffeine in a cup of coffee depends on the strength of the brew, but the alkaloid is generally present at a concentration of about 100–150 mg per cup. (Decaffeinated coffee, of course, contains virtually none.) The lethal dose for an adult is estimated at about 10 g, taken orally. That amounts to 70 to 100 cups of brewed coffee at one sitting, a quantity that would generate plenty of discomfort long before the fatal dose is approached.

Nicotine, on the other hand, is a lethal substance that's used as a powerful agricultural insecticide. Absorbing less than 50 mg of nicotine can kill an adult in a few minutes. The high temperatures and rapidly moving air stream that accompany smoking oxidize most of the nicotine of tobacco, converting this alkaloid to less toxic products. Without this oxidation, no cigarette smoker could possibly last long enough to develop smoking into a habit.

Several generations ago, cigarette smoking was generally considered fashionable and sophisticated. With growing recognition of the health hazards posed by nicotine and other components of cigarette smoke—both to smokers themselves and to those exposed to secondhand smoke—smoking in public places has been severely restricted or prohibited entirely.

The lethal effects of smoking aren't due exclusively to nicotine. Cigarette smoke is rich in toxic carbon monoxide (CO), which decreases the blood's ability to carry oxygen. The smoke also contains tarry substances capable of generating cancer. According to the U.S. Centers for Disease Control, an estimated 400,000 people die each year in the United States from smoking-related causes, including cardiovascular diseases and cancers. Smoking during pregnancy is particularly dangerous to the fetus and can lead to spontaneous abortions, low birth weights, birth defects, and infant deaths.

Opioid narcotics Whether we consider it a medicine or a drug, the essence of a narcotic is clear. A narcotic—a word derived from the Greek *narkotikos*, meaning numbing or stupefying—is a chemical that dulls the mind, induces sleep, and generally numbs the senses.

The use of narcotics is an ancient activity. The Sumerians, one of the earliest civilizations, were probably familiar with the narcotic effects of opium, the dried sap of the poppy, some 6000 years ago. (The Sumerian term for the flower can be translated as "joy plant.") The first clear reference to the sap of the poppy appeared in Greek writings from about 300 BCE. Later, both Arabian and Oriental healers used its medical powers. Thomas Sydenham, the 17th-century English physician known as the British Hippocrates, wrote of the substance's powers to ease pain:

> *Among the remedies which it has pleased Almighty God to give to man to relieve his sufferings, none is so universal and so efficacious as opium.*

Common alkaloids • Figure 12.14

The alkaloids caffeine and nicotine are both produced by plants.

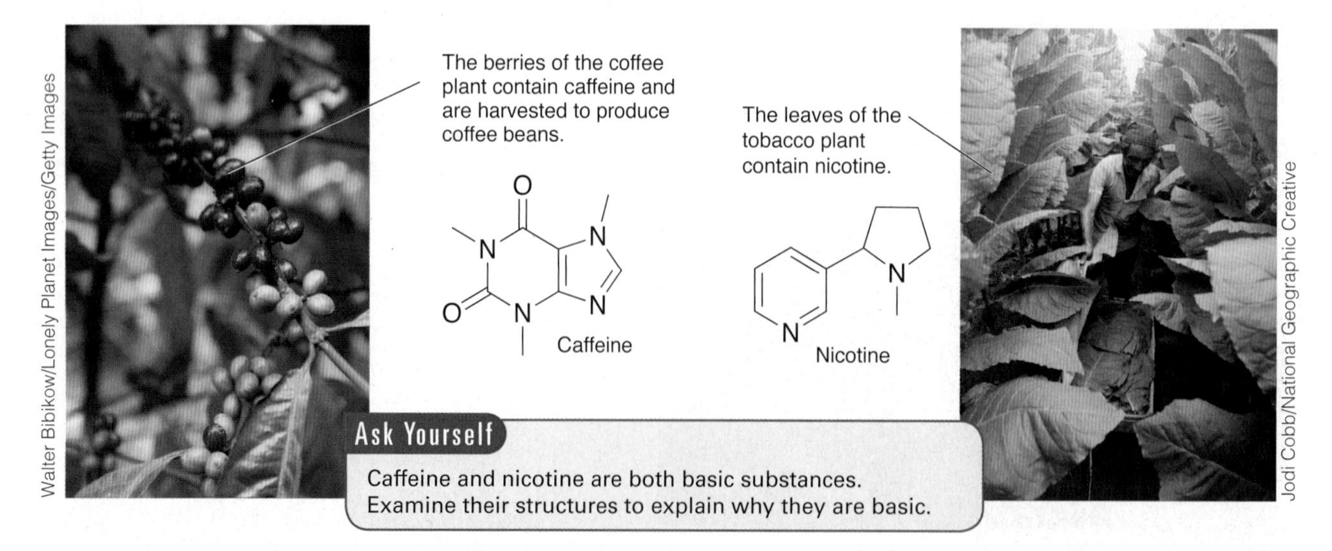

Walter Bibikow/Lonely Planet Images/Getty Images
The berries of the coffee plant contain caffeine and are harvested to produce coffee beans.
Caffeine
The leaves of the tobacco plant contain nicotine.
Nicotine
Jodi Cobb/National Geographic Creative
Ask Yourself
Caffeine and nicotine are both basic substances. Examine their structures to explain why they are basic.

Narcotics found in opium • Figure 12.15

The opium poppy produces various alkaloids, including codeine and morphine.

farmer images/Moment/Getty Images

Here, a milky fluid called latex drips out of incisions made to the seed pod of the opium poppy. The dried liquid yields opium, which contains various alkaloids, including codeine and morphine.

Codeine

Morphine

Ask Yourself

1. In what way does the molecular structure of codeine differ from the molecular structure of morphine?
2. What structural component of codeine and morphine makes them alkaloids?

Opium's properties arise from several of its alkaloid constituents. In the dried opium itself, alkaloids of all sorts make up about a quarter of the total weight. Morphine, the first alkaloid of any kind ever to be isolated, leads the list at about 10%. In 1803, Friedrich Sertürner, a German pharmacist, obtained this alkaloid in pure form, described it, and named it for Morpheus, the Roman god of dreams. Morphine is a powerful narcotic, and for many years it was a valuable tool for physicians. It's a potent analgesic as well as a cough suppressant. Morphine produces a variety of psychological responses, including apathy and euphoria. Another opium alkaloid is codeine, isolated in 1832 by a French pharmacist, Pierre-Jean Robiquet. Codeine is a less potent analgesic than morphine, yet it's one of the most powerful cough suppressants known.

Although they differ in their specific actions on the body and in their potency, these and the other opium alkaloids share two important characteristics: They are all highly addictive, and they are all among the most powerful constipating agents known, with substantial value in controlling diarrhea. **Figure 12.15** shows their molecular structures.

Semisynthetic opioids The medical value of morphine in vanquishing pain is counterbalanced by its addictive effect. Just as the search for better, less toxic analgesics through molecular modifications of acetanilide led to the *p*-aminophenols (Section 12.2), a search for nonaddictive molecular modifications of morphine soon produced a promising semisynthetic substance. In 1898, the diacetylation of morphine in the chemical laboratories of a German dye manufacturer led to diacetylmorphine, through a chemical transformation very much like the acetylation of salicylic acid to aspirin (**Figure 12.16**). Immediately on its introduction into medicine at the turn of the century, this diacetylated molecule proved to be a much more powerful narcotic and cough suppressant

Heroin semisynthesis • Figure 12.16

Heroin (diacetylmorphine) is prepared easily through the reaction of one molecule of morphine with two molecules of acetic anhydride.

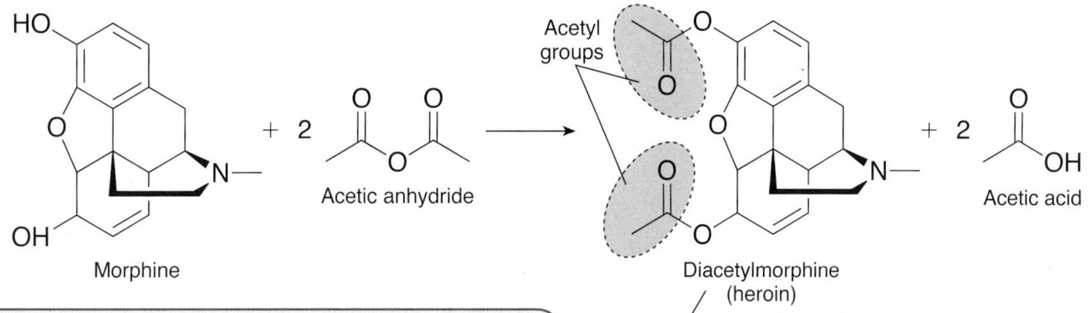

Morphine + 2 Acetic anhydride → Diacetylmorphine (heroin) + 2 Acetic acid

Acetyl groups

This chemical name tells us that two acetyl groups have been added to morphine.

Ask Yourself

1. What functional group is present in morphine but not in heroin?
2. Why is heroin considered a semisynthetic product?

than morphine itself. It was so powerful, in fact, and its effective doses were so small, that it gave promise, at first, of being nonaddictive.

The promise fell through, badly. Diacetylmorphine proved to be one of the most addictive drugs known, so powerfully addictive that its use, possession, manufacture, and importation into the United States and several other countries have been banned by law. It's now known more commonly by its early trade name, heroin, dating from the time of its medical trials and the "heroic" feelings its users reported.

A bit of practical chemistry comes into play in enforcing the ban on the manufacture and distribution of heroin. One of the by-products of the acetylation of morphine with acetic anhydride is acetic acid, the major organic component of vinegar. Drug detector dogs are trained to be alert to the vinegar-like odor of heroin due to the residual acetic acid.

In addition to heroin, several other semisynthetic opioid narcotics have been developed (**Figure 12.17**). Many of these have important medical uses for the treatment of pain, but they also show a strong potential for abuse and addiction. One of these drugs, Oxycontin, an extended-release pill form of oxycodone, was designed to prevent abuse. When swallowed, its contents are released over the course of several hours, thus preventing the kind of immediate high craved by addicts. Ironically, this medicine soon became widely abused because it could be easily crushed into a powder and snorted, thus delivering its large dose of opiate almost instantly. In 2010 the pill was reformulated to prevent this sort of tampering.

Synthetic opioids A variety of synthetic narcotics have been developed that act on the same opioid receptors in the brain as the natural and semisynthetic opioids, yet do not share the same core, morphine-like structure as these older opioids. These newer compounds include fentanyl, used medically as an anesthetic and analgesic, with a potency 80 times the strength of morphine; and methadone, used both as an analgesic in its own right and as a treatment for heroin addiction. Although it is itself addictive, methadone doesn't provide the euphoria or other psychological effects of heroin. Heroin addicts can avoid the crippling effects of withdrawal through a methadone maintenance program that permits them to remain productive members of society.

Alkaloid stimulants Still another plant, the South American coca bush, produces the alkaloid cocaine, which is used medically as a topical or local anesthetic. Cocaine resembles morphine in its medical value as a local anesthetic, but cocaine also has the ability to produce a fast and powerful addiction. Beyond its therapeutic uses, cocaine produces a great sense of well-being and delusions of immense power. However, these sensations are followed by a depression that leads the user to crave another jolt of euphoria, followed by the same cycle of euphoria, depression, and a craving for more.

Morphine and heroin produce a truly physical addiction in that physiological symptoms—watery eyes and nose, sweating, goose flesh, and dilated pupils, for example—occur during withdrawal. Cocaine's addiction seems to be purely psychological, without physical symptoms upon withdrawal, but its psychological addiction is nonetheless as real and as powerful as any physical addiction.

Sigmund Freud, founder of psychoanalysis, played a major role in the discovery of cocaine's addictive and anesthetic potential. A few years after the compound's isolation from coca leaves, Freud began studying it as an aid to his treatment of patients. He soon persuaded a fellow physician who was addicted to morphine, Karl Koller, to use cocaine as an aid in breaking his morphine habit. Koller

Semisynthetic opioids • Figure 12.17

The codeine derivatives, oxycodone and hydrocodone, and the morphine derivative, oxymorphone, are powerful analgesics. They are sometimes combined with aspirin or acetaminophen in pill form, as in Percocet and Vicodin.

Put It Together Review Figure 12.15 and answer the questions.

1. How are oxycodone and hydrocodone structurally different from codeine?

2. How is oxymorphone structurally different from morphine?

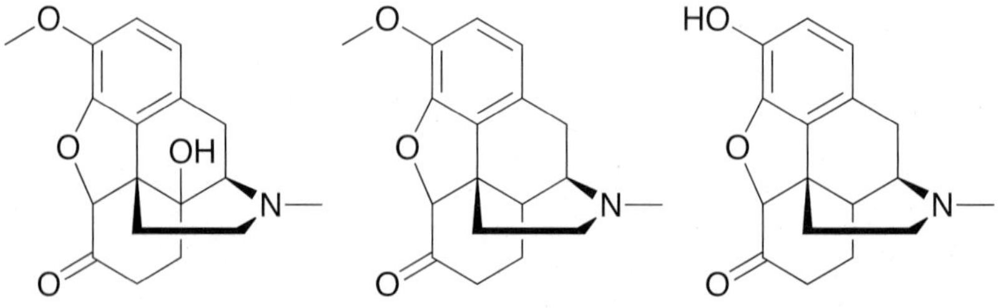

Oxycodone (Oxycontin; Percocet)

Hydrocodone (Vicodin)

Oxymorphone (Opana)

Cocaine • Figure 12.18

The coca plant contains cocaine in its basic form, which can be treated with acid to produce cocaine hydrochloride, a highly addictive white powder that is often inhaled through snorting. Treatment of this acid form with base yields "free base" or crack, which can be inhaled through smoking and which is faster acting and even more addictive.

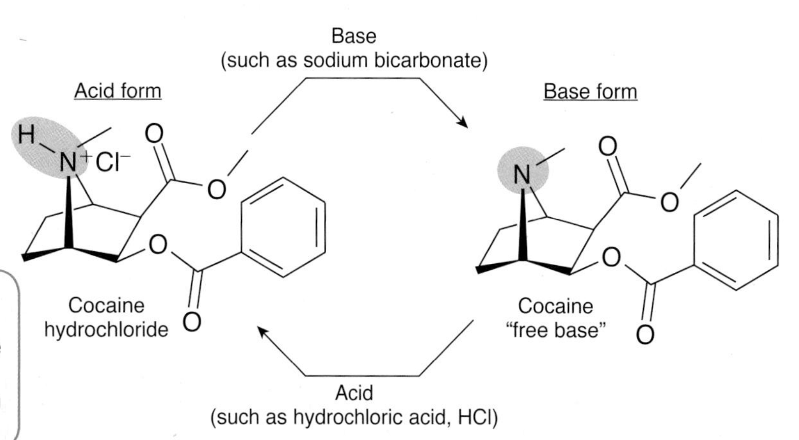

Put It Together Review Figure 12.16 and answer the questions.
1. What product would be formed on treating morphine with hydrochloric acid?
2. Is this product predicted to be more or less soluble in water than morphine?

succeeded, at the cost of becoming addicted to cocaine instead. During his experimentation with cocaine, Koller recognized its properties as an anesthetic and, in 1884, began using it to anesthetize his patients' eyes during medical procedures. Within the year, another physician began using the drug in dentistry as the first local dental anesthetic.

Cocaine provides an apt example of the connection between chemical properties and the use or abuse of a substance. Chemically, cocaine is a nitrogen-containing base, with chemical properties similar to those of other amines we have examined. Like other amines, cocaine reacts with acids, including hydrochloric acid, to form salts. Extraction of the cocaine in coca leaves and treatment with acid yields cocaine hydrochloride, a salt with physical properties similar to those of common sodium chloride. It's readily soluble in water and fairly stable toward heat. It doesn't vaporize readily. On reaction with a base, cocaine hydrochloride is converted to cocaine itself, the pure base, also known as "free base" (**Figure 12.18**).

This basic form of cocaine has far different physical and chemical properties from the hydrochloride. During the chemical conversion of cocaine hydrochloride (the salt) to cocaine (the base), the generated base forms as a white solid sheet that cracks into large lumps or "rocks." This formation of "rocks" gives the product the name crack cocaine or simply crack. Unlike the crystalline salt, the free base readily vaporizes. Inhaling cocaine vapors quickly produces a sharper, more intense sensation than that produced by the hydrochloride salt. The fall into depression that follows is also steeper.

Other Abused Drugs

Drugs can also affect perceptions other than pain. Some substances with no medical value whatever can stimulate hallucinations and can induce psychological states so powerful that they change the perception of reality itself. These substances are known as **hallucinogens**.

Hallucinogens Two hallucinogens with particularly interesting histories, mescaline and lysergic acid diethylamide (LSD), are shown in **Figure 12.19**.

Ask Yourself
1. Molecules of both mescaline and LSD contain a benzene ring. What other functional group is common to both?
2. The amide functional group contains a nitrogen atom bonded to a carbonyl group. Identify the amide group in the LSD molecule.

LSD and mescaline • Figure 12.19

LSD and mescaline are highly potent hallucinogens. Note that each compound contains a phenylethylamine fragment, in which an amine nitrogen is connected through two carbons to a benzene ring. This structural feature is common to a variety of psychoactive substances.

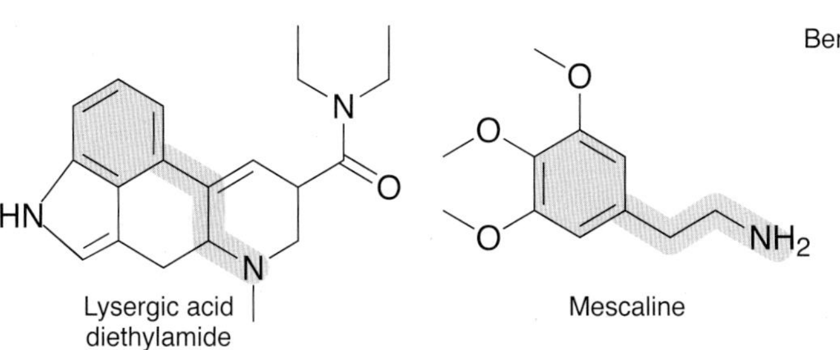

Amphetamine and structurally related drugs • Figure 12.20

Each of these compounds contains the phenylethylamine structure.

Amphetamine

Methamphetamine, also known as "meth" or "speed"

Methylenedioxymethamphetamine (MDMA), also known as "ecstasy"

Ask Yourself

1. How is the structure of methamphetamine different from that of amphetamine?
2. How is the structure of MDMA different from that of methamphetamine?

Mescaline occurs naturally in the tiny, peyote cactus native to southwest Texas and Mexico. Eating this plant produces distortions of reality and states of deep meditation that have been part of longstanding religious rituals of some Native American groups of the Southwest. Today the drug is generally illegal in the United States, although in some regions its use is permitted for religious services of the Native American Church, which serves about a quarter of a million followers.

Unlike mescaline, LSD came first from the chemical laboratory. The drug was initially prepared in 1938 by Albert Hofmann and a co-worker at the Swiss pharmaceutical firm, Sandoz. They set it aside after a few tests indicated that it was a thoroughly uninteresting substance. Five years later, while studying the chemistry of lysergic acid, one of the alkaloids of the ergot fungus that grows on rye and other cereal grains, Hofmann prepared a bit more of the amide, a few milligrams. Shortly afterward

A DEEPER LOOK Optical isomers and Adderall

Carbons bonded to four different groups are capable of forming a type of isomer we have not seen before, one that depends on the particular arrangement of these four groups around the carbon atom. Any carbon bonded to four groups lies in the center of a tetrahedron, a four-sided structure resembling a pyramid. With the carbon at its center, each of the four substituents lies at one of the four corners of the tetrahedron. If each of these four substituents is different, such as in the generalized molecule CWXYZ, we can represent this structure as shown in Figure a.

We'll examine the unusual behavior of the CWXYZ carbon through the amphetamine molecule (Figure b), which is a component of the drug Adderall. Here, the central carbon in amphetamine is bonded to four different groups. ("Ph" is shorthand for a phenyl group, which is simply a benzene ring.)

The specific orientation in space of these four groups surrounding the central carbon gives rise to two possibilities, called *optical isomers*, as shown in Figure c.

Optical isomers share a property also found in your right and left hands: They are mirror images of each other, but they cannot be superimposed onto one another. We consider two structures to be superimposable if we can merge them in space so that each and every point on one of the structures coincides exactly with its equivalent point on the other structure. If such point-for-point merging in space is not possible, the two are not superimposable. Thus the amphetamine isomers shown in Figure c are nonsuperimposable.

Molecules with this property of handedness are called **chiral** (pronounced KY-rul), from a Greek word meaning

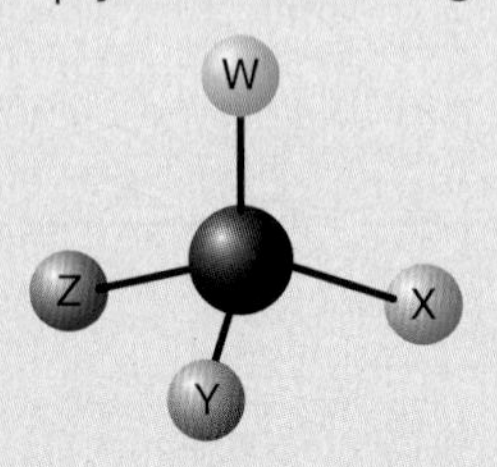

Figure a. A tetrahedral carbon atom bonded to four different groups: W, X, Y, and Z.

$$PhCH_2-\underset{CH_3}{\overset{NH_2}{C}}-H$$

Figure b. Amphetamine.

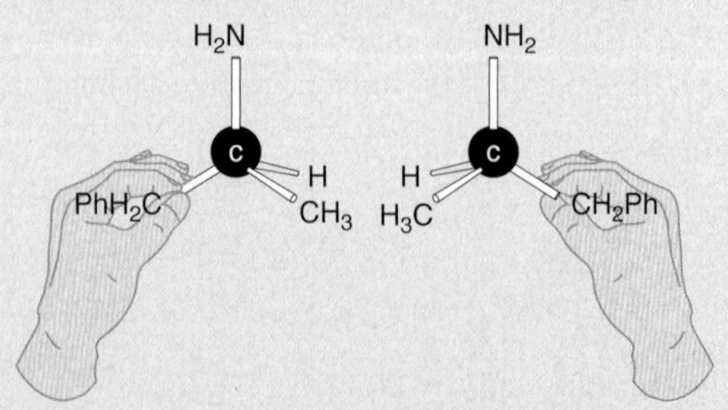

Figure c. The two optical isomers of amphetamine.

he developed a restlessness and dizziness followed by a fantasy-filled delirium, including what he later called an, "intense kaleidoscopic play of colors."

In tracing the source of the disturbance, Hofmann focused on the diethylamide of lysergic acid that he had prepared on the day of the attack. To test his suspicions, he swallowed a quarter of a milligram, 0.00025 g, of the substance. The result was more than six hours of distortions of vision, space, and time; alternating restlessness and paralysis; dry throat; shouting; babbling; fear of choking; and a sense of existing outside of his body. The worst of the symptoms ended after six hours, but the distortions of shapes and colors continued throughout the day.

LSD gained its greatest notoriety during the countercultural movement of the 1960s, but its current incidence of use is much lower than that of many other types of drugs. The federal government lists LSD and mescaline as "Schedule I" controlled substances, a category of drugs that includes heroin and that is subject to the most severe penalties for possession and use.

Amphetamines In addition to LSD and mescaline, the opioid narcotics and many other compounds that significantly alter our perceptions are composed of molecules containing a phenylethylamine segment. Molecules containing this structural fragment often affect the way our nerves carry their messages to the brain. Other substances with this unit include the amphetamines and structurally related compounds (**Figure 12.20**).

Amphetamines raise the pulse rate and blood pressure, reduce fatigue and appetite, and temporarily suspend the desire for sleep. The drug Adderall—prescribed to treat attention/deficit hyperactivity disorder (ADHD), but also widely abused—is a mixture of amphetamine isomers (see *A Deeper Look*).

Amphetamines carry a great potential for abuse and can produce or contribute to paranoia and mental illness. The powerful stimulant methamphetamine, commonly known as *speed*, is particularly susceptible to abuse and addiction. The drug MDMA, also called *ecstasy* (see Figure 12.18), is similar in structure to methamphetamine, but its effects

hand, and are widely represented in nature. For example, all but one of the 20 naturally occurring α-amino acids are chiral and share a common orientation about a central chiral carbon atom (see Table 5.1). The exception, glycine, $H_2NCH_2CO_2H$, is not chiral because it does not have a chiral carbon in its structure. That is, neither of its carbons is bonded to four different substituents.

Many compounds used as medicines are chiral. The biological effects of one optical isomer of a compound can be quite different from that of its mirror-image isomer. Sometimes, the side effects caused by one optical isomer of a compound may be so serious as to prevent its use as a drug.

Many drugs are sold as either an equal mixture of a compound's optical isomers, called a **racemic mixture**, or as a single optical isomer. For example, omprezaole (Prilosec), a heartburn drug, is a racemic mixture, and esomeprazole (Nexium), a newer version of this product, is a single optical isomer. Adderall is a somewhat unusual medication in that it contains a mixture of unequal amounts of optical isomers (Figure d).

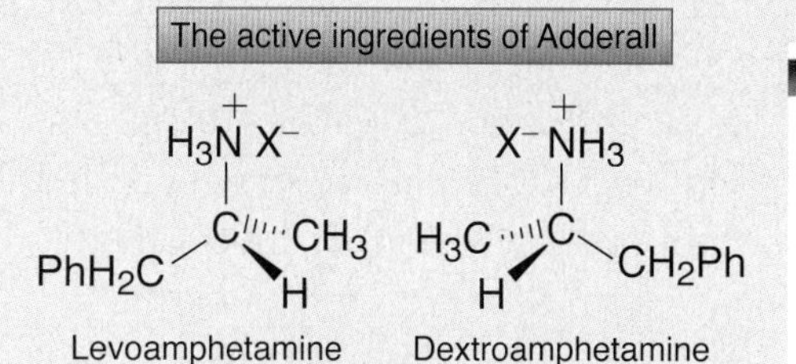
The active ingredients of Adderall
Levoamphetamine
Dextroamphetamine
The positive charge on each nitrogen is balanced by the negative charge of anions of various forms, indicated as X⁻

PhotoAlto/Sigrid OlssonGetty Images

Figure d. Adderall contains a mixture of the optical isomers of amphetamine salts shown here, with the dextroamphetamine isomer predominant. The prefix dextro- derives from the Latin *dextra* for right, and levo- comes from the Latin laevus for left. Adderall is prescribed to increase alertness and focus in those with ADHD. These effects, however, have also led to its widespread abuse as a study aid.

Put It Together Review Figure 12.18 and answer the question. How could you convert the compounds shown in Figure d to those shown in Figure c?

are quite different. It acts as both a stimulant and a hallucinogen, causing feelings of well-being and emotional warmth toward others, as well as distortions of time and perception. The short-term physical effects of MDMA include elevated body temperatures and chills. MDMA also temporarily boosts serotonin levels in the brain, and as these levels drop in the days following MDMA use, they can often produce feelings of anxiety, irritability, or sadness.

CONCEPT CHECK STOP

1. **What** is the name of the compound, formed from the oxidation of alcohol in the body, that is associated with some of alcohol's unpleasant side effects?
2. **Which** functional group is shared by all alkaloids?
3. **What** is the structure of phenylethylamine?

12.4 Chemistry of Genetics

LEARNING OBJECTIVES

1. **Describe** the structure of DNA and its role in genetics.
2. **Explain** how mutations in genes can cause disease.
3. **List** several applications of genetic technologies.

Not all diseases are caused by pathogens, such as bacteria and viruses. Some diseases are genetic in nature, meaning they are caused by inherited defects in molecules that have important functions in our bodies. The approach to treating these kinds of diseases is different from what we have seen so far in this chapter, and is at the forefront of biochemical research. Many of the techniques used to develop treatments for these kinds of illnesses are also used in other areas, including the modification of plant crops and farm animals to increase their productivity or their resistance to pests. Some of these advances have raised ethical issues that have become the subject of public debate. To understand some of these issues, we first examine the chemistry of genetics.

The Molecular Basis of Heredity

People have long observed that the offspring of plants and of animals bear a resemblance to their parents. Such observations have formed the basis of breeding methods that seek to optimize desirable traits of domesticated plants and animals. For instance, years of selective breeding have led to the many different kinds of dogs we see and to the different varieties of apples available to us. For most of human history, however, the scientific basis of heredity has remained largely unknown.

Our understanding of these issues was greatly aided by the work of Gregor Mendel (1822–1884), an Austrian monk who studied the effects of cross-breeding on the traits of garden peas. Mendel developed statistical models to explain how individual traits, such as flower color, are transmitted from a plant to its offspring. He found that some of these traits, referred to as **dominant**, are expressed in progeny more frequently than others, referred to as **recessive**. Mendel's research identified some heritable factors within plants that are invisible to the naked eye, but that govern how these traits are passed from generation to generation. We now refer to observable traits of an organism as **phenotypes**, and we understand that they originate in a genetic feature or structure carried by a parent, called a **genotype**. Although Mendel did not know the chemical nature of this genotype, his writings opened the path to our current understanding of the molecular basis of heredity.

We now recognize that all organisms—both plants and animals—resemble their parents because each generation transfers to its offspring its own set of biochemical information. This information is carried from one generation to the next by a molecular messenger called DNA, which exists within the cells of all organisms.

Cell structure and replication The **cell** is the fundamental structural unit of all living things. With a few exceptions, such as red blood cells and the cells of simple organisms known as prokaryotes, each cell is surrounded by a cellular membrane and contains a well-defined **nucleus** within it. Structures within the **cytoplasm**—the fluid inside the cell but outside the cell's nucleus—do most of the work of the cell, including photosynthesis (in plants) and cellular respiration (in both plants and animals).

The cell cycle and mitosis • Figure 12.21

This type of cellular reproduction occurs in three phases—interphase, mitosis, and cytokinesis—and yields identical copies of the parent cell.

a. Relative duration of each stage of the cell cycle

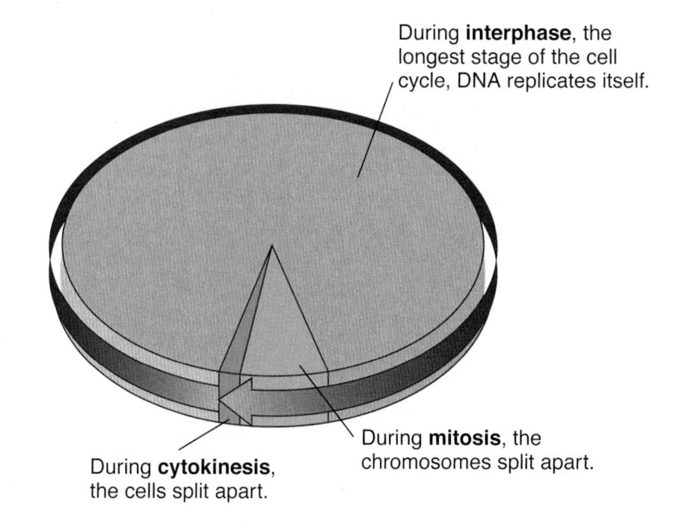

During **interphase**, the longest stage of the cell cycle, DNA replicates itself.
During **mitosis**, the chromosomes split apart.
During **cytokinesis**, the cells split apart.

b. Visualization of the stages of the cell cycle

1. Interphase.
DNA, which exists as threads called **chromatin**, forms a copy of itself.

2. Mitosis.
The chromatin condenses into **chromosomes.** The sister chromatids of each chromosome split apart and migrate to opposite ends or poles of the cell. (Here we've simplified the depiction of mitosis, which actually occurs in four stages.)

3. Cytokinesis
The cells divide, yielding a new cell identical to the parent.

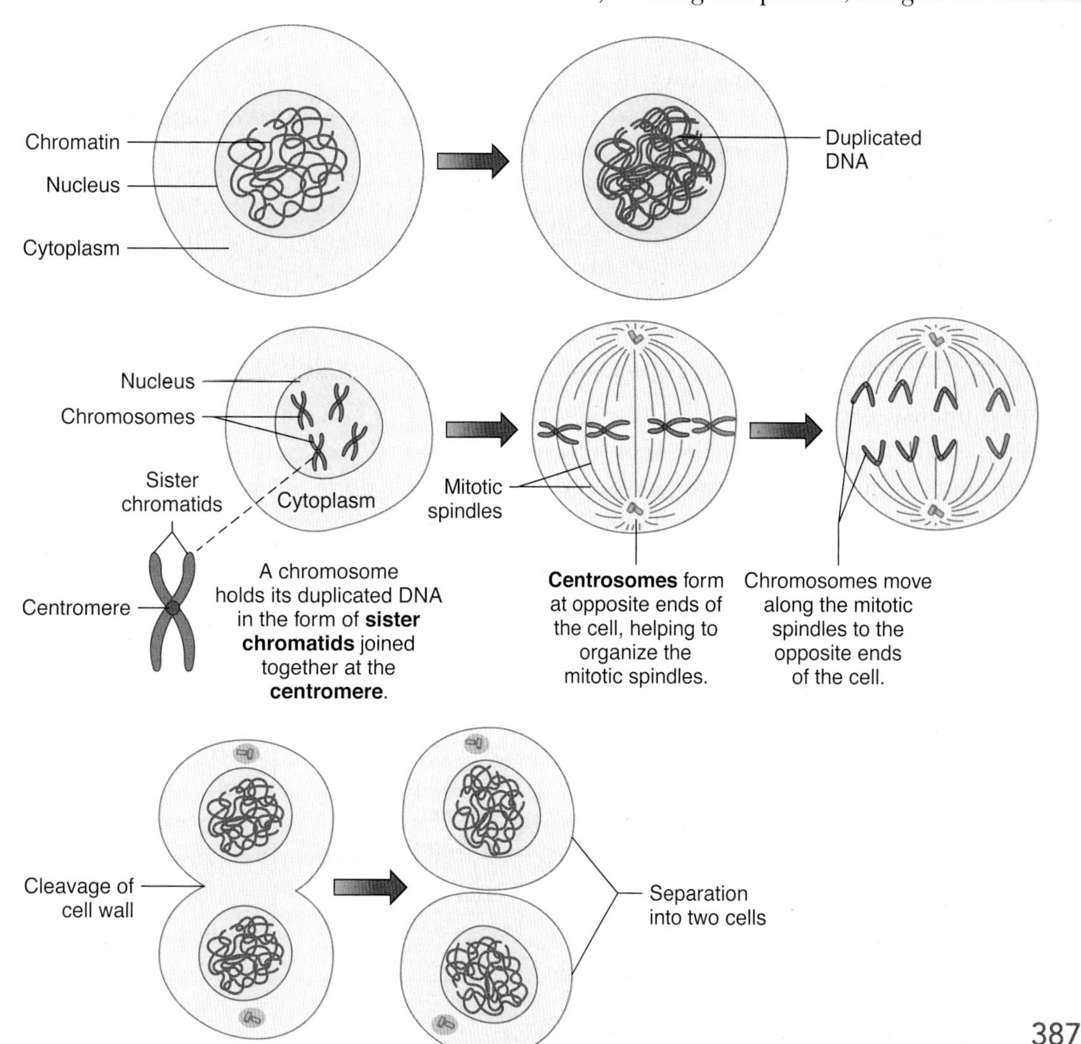

Our interest here lies in the nucleus, which carries the organism's genetic information. As long as the cell is not in the process of dividing, the genetic material lies diffused within the nucleus as thread-like filaments called **chromatin**. But as the cell prepares to divide, these filaments coalesce into more highly ordered structures called **chromosomes**. The term chromosome (from the Greek *khroma*, for color, and *soma*, for body) reflects the ability of these structures to be stained by certain dyes. Except for cells involved in reproduction, all the cells of each organism of any given species, called somatic cells, hold the same number of chromosomes, a number that varies from one species to another. For humans, the number of chromosomes in somatic cells is 46; for rice, the number is 24; for cats, 38; for elephants, 56; and for dogs, 78. Because chromosomes commonly occur in pairs, the 46 human chromosomes of our somatic cells occur as 23 pairs of chromosomes.

In cellular reproduction involving **mitosis**, a cell divides to produce two identical copies of itself (**Figure 12.21**). During this process, the genetic material

Chromosomes and DNA • Figure 12.22

DNA coils around specialized proteins called *histones*. This enables the long DNA molecule to form the highly compact structure of a chromosome. Each sister chromatid of a chromosome contains identical DNA.

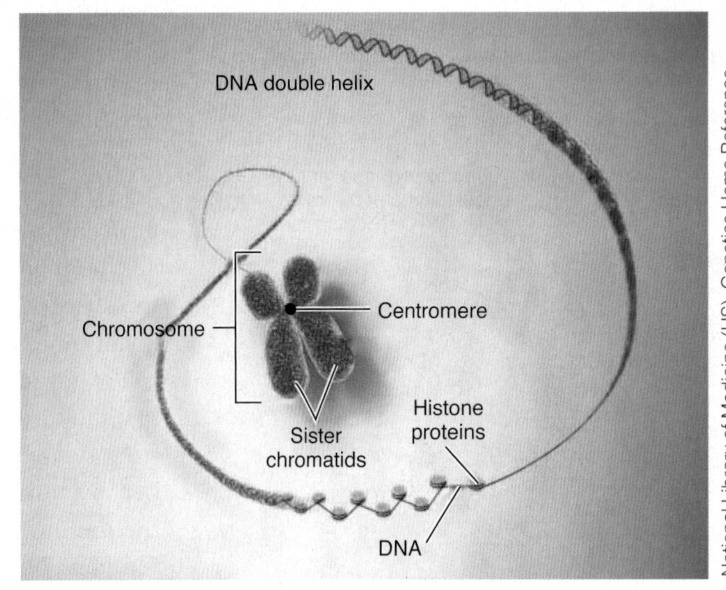

National Library of Medicine (US). Genetics Home Reference [Internet]. Bethesda (MD): The Library; 2014 Aug 4. [Illustration] Gene; [cited 2014 Aug 6]; [about 1 screen]. Available from: http://ghr.nlm.nih.gov/handbook/illustrations/geneinchromosome

within a cell duplicates itself, with each of the two identical sets of genetic material going into each of the two new cells. Where we previously had a single cell, we now have two cells, identical to each other and to the original cell. This form of cellular reproduction represents the mode by which most body cells, including those of the skin and the blood, divide.

Meiosis, the kind of division important in sexual reproduction, represents a more complex process. Here half the chromosomes (one from each pair) move to one of the new cells, and half (the other member of each pair) go to the other. This kind of division occurs in the formation of sperm and egg cells. To produce a cell containing the full complement of chromosomes (46 in the case of humans), two of these cells—one sperm cell and one egg, each containing half the required number of chromosomes—must unite. In human reproduction, this produces again the characteristic 23 pairs of chromosomes. Half the new set of 46 chromosomes (and half the genetic information) comes from one parent, half from the other. Unlike mitosis, meiosis, with its resultant mixture of the genetic properties of two parents, produces something new rather than more of the same.

DNA structure and replication Chromosomes consist of very long, tightly coiled molecules of **deoxyribonucleic acid (DNA)**, a compound found almost exclusively in the nuclei of our cells, hence the *nucleic* portion of the term (**Figure 12.22**). These DNA molecules carry all the genetic information of every living organism. Moreover, the DNA in virtually every cell of any individual organism is identical with the DNA in every other cell of that same organism.

The DNA molecule in a biological chromosome consists of three chemical segments:

- a series of four cyclic amine bases—adenine, cytosine, guanine, and thymine—that are bonded to a backbone consisting of:
- 2-deoxyribose, a carbohydrate, and
- a structural portion of phosphoric acid, H_3PO_4, called a phosphate group.

Combining one of the four cyclic amine bases with 2-deoxyribose and a phosphate group produces a **nucleotide** (**Figure 12.23a**), the repeating structural unit of DNA. A long sequence of one nucleotide unit after another, each bearing one of the bases of **Figure 12.23b**, produces the primary structure of a DNA molecule (**Figure 12.23c**). To sum up, the DNA chain is simply a set of alternating 2-deoxyribose and phosphate units, with four amines—adenine, cytosine, guanine and thymine—strung out as substituents along the chain.

As we saw in Chapter 5, the primary structure of a protein—that is, the sequence of amino acids along the polypeptide chain—tells only part of the chemical story; proteins organize into higher-order structures. Similarly,

deoxyribonucleic acid (DNA) The molecule within the cell's nucleus that carries the genetic blueprint of the entire organism.

nucleotide The repeating structural unit of DNA.

The components and primary structure of DNA • Figure 12.23

The repeating structural unit of DNA is the nucleotide (a), which contains any of the four amine bases shown in b. Nucleotides connect as a chain to form the primary structure of DNA (c), which is also shown schematically in d.

a. The structure of a typical nucleotide.

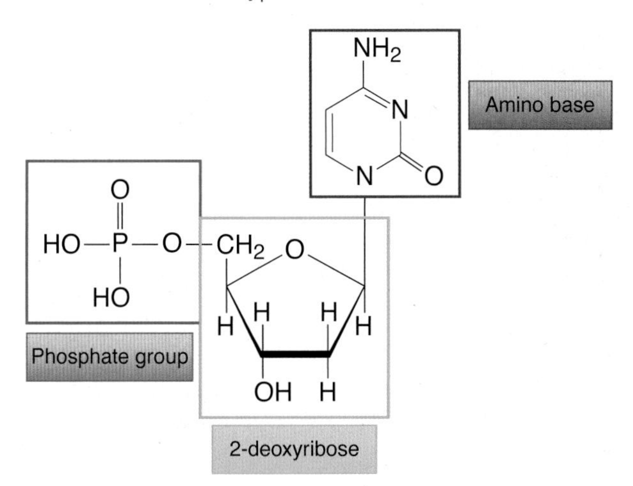

b. The amine bases in DNA. The squiggle lines show where the 2-deoxyribose units attach to each of these bases within DNA.

Adenine (A)

Thymine (T)

Guanine (G)

Cytosine (C)

c. A segment of the DNA chain. This strand can run hundreds of thousands of nucleotides long.

Adenine

Thymine

Guanine

Cytosine

d. A schematic representation of the DNA strand shown in Figure c.

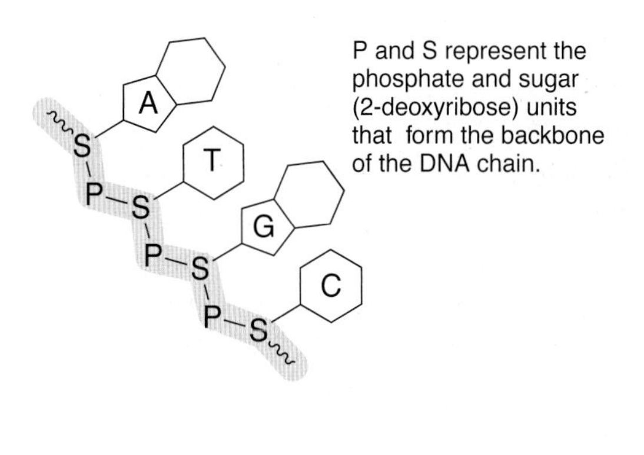

P and S represent the phosphate and sugar (2-deoxyribose) units that form the backbone of the DNA chain.
A
T
G
C

Ask Yourself

Which base is represented in the nucleotide of part a?

The DNA double helix • Figure 12.24

The two DNA strands that form the double helix are complementary, which means that every adenine of one strand lies opposite a thymine of the other strand; every cytosine of one strand lies opposite a guanine of the other strand. The hydrogen bonds that hold the strands together form between these base pairs, which occupy the central core of the double helix.

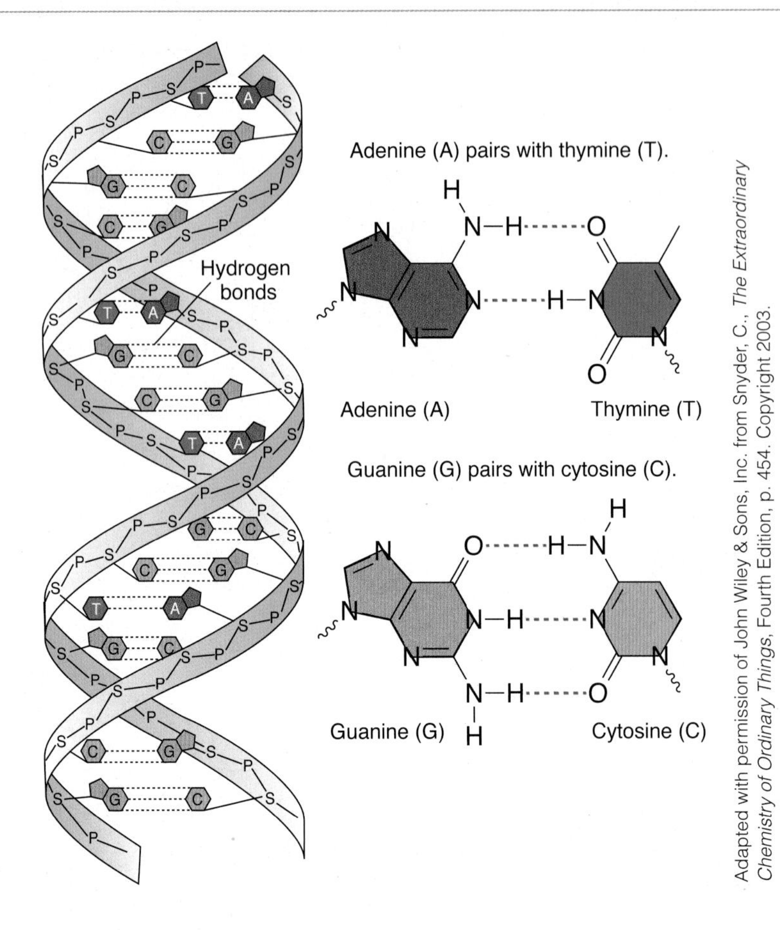

Ask Yourself

1. Why are the two DNA strands of the double helix regarded as complementary rather than identical?
2. If the sequence of bases along a segment of one strand of DNA is GCGGATGAT, what is the sequence of bases along its complementary segment?

the primary structure of DNA—the sequence of bases along the DNA backbone—reveals only part of the picture. DNA commonly exists as a double helix (**Figure 12.24**). This discovery was made in 1953 by a pair of biologists, the American James Watson and the Englishman Francis Crick, and has led to rapid advances in our understanding of the chemical basis of heredity. Working in Cambridge, England, Watson and Crick determined the structure of DNA based on X-ray studies of DNA crystals carried out in London by Rosalind Franklin, a biophysicist. In 1962, Watson and Crick shared the Nobel Prize in medicine with Maurice Wilkins, who had worked with Franklin. Because Nobel Prizes may not be awarded posthumously, the name of Rosalind Franklin, who died in 1958, does not appear on the Prize.

When a cell divides, replicas of its DNA form as the double helix uncoils and a new complementary strand forms adjacent to each one of the original strands (**Figure 12.25**). Each of the two new cells receives one of the newly formed double helixes. This accounts for the identity of the DNA in each cell of an organism.

DNA and genes An estimated 6 billion DNA bases (or 3 billion base pairs) exist in each human cell except for those involved in reproduction, which contain half that amount. Certain segments along the length of the DNA constitute **genes**, or chemical blueprints of heredity. Genes enable the cell to form the polypeptides or proteins—such as the enzymes, hormones, and structural and regulatory proteins—that determine the form, appearance, and functions of an organism.

Each gene encodes for a specific protein, so the sequence of DNA bases within each gene determines the sequence of the amino acids that constitute its corresponding protein. The actual synthesis of proteins takes place outside the cell nucleus, in the cytoplasm. As a result, the information contained in the DNA molecule, which remains confined to the nucleus, must be transferred out of the nucleus, into the cytoplasm. The transfer begins within the nucleus with **transcription**, a process in which the information

gene A segment of DNA that carries the information needed to create a specific polypeptide or protein.

DNA replication • Figure 12.25

During DNA replication, enzymes aid in the uncoiling of the double helix and the formation of new complementary strands.

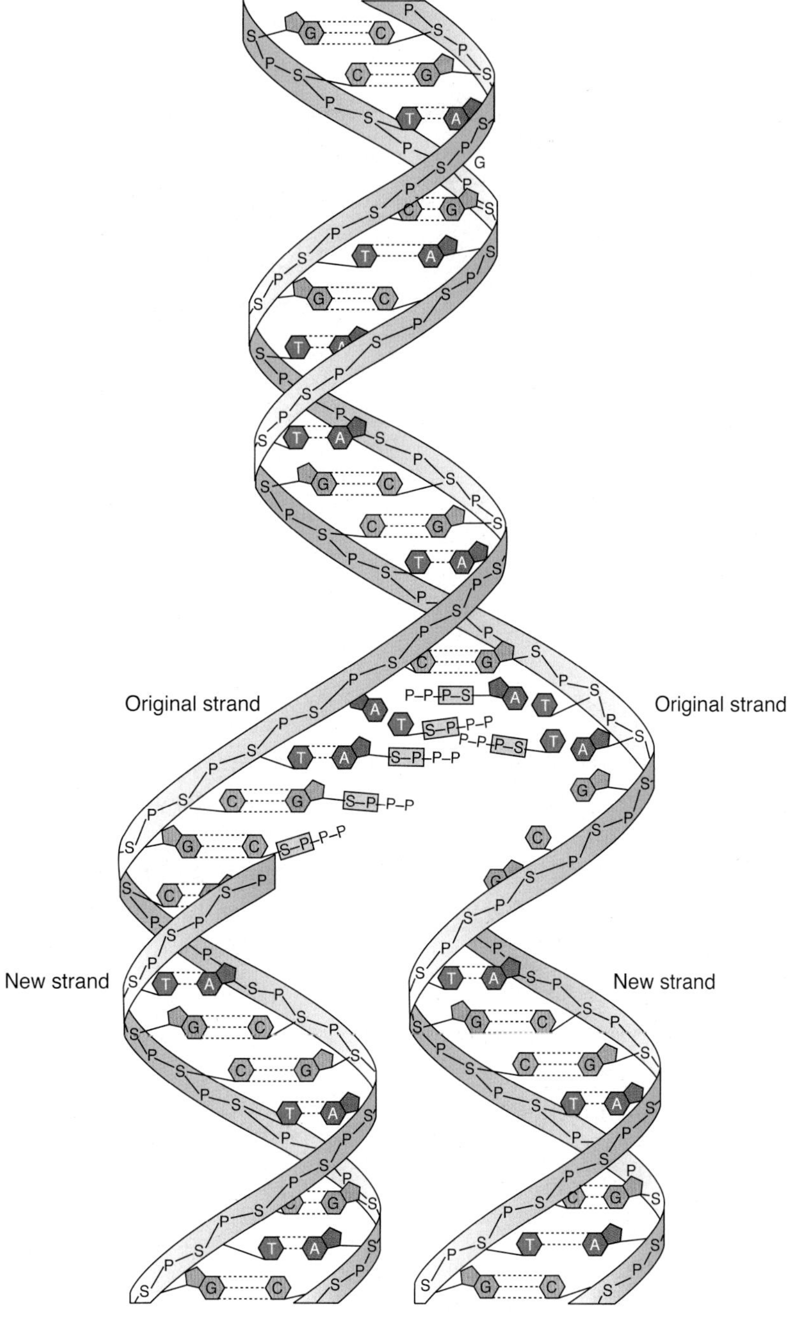

Ask Yourself

What links need to be broken when double-stranded DNA breaks apart into two single strands?

Differences in the components of DNA and RNA • Figure 12.26

RNA incorporates a different sugar unit (a) and base unit (b) than those contained in DNA.

a. The ribose of RNA and the 2-deoxyribose of DNA differ only in the groups attached at carbon #2 in each structure. Ribose has an –OH group at this position, 2-deoxyribose does not.

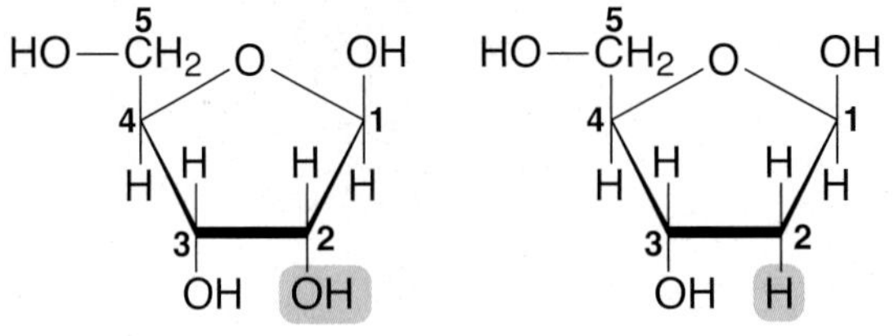

Ribose (in RNA)

2-deoxyribose (in DNA)

b. The uracil of RNA replaces the thymine of DNA.

Uracil (in RNA)

Thymine (in DNA)

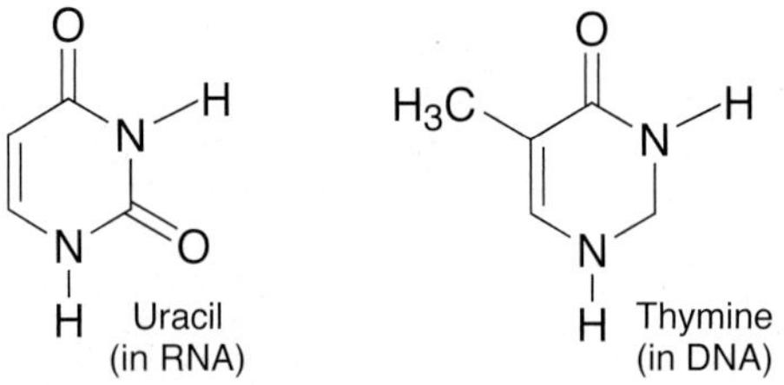

ribonucleic acid (RNA) A type of nucleic acid that acts as an intermediary in the production of proteins within an organism.

contained on the DNA molecule is transcribed, in portions, onto another kind of molecule, called **ribonucleic acid (RNA)**. RNA, formed by biochemical reactions within the nucleus, differs from DNA in four important ways:

- RNA molecules can pass from the nucleus into the cytoplasm.
- Ribose (rather than 2-deoxyribose) forms the carbohydrate ring of the RNA chain (**Figure 12.26a**).
- In RNA the amine uracil replaces the thymine of DNA (**Figure 12.26b**).
- RNA molecules are much shorter than DNA; each RNA carries only a small segment of the genetic information stored on the entire strand of DNA.

RNA occurs in two primary forms, each with a specific function. One form, called **mRNA** (for messenger RNA), carries information stored in DNA (the message) from the nucleus to the cytoplasm, where protein synthesis occurs. For the protein synthesis itself, another form, called **tRNA** (transfer RNA), finds and transports (transfers) each required amino acid to the site where the peptide bonds are formed under the direction of the mRNA. This step, called **translation**, can be likened to translating the genetic message from a language made up of RNA bases to a language of amino acids. This peptide synthesis takes place within the cytoplasm, in molecular structures known as **ribosomes**.

The genetic code The sequence of the amine bases strung out along the RNA molecule determines the primary structure of the polypeptide it forms. Each set of three sequential amines on RNA either identifies one particular amino acid that fits into the polypeptide chain or acts as a signal for the chain-forming process itself, such as "start" or "stop." To indicate this coding function, sets of three successive amine bases are called **codons**. The correspondence between any particular RNA codon and the specific amino acid or function it represents is known as the **genetic code** (**Table 12.1**).

With four different bases arranged in codons of three bases each, a total of 64 different sequences (codons) is available. This allows for some redundancy in specifying the 20 amino acids. For example, with adenine, cytosine, guanine, and uracil represented by A, C, G, and U, respectively, each of the following four codons represents the simplest of the amino acids, glycine: GGA, GGC, GGG, and GGU. The signal to terminate the protein chain is given by each of the codons UAA, UAG, and UGA. **Figure 12.27** (on the next page) shows how the sequence of codons along a strand of mRNA serves as the template for the formation of a specific protein, assembled one amino acid at a time, in sequence.

To summarize the entire process, genetic information is stored in a sequence of four different amine bases in molecules of DNA, which lie within the nucleus of the cell. The information is transcribed, in portions, into a sequence of bases of newly formed mRNA molecules. Three of these bases on the mRNA are identical with those on DNA; one is different. This mRNA leaves the nucleus for the ribosomes of the cytoplasm and there serves as a guide for joining individual amino acids, brought to it by tRNA, into useful polypeptides.

The human genome Given the importance of our genetic information, an enormous amount of time and

The genetic code Table 12.1

Amino Acid	Codon[a]	Amino Acid	Codon[a]
Ala	GCA	Lys	AAA
	GCC		AAG
	GCG	Met	AUG
	GCU	Phe	UUU
Arg	AGA		UUC
	AGG	Pro	CCA
	CGA		CCC
	CGC		CCG
	CGG		CCU
	CGU	Ser	AGC
Asn	AAC		AGU
	AAU		UCA
Asp	GAC		UCC
	GAU		UCG
Cys	UGC		UCU
	UGU	Thr	ACA
Gln	CAA		ACC
	CAG		ACG
Glu	GAA		ACU
	GAG	Trp	UGG
Gly	GGA	Tyr	UAC
	GGC		UAU
	GGG	Val	GUA
	GGU		GUG
His	CAC		GUC
	CAU		GUU
Ile	AUA	Initiate peptide chain	AUG
	AUC		GUG
	AUU	Terminate peptide chain	UAA
Leu	CUA		UAG
	CUC		UGA
	CUG		
	CUU		
	UUA		
	UUG		

[a]A = adenine; C = cytosine; G = guanine; U = uracil

Interpret the Data

What amino acid does the codon CAC correspond to? What other codon corresponds to this amino acid?

Protein synthesis • Figure 12.27

Protein synthesis, known as **translation**, occurs within bodies of the cell called ribosomes. The sequence of bases along the strand of mRNA determines the sequence of amino acids in the newly formed protein.

1 Here, the amino acids methionine (Met) and proline (Pro) serve as the first two units of a growing polypeptide chain. The third unit in this chain, aspartic acid (Asp), is being transported to the site of protein synthesis by a molecule of tRNA.

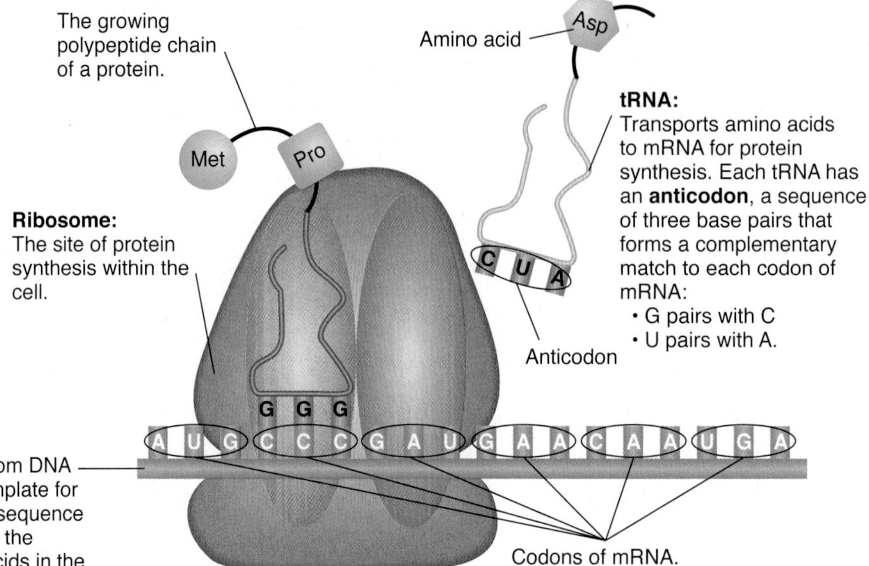

The growing polypeptide chain of a protein.
Amino acid
Asp
Met
Pro
tRNA:
Transports amino acids to mRNA for protein synthesis. Each tRNA has an **anticodon**, a sequence of three base pairs that forms a complementary match to each codon of mRNA:
• G pairs with C
• U pairs with A.
Ribosome:
The site of protein synthesis within the cell.
C U A
Anticodon
G G G
A U G C C C G A U G A A C A A U G A
mRNA:
Carries information from DNA and serves as the template for protein synthesis. Its sequence of codons determines the sequence of amino acids in the growing polypeptide (protein) chain.
Codons of mRNA.

2 The tRNA bearing the amino acid, Asp, docks onto an mRNA codon for Asp.

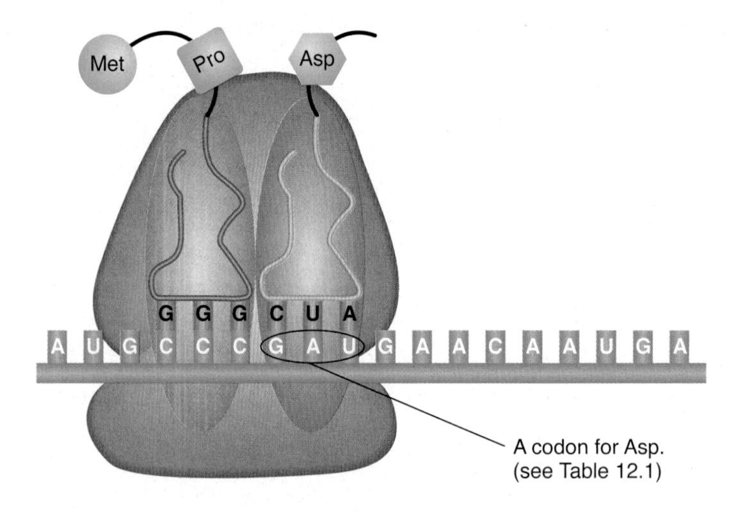

3 A peptide link forms between the amino acids Asp and Pro, thus lengthening the protein chain.

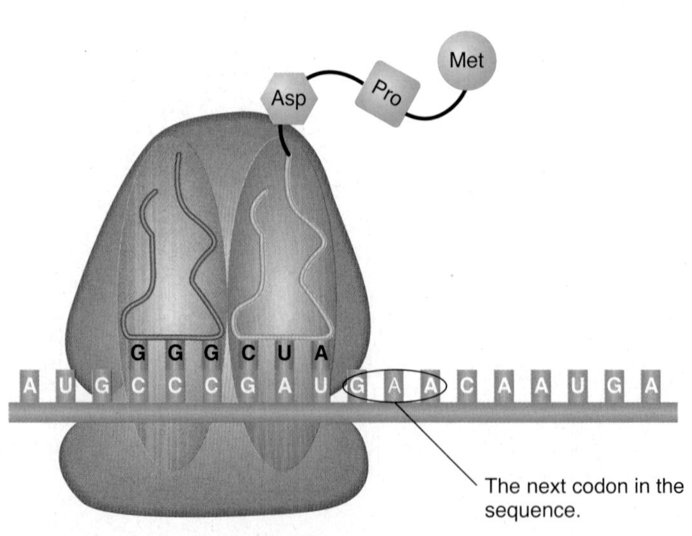

Interpret the Data

Refer to Table 12.1 to predict the fourth, fifth, and sixth amino acids in the growing polypeptide chain shown here.

effort has gone into determining the precise sequence of genes and their bases in our DNA. This sequence for any particular species or individual is called a **genome**. The sequence responsible for our human characteristics represents the human genome. The Human Genome Project, initiated in 1990 and completed in 2003, revealed that our genome consists of approximately 20,500 genes. About 3000 amine bases make up the average gene, but this number can vary widely. For instance, one of the largest genes, the one responsible for the protein dystrophin, contains 2.2 million bases. (Dystrophin is required for healthy muscle tissue. Defects in dystrophin formation are associated with the disease known as muscular dystrophy.)

Through the work of the Human Genome Project and other studies, we now know that a huge portion of the DNA molecule itself, slightly over 98%, has sequences of bases that do not encode for protein formation. This noncoding DNA lies between genes as well as within genes. Historically, much of it has been called junk DNA, with no apparent use. We now know, however, that some of these segments provide regulatory and other useful functions.

Although each of us is an individual, different from anyone else now alive or who has ever lived, we are all members of the human species. You might wonder, then, how much of your DNA belongs to you alone, making you the individual you are, and how much you share with all other members of our species. The answer is that at least 99.6% of the base sequences of our DNA is shared by all other humans. Therefore, less than half of 1% of our genome makes us the individuals we are. Moreover, the analysis of genomes of other species reveals that many genes are closely related across species. This means that genes found within one species are often quite similar to those existing in others. For instance, as many 70% of the genes in yeast bear a striking similarity to genes found in humans. As a result, biomedical research involving organisms carrying genes similar to those found in humans can produce results useful to our own lives.

Genetic Factors in Health and Disease

Just as you inherit genes that confer certain physical traits, such as hair and eye color, you also inherit genes that may confer a low risk or high risk for developing certain diseases or disorders. For example, sickle cell anemia, which affects the blood, and cystic fibrosis, which affects the lungs, are **single-gene disorders**. That is, they result from a defect or mutation in a single gene passed from parent to child. However, most diseases with a genetic component are **multifactorial disorders**, meaning that a combination of various genetic and environmental factors contribute to their development. Heart disease, diabetes, and various forms of cancer fall within this class of multifactorial disorders. Certain behavioral and psychiatric disorders do as well, including alcoholism and certain forms of mental illness.

Sometimes lifestyle choices can help reduce the risks of developing multifactorial disorders. For instance, those who don't smoke, and who regularly exercise and maintain a healthy diet, appear to be at lower risk of developing cardiovascular disease and certain cancers. However, since single-gene disorders, such as sickle cell anemia, are inherited at birth, treating and managing these conditions is often the only course of action. (We'll discuss shortly an alternative approach to treating certain genetic disorders, known as gene therapy.)

A minor mutation in a gene can have serious consequences. For instance, consider the protein hemoglobin, the complex molecule in red blood cells that transports oxygen throughout the body. Hemoglobin consists of two pairs of polypeptide chains. Sickle cell anemia, which occurs principally in people of West African ancestry, results from a genetic error that replaces one amino acid for another at a precise point within one of the pairs of the polypeptide chains of hemoglobin (**Figure 12.28** on the next page.) As a result of this small change in primary structure, the hydrogen bonds that form the higher structures of hemoglobin produce an abnormal molecular structure, causing hemoglobin molecules to aggregate or clump together. This forces the red blood cells into a characteristic sickle shape and impairs their ability to carry oxygen.

Genetic Technologies

The disovery of the structure of DNA over 60 years ago set the stage for various innovations in the field of genetics. Such advances include the production of genetically modified organisms (GMOs), genetic testing, and DNA fingerprinting. Advanced laboratory techniques have enabled another innovation, cloning, to become a reality, and advances in computing and associated technologies have enabled the sequencing and analysis of the DNA of entire organisms, a technique known as **whole-genome analysis**. In this section we'll explore the meaning of some of these terms as well as concerns raised by some of these applications.

Cloning Reproduction in humans and other higher animals begins when a sperm cell, with its set of the male organism's DNA, enters an egg cell, with its set of the female's DNA. The two sets of DNA combine to form the full complement of DNA that characterizes the offspring. As the fertilized egg begins dividing, each of the newly formed cells of the developing embryo carries this same complete set of DNA. As a result, virtually every cell of the new organism has this same complete set of DNA that was created in the newly fertilized egg and that characterizes the organism.

It's possible to carry out this reproductive process in an entirely new and different way. In **nuclear-transfer cloning**, the nucleus of a female donor's egg cell, with its half set of chromosomes, is removed and replaced by a nucleus extracted from a skin cell or other body-cell of another adult animal. This nucleus has a full set of the adult organism's chromosomes. The egg now contains a nucleus with a full complement of chromosomes, all of which came from the adult donor organism. Genetically, the egg is now the equivalent of a freshly fertilized egg, but with all of its genetic information coming from a single existing organism rather than from two parents.

The egg is now induced to begin dividing. When it reaches an early embryonic stage of about 150 cells, called a **blastocyst**, it is implanted into the reproductive system of a female, a surrogate mother. Here it develops into a fetus and in time is born as a new infant, genetically identical to the adult whose DNA was inserted into the egg. The adult individual that served as the source of the DNA has been cloned, a process referred to as **reproductive cloning**. By contrast, **therapeutic cloning** involves the same nuclear-transfer technique, but instead of implanting the resulting blastocyst in the uterus of a

Sickle cell anemia • Figure 12.28

This genetic disease affecting the blood results from a single mutation within the gene encoding hemoglobin.

a. Hemoglobin, shown here in its normal higher-order structure, consists of four polypeptide chains: two α-chains, each with 141 amino acids and two β-chains, each with 146 amino acids. The square structures (in red) contain embedded iron atoms.

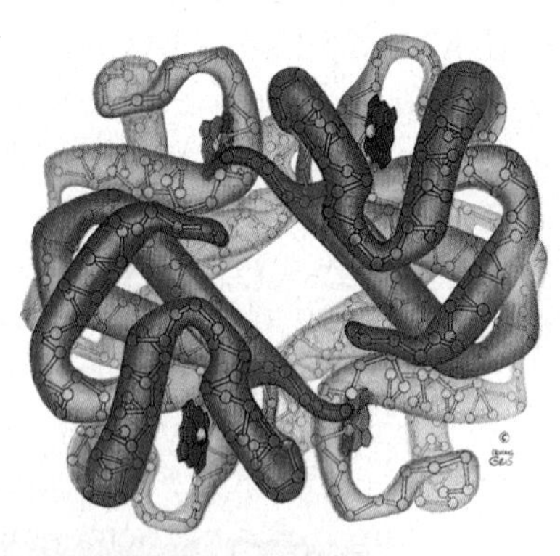

Reprinted with permission of John Wiley & Sons, Inc. from Snyder, C., *The Extraordinary Chemistry of Ordinary Things*, Fourth Edition, p. 434. Copyright 2003.

b. Sickle cell anemia results from a single-base mutation in the gene encoding the β-chains of hemoglobin. This error, which replaces an adenine (A) with a thymine (T), causes the required glutamic acid to be replaced by valine at a single position in each of the β-chains.

	Normal	Sickle Cell
DNA	...G–A–G...	...G–T–G...
mRNA	...G–A–G...	...G–U–G...
Amino acid	$H_2N—CH(CH_2CH_2CO_2H)—CO_2H$ Glutamic acid	$H_2N—CH(CH(CH_3)_2)—CO_2H$ Valine

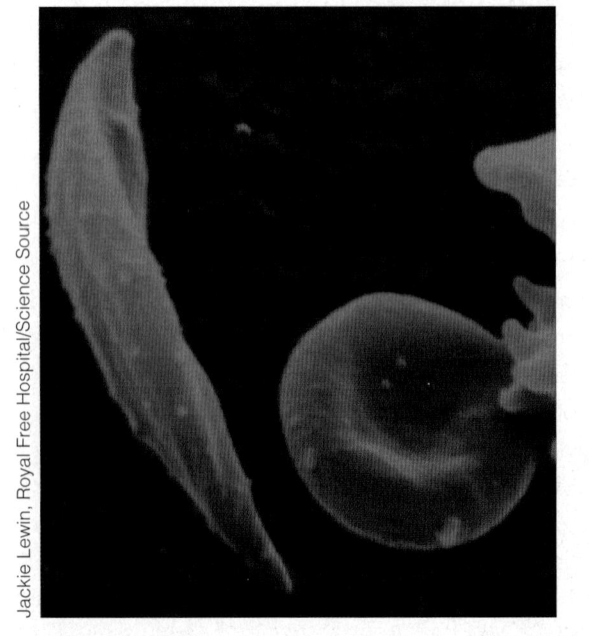

Jackie Lewin, Royal Free Hospital/Science Source

c. Sickle cell anemia is characterized by deformed or sickle-shaped red blood cells (left) and impaired blood function. A normal, circular-shaped red blood cell is seen at right for comparison.

Ask Yourself

1. What would be the expected result if the DNA mutation in question were from G-A-G to G-A-A, instead of from G-A-G to G-T-G? (Refer to Table 12.1 to answer the question.)
2. Compare the side chains of glutamic acid and valine, shown in red. Which is polar? Which is nonpolar?

Reproductive cloning • Figure 12.29

CC, at right, is the genetic clone of Rainbow, at left, but does not share the same markings.

AP Photo/Pat Sullivan

Ask Yourself

Do Rainbow and CC share the same: **(a)** genotype? **(b)** phenotype?

surrogate mother, **stem cells** are extracted from the blastocyst. These embryonic stem cells can differentiate into any number of different cell types, such as blood cells, neurons, or cells of specific organs. This opens up the possibility that stem cells thus produced from an individual could be used to restore diseased or damaged tissue within the same person.

Cloning is a tricky process, with a high failure rate. The first reported reproductive cloning of an adult mammal came in 1997, when Ian Wilmut, of the Roslin Institute in Scotland, announced that he and his colleagues had successfully cloned a sheep, producing a lamb they named Dolly. The success came after 277 previous failures by the same team of scientists.

Following Dolly, cows, goats, mice, pigs, sheep, horses, dogs, and other animals have also been cloned. In 2001, a cloned cat, called CC (for Carbon Copy), was born, the first success in a total of 87 failed attempts. Although CC is genetically identical to the adult DNA donor, a 2-year old female named Rainbow, the kitten's markings are different from those of the donor (**Figure 12.29**). This difference in fur patterns illustrates an important and more generalizable point: Reproductive cloning produces a new organism with the same genetic information as the cloned individual, but not a replica. Although CC and Rainbow are genetically identical, factors other than their (identical) DNA influence at least some aspects of their physical appearance. These include random events that may occur during the development of the fetus, such as variations in blood flow to the fetus.

Beyond the cloning of pets and livestock lurks the possibility of cloning humans. Pairs of humans with mutually identical DNA—identical twins—are well known. Yet despite having developed under nearly identical fetal conditions and having been raised in nearly identical environments, each twin still shows distinguishing physical and other characteristics. Similarly, although science might someday be able to clone any one of us, just as it has already cloned several other animals, it's unlikely that there will ever be anyone else quite like any one of us. Clearly, the string of bases of a DNA molecule does not tell the entire story of any plant or animal.

In 2005, the United Nations issued a nonbinding declaration in opposition to the practice of human cloning, based on the various ethical, social, and medical concerns it raises. Various countries, including the United States, Canada, Australia, and those of the European Union, have enacted laws banning the practice as well.

Recombinant DNA Prior to the introduction of **recombinant DNA** technology, therapeutic proteins, such as the insulin required to treat some forms of diabetes, were commonly obtained from the organs of slaughtered animals. For instance, insulin was traditionally obtained from the pancreases of pigs, cattle, and horses. For those suffering from hemophilia, the only viable source of the needed blood-clotting factors was the blood plasma of human donors, which carried the risk of viral infection.

recombinant DNA Genetic material produced by introducing DNA sequences of one organism into the DNA of another organism.

The biochemical manipulation of DNA has produced entirely new ways of obtaining these important therapeutic proteins, also known as **biologics**. For example, human genes that generate biologics can be introduced into the cells of the common bacterium, *E. coli*. This organism, genetically modified to produce these proteins, can be cultured on a large scale, allowing for almost limitless quantities of these medicines to be isolated and purified. In 1982, Humulin—short for human recombinant insulin—was the first commercial product manufactured through these methods. A variety of biologics are now

PROCESS DIAGRAM

Gene splicing • Figure 12.30

Genes from humans and other organisms can be spliced into bacterial plasmids and utilized by the bacterial host.

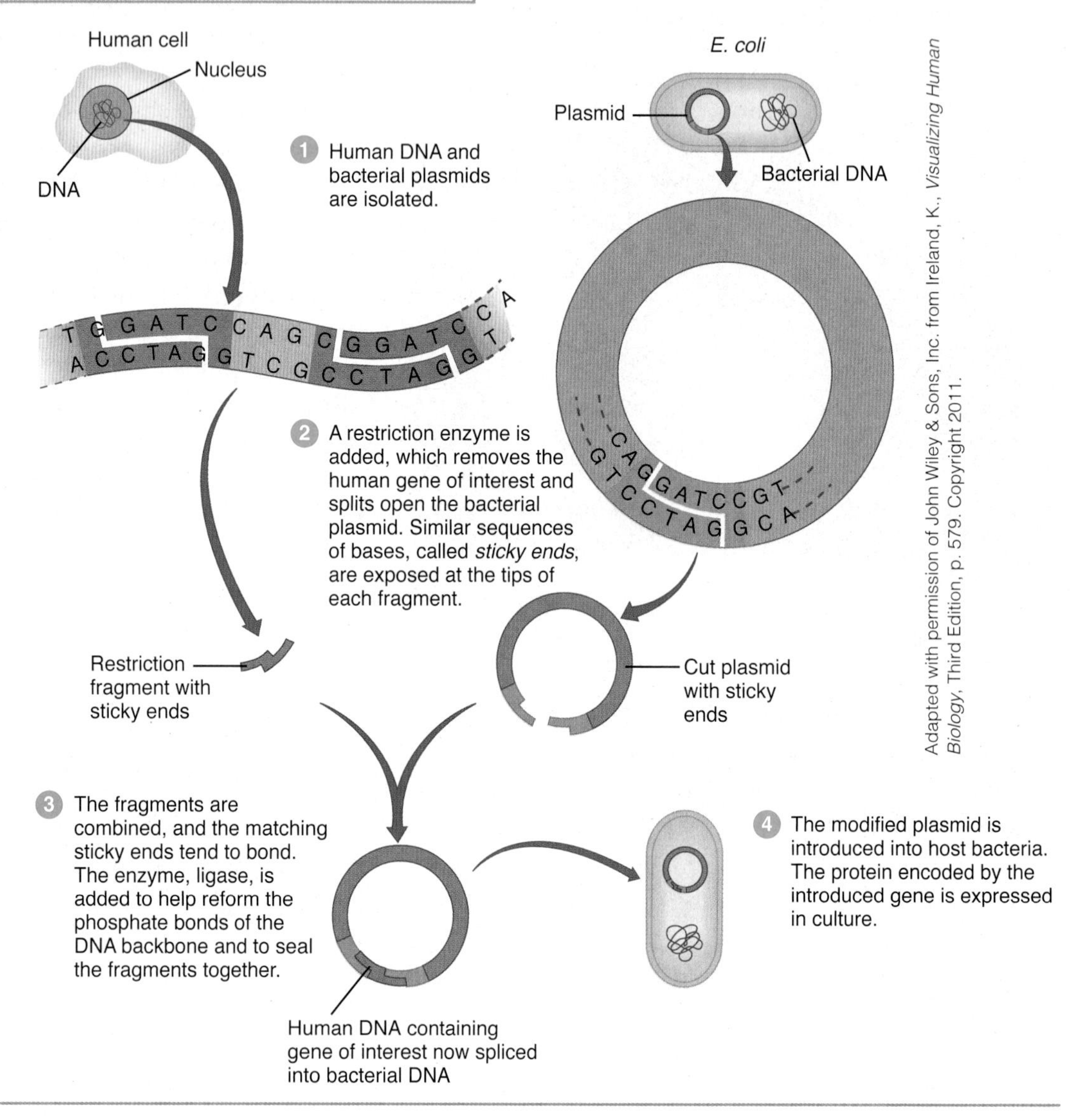

Human cell
Nucleus
DNA
E. coli
Plasmid
Bacterial DNA
1 Human DNA and bacterial plasmids are isolated.
T G G A T C C A G C G G A T C C A
A C C T A G G T C G C C T A G G T
2 A restriction enzyme is added, which removes the human gene of interest and splits open the bacterial plasmid. Similar sequences of bases, called sticky ends, are exposed at the tips of each fragment.
C A G G A T C C G T
G T C C T A G G C A
Restriction fragment with sticky ends
Cut plasmid with sticky ends
3 The fragments are combined, and the matching sticky ends tend to bond. The enzyme, ligase, is added to help reform the phosphate bonds of the DNA backbone and to seal the fragments together.
4 The modified plasmid is introduced into host bacteria. The protein encoded by the introduced gene is expressed in culture.
Human DNA containing gene of interest now spliced into bacterial DNA

obtained using recombinant DNA (**Table 12.2**). In addition to *E. coli*, other genetically modified host organisms, including yeast, are also used.

Recall that DNA commonly exists as a very long molecular strand, and genes occur in sequences along this strand. (Human DNA, if stretched out in a linear string rather than as a compacted helix, would reach an estimated 2 meters in length!) To create recombinant DNA, a gene located on the DNA of one organism is removed and spliced into the DNA of another. In this process, biotechnology tools, called **restriction enzymes**, act like molecular scissors, cleaving DNA strands at precise points. Other enzymes known as **ligases** help seal and repair the newly spliced DNA sequences. In the case of bacterial hosts such as *E. coli*, the transferred DNA is inserted into **plasmids**, which are circular pieces of DNA located outside the nucleus of bacterial cells (**Figure 12.30**).

> **restriction enzyme** An enzyme that cleaves DNA at a specific location along its strand.

Genetically modified crops The genetic modification of agricultural plants through recombinant DNA technology is now quite common. Unless you maintain a diet consisting entirely of organic foods—which by definition cannot be derived from **genetically modified organisms**, or GMOs—it's likely that you consume food ingredients, such as soybean oil and corn syrup, derived from GMOs. Currently, about 90% of the corn, soybean, and cotton grown in the United States is genetically modified.

> **genetically modified organism (GMO)** An organism whose genetic material has been altered through genetic engineering.

Biologics produced through recombinant DNA technology Table 12.2

Trade Name (generic name)	Activity and Use
Humulin (human insulin)	Moderates glucose levels in the treatment of diabetes.
Epogen (erythropoietin)	Increases red blood cell production; treats anemia.
Humatrope (human growth hormone)	Stimulates growth in humans.
Factor VIII (blood-clotting protein)	Treats hemophilia.
Activase (tissue plasminogen activator)	Decomposes blood clots; treats stroke and heart attack.
Bovine somatotropin (bST)	Increases milk production in dairy cattle.

A large majority of these genetically modified or GM crops now incorporate one or both of the following genes:

- A gene that confers tolerance to an herbicide called *glyphosate*, allowing this chemical to be sprayed on fields of these crops so that it selectively kills weeds (**Figure 12.31**);
- A gene originating in the bacterium known as *Bacillus thuringensis*, or Bt, that encodes for an insecticidal protein. Crops engineered to contain this gene produce the insecticidal protein and are thereby immune to a variety of insect pests.

GM crops such as these can provide benefits to farmers in terms of higher yields and a reduced need to apply herbicides and pesticides. However, a variety of concerns have been raised over the use of genetically modified crops. For instance, herbicide-resistant genes can be carried by pollen and can potentially cross-pollinate weeds, thereby conferring this trait to the very types of plants that we do not want to tolerate herbicides. Protocols for the use of genetically modified seed in the United States require farmers to create buffer zones around fields containing GM plants in order to block the migration of their pollen. Another concern is that GM crops could create proteins that cause allergic reactions in some consumers. For example, in 1998 a genetically modified corn called StarLink was approved by the Environmental Protection Agency (EPA) for use in animal feed only, because of a very slight but real risk of allergic reactions. The product was removed from commercial use in 2000 when it was discovered that some of this GM corn had entered the human food supply as a component of taco shells and other products.

Critics of GM foods advocate for greater regulatory oversight of these products and for widespread labeling of foods containing GMOs. Although GM foods currently approved for human consumption appear to be as safe as their non-GM counterparts and appear to offer the same nutritional benefits, the long-term health effects of consuming these products are not yet known.

Genes and medicine Understanding the role genes play in various diseases is currently a matter of widespread interest. As the cost of sequencing the whole genome of an individual continues to drop, this process may eventually become a standard part of health care. **Personalized**

Genetically engineered plants • Figure 12.31

Many crops, including soybean and corn, have been genetically engineered to contain a gene conferring resistance to the herbicide glyphosate.

a. Glyphosate, sold under the trade name Roundup, is a widely used herbicide that shows low toxicity in animals and undergoes biodegradation or natural breakdown in the environment. Various functional groups present in glyphosate are shown.

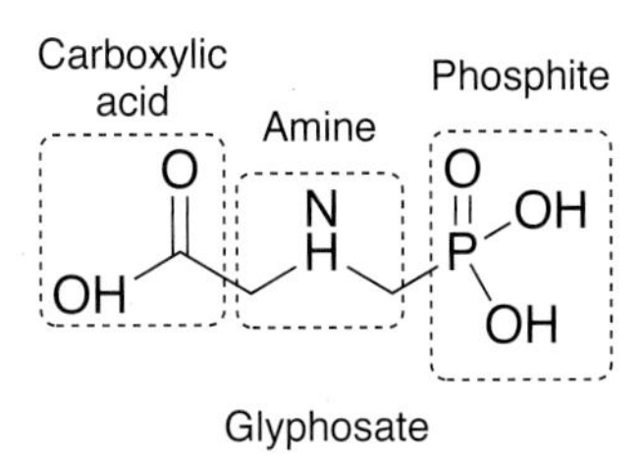

b. Researchers collect plant and insect specimens in a glyphosate-resistant soybean test field. The widespread use of glyphosate has led to an increase in weeds resistant to this herbicide.

Bloomberg via Getty Images

DID YOU KNOW?

Genetically engineered plants and animals can yield medications and other valuable materials

In a process called pharmaceutical farming, sometimes referred to as pharming, plants and animals are genetically engineered to produce protein therapeutics—in some sense serving as chemical factories. This may seem like science fiction, but it has already resulted in FDA-approved drugs for human use. For example, in 2009, an anticoagulant drug expressed in the milk of genetically engineered goats was approved for human use. This compound is used to prevent the formation of life-threatening blood clots in individuals with a rare genetic disorder. In 2012, a fat-metabolizing enzyme produced in the cells of genetically engineered carrots received FDA approval. This compound is used to treat a genetic disorder called Gaucher's disease, whose sufferers produce insufficient amounts of this enzyme in their own bodies.

Pharming offers a potentially lower-cost and more flexible alternative to conventional recombinant protein methods, which rely on cultured bacterial, yeast, or mammalian cells. For instance, if you want to increase the production of a recombinant protein with pharming, you simply breed more of the drug-producing plants or animals (**Figure a**). Like most genetic technologies, however, pharming raises ethical questions and environmental concerns. So far, the technique does not appear to have harmed the animals involved, but caution needs to be exercised to protect animal welfare. Another concern is that genetically engineered organisms (GMOs), whether plant or animal, could cross-breed with their non-GMO counterparts, thus disseminating nonnative genes into the wild.

Organisms can be genetically engineered to produce a variety of compounds they do not normally make, or to increase the yields of substances they already do generate. A promising area of research is in the genetic modification and cultivation of bacteria that produce hydrocarbons, and of algae that create high levels of oils (**Figure b**). If these processes become commercially successful, the biofuels they create will serve as environmentally friendly alternatives to fossil fuels.

Joe Proudman/UC Davis

a. Dr. James Murray of the University of California, Davis, leads a team in developing genetically engineered goats to express high levels of the enzyme human lysozyme in their milk. Variants of this enzyme occur naturally in the milk of all mammals. Human lysozyme has been shown to kill bacteria associated with childhood diarrhea, a serious cause of death among children in the developing world.

Bloomberg via Getty Images

b. Various companies are developing methods to increase the yield of oils from cultured algae.

Think Critically If lysozyme-rich goat milk is determined to be safe for human consumption and effective in treating childhood diarrhea, would you consider it to be a food, a medicine, or some combination of the two? Does your answer change if this milk is shown to offer protection against the development of this condition in an otherwise healthy individual?

medicine offers the possibility of using information about a person's genotype to identify for each patient those specific medicines or other treatments that are shown to be highly effective in other individuals with the same genetic profile. Historically, the clinical testing of drugs has involved large, heterogeneous groups of people, but we now know that some subgroups of individuals respond either more or less favorably to a medicine because of genetic factors they have in common.

In cancer therapy, for example, the genotyping of certain tumor cells has become routine. Breast cancer associated with the BRCA (pronounced *brak-uh*) gene has been found to be less likely to respond to some hormone-based therapies commonly used against breast cancer. As a result, other approaches are now used to treat it. Breast cancer that expresses the HER2 gene, which is associated with certain skin growth factors, has been shown to be particularly aggressive. However, these tumor cells often respond to treatment with a protein-based drug, Herceptin, which is one of many products manufactured through recombinant DNA technology.

Gene therapy offers the potential for treating diseases caused by a mutation in a single gene, especially if the condition expresses itself in a single type of body tissue or organ. In this approach, DNA containing the corrected gene is combined at the molecular level within a **vector**, such as a virus, that can deliver or transfer this gene into the patient's affected cells. With a sucessful transfer of the required gene, the diseased tissue becomes able to produce the required protein. Although a large number of clinical trials involving gene therapy have been conducted since 1990, safety and other concerns delayed the first regulatory approval for the use of this type of therapy in humans until 2012. Biotechnology innovations such as these can lead to a future in which the analysis and manipulation of DNA will become a more common presence in health care and related fields (see *Did You Know?*).

CONCEPT CHECK

1. **How** does the genetic code within a segment of DNA result in the formation of a specific protein?
2. **What** is the chemical basis for sickle cell anemia?
3. **How** is reproductive cloning different from therapeutic cloning?

Summary

1 Nonprescription Medicines 366

- ***How does aspirin work, and what are non-steroidal anti-inflammatory drugs (NSAIDs)?***
  Aspirin (acetylsalicylic acid) and certain other pain relievers, including ibuprofen, belong to a class of compounds called NSAIDs, short for **n**on-**s**teroidal **a**nti-inflammatory **d**rugs. NSAIDs act as **analgesics** (to relieve pain), **antipyretics** (to lower fever), and **anti-inflammatory agents** (to reduce inflammation). These medicines, which do not have steroidal structures, inhibit an enzyme involved in the production of **prostaglandins**—compounds involved in the sensation of pain and its transmission along the nervous system, in the generation of fevers, and in the swelling of inflammation. In a common mechanism of **enzyme inhibition** (shown here), enzyme inhibitors (I) block the active site of an enzyme, thus preventing a substrate molecule (S) from reacting to form product.

Figure 12.3 • Enzyme inhibition

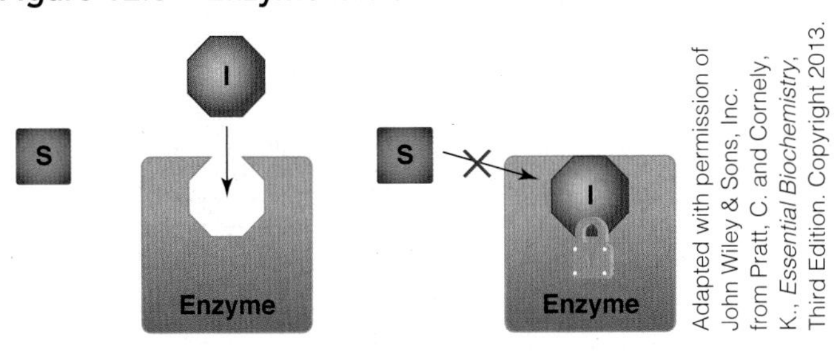

- ***How do decongestants and antihistamines work?***
  Decongestants help relieve swelling of the nasal passages by restricting blood flow to the sinuses. The most commonly used decongestant is phenylephrine, which falls within the class of compounds known as phenylethylamines. **Antihistamines** can help reduce allergic symptoms. They operate by blocking the action of a chemical, known as histamine, that triggers inflammation within our bodies.

2 Prescription Medicines 370

- ***How are medications developed?***
  Compounds that show therapeutic potential, called **drug candidates**, first undergo preclinical testing in the laboratory and in animals to assess their efficacy, toxicity, and other factors. Based on these results, only the most promising compounds may then undergo testing in humans, known as **clinical trials**. To minimize the risk of bias, clinical trials are commonly carried out **double-blinded** and **placebo-controlled**, meaning that during the trial neither the investigators

having patient contact nor the participants know whether any participant receives the medicine or a **placebo**, an inert substance that looks identical to an actual medicine. Clinical trial data are analyzed to determine the efficacy and side effects of the medicine as compared to placebo.

- ***What are drug analogs?***
Drug analogs share the same basic molecular framework as the comparable drugs, but the drug analogs have small structural differences from the drugs themselves, which may result in different therapeutic properties. For example, penicillin-type antibiotics include amoxicillin and other compounds that share a common structure with penicillin, but that show slightly different benefits against certain classes of bacteria.

- ***What are steroids?***
Steroids constitute a large class of organic compounds that all share a common four-ring structure. Hundreds of steroids are known, including those occurring naturally in animals and plants and those produced synthetically. The most common steroidal medicines are the **corticosteroids**, which reduce inflammation. Another major group of steroids includes the sex hormones estrogen, progesterone, and testosterone (shown here). **Anabolic steroids** are synthetic compounds that act on the body similarly to testosterone.

Figure 12.10 • Steroidal structure

3 Recreational, Illicit, and Abused Drugs 377

- ***How is alcohol metabolized by the body?***
Within the body, alcohol is metabolized by the liver through a series of **oxidation reactions** facilitated by enzymes. Initially, ethanol (CH_3CH_2OH) is oxidized to acetaldehyde (CH_3CHO); then, acetaldehyde is oxidized to the acetate ion ($CH_3CO_2^-$). The build-up of acetaldehyde within the blood is responsible for some of the unpleasant short-term side effects associated with alcohol consumption.

- ***What are alkaloids?***
Alkaloids constitute a large class of compounds of plant origin, often bitter, that contain the basic amine group (such as shown here). Naturally occuring alkaloid drugs include the stimulants caffeine, nicotine, and cocaine, the opioid narcotic morphine, and the hallucinogen mescaline. Semisynthetic derivatives of alkaloids exist as well, such as heroin, which is derived from morphine. **Semisynthetic drugs** are made by chemical modification of a natural product.

Figure 12.14 • Common alkaloids

Caffeine

Nicotine

- ***What are phenylethylamines?***
Many mind-altering substances, including LSD, mescaline, the opioid narcotics, and the amphetamines and their derivatives, contain a phenylethylamine segment. Molecules containing this structural fragment often affect the way our nerves carry their messages to the brain. Phenylethylamine itself has the structural formula C_6H_5-CH_2-CH_2-NH_2, with a benzene ring attached through a two-carbon atom chain to an amine group.

4 Chemistry of Genetics 386

- ***What is the structure of DNA, and what is its role in heredity?***
The primary structure of DNA consists of a long sequence of **nucleotides**, each of which has a phosphate group connected to a sugar unit bound to one of four amine bases—adenine (A), cytosine (C), guanine (G), and thymine (T). DNA exists as a double helix (as shown here), with the bases of complementary DNA strands forming hydrogen bonds to one another (A pairs with T and G pairs with C). When a cell divides, replicas of its DNA form as the double helix uncoils and a complementary strand forms adjacent to each one of the original strands.

Figure 12.24 • The DNA double helix

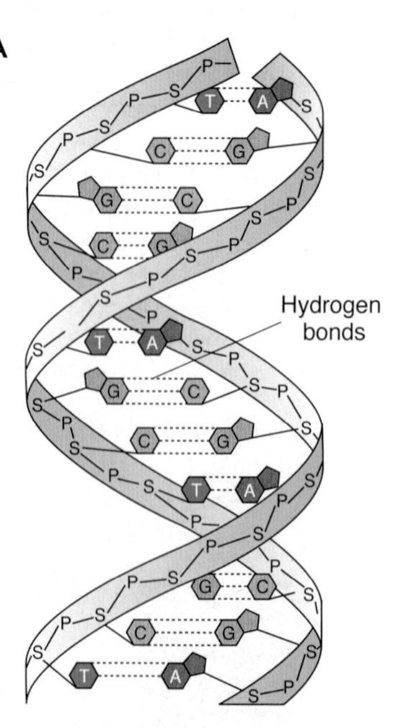

A **gene** is a sequence of bases along the DNA strand that encodes for the sequence of amino acids in a particular protein. The **transcription** of DNA produces a complementary strand of messenger RNA (mRNA). A **codon** is a three-base sequence of mRNA that encodes for a specific amino acid. The sequence of codons along the mRNA chain serves as a template for protein synthesis. In **translation**, each sequential amino acid of the growing protein chain is transported to its corresponding codon of mRNA by a molecule of transfer RNA (tRNA).

- ***What is recombinant DNA technology?***
Recombinant DNA is produced by splicing together DNA sequences from two or more organisms. This techology produces **genetically modified organisms (GMOs)**, such as agricultural crops designed to be resistant to herbicides, and bacteria and other organisms designed to produce protein-based medications. These medications include insulin, used for treating diabetes, and erythropoeitin, used for treating anemia, a deficiency of red blood cells.

Key Terms

- alkaloid 379
- antibiotics 372
- deoxyribonucleic acid (DNA) 388
- enzyme inhibition 369
- gene 390
- generic drugs 372
- genetically modified organism (GMO) 399
- natural product 366
- NSAID 367
- nucleotide 388
- placebo 372
- recombinant DNA 397
- restriction enzyme 398
- ribonucleic acid (RNA) 392
- semisynthetic 366

What is happening in this picture?

DNA fingerprinting analysis is a technique used to identify individuals according to certain characteristics of their DNA. It is widely used in law enforcement, paternity testing, and other forensic applications. This technique can also be applied to animal species other than humans. Here, a pet owner swabs the inner cheek of her dog to collect a sample of cells to be used for DNA fingerprinting analysis.

Canine DNA evidence has been deemed admissible in criminal trials, as can human DNA. A 2009 murder in London involving a vicious attack by a dog was prosecuted in part based on canine DNA evidence, which showed that there was less than a one-in-a-billion chance that the alleged perpetrator's dog was not involved in the attack.

Think Critically

1. DNA fingerprinting involves the analysis of repeating patterns of DNA bases in *noncoding* regions of DNA. What does the term noncoding here mean?
2. Is DNA present in skin cells? Based on this image, do you think it is possible that the sample could become cross-contaminated with human DNA?
3. If somatic cells of a human and a dog become mixed, is there a way to determine the origin of each based on the number of chromosomes in each type? Explain.

Exercises

Review

1. Explain, identify, describe, or define each of the following terms:
   (a) acetaminophen
   (b) amphetamines
   (c) acetylsalicylic acid
   (d) alkaloids
   (e) analgesic
2. Identify a physiologically active substance that comes from each of the following:
   (a) the bark of the willow tree
   (b) the tobacco plant
   (c) the latex of the opium poppy
   (d) the cannabis plant
   (e) the peyote cactus
   (f) the coca plant
3. Describe the connection or distinction between the items in the following pairs:
   (a) salicylic acid and aspirin
   (b) NSAIDs and prostaglandins
   (c) morphine and heroin
   (d) methamphetamine and MDMA
   (e) the test subjects of preclinical trials and those of clinical trials
   (f) generic drugs and brand-name drugs
4. (a) Identify the phenylethylamine structure within each of the following compounds:

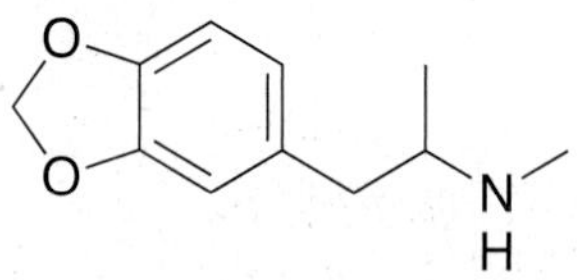

Methylenedioxymethamphetamine (MDMA)

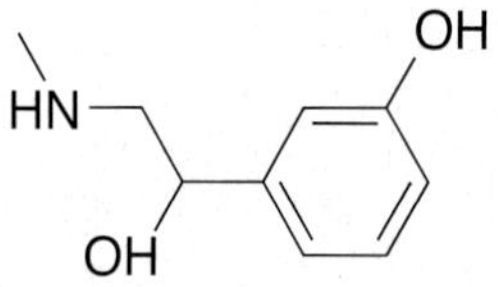

Phenylephrine

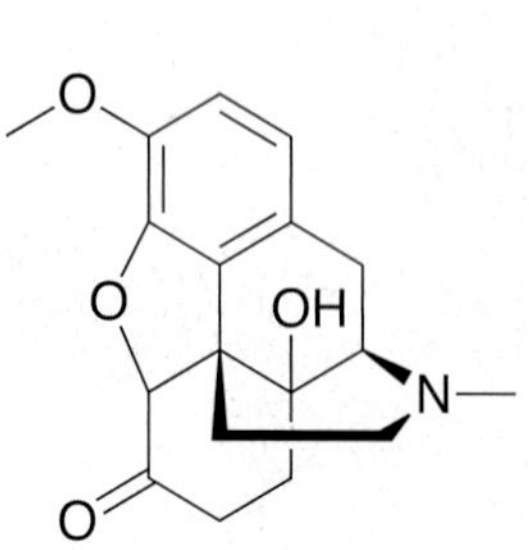

Oxycodone

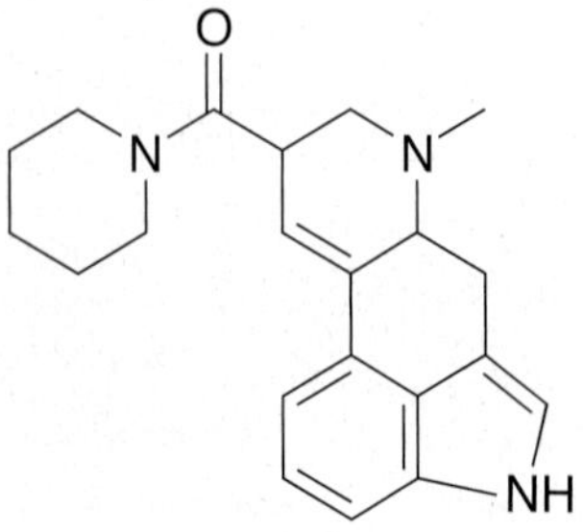

Lysergic acid diethylamide (LSD)

   (b) Identify the uses or effects of each compound in part a.

5. Explain the role of acetylation as it pertains to the semisynthesis of aspirin and heroin.
6. What's the difference between an analgesic medication and an antipyretic medication?
7. What symptoms are alleviated by aspirin and ibuprofen, but not by acetaminophen?
8. (a) What does the acronym NSAID stand for?
   (b) Identify three symptoms of illness or injury that NSAIDs relieve.
9. (a) Identify the ester functional groups in cocaine, shown here.
   (b) Does this molecule represent the acidic or basic form of the drug?
   (c) What functional group common to all alkaloids is present in this molecule?

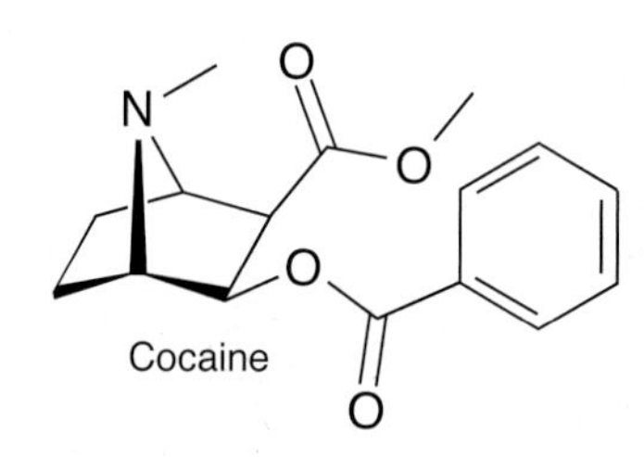

Cocaine

10. (a) Electronic cigarettes, or e-cigarettes, contain no tobacco but rather are devices that use a heating element to vaporize the active ingredient present in tobacco. Name the alkaloid that is vaporized.
    (b) Since e-cigarettes do not burn tobacco, harmful combustion products normally found in cigarette smoke are not formed. Identify one such compound that is absent from the vapors of e-cigarettes.
11. Name and give the structure of a compound you could use to convert an alcohol or a phenol into an ester.
12. (a) Identify the phenol functional group present in morphine and THC, which are shown here.
    (b) Identify the plant sources of each of these compounds.
    (c) What functional group is produced upon treating each of these compounds with acetic anhydride?

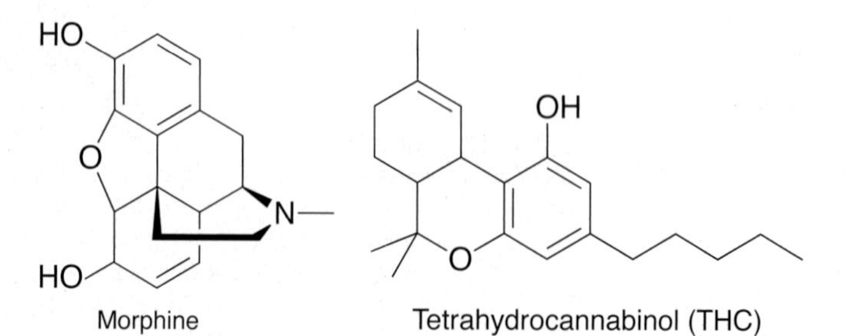

Morphine

Tetrahydrocannabinol (THC)

13. Why is diacetylmorphine a particularly dangerous chemical?

14. (a) In what way are methadone and heroin alike in their effects on humans?

(b) In what way are they different?

15. (a) Name two alkaloids that are used often, legally, and in large quantities throughout most of the world, even though they can be addictive.

(b) In what form is each consumed?

16. Describe or identify the kinds of cells produced by:

(a) mitosis;

(b) meiosis.

17. In what biological structures are chromosomes located?

18. What four amine bases occur in DNA? In RNA?

19. What holds together the two strands of the DNA double helix?

20. (a) Identify the two DNA bases shown here.

(b) What is represented by the blue dotted lines?

(c) In DNA, what carbohydrate group is bonded to each base?

(d) What functional groups are responsible for the basic properties of DNA bases?

(e) Which one of the two bases shown here does not occur in RNA? What is the name of the base that replaces it in RNA?

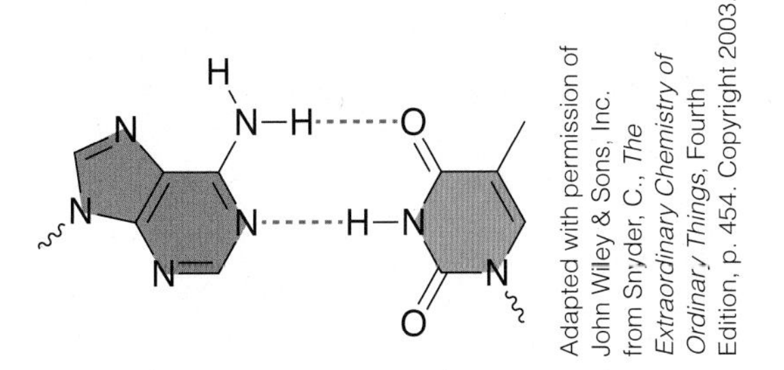

21. What four sequences of amine bases on mRNA form the codons for valine (Val)?

22. What is the genetic code?

23. (a) Provide labels for the following components of an animal cell.

(b) Identify the sister chromatids in the image.

(c) Has the DNA contained within features of the image shown here already replicated? Explain.

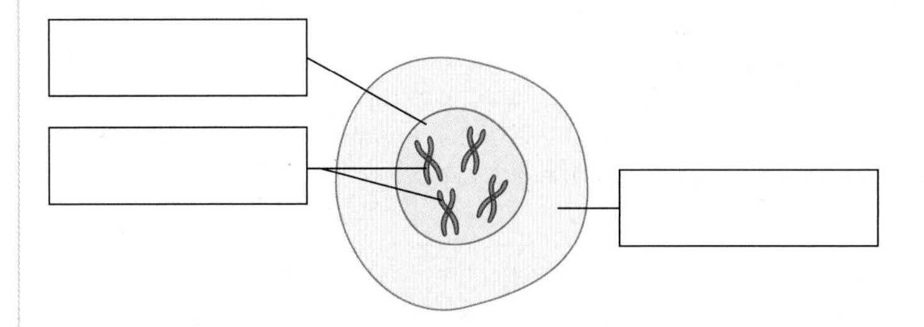

24. Explain or describe the difference between:

(a) a genotype and a phenotype

(b) a gene and a genome

(c) chromatin and a chromosome

(d) DNA and RNA

(e) a nucleotide and a DNA base

(f) mRNA and tRNA

(g) meiosis and mitosis

(h) a codon and an amino acid

(i) an amino acid and a polypeptide

(j) reproductive cloning and therapeutic cloning

(k) a single-gene disorder and a multifactorial disorder

(l) a single-gene disorder and gene therapy

(m) a plasmid and a restriction enzyme

25. Is red hair classified as a phenotype or a genotype? What term do we give to the carrier of the genetic information that transmits the characteristic of red hair from parent to child?

26. (a) What do the groups P and S represent in the segment of DNA shown here? (b) What is the sequence of bases of the mRNA strand that is complementary to the segment of DNA shown here?

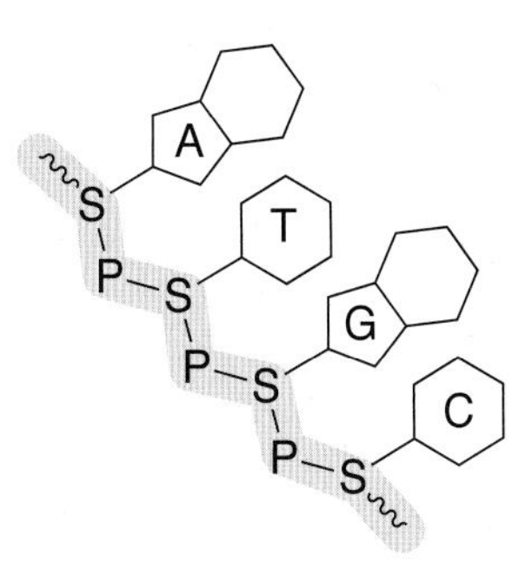

27. Explain, identify, describe, or define each of the following terms:
   (a) plasmid
   (b) restriction enzyme
   (c) glyphosate
   (d) recombinant protein
   (e) transgenic

28. What was the first adult animal to be cloned successfully? When and where did this occur, and how many unsuccessful attempts preceded the first success?

Think

29. Under what conditions or circumstances is it dangerous, and possibly fatal, to use aspirin?

30. Why does heroin often exhibit a slight scent of vinegar?

31. Aspirin tablets kept for many months, especially in hot, humid climates, often smell of vinegar. Why?

32. Draw the portion of their molecular structures that is common to all the salicylates.

33. Do you think that the use of all mood-altering or perception-altering drugs should be prohibited except for medical purposes? Explain your reasoning.

34. The consumption of a combination of alcohol and energy drinks, which contain caffeine and other stimulants, can be hazardous. What opposing effects on the body do these substances cause? What risks are associated with this practice?

35. (a) Do you think that a person who drinks several cups of decaffeinated coffee, thinking mistakenly that it is caffeine-containing coffee, might show signs of agitation or sleeplessness as a result?
   (b) If this would occur in some cases, to what would you attribute the phenomenon?

36. Why is the involvement of an independent third party necessary as part of a double-blind, placebo-controlled clinical trial?

37. Certain dog breeds, including Chihuahuas and Great Danes, as well as a variety of domesticated crops, including corn, have developed through specialized breeding over many generations. Do you consider these examples to be genetically modified organisms since they required human intervention to exist?

38. Antibiotics were initially touted as wonder drugs. What observations led to a more tempered view of antibiotics?

39. How is the resistance of weeds to widely used herbicides similar to antibiotic resistance?

40. Is DNA a protein? Explain.

41. In one of their early models of the double helix, Watson and Crick assumed that the amine substituents on each DNA molecule pointed outward, away from the central axis. They later discarded this model in favor of the one in which the amine substituents point inward, toward the central axis. Why do inward-pointing amine bases provide a more stable double helix?

42. A rabbit can run from a hungry fox, but a plant can't run from a hungry insect. Describe what defense(s) a plant might possess against a predatory insect.

43. (a) What function does cell division by meiosis serve in the reproduction of an organism?
   (b) In this form of reproduction, why can meiosis be considered superior to mitosis?

44. Why does reproductive cloning produce a being that is genetically identical but not an exact replica of the cloned organism?

45. Human cloning is a controversial topic. Suggest some ethical considerations in favor of human cloning. Suggest some in opposition to human cloning.

46. Give your own answers to the following questions and explain your reason for each:
   (a) Would you wish to have a pet of yours cloned?
   (b) Would you wish to have yourself cloned?
   (c) Should governments control animal and/or human cloning, or should cloning be left to the judgments of individuals, physicians, and scientists?

47. Suggest some arguments in favor of the use of recombinant DNA technology. Suggest some arguments opposing its use.

Calculate

48. Show that the four amines of RNA, taken three at a time, give 64 different sequences (codons).

49. How many different nucleotides can a strand of DNA contain? Explain.

50. One of the largest human genes known is responsible for the formation of the protein CNTNAP-2, which is involved in cell-to-cell interactions in the central nervous system. This gene contains 2.3 million bases. How many codons does this represent?

51. An estimated 3.1 billion base pairs exist in the DNA of each human cell. The thymine–adenine pair is connected by two hydrogen bonds, and the cytosine–guanine pair by three. Assuming that these two base pairs are present in equal numbers in the DNA of each human cell, calculate the total number of hydrogen bonds in the DNA of each human cell.

52. DNA fingerprinting analysis, used by law enforcement, measures variations in short tandem sequences (STRs) exhibited by individuals. An STR is a repeating sequence of DNA bases found at a particular **locus**, or site on a chromosome. An individual exhibits a certain STR type at a given locus. Of the 13 standardized STR loci established by the FBI for DNA fingerprinting, the probability that two randomly selected individuals share the same STR type at a given locus is given by the table below.

For instance, the chances that two randomly selected people share the same STR type at locus #1 is 0.075 or 7.5%. For #2, the value is 0.063 or 6.3%. The chance that these individuals share the same STR type at both loci #1 and #2 is $(0.075 \times 0.063) = 0.0047$, which corresponds to a 1 in 211 chance. [We get the number 211 by dividing 1 by 0.0047.] What is the chance that two randomly selected individuals share the same STR type at all 13 loci?

Table for Question 52

STR Locus	#1	#2	#3	#4	#5	#6	#7	#8	#9	#10	#11	#12	#13
Probability	.075	.063	.036	.081	.195	.112	.158	.085	.065	.067	.039	.028	.089

13

Plastics, Pollution, and Sustainability

When you discard trash in a garbage can or recycle bin, where exactly does it go? Slightly more than half of our municipal waste is sent to landfills (and a small percentage is incinerated), but the paper, glass, metal, and plastic materials of trash are increasingly reclaimed and recycled, forming partial or whole components of new products. For example, the disposable plastic water bottle you toss into a recycle bin one day could go on to form the polyester fibers of new clothing or carpeting. As shown here, the official uniforms worn by members of the U.S. Olympic basketball team are constructed largely from recycled polyester fibers.

In this chapter we'll explore the chemistry underlying the synthetic fibers and plastics we use every day, materials composed of compounds called *polymers*. We'll learn about the many valuable uses of these materials, and about some of their environmental costs. For example, throughout the world, discarding plastics and other waste materials and chemicals into the environment not only increases pollution but can have other wide-ranging impacts. We'll examine these effects and see how various approaches designed to make polymers degradable, reduce pollution, and conserve resources offer hope for a more sustainable future.

Christian Petersen/Getty Images

CHAPTER OUTLINE

13.1 Polymer Uses and Structures

LEARNING OBJECTIVES

1. **Distinguish** between the terms *plastic* and *polymer*.
2. **Describe** different types of synthetic polymers and their chemical structures.

Plastics, which belong to the chemical class known as polymers, constitute some of the most common materials of our everyday lives. The manufacture of plastics and their conversion into consumer products has become one of the largest industries in the world today. In this section we'll explore various types and uses of polymers and plastics, as well as their underlying chemical structures.

Plastics and Society

Think for a moment of the various plastic items you use or come in contact with on a daily basis. These may include plastic water or soft-drink bottles; plastic cups, utensils and take-out food containers; plastic bottles and containers for shampoos, laundry detergents, cosmetics, and other personal care items; plastic credit cards, pens, toothbrushes, and razors; plastic shopping bags and garbage bags; and the plastic components of televisions, laptops, smartphones, and other electronic devices. Other common uses of plastics or polymeric materials include components of cars, household appliances, and furniture; components of paints, coatings, and adhesives; and the textiles used in clothing, carpeting, and upholstery.

Plastics are highly versatile and exhibit a range of properties, from firm and rigid to soft and flexible. They are also relatively light and inexpensive, and can be formed into an almost limitless variety of shapes and objects (**Figure 13.1a**). Plastics are often quite durable and resistant to decomposition, but this property can also be a liability, especially when these materials are discarded into the environment.

Worldwide, about 300 million tons of various types of plastics are produced each year. **Figure 13.1b** summarizes the major uses of these materials, which include packaging, building and construction materials, electrical insulation, and housewares.

The word "plastic" derives from the Greek *plastikos*, meaning a material that can be molded and shaped. Chemically, commercial plastics are **polymers**, which are very large molecules, or **macromolecules**, made

polymer A very large molecule formed by the repeated combination of much smaller molecules.

Uses of plastics • Figure 13.1

Plastic items are produced in a virtually unlimited variety of designs and shapes, some of which are now made through the technology of 3D printing (**a**) Plastics are highly versatile materials with a wide variety of uses (**b**).

a. Digital designs can now be converted into three-dimensional plastic objects quite easily using 3D printers. The plastic components of these table lamps were produced by 3D printing.

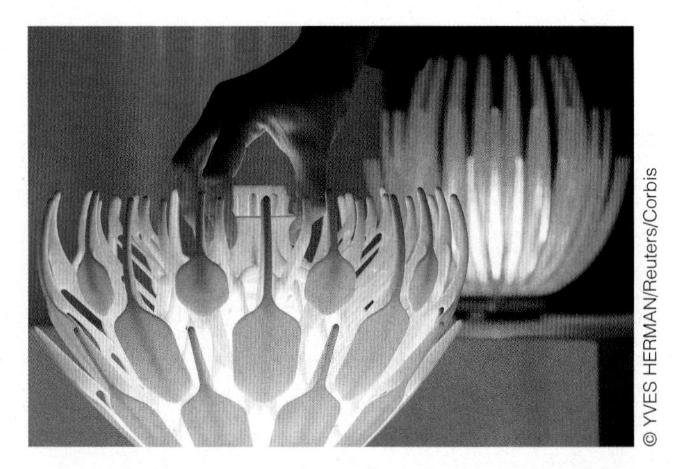

© YVES HERMAN/Reuters/Corbis

b. Plastics are used largely for packaging, but they have a wide variety of other applications, including building materials, electrical insulation, and transportation.

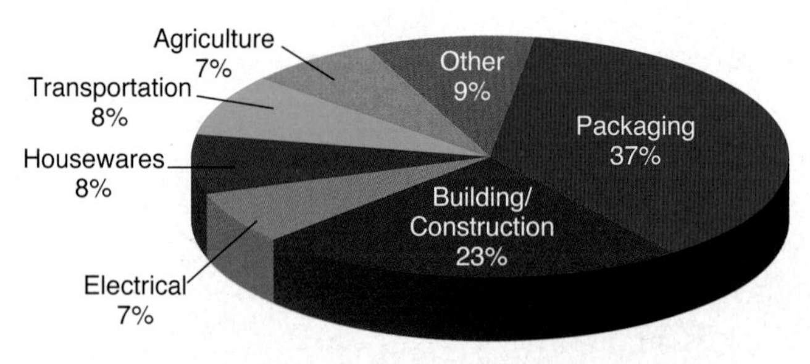

Polyethylene, an addition polymer • Figure 13.2

Thousands of ethylene units link into a chain to produce polyethylene. This process can be likened to forming a human chain. Typically, a free radical (represented as R•) initiates the process.

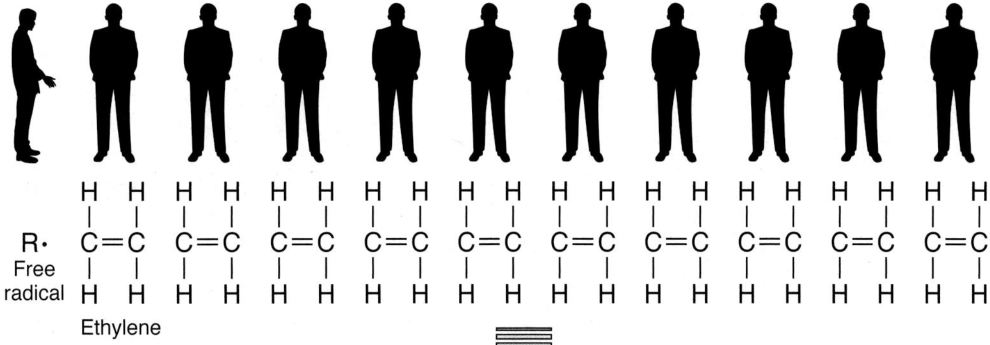

A visual analogy for the formation of polyethylene.

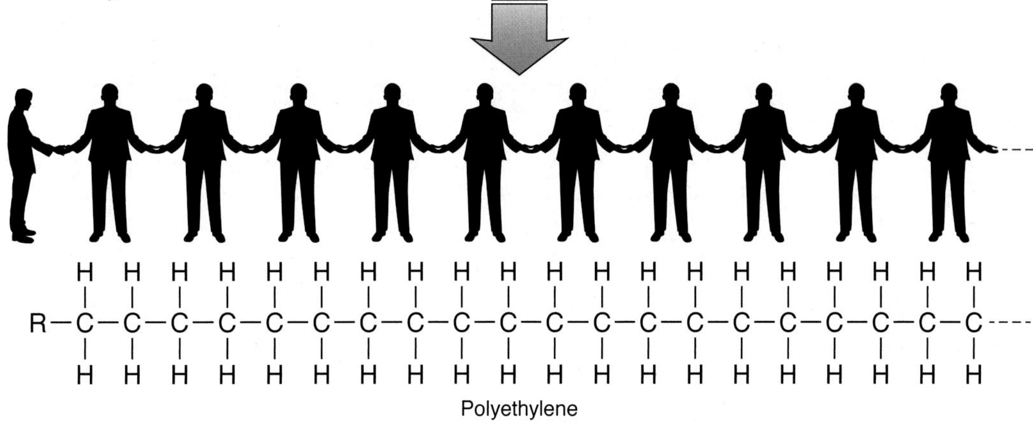

up of a large number of linked **monomers**. A typical synthetic polymer molecule contains thousands of monomers linked together in a chain, thus producing a compound with a very high molecular mass. For instance, the polyethylene in bottles and jugs commonly used for liquid laundry detergent consists of polymeric chains with an average molecular mass exceeding 280,000 atomic mass units (u). By contrast, each ethylene unit (C_2H_4), which is the monomer that forms the repeating links of these polyethylene chains, has a molecular mass of only 28 u. This means that more than 10,000 ethylene units may link together to form the typical polyethylene chains of these plastic jugs.

monomer A small molecule, many of which link together to form a polymer.

Almost all commercial plastics are **synthetic polymers**, produced in the laboratory or factory from one or more different types of monomers. For instance, polyethylene is manufactured from the monomer ethylene, a gas, and the ethylene itself is chemically derived from petroleum. In fact, almost all synthetic polymers are manufactured from monomers obtained from petroleum.

Polymers produced by living organisms, or **biopolymers**, are also common. These include the polysaccharide chains of starch and cellulose (Chapter 5), each consisting of repeating glucose units; the polypeptide chains of proteins (Chapter 5), made up of linked amino acids; and the strands of DNA molecules (Chapter 12), formed from linked nucleotide units. Plastics are synthetic polymers, but biopolymers—such as proteins and DNA—are not plastics.

Polymer Formation

Monomers can link together to form polymers by either of two general methods: addition polymerization, which produces addition polymers; and condensation polymerization, which produces condensation polymers. We'll now examine each of these categories.

Addition polymers Imagine a long line of people standing side-by-side. As each person clasps the hand of the person on each side, a human chain forms **Figure 13.2**. We can liken this process to the formation of **addition polymers**. In this case, an **unsaturated** monomer—a small molecule containing a double or triple bond—uses the two electrons of one of its multiple bonds to covalently link with its neighboring monomers. One of the electrons links to the neighbor on one side, and the other electron

addition polymer A polymer in which all atoms originally present in the monomer units are retained.

Polyethylene, the process of formation • Figure 13.3

The formation of polyethylene proceeds as follows. Each red arrow represents the flow of a single electron.

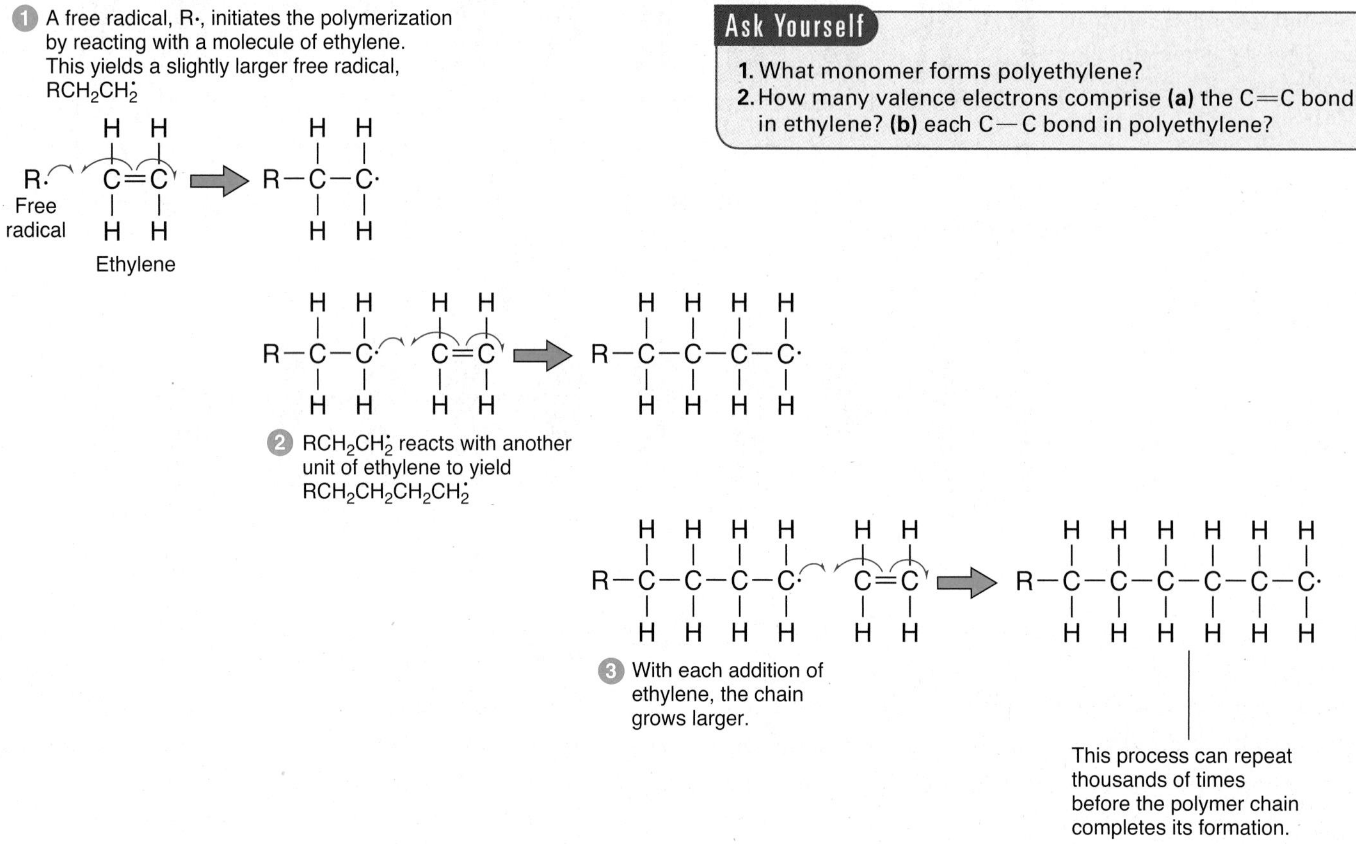

1 A free radical, R·, initiates the polymerization by reacting with a molecule of ethylene. This yields a slightly larger free radical, $RCH_2CH_2^{\cdot}$

Free radical

Ethylene

2 $RCH_2CH_2^{\cdot}$ reacts with another unit of ethylene to yield $RCH_2CH_2CH_2CH_2^{\cdot}$

3 With each addition of ethylene, the chain grows larger.

This process can repeat thousands of times before the polymer chain completes its formation.

Ask Yourself

1. What monomer forms polyethylene?
2. How many valence electrons comprise **(a)** the C=C bond in ethylene? **(b)** each C—C bond in polyethylene?

links to the neighbor on the other side. As a very large number of monomers link together to form a long chain, a polymer is created. **Figure 13.3** illustrates this process for the formation of polyethylene.

Polyethylene and a variety of closely related molecules are the most widely manufactured polymers. Each of these compounds is formed through addition polymerization of ethylene or one of its derivatives (**Table 13.1**). For example, polymerization of the ethylene derivative tetrafluoroethylene produces polytetrafluoroethylene (more commonly known as Teflon), which is used as an antistick coating on cookware. Since the polymer chains of polyethylene and its derivatives consist of a repeating structural unit, we can represent the structure of each of these polymers with the following general notation or shorthand: $-[A]_n-$, where A represents the repeating unit and the subscript n represents the number of times these units are linked together in a given polymer chain, a value that often exceeds several thousand.

Condensation polymers In condensation polymerization, the creation of each new link in the growing polymer chain is accompanied by the generation and release of a small molecule, such as water or alcohol. [The general term *condensation reaction* probably originated in early observations of water or similar liquids forming droplets of condensation on the sides of flasks during this type of reaction.]

The polypeptide chains of proteins, discussed in Chapter 5, are naturally occurring **condensation polymers**. As each amino acid forms a peptide link in the growing protein chain, a molecule of water is formed and released. Common synthetic condensation polymers include polyethylene terephthalate (pronounced teref–THAL–ate), or PET, a clear plastic widely used to manufacture bottles for water and soft drinks; and nylon, formed into high-strength fibers used in fabrics and ropes. PET is a *polyester* because each linkage formed between monomers in

condensation polymer A polymer in which the formation of each link is accompanied by the release of a small molecule.

Polyethylene and its derivatives Table 13.1

Polymerization of...	... produces...	... with the following uses.
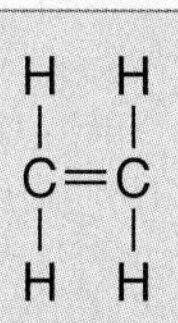 Ethylene	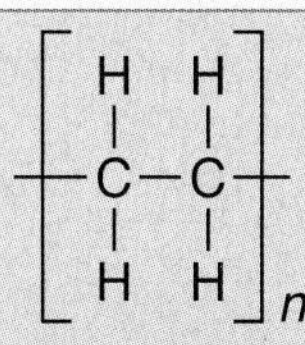 Polyethylene (PE)	Milk and juice containers, grocery and produce bags, detergent and motor oil jugs, and toys.
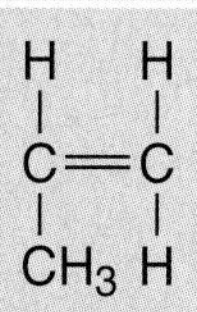 Propylene	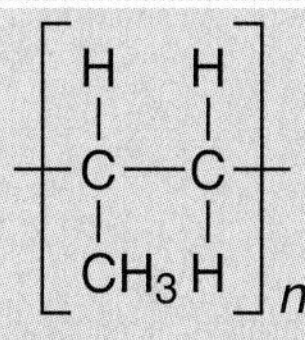 Polypropylene (PP)	Food-storage containers, plastic yogurt cups, bottle caps, drinking straws, rope, carpeting, upholstery, and plastic automotive parts.
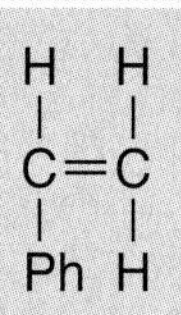 Styrene[a]	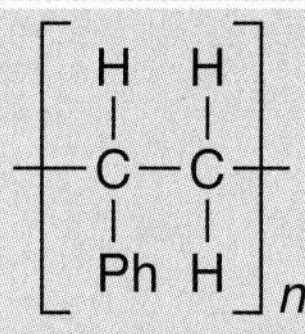 Polystyrene (PS)	Styrofoam cups, utensils, insulated coolers, and packing peanuts.
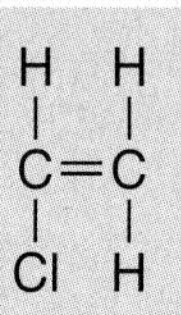 Vinyl chloride	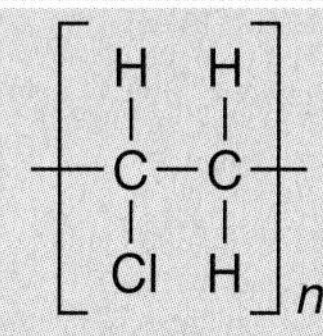 Polyvinylchloride (PVC)	Water and sewer pipes, electrical insulation, outdoor furniture, and vinyl siding and flooring.
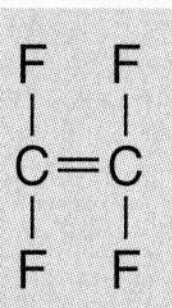 Tetrafluoroethylene	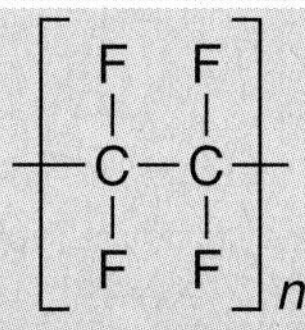 Polytetrafluoroethylene (PTFE, or Teflon)	Nonstick cookware coating, anticracking component of nail polish, stain-resistant coating on carpets and upholstery, and coating on windshield-wiper blades.
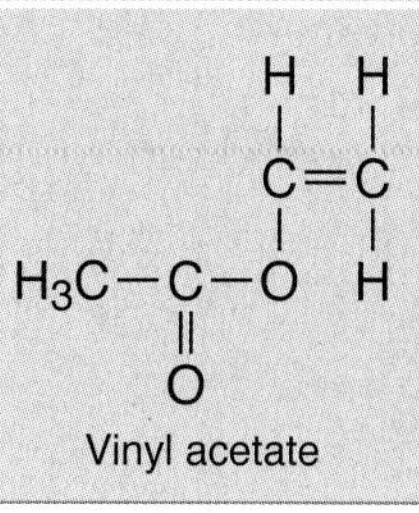 Vinyl acetate	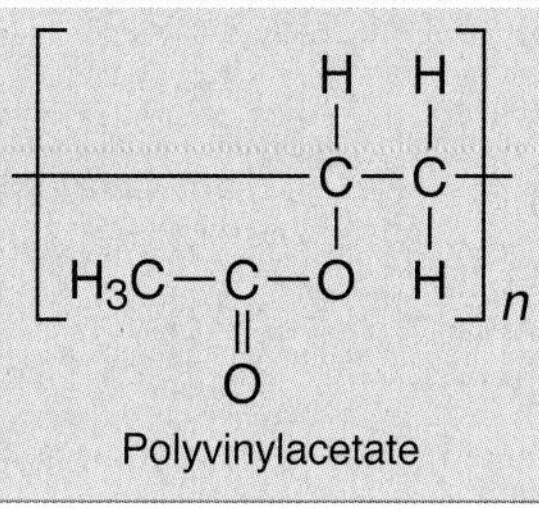 Polyvinylacetate	Component of white glue, carpenter's glue, and latex paint.

[a] "Ph" represents a *phenyl* group (C_6H_5) or benzene ring substitutent.

Ask Yourself

How is ethylene **(a)** structurally similar to, and **(b)** structurally different from, each of the other monomers in Table 13.1?

its polymer chains is an ester group. Nylon is a *polyamide*, with an amide (or peptide) group forming each linkage between the monomers of its chains (**Figure 13.4**).

Regardless of how polymers may form, whether through addition or condensation reactions, they can also be categorized in either of two ways, based on the repeating pattern of their monomer links. **Copolymers** consist of chains produced by two or more different kinds of monomers. For example, polyethylene terephthalate (PET) and nylon are copolymers because in each case, they are produced from two different monomers. Each exhibits the general pattern of A-B-A-B-A-B . . . along their chains, where A and B represent each monomer type. Copolymers such as these can be represented more simply by the general form $-[A\text{–}B]_n-$ where *n* indicates the number of times this unit within the bracket repeats in the polymer chain.

Chemistry InSight Condensation polymers • Figure 13.4

A small molecule, such as water, is released with the formation of each link in a condensation polymer chain. Synthetic condensation polymers include polyethylene terephthalate (a) and nylon (b).

a. Formation of polyethylene terephthalate (PET), a polyester.

Recall from Section 11.1 that an ester can be formed by the reaction of a carboxylic acid and an alcohol, with the release of water, as shown here.

$H_3C-C(=O)-OH + H-O-CH_2-CH_3 \rightarrow H_3C-C(=O)-O-CH_2-CH_3 + H_2O$

Acetic acid, a carboxylic acid

Ethanol, an alcohol

Ester link

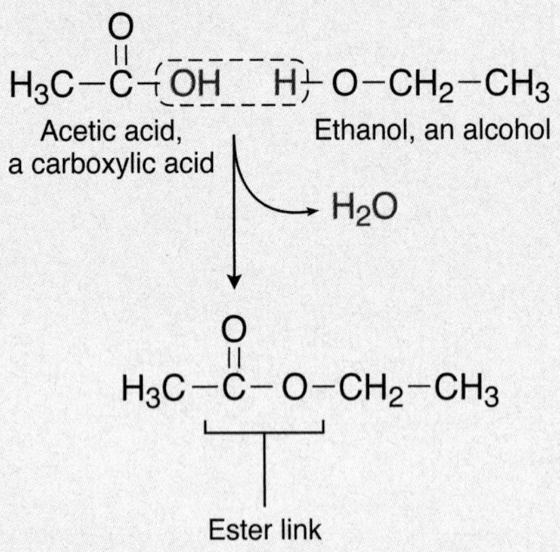

Similarly, a polyester can be formed from a dicarboxylic acid (a molecule containing two carboxylic acid groups), and a diol (a molecule containing two alcohol groups). For example, polyethylene terephthalate (PET) is a polyester formed from the dicarboxylic acid, *terephthalic acid*, and the diol, *ethylene glycol*, as shown here.

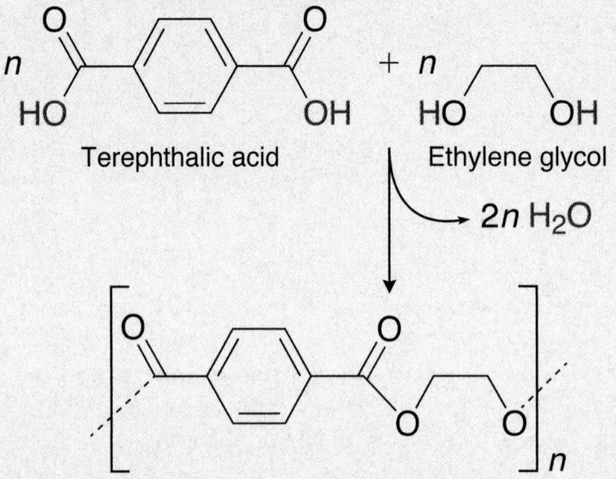

n HO OH + n HO OH → 2n H_2O

Terephthalic acid

Ethylene glycol

Polyethylene terephthalate (PET)

A PET water bottle.

filonmar/iStockphoto

b. Formation of nylon, a polyamide.

Recall from Section 5.4 that a peptide link (also known as an amide group), forms by the reaction of carboxylic acid and an amine, with the release of water, as shown here.

$H_2N-CH_2-C(=O)-OH + H-NH-CH(CH_3)-C(=O)-OH \rightarrow H_2N-CH_2-C(=O)-NH-CH(CH_3)-C(=O)-OH + H_2O$

Peptide (amide) link

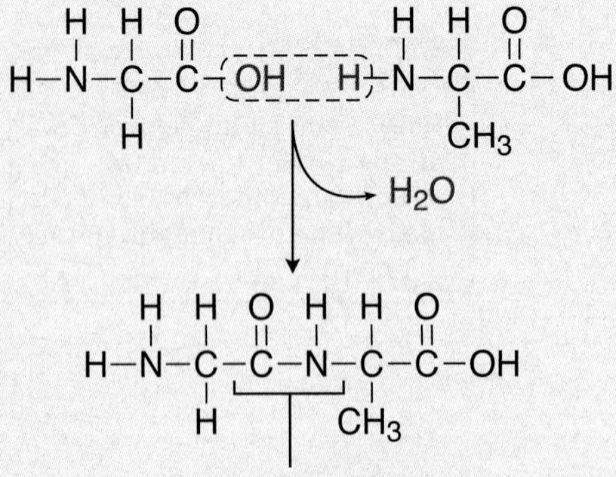

Similarly, a polyamide can be formed from a dicarboxylic acid and a diamine. For example, nylon is a polyamide formed from the dicarboxylic acid, *adipic acid*, and the diamine, *1,6-hexanediamine*, as shown here.

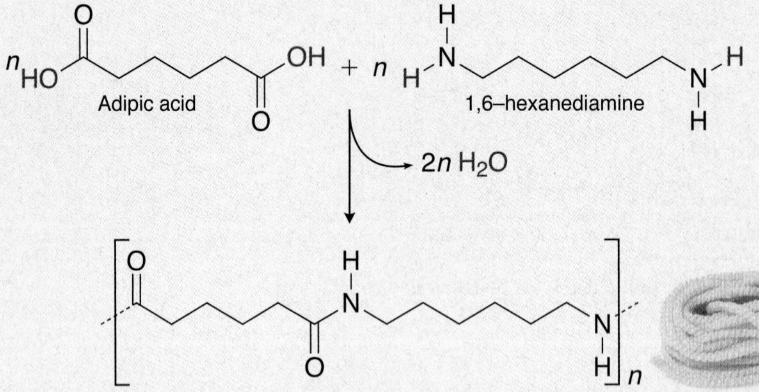

n HO OH + n H N H → 2n H_2O

Adipic acid

1,6–hexanediamine

Nylon

A nylon rope.

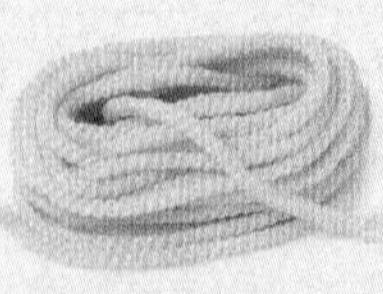

Evgeny Tomeev/ Shutterstock

Ask Yourself

1. Identify the ester link in the structure of polyethylene terephthalate shown in part a.
2. Identify the amide link in the structure of nylon shown in part b.
3. What do the dashed lines in each of the polymer structures indicate?
4. The nylon molecule shown in part b is also known as nylon 6,6 because each of its constituent monomers contains six carbon atoms. What is the structure of nylon 4,4?

WHAT A CHEMIST SEES

Low-density and high-density polyethylene

Differences in the physical properties of low-density polyethylene (LDPE) and high-density polyethylene (HDPE) derive from differences in their molecular structures. LDPE is more flexible (a); HDPE is more rigid (b). Artwork: Adapted with permission of John Wiley & Sons, Inc. from Snyder, C., *The Extraordinary Chemistry of Ordinary Things*, Fourth Edition, p. 543. Copyright 2003.

a. The disorganized and branched strands of LDPE produce a more flexible material. Uses include plastic food wrap and the plastic squeeze bottles shown here.

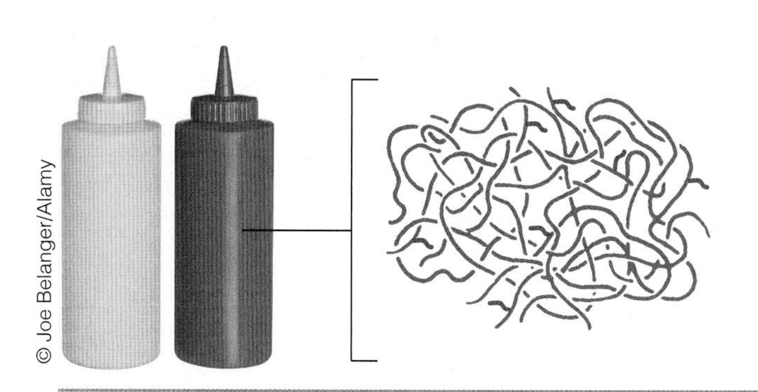

b. The linear, more organized strands of HDPE produce a stronger material. A wide variety of uses include beverage bottles and containers for consumer products, such as laundry detergent and motor oil.

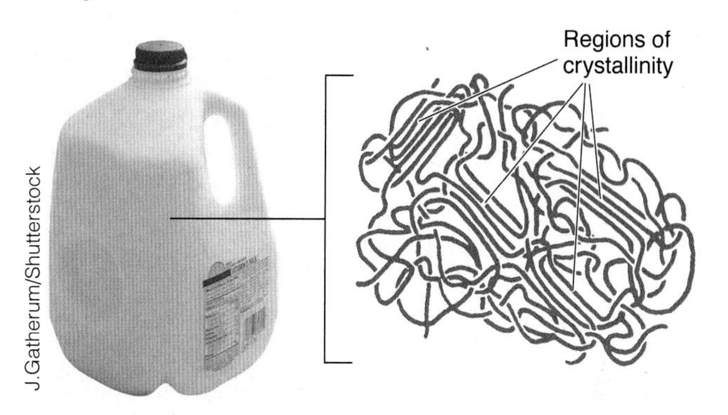

Homopolymers, by contrast, consist of chains produced from a single type of monomer. Polyethylene and its derivatives (Table 13.1) are homopolymers, because in each case, a single type of monomer forms the polymer chain. Homopolymer chains exhibit the pattern A-A-A-A-A . . . and can be represented more simply as $-[A]_n-$.

Notice that any particular polymer can occupy several different categories. For instance, both condensation and addition polymerization can produce homopolymers and copolymers.

Properties of Polymers

Polymers exhibit a wide range of properties. Some are soft and pliable, such as rubber. Others are so hard and durable—such as Kevlar—that they're used for bulletproofing. The physical properties of a given polymer depend on:

- the structure of the monomer(s) from which the polymer forms;
- the average length of the polymer chains;
- the architecture of the polymer chains—whether linear or branched, for example—and the presence of attractions or bonds between the chains;
- the presence of additives, such as plasticizers, that may affect the flexibility and other properties of the material.

In addition, the physical properties of a piece of plastic may depend on the shape or form into which the plastic is molded or extruded. For example, many polymers are first produced as pellets, and then melted and converted into a variety of shapes. These include the long, flexible fibers of textiles; the thin films used for coatings and bags; the tubes used for piping; the solid foams of insulation and packing materials; and the dense solids used for machinery parts.

We'll explore some of the physical properties of polymers with the example of polyethylene, the most widely produced plastic, accounting for over 36% of the 53 million tons of plastic made annually in the United States. Polyethylene exists in two principal forms: low-density polyethylene (LDPE) and high-density polyethylene (HDPE). The average density of LDPE is 0.92 g/cm^3, and that of HDPE is 0.95 g/cm^3. These differences are due to structural variations between the two forms.

The polymer chains of LDPE contain branches. At the molecular level, this produces a more disordered network of polymer strands, like a loose pile of yarn. By contrast, the polymer chains of HDPE are linear, without branches. As a result, molecular strands of HDPE tend to be more ordered as compared to the tangled chains of LDPE. This allows for pockets of aligned molecules, or crystallinities, to exist within the bulk of HDPE (see *What a Chemist Sees*). Because of this molecular ordering,

Rubber • Figure 13.5

The polymer *polyisoprene* is formed from the monomer *isoprene* (a). This polymer is concentrated in the sap of certain trees (b). This sap is harvested and processed into rubber.

a. The addition polymer, polyisoprene, is formed from isoprene. Each curved, red arrow indicates the flow of two electrons in forming each new bond of the resulting polymer. (Though this flow is pointing to the right, we could have equally shown it moving to the left.)

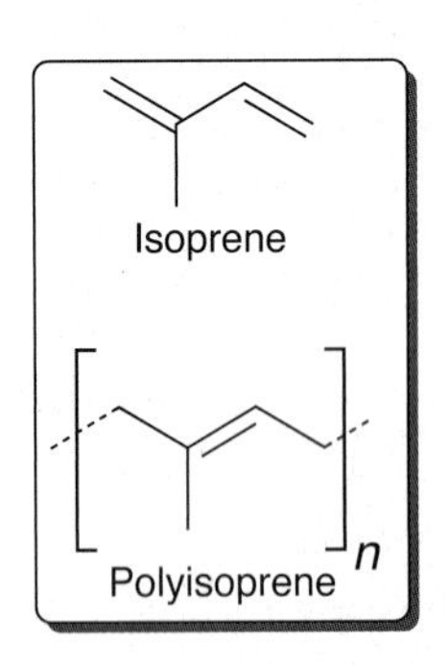

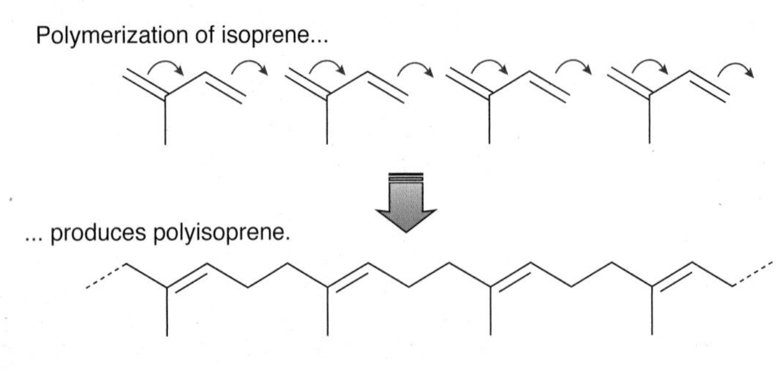

b. Here, the rubber tree (*Hevea brasiliensis*) is tapped for its latex, which is rich in *polyisoprene*. This collected sap is processed into rubber.

Ask Yourself

1. Why is polyisoprene classified as an addition polymer rather than a condensation polymer?
2. a. How many carbon atoms exist within one isoprene molecule? **b.** How many carbon atoms exist within a polyisoprene chain made from 1000 isoprene units?

HDPE is denser and more durable than LDPE and exhibits a higher melting point. LDPE, by contrast, is less durable, with a lower melting point. [Similar trends are observed in the melting points of saturated and unsaturated fats, as discussed in Chapter 5. Saturated fats melt at higher temperatures than unsaturated fats because in the solid state, saturated fat molecules can align with one another more readily, thus generating stronger intermolecular forces.]

Plastics can be categorized in either of two ways, based on their response to heat. Most plastics, including polyethylene and its derivatives, soften as they become warm. You may have noticed this effect if you've ever left a plastic object out in the sun or in a hot car. Plastics that soften in response to heat (and harden when cooled) are called **thermoplastics**. By contrast, **thermosetting plastics**, or **thermosets**, are soft and moldable when first prepared, but on heating they "cure" or harden permanently. If exposed to high enough temperatures, thermosets may decompose, but they won't soften again. Examples of thermosets include melamine, used in plastic kitchenware; the Bakelite of heat-tolerant pot handles; epoxy resins that are components of adhesives and coatings; and polyurethane foams, used in furniture and insulation.

To summarize, thermoplastics behave like fats in their response to heat. They soften and eventually melt when heated, and they become firmer when cooled. Thermosets, on the other hand, behave more like eggs; heating hardens them and produces irreversible changes.

Elastomers, which are elastic polymers, produce yet another kind of material, one that returns to its original shape after it's stretched or squeezed. These substances include synthetic polymers, such as polyurethanes, that are used in seat cushions, and natural polymers, such as rubber, the major component of tires. The term *rubber* derives from the accidental discovery of its ability to rub out pencil marks. Chemically, rubber is the polymer *polyisoprene*, found in the latex or sap of certain trees (**Figure 13.5**).

CONCEPT CHECK STOP

1. **Why** are all commercial plastics polymers, but not all polymers are plastics?
2. **How** does addition polymerization differ from condensation polymerization?
3. **Why** are thermosets often used in high-temperature applications?

13.2 The Development and Future of Polymers

LEARNING OBJECTIVES

1. **Describe** how various plastics were discovered.
2. **Summarize** efforts to improve the sustainability of plastics.

Having reviewed how polymers form and described some of their structures and properties, we'll now examine the development and the advanced applications of several types of polymers. Then we'll explore efforts to lessen the adverse environmental impacts of plastics by making them biodegradable.

Polymer Discovery

The first artificial polymers to be made were **semisynthetic**, produced by chemical modification of naturally occurring polymers, such as the cellulose of cotton. These earliest semisynthetic polymers include nitrocellulose (also known as guncotton) and celluloid, both of which originated in the mid-19th century.

In 1845, Swiss chemist Christian Schoenbein invented guncotton when he accidentally spilled a mixture of nitric acid and sulfuric acid in his kitchen and then wiped up the liquids with a cotton apron. After rinsing the apron in water and hanging it up to dry near a warm stove, it unexpectedly burst into flames, producing very little smoke. We now recognize that the guncotton (or nitrocellulose) Schoenbein discovered is produced by the reaction of cellulose with nitric acid, which converts some of the hydroxyl groups (–OH) of cellulose into nitrate groups ($–O–NO_2$). (The sulfuric acid catalyzes this reaction.) As a smokeless explosive, guncotton—also known as smokeless powder—proved to be far superior to the very smoky gunpowder used at the time.

The discovery of celluloid in the late 1860s originated in the need to find a suitable substitute for ivory, an expensive material derived from elephant tusks. Even at that time, ivory was overexploited due to its widespread use in carved ornaments and other objects. The elephants from which it was obtained were rampantly slaughtered in the process. In 1869, the American John Wesley Hyatt produced a plastic that he called celluloid. When melted and then allowed to harden, this new plastic formed a hard material resembling ivory. He generated this new material by combining camphor, a pungent, waxy compound obtained from the camphor tree, with a lightly nitrated form of nitrocellulose. Hyatt patented his idea and established a company to manufacture celluloid. His company (and later others) introduced a host of successful products made from celluloid, including dentures, photographic film, combs, brush handles, and ping pong balls. Movie film was also once made of celluloid, but this highly flammable material often ignited from the heat of the projector.

Rubber and its derivatives Rubber is highly valued for its elasticity and bounce. But if you if heat a piece of pure, natural rubber, you'll find that as it warms it becomes soft and soggy, and loses its resilience. This happens because at high temperatures the chains of the polyisoprene molecules of rubber lose their attractions for one another and slide past each other too readily (**Figure 13.6a**).

In 1839, the American inventor Charles Goodyear (whose last name is still associated with a well-known tire

Vulcanized rubber • Figure 13.6

Natural rubber (a) contains weak associations between polymer chains. Vulcanized rubber (b) contains sulfur cross-links between polymer chains.

Adapted with permission of John Wiley & Sons, Inc. from Snyder, C., *The Extraordinary Chemistry of Ordinary Things*, Fourth Edition, p. 543. Copyright 2003.

a. The intertwined polyisoprene chains of natural rubber can easily slip past one when the material is heated, so the material does not return to its original shape when stretched.

b. Vulcanized rubber retains its elasticity when heated because the sulfur cross-links help prevent the chains from slipping past one another when stretched.

Sulfur cross-links

Ask Yourself

Is vulcanized rubber a semisynthetic polymer? Explain.

Synthetic rubber • Figure 13.7

Styrene-butadiene copolymer (a) is the most common synthetic rubber. Another important synthetic rubber is Neoprene (b).

Ask Yourself

1. Why is Neoprene a homopolymer?
2. How does the structure of chloroprene differ from that of isoprene, shown in Figure 13.5?

a. The synthetic rubber, styrene-butadiene, is highly durable and widely used in tire treads and automotive belts and hoses.

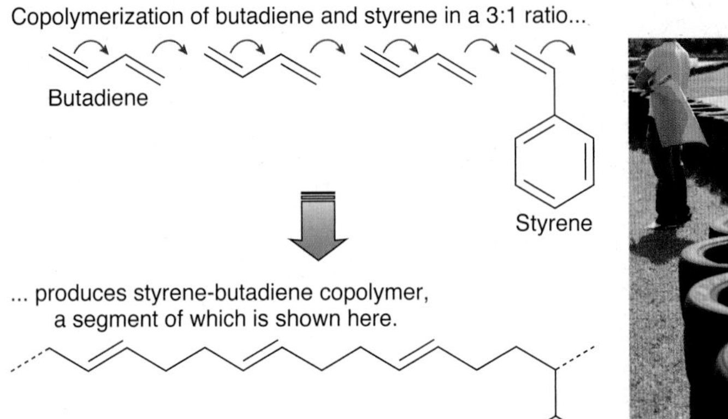

Copolymerization of butadiene and styrene in a 3:1 ratio...
Butadiene
Styrene
... produces styrene-butadiene copolymer, a segment of which is shown here.

CÃ©cileDÃ©gremont/Photononstop/ Getty Images

b. The synthetic rubber, Neoprene, resembles vulcanized rubber except that it is much more resistant to heat and exposure to hydrocarbon solvents such as gasoline, greases, and oils. It's often used in automotive and gasoline hoses. Neoprene also provides good thermal insulation and is used in wetsuits.

Polymerization of chloroprene...

Cl Cl Cl Cl

... produces polychloroprene, also known as Neoprene.

Cl Cl Cl Cl

A Neoprene wetsuit.

Fotosearch/Getty Images

company) accidentally discovered a form of rubber that retains its elasticity even when hot. Goodyear had inadvertently dropped a mixture of crude rubber and sulfur onto a hot stove. When the mixture cooled slightly, he found that it was nicely elastic, even though still warm. Though neither he nor his contemporaries could explain the basis for this result—the chemical structure of rubber itself was not even known at the time—we now know that when sulfur is heated with rubber, the sulfur forms **cross-links** between the polyisoprene chains of the rubber (**Figure 13.6b**). With this cross-linking, the polymer chains become loosely bound to each other in a three-dimensional lattice. The sulfur links keep the long molecules from slipping past their neighbors at high temperatures and thereby keep the rubber resilient. This process of strengthening rubber by heating it with sulfur is known as **vulcanization**, for the Roman god of fire, Vulcan. Though Goodyear proved unsuccessful in business and died poor and in debt, almost all natural rubber still produced today is vulcanized. This strengthened rubber is used in automobile and truck tires, rubber gaskets and seals, shoe soles, and other products.

Several synthetic elastomers now supplement or replace rubber in various consumer applications, including tire treads, engine hoses, and drive belts. The most common synthetic elastomer, styrene-butadiene rubber, results from the copolymerization of a mixture of 25% styrene and 75% butadiene, and serves as a good substitute for natural rubber in the manufacture of various consumer goods (**Figure 13.7a**). Another commercially important synthetic rubber is Neoprene, a homopolymer produced from chloroprene, a monomer resembling isoprene (**Figure 13.7b**). Neoprene was invented in 1931 by the American chemist Wallace Carothers, who also invented nylon a few years later.

Other synthetic polymers Although guncotton, celluloid, and vulcanized rubber were among the first commercially successful polymeric materials, they are all semisynthetic, made from chemical modification of naturally occurring polymers. Production of the first truly synthetic polymer was reported in 1909 by the Belgian-American chemist Leo Baekeland. While attempting to develop a substitute for shellac (a resinous substance produced by certain beetles), Baekeland discovered a synthetic, moldable substance that quickly solidified. This substance, which he called Bakelite (pronounced BAKE-a-lite), maintains its hardness even on heating. Baekeland soon founded a company to manufacture the material and became quite wealthy.

Bakelite is noted for its resistance to heat and electricity and is used in many products, including the black, heat-resistant handles on pots and pans. Though the structure of Bakelite is quite complex and beyond our discussion, it is a condensation polymer formed from the combination of phenol and formaldehyde, with loss of water (**Figure 13.8**).

Bakelite • Figure 13.8

Bakelite was the first synthetic polymer to be prepared. Though no longer widely used, it created an entire class of polymers called *phenolic resins*, used today as adhesives in the manufacture of plywood, fiberboard, and electronic printed circuit boards.

a. Some of the earlier uses of Bakelite included the plastic housings of radios and telephones, as well as in ornaments and jewelry. Noted for its heat resistance, Bakelite found widespread use in pot and pan handles.

Eric Tadsen/E+/Getty Images

b. Bakelite is prepared by mixing phenol and formaldehyde. This reaction produces a network of phenol rings connected by CH_2 links. (The loss of water accompanies the formation of each link.) A simplified structure of the resulting polymer is shown here.

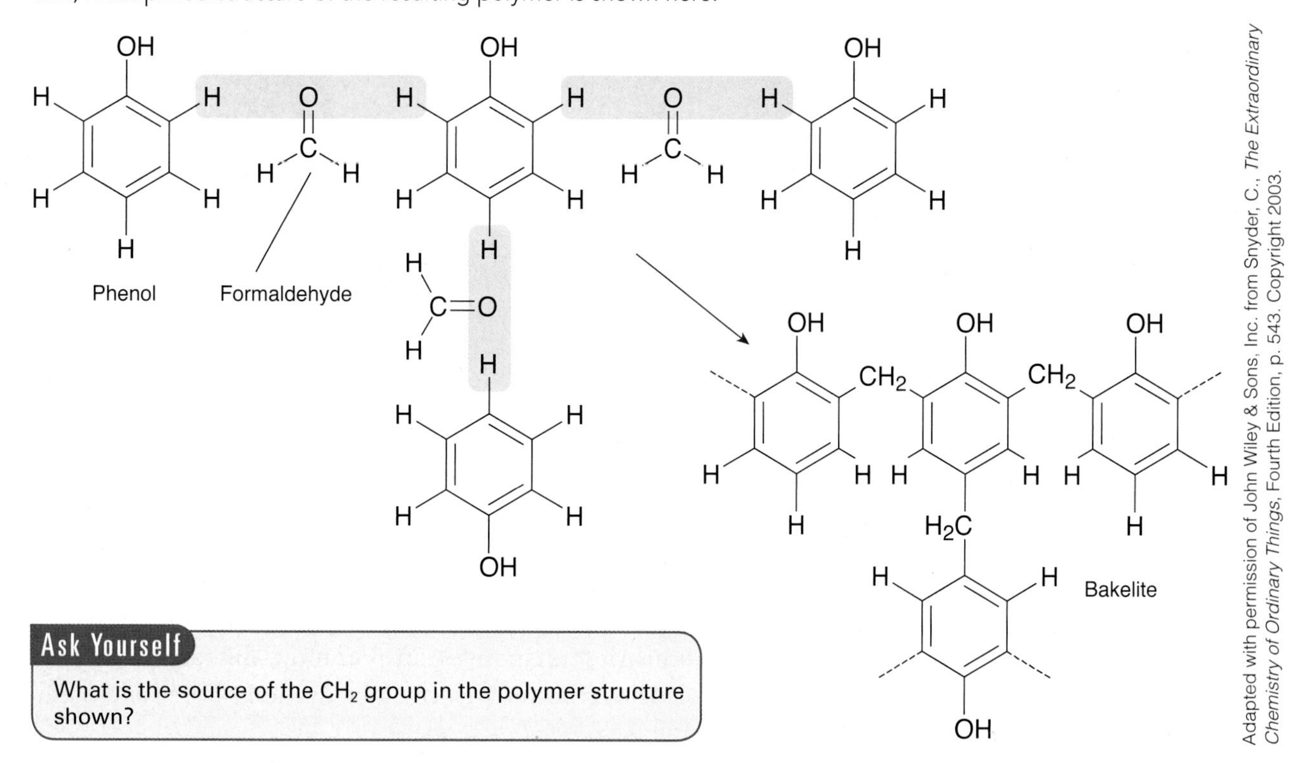

Adapted with permission of John Wiley & Sons, Inc. from Snyder, C., *The Extraordinary Chemistry of Ordinary Things*, Fourth Edition, p. 543. Copyright 2003.

Ask Yourself

What is the source of the CH_2 group in the polymer structure shown?

Key developments in the history of synthetic polymers • Figure 13.9

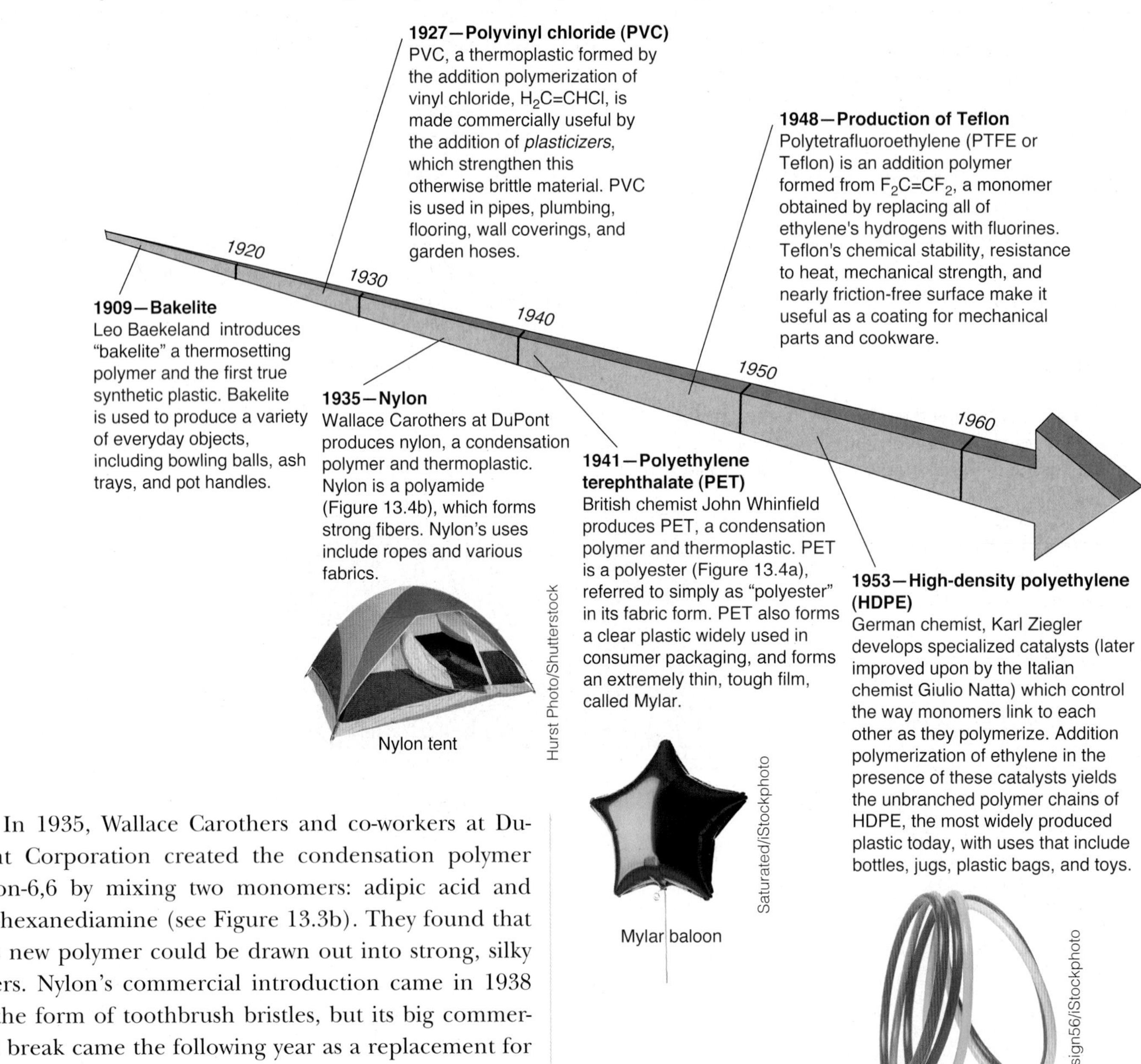

1920
1930
1940
1950
1960
1909—Bakelite
Leo Baekeland introduces "bakelite" a thermosetting polymer and the first true synthetic plastic. Bakelite is used to produce a variety of everyday objects, including bowling balls, ash trays, and pot handles.
1927—Polyvinyl chloride (PVC)
PVC, a thermoplastic formed by the addition polymerization of vinyl chloride, $H_2C{=}CHCl$, is made commercially useful by the addition of *plasticizers*, which strengthen this otherwise brittle material. PVC is used in pipes, plumbing, flooring, wall coverings, and garden hoses.
1935—Nylon
Wallace Carothers at DuPont produces nylon, a condensation polymer and thermoplastic. Nylon is a polyamide (Figure 13.4b), which forms strong fibers. Nylon's uses include ropes and various fabrics.
Nylon tent
Hurst Photo/Shutterstock
1941—Polyethylene terephthalate (PET)
British chemist John Whinfield produces PET, a condensation polymer and thermoplastic. PET is a polyester (Figure 13.4a), referred to simply as "polyester" in its fabric form. PET also forms a clear plastic widely used in consumer packaging, and forms an extremely thin, tough film, called Mylar.
Mylar baloon
Saturated/iStockphoto
1948—Production of Teflon
Polytetrafluoroethylene (PTFE or Teflon) is an addition polymer formed from $F_2C{=}CF_2$, a monomer obtained by replacing all of ethylene's hydrogens with fluorines. Teflon's chemical stability, resistance to heat, mechanical strength, and nearly friction-free surface make it useful as a coating for mechanical parts and cookware.
1953—High-density polyethylene (HDPE)
German chemist, Karl Ziegler develops specialized catalysts (later improved upon by the Italian chemist Giulio Natta) which control the way monomers link to each other as they polymerize. Addition polymerization of ethylene in the presence of these catalysts yields the unbranched polymer chains of HDPE, the most widely produced plastic today, with uses that include bottles, jugs, plastic bags, and toys.
HDPE hula hoops
design56/iStockphoto

In 1935, Wallace Carothers and co-workers at DuPont Corporation created the condensation polymer nylon-6,6 by mixing two monomers: adipic acid and 1,6-hexanediamine (see Figure 13.3b). They found that this new polymer could be drawn out into strong, silky fibers. Nylon's commercial introduction came in 1938 in the form of toothbrush bristles, but its big commercial break came the following year as a replacement for the silk of stockings and other clothing. Soon thereafter, during World War II, the U.S. government used most of the nation's limited supplies of nylon for making parachutes, ropes, and other military supplies. Not until the early 1950s was there sufficient production capacity to fill the popular demand for "nylons," as nylon stockings came to be known, and to provide enough of the plastic for other uses. Today, nylon is used in ropes, fishing lines, synthetic fabrics, and industrial parts. Ripstop nylon, a fabric resistant to tearing, is used in parachutes, outdoor clothing, and camping equipment, such as tents and sleeping bags.

The period from the late 1920s to the 1950s marked an especially prolific time in the discovery of commercially useful polymers. Many of the common plastics used today, including polyethylene, polyethylene terephthalate (PET), polystyrene, and nylon, were discovered during this time. **Figure 13.9** summarizes some of these developments. Today, advances in polymer science continue to lead to valuable new materials with novel properties and applications. Advanced polymers find uses in a variety of industries, including transportation, military and police security, consumer electronics, fashion, sports, and medicine (see *What a Chemist Sees*).

WHAT A CHEMIST SEES

Advanced Polymers

Polymers lend themselves to a variety of applications, due to their wide-ranging properties. The properties of a given polymer derive in part from its molecular structure.

The outer shell of this racing helmet is made from **Kevlar**, an extremely strong, yet light material also used in military and police gear, such as bulletproof vests and helmets.

The visor of this helmet and the front windshield are made of **polycarbonate** plastic, valued for its impact-resistance. This material is also used for eyeglass lenses as well as in the construction of bulletproof glass.

AP Photo/Nick Wass

Race car drivers and firefighters wear fire-resistant body suits made of **Nomex**.

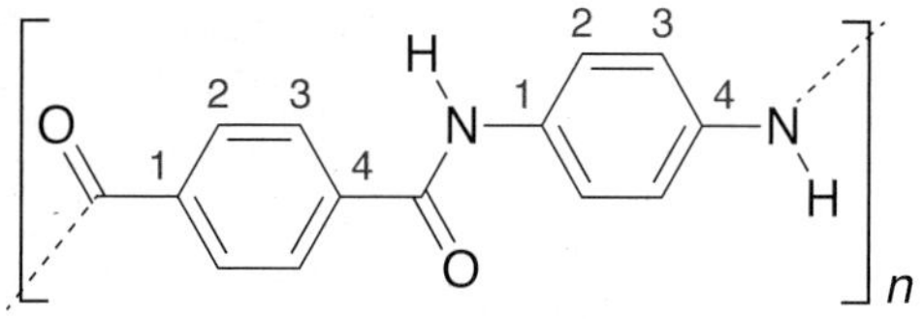

Kevlar

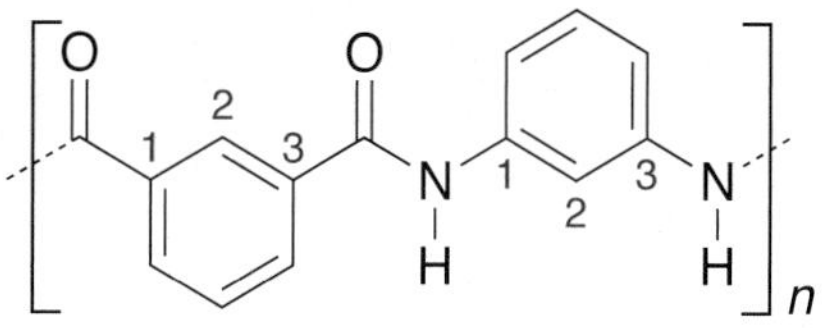

Nomex

Both Kevlar and Nomex are polyamides. They differ from one another in the positions at which groups attach to the aromatic rings. In Kevlar the attachment pattern is 1,4-; in Nomex the pattern is 1,3-. Kevlar owes its strength, in part, to the large number of hydrogen bonds that form between its chains, as shown here.

Hydrogen bonds between Kevlar chains

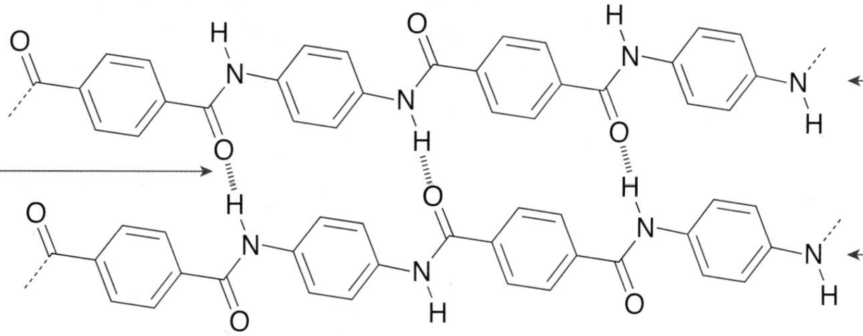

Kevlar chains

Put It Together Review Figure 13.4b and answer the questions.

1. Identify the amide linkages in Kevlar and Nomex.
2. What diamine monomer is used to produce **(a)** Kevlar; **(b)** Nomex?

Recyclable plastics and their uses • Figure 13.10

The plastic used in the creation of commercial goods is often identified with a resin code number to aid in recycling and incorporation into new products (a). Under certain circumstances, even mixed plastic wastes can be converted into useful products (b).

a. Resin codes and uses of recycled plastics

Resin Code	1	2	3	4	5	6
Name	Polyethylene terephthalate **PET**	High-density polyethylene **HDPE**	Polyvinyl chloride **PVC**	Low-density polyethylene **LDPE**	Polypropylene **PP**	Polystyrene **PS**
Can be recycled and made into...	Water and soft-drink bottles, polyester fabrics, and paint brushes.	Detergent and cleaner jugs, plastic lumber, and trash cans.	Pipes and hoses, floor mats and mud flaps.	Garbage bags, plastic lumber, and toys.	Fiber insulation for coats and sleeping bags, buckets, and tool handles.	Packaging materials and building insulation.

b. Chilean artist Rodrigo Alonso creates novel molded stools from mixed plastic waste.

Rodrigo Alonso Schramm (ralonso.com)

Plastics and Sustainability

We have discussed many of the valuable applications of plastics, but one advantage that is sometimes overlooked is that these materials can provide environmental benefits. For example, due to their lighter weights, plastics are often preferred as packaging materials over glass, paper or metal, which saves on fuel costs in the transportation of packaged goods. Also, because many plastics exhibit high mechanical strength in addition to their light weight or low density, plastics and plastic composite materials are increasingly used to replace metal parts in the construction of cars, trucks, and airplanes. These materials help reduce weight and improve fuel efficiency. The Boeing 787 Dreamliner aircraft, for example, incorporates carbon-fiber reinforced polymers for much of its primary construction. This material is both lighter and stronger than the aluminum typically used for the construction of an airliner's fuselage and wings.

Despite the various benefits of plastics, their widespread use also carries significant environmental costs. For instance, most plastics are produced from nonrenewable resources, such as petroleum or natural gas. As a result, manufacturing plastic from these resources is not sustainable. In addition, since most plastic waste does not decompose readily, large quantities of waste plastic remain within the environment. In 2011, the United States produced about 32 million tons of plastic municipal waste, only about 10% of which was recycled. Generally, about 80% of our plastic waste ends up in landfills, where it simply accumulates along with other trash. Incineration, an alternative approach used with about 10% of our plastic trash, is not only expensive but potentially hazardous as well. Some plastics, including polyvinyl chloride (PVC), produce irritating or toxic gases when they burn.

Globally, an additional concern results from the simple discharge into the environment of much plastic waste (along with other untreated garbage). By some estimates, as much as seven million tons of the world's plastic waste ends up in the oceans each year, causing various hazards to wildlife. Much of this floating debris becomes trapped within rotating ocean current systems called gyres (pronounced with a soft *g* as in *gyration*). This has produced an accumulation of marine debris known as the Great Eastern Pacific garbage patch, which occupies a region of the ocean's surface located between California and Hawaii. The size of this patch is difficult to determine, but it may be well over 100,000 square kilometers. Because much of the plastic accumulating in the world's oceans degenerates into small pieces, the waste plastic of oceanic garbage patches can be easily ingested by fish and birds.

Various approaches offer hope for reducing the environmental impacts of plastics. Here, the three R's of resource conservation—reduce, reuse, and recycle—apply to plastics as they do to other recyclable materials. To reduce plastic consumption, manufacturers can decrease the amount of plastic used for product packaging and can substitute recycled plastic for virgin plastic. Relatively small changes in consumer behavior carried out on a large scale can also help reduce plastic consumption. For instance, consumers can use refillable water bottles (in regions of the world where tap water is safe for drinking) in order to cut back on bottled water use. Also, employing reusable shopping bags can significantly decrease the use of disposable bags.

Recycling Making waste plastics into new products, much as paper, glass, and metal wastes are recovered and incorporated into new and useful items, has proven important in conserving resources. Here consumers can play an important role by participating in available recycling programs, both at home and at work, and by purchasing products made with recycled contents.

Recyclable plastics are assigned a "resin code number" to aid consumers in classifying them for recycling. Numbers 1 through 6 correspond to the most common recyclable plastics, and number 7 includes all other types of plastics. **Figure 13.10** summarizes these codes and identifies some of the products made from recycled plastics. Currently, polyethylene terephthalate (PET) and high-density polyethylene (HDPE) are the most frequently recycled plastics. According to the U.S. Environmental Protection Agency (EPA), the recovery rate for recycling these plastics was about 30% for each in 2011.

Increasingly, municipalities use **single-stream recycling**, in which all recyclable materials (paper, plastic, glass, and metal) are commingled for collection. Under this system, these four materials are first separated from one another at waste recovery facilities and processed separately. The plastics are first sorted by polymer type, then cleaned and shredded. The shredded flakes may undergo additional sorting by immersion in flotation tanks, which separate polymers according to their densities [see WileyPlus for a demonstration you can carry out involving this process]. Once dried, the flakes are melted and formed into granules that are sold to manufacturers and incorporated into new plastic products. One such product is recycled plastic lumber (RPL), which is used to make picnic tables, decking, and landscape structures. RPL offers advantages over wooden lumber in that it can be molded into various shapes and is more durable and resistant to rot.

In spite of the successes of recycling programs, challenges remain. Since rules for the types of plastics accepted for recycling can vary from one community to another, confusion remains among some consumers as to which plastics are eligible to be recycled. Another problem is that some plastic products don't readily lend themselves to collection for recycling because of their mixed composition. The plastic liners of baby diapers illustrate this problem.

Bioplastics, which derive from renewable resources, such as plant materials, and which are often biodegradable, offer potential solutions to reducing the environmental costs of plastic use. Bioplastics present an alternative to conventional plastics, whose raw materials typically derive from nonrenewable resources, such as petroleum or natural gas. One of the more common of the variety of bioplastics available is polylactic acid, or PLA, which is produced from the starch derived from corn or other plants (**Figure 13.11a**). PLA is a clear thermoplastic, used to make deli-food containers, cups, plates, utensils, and other objects.

The advantages of PLA are that it derives from a renewable resource, is biodegradable, and can replace polyethylene terephthalate (PET) in certain applications. However, in order for waste PLA plastic to decompose as intended, it must be sent to industrial-scale composting facilities. The environmental conditions provided by these facilities include high temperatures (in excess of 140 degrees Farenheit, generated by the action of microorganisms); high humidity; and adequate exposure to oxygen. Even under these conditions, it still may take several months for PLA to biodegrade.

One concern is that only very small amounts of PLA waste are directed to these composting facilities. Most PLA waste is discarded or combined with other recycled plastic waste. This presents additional concerns because PLA plastic can easily be mistaken for PET plastic in materials recovery facilities, where recyclable materials are collected and sorted for further processing. This can be a costly nuisance to recyclers because contaminating PET waste with even small amounts of PLA can render the PET unsuitable for recycling. In addition, PLA waste sent to landfills, as if it were conventional plastic waste, simply accumulates without decomposing effectively because of the unfavorable environmental conditions.

Bioplastics • Figure 13.11

a. Polylactic acid (PLA) can be produced in a multistep process starting from plant starch. PLA produced in the United States uses corn as a source of this starch, although other plants rich in starch or sugars, such as potato and sugarcane, can be used as well. PLA is biodegradable under certain controlled composting conditions.

Corn → Starch → Lactic acid → → Polylactic acid (PLA)

b. The polyethyethylene terephthalate (PET) that forms these bottles is made from the combination of plant-derived ethylene glycol and petroleum-derived terephthalic acid.

Ken Karp

Think Critically
1. Is PLA a polyester, a homopolymer, and/or a thermoplastic? Explain.
2. What potential benefits do **(a)** PLA and **(b)** PET bioplastics provide over fossil-fuel derived PET?

Another type of bioplastic is PET, generated from renewable resources rather than from fossil fuels. Recall from Figure 13.4 that PET is made from the monomers terephthalic acid and ethylene glycol. Both of these compounds are traditionally derived from petroleum, but methods have recently been developed to produce ethylene glycol from plant starch. This plant-derived ethylene glycol is currently used to produce PET used in food packaging (**Figure 13.11b**). Research is currently underway to devise a suitable method for producing terephthalic acid from plant starch as well. Successful research will allow a fully plant-based PET to be produced. Chemically, PET derived from renewable resources is identical with PET produced from petroleum, so it can be recycled in the same way. However, it is no more biodegradable than petroleum-derived PET.

Some opponents of bioplastics have expressed concern that the use of these materials in food packaging may amount to **greenwashing**—making products appear more environmentally beneficial than they truly are. Their argument is that cultivating crops and processing them into the monomers of bioplastics requires a significant amount of energy, which is often supplied by fossil fuels. Opponents also claim that the benefits of certain products, such as bottled water, do not justify their environmental costs, no matter how "green" the packaging. Proponents of bioplastics argue that the use of these materials represents a positive step since, on balance, fewer nonrenewable resources are used to make bioplastics than conventional plastics.

CONCEPT CHECK STOP

1. **What** is vulcanized rubber, and how does it differ from natural (latex) rubber?
2. **What** are resin identification codes, and how are they used?

13.3 Pollution and Wastes

LEARNING OBJECTIVES

1. **Provide** a working definition of pollution.
2. **Describe** types of pollution and methods used to control the release of pollutants.
3. **Differentiate** between municipal and hazardous wastes.

We've seen how plastics benefit our lives as well as contribute to pollution. Disposal of plastics in the environment is a major concern, but it's only part of the broader picture of waste disposal and environmental pollution. We now examine this broader issue, especially the pollution produced by waste that is generated from transportation and industrial sources. We'll also examine methods of reducing pollution and managing hazardous waste.

Defining Pollution

When we discuss releasing waste chemicals and similar materials into the environment, the word "pollution" naturally comes to mind. But "pollution" can have various meanings and can be hard to define precisely.

From a legal standpoint, pollution is often defined in terms of the limits of specific contaminants permitted by law to exist in a given environment. For instance, the Safe Drinking Water Act (discussed in Chapter 7) specifies pollution of drinking water in quantitative terms. If municipal or tap water contains contaminants at levels in excess of those specified by this Act, the water is considered polluted.

On a personal level, what constitutes pollution can be more subjective. For instance, municipal drinking water is often fluoridated at a level of 1 part per million (ppm) to provide protection against tooth decay. From a public health standpoint, evidence indicates that the benefits of this treatment outweigh its risks, but for those who have opposing views, any addition of fluorides to drinking water may be perceived as an unacceptable contamination of the water. As another example, the World Health Organization recommends that sodium levels in drinking water not exceed 200 ppm. According to this guideline, water containing sodium levels far below this 200 ppm level could be said to be free of any sodium pollution, even though you may still consider this water too salty for your own tolerance or preference. (You can explore your own sensitivity to the taste of salt in water with the blind taste-test described in Section 7.4.)

We often associate pollution with human activities, but pollution can be caused by nature as well. For instance, volcanic eruptions can emit massive amounts of pollutants into the air. The 1991 eruption of Mount Pinatubo in the Philippines released a host of contaminants into the atmosphere, including an estimated 20 million tons of sulfur dioxide within a single day. Even lightning produces pollutants, in the form of nitrogen oxides (NO_x) and ozone (O_3), an irritating, toxic gas.

What constitutes pollution depends not only on the contaminant in question and its concentration, but also on the location of the substance. For instance, in the troposphere—the lowest region of the atmosphere and the one that contains the air we breathe—ozone is a pollutant due to the various respiratory problems it can cause. However, in the stratosphere, many kilometers higher, ozone is a life-saving gas that shields us from the sun's harmful ultraviolet radiation.

Clearly, we need a suitable working definition of pollution. For our discussion, we can say that **pollution** occurs when an *excess* of a substance generated either by human activity or by nature is present in the wrong environmental location. It's not precise, but we'll find that it works well. For example, when we consider that our land, water, and air are unpolluted, we do not imply that these natural resources are completely free of contaminants, but rather that the degree of contamination is within acceptable levels (**Figure 13.12**).

Environmental purity • Figure 13.12

Even pristine air, rain, and bodies of fresh water are not pure in the chemical sense. Each is naturally contaminated by other substances.

Air can contain ozone, produced from lightning; and volatile organic compounds, generated from plants and animals.

Rain absorbs carbon dioxide naturally present in the air, to form carbonic acid.

Fresh water from wells, springs, and ponds contains naturally dissolved minerals drawn from Earth's crust, as well as microorganisms.

John Burcham/National Geographic/Getty Images

Ask Yourself

Denali National Park in Alaska, seen in this image, is remote from any industrial sources of pollution. Is this environment free of contaminants?

Types of Pollution

Humans have been releasing chemicals into the environment, at least on an industrial scale, for about two hundred years. Compared to the age of Earth, which is about 4.6 billion years according to radioisotopic dating techniques, this **anthropogenic** (or human-generated) pollution is an exceedingly recent occurrence. For instance, if Earth's history could be compressed into a 24-hour day, industrial pollution would have existed only within the last three milliseconds of the day. As recent as this is (on a geological time scale), industrial pollution has generated significant impacts on our environment. In the following sections, we'll explore the chemistry of some of the major pollutants of the air, water, and land, and discuss various methods of reducing pollution.

Air pollution Each year the population of the world consumes roughly 8×10^{16} kcal of energy in all its forms. That's enough energy to heat 8×10^{14} kg [about 8×10^{11} (800,000,000,000) tons] of water from its freezing point to its boiling point. A major portion of all this energy comes from the combustion of fossil fuels (Chapter 4). Burning the hydrocarbons of these fuels generates the greenhouse gas carbon dioxide, which contributes to climate change, as well as adding much water vapor to the atmosphere. But fossil fuels rarely consist of hydrocarbons alone. Petroleum and coal usually contain impurities, including small amounts of sulfur compounds. Most of the sulfur within petroleum is incorporated into organosulfur compounds (organic compounds containing sulfur). Most of the sulfur of coal is combined with iron in the mineral pyrite, FeS_2.

Although the processing of raw petroleum, coal, and natural gas removes many of their impurities, eliminating every trace of all impurities becomes prohibitively expensive. When we burn coal, gasoline, and similar fuels, we oxidize not only their hydrocarbons but their impurities as well. The sulfur of the pyrite within coal, for example, oxidizes to *sulfur dioxide*, SO_2, an irritating gas with an acrid odor. The reaction for this oxidation is

$$\underset{\text{Pyrite, a sulfur-containig inpurity in coal}}{4\,FeS_2} + \underset{\text{Oxygen}}{11\,O_2} \longrightarrow \underset{\text{Iron oxide}}{2\,Fe_2O_3} + \underset{\text{Sulfur dioxide}}{8\,SO_2}$$

Sulfur dioxide that forms through the burning of coal and petroleum products enters the air as a **primary air pollutant**. Once sulfur dioxide enters the atmosphere and remains in contact with the oxygen of the air, it can oxidize further to *sulfur trioxide*, SO_3, a highly acrid gas and a **secondary air pollutant**.

primary air pollutant An air pollutant generated as the direct result of a specific activity.

secondary air pollutant A pollutant formed by the further reaction of a primary air pollutant.

Primary and secondary air pollutants • Figure 13.13

Primary air pollutants are emitted directly from either human or natural sources. Secondary air pollutants result from chemical reactions involving primary air pollutants.

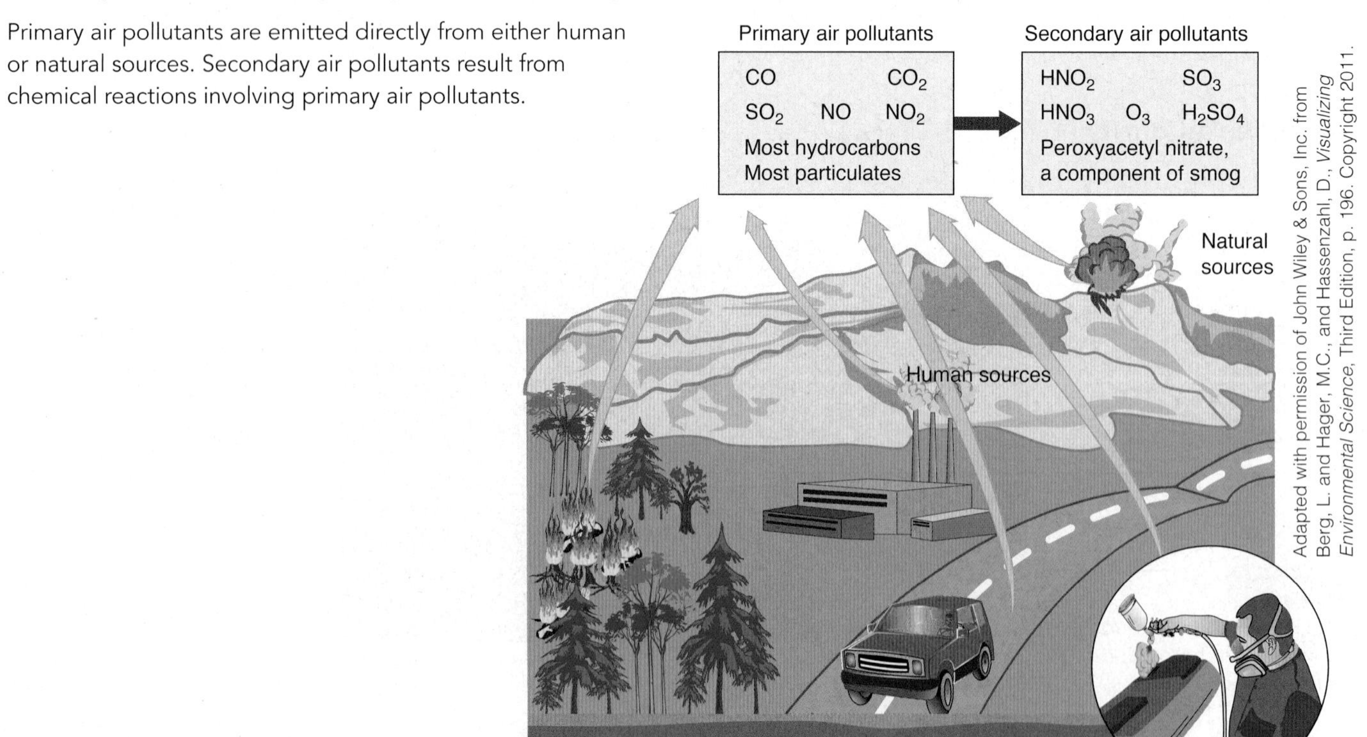

This sulfur trioxide can combine with moisture (H_2O) in the air to form sulfuric acid, H_2SO_4, a major component of acid rain (Section 8.4) and another secondary air pollutant. **Figure 13.13** shows the generation of primary and secondary air pollutants.

Coal, and to a lesser extent petroleum, also contain trace amounts of heavy metals, such as mercury, which are released to the atmosphere during combustion. Coal-burning power plants are the largest anthropogenic source of mercury emissions in the United States. These emissions eventually settle onto land and water, where the mercury is converted by bacteria into *methylmercury*, CH_3Hg^+, an ion formed from the bonding of mercury to a *methyl* (CH_3) group. Once formed, methylmercury may enter and move up the food chain (**Figure 13.14**). The **bioaccumulation** of this substance in certain species of fish, including tuna, has led to recommendations that young children

bioaccumulation A process whereby pollutants accumulate in the tissues of animals and plants.

Mercury bioaccumulation • Figure 13.14

Emissions of mercury from coal-fired power plants, waste incinerators, and other sources deposit onto land and water. Bacteria convert mercury into methylmercury, which accumulates in the aquatic food chain.

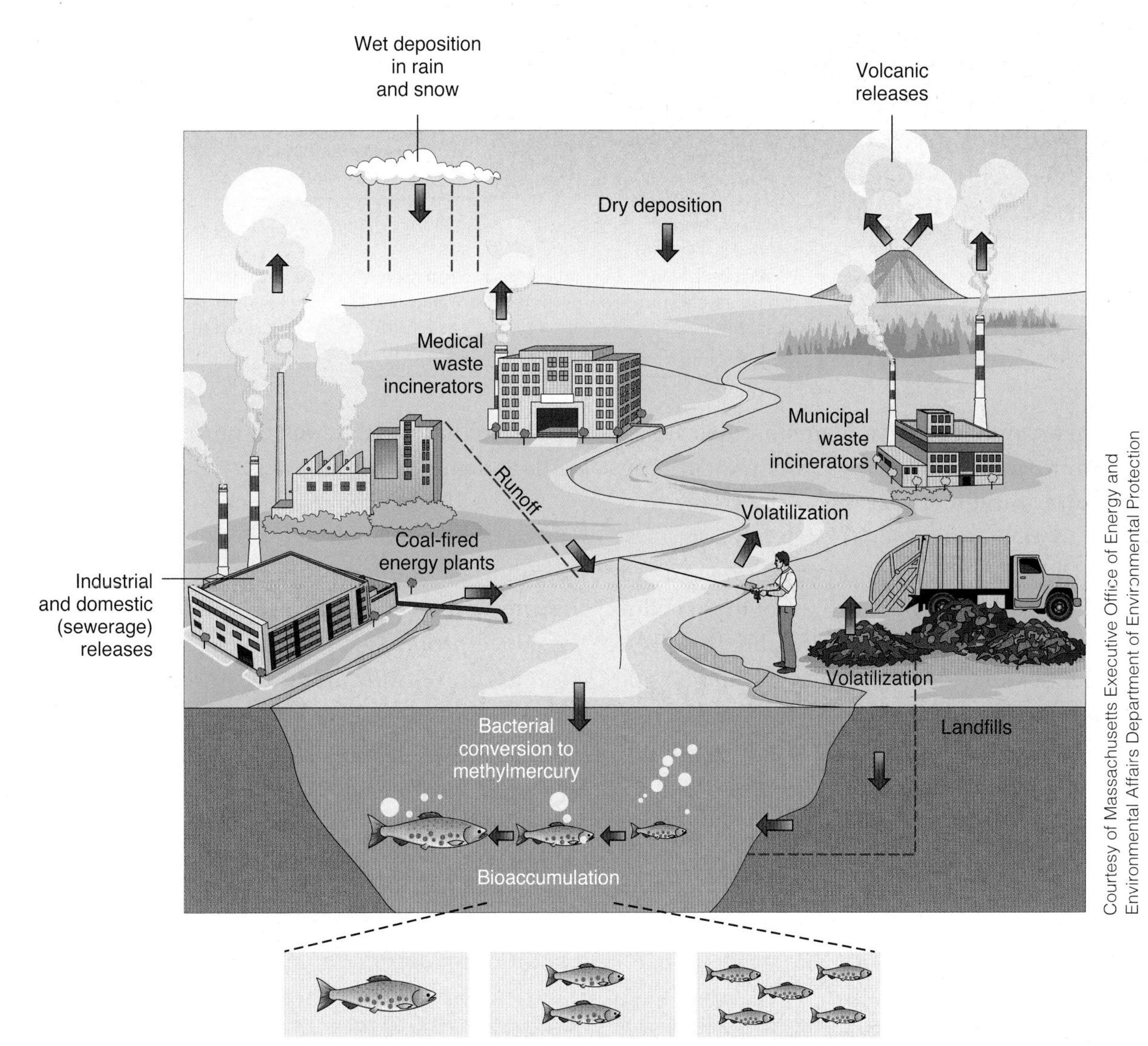

PROCESS DIAGRAM

Photochemical smog • Figure 13.15

Photochemical smog is a noxious mixture of ozone, nitrogen dioxide, hydrocarbons, and other pollutants, sometimes seen as a brownish haze over cities (a). This type of pollution is initiated by the decomposition of ground-level nitrogen dioxide through the action of sunlight (b).

a. Photochemical smog over Los Angeles.

Damir Frkovic/Radius Images/Getty Images

b. Nitrogen dioxide can absorb ultraviolet radiation to release an oxygen atom. The resulting oxygen atom can react with an oxygen molecule to form ozone.

Think Critically

1. Why does photochemical smog tend to be worse in cities with highly congested traffic?

2. Write a balanced equation for the reaction of ozone with nitric oxide to form an oxygen molecule and nitrogen dioxide. (Normally, this reaction reduces ground-level ozone. However, when other air pollutants, such as hydrocarbons, are present, this reaction is inhibited, leading to an increase of ozone concentration.)

as well as pregnant and nursing women limit their intake of foods known to contain higher levels of mercury. Significant exposure to mercury has been shown to cause birth defects and impair neurological development.

Nitrogen oxides, NO_x, form another group of air pollutants. Unlike sulfur oxides, the nitrogen of these pollutants comes principally from the air itself rather than from impurities in fuels. Molecules of the nitrogen and oxygen of the air combine at very high temperatures (such as those produced by lightning and those found within the combustion chambers of internal combustion engines) to form *nitric oxide*, NO, a colorless gas:

$$\underset{\text{Nitrogen}}{N_2} + \underset{\text{Oxygen}}{O_2} \longrightarrow \underset{\text{Nitric oxide}}{2\,NO}$$

Once in the atmosphere, nitric oxide reacts with additional oxygen to form *nitrogen dioxide*, NO_2, a red-brown, toxic gas that causes irritation to the eyes and the respiratory system. Further reaction of nitrogen dioxide with atmospheric oxygen and water produces still other oxides of nitrogen, as well as nitric acid, HNO_3, another component of acid rain.

In addition, the energetic ultraviolet radiation of sunlight can dislodge one of nitrogen dioxide's oxygen atoms. The resulting, liberated oxygen atom is highly reactive and can combine with diatomic oxygen, O_2, to produce ozone, O_3. The newly generated oxygen atom can also react with unburned hydrocarbons of auto exhausts to produce a variety of new pollutants. The complex combination of all of these pollutants—nitrogen dioxide, ozone, unburned hydrocarbons from engine exhaust, and other chemicals resulting from the initial reaction of nitrogen dioxide with sunlight—is known as **photochemical smog**. We see it as the brown haze that sometimes forms over cities (**Figure 13.15**). Exposure to this smog can cause irritation of the eyes and difficulties breathing.

photochemical smog A complex combination of pollutants resulting from the interaction of sunlight with nitrogen dioxide and subsequent reactions involving atmospheric oxygen and hydrocarbon pollutants.

Still another pollutant, one that we can't see and that produces no sense of irritation, is *carbon monoxide*, CO. This is known as a silent killer because it is odorless, tasteless, and invisible, but can enter the lungs and be absorbed into the bloodstream even more effectively than oxygen, leading to death by carbon monoxide poisoning. A product of the incomplete combustion of carbon or organic compounds, such as the hydrocarbons of gasoline, CO is primarily a pollutant of cities and usually fluctuates with the flow of traffic.

These pollutants aren't the only ones affecting the quality of the air we breathe. Liquid organic compounds often become air pollutants simply by evaporating. Those that evaporate easily are called **volatile organic compounds (VOCs)**. (A volatile substance is one that evaporates readily.) VOCs can be found in a variety of everyday products, including gasoline, paints, paint thinners, varnishes, roofing tar, degreasers, and epoxy glues. These volatile compounds are also used in cosmetic products, such as nail polishes and nail polish removers, hairsprays, perfumes, colognes, and after-shave lotions. VOCs give rise to the odors of the interiors of new cars (new car smell) and to the lingering scent of freshly installed carpeting.

volatile organic compounds (VOCs) Organic (carbon-based) compounds with relatively low boiling points (< 250°C) and a propensity to evaporate.

Exposure to VOCs can pose health concerns, especially in closed locations with limited ventilation. Recent environmental regulations have led to restrictions on the content of VOCs in various products, including paints, in order to reduce hazards from inhalation (**Figure 13.16**).

CFCs and the ozone hole In Section 6.2 we briefly discussed a class of volatile organic compounds called **chlorofluorocarbons (CFCs)**. These substances are now banned, but were formerly used as aerosol propellants in spray cans; as refrigerants in air conditioners and refrigerators; as foam-producing agents in the manufacture of certain plastics and insulation; and as solvents for cleaning circuit boards and other electronic parts. (In the peak year for production of CFCs, 1986, over a billion kilograms (about 1.25 million tons) of these chemicals were produced.) Ironically, the very desirable properties of CFCs—low chemical reactivity, low flammability, low toxicity, and low cost—proved to be a major environmental liability. When released into the atmosphere, these pollutants can drift upward into the stratosphere and remain there for decades, causing harm to the **ozone layer**, a band of gas surrounding Earth some 20 to 30 km (about 15 miles) above Earth's surface. This stratospheric ozone protects us from the sun's ultraviolet (UV) radiation, which is invisible to the naked eye but contains far more energy than the accompanying infrared radiation (heat) and light of the sun. Short-term exposure to UV radiation can produce sunburns, while long-term exposure can cause cataracts, premature aging of the skin, and skin cancer (see Section 11.2).

Low VOC paints • Figure 13.16

All paints contain solvents designed to evaporate as the paint dries and binds to surfaces. Paints low in volatile organic compounds (VOCs) are typically water-based and contain nonvolatile, polymeric components, so as to reduce levels of unwanted emissions.

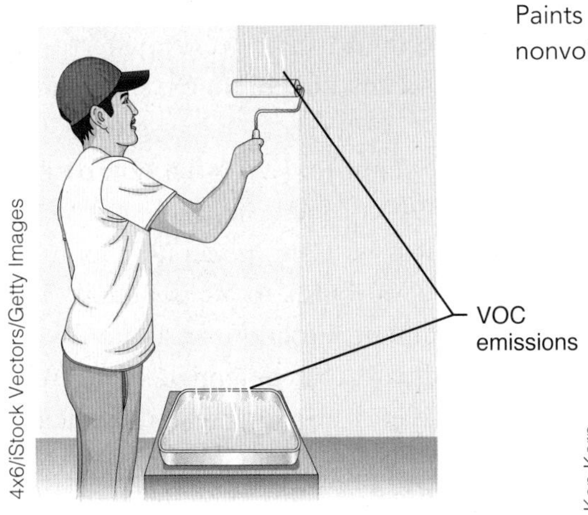

4x6/iStock Vectors/Getty Images

VOC emissions

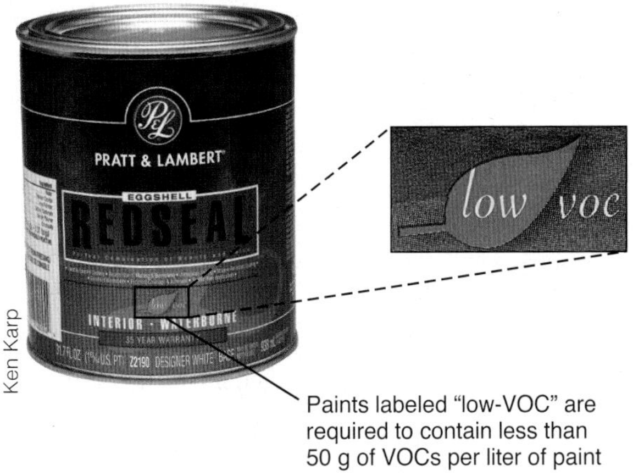

Ken Karp

Paints labeled "low-VOC" are required to contain less than 50 g of VOCs per liter of paint

Protective effects of stratospheric ozone • Figure 13.17

The ozone layer protects Earth's surface from incoming ultraviolet radiation (a). Ozone molecules dissociate during this process, but reform rapidly (b), establishing stable concentrations of ozone in the stratosphere.

a. Stratospheric ozone normally absorbs about 99% of the incoming ultraviolet radiation of the sun.

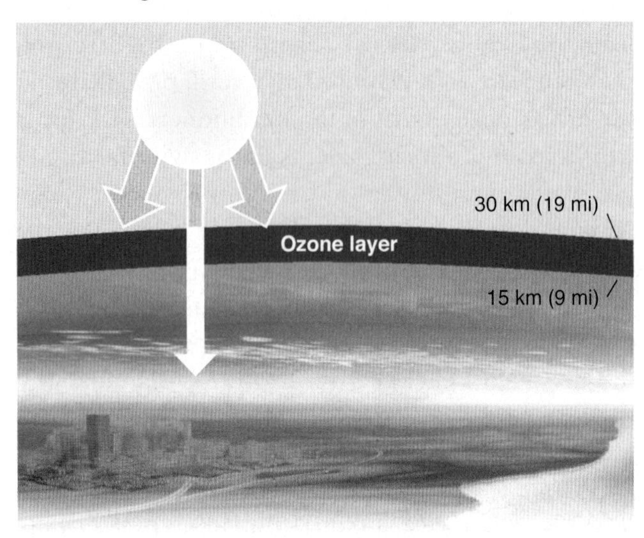

Adapted with permission of John Wiley & Sons, Inc. from Berg, L. and Hager, M.C., and Hassenzahl, D., *Visualizing Environmental Science*, Third Edition, p. 231. Copyright 2011.

b. Ozone–oxygen cycling in the ozone layer.

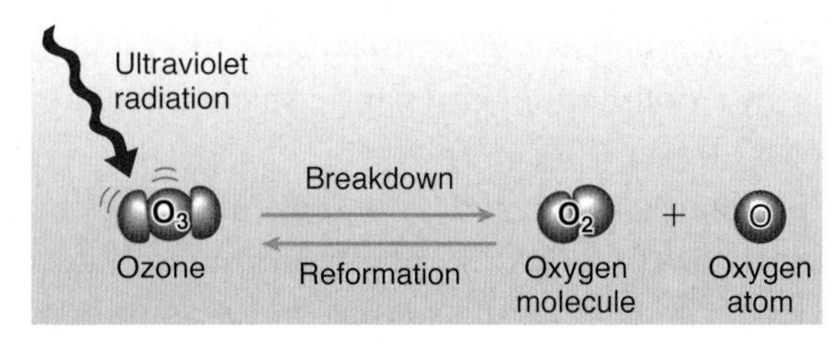

Adapted with permission of John Wiley & Sons, Inc. from Snyder, C., *The Extraordinary Chemistry of Ordinary Things*, Fourth Edition, p. 347. Copyright 2003.

When an ozone molecule, O_3, within the ozone layer absorbs UV radiation, it dissociates into an oxygen molecule, O_2, and an oxygen atom, O. This oxygen atom is highly reactive and combines with a nearby oxygen molecule to reform ozone (**Figure 13.17**). Left undisturbed over millennia, this cycling between the destruction and formation of ozone molecules has produced a reasonably stable concentration of ozone that has allowed surface life to evolve and thrive under its invisible shield.

In the upper levels of the atmosphere, CFCs (and related volatile organic compounds containing chlorine and/or bromine atoms in their structures) absorb the sun's ultraviolet radiation and decompose, releasing chlorine and bromine atoms in the process. These chlorine and bromine atoms act as catalysts in the conversion of ozone molecules, O_3, into oxygen molecules, O_2, thus decreasing stratospheric ozone levels (**Figure 13.18a**). This ozone-destructive potential of atmospheric CFCs was recognized as early as 1974 by the chemist F. Sherwood Rowland at the University of California, Irvine, and his co-worker, Mario J. Molina. This discovery earned Rowland and Molina a Nobel Prize in Chemistry in 1995, which they shared with Paul Crutzen of Germany, who investigated other catalytic, ozone-depleting processes.

The strongest evidence for the erosion of the ozone layer lies in the discovery of a sharp drop every autumn in the density of the part of the ozone layer that lies over the Antarctic (**Figure 13.18b**). In 2006, this annual drop in ozone density, known as the **ozone hole**, reached a record size, covering an area almost equal to all of North America. Although ozone depletion over the Arctic occurs to a much smaller extent than over Antarctica, the loss of stratospheric ozone over the Arctic briefly reached a record level of its own in 2011.

In 1987, more than 140 nations signed a treaty known as the **Montreal Protocol** (and its later revisions), to end or severely restrict the production of CFCs and related compounds in order to protect the ozone layer. The ozone hole appears to be stabilizing and on a slow path to recovery, which may be fully complete in another 50 years.

The combustion of fossil fuels to generate energy remains by far the major contributor to air pollution. The United Nations Environment Program and the World Health Organization estimate that potentially dangerous levels of air pollutants of every category threaten the health of over 1 billion people throughout the world. Of these, well over half face hazardous levels of sulfur dioxide. **Table 13.2** lists major sources of air pollutants in the United States.

CFCs and the ozone hole • Figure 13.18

Chlorofluorocarbons (CFCs) and related compounds can destroy ozone through a series of reactions (a). Thinning of stratospheric ozone occurs most intensely over the Antarctic (b).

a. CFCs can absorb UV radiation and release free chlorine atoms, which are highly reactive. These chlorine atoms act as catalysts in the conversion of ozone to oxygen molecules.

b. Visualizations of the annual ozone hole over the Antarctic, which peaks each year in September. Though this hole is much larger now than in the 1970s, it appears to be stabilizing due to restrictions on CFC use.

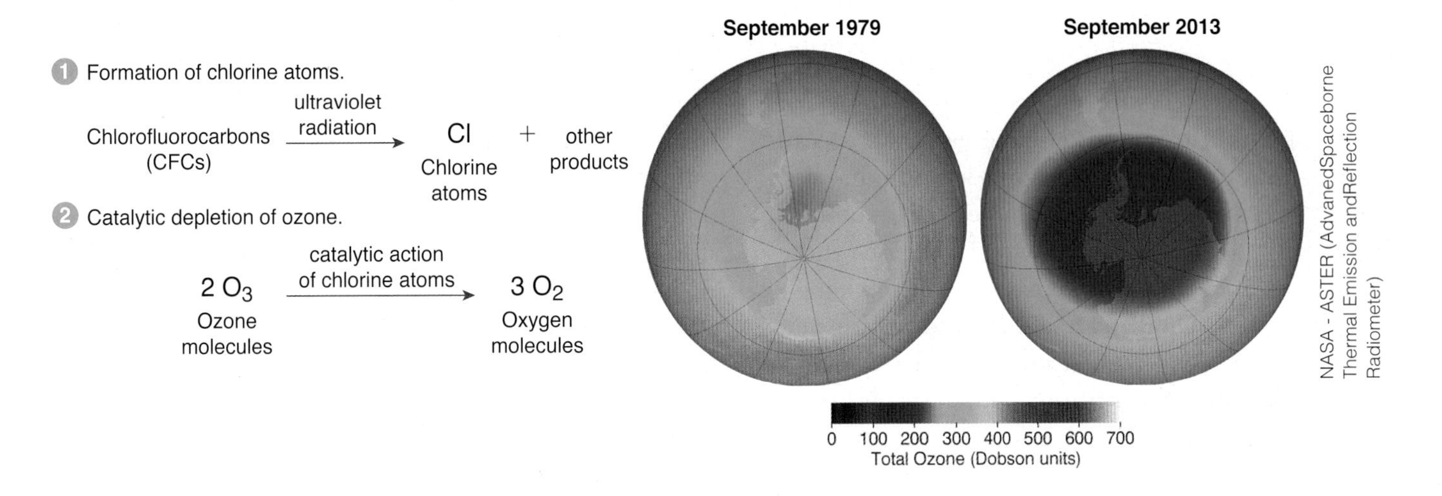

Major sources of air pollution in the United States, 2008 Table 13.2

Source	Millions of Tons				
	Carbon monoxide	**Nitrogen oxides**	**Particulate matter**	**Sulfur dioxide**	**Volatile Organic Compounds**
Fuel combustion, stationary sources	5.3	5.6	2.3	9.9	1.4
Fuel combustion, mobile sources	56.9	9.5	0.9	0.5	6.0
Industrial processes other than fuel combustion	3.8	1.0	2.2	1.0	7.1
Miscellaneous[a]	11.7	0.3	14.3	0.1	1.3
Total emissions	77.7	16.3	19.7	11.4	15.9

[a]Includes agricultural and natural sources
Source: U.S. Statistical Abstracts, 2012.

Interpret the Data

Stationary sources of fuel combustion include power plants and homes. Mobile sources of fuel combustion include cars. Which of these two sources—stationary or mobile—is responsible for more **(a)** CO emissions; **(b)** SO_2 emissions?

Strategies for reducing air pollution What can be done about reducing air pollution? Several options are available, including:

- using alternative energy sources;
- conserving energy through improving the efficiency of power plants and vehicles;
- converting from coal to natural gas, which is less polluting;
- removing pollutants from the exhausts of vehicles and power plants.

Catalytic converters in cars (Section 4.3) effectively reduce pollution from automobile exhaust. Because coal-fired power plants play a significant role in energy production, improving the efficiency of coal combustion and removing pollutants from the exhaust gases of these power plants are important approaches to reducing air pollution as well. **Fluidized-bed combustion** is a newer technology in which crushed coal particles are mixed with limestone particles within rapidly flowing air during the combustion process. This technology produces more heat from a given quantity of coal, thus reducing the total amount of coal needed to produce a desired quantity of energy and therefore reducing as well the amount of carbon dioxide produced in the emissions. The limestone particles, consisting of calcium carbonate ($CaCO_3$), help reduce sulfur dioxide emissions as well. (We'll discuss this **desulfurization** process in more detail shortly.)

Several other approaches also aim to reduce pollutants in industrial exhausts, including electrostatic precipitation and scrubbing. Electrostatic precipitation removes particulate matter (PM), consisting of small particles of liquids or solids dispersed in smoke. As shown in **Figure 13.19a**, exhaust gases pass between two charged vertical plates or electrodes. The particles pick up electrons supplied by the negative electrode and move to the more positive one. There the liquid particles accumulate and flow to the bottom of the collector. With agitation, the solid particles drop off. Scrubbers operate by passing exhaust gases through water, often present as a fine spray (**Figure 3.19b**). Scrubbers that force the gases through a slurry of calcium carbonate, $CaCO_3$, or magnesium hydroxide, $Mg(OH)_2$, are especially useful for removing sulfur dioxide, SO_2. This desulfurization is widely used to reduce sulfur emissions in coal-fired power plants. Carried out with magnesium hydroxide, this process converts sulfur dioxide, a gas, into magnesium sulfite, a solid,

Reducing pollution from industrial emissions • Figure 13.19

Electrostatic precipitators (a) remove particulate matter from exhaust. Scrubbers (b) can remove particulate matter and some gaseous pollutants.

a. Schematic cutaway of an electrostatic precipitator.

Central electrode carries negative charge.
Grounded cylinder walls carry positive charge.
3 Particulate-free exhaust gases leave.
2 Particulates matter acquire negative charges from central electrode and move to positively charged walls.
1 Entering exhaust gases contain particulate matter.

Adapted with permission of John Wiley & Sons, Inc. from Snyder, C., *The Extraordinary Chemistry of Ordinary Things*, Fourth Edition, p. 343. Copyright 2003.

b. Schematic cutaway of a scrubber.

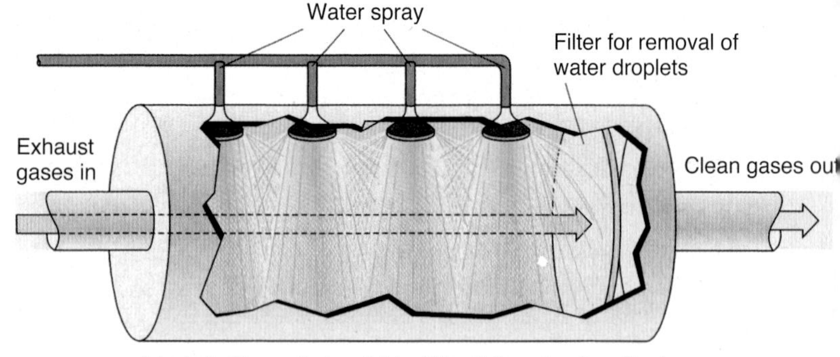

Adapted with permission of John Wiley & Sons, Inc. from Snyder, C., *The Extraordinary Chemistry of Ordinary Things*, Fourth Edition, p. 344. Copyright 2003.

which settles within the base of the scrubber and can be removed.

$$\underset{\text{Sulfur dioxide}}{SO_2} + \underset{\text{Magnesium hydroxide}}{Mg(OH)_2} \longrightarrow \underset{\text{Magnesium sulfite}}{MgSO_3} + \underset{\text{Water}}{H_2O}$$

Water pollution Several forms of pollution affect our water supplies, including biological, thermal, sedimentary, and chemical. **Biological contamination** results from the presence of disease-causing and life-threatening microorganisms, especially in drinking water. To remove these microorganisms, municipal water systems in the United States, which provide water to more than 90% of the population, are generally disinfected with chlorine gas (Cl_2) or ozone (O_3).

Thermal pollution raises the temperature of bodies of water. This typically occurs by the transfer of excess heat generated by industrial processes (such as nuclear power generation) to the water. Gases, including the oxygen needed by both land and aquatic animals, are less soluble in warm than in cold water. Thus a rise in the temperature of a body of water can deprive fish and other aquatic animals of needed oxygen.

Sedimentary pollution results from the accumulation of suspended particles within a body of water. These can be particles of soil washed into the water by the runoff of rainwater from the land, or particles of insoluble organic and inorganic chemicals carried along by runoff or by waste water. These sediments pollute in many ways, including the simple blocking of sunlight. A drop in the amount of light that penetrates into a body of water interferes with photosynthesis by aquatic plants and decreases the ability of aquatic animals to see and find food. Sediments also carry absorbed chemical and biological pollutants with them. Worldwide, sediments account for the greatest mass of pollutants and generate the greatest amount of water pollution.

Chemical pollution is caused by the release of harmful or undesirable chemicals. These pollutants that find their way into Earth's waters are a varied group of substances, including agricultural chemicals, such as fertilizers and pesticides; household chemicals, such as paints and solvents; industrial pollutants, including factory and mining waste; and petroleum from major oil spills. We'll discuss some of these kinds of hazardous wastes shortly.

Not all forms of pollution are necessarily harmful to all living things. For instance, the nitrates (NO_3^- ions) and phosphates (PO_4^{3-} ions) of agricultural fertilizers are effective plant nutrients. When these chemicals enter the waters of still lakes and slow-moving streams, they produce a rapid growth of surface plant life, especially algae, which can produce mats that cover the surface, sealing off the rest of the water from the oxygen of the air. Deprived of dissolved oxygen, fish and other aquatic animals virtually disappear from these waters. This form of water pollution, which operates through selective stimulation of plant life at the expense of animals, is called **eutrophication** (Section 11.1), from Greek words meaning "well nourished."

Land pollution Pollution of land can occur when solid or liquid wastes are improperly disposed or released onto land. This type of pollution can result from agricultural activities, such as the use of fertilizers, pesticides, and herbicides; from mining and industrial activities, such as the improper release or disposal of hazardous wastes; and from municipal activities, such as the disposal of wastes from homes, businesses, and public institutions.

Many substances can move easily and quickly from one part of the environment to another, so sorting out pollutants according to the region of the environment that they impact isn't a simple matter. What we recognize as a land pollutant one day can be an air or a water pollutant the next. We've already seen how an air pollutant can be transformed into a water pollutant as the sulfur oxides of air pollution come back to damage Earth and its waters through the medium of acid rain.

Solid and Hazardous Waste

The generation of various types of trash or material waste through the manufacturing and consumption of consumer goods is an unavoidable aspect of modern life, especially in nations with large populations and/or substantial industrial activity. In the next sections, we'll explore how society deals with and manages the ever-increasing amounts of waste materials produced. As we noted earlier, some of the more enduring parts of this waste consist of discarded plastics.

Municipal solid waste The population of the United States generates on average about 4.4 pounds (2.0 kg) of **municipal solid waste** per person per day, a value greater than that of any other country. Various reasons account for this, including our

municipal solid waste
Solid materials discarded by homes, businesses, and public institutions.

Solid waste • Figure 13.20

Most solid waste produced in the United States results from mining activities. Municipal solid waste represents less than 2% of the total solid waste generated. Source: EPA

a. Sources of total solid waste, by weight

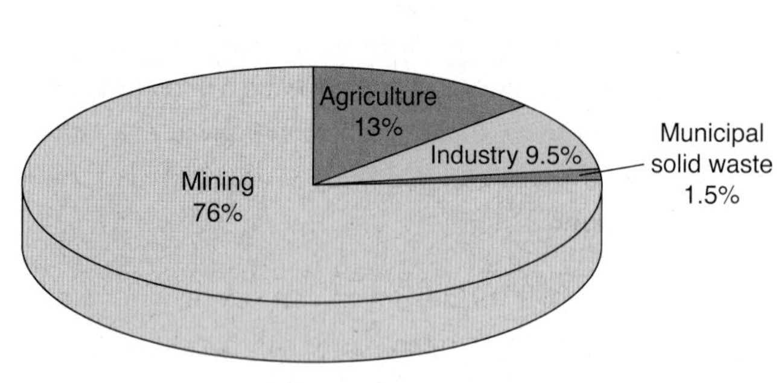

b. Composition of municipal solid waste, by weight, 2011

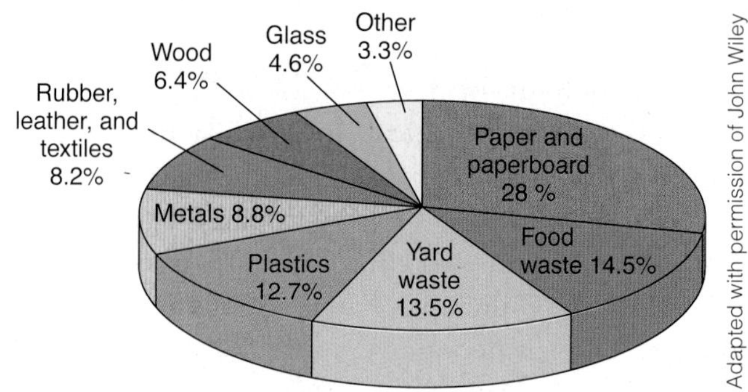

Adapted with permission of John Wiley & Sons, Inc. from Berg, L. and Hager, M.C., and Hassenzahl, D., *Visualizing Environmental Science*, Third Edition, p. 397. Copyright 2011.

relatively high standard of living. The more material substances we buy, the more products we discard that might otherwise be reused, repaired, or recycled in less prosperous nations. Also, land resources in the United States are still relatively abundant, allowing us to build more landfills to store our waste.

The materials of municipal solid waste include paper and paperboard, plastics, metals, glass, food waste, and lawn and yard trimmings. According to the EPA, the United States generated 250 million tons of this waste in 2011. As much trash as this amounts to, it still represents less than 2% of the total solid waste generated annually. The vast majority of this national waste originates in mining, industrial, and agricultural activities (**Figure 13.20**).

In 2011, 54% of our municipal waste was sent to landfills, 12% was incinerated, and 34% was either composted or recycled. We'll now examine each of these methods for handling municipal waste.

Sanitary landfills are large disposal sites in which garbage is compacted and covered each day by a thin layer of soil. Sanitary landfills offer the following improvements over open garbage dumps, which have been largely phased out in the United States:

- since the garbage in sanitary landfills is covered with soil each day, this reduces pest and odor problems;
- since sanitary landfills employ containment liner systems underneath the garbage, this prevents noxious liquids from seeping into groundwater;
- since large quantities of methane gas are naturally generated by landfills through the decomposition of organic matter, most sanitary landfills employ systems to capture this gas. Often the gas is simply burned off, but a growing number of landfills capture the gas and use it to generate electricity.

In spite of the benefits of modern landfills, they do not represent sustainable solutions for waste management. For example, landfills can fill up, and often communities oppose the building of new ones, especially near homes and businesses. Furthermore, locating landfills far from population centers is not ideal because of the added costs of transporting garbage long distances.

Incinerating combustible trash, such as paper, plastic, and old tires, provides both benefits and risks. On the positive side:

- incineration reduces the volume of solid waste by about 90% (the ash remaining afterward comprises about 10% of the original volume);
- because trash is being incinerated, less of it is directed to landfills;
- the heat produced by incineration can be used to generate electricity.

The disadvantages of incineration include:

- the release into the atmosphere of various pollutants, including carbon monoxide, heavy metals, particulate matter, and other hazardous agents;
- the need to dispose of the ash resulting from incineration into specialized landfills, since this ash is classified as hazardous waste;
- the expense of installing pollution-control devices on incinerators.

Composting and recycling are generally preferred over other methods of handling municipal solid waste. Recycling conserves resources, and both composting and recycling can greatly reduce the amount of garbage sent to landfills or incinerated. Composting food and garden waste yields nutrient-rich organic material that can be used as a natural fertilizer and for other landscaping purposes.

Recycling is a cyclical process and involves far more than simply separating recyclable materials from other waste. In order for recycling to work, not only must recyclable materials be collected, sorted, and processed, but manufacturers must be willing to incorporate these materials into their products, and consumers must be willing to purchase products made with recycled content. The United States currently recycles an average of about 60% of its paper and paperboard waste (including corrugated boxes); 30% of its metal waste; 25% of its glass waste; and about 10% of its plastic waste. Most state laws require the recycling of certain hazardous items, such as lead–acid car batteries, which have a reclamation rate of 96%.

Hazardous waste About 1% of the total waste generated in the United States is classified as **hazardous waste**. This includes radioactive materials; various types of industrial and agricultural wastes; medical wastes; and a variety of products we use in our homes, such as paints, solvents, motor oils and antifreeze, lawn and garden chemicals, electronic devices, and certain types of batteries. Hazardous wastes can be solids, liquids, or gases; they include corrosive chemicals (such as battery acid) and other substances that present a high risk of toxicity.

> **hazardous waste** Chemical waste that may endanger public health or wildlife.

The improper dumping or release of hazardous industrial wastes is a continuing concern, especially in nations with lax environmental oversight or controls. But even in countries such as the United States, with relatively strong regulatory oversight, there is a legacy of the improper disposal of various forms of hazardous waste carried out decades ago. In 1980, the United States created the **Superfund Act**, a federal program designed to clean up uncontrolled hazardous waste disposal sites. This legislation was enacted partly in response to the discovery of a hazardous waste dump near Niagara Falls, New York. Here, in the 1940s and early 1950s, a local chemical company had buried some 22,000 tons of toxic chemicals in a 10-square-block area known as Love Canal. The chemicals, most of them in sealed drums, were covered with topsoil and the land was donated to the local board of education. Years later a small community and a school were built on the site. By 1976, the chemicals had begun leaking out of the corroded drums and eventually contaminated the entire area. By 1980, the resulting harm to the people living there (including high incidences of miscarriages and birth defects) resulted in the relocation of more than 230 families, and the land was declared a federal disaster area under the Superfund Act (**Figure 13.21**). By 1990, after nearly a decade of

Love Canal • Figure 13.21

The homes shown here, as they existed in 1980, were evacuated and demolished in order to remove toxic waste that had been buried there decades earlier. Over 1000 hazardous-waste disposal sites posing the greatest risk to public health have been placed on the Superfund National Priorities list in order to carry out necessary clean-up efforts.

Joe Traver/Contributor/Hulton Archive/Getty Images

Persistent organic pollutants • Figure 13.22

Some of the more hazardous organic compounds known to resist degradation in the environment include chlorinated pesticides (a), dioxins (b), and PCBs (c). Accumulation of these compounds in the body has been linked to reproductive and developmental effects, as well as certain cancers.

a. Chlorinated pesticides, such as DDT and chlordane, contain chlorine atoms bonded to carbon atoms.

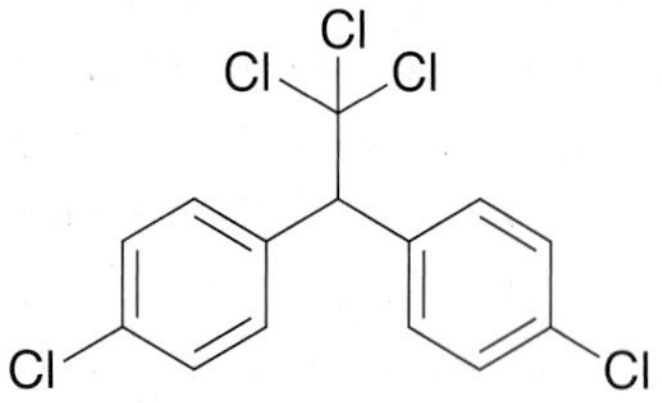

Dichloro**d**iphenyl**t**richloroethane (DDT)

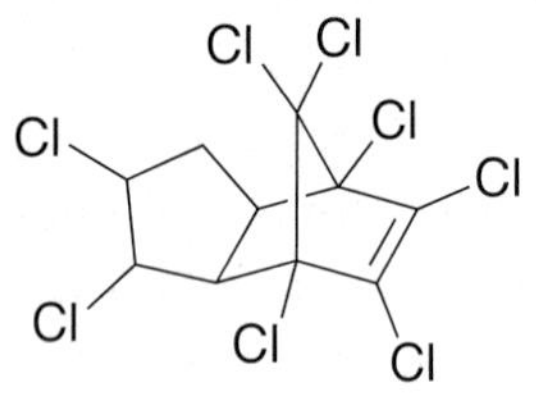

Chlordane

b. Dioxins, such as tetrachlorodioxin, represent a large class of structurally related compounds. Dioxins form as by-products of combustion and are released by both human activities (such as the burning of backyard, municipal, and medical wastes) and by natural causes, such as forest fires.

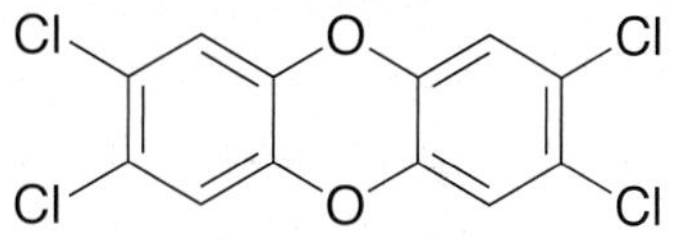

Tetrachlorodioxin

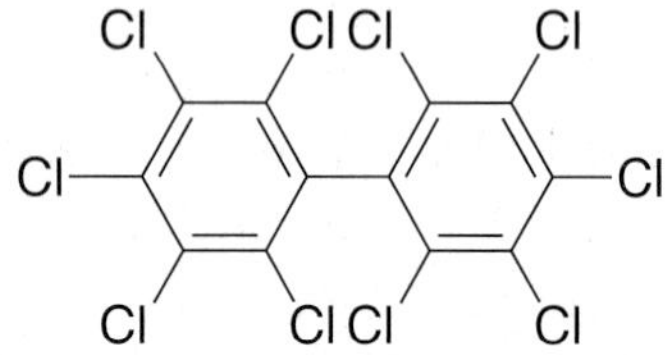

A **poly**c**hlorinated** **b**iphenyl (PCB)

c. PCBs include a large class of chlorinated compounds similar in structure to the compound shown here. Prior to the EPA's ban on the manufacturing of PCBs in 1979, these liquid compounds were widely used in electrical transformers, hydraulic systems, electrical insulation, and other applications. The historical dumping of PCBs into the environment continues to present concerns because of their toxicity and resistance to degradation.

Ask Yourself

1. Which three elements occur in all of the compounds shown?
2. What structural feature is shared by all of the compounds shown except for chlordane?

remediation, the EPA declared much of the Love Canal area safe for resettlement.

Environmental damage operates on many scales of both time and quantity. The daily release of small amounts of hazardous chemicals by commercial entities rarely reaches catastrophic proportions, but these substances can accumulate in the environment over time and pose serious risks to human health and to wildlife. One such class of hazardous substances are called **persistent organic pollutants** (POPs). These are organic compounds that resist degradation in the environment and can undergo bioaccumulation in the food chain.

The first of the POPs to gain notoriety was DDT (**d**ichloro**d**iphenyl**t**richloroethane), a synthetic insecticide widely used in the United States (and elsewhere) beginning in the 1940s and initially thought to be perfectly safe. DDT shares features common to various POPs, including:

- resistance to degradation in the environment;
- existence as a chlorinated organic compound;
- high solubility in fat, including the fat of milk and the body fat of animals;
- interference with the reproduction of birds, fish, and other animals.

The effects of DDT and its potential for harm were described by the American biologist Rachel Carson in her 1962 book *Silent Spring*, a publication credited with spurring the modern environmental movement. Partly as a consequence of this warning, the use of DDT as an insecticide was severely restricted in the United States in 1973 and was phased out by most industrial nations by the 1980s. In 2001, a United Nations treaty signed in Stockholm, Sweden, and known as the **Stockholm Convention**, bans the production, use, and environmental release of several chemicals known to be persistent organic pollutants. **Figure 13.22** identifies a few of these compounds.

Periodically, we learn about industrial accidents that release large quantities of hazardous chemicals into the environment. One of the most tragic of these occurred in 1984 in Bhopal, India. This incident involved an explosion at a Union Carbide plant that manufactured methyl isocyanate, $CH_3{-}N{=}C{=}O$, a deadly compound used in the production of pesticides. The explosion released a cloud containing an estimated 23,000 kg (50,000 lb) of methyl isocyanate. As the cloud drifted over Bhopal, it killed 2000 people and injured another 200,000, all within a few hours. In response to this incident, most of the global chemical industry has pledged to adopt practices to improve factory safety, to increase dialog with the communities in which it operates, and to improve its environmental record, independent of any legal requirements to do so.

Due to the enforcement of various state and federal laws concerning the disposal of hazardous wastes, companies that produce these wastes are generally well aware of their responsibilities to dispose of them properly. The preferred method of dealing with any type of waste, whether hazardous or not, is **source reduction**, which means decreasing the amount of waste initially produced. In the case of reducing hazardous waste, **green chemistry** can play an important role. For example, chlorinated solvents are often used in the manufactuing of electronics and pharmaceuticals, and for dry cleaning. Sometimes these chlorinated solvents can be replaced by less hazardous solvents, such as supercritical carbon dioxide (Section 6.2). Another example involves the bleaching of wood pulp in the manufacturing of paper. This whitening process was traditionally carried out with chlorine-based bleaches but is now largely accomplished with less hazardous, oxygen-based bleaches.

> **green chemistry** Chemical practices that aim to conserve resources and reduce the generation of waste and toxic substances.

Many of the consumer products we've discussed—including biofuels (Chapter 4), "green" cleaners (Chapter 11), and biodegradable polymers (Chapter 12)—utilize green chemistry. But those that do are exceptions rather than the norm. Concerns about: (1) potential health impacts of our exposure to the variety of chemicals in consumer products, and (2) waste materials generated by the manufacture of these substances, have led to growing support for stricter oversight in their regulation. The Toxic Substances Control Act (TSCA), enacted in 1976, requires the EPA to protect consumers and the environment from unreasonable risks associated with the manufacture and use of various substances. However, the TSCA has been criticized as inadequate by many, including the U.S. Government Accountability Office. Because of the insuffiency of safety data relating to thousands of chemicals used in the manufacture of many of our commercial goods, the EPA may be unable to use the provisions of the TSCA to ban or limit the production of numerous potentially hazardous substances. Pressure for regulatory reform, including the Safe Chemicals Act, proposed to Congress in 2013, points to a future in which principles of green chemistry play a greater role.

Our daily concerns with potentially hazardous chemical wastes arise from ordinary things like exhausted but not quite empty pesticide spray cans, small amounts of house paint remaining from renovations, used motor oil, partly empty containers of automotive fluids, and leftover garden and pool chemicals. Although it may be tempting to pour these wastes down the kitchen drain, into a toilet, down a storm drain, or even onto the ground itself, that's often a certain route into the groundwater. Combining these hazardous wastes with normal household trash can be equally problematic. To handle these types of household wastes, most municipalities have hazardous or special waste programs, which often include drop-off sites for the collection of these residual substances.

Electronic wastes (or **e-waste**), such as older, unwanted cell phones, computing devices, and TVs, can contain small amounts of hazardous substances, including lead, mercury, and cadmium. The disposal of these devices also represents a substantial waste of resources because of the various valuable metals and other materials they may contain. Several states and organizations have adopted programs to help consumers recycle (or **e-cycle**) these products. According to the EPA, recyclers in the United States recover more than 100 million pounds of useful materials from electronics each year.

CONCEPT CHECK

STOP

1. **Why** is it necessary to specify the location of a contaminant in the environment in defining pollution?
2. **What** technique is used to remove sulfur from exhaust gases in coal-fired power plants, and what environmental benefits does this desulfurization provide?
3. **What** types of (a) municipal solid waste and (b) hazardous waste are generated by households?

Summary

1 Polymer Uses and Structures 410

- ***How are plastics used by society?***
Plastics are inexpensive, lightweight, and generally durable materials widely used by society. Their largest use is in packaging materials, but other important uses include building materials, electrical insulation, transportation applications, housewares, and other consumer goods. Plastics exhibit a range of properties and can be molded into a variety of shapes, both solid and hollow. They can also be formed into thin sheets, stretched into long fibers, and permeated with gases to form spongy or insulating solids.

- ***How are synthetic polymers classified?***
All plastics are synthetic **polymers**, which are **macromolecules** formed from the repeated combination of a very large number of small units, called **monomers**. These monomers form either **addition polymers**, compounds in which the monomers add to each other without loss of any of the atoms originally present in the monomer units; or **condensation polymers**, compounds in which each link formed in the polymer chain is accompanied by the release of a small molecule, such as water, from the monomers involved in forming the link. Polymers that are made up of only one type of monomer are **homopolymers** (such as polyethylene, shown here); those that are made up of two or more types of monomers are **copolymers**. Polymers that soften in response to heat are **thermoplastics**; those that do not are **thermosets**. The most widely produced plastic is polyethylene, an addition polymer of ethylene. Polyethyethylene terephthalate (PET) and nylon are common condensation polymers.

Polyethylene and its derivatives Table 13.1

Polymerization of...	... produces...
$H_2C{=}CH_2$ Ethylene	$-\!\left[CH_2-CH_2\right]_n\!-$ Polyethylene (PE)

2 The Development and Future of Polymers 417

- ***How and when were different types of polymers discovered or invented?***
The first polymeric materials made by people were semisynthetic, produced by modification of **biopolymers**. These discoveries, all of which involved a degree of serendipity, include vulcanized rubber (1839, illustrated here), guncotton (1845), and celluloid (1869). Bakelite (1909), a condensation and thermosetting polymer formed from phenol and formaldehyde, was the first synthetic polymer to be produced. The field of modern polymer science was initiated in the 1920s. Many of the polymers used today—including nylon, polyethylene terephthalate (PET), high-density polyethylene (HDPE), and Teflon—were first produced in the period from the late 1920s to the early 1950s. Countless other valuable polymers have been synthesized since, including Kevlar, used as bulletproofing material, and Nomex, used in fire-resistant clothing.

Figure 13.6 • Vulcanized rubber

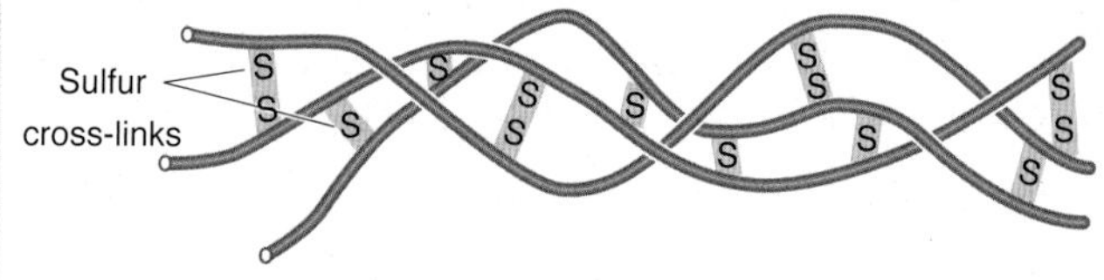

- ***How are plastics recycled?***
Various recyclable plastics are assigned resin code numbers. Numbers 1 through 6 correspond to the most common recyclable plastics, and number 7 includes all others. For **single-stream recycling**, in which recyclable materials are commingled, waste plastics are separated from other waste, then sorted by polymer type, cleaned, and shredded. The shredded flakes may then be sorted in flotation tanks that separate plastics according to their different densities. Once dried, the flakes are melted and then formed into solid granules. This material can be formulated into a variety of new plastic products.

3 Pollution and Wastes 424

- ***How is pollution defined?***
The term *pollution* can be defined in different ways. From a legal standpoint, pollution is often defined in terms of the limits of specific contaminants permitted by law to exist in a given environment. In our discussion, we say that pollution occurs when an excess of a substance generated by human activity or by nature is present in the wrong environmental location.

- ***What are the major causes and types of pollution?***
The combustion of fossil fuels is the major cause of air pollution, but manufacturing contributes as well. Fossil fuel combustion releases **primary air pollutants** into the atmosphere, such as sulfur dioxide, which can undergo subsequent reactions to form **secondary air pollutants**, such as sulfuric acid, a component of acid rain. Other air pollutants of concern include carbon monoxide, nitrogen oxides (shown here, involved in the generation of photochemical smog), particulate

Figure 13.15 • Photochemical smog

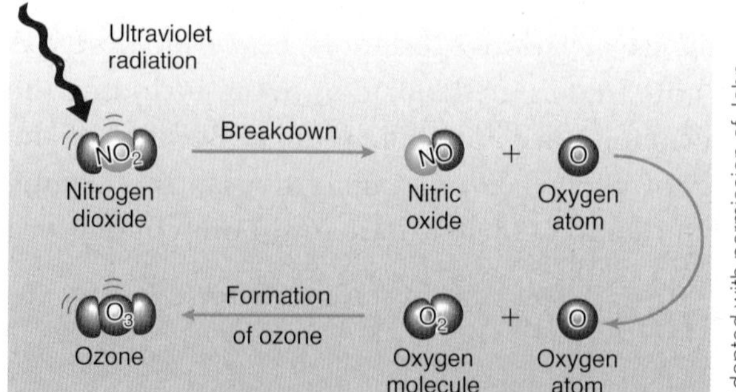

matter, and **volatile organic compounds**. Water pollution may result from biological agents, sediments, thermal waste, and/or chemical agents.

- ***What constitutes hazardous waste?***
  Chemical waste that endangers public health or wildlife constitutes hazardous waste. This can result from the improper disposal of hazardous chemicals, as exemplified by the burial of industrial wastes at Love Canal, which helped galvanize support for more stringent environmental laws. Chlorinated organic compounds that resist degradation in the environment represent a class of **persistent organic pollutants**. Most hazardous waste is generated by industrial, agricultural, and mining activities, but consumer and domestic activities, such as the disposal of used motor oils, leftover paints, and lawn chemicals, can contribute to hazardous waste as well.

Key Terms

- addition polymer 411
- bioaccumulation 427
- condensation polymer 412
- green chemistry 437
- hazardous waste 435
- monomer 411
- municipal solid waste 433
- photochemical smog 428
- polymer 410
- primary air pollutant 426
- secondary air pollutant 426
- volatile organic compound (VOC) 429

What is happening in this picture?

Nylon 6,6 can be made in the laboratory by combining:

- an aqueous or water solution of 1,6-hexanediamine, and
- a hexane solution of adipoyl chloride.

Because water is polar, and the hydrocarbon, hexane (C_6H_{14}), is nonpolar, these two solutions form two separate layers or phases (much like water and oil) when they are mixed together.

The condensation reaction that produces nylon 6,6 occurs at the boundary or interface between these two solutions. Here, 1,6-hexanediamine reacts with adipoyl chloride. The newly created nylon fibers can be drawn out into a long, continuous silky thread by twirling them onto the end of a glass rod, as seen here.

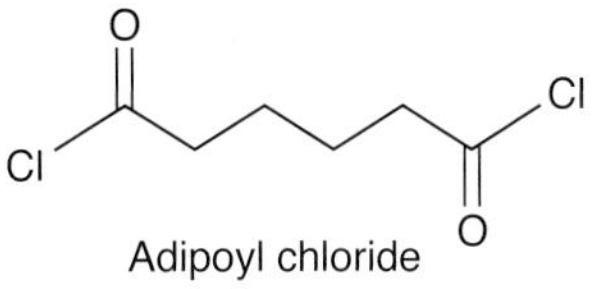

Adipoyl chloride

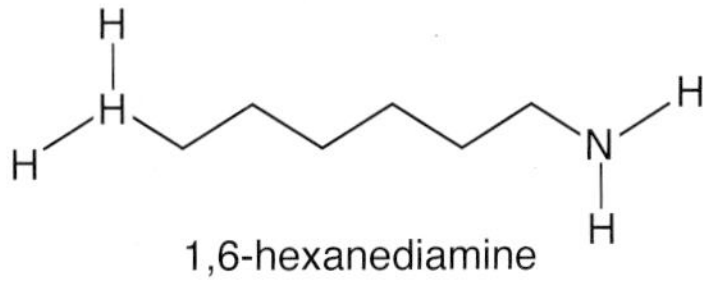

1,6-hexanediamine

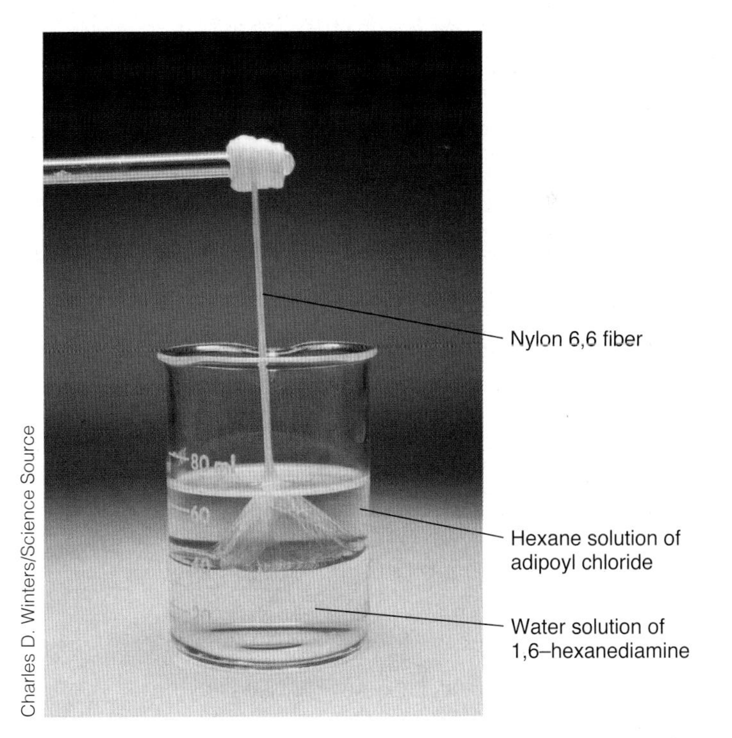

Charles D. Winters/Science Source

Think Critically **1.** What does the term *condensation reaction,* used in the above passage, mean?
2. In Figure 13.4b, we show that nylon 6,6 can form from the reaction of adipic acid with1,6-hexanediamine, releasing water in the process. What small molecule is released in the reaction shown here?

Exercises

Review

1. Differentiate between the terms within each set:
   a. monomer; polymer
   b. homopolymer; copolymer
   c. addition polymer; condensation polymer
   d. thermoplastic; thermoset
   e. biopolymer; semisynthetic polymer; synthetic polymer
   f. polyamide; polyester
   g. primary air pollutant; secondary air pollutant

2. Use the following terms to classify each of the substances listed below. (As an example, the answer for part a has been given.)

   homopolymer (HOM); copolymer (COP); addition polymer (ADD); condensation polymer (CON); thermoplastic (TP); thermoset (TS); elastomer (EL) biopolymer (BP); semisynthetic polymer (SSP); synthetic polymer (SP)

   a. polyethylene. Answer: HOM, ADD, TP, SP
   b. polystyrene
   c. nylon
   d. vulcanized rubber
   e. Bakelite
   f. polylactic acid
   g. Neoprene
   h. polyethylene terephthalate

3. Name the polymer each of the following pairs of compounds form and describe its use or importance:
   a. adipic acid and 1,6-hexanediamine
   b. phenol and formaldehyde
   c. styrene and butadiene
   d. ethylene glycol and terephthalic acid

4. Provide the structure(s) of the monomer(s) used to make each of the following plastics, and describe a product in which each is used.
   a. Mylar
   b. Teflon
   c. Nomex
   d. Kevlar

5. State how high-density and low-density polyethylene differ, if at all, in the following:
   a. the monomer from which they form
   b. the structures of their polymeric molecules
   c. the forms or shapes that their molecules take within the bulk plastic
   d. the physical properties of the bulk plastic

6. Give three examples of polymers that we eat.

7. What commercial plastic is produced in the largest quantities each year?

8. Identify five polymers discussed in the chapter that contain phenyl groups or benzene rings in their structures.

9. Identify each of the following as a polyamide or a polyester:
   a. nylon
   b. a polypeptide
   c. polylactic acid (PLA)
   d. polyethylene terephthalate (PET)
   e. Nomex
   f. Kevlar

10. a. Identify two principal environmental concerns about the continued use of plastics.
    b. In what two ways can science and technology help alleviate these concerns?

11. Identify a commercial plastic that contains:
    a. chlorine
    b. fluorine
    c. nitrogen
    d. acetate groups.

12. What causes plastic waste to accumulate in certain regions of the world's oceans?

13. a. Identify the recyclable polymer corresponding to each of the following symbols.

    b. What does the #7 symbol of recyclable plastics represent?

14. What natural phenomenon causes the degradation of polylactic acid (PLA)?

15. Give two examples of serendipity in the discovery or development of a polymer.

16. Review the figure below and answer the following questions:
    a. Identify three combustion sources of mercury emissions.
    b. What is a natural source of mercury emissions that is also responsible for sulfur dioxide emissions?
    c. How does mercury convert from being an air pollutant to being a water pollutant?
    d. How does mercury convert from being a land pollutant to being a water pollutant?
    e. What does volatilization refer to in this figure, and what other types of pollutants mentioned in the chapter enter the air through volatilization?
    f. How does mercury enter the food chain?

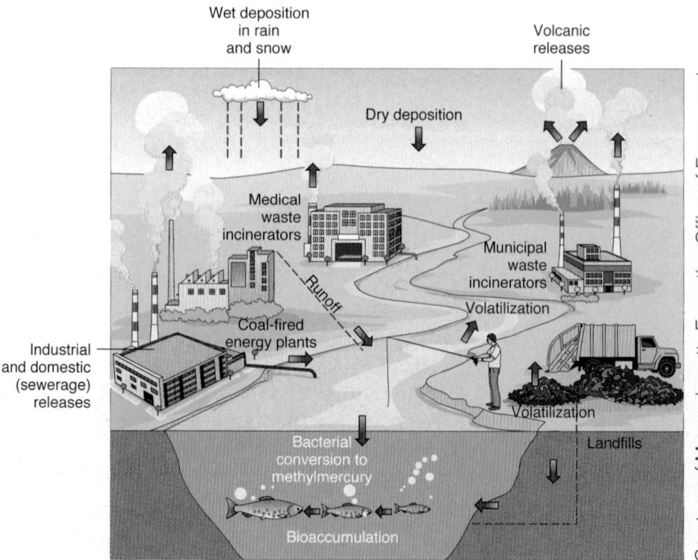

Courtesy of Massachusetts Executive Office of Energy and Environmental Affairs Department of Environmental Protection

17. Identify two primary air pollutants and two secondary air pollutants containing nitrogen and describe their origins or sources.

18. An element that is found among the impurities of both coal and fuels derived from petroleum is a major contributor to air pollution. What is that element?

19. Write the molecular formulas of the following:

(a) pyrite

(b) nitrogen dioxide

(c) nitric oxide

(d) calcium carbonate

(e) magnesium sulfite

20. Describe how each of the following removes pollutants from exhaust gases:

a. scrubbing

b. electrostatic precipitation

21. What are the environmental benefits of the fluidized-bed combustion of coal?

22. Write balanced reactions for:

a. conversion of sulfur dioxide to sulfur trioxide

b. reaction of sulfur dioxide with magnesium hydroxide

c. formation of ozone from a sequence of reactions initiated by the action of sunlight on nitrogen dioxide

d. the reaction of sulfur trioxide with water to produce sulfuric acid.

23. Higher levels of nitric oxide and ozone can be found in the atmosphere after thunderstorms. Why? What is their origin?

24. The burning of coal containing pyrite impurities contributes to the presence of sulfuric acid in the atmosphere. Write a single balanced equation for the conversion of the pyrite impurity of coal into atmospheric sulfuric acid. Assume that water vapor is present in the atmosphere.

25. Worldwide, what is the most widespread form of water pollution?

26. Which two chemicals are most commonly used to disinfect municipal water supplies?

27. Name and describe four types of water pollution. Explain how each damages aquatic life.

28. Give the names and molecular formulas of two acids found in acid rain and discussed in this chapter.

29. How can a substance that nourishes and sustains certain forms of aquatic life also act as a pollutant?

30. What does the term VOCs mean? How might you be exposed to VOCs while pumping gas or painting a house?

31. a. What substances were banned under (i) the Montreal Protocol; and (ii) the Stockholm Convention?

b. Why were these substances banned in each case?

32. For each of the following, identify or describe the hazardous substance involved and describe how it entered the environment:

a. Bhopal

b. Love Canal

Think

33. Why are thermoplastics recyclable but thermosets nonrecyclable?

34. The nitric acid in Schoenbein's accidental invention of guncotton reacted with free hydroxyl groups on each of the monomeric glucose links of the polymer cellulose. How many free hydroxyl groups does each of these monomeric units have available for this reaction?

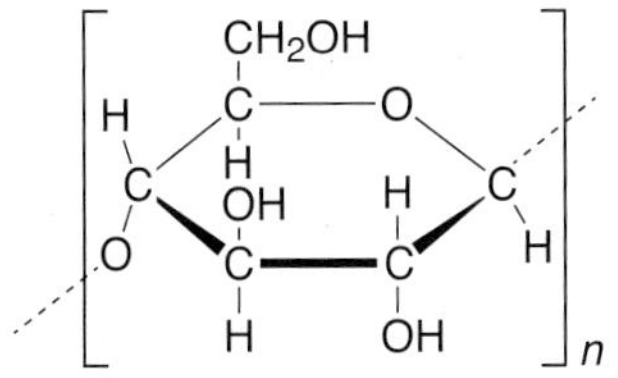

Cellulose

35. Figure 13.7 shows a portion of the polymer that forms on polymerization of a mixture of 75% butadiene and 25% styrene. Draw a similar segment of the polymer you would expect to form from a mixture of 25% butadiene and 75% styrene.

36. What's the difference between a plastic and a polymer?

37. Why was the discovery of vulcanization so important to the development of the commercial rubber industry?

38. Give examples of:

a. a natural condensation homopolymer

b. a natural condensation copolymer

c. a natural addition homopolymer

39. Give examples of

a. a synthetic addition homopolymer

b. a synthetic addition copolymer

c. a synthetic condensation copolymer

40. Terephthalic acid is a dicarboxylic acid used as a monomer in the manufacture of a condensation polymer. What is another dicarboyxlic acid used successfully in the manufacture of a condensation polymer?

41. What are three factors, operating at the molecular level, that affect the bulk properties of a plastic?

42. What name would you give the plastic made from the addition polymer of $CH_2{=}CH{-}Br$?

43. Hydrogen bonds between polymer strands of many kinds of polyamides, such as Kevlar (shown here), helps gives these polymers their strength. Show the hydrogen bonds that can exist between segments of two strands of nylon 6,6.

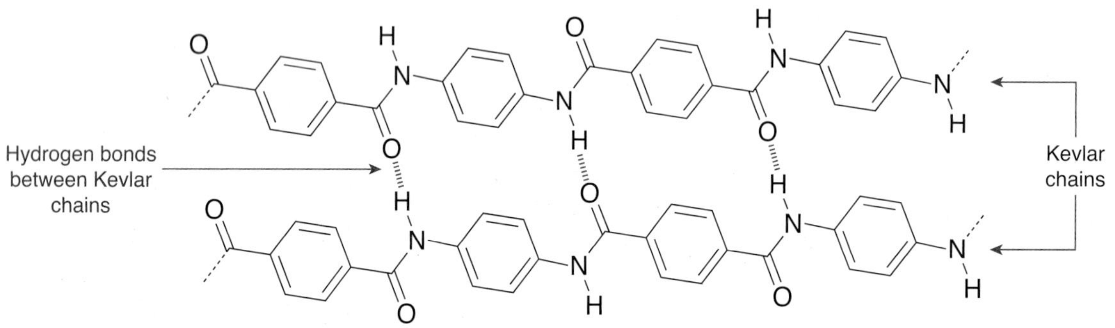

Hydrogen bonds between Kevlar chains

Kevlar chains

44. What polymer would you use if you wanted to manufacture the following items:

a. a coating on windshield wipers that makes them glide over the windshield more easily

b. a shiny, strong wrapper for boxes of perfume that would not let the perfume's odor escape into a store or a room

c. an inexpensive, imitation silk fabric

d. the handle of a frying pan

e. a flexible hose for transferring gasoline

45. Molecules of low-density polyethylene can be described by the shorthand form $-(CH_2)-_n$, but this does not adequately describe the polymer. Explain why it does not.

46. Write the molecular structure of Neoprene in the form $-[A]_n-$.

47. In what important way are the molecular structures of nylon polymers and protein polymers similar to each other? In what ways are they different?

48. High-density polyethylene is used for manufacturing large containers for liquid laundry detergent, while low-density polyethylene is used for trash bags. What consequences would result from using high-density polyethylene for trash bags and low-density polyethylene for containers of liquid detergents?

49. If you cool a racketball to −195°C with liquid nitrogen and then throw it against a wall, the ball will shatter instead of bouncing back. Yet if you let the cold ball warm to room temperature, it will once again bounce normally. This cycle of cooling and heating, with corresponding changes in the properties of the tennis ball, can be repeated almost indefinitely. Does the elastomer that forms the ball resemble a thermoplastic or a thermoset in its behavior?

50. Describe the environmental problems generated by the production of synthetic plastics, their possible solutions, and the challenges associated with each possible solution.

51. In an analogy to polypeptides and proteins, the primary structure of a polymer is represented by $-(\text{-}X\text{-})-_n$. Which polymers discussed in this chapter, if any, would you expect to show a secondary structure? Explain.

52. Does DNA represent a homopolymer or a copolymer? Explain.

53. a. Describe how household trash is disposed of by the community in which you currently live. Is it placed in a landfill? Incinerated? Transferred to a distant location? Disposed of in some other way?
b. What plastics are accepted for recycling in your community? (You may use the Internet to find your answer.)

54. The sulfur of sulfur oxides originates in impurities of the fossil fuels we burn to produce energy. What is the principal source of the nitrogen of the nitric oxides that pollute the air?

55. Teflon is useful as a coating in pots and pans because foods do not tend to stick to the Teflon surface. Why, then, does the Teflon coating stick firmly to the metal surface of the pot or pan? How is the Teflon coating applied so that it adheres strongly to the metal surface? (You may use the Internet to find your answer.)

56. Would adding fresh water to polluted water reduce the concentration of contaminants sufficiently to serve as an effective method for reducing pollution? Explain.

57. Seawater is too salty to drink. Is seawater polluted by salt?

58. How could energy conservation help reduce air pollution?

59. Assuming that our use of energy does not decrease, what would you recommend as the best way to reduce air pollution due to carbon monoxide?

60. Under what circumstance is ozone beneficial to life? In what circumstances is it harmful?

61. CFCs generally exhibit low chemical reactivities, low or nonexistent flammability, and low toxicity. Using the definition of pollution presented in this chapter, explain why CFCs might be considered to be pollutants.

62. Describe the connection between air pollution and the generation of energy.

63. Explain why the use of natural gas as a fuel is less likely to lead to acid rain than the use of other fossil fuels, such as petroleum or coal.

64. Discuss the environmental impact of replacing electricity-generating plants that burn coal or petroleum with electricity-generating plants that use nuclear power. Refer to Chapter 9.

65. What does the term *environmental persistence* mean when it is applied to insecticides and herbicides, and why may this be a concern?

66. Give an example of a substance, other than water or gasoline, that is useful or beneficial to us in one place but that acts as a pollutant in another.

67. Suppose you are a saltwater fish, one that lives its entire life in the ocean but cannot survive in fresh water. Would you use the term *polluted* to describe an environment of fresh water, suitable for humans to drink?

68. Describe the most recent observations of the ozone hole in both Arctic and Antarctic atmospheres. (You may use the Internet to provide your answer.)

69. Now that CFC use has been sharply curtailed in order to protect the ozone layer, what chemicals have replaced them as refrigerants? (You may use the Internet to provide your answer.)

70. Identify the chemical known as tetrachlorodioxin and describe how it is related to the topics of this chapter.

71. What element, absent from natural rubber, is present in: a. Neoprene? b. vulcanized rubber?

72. What are the chemical names of the three reagents needed for the production of guncotton?

73. Plastic makes up about 10% of the mass of our discarded trash but about 25% of its volume. What is the likely reason for the difference between these two figures?

74. Which method of removing air pollutants from exhaust gases—electrostatic precipitation or scrubbing—operates by producing chemical changes in the pollutants?

75. One of the world's worst environmental catastrophes occurred in Minamata, Japan, in the middle of the past century. Industrial wastes containing tons of mercury were poured into the bay at Minamata. Consumption of locally caught fish resulted in the deaths of hundreds of people and severe mental and physical disorders in thousands of others. How does mercury released into the environment enter and move up the food chain?

76. Swordfish, like tuna, are large, predatory fish and are classified as fatty or oily fish—those with higher fat content. How would these factors help explain the higher levels of mercury found in swordfish as compared to many other types of fish?

Calculate

77. The polyethylene terephthalate that makes up the bottles shown here is made from plant-derived ethylene glycol and petroleum-derived terephthalic acid. This plastic contains 31% plant-derived content. Explain. [*Hint*: You will need to use the atomic masses of carbon, hydrogen, and oxygen, listed in the periodic table, to answer this question.]

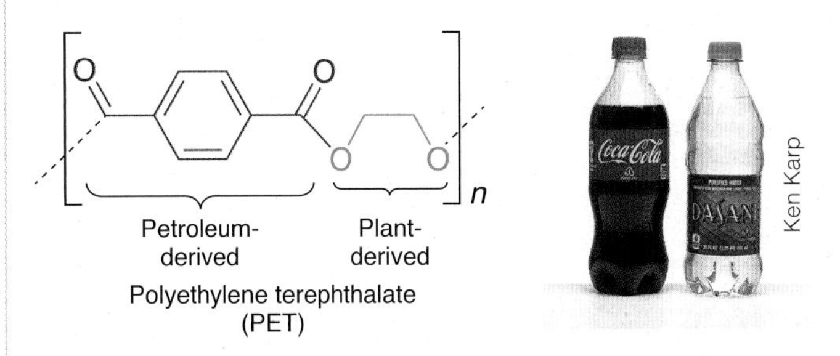

78. List the societal function or human activity that contributes the greatest combined weight of:
    a. sulfur and nitrogen oxides
    b. carbon monoxide
    c. VOCs to the pollution of the atmosphere (Refer to Table 13.2.)

79. a. Write a single, balanced equation for the conversion of atmospheric nitrogen and atmospheric oxygen into nitrogen dioxide under the influence of the high temperatures of lightning or of combustion within the internal combustion engine.
    b. Write a single, balanced equation for the formation of ozone and nitric oxide from a mixture of nitrogen dioxide and atmospheric oxygen.
    c. Now write a single, balanced equation for the formation of ozone from the high-temperature reaction of atmospheric nitrogen and atmospheric oxygen.

80. Use the chart provided to answer the following questions.
    a. About 57% (by weight) of the yard waste generated in 2011 was recovered for composting and mulching. How many tons of this waste were kept out of incineration facilities and landfills as a result of this recovery?
    b. Of the total plastic waste generated in 2011, 8.3% of this material by weight was recovered for recycling, a value also known as the recovery rate for plastic. By contrast, the recovery rate for paper and paperboard in the same year was 65.6%. How many tons of (i) plastic; (ii) paper and paperboard; were not recycled in 2011?

Composition of municipal solid waste generated in the U.S. in 2011, as a percentage by weight.

(250 million tons of this waste was generated in total in 2011.)

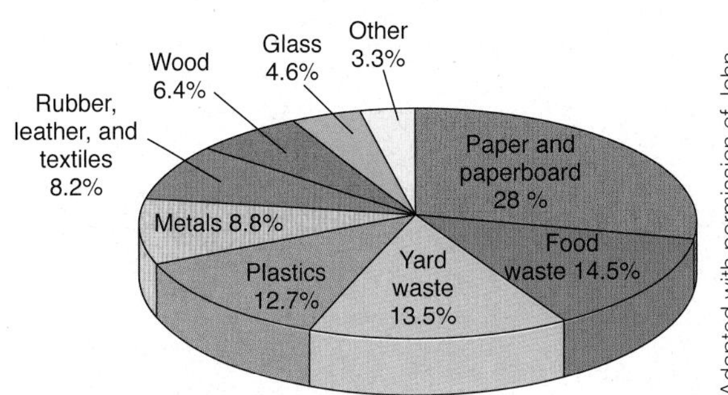

Adapted with permission of John Wiley & Sons, Inc. from Berg, L. and Hager, M.C., and Hassenzahl, D., *Visualizing Environmental Science*, Third Edition, p. 397. Copyright 2011.

14

Micronutrients, Food Additives, and Food Safety

In Chapter 5 we saw that foods consist largely of macronutrients: the fats and oils, proteins, and carbohydrates that provide us with the material structures and energy of our bodies. In this chapter we examine the much smaller components of foods that do not provide us directly with bodily substance or energy, but that are nonetheless vital to our well-being (i.e., vitamins and minerals), or that contribute to the taste, texture, or stability of foods (i.e., food additives). Some of these additives occur in **fuctional foods**, substances that may offer additional health benefits.

We also look at the safety of the foods and liquids that we eat and drink. We examine whether poisons exist in our food supply—poisons that can make us sick or even kill us—put there either by nature or by people, individually and commercially, as we prepare, process, cook, store, and modify foods. Produce that is stored for extended periods, for example, can harbor **aflatoxin**, a fungal toxin; the skins of potatoes and the leaves and stems of tomatoes naturally contain the toxin **solanine**; and **thiosulfates**, which are toxic to dogs and cats, are components of onions. We conclude the chapter with the question, "Is our food safe to eat?" and with an examination of what we mean when we use the word "safe."

WIN-Initiative/Neleman/Riser/Getty Image

CHAPTER OUTLINE

14.1 Micronutrients

LEARNING OBJECTIVES

1. **Define** vitamins and describe their classification and roles in health.
2. **Define** minerals and describe their classification and roles in health.

In Chapter 5 we studied the macronutrients of food: its fat, carbohydrates, and proteins. But food contains small quantities of other essential natural products as well, known as **micronutrients**. Obtaining adequate amounts of these micronutrients, which include vitamins and minerals, is essential for good health. In this section, we explore the chemistry of micronutrients, learn about their dietary sources, and examine the health consequences of micronutrient deficiencies.

micronutrients Dietary substances needed in trace amounts for proper health.

Vitamins

Vitamins are organic compounds that form in small amounts within the plants and animals we eat as food, and that are essential for life and good health. The word "vitamin" came into our language in 1912, devised by the Polish biochemist (later, American citizen) Casimir Funk. He discovered these organic substances and believed that they were all amines (Section 12.2) that are vital to our health, or "vital amines." Hence he named them "vitamines," later shortened to our current "vitamins." We now know that not all vitamins are amines, but we also know that they are, indeed, vital to our health.

vitamins Organic compounds required in small amounts for proper health.

Vitamins Table 14.1

Chemical Name (and Letter Designation)	Dietary Source	Deficiency Symptoms or Disease
Water-Soluble Vitamins		
Ascorbic acid (vitamin C)	Fruits, especially citrus; many vegetables	Scurvy; degeneration of tissues
Biotin	Various foods	Rare; nausea, loss of appetite
Cobalamin (vitamin B_{12})	All foods of animal origin	Pernicious (fatal) anemia
Folic acid	Meat; green, leafy vegetables	Anemia
Niacin, or its equivalent in tryptophan (vitamin B_4)	Meat; legumes; grains	Pellagra; skin, digestive, and nervous system disorders; depression
Pantothenic acid (vitamin B_3)	Widely distributed; occurs in virtually all foods	Defects in metabolism
Pyridoxine, Pyridoxal, Pyridoxamine (Forms of vitamin B_6)	Various foods; widely distributed	Deficiency is rare; results in defects in amino acid metabolism with various symptoms
Riboflavin (vitamin B_2)	Meat, especially animal organs; milk and dairy products; green vegetables	Skin disorders
Thiamine (vitamin B_1)	Pork; animal organs; whole grains; nuts; legumes	Beriberi; muscular weakness; paralysis
Fat-Soluble Vitamins		
Cholecalciferol (vitamin D_3)	Liver and liver oils; fatty fish (such as salmon); fortified milk	Rickets; malformation of the bones; osteomalacia (adult counterpart of rickets)
Retinol (vitamin A)	Liver and liver oils; carrots and other deeply colored vegetables	Night blindness; degenerative diseases of the eyes leading to total blindness
α-Tocopherol (vitamin E)	Various foods, especially grain oils	Deficiency disease unknown in humans
Vitamin K	Plants and vegetables; produced by intestinal bacteria and absorbed through the intestinal wall	Deficiency disease unknown in adults; needed for blood clotting

Adapted with permission of John Wiley & Sons, Inc. from Snyder, C., *The Extraordinary Chemistry of Ordinary Things*, Fourth Edition, p. 466. Copyright 2003.

Vitamin C • Figure 14.1

The structure of vitamin C, ascorbic acid, indicates that it is water-soluble.

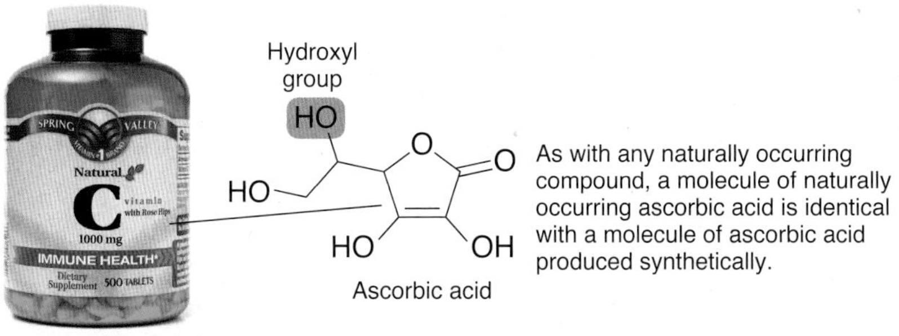

Hydroxyl group

HO

HO

O

O

HO OH

Ascorbic acid

As with any naturally occurring compound, a molecule of naturally occurring ascorbic acid is identical with a molecule of ascorbic acid produced synthetically.

smartstock/iStockphoto

Think Critically

1. How many hydroxyl groups are present in ascorbic acid?
2. Most acids present in foods are carboxylic acids (Section 8.3). Is ascorbic acid a carboxylic acid? Explain.
3. What structural features of ascorbic acid help make this compound water-soluble? (You may find it helpful to refer to the discussion of solubility in Section 7.1.)

Like all chemicals, each vitamin has a specific chemical name. In addition, many vitamins are associated with a specific capital letter of the alphabet, and some carry a numerical subscript to that letter. Ascorbic acid, for example, a vitamin that protects us from a disease called scurvy, is also known as Vitamin C. Retinol, a vitamin needed for healthy eyes, is referred to as Vitamin A. Among the vitamins that carry subscripts is a large set of the B vitamins, known collectively as the B complex:

Thiamine (B_1), riboflavin (B_2), pantothenic acid (B_3), niacin (B_4), pyridoxine (B_6), and cobalamin (B_{12}).

Vitamins can be divided into two important groups: the **fat-soluble vitamins** and the **water-soluble vitamins**. Fat-soluble vitamins are much more soluble in fats, hydrocarbons, and similar solvents than in water. Water-soluble vitamins exhibit the opposite property. Vitamins A, D, E, and K constitute the fat-soluble class, and the B complex and vitamin C are water-soluble. This classification (fat-soluble vs. water-soluble) doesn't tell us much about what kinds of food contain any particular vitamin. Some fatty foods can be good sources of the water-soluble vitamins, and moist, green, leafy vegetables often provide ample supplies of fat-soluble vitamins. **Table 14.1** presents the best-known vitamins and brief descriptions of their dietary sources and biological properties.

A full description of the chemistry and the nutritional importance of all the vitamins is beyond the scope of this discussion. Instead, we'll look closely at three of the better-known vitamins: the water-soluble vitamin C and the fat-soluble vitamins A and D.

Vitamin C, a water-soluble vitamin Vitamin C, the best known of all the vitamins, is one of the most widely used dietary supplements. Its chemical name, *ascorbic acid* (**Figure 14.1**), represents two properties of the substance, one chemical, the other biological. First, Vitamin C is an acid, although it's not a carboxylic acid. Second, the term **ascorbic** reflects its ability to protect against the disease **scurvy**. (The words **scurvy** and **ascorbic** come from the Latin word for this disease, *scorbutus*.)

The symptoms of scurvy can develop from a dietary deficiency of vitamin C. Humans require this vitamin for the formation of bonds that hold together strands of the collagen of our connective tissues (Section 5.4). Tough fibers of this protein are important components of our skin, muscles, blood vessels, scar tissue, and similar bodily structures. The gums in particular are rich in blood vessels and are subject to wear and abrasion through eating and the brushing of teeth. Thus, the bleeding of gum tissue, weakened by deterioration of its collagen, can be the first visible symptom of scurvy. As the disease progresses, the gums weaken and decay, the teeth fall out, and the body bruises easily.

Until a few hundred years ago, scurvy was the scourge of sailors, explorers, and those on long military expeditions—people who had to survive for months or years on stored provisions. These provisions lacked the fresh fruits and vegetables that are rich in vitamin C. Although some Dutch and English seamen of the 1500s knew of the value of fresh fruit and lemon juice in preserving sailors' health on long voyages, it wasn't until 1795 that the British navy officially made lemons part of the required stores of its fleet at sea.

Vitamin A, a fat-soluble vitamin Vitamin A maintains the health of eyes, skin, and mucous membranes and is particularly important for good vision in dim light. A deficiency can produce "night blindness," an inability to see in the dark. More severe deficiencies can cause eventual blindness.

Formation of vitamin A • Figure 14.2

Human metabolism converts one molecule of β-carotene into two molecules of vitamin A, also known as retinol.

β-carotene

Chemical reactions within the body

Retinol

OH

HO

Retinol

Think Critically

1. What chemical group is present in retinol but not in β-carotene?

2. What structural features of β-carotene and retinol help make these compounds fat-soluble?

(To answer this question, you may find it helpful to review Section 7.1 on factors that affect solubility.)

Foods rich in β-carotene include root vegetables, such as sweet potatoes and carrots; melons and squashes, such as cantaloupe and pumpkin; green leafy vegetables, such as spinach and kale; and certain fruits, such as apricots, mangos, and papayas.

Wally Eberhart/The Image Bank/Getty Images

As one of the fat-soluble vitamins, vitamin A accumulates in the body's fat cells, particularly in the liver. A moderate excess of vitamin A in our bodies normally does no harm. We simply store it, largely in our livers, for use later, when it might be in short supply. A large excess, though, can overwhelm the body's storage capacity and can generate toxic symptoms, including nausea, blurred vision, loss of hair, and worse, including death. Moreover, even though vitamin A is required for the normal development of an embryo, excesses taken in early stages of pregnancy–and even shortly before conception, since the vitamin can be stored–can produce serious birth defects.

Toxic doses of the vitamin can result from overzealous use of vitamin supplements and from eating certain rare and bizarre foods that are particularly rich in vitamin A. Polar bear livers, for example, contain unusually large quantities of the substance; a 3-oz portion of polar bear liver provides about 200 times the currently recommended daily intake for an adult. There have been reports of extreme toxicity and death among early Arctic explorers who ate large quantities of these livers.

Although animal livers can store very large concentrations of vitamin A, plants don't contain any. Carrots, which are supposed to be a very good source of the nutrient (and indeed are), have none whatever. This strange-sounding contradiction makes more sense when we examine the chemistry of the vitamin.

Our bodies can use vitamin A (known chemically as *retinol*) itself, or we can easily convert several chemical structures closely related to retinol into the vitamin, or even use these closely related compounds directly as retinol's equivalents. For this reason, vitamin levels in foods or supplements are sometimes stated in Retinol Equivalents (REs) of the vitamin, or in International Units (IUs). REs and IUs represent not only the quantity of the vitamin itself, but the total amount of the vitamin and all its physiological equivalents that can be converted into or used as the vitamin. (One RE of vitamin A is equivalent to 1 μg—one microgram or 10^{-6} g—of retinol. One IU of vitamin A represents 0.3 μg of retinol. A little arithmetic shows that 3.33 IUs of retinol constitute one RE.)

Although carrots contain no retinol, they do provide large amounts of β-carotene, a deeply colored material that we easily convert into vitamin A (**Figure 14.2**). With enough of the vitamin stored in our liver, our body stops converting the β-carotene into retinol. As a result, we can't poison ourselves by overeating carrots and other deeply colored red, orange, and yellow fruits and vegetables, all of which provide plenty of β-carotene.

When we consume more β-carotene than we need, some of it reaches our surface tissues, where it's stored and begins imparting its own color to the skin and eyes. There are reports of people whose skin acquires a yellow–orange tint from consuming enormous quantities of carrots and/or tomato juice (which is also rich in carotenes) over a long period. It's a harmless and reversible condition that disappears when carrots and tomatoes are removed from the diet.

Vitamin D, a fat-soluble vitamin Vitamin D promotes the absorption of food-borne calcium and phosphorus through our intestinal wall and into the bloodstream, thereby providing the raw materials for forming and maintaining healthy bones. Without vitamin D, children's bones develop poorly, resulting in the severely bowed legs and other skeletal deformations that characterize the condition known as rickets. Fortifying milk with added vitamin D makes sense because milk is rich in calcium, and the fat-soluble vitamin dissolves readily in milk's fatty components.

However, there's another side to vitamin D. Both rickets and vitamin D itself may be largely the products of the way civilization has developed in the northern and southern latitudes. For some of us, the chemical may not be a vitamin at all because vitamin D, unlike other vitamins, can form in our own bodies. Under the right conditions, we generate vitamin D in our own skin through the action of the sun's ultraviolet rays.

Furthermore, vitamin D isn't even a single organic compound. The term applies to a set of very closely related molecular structures, differentiated from each other by subscripts. There are vitamins D_1, D_2, D_3, and so on, all with the same physiological function. As ultraviolet radiation strikes our skin, it converts a bodily substance known as *7-dehydrocholesterol* into the form of the vitamin known as vitamin D_3, or *cholecalciferol* (**Figure 14.3**).

With plenty of exposure of our skin to sunshine, lots of cholecalciferol would be in plentiful supply in all our bodies, rickets would be virtually unknown, and vitamin D wouldn't be a vitamin at all. But with the limited sunshine, copious clothing, and indoor living and working conditions of the more extreme latitudes, the chemical becomes scarce. This could be especially dangerous for children, whose bones are still growing. Under these conditions, rickets would be common if it weren't for the addition of vitamin D to children's diets, usually in the form of fortified dairy products.

Minerals

You might think of minerals as hard, sometimes metallic or crystalline substances we obtain from the ground, ranging in value from lead to gold, or to gems such as opals or rubies. Nutritionally, though, the term **minerals** refers to all the nutritionally important chemical elements of our foods, except for the carbon, hydrogen, nitrogen, and oxygen that come to us with the macronutrients. Of all the minerals, calcium is the most abundant in our bodies, ranking fifth among all of our atoms, just behind hydrogen, oxygen, carbon, and nitrogen.

minerals Elements needed by the body in small amounts for proper health.

Two significant features distinguish minerals from vitamins: their structure and their origin. Vitamins are chemical compounds, composed of molecules in which atoms of various elements are held together by covalent bonds. However, all the minerals of food consist of atoms or ions of individual elements. The molecules of vitamins are formed biochemically by the bonding together of various elements within the structures of the plants or animals we use as food; minerals must be absorbed by plants from the soil in which they grow, or ingested by animals as they eat the plants that have absorbed the minerals. In essence, vitamins are synthesized by the living things we eventually eat; minerals can't be created by living things, but (ultimately) must be absorbed from the earth in which they exist.

Major minerals Between 1.5 and 2.0% of your weight comes from calcium—in the form of the Ca^{2+} ion—in your body, and more than 99% of all this calcium lies in your bones and teeth. This means that if you weigh, for example, 70 kg (154 lb), you're carrying around some 1.0–1.5 kg (roughly 2–3 lb) of calcium in your skeleton and in your mouth.

Formation of vitamin D • Figure 14.3

We naturally produce vitamin D when our skin is exposed to the ultraviolet radiation of sunlight.

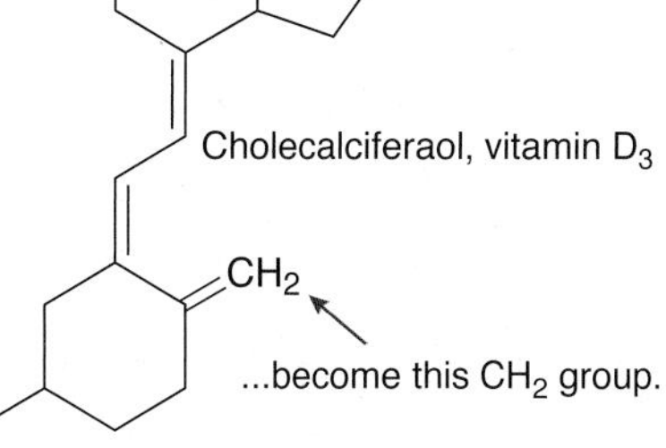

7-dehydrocholesterol, a precursor to vitamin D_3, which occurs in the skin.

Sunscreens offer benefits in protecting the skin against sun damage, but they also decrease vitamin D synthesis.

Uwe Krejci/Digital Vision/Getty Images

Ask Yourself

1. What structural differences exist between 7-dehydrocholesterol and vitamin D_3?
2. Why does wearing sunscreen decrease vitamin D synthesis?

Adapted with permission of John Wiley & Sons, Inc. from Snyder, C., *The Extraordinary Chemistry of Ordinary Things*, Fourth Edition, p. 469. Copyright 2003.

major minerals Minerals required in the diet in amounts greater than 100 mg per day or present in the body in amounts greater than 0.01% of body mass.

Next in rank of the **major minerals** comes the element phosphorus. Although calcium is the major mineral of our bones and teeth, it takes about half a gram of phosphorus to pack each gram of calcium firmly into the lattices of your bones and teeth and to hold it there. Add to this the smaller amounts of phosphorus in our soft tissues, and a 70-kg person carries around a total of almost a kilogram (about 2 lb) of this element.

We have smaller amounts of potassium, chlorine, sodium, and magnesium inside us, in that order. These elements form the major ions of the fluids in and around cells. (The fluids within the cells are *intracellular*; those outside are *extracellular*.) Potassium ion (K^+), the principal cation inside the cells, governs the activities of the cellular enzymes. Sodium ion (Na^+), which is the dominant cation outside the cells, keeps the water content of the intracellular and extracellular fluids in a healthy balance and thereby helps maintain proper fluid levels in the body. Both sodium and potassium cations regulate the distribution of hydrogen ions throughout the body and thereby keep the acidities of various bodily fluids within their normal ranges. They also play a critical role in the transmission of nerve signals within the body.

Along with these alkali metal cations, chloride anions (Cl^-) also help regulate fluid balances and, in combination with protons, provide the hydrochloric acid in gastric juices. They also serve to balance the electrical charges of the various bodily cations.

Magnesium—in the form of the Mg^{2+} ion within the body—plays several secondary roles. It's second to calcium as the hard mineral of bones and teeth, and second to potassium in regulating biochemical activities within the cells. Magnesium helps control the formation of proteins inside the cells and the transmission of electrical signals from cell to cell.

Another major mineral, sulfur, is a component of the amino acids methionine and cysteine and hence is found within various proteins of the body. Sulfur is also present within other biomolecules, including one essential for the metabolism of all foods and another that helps rid the body of toxins. Sulfur also occurs as part of the sulfate ion (SO_4^{2-}), one of several anions in the blood.

Trace minerals Together, just 11 elements–the carbon, hydrogen, nitrogen, oxygen, and sulfur of the macronutrients and of water, and the calcium, phosphorus, potassium, chlorine, sodium, and magnesium of the micronutrients–compose well over 99% of the mass of all living matter. All the remaining elements necessary for good health constitute the category of **trace minerals**. Generally, we consider trace minerals to be those needed

trace minerals Minerals required in the diet in amounts of less than 100 mg per day.

Dietary sources of minerals Table 14.2

Major Minerals	Source
Calcium, Ca	Milk and dairy products such as cheese and ice cream; fish, such as sardines, eaten with their bones; broccoli and other dark green vegetables; legumes
Chlorine, Cl	Table salt
Magnesium, Mg	Whole grains; nuts; green vegetables; seafood
Phosphorus, P	Nearly all foods
Potassium, K	Nearly all foods, especially meat, dairy products, and fruit
Sodium, Na	Table salt
Sulfur, S	Dietary proteins of meat, eggs, dairy products, grains, legumes
Minor Minerals	**Source**
Chromium, Cr	Brewer's yeast; meat (except fish); whole grains
Cobalt, Co	Most animal products, including meta, milk, and eggs
Copper, Cu	Liver and kidneys; shellfish; nuts; raisins; dried legumes; drinking water in some areas
Fluorine, F	Drinking water (in some regions); tea; fish, eaten with their bones
Iodine, I	Iodized salt; seafood; bread
Iron, Fe	Liver and other red meats; raisins; dried apricots; whole grains; legumes; oysters
Manganese, Mn	Nuts; whole grains; leafy green vegetables; dried fruits; roots and stalks of vegetables
Molybdenum, Mo	Animal organs; cereals; legumes
Selenium, Se	Various foods, including Brazil nuts, grains, meat, and seafood
Zinc, Zn	Meat; eggs; seafood (particularly oysters); dairy products; whole grains

WHAT A CHEMIST SEES

Dietary Minerals and the Periodic Table

Dietary minerals are atoms or ions of elements that are necessary in small amounts for good health.

Major minerals collectively account for about 3% of the mass of our bodies. These minerals include the Group 1 elements, sodium (Na) and potassium (K); the Group 2 elements, magnesium (Mg) and calcium (Ca); and the nonmetals, phosphorus (P), sulfur (S), and chlorine (Cl). Calcium is the most abundant major mineral, followed by phosphorus.

Trace minerals are present in far lower quantities within the body than the major minerals. These elements include the Period 4 transition metals, vanadium (V) through zinc (Zn); the Period 5 transition metal, molybdenum (Mo); and the nonmetals, selenium (Se), fluorine (F), and iodine (I). Iron is the most abundant trace element within the body, yet constitutes only 0.005% of our body mass.

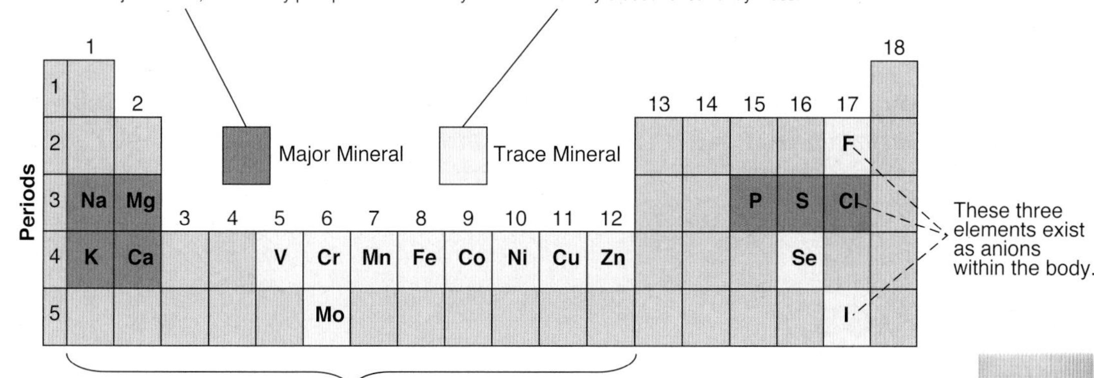

These elements exist as cations within the body.

Since almost all dietary minerals exist as ions, multimineral supplements, such as that shown here, provide these minerals in the form of ionic compounds.

Phosphorus (as Dicalcium Phosphate)
Iodine (as Potassium Iodide)
Magnesium (as Magnesium Oxide)
Zinc (as Zinc Oxide)
Selenium (as Sodium Selenate)
Copper (as Cupric Sulfate)
Manganese (as Manganese Sulfate)
Chromium (as Chromium Picolinate)
Molybdenum (as Sodium Molybdate)
Chloride (as Potassium Chloride)
Potassium (as Potassium Chloride)

Ken Karp

Ask Yourself

Just four elements—hydrogen, carbon, nitrogen, and oxygen— make up 96% of the mass of our bodies. Where are these elements located in the periodic table?

in our diets at levels of less than a tenth of a gram (less than 100 mg) per day.

Among the more important trace minerals are iron, fluorine, zinc, copper, selenium, manganese, iodine, molybdenum, chromium, and cobalt, arranged here in decreasing order of their presence in an adult body. They function in various ways, with most of them incorporated into the structures of enzymes, hormones, and related biologically important molecules.

Iron forms a critical part of the hemoglobin molecule in red blood cells. The iron of hemoglobin bonds to oxygen molecules acquired as the blood passes through the lungs. As the blood circulates throughout the body, it dispenses this oxygen to all the body's cells. Anemia, a condition characterized by fatigue, results from a deficiency of iron and the resulting inability of the blood to carry sufficient oxygen to the cells.

Fluorine, acting as the fluoride anion, helps harden the enamel of teeth, making them resistant to decay. Many communities add fluoride salts to drinking water to protect the teeth of children. The fluoride anion, often as sodium fluoride, is also an ingredient of most anticavity toothpastes.

Other trace minerals include zinc, which promotes growth, the healing of wounds, and the development of male sex glands. Iodine is important to the proper development and operation of the thyroid gland in the neck. A deficiency of iodine can lead to enlargement of the thyroid gland, a condition known as a goiter, and has been implicated in learning deficiencies. A single cobalt atom is incorporated into every molecule of vitamin B_{12}.

Still other trace minerals play various roles in the body. Deficiencies of copper can produce symptoms ranging from changes in hair color to anemia and bone disease, and may impair the immune system. Manganese is essential for healthy bones, a well-functioning nervous system, and reproduction. Chromium plays an important part in the metabolism of glucose.

Table 14.2 summarizes some of the more common dietary sources of minerals. *What a Chemist Sees* summarizes some of the distinctions between the major minerals and the trace minerals, and their locations within the periodic table.

Eating a well-balanced diet and a variety of foods is the best way to ensure that you are obtaining sufficient micronutrients. Generally, healthy adults who maintain a good diet do not need to take vitamin and mineral supplements. Nevertheless, these supplements may be recommended for certain groups of individuals, including:

- those on calorie-restricted diets (<1600 Calories/day)
- vegans (who may lack sufficient vitamin B_{12}, since this vitamin does not occur in plants)
- those who abstain from eating dairy products (and who may therefore lack sufficient calcium and vitamin D)
- women of child-bearing age and pregnant women (who should consume at least 400 mg of folic acid daily), and
- older adults

CONCEPT CHECK STOP

1. **Why** are carrots considered a good source of vitamin A even though they do not contain any of this substance?
2. **What** is the most abundant (a) major mineral and (b) trace mineral within the body?

14.2 Food Additives

LEARNING OBJECTIVES

1. **State** the reasons that prompted the regulation of food additives in the United States.
2. **Summarize** the legal definition of a food additive.
3. **Describe** the classifications and functions of food additives.

We'll now turn from the vitamins and minerals naturally present in our foods to an examination of **food additives**. These are substances we add to our foods as we prepare them, as we eat them, and as commercial manufacturers process and package them. They serve a variety of functions and are closely regulated by the federal government.

food additives Chemicals added to foods to help preserve them and to make them more nutritious and more appealing.

The Need for Federal Supervision

Additives of one kind or another have been used since prehistoric times as cooks smoked meat and added salt in order to preserve meat against decay. Food dyes were used by the Egyptians as early as 3500 years ago in order to improve the appearance of food. Adding herbs, spices, and sweeteners to improve taste is also an ancient practice.

Another ancient practice, especially among the manufacturers and vendors of food, was the use of additives to make spoiled food more palatable. Covering the foul tastes and odors of decay in food was the principal function of spices in the Middle Ages. Indirectly, the lack of good methods for preserving food in those times led to the opening of trade routes to the Orient and to the European discovery of the Western Hemisphere. It was at least partly the search for foreign spices that led Marco Polo eastward to the Orient and Columbus westward to the New World.

With the development of chemical technology and food processing late in the 19th century, the use of chemical additives grew enormously and indiscriminately. By 1886, the U.S. Patent Office had issued its first patent for a food additive—a combination of sodium chloride and calcium phosphate designed for use as a food seasoning. As the use of additives grew, the variety of the chemicals used as additives also increased, and so did their hazards.

We now have governmental regulations designed to manage the addition of chemicals to our foods. That wasn't always the case. In the United States of the late 1800s, the uninformed and widespread use of chemicals as food additives led to frequent and sometimes fatal outbreaks of illness and, eventually, to detailed oversight by the federal government. The first step, taken in 1902, was the formation of a group within the U.S. Department of Agriculture to examine both the usefulness and the dangers of food additives. Headed by Dr. Harvey Wiley, an American chemist, this group involved a dozen healthy young volunteers who actually ate the additives under investigation, along with their regular meals, while Wiley watched for signs of illness.

Wiley's investigations and his vigorous public support of effective control of food additives, coupled with public indignation at the filth of slaughterhouses and meat products depicted in Upton Sinclair's novel *The Jungle* (1906), stimulated Congress to pass both the Meat Inspection Act of 1906 and the Pure Food and Drug Act of 1906. Since 1906, the enforcement of the Pure Food and Drug Act and its subsequent revisions has been the responsibility of organizations bearing a variety of names and titles. In 1931, Congress created the Food and Drug Administration (FDA) to ensure the safety of the chemicals added to the food and drugs sold to the public. The FDA is now part of the U.S. Department of Health and Human Services, and the Pure Food and Drug Act is now known as the Federal Food, Drug, and Cosmetic Act (or, for brevity, the FD&C Act).

Defining Food Additives

Chapter IV of the FD&C Act covers food additives and defines a food additive as any substance added to a food, directly or indirectly, except for:

- those shown by scientific studies to be safe, at least under the conditions of their use in foods;
- those used as additives before January 1, 1958 (the date this part of the Act became effective), and shown to be safe either by scientific studies or by common experience;
- pesticides;
- color additives;
- substances approved by earlier acts of Congress;
- new animal drugs; and
- ingredients of dietary supplements.

To work backward through this list of exemptions, the bottom five are defined and regulated either in other laws or in some other section of this Act, so they aren't actually considered to be food additives by this particular section of the law.

The second exemption results from a combination of expediency and common sense. By January 1, 1958, the date this particular section of the Act went into effect, so many substances had been used as food components regularly and for so long that testing all of them would have been entirely impractical. Instead, panels of scientists evaluated the safety of these materials that had been in common and long-term use. Those that were generally recognized by the panelists to be safe, on the basis of common experience (but without any laboratory testing), were compiled into a list of substances **G**enerally **R**ecognized **A**s **S**afe, called the **GRAS list**. In effect, the GRAS list constitutes the second category of exemptions. (Legally, GRAS substances aren't food additives either.)

GRAS list A list of chemical substances generally regarded as safe by the FDA.

This GRAS list is by no means fixed and unchangeable, nor are all the substances on the list necessarily free of hazard. They are simply generally recognized as safe, and they are reviewed periodically to determine their continued suitability for GRAS status. **Table 14.3** contains

Examples of GRAS substances Table 14.3

Category	Substance	Structure or Chemical Formula
Flavoring agents	Acetaldehyde	$CH_3-C(=O)-H$
Flavoring agents	Anise	
Flavoring agents	Cinnamon	
Flavoring agents	Ethyl acetate	$CH_3-C(=O)-O-CH_2-CH_3$
Anticaking agent	Aluminum calcium silicate	$CaAl_2(SiO_4)_2$ and $Ca_2Al_2SiO_7$
Preservatives	Sodium metabisulfite	$Na_2S_2O_5$
Preservatives	Sorbic acid	$CH_3-CH=CH-CH=CH-CO_2H$
Dietary Supplements	Ascorbic acid	$C_6H_8O_6$
Dietary Supplements	Calcium phosphate	$Ca_3(PO_4)_2$
Dietary Supplements	Ferrous sulfate	$FeSO_4$
Dietary Supplements	Linoleic acid	$CH_3-(CH_2)_4-CH=CH-CH_2-CH=CH-(CH_2)_7-CO_2H$
Dietary Supplements	Zinc oxide	ZnO
Sequestraits	Dipostassium hydrogen phosphate	K_2HPO_4
Sequestraits	Trisodium citrate	$HO-C(CH_2\text{-}CO_2Na)_2-CO_2Na$

Adapted with permission of John Wiley & Sons, Inc. from Snyder, C., *The Extraordinary Chemistry of Ordinary Things*, Fourth Edition, p. 481. Copyright 2003.

Representative substances used to increase the appeal of foods Table 14.4		
Substance	**Structure or Chemical Formula**	**Function**
β-Carotene	$C_{40}H_{56}$	Colorant
Ethyl acetate	$CH_3-C(=O)-O-CH_2-CH_3$	Flavoring and fragrance
Ferric oxide	Fe_2O_3	Colorant
Glucose	$C_6H_{12}O_6$	Sweetener
Monosodium glutamate, MSG	$HO_2C-CH_2-CH_2-CH(NH_2)-CO_2Na$	Flavor enhancer
Paprika		Flavoring and colorant
Sucrose	$C_{12}H_{22}O_{11}$	Sweetener
Titanium dioxide	TiO_2	Colorant

examples of chemicals and other substances currently on the list, together with brief descriptions of their functions in food. We'll explore some of these functions in greater depth shortly.

The remaining exemption, the first one of the series, requires that newly developed chemicals, which are by no means GRAS, be shown by scientific studies to be safe before they can be used in foods. Once they are demonstrated to be safe and may legally be added to food, they're no longer legally defined as food additives.

There are other good and workable (and certainly simpler) definitions of food additives, aside from this legal one of the Federal Food, Drug, and Cosmetic Act. A more practical definition holds that a food additive is simply **anything intentionally added to a food to produce a specific, beneficial result, regardless of its legal status**. The importance of the extended, legal definition of the Act comes from the force of U.S. federal law behind it. As a result, it has the close attention of anyone who processes food for sale to the public.

Functions of Additives

We use additives simply to improve or maintain the quality of our foods. For many of us, adding some combination of salt, sweeteners such as sucrose (table sugar) or an artificial sweetener, and seasonings like mustard, pepper, and other condiments can improve the taste of an otherwise bland dish. Food colorings, another class of additives, enhance the appearance of foods by converting a pallid frosting, for example, into the colorful decorations of a birthday cake.

Other additives function in less obvious ways. Adding vitamin D to milk doesn't add any appeal to the milk through taste, odor, or color, but it does increase the nutritional value of milk by improving the absorption of calcium into the body. Ascorbic acid (vitamin C), on the other hand, not only increases the nutritional qualities of food but also protects other food components from air-oxidation. The ascorbic acid is itself so easily oxidized that it reacts preferentially with atmospheric oxygen, thereby protecting other food components from similar oxidation. As a food additive, then, ascorbic acid acts both as a nutrient and as a preservative (specifically, an antioxidant).

Although additives serve a large variety of purposes in foods, we can place them into four major groups, according to their function. Generally, chemicals are introduced into foods in order to make them more appealing or nutritious, to preserve freshness, to make them easier to process, and to keep them stable during storage.

These categories aren't exclusive, and they even overlap a bit. For example, we've just seen that through its own preferential oxidation, ascorbic acid can act both as a preservative and as a nutrient. In this same sense, any one chemical additive can function in two or more of these categories. **Tables 14.4–14.7** present some of the substances added to food to perform each of these four functions.

Representative substances used to improve nutrition Table 14.5		
Substance	**Chemical Formula**	**Function or Classification**
Ascorbic acid	$C_6H_8O_6$	Vitamin C
β-Carotene	$C_{40}H_{56}$	Provitamin A[a]
Ferrous sulfate	$FeSO_4$	Mineral
Potassium iodide	KI	Prevents goiter
Riboflavin	$C_{17}H_{20}N_4O_6$	Vitamin B_2
Zinc sulfate	$ZnSO_4$	Mineral

[a]A provitamin is a substance that can be converted easily into a vitamin.

Representative substances used as preservatives Table 14.6

Substance	Structure or Chemical Formula	Function
Ascorbic acid	$C_6H_8O_6$	Antioxidant and antimicrobial
Butylated hydroxyanisole, BHA	OH, $C(CH_3)_3$, OCH_3 (benzene ring); OH, $C(CH_3)_3$, OCH_3 (benzene ring)	Antioxidant
Butylated hydroxytoluene, BHT	OH, $CH_3-C(CH_3)_2-$, $-C(CH_3)_2-CH_3$, CH_3 (benzene ring)	Antioxidant
Calcium propionate	$(CH_3-CH_2-CO_2)^- Ca^{2+}\ ^-(O_2C-CH_2-CH_3)$	Inhibits growth of molds and other microorganisms
EDTA (ethylenediaminetetraacetic acid)	$C_{10}H_{16}N_2O_8$ (see Figure 14.4)	Antioxidant
Sodium benzoate	$CO_2^- Na^+$ (benzene ring)	Inhibits growth of microorganisms in acidic foods
Sodium nitrite	$NaNO_2$	Inhibits growth of microorganisms in meat
Sorbic acid and its salts	$CH_3-CH=CH-CH=CH-CO_2H$	Inhibits growth of molds and yeast, especially in cheese

Adapted with permission of John Wiley & Sons, Inc. from Snyder, C., *The Extraordinary Chemistry of Ordinary Things*, Fourth Edition, p. 485. Copyright 2003.

Representative substances used in processing and to maintain stability Table 14.7

Substance	Structure or Chemical Formula	Function
Acetic acid	CH_3-CO_2H	Control of PH
Calcium silicates	$CaSiO_3$, Ca_2SiO_4, Ca_3SiO_5	Anticaking agents
Glycerine	$CH_2-CH-CH_2$ OH OH OH	Humectant
Glyceryl monostearate	$CH_2\text{-}O_2C-(CH_2)_{16}-CH_3$, / $CH\text{-}OH$ / CH_2-OH $CH_2\text{-}OH$ / $CH-O_2C-(CH_2)_{16}-CH_3$ / $CH_2\text{-}OH$	Humectant
Gum arabic	(gummy plant fluid)	Thickener, texturizer
Mono-and diglycerides	$CH_2\text{-}O_2C-R$ / $CH-OH$ / CH_2-OH $CH_2\text{-}O_2C-R$ / $CH\text{-}O_2C-R$ / CH_2-OH	Emulsifier
Phosphoric acid and its salts	H_3PO_4, NaH_2PO_4, Na_2HPO_4, Na_3PO_4	Control of PH
Silicon dioxide	SiO_2	Anticaking agents
Xanthan gum	Complex polysaccharide from corn fermentation	Emulsifier, thickener

Adapted with permission of John Wiley & Sons, Inc. from Snyder, C., *The Extraordinary Chemistry of Ordinary Things*, Fourth Edition, p. 488. Copyright 2003.

Chemistry InSight Food additives • Figure 14.4

Chemicals added to foods perform a variety of functions, including increasing their appeal, nutritional content, freshness, and stability. The examples of food additives shown here appear on the FDA's GRAS (Generally regarded as safe) list.

a. The sequestrant, EDTA.

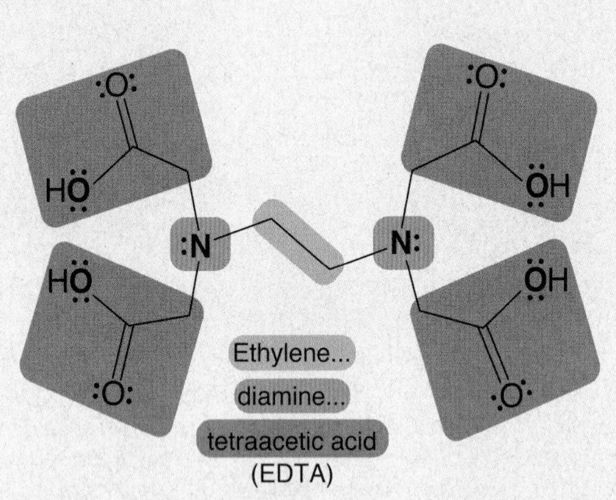

EDTA consists of an ethylene core ($-CH_2CH_2-$) to which two amine nitrogens are attached. Each of these amine nitrogens is bonded to two groups derived from acetic acid, for a total of four acidic structural units. Note that six of the atoms of the EDTA (shown in bold) can use their lone pair electrons to form covalent bonds to metal ions.

Adapted with permission of John Wiley & Sons, Inc. from Snyder, C., *The Extraordinary Chemistry of Ordinary Things*, Fourth Edition, p. 487. Copyright 2003.

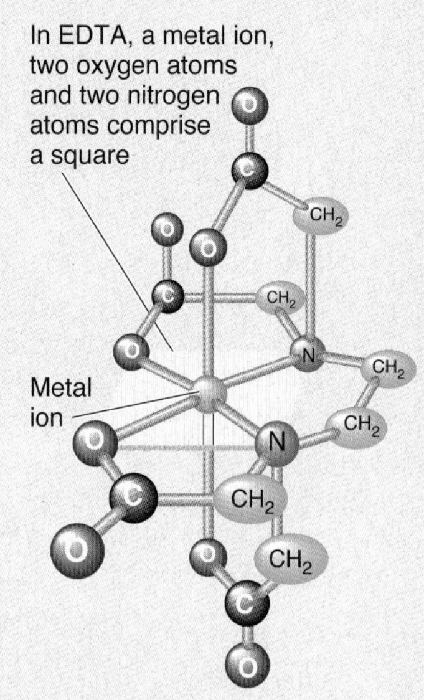

When added to foods, the EDTA molecules wrap around and bond to extraneous metal ions that may be present. This sequesters or traps metals ions as if in molecular cages, thereby preventing these ions from catalyzing the oxidation of fats and other components of foods. Sequestrants help prevent rancidity and preserve food quality.

Table 14.6 shows some chemical **preservatives**, which are compounds that help keep foods fresh and retard the process of spoilage and deterioration. This group of compounds include:

- **antioxidants**, which protect against oxidation by air, and
- **antimicrobials**, which protect against spoilage by bacteria and mold.

We've already seen that ascorbic acid, by oxidizing preferentially, can slow the reaction between food chemicals and atmospheric oxygen. Other widely used antioxidant additives include BHA and BHT, both of which contain the phenol group. The additives that guard food against the growth of bacteria and other microorganisms not only preserve the taste and appearance of food, but also protect us against microbiologically generated toxins. Labels on breads and other baked goods often list calcium propionate among their ingredients. Propionic acid occurs naturally in cheeses and other dairy products, and both this acid and its salts (such as calcium propionate) effectively inhibit the growth of mold, a major cause of the spoilage of both dairy products and baked goods. Packaged fruit drinks, soft drinks, and other processed foods often contain added sodium benzoate, the sodium salt of benzoic acid, a carboxylic acid. When present in amounts a little less than 0.1%, the sodium benzoate serves as an effective antimicrobial agent.

Some additives protect food against spoilage through indirect chemical action. One of these, EDTA or ethylene*d*iamine*t*etraacetic *a*cid, acts as an antioxidant (indirectly) by interacting with metal ions present in foods. Trace quantities of the ions of various metals—aluminum, iron, and zinc, for example—easily enter our processed foods as they go through their many stages

b. The humectant, propylene glycol.

Humectants, such as propylene glycol, help retain moisture through *hydrogen bonding* to water molecules. These coconut flakes contain propylene glycol to keep them moist.

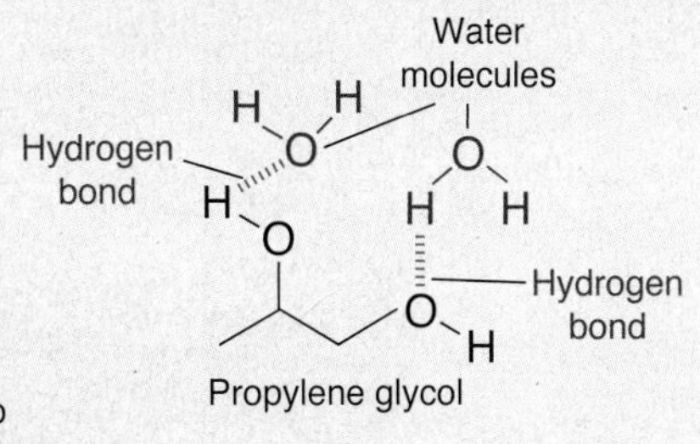

c. The emulsifiers, mono- and diglycerides.

Ingredients
MADE FROM ROASTED PEANUTS AND SUGAR, CONTAINS 2 PERCENT OR LESS OF: MOLASSES, FULLY HYDROGENATED VEGETABLE OILS (RAPESEED AND SOYBEAN), **MONO- AND DIGLYCERIDES** AND SALT.

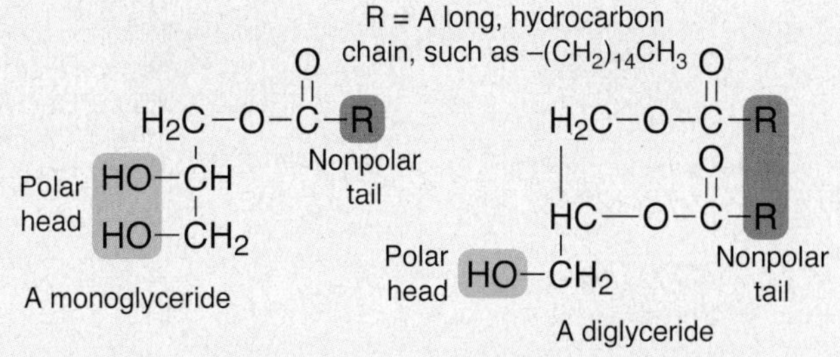

Emulsifiers, such as mono- and diglycerides contain both polar (water-attracting) groups and nonpolar (oil-attracting) groups, enabling two liquids that normally do not mix together, such as oil and water, to form a stable mixture or *emulsion*. Added to peanut butter, emulsifiers help prevent the peanut solids and oils from separating.

Think Critically Oxalates, derivatives of oxalic acid found in spinach, chocolate, and other foods, are known to interfere with the body's absorption of certain minerals, such as the ions of calcium and iron. Compare the structure of oxalic acid to EDTA and suggest a mechanism by which this interference occurs.

Oxalic acid

of preparation in metal equipment. Although these exceedingly small quantities of metal ions don't affect us or our foods directly, they can very effectively catalyze the air oxidation of many of the compounds that are present in food and thereby lead to spoilage. EDTA acts as a **sequestrant**, tightly binding with traces of metal ions within some foods to protect against this catalytic oxidation (**Figure 14.4a**).

Table 14.7 (on page 455) lists some processing aids and stabilizers—substances that help prevent undesirable changes in the appearance or physical characteristics of processed foods while they're being stored. Among these are **humectants**, which are compounds that help attract and retain moisture. Typical humectants include propylene glycol, which keeps shredded coconut moist, and glyceryl monostearate, which softens marshmallows (**Figure 14.4b**). **Anticaking agents**, such as silicon dioxide and the calcium silicates, keep table salt, baking powder, and other finely powdered food substances dry and free-flowing. **Emulsifiers** help blend oil and water into a stable, homogeneous mixture. Examples of emulsifiers include xanthan gum, which blends the oils and water of salad dressings into a smooth mixture; and mono- and diglycerides, which help peanut butter maintain a creamy smoothness during storage. Chemically, emulsifiers are surfactants (Section 11.1) that make it possible to mix two mutually insoluble liquids, such as oil and water (**Figure 14.4c**).

Consumers increasingly look to their diets to supply health benefits in addition to basic nutrition. Foods that contain substances capable of providing health benefits beyond conventional nutrition are known as **functional foods**. Chemicals providing these additional health benefits can occur naturally, such as the antioxidants carried

Functional foods • Figure 14.5

A variety of processed foods contain additives that may provide health benefits.

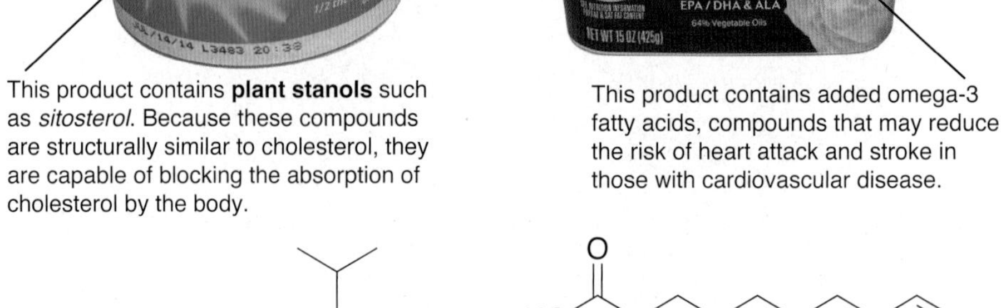

This product contains **plant stanols** such as *sitosterol*. Because these compounds are structurally similar to cholesterol, they are capable of blocking the absorption of cholesterol by the body.

This product contains added omega-3 fatty acids, compounds that may reduce the risk of heart attack and stroke in those with cardiovascular disease.

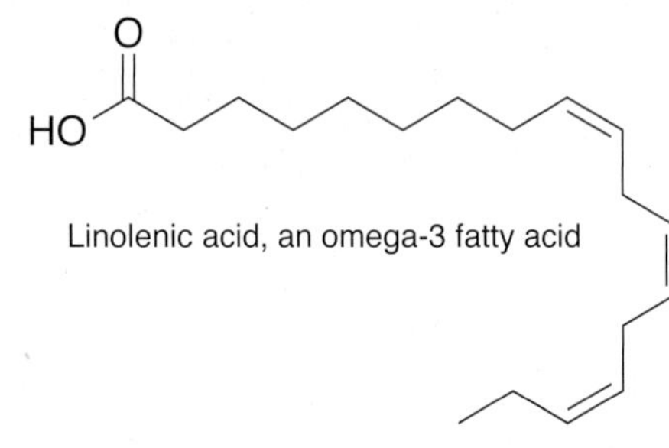

Linolenic acid, an omega-3 fatty acid

HO

Sitosterol

by many fruits and vegetables, or they can be added to foods during manufacture or processing (**Figure 14.5**). Although the U.S. Food and Drug Administration (FDA) has not yet formally recognized the term **functional food**, the agency does allow certain health-related claims to appear on food packaging, provided that these claims have sufficient scientific support.

CONCEPT CHECK — STOP

1. **What** federal agency is responsible for ensuring the safety of the chemicals added to the food and drugs sold in the United States?
2. **What** is the purpose of the GRAS list?
3. **What** functions does ascorbic acid serve as a food additive?

14.3 Food Safety

LEARNING OBJECTIVES

1. **Explain** why the toxicity of any substance is dosage-dependent.
2. **Describe** why identifying a substance as "safe" involves some acceptability of risk.

We all know that some of the things we eat or drink can make us sick, or can even poison us. But what we might not realize is that a poison doesn't have to be a rare, exotic, complex, or synthetic substance. A poison can be a natural component of foods, such as the caffeine of coffee, the tannins of tea, or the oxalic acid of rhubarb and spinach.

Toxicity of water • Figure 14.6

Substances that are normally benign or even necessary for life, such as water, can cause harm if consumed in gross excess.

a. Taken in moderation, water is essential to health and life.

fotum/iStockphoto

b. Swallowed in massive amounts, frequently, water can produce illness or death.

hkuchera/iStockphoto

Or it can be something we routinely add to meals, like the table salt many of us sprinkle on food every day.

Even something as simple and common as water—a liquid that makes up over half our weight and that's necessary for health and for the very existence of life itself—can make us sick or even kill us if we drink too much of it, too quickly. If drinking excessive amounts of water over a short period of time can cause illness and death, should we consider water as a poison, even though moderate amounts are necessary for health and life?

In this section we'll examine this and related questions as we discuss what makes a substance a poison and when we can consider something as safe. As we do this, we'll look at some of the most common hazardous substances of our everyday lives and some of the most powerful poisons.

Toxicity

The lethal effect of drinking excessively large amounts of water over short periods of time can result from large, rapid changes in the composition of the body's fluids and the concentrations of the body's **electrolytes**—especially a drastic drop in the level of sodium ions in the blood. This can produce a potentially fatal condition known as water intoxication or **hyponatremia**.

electrolytes The cations and anions that constitute the ionic compounds of our blood and other bodily fluids.

Among the many reported cases of fatal water intoxication was that of a 44-year-old man in England who died of a heart attack in 2008 after drinking huge quantities of water, approximately 17 pints over 8 hours a day for three consecutive days, in order to relieve a painful condition of his gums. Similarly, a 29-year-old Florida woman, convinced that poisons were accumulating in her body, died after drinking as many as 4 gallons of water a day for several days to flush these imaginary poisons out of her system. Marathon runners sometimes suffer nonfatal seizures from sharp drops in sodium concentrations within their bodies when they drink large amounts of water, rapidly, after they run.

Clearly, drinking excessive amounts of water, rapidly and frequently, can produce undesirable effects, ranging from discomfort to death. Should we then say that water *is* a poison, or simply that water *can be* a poison under certain conditions? We'll keep this question in mind as we continue (**Figure 14.6**).

As we proceed, we'll consider two types of harmful substances, poisons and toxins. Although these two terms are often used interchangeably, there is a subtle distinction between them. A **poison** is generally any chemical substance that can cause illness or death when it enters the body by eating, drinking, inhalation, or injection. In many mystery stories, arsenic is a common poison. In everyday life, lead and sodium cyanide are widely recognized as poisons. A **toxin** is a poison of biological origin, produced by a plant or an animal. It might enter the body if you eat a spoiled food, are infected by a virulent disease, are stung or bitten by an insect or a snake, or have a cut that becomes infected.

poison A chemical that enters our body and causes illness or death.

toxin A poison of biological origin.

The most sensible way to measure the virulence of a poison or a toxin is to determine the amount needed

The LD_{50} for aspirin • Figure 14.7

The lethal dosage of aspirin (and other substances) is defined as that which is lethal to 50% of a population of animal subjects, such as mice.

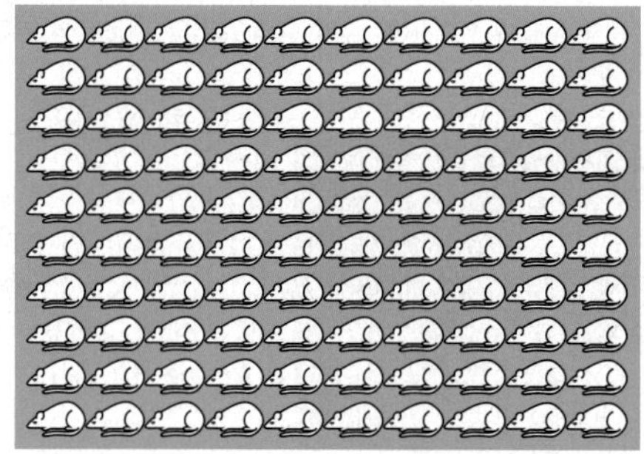

Feeding aspirin to a large group of mice at the level of 1.5 g of aspirin per kg of mouse . . .

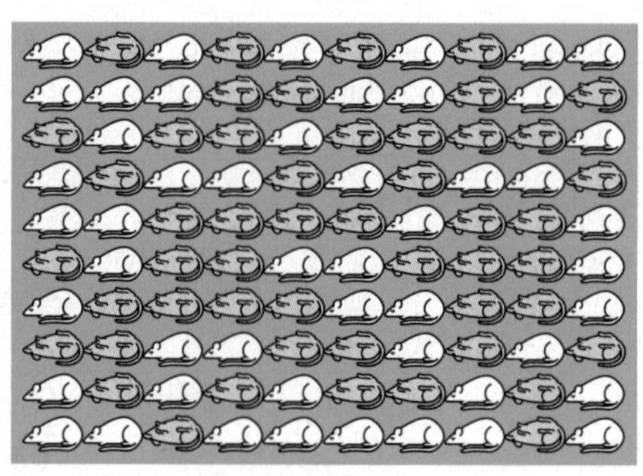

. . . kills half of them.

actually to kill a living thing: the lethal dose. There are some problems with this approach, though, even aside from any ethical issues involved in deliberately killing an animal to measure a chemical's lethal strength. As a practical matter, simply feeding the substance in question to a laboratory animal and determining the amount needed to kill the creature doesn't give a satisfactory measure of the lethal dose.

Individual animals of the same species, even of the same litter, can show different responses to identical stresses and to identical poisons. Some unusually sturdy individuals, for example, can withstand relatively large amounts of any particular toxin, while others, more susceptible, succumb to traces. To nullify the effects of these relatively rare individuals, it's customary to apply the test substance to large groups of animals, with each group containing a large, statistically realistic spectrum of individual susceptibilities. One useful measure that comes from studies of this sort is the amount of a chemical that kills exactly half of a large population of animals, usually within a week. It's known as the **LD_{50}** or **L**ethal **D**ose for 50% of the group.

> **LD_{50}** The dose of a substance that kills half of the animals tested.

Since a chemical's capacity to cause harm depends partly on its concentration in the animal's body, it's common practice to report the LD_{50} of a substance in terms of the weight of the poison per unit weight of the test animal. Moreover, because the method used for administering the substance often affects the results, that is usually included as well. For example, the LD_{50} for aspirin fed orally to mice and rats is 1.5 g/kg. This means that feeding aspirin orally to a large group of mice or rats, at the level of 1.5 g of aspirin per kilogram of animal, kills half of the population (**Figure 14.7**). From still another viewpoint, we could say there's a 50% chance that 1.5 g of aspirin per kilogram of body weight will kill any particular mouse or rat that eats it.

In contrast, the LD_{50} of sodium cyanide is only 15 mg/kg, fed orally to rats. Since LD_{50}s serve as a good guide to the potency of poisons, we can conclude that sodium cyanide is about 100 times as lethal as aspirin, at least when taken orally by rats. Notice that the smaller the value of the LD_{50}, the less it takes to kill and the more toxic the substance is. **Table 14.8** lists the LD_{50}s of some more familiar chemicals, including the caffeine of coffee and tea, the nicotine of tobacco, the glucose of food sweeteners, and the sodium chloride of table salt. (Although we've seen that huge quantities of water can cause illness and death when taken into the body over short periods of time, the toxicity of water is too low to be determined accurately as an LD_{50}.)

Although these LD_{50}s serve as a useful guide to a poison's lethal power, they have to be interpreted with a bit of caution. Different animal species often respond differently to a particular chemical and, as we noted earlier, the method of introducing the substance into the animal influences the results. Even for the same species, different methods of administration—orally, by injection under the skin (subcutaneously), directly into the abdominal cavity (intraperitoneally), or directly into the bloodstream (intravenously)—can produce different values for the LD_{50}. For example, the LD_{50} for nicotine administered to mice is 230 mg/kg orally, 9.5 mg/kg intraperitoneally, and only 0.3 mg/kg intravenously.

Considering that anything can cause harm if it's used in excess (a word whose meaning varies with the material we're talking about), it's clear that to be realistic we've got to put everything around us into the same category. This includes chemicals ranging from cyanide and the most lethal biological toxins to the caffeine, nicotine, glucose, and sodium chloride we mentioned above. It even includes water and our food's macronutrients, micronutrients, and chemical additives. Each one of these has the potential to cause harm.

Understanding this, we're now in a position to answer one of the questions that began this section: "Are there actual poisons in our foods?" The answer must be, "Yes, there are." But since everything presents a potential hazard, both the question and its answer have lost most of their meaning. The significant question now is whether our foods, with all their chemicals, are actually safe. The answer to this one isn't quite as clear-cut.

Safety, A Matter of Personal and Societal Judgment

Recognizing that there are, indeed, potentially harmful substances in our foods—poisons—we can now turn to the question "Are our foods safe?" A reasonable answer requires a reasonable definition of "safe." Often, when we say that something is "safe" we imply that it is free of danger, hazard, or harm. But no substance is inherently and totally free of danger, hazard, or harm. The harm any substance can do depends on:

- its chemical characteristics
- how much of it we use
- how we use it
- how susceptible to its hazards we are as humans
- how susceptible we are as individuals

For a more realistic test of safety, we can ask whether, knowing the risks involved in using a chemical (or in anything else for that matter), we are willing to accept those risks. In this sense **safety**, as defined by William Lowrance (a contemporary American organic chemist and consultant on health ethics and policy), is **the acceptability of risk**. It's a realistic and practical definition of a subtle and sometimes ambiguous term, and it provides us with a convenient way to examine the safety of our food supply and of chemicals in general. With this definition in mind, we can look to our own (informed) judgment in deciding matters of safety. In short, everything in life involves a risk of one sort or another. Those risks we find acceptable are the ones we, ourselves, define as safe. Those we find unacceptable we define as unsafe. All of us, collectively, make similar judgments about the risks we are willing to take as a society.

Viewing safety as a matter of acceptable risk shines a somewhat clearer light on any attempts to prove that something is indeed safe. It is simply impossible to prove that anything is safe in the sense that it presents no hazard whatever to anyone, at any time, in any possible circumstances. We might think that if we tried to demonstrate in 10,000 different ways that a certain chemical, for example, is harmful to mice and failed each time, we might have shown it to be safe. Yet all this is negative evidence, which proves nothing at all. There's always the possibility that one ingeniously designed additional test, a 10,001st experiment, might show that the chemical does, indeed, do harm to the mice.

LD_{50}s of some familiar substances Table 14.8

Substance	Animal[a]	LD_{50}
Acetaminophen (an analgesic in medications such as Excedrin and Tylenol)	Mice	0.34 g/kg
Acetic acid (major organic component of vinegar)	Rats	3.53 g/kg
Arsenic trioxide (poison of mystery stories)	Rats	0.015 g/kg
Aspirin	Mice, rats	1.5 g/kg
BHA (antioxidant food additive)	Mice	2 g/kg
BHT (antioxidant food additive)	Mice	1 g/kg
Caffeine	Mice	0.13 g/kg
Citric acid	Rats (abdominal injection)	0.98 g/kg
Ethyl alcohol	Rats	13 mL/kg
Glucose	Rabbits (intravenous)	35 g/kg
Niacin (vitamin B_4)	Rats (injection under skin)	5 g/kg
Nicotine	Mice	0.23 g/kg
Sodium chloride	Rats	3.75 g/kg
Thiamine hydrochloride (vitamin B_1)	Mice	8.2 g/kg
Trisodium phosphate (pH-adjusting food additive)	Rats	7.4 g/kg

[a]The substance is administered orally unless noted otherwise.

Interpret the Data

Which substance listed in this table shows the greatest potency as a poison?

With safety defined as the acceptability of risk, the question of which risks are acceptable and which are not falls on us as individuals and also as members of a larger society, one represented by our elected government and its laws. As individuals, for example, we're free to choose whether or not to use aspirin to relieve headaches and other minor pains, inflammations, and fevers. The choice is ours, individually, and the judgment of safety rests with each of us as individuals. In 1970, though, with the passage of the Poison Prevention Packaging Act, Congress placed a larger, societal judgment on free access to aspirin. Aspirin, they said in effect, is safe enough to be handled freely by anyone who can open a simple but obstinate twist-and-snap cap, but it is unsafe for those very young children who can't. That judgment has become society's assessment of the safety of aspirin.

It also serves as a model for our decisions about the safety of chemicals in foods. Each of us makes daily decisions about what to eat, but our representatives at the centers of government decide what chemicals are safe enough to be used as additives in commercial foods and in what quantities these additives may be used. The executive branch of the government operates through the Food and Drug Administration (the FDA) and many other agencies to implement legislative judgments of the safety of various chemicals. In addition to the FDA, these other agencies include the:

- Environmental Protection Agency (EPA), which regulates chemical pesticides, among other matters affecting the environment
- Occupational Safety and Health Administration (OSHA), which is concerned with exposure to chemicals in the workplace
- Bureau of Alcohol, Tobacco, Firearms, and Explosives (ATF), a division of the Department of Justice that has jurisdiction over beer, wine, liquor, and tobacco
- U.S. Public Health Service (USPHS), which, among its other activities, investigates outbreaks of illness due to food spoilage

Are natural foods free of chemicals? With the variety of chemicals that might be added to our processed foods, it would seem that eating natural foods, such as fresh fruits and vegetables, would expose us to far fewer hazardous chemicals than those present in processed foods. That's not necessarily the case. We know, for example, that all commercially processed foods must carry ingredients labels—lists of the chemicals added to them during processing—with the ingredients presented in decreasing order of their abundance. But what might our response be if *all* foods, both processed and natural, carried these ingredients labels?

Imagine what we might find if an orange growing on a tree or a potato freshly dug out of the ground carried a stamp showing all the chemicals it contained. To learn what's in unprocessed foods like oranges freshly off the trees or potatoes freshly out of the ground, or in apples, steaks, tomatoes, and the like, we're forced to turn to the chemists who analyze them and study their components. Without exception, the lists of chemicals isolated from foods grown on the farm and in the orchard and sold with little or no processing far exceed those appearing on the ingredients panels of processed foods. The oil of an orange, for example, contains more than 40 different chemicals, grouped by chemists into categories such as alcohols, aldehydes, esters, hydrocarbons, and ketones. A potato yields some 150 different compounds, each of which can be synthesized in a chemical laboratory or poured from a bottle on the chemist's shelf. Each is a chemical put there by nature, synthesized by the plant itself as it grows.

The mango, a particularly tasty tropical fruit, offers an excellent example of what an ingredients label on a piece of a natural, unprocessed fruit might look like. Mangoes are prized delicacies throughout much of the world, especially in India and the Far East. In terms of total tonnage consumed, the mango is the most popular of all fruits. **Figure 14.8** presents a partial list of the chemicals responsible for the flavor of a typical mango. It represents only a small fragment of all the chemicals that make up the fruit. Very few of them have been examined sufficiently for use as food additives.

Is the mango safe to eat? By any reasonable standard, and based on the masses of people throughout the world who eat mangoes daily without ill effects, the answer must be yes. (Of course, the answer could be no for those individuals who may be allergic to mangoes, for those with diabetes who may have to avoid the mango's sugars, and for others who may be sensitive to individual chemicals within the fruit.) Yet since its chemicals have not been tested for safety and have not been approved for use in foods, an identical mango that might somehow be manufactured in a food processing plant could not be sold legally as a food.

Natural toxins in foods The example of mangoes can be repeated with all of our other foods, many times over and sometimes with sinister implications. Many of the compounds that form in plants and that become part of our food supply are very effective insect poisons, rivaling commercial insecticides in toxicity (certainly to insects and perhaps to humans) and overwhelming them in number. With some reflection, this shouldn't surprise us.

Chemicals in a typical mango • Figure 14.8

The compounds shown below comprise a partial ingredient list of the essential oil of a mango. These and similar compounds exist within mangos whether the fruit is grown conventionally or organically.

RedHelga/iStockphoto
Ask Yourself
Identify the two- or three-letter suffix common to the compounds within each category of this table.

Alcohol	Aldehyde	Alkene	Ester	Ketone
1–butanol	5–methylfurfural	3–carene	ethyl 2–butenoate	2,3-pentanedione
1–hexanol	ethanal	α–humulene	ethyl acetate	4–methylacetophenone
1–methyl–1–butanol	furfural	α–terpinene	ethyl butanoate	5–butyldihydro–3H–2–furanone
carveol	geranial	α–terpinolene	ethyl decanoate	6–pentyltetrahydro–2H–2–pyranone
cis–3–hexene–1–ol	hexadecanal	α–thujene	ethyl hexadecanoate	dihydro–5–hexyl–3H–2–furanone
ethanol	hexanal	β–caryophyllene	ethyl octanoate	dihydro–5–octyl–3H–2–furanone

Adapted with permission of John Wiley & Sons, Inc. from Snyder, C., The Extraordinary Chemistry of Ordinary Things, Fourth Edition, p. 511. Copyright 2003.

With eons of evolution behind them, it's entirely reasonable that plants should develop powerful toxins as part of their chemical armory against insects and other predators. Unlike animals, plants can't run from their enemies; they can only poison them.

The armaments of some plants include chemicals that are not only toxic in the more common sense of the word, but are **carcinogens** as well. Safrole, for example, makes up about 85% of oil of sassafras, which comes from the bark around the root of the sassafras tree and is also a minor component of cocoa, black pepper, and spices and herbs such as mace, nutmeg, and Japanese wild ginger. Safrole produces the taste of sassafras tea and was at one time used as the principal flavoring ingredient in the manufacture of root beer.

> **carcinogen** A cancer-causing substance.

Both the oil of sassafras and safrole itself have been banned from use as food additives. Safrole produces liver cancer in mice and has been listed as a carcinogen by the Environmental Protection Agency. The FDA judges both safrole and the oil to be too dangerous for use in foods. **Figure 14.9** shows the structures of safrole and several additional plant substances that produce tumors in mice, and **Table 14.9** on the next page, lists these and related substances found naturally in foods or produced during cooking.

The costs of evaluating the **carcinogenicity**, or cancer-causing potential, of various substances and ethical

Plant substances that produce tumors • Figure 14.9

Some natural substances found in foods are considered too dangerous to be used as food additives.

Safrole, a major component of oil of sassafras.

Estragole, a major component of oil of tarragon and occurring also in the herbs basil and fennel.

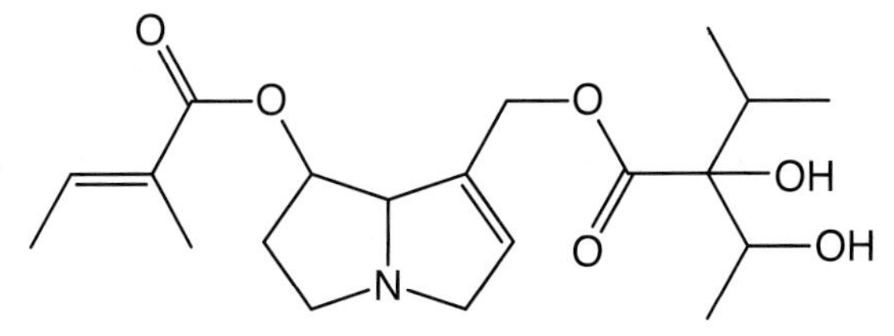
O
O
O
O
OH
OH
O
N

Symphytine, a component of the comfrey plant and one of its natural defenses. (The comfrey plant is used for brewing herbal teas.)

Allyl isothiocyanate, a pungent, irritating oil that occurs in brown mustard as well as horseradish and garlic and is also known as mustard oil.

Adapted with permission of John Wiley & Sons, Inc. from Snyder, C., *The Extraordinary Chemistry of Ordinary Things*, Fourth Edition, p. 512. Copyright 2003.

concerns about the use of large numbers of laboratory animals to do so have spurred the development of faster, less expensive, and more humane tests. One such procedure is called the **Ames test**, developed by Bruce Ames of the University of California, Berkeley (**Figure 14.10**). This laboratory test operates on the assumptions that both cancer and mutations begin with genetic damage of some sort and that a **chemical mutagen** stands a very good chance of being a carcinogen as well. The test uses a strain of *Salmonella* bacteria that is modified to make it very sensitive to genetic damage by chemical means and therefore sensitive to chemically induced mutation. What's more, the bacteria are altered further so that the amino acid histidine, which is normally a nonessential dietary amino acid for the bacteria, becomes an essential amino acid (Section 5.5). As a result of its newly acquired dependence on histidine as an essential amino acid, a colony of this bacteria won't grow in the absence of this amino acid. But exposing it to a mutagen that switches it back to its more common form allows it to synthesize its own histidine and to flourish.

Ames test A test that determines the mutagenic potential of a chemical substance.

chemical mutagen A chemical substance that produces a genetic mutation in an organism.

Because of a substantial overlap between the list of chemicals known to produce mutations and those known to cause cancer—about 90% of the chemicals on each list appear on the other—the Ames test and similar techniques can serve as useful, rapid, and inexpensive screening tools for identifying chemical carcinogens. Yet laboratory animals, which stand far closer to humans in their genetic makeup than bacteria do, still offer the best and most reliable estimate of chemical dangers to humans. In this respect, questions of ethical and economical values in animal testing must be balanced against the reliability of the animal experiments in protecting humans against chemical harm.

The roster of toxic, carcinogenic, or otherwise hazardous chemicals in our natural food supply continues, seemingly without end. Yet we do accept the hazards of our natural food supply, those that exist through the action of nature even in the absence of any processing or chemical additives. Despite the presence of an enormous variety of naturally occurring chemicals that may cause harm to us, we ordinarily consider the fruits, vegetables, grains, legumes, dairy products, eggs, meats, poultry, fish, and other foods that we eat in our daily meals to be safe, especially when we eat each in some measure of moderation.

This word *moderation* holds the key to our own defenses against food toxins. For most ordinary foods, normal levels of consumption lie far below anything that might produce an acute illness. Although the myristicin of nutmeg, eaten in very large quantities at one sitting, can produce hallucinations and liver damage, the amount of nutmeg ordinarily used in even the most heavily spiced meal runs about 1 or 2% of the quantity needed to produce anything more notable than flavorful food. To suffer cardiovascular damage from the glycyrrhizic acid of licorice seems to require eating perhaps 100 g of the candy each day for several days. Caffeine, with its LD_{50} of about 130 mg/kg in mice, is the major physiologically active compound of coffee and a powerful stimulant of the central nervous

Toxic chemicals occurring naturally in foods or produced through cooking Table 14.9

Food Substance	Source	Potential Hazard
Allyl isothiocynate	Brown mustard, horseradish, garlic	Tumors
Benzo(a)pyrene	Smoked and broiled meat	Gastrointestinal cancer
Cyanides	Oil of bitter almond, cashew nuts, lima beans	General toxicity
Dimethylnitrosamine	Cooked bacon	Cancer
Estragole	Basil, fennel, oil of tarragon	Tumors
Glycyrrhizic acid	Licorice root	Hypertension and cardiovascular damage
Hydrazines	Raw mushrooms	Cancer
Myristicin	Black pepper, carrots, celery, dill, mace, nutmeg, parsley	Hallucinations
Oxalic acid	Rhubarb, spinach	Kidney damage
Saxitoxin	Shellfish	Paralysis
Symphytine	Comfrey plant	Tumors
Tannic acid and related tannins	Black teas, coffee, cocoa	Cancer of the mouth and throat
Tetrodotoxin	Pufferfish	Paralysis

Adapted with permission of John Wiley & Sons, Inc. from Snyder, C., *The Extraordinary Chemistry of Ordinary Things*, Fourth Edition, p. 513. Copyright 2003.

The Ames test • Figure 14.10

This test measures the potential of a chemical substance to act as a mutagen. Generally, the more likely it is that a substance causes genetic mutations or changes in the DNA of an organism, the more likely that it also acts as a carcinogen.

1. *Salmonella* bacteria that have been altered to require histidine to grow are put in a medium that contains all the needed nutrients for growth, except histidine, as well as an extract taken from rat livers. Using the liver extract helps identify chemicals that may not be mutagens themselves but that are converted into other, mutagenic substances as the liver enzymes detoxify them.

2. The chemical under examination is added to the test dish but not to the control dish.

3. The dishes are incubated for 12-24 hours. Rapid growth of the bacterial colony in the test dish (as compared to the control) indicates that the added chemical (or whatever it's being metabolized to) is a mutagen and causes the bacteria to revert to the form that does not require histidine for growth.

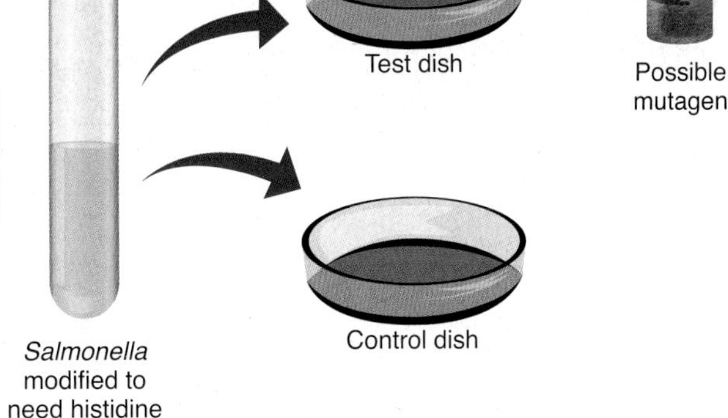

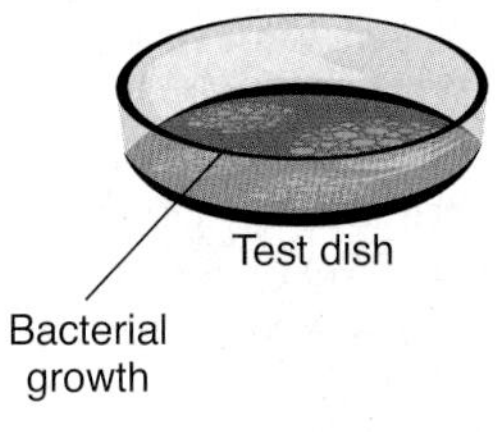

What would you conclude if the test dish does not show signs of bacterial growth after incubation?

system. At 100–150 mg of caffeine to a cup of coffee, it would take about 70 cups of coffee at one sitting to reach the mouse's LD_{50} in a 70-kg human. Surely other, urgent problems would arise long before the approach of a lethal dose.

The body's most effective shield against molecular poisons lies in the operation of the liver. This organ can call up a great variety of metabolic reactions to change the chemical structures of poisons, usually rendering them harmless. As long as a healthy liver isn't overpowered by a massive dose of any one toxin, it can effectively protect the rest of the body. By using a variety of metabolic defenses, the liver far more easily disposes of, say, a tenth of a gram of each of 10 different toxins than the same total weight, 1 g, of a single toxin. Although small amounts of the 10 poisons might be metabolized effectively by 10 different enzymatic reactions, a large dose of a single poison could overwhelm the only biochemical path available for its removal. In this sense a varied diet, which implies eating foods containing a variety of natural toxins, each present in exceedingly small amounts, is far less hazardous than a diet that concentrates on large quantities of only one or two foods.

We've seen that all chemicals present hazards. Poisons and carcinogens are indeed present in foods, and they occur in natural, unprocessed foods as well as in manufactured or processed foods. Various governmental agencies have the power to protect us from excessive, known chemical risks, but our own judgment of the acceptability of risks, both as individuals and as members of society, ultimately determines the issue of safety.

We've seen, finally, that the idea of absolute safety is a phantasm. To the extent that we are well informed and that our judgments are sound, we can weigh the very real benefits that chemicals provide and balance them against the very real risks of their use. In this way, and only in this way, can we ensure that our food, while never free of hazards, is indeed safe.

CONCEPT CHECK

1. **Which** two additional items of information must be reported along with an LD_{50} of a substance in order to give precise meaning to the value?
2. **Why** does considering a substance to be "safe" imply accepting some level of risk?

Summary

1 Micronutrients 446

- ***What are vitamins and minerals, and how are they classified?***

  Vitamins are organic compounds essential for good health that occur in small amounts in many of the foods we eat. **Minerals** are atoms or ions of individual elements needed by the body in small amounts for good health. Vitamins are classified based on their solubilities in water and fat or oil. Water-soluble vitamins include the B vitamins and vitamin C. Fat-soluble vitamins include vitamins A, D, and E. **Major minerals**, including calcium and phosphorus (shown in the table here), are required in the diet at levels above 100 mg/day. **Trace minerals**, including iron, are required in the diet at levels below 100 mg/day.

What a Chemist Sees • Dietary minerals and the periodic table

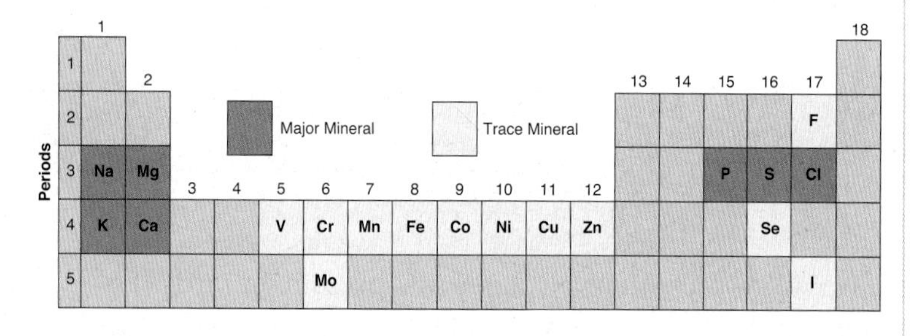

- ***What are the roles of vitamins and minerals in diet and health?***

  Vitamins and minerals help regulate various activities within the body. A prolonged dietary deficiency of any vitamin can lead to disease. For example, *scurvy*, a disease characterized by bleeding gums, results from vitamin C deficiency; night-blindness and other vision impairment can result from vitamin A deficiency; and *rickets*, characterized by poor bone development in children, results from vitamin D deficiency. Some major minerals, such as calcium and phosphorus, serve structural functions within bones and teeth; other major minerals, such as sodium, potassium, and chloride ions, act as *electrolytes,* helping to maintain proper fluid balance. Trace minerals participate in protein functions.

2 Food Additives 452

- ***How are food additives regulated?***

  Food additives are regulated by the FDA under the Federal Food, Drug, and Cosmetic Act (or FD&C Act). Legally, additives that fall under the FDA's **GRAS** (or generally regarded as safe) **list**, are not considered food additives for purposes of regulation. However, a more practical definition holds that a food additive is simply anything intentionally added to a food to produce a specific, beneficial result, regardless of its legal status.

- ***What are the functions of food additives?***

  Food additives are used to improve or maintain the quality of foods. Additives serve a large variety of functions in foods: adding to their appeal and nutritional value, preserving their freshness, and making them easier to process and more stable during storage. Any single chemical additive can function in multiple ways. **Antioxidants** help preserve foods by preventing oxidation; **antimicrobials** protect against spoilage by bacteria and mold; **sequestrants** (such as EDTA, shown here) bind to trace metals in foods, thereby preventing them from catalyzing oxidations; **humectants** help foods retain moisture, and **emulsifiers** help blend mixtures that would otherwise separate into two phases.

Figure 14.4 • Food additives

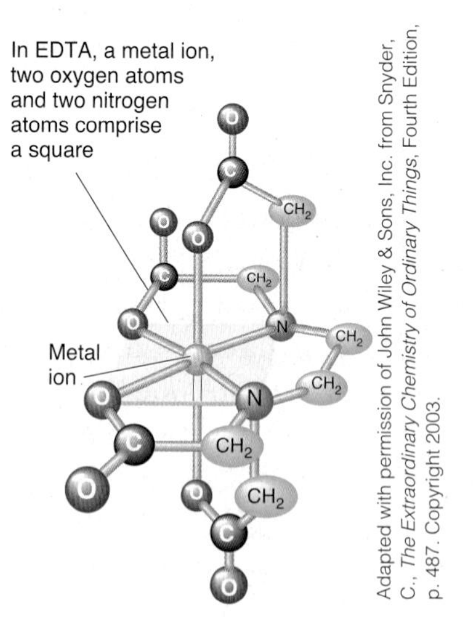

3 Food Safety 458

- ***How is toxicity measured?***

  Virtually any substance, including water, can be toxic if taken in large enough doses over short time periods. One useful measure of toxicity involves determining the amount of a chemical that is lethal to exactly half of a large population of test animals, as illustrated here. This dosage is known as the **LD_{50}**, which is typically reported in terms of the weight of the poison per unit weight of the test animal. Because the method used for administering the substance often affects the results, it is usually included in the data.

Figure 14.7 • LD_{50} for aspirin

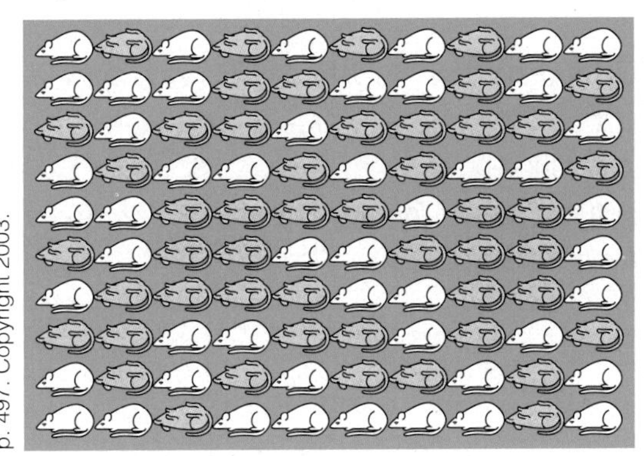

The LD_{50} for aspirin administered orally in mice is 1.5g/kg

- ***What does it mean to label a chemical substance as "safe"?***

  No chemical substance is inherently free of hazards. Labeling a chemical substance as "safe"—either through federal regulation or personal choice—implies a willingness to accept some level of risk.

- ***Are chemicals naturally present in foods?***

  A large number of chemicals are naturally present in foods but are not normally a cause for concern. Certain plant foods naturally contain toxins as part of their chemical defense against insects and other predators. Consuming foods in moderation generally protects us against food toxins, which are typically metabolized by the liver. Toxins or poisons can often be identified in the laboratory with procedures such as the **Ames test,** which assesses the mutagenic potential of a substance.

Key Terms

- Ames test 464
- carcinogen 463
- chemical mutagen 464
- electrolytes 459
- food additives 452
- GRAS list 453
- LD_{50} 460
- major minerals 450
- micronutrients 446
- minerals 449
- poison 459
- toxin 459
- trace minerals 450
- vitamin 446

What is happening in this picture?

Here, children in northern Russia, above the Arctic Circle, are being exposed to ultraviolet radiation to help prevent vitamin D deficiency. All are wearing eye masks to protect their vision from this radiation.

Global Locator
Murmansk, Russia

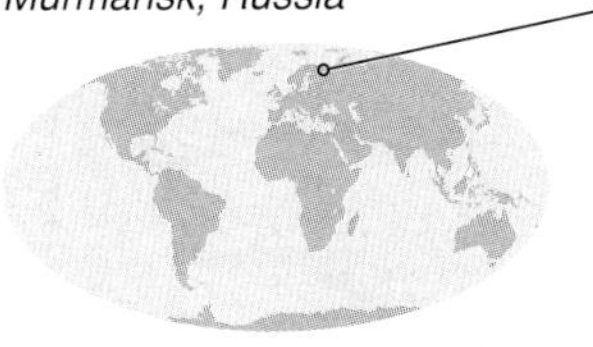

Think Critically

1. Why are these children at risk for vitamin D deficiency?
2. How does this treatment help prevent vitamin D deficiency?
3. What alternative method could be used to prevent this vitamin deficiency?

Exercises

Review

1. Fill in the blanks of the following statement with the words and phrases provided.

The __________ and minerals of our diet constitute our __________. Except for carbon, hydrogen, nitrogen, and oxygen (and, by some definitions, sulfur), the elements of our foodstuffs make up the class of __________. Of these, __________ is the major mineral of our body, located largely in our __________. Sodium and __________ are the major cations of our bodily fluids, with __________ ions as the major anion. Among our vitamins (conveniently divided into the two categories of the __________ A, D, E, and K and the __________ B and C) are vitamin C, which is known chemically as __________ and which protects against __________; vitamin D, which protects against __________ and vitamin A, which protects against a vision defect known as __________. Much of our vitamin A comes from the __________ of our yellow, orange, and green vegetables.

ascorbic acid
minerals
bones and teeth
micronutrients
calcium
night blindness
β-carotenes
rickets
chloride
fat soluble
vitamins
potassium
water soluble
scurvy

2. Fill in the blanks of the following statement with the words and phrases provided.

In its most general sense, a __________ is any substance deliberately added to a food to help __________. Among the additives that make our foods __________ are __________, which adds its flavor to foods; glucose and sucrose, which serve as __________; and ferric oxide and __________, which add __________. Increasing the __________ are additives such as __________, which helps prevent goiter formation. Several additives, including ascorbic acid and __________, help prevent the air oxidation of foods. __________ protects against oxidation by combining with traces of __________ and reducing their ability to __________ the process. __________ are surfactants that help __________ foods and keep them from separating during storage.

BHA and BHT
catalyze
metal ions
EDTA
monosodium glutamate
emulsify
mono- and diglycerides
nutritional value
more appealing to the senses
titanium dioxide
color
potassium iodide
food additive
preserve it or make it more nutritious or appealing
food sweeteners

3. Explain, define, or describe the significance of each of the following terms:

a. cholecalciferol
b. GRAS
c. collagen
d. retinol
e. retinol equivalent
f. FDA
g. sodium benzoate
h. goiter
i. trace minerals

4. What is the difference between a macronutrient and a micronutrient? Name two of each.

5. Describe the major function of

a. phosphorus in the bones
b. sodium in the extracellular fluid
c. fluoride in the teeth
d. chloride in the gastric juices and
e. cobalt in the body

6. Name a vitamin that plays an important role in:

(a) strengthening collagen fibers
(b) the absorption of calcium

7. Name two good sources of:

a. vitamin A
b. vitamin C
c. vitamin D

8. What is meant by the IU and the RE of vitamin A, and what is the advantage of describing the recommended daily dosage of vitamin A in terms of its IUs and REs?

9. a. What molecule is depicted in this figure?

b. How many atoms of this molecule bind to the ion shown at its center?

c. What type of element does this ion represent?

d. How does this molecule function as a food additive?

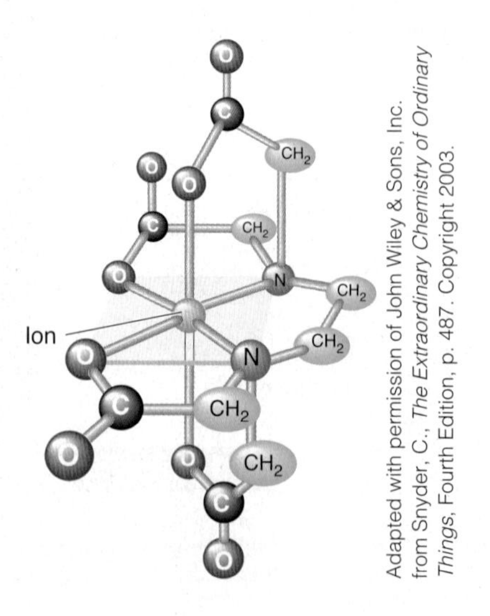

Adapted with permission of John Wiley & Sons, Inc. from Snyder, C., *The Extraordinary Chemistry of Ordinary Things*, Fourth Edition, p. 487. Copyright 2003.

10. a. What is the function of glycerine as a food additive?

b. By what mechanism does glycerine carry out this function?

$$\begin{array}{ccccc} CH_2 & - & CH & - & CH_2 \\ | & & | & & | \\ OH & & OH & & OH \end{array}$$

Glycerine

11. a. Which of the following elements are major minerals, and which are trace minerals?

b. Identify whether each of the following exist within the body as neutral atoms, anions, or cations:

i. the metallic elements shown in this figure

ii. the Group 17 elements shown in this figure

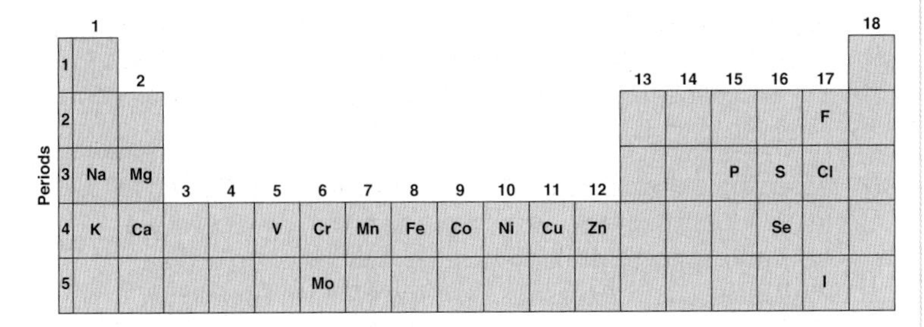

12. Why are bleeding gums one of the first symptoms of scurvy?

13. What are four major benefits a chemical can provide when it is added to a food?

14. Describe at least one function each of the following chemicals provides when it is used as a food additive:

a. potassium iodide

b. iron oxides

c. calcium propionate

d. mono- and diglycerides

e. sodium nitrite

15. Into which of the four major groups of functions of food additives does each of the following substances fall:

a. vitamin D

b. potassium iodide

c. pepper

d. sugar

e. food coloring

16. What is the function of a humectant? Name a chemical used as a humectant and identify a food to which it is added.

17. What two functions does ascorbic acid serve when it is used as a food additive?

18. What connection existed between the lack of effective methods of food preservation and the travels of Marco Polo and Columbus?

19. Legally, food colorants are not food additives. Explain why not.

20. Name the contribution each of the following make to the subject matter of this chapter:

a. William Lowrance

b. Bruce Ames

c. Harvey Wiley

21. Name the food or edible substance in which each of the following chemicals occur:

a. lactose

b. glycyrrhizic acid

c. myristicin

d. oxalic acid

22. Name the governmental agency responsible for:

a. investigating outbreaks of illness caused by food spoilage

b. the chemicals of beer, wine, liquor, and tobacco

c. chemical pesticides

d. exposure to chemicals in the workplace

e. chemicals used as food additives

23. What is the difference between a mutagen and a carcinogen?

24. On what factors does the harm that any particular substance can do to us depend?

25. List the natural sources of each of the following chemicals:

a. caffeine

b. symphytine

c. safrole

26. What characteristic or property of a chemical does the Ames test reveal?

Think

27. Which of our macronutrients would you expect to be a major source of dietary sulfur?

28. a. What advantage is there in taking β-carotene rather than retinol as a food supplement?

b. Why is the vitamin A content of food better described in terms of International Units and Retinol Equivalents than in terms of actual weight?

29. Would a chemical analysis of the soil of a vegetable garden reveal:

a. what minerals we might find in the vegetables that come from the garden?

b. what vitamins we might find in the vegetables that come from the garden?

Explain.

30. Would you expect the vitamin C contained in an orange grown in soil of poor quality to be inferior to the vitamin C contained in an orange grown in high-quality soil? Explain.

31. In what chemical or physical way does the vitamin C produced by chemical reactions carried out in laboratory equipment differ from the vitamin C obtained from fruits and vegetables?

32. To provide fresh meat on long voyages, sailors of past centuries took with them live cattle, which they slaughtered as they needed meat. With no supplies of fresh fruit and vegetables, the sailors often succumbed to scurvy. Why didn't the live cattle also develop scurvy? (You may use the Internet to find your answer.)

33. People living in the far northern latitudes receive little or no sunshine during the winter, wear a good deal of clothing for protection against the cold, and lack readily available vitamin supplements and fortified milk. Where does their vitamin D come from?

34. In cooking fresh vegetables in water, it's advisable to cook them for a short period in the minimum amount of water required or to steam them. Why does this make good nutritional and good chemical sense?

35. A provitamin is a chemical substance that is converted into a vitamin within the body. For example, panthenol is converted into pantothenic acid (vitamin B_5) by the action of enzymes within the body, so panthenol is also known as provitamin B_5.

a. How is the structure of pantothenic acid different from that of panthenol?

b. Provide two other examples from the chapter of the conversion of a precursor molecule or provitamin into a vitamin.

37. Mono- and diglycerides act as effective emulsifiers, but triglycerides (Section 5.2) do not. Explain.

38. Why were spices one of the most widely used food additives of the Middle Ages?

39. One definition regards a food additive as "anything intentionally added to a food in order to produce a specific, beneficial result," regardless of its legal status. In what way is this definition:

a. more useful than the legal definition?

b. less useful than the legal definition?

40. Suppose someone began marketing packages of refined sugar that's enriched with vitamins A, the B complex, C, and D, and that had added iron, phosphorus and zinc. Would you consider this a health food or a junk food? Explain.

41. Many chemicals used as food additives have never been subjected to scientific laboratory tests for safety, and yet their use is perfectly legal. What do all these legal, yet untested, additives have in common?

42. Assuming that it's desirable to test chemicals that are added to our foods to determine their effects on us, do you believe it is better to test them on laboratory animals or on human volunteers, such as those of Dr. Wiley's Poison Squad (Section 14.2)? Under what conditions, if any, would you volunteer to participate in such a test?

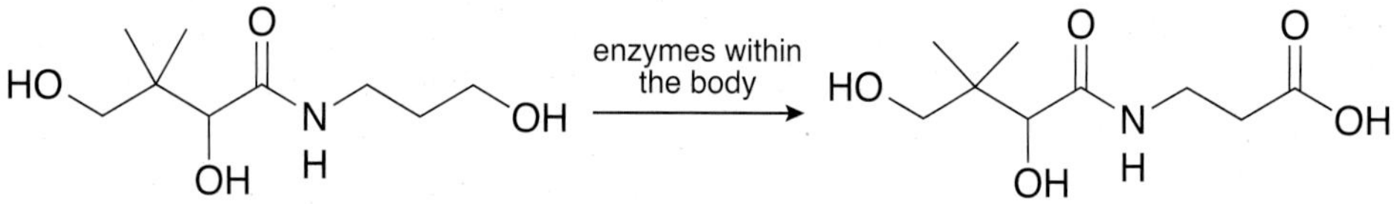

Panthenol (provitamin B_5) $\xrightarrow{\text{enzymes within the body}}$ Pantothenic acid (vitamin B_5)

36. The two isomers of gyceryl monostearate, shown here, can function as humectants as well as emulsifiers. Explain.

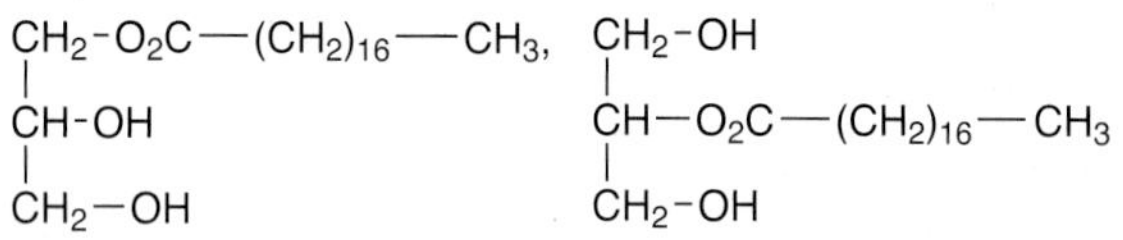

$$\begin{array}{ll} CH_2\text{-}O_2C\text{—}(CH_2)_{16}\text{—}CH_3, & CH_2\text{-}OH \\ | & | \\ CH\text{-}OH & CH\text{—}O_2C\text{—}(CH_2)_{16}\text{—}CH_3 \\ | & | \\ CH_2\text{—}OH & CH_2\text{-}OH \end{array}$$

Glyceryl monostearate

43. With "safety" defined as the acceptability of risk, name three activities you would consider to be unsafe.

44. Give an example of how the hazard presented by a substance can depend on the method by which it is introduced into the animal's body.

45. Suppose that laboratory tests on a newly discovered chemical showed that it produced absolutely no effects on any animal tested, no matter how or at what level it was administered. Would you consider this new chemical to be "safe"? Explain.

46. A statement sometimes used about the hazards of medicines is: "The poison is in the dosage." Explain what this means.

47. Which of the following does our society (as represented by its laws) consider to be safe?

    a. use of ethylenediaminetetraacetic acid (EDTA) as a food additive

    b. cigarette smoking by adults in the privacy of their own homes

    c. easy access to aspirin by children under age five

48. Describe your own thoughts about the safety of each of the following:

    a. sodium chloride

    b. aspirin

    c. ethanol

    d. caffeine

    e. nicotine

49. If we can say that anything is hazardous if it is used in excess, can we also say that everything is safe if used in very small amounts? Explain.

50. a. What conclusions, if any, could you draw from a positive Ames test (with bacterial growth) in an examination that did not contain added liver extract? b. What if this same examination produced a negative result?

51. Why is the quantity of a chemical that is lethal to 50% of a population a better measure of its hazard than the quantity that is lethal to 100% of a population?

52. Vitamin-fortified water is a poor source of vitamins A and D. Explain.

Calculate

53. Using the data from the section on major minerals, calculate the approximate ratio of calcium atoms to phosphorous atoms in bones and teeth.

54. Assuming that an average carrot has a mass of 72 g, and provides us with 8000 IUs of vitamin A, determine the mass-percent of retinol equivalent in an average carrot.

55. How many REs of retinol do 5 IUs of retinol represent? How many IUs of retinol does 1 μg of retinol represent?

56. Which is more toxic to rodents when administered orally:

    a. arsenic trioxide or sodium chloride?

    b. aspirin or trisodium phosphate?

    c. caffeine or nicotine?

    d. acetaminophen or BHA?

57. What three poisons of Table 14.8 are the most powerful when administered orally to mice or rats? What three are the least powerful when administered in the same way?

58. Commercial aspirin contains 325 mg of aspirin (acetylsalicylic acid) per tablet. Assuming that the LD_{50} for aspirin in mice and rats applies equally well to humans, how many aspirin tablets, taken all at once, would produce a 50% chance of a lethal dose of aspirin in a 70-kg person?

59. Bottled fruit punch often contains 0.1% of sodium benzoate as a preservative. Studies provide a value of 4 g/kg for the LD_{50} of sodium benzoate, orally in rats. Assuming that humans respond to this chemical as rats do, and assuming that the density of a commercially bottled fruit punch is 1.0 g/mL, how many liters of this beverage would a 70-kg person have to drink, all at the same time, to produce a 50% chance of a lethal dose of sodium benzoate?

60. Arrange the following substances in order of their ability to cause illness or death (from the most hazardous or lethal to the least): aspirin, sodium chloride, sodium cyanide, water.

Appendix A

SI Units and Conversion Factors

Length			Mass			Volume		
SI unit: meter (m)			**SI unit: kilogram (kg)**			**SI unit: cubic meter (m^3)**		
1 meter	=	1000 millimeters	1 kilogram	=	1000 grams	1 liter	=	1000 milliliters
	=	1.0936 yards		=	2.20 pounds		=	10^{-3} m^3
1 centimeter	=	0.3937 inch	1 gram	=	1000 milligrams		=	1 dm^3
1 inch	=	2.54 centimeters (exactly)	1 pound	=	453.59 grams		=	1.0567 quarts
				=	0.45359 kilogram	1 gallon	=	4 quarts
1 kilometer	=	0.62137 mile		=	16 ounces		=	8 pints
1 mile	=	5280 feet	1 ton	=	2000 pounds		=	3.785 liters
	=	1.609 kilometers		=	907.185 kilograms	1 quart	=	32 fluid ounces
			1 ounce	=	28.3 grams		=	0.946 liter
1 angstrom	=	10^{-10} meter	1 atomic mass unit	=	1.6606×10^{-27} kilograms		=	4 cups
						1 fluid ounce	=	29.6 mL

Temperature			Energy			Pressure		
SI unit: kelvin (K)			**SI unit: joule (J)**			**SI unit: pascal (Pa)**		
0 K	=	−273.15°C	1 joule	=	1 kg m^2/s^2	1 pascal	=	1 kg/(ms^2)
	=	−459.67°F		=	0.23901 calorie	1 atmosphere	=	101.325 kilopascals
K	=	°C + 273.15	1 calorie	=	4.184 joules		=	760 torr
°C	=	$\frac{(°F - 32)}{1.8}$					=	760 mm Hg
°C	=	$\frac{5}{9}(°F - 32)$					=	14.70 pounds per square inch (psi)
°F	=	1.8(°C) + 32						

Appendix B

Answers to In-Chapter Questions

Chapter 1

Figure 1.1

Chemical expertise is required in the refining of crude oil into its component fractions and converting these fraction into various products, such as transportation and heating fuels and paving materials.

What a Chemist Sees

Deployment of air bags; combustion of the hydrocarbons of gasoline; modification of exhaust gases that pass through the catalytic converter; use of the battery.

Section 1.1 Concept Check

1. Combustion of fossil fuels provides electricity that powers electric heaters and air conditioners. Combustion of heating fuels powers home furnaces.
2. (a) polymers and dyes; (b) surfactants; (c) macronutrients, micronutrients; (d) analgesics, antipyretics, antibiotics and statins, among others

Figure 1.3

1. An increase in the resistance of the plastic to breaking or shattering, and an ability of the resin to remain in contact with food for long periods of time without changing the food's flavor or consistency.
2. There is a risk that BPA may pass from the container or liner into the stored food or drink, and then into our bodies as we eat or drink the contaminated contents.

Section 1.2 Concept Check

1. The poison is in the dosage, meaning that virtually any substance ingested can be poisonous if taken in large enough quantities.
2. (a) Caffeine provides a stimulation to the central nervous system that many of us find pleasant and beneficial. Yet, in large amounts caffeine can be lethal. (b) Aspirin reduces pain, fever, and inflammation and can provide a measure of protection against heart attacks. However, excessive use of aspirin can produce susceptibility to bruising, gastrointestinal bleeding, and strokes, and an overdose of aspirin can kill.
3. BPA is found in certain plastic bottles and in the liners for food cans and can leach from these containers into beverages or food.

Section 1.3 Concept Check

1. Those in more highly developed countries – about a fifth of the world's population – consume well over half the world's energy and materials. Improvements in standards of living throughout the world will continue to drive growth in demand for energy and goods.
2. Nonrenewable resources include natural gas, coal, petroleum, and other mineral resources. Renewable resources include trees, food crops, and fresh water (if restored by rain).
3. Chemists are developing improved materials for use in solar panels and in devices that produce hydrogen gas from water and sunlight.

Figure 1.9

1. No, a theory gains footing not because it is proven correct, but rather because a large set of confirmed observations support it and none can disprove it.
2. In formulating questions and hypotheses, and in carrying out experiments and interpreting findings.

Section 1.4 Concept Check

1. Pharmaceutical/Medical: drug discovery; Agricultural: developing new chemicals to improve crop yields; Environmental: monitoring the environment for pollutants; Energy/Materials: developing new energy technologies; Criminal Justice: developing new methods of analyzing contraband.
2. High-throughput screening screen can help identify potential drug candidates from libraries of thousands of compounds. Here, chemistry plays a central role in bridging between various disciplines, including engineering, robotics and the life sciences.
3. Observing the world and formulating questions; developing hypotheses; carrying out experiments; interpreting findings; communicating findings to others.

Did You Know?

1. To maintain scientific objectivity.
2. Detector dogs alerting to the presence of cocaine.
3. A drug-detector dog should alert to currency that is recently tainted by illicit cocaine, but not to currency in general circulation. Since these dogs are trained to detect methyl benzoate, the signature odor of cocaine, it's important to know the rate at which this odor dissipates from tainted currency.

$$\frac{600\ \cancel{\text{miles}}}{1\ \cancel{\text{hour}}} \times \frac{5280\ \cancel{\text{feet}}}{1\ \cancel{\text{mile}}} \times \frac{1\ \text{meter}}{3.28\ \cancel{\text{feet}}} \times \frac{1\ \cancel{\text{hour}}}{60\ \cancel{\text{minutes}}} \times \frac{1\ \cancel{\text{minute}}}{60\ \text{seconds}} = 268\ \text{meters per second}$$

Know Before You Go

1. decimeter (left); centimeter (right).
2. \$1.82/lb
3. 16.8 km/L
4. 30 mL

Section 1.5 Concept Check

1. Prefixes for units are based on powers of ten.
2. meters, seconds.
3. The prefix modifies the unit. For example, a kilometer is 1000 meters and a millimeters is $1/1000^{th}$ of a meter.

What is happening in this picture?

Chlorine poses both benefits and risks. Used in proper amounts, chlorine serves as an effective disinfectant. However, this label found on storage tanks of chlorine serves as a warning that acute ingestion of chlorine can be fatal. As with any substance, toxicity depends on the level and manner in which one is exposed to the substance.

Chapter 2

Figure 2.1

Yes.

Figure 2.3

Figure 2.4

They are solids.

Figure 2.7

The alpha particles would be expected to have a very low likelihood of colliding with the nucleus if the nucleus is miniscule as compared to the size of the atom. This is what is observed.

Figure 2.8

Both atomic models show a nucleus at the center surrounded by electrons.

Figure 2.9

The red line arises from an electron transition from the $n = 3$ to $n = 2$ quantum level.

A Deeper Look

We divide the maximum number of electrons the shell can hold by two since each orbital can hold two electrons.
$32/2 = 16$, which corresponds to answer (c).

Figure 2.11

Approximately 100,000 to 1.

2.1 Concept Check

1. Dalton built on Democritus' earlier notion that matter is made up of indivisible particles, or atoms. Dalton saw each element as representing a unique type of atom, and that atoms of different elements can bond together to form *compounds.*
2. In the Rutherford and Bohr models, the atom has a central, dense, positively-charged nucleus, with negatively-charged electrons outside and surrounding the nucleus. In the Thomson model, negatively charged electrons are embedded within a diffuse "pudding" of positive charge.
3. The diameter of a hydrogen atom is about 100,000 times the size of its nucleus.

Figure 2.12

Because it contains one proton and one electron. The positive charge of the proton balances the negative charge of the electron.

2.2 Concept Check

1. The three subatomic particles are the electron, proton, and neutron. Protons and neutrons can be found in the nucleus.
2. An element is defined by the number of protons in the nucleus, or atomic number.
3. To determine the number neutrons in an atom, we subtract the atomic number from the mass number.

Figure 2.15

1. weight
2. mass

Figure 2.16

No, it takes more force to bring the filled shopping cart to a stop.

Know Before You Go

$[(8 \times 6) + (92 \times 7)] \div 100 = 6.92$ u

2.3 Concept Check

1. Isotopes of a given element differ in the number of neutrons and hence mass number.
2. Weight is dependent on gravitational force, which can vary depending on location in the universe.
3. A proton or neutron has about 1900 times the mass of an electron.

Figure 2.20

1. Most elements are metals.
2. The period containing potassium is the row beginning K, Ca, Sc, Ti, etc.
3. The group containing phosphorus is the column labeled 5A or 15. From top to bottom it reads N, P, As, etc.
4. Gold is Au, silicon is Si, germanium is Ge, and helium is He.
5. Bromine and mercury.

Figure 2.21

(a) Water (b) Nitrogen, N_2.

2.4 Concept Check

1. The neighbor to the left has one less proton and the neighbor to the right has one more proton.
2. For each new element in the horizontal series we increase the atomic number by one. Elements in a given column have the same number of electrons in their outermost quantum shell.
3. Elements in a given column tend to have similar chemical properties. For instance, elements in the column that includes He and Ne, are generally unreactive, whereas elements in the column that includes Li and Na (with the exception of H) reactive vigorously with water.

What is happening in this picture?

1. Group 14
2. Group 18. It is an inert gas, lacking chemical reactivity with most elements, including silicon.

Chapter 3

Figure 3.1

1. (a) one (b) two (c) seven (d) eight
2. (a) two (b) eight

Figure 3.3

halogen

3.1 Concept Check

1. They are in the same column of the periodic table.
2. Seven
3. $:\dot{\ddot{Se}}:$

Figure 3.4

1. Noble gases
2. (a) 11 protons, 10 electrons (b) 17 protons, 18 electrons

Figure 3.5

No, because it is formed from a single element.

Figure 3.8

1. Two
2. In part **a**, the bonding electrons are equally shared between atoms. In part **b**, the bonding electrons are also shared but reside closer to one atom, the chlorine.
3. The bond in part **c** is ionic, formed from the attraction of oppositely charged ions. The bonds in parts **a** and **b** are covalent, formed through shared electrons.

Figure 3.12

1. Nonpolar
2. Two

Figure 3.14

1. Gas; solid
2. Solid; gas

3.2 Concept Check

1. According to Figure 3.4, when a sodium atom loses its lone electron from the third quantum shell, the resulting sodium cation holds eight valence in the second quantum shell. When a chlorine atom gains an electron, the resulting chloride anion holds eight electrons in its valence shell.
2. Covalent bonds are characterized by shared electrons and ionic bonds result from electron transfer.
3. Because the more electronegative element of a polar covalent bond has a greater attraction to the shared electrons.
4. Ionization involved transfer of electrons. Since the bonding electrons of a compound's polar covalent bond are closer to one atom, this may make the electrons more prone to transfer to that atom and cause ionization.
5. Because the large number of dissolved Na^+ and Cl^- ions are able to carry electrical current through the solution.

Know Before You Go

1. (a) Magnesium iodide; 278.1 u (b) Beryllium sulfide; 41.08 u (c) Potassium oxide; 94.20 u
2. +4
3. (a) AgBr; Silver bromide (b) SrS; Strontium sulfide (c) K_3N; Potassium nitride (d) CaO; Calcium oxide (e) Na_2S; Sodium sulfide

3.3 Concept Check

1. KBr; $MgCl_2$; $CaCl_2$
2. The magnitude of charge on the cation and on the anion.
3. To determine the formula mass of an ionic compound, sum up the atomic masses of all the atoms in its chemical formula.

Figure 3.22

1. Six and one, respectively.
2. (a) Eight (b) Four (c) In lone pairs on the oxygen atom.

Figure 3.23

1. Four
2. Eight
3. Ten

Figure 3.24

1. Six, five, and four, respectively.
2.

Compound	(a)	(b)
Water	4	2
Ammonia	2	3
Methane	0	4

3.4 Concept Check

1. H_2O exists as a molecule, a discrete, electrically neutral assembly of atoms held together by covalent bonds. Na_2O exists as an orderly arrange of ions held together by ionic bonds within a crystal lattice.
2. Molecular formula: $C_{21}H_{28}O_5$; Molecular mass: 360.4 u
3. Because it is made up of Na^+ and Cl^- ions held together by ionic bonds with a crystal lattice.

Figure 3.26

Compound	C—H bonds	C—C bonds
Methane	4	0
Ethane	6	1
Propane	8	2

Table 3.5

1. Hexane
2. $C_{16}H_{34}$

Figure 3.29

1.

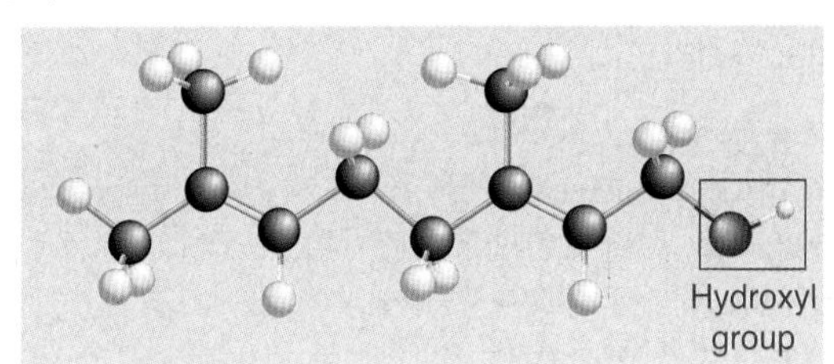

2. (a) Two (b) Two (c) Three (d) Three

Know Before You Go

4. (a) Eight (b) C_8H_{18}
5. Octane
6.

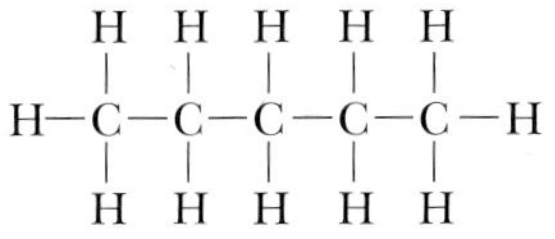

```
   H  H  H  H  H
   |  |  |  |  |
H—C—C—C—C—C—H
   |  |  |  |  |
   H  H  H  H  H
```

Pentane

```
       H
     H | H
   H  \C/ H  H  H
   |   |  |  |  |
H—C—C—C—C—C—H
   |   |  |  |  |
   H   H  H  H  H
```

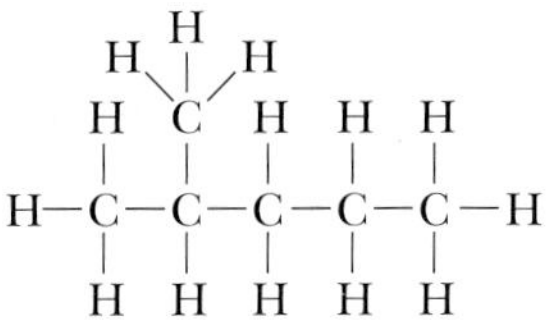

2-methylbutane

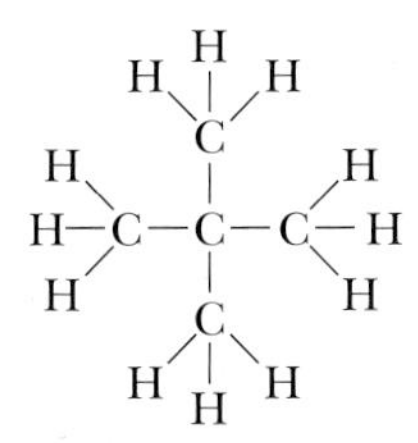

2,2-dimethylpropane

3.5 Concept Check

1. A covalent compound containing carbon. Foods produced without the use of synthetic fertilizers or pesticides, or through the use of genetic engineering.
2. Isomers

What is happening in this picture?

1. (a) Al_2O_3 (b) CaO (c) Na_2O
2. Ionic **3.** Fe^{3+}

Chapter 4

Figure 4.1

a) At the top of the arc b) At the lowest point in the arc.

Figure 4.2

(a) First (b) Second

Section 4.1 Concept Check

1. Joule
2. Residential; commercial; transportation; industrial.

Figure 4.5 Ask Yourself

Renewable energy sources, such as solar, hydroelectric, and wind power.

Figure 4.5 Interpret the Data

a) oil and coal; b) oil and coal

Figure 4.6

840 million gallons; About 50%

Section 4.2 Concept Check

1. (a) heating and electricity production (b) transportation and heating fuels (c) electricity generation
2. About 80%
3. CO_2 gas and water vapor

Figure 4.8

a) compression and power strokes; b) intake and exhaust strokes

Figure 4.10

The compounds evaporate at different rates. More volatile components evaporate faster so the fragrance of the cologne will initially reflect the odors of these compounds. After a period of time, the odors of the less volatile components will become more apparent.

Figure 4.11

a) At the top; b) At the bottom.

Figure 4.12

Catalytic cracking converts larger hydrocarbons, which are less suitable for use in gasoline, to smaller hydrocarbons, which are more useful in gasoline.

Figure 4.13

1. No, it exhibits the same knocking properties as mixture of 87% "octane" and 13% heptane would exhibit, but like all gasoline it contains over one hundred different hydrocarbons plus additives.
2. 7% heptane; 93% 2,2,4-trimethylpentane

Figure 4.14

1. No, because unburned hydrocarbons and carbon monoxide are converted into carbon dioxide through the action of catalytic converters.
2. Oxygen

Figure 4.15

a) Both compounds have at least one carbon-oxygen bond;
b) MTBE is an ether, with oxygen bonded to two carbon groups; Ethanol has oxygen bonded to one hydrogen and one carbon.

Section 4.3 Concept Check

1. Octane rating, which is improved, for example, by higher concentrations of branched hydrocarbons.
2. Crude oil is heated in order to vaporize it and allow the separation of its components.
3. Catalytic converters oxidize unburned hydrocarbons into carbon dioxide and water vapor. Fuel oxygenates improve the efficiency of hydrocarbon combustion and thereby decrease the level of hydrocarbon emissions.

Figure 4.16

Carbon dioxide

Figure 4.17

i) decreasing; ii) increasing; iii) increasing

Section 4.4 Concept Check

1. Photosynthesis combines carbon dioxide and water to form organic compounds and oxygen. Cellular respiration combines organic compounds and oxygen to form carbon dioxide and water.
2. Fossil fuel formation is infinitesimally slower than the rate by which we are burning these fuels. As a result, CO_2 from the combustion of fossil fuels tends to accumulate in the atmosphere as well as in the oceans.

Section 4.5 Concept Check

1. Greenhouse gases absorb infrared energy (through molecular vibrations) and re-radiate a portion of this energy back towards the earth's surface thereby warming the atmosphere.
2. Roughly 17°C.
3. A number of countries throughout the world have made commitments to reduce greenhouse gas emissions, but no global commitment has as yet been achieved. Approaches include: energy conservation (such as increasing the fuel economy of cars), the development of renewable energies, and switching to greater use of natural gas over other fossil fuels.

Figure 4.25

No, the bus releases water vapor, a product of the reaction of hydrogen and oxygen in the fuel cell. Though benign, water vapor is still an emission.

Did you know?

Advantages: Derived from a renewable resource. Domestically produced, so reduces demand for imported oil. Cleaner burning; emits fewer harmful emissions than gasoline.

Disadvantages: Reduced fuel economy, about 25% less than conventional gasoline; Production of which requires fossil fuels for energy; Limited amount available as compared to the enormous demand for liquid transportation fuels.

Section 4.6 Concept Check

1. Costs and convenience.
2. Biofuels, which are derived from plants, have potentially a smaller carbon footprint than fossil fuels since the CO_2 emissions from the burning of biofuels are partially offset by the CO_2 that had been previously removed from the atmosphere as the plants grew.

What is happening in this picture?

1. No, because the carbon in the plants was drawn from the atmosphere.
2. It would increase atmospheric greenhouse gases, because fewer plants are available to sequester CO_2 from the atmosphere.
3. Methane, CH_4.

Chapter 5

Figure 5.1

1. 100 calories
2. No, because the Calorie is defined with respect to liquid water. Water boils at 100°C so would vaporize.

Know Before You Go

1. 17.5 kcal

Figure 5.2

1. 4.2 kJ per kcal
2. kcal

Figure 5.3

28 hr

Figure 5.5

1544 Cal

Know Before You Go

3. (a) 8.4×10^{12} Cal; (b) 8.4×10^{12} Cal; (c) 1.6×10^{13} Cal

Know Before You Go

4. 52,500 Cal

Section 5.1 Concept Check

1. (a) 1 kcal (b) 1 Cal (b) 4184 J
2. They provide more than double the number of Calories per gram as either carbohydrates or protein.
3. Long-term calorie storage is in the form of fat (adipose tissue).

Figure 5.9

It would coagulate or freeze into a waxy substance. On warming, it would return to its liquid state.

Figure 5.7

Yes, each has 18 carbons and is also referred to as a C-18 fatty acid.

Figure 5.10

Corn oil would be expected to have a larger iodine number since it has a much greater percentage of unsaturated fats.

Figure 5.13

They are all in the *cis* configuration.

Figure 5.14

1. Foods derived from plants
2. Peanut butter

Section 5.2 Concept Check

1. A triglyceride is an ester of glycerol and three fatty acids. A fatty acid is a type of carboxylic acid.
2. Oleic acid is mono-unsaturated, meaning it has a C=C bond. Stearic acid is saturated, meaning it has no C=C bonds.
3. Animal fats tend to be higher in saturated fat. Consumption of saturated fat is associated with increased serum cholesterol levels.
4. Partial hydrogenation increases levels of saturated and trans fats, both of which increase LDLs and decrease HDLs, raising the risk of atherosclerosis.

Figure 5.15

1. H_2O
2. Yes, they are isomers because they have the same molecular formula ($C_{12}H_{22}O_{11}$), but have different structures.

Figure 5.18

Sucrase speeds up the reaction because it allows sucrose to react more rapidly with water. In step 4, the enzyme's active site is freed up to participate in another catalytic cycle.

Figure 5.19

Glucose

A Deeper Look

It becomes an alcohol functional group C—OH.

Figure 5.20

Glycogen. Both starch and glycogen are polymers of α-glucose.

Figure 5.21

Glucose and galactose

Section 5.3 Concept Check

1. cellulose (a polysaccharide) > maltose (a disaccharide) > fructose (a monosaccharide)
2. water (H_2O)

3. We lack enzymes to break down the cellulose present in grass and hay.

Table 5.1

1. Alanine, isoleucine, leucine, phenylalanine, proline, valine
2. Serine, threonine, tyrosine

What a Chemist Sees

Irreversible

Figure 5.28

Yes, peanut butter is a legume, and bread is made from wheat, a grain.

Section 5.4 Concept Check

1. H_2O
2. The sequence of amino acids in the peptide chain
3. *Quality* of protein refers to how well the protein provides essential amino acids, which must come from diet.
4. The heat disrupts attractive forces that maintain the protein's higher-order structure.

Did You Know?

Ask Yourself

In olestra, fatty-acid groups are linked to the $-OH$ groups in sucrose. In sucralose, three of the $-OH$ groups in sucrose are replaced by Cl (chlorine) atoms.

Put It Together

The Asp-Phe dipeptide has a free carboxylic acid group on Phe (phenylalanine). In aspartame this carboxylic acid group is capped with a $-CH_3$ group from methanol.

What is happening in this picture?

1. Arginine has an amino group in its side chain but methionine does not.
2. Lysine is an essential amino acid, whereas arginine is a nonessential amino acid.

Chapter 6

Figure 6.1

(a) solids (b) solids, liquids

Figure 6.3

Body heat increases the rate at which the components in perfume in colognes evaporate from the skin, and you can readily detect the resulting vapors.

Figure 6.4

Glycerine experiences greater hydrogen bonding than ethylene glycol because glycerine contains three polar O—H bonds whereas ethylene glycol contains two. The stronger intermolecular attractions among glycerine molecules give rise glycerine's higher boiling point.

Figure 6.5

1. Ethanol boils at a higher temperature because the intermolecular forces among ethanol molecules (hydrogen bonds) are much stronger than those among ethane molecules (dispersion forces).
2. Since mineral oil exists as a liquid at room temperature its melting point lies lower than room temperature, and since petroleum jelly exists as a semi-solid at room temperature its melting point lies above room temperature. Melting temperatures of hydrocarbons generally increase with molecular size, so the average size of the hydrocarbon molecules in mineral oil are expected to be smaller than the average size of those in petroleum jelly.

Table 6.1

1. (a) Oxygen, nitrogen, propane, ammonia (b) Ethanol, acetone, water, acetic acid (c) Sucrose, sodium chloride, gold.
2. The intermolecular forces in liquid propane are much weaker than those of water, because propane boils at a much lower temperature.

Know Before You Go

1. Water vapor or humidity condenses on cold surface of the container.

Did You Know?

Yes, an oxygen atom of a hydroxyl group from cellulose can form a hydrogen bond with a hydrogen atom of water, as shown here.

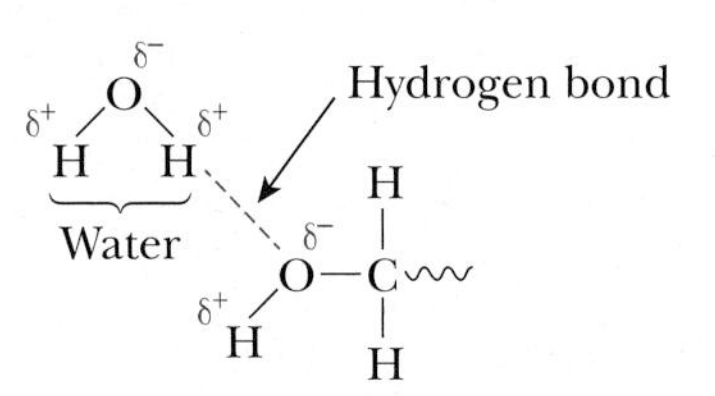

Table 6.2

Since ice is less dense than liquid water, fewer H_2O molecules occupy a given volume of space in ice as compared to in water. The average distance between H_2O molecules in ice is greater than that found in water.

Know Before You Go

3. The block of aluminum.

Section 6.1 Concept Check

1. The kinetic energy increases with temperature.
2. Because the average distance between chemical particles is far greater in the gaseous state than in the liquid or solid states.

Figure 6.10

The density of air is higher at sea level than at very high altitudes.

Figure 6.11

Since the pressure gauge reads 28 psi, the tire's internal pressure is (14.7 + 28) psi = 42.7 psi

Figure 6.12

Yes, the molecules possess translational energy and are constantly in motion.

Figure 6.14

The balloon should appear to inflate, returning to its original size as it warms.

Figure 6.15

40 °C

Figure 6.16

1. The dashed line represents the predicted volume of the sample as it is cooled towards absolute zero *if it were to behave according to Charles's Law.* However, since the sample condenses to a liquid at the temperature at which the dashed line begins (77 K), at temperatures below this, we do not observe the predicted decrease in volume.
2. 4 K, −269 °C

Know Before You Go

4. 99.5 °C
5. −107 °C
6. 1.3 atm

Figure 6.19

1. After it passes through the expansion valve. Heat is absorbed by the refrigerant.
2. After it passes through the condenser. Heat is released by the refrigerant.

A Deeper Look

The sample would melt, turning from a solid to a liquid. No, it would not be appropriate to call it dry ice under these circumstances since it produces a liquid.

Section 6.2 Concept Check

1. Nitrogen and argon
2. According to Charles's Law its volume would increase as it is heated if the pressure remains constant.
3. Expansion of a gas lowers the kinetic energy of its particles.

Figure 6.20

In the reactant, H_2O_2, each oxygen atom forms a covalent bond with one oxygen atom and with one hydrogen atom. In the product, O_2, the oxygen atoms form a double covalent bond, and in the product, H_2O, the oxygen atom is covalently bonded to two hydrogen atoms.

Know before you go

7. $2\ C_2H_2 + 5\ O_2 \longrightarrow 4\ CO_2 + 2\ H_2O$
8. $2\ NaN_3 \longrightarrow 3\ N_2 + 2\ Na$

Section 6.3 Concept Check

1. One or more new chemical species are formed.
2. The law of conservation of matter.

Figure 6.25

1. 5.6 mol H_2O
2. 3.3×10^{24} H_2O molecules

Know Before You Go

9. a. They have the same number of atoms; b. One mole of lead.
10. 50 g of H_2O represents a greater number of moles.
11. a. 0.61 mol Al; b. 3.7×10^{23} Al atoms

Figure 6.26

1. 2 dozen
2. 2 dozen

Know Before You Go

12. 31 mol O_2.
13. a. $C_6H_{12}O_6 \longrightarrow 2\ C_2H_6O + 2\ CO_2$; b. 16 mol CO_2

Figure 6.27

Two C—C bonds and eight C—H bonds

Know Before You Go

14. 27.5 g CO_2
15. 426 kcal

Section 6.4 Concept Check

1. Divide its mass by its molar mass.
2. The coefficients in a balanced equation represent the ratio of individual chemical particles (such as molecules) involved in a reaction, as well as the ratio of moles of these particles.
3. The amount of heat released from the combustion of one mole of the hydrocarbon.

What is happening in this picture?

Both types of changes are involved. A chemical reaction between resin and hardener produces a polymer product. Evaporation of solvents constitutes physical changes.

Chapter 7

Figure 7.1

1.

Solution	Solvent	Physical state of solvent	Physical state of solution
Air	nitrogen	gas	gas
Champagne	water	liquid	liquid
18K gold	gold	solid	solid

In each case, the physical state of the solvent determines the physical state of the resulting solution.

2. a) Oxygen; b) Solutes in gold alloys include copper, silver, nickel and other metals.
3. Various correct answers apply. Examples include: a) sugar or salt in water; b) ethanol in water; c) carbon dioxide in water; d) copper in gold

Figure 7.3

1. Hydrogen bonding
2. Electron poor hydrogen atoms of water can form hydrogen bonds with electron rich oxygen atoms of O—H groups of glucose. Electron rich oxygen atoms of water can form hydrogen bonds with electron poor hydrogen atoms of O—H groups of glucose.

Figure 7.5

1. Sugar is more soluble in hot water. As the hot, saturated solution cools, the solution temporarily holds more dissolved sugar than it can hold. As a result, crystals begin to form.
2. There is no change in the chemical species present. For example, sucrose ($C_{12}H_{22}O_{11}$) dissolves in water to produce the supersaturated sugar solution. After crystallization, sucrose and water are still present.

Figure 7.8

Egg yolk contains lecithin, an emulsifier.

Section 7.1 Concept Check

1. The physical state of the solvent determines the physical state of the resulting solution.
2. Opposite charges attract. The oxygen atom of water carries a partial negative charge, allowing it to associate with cations and with atoms on polar molecules bearing partial positive charges. The hydrogen atoms of water carry partial positive charges, allowing them to associate with anions and with atoms on polar molecules bearing partial negative charges.
3. The dispersed chemical particles in suspensions are larger than dissolved particles in solutions.

Figure 7.10

1. 10:4 and 20:8.
2. Henry's Law predicts that at a given temperature, the concentration of a gas dissolved a liquid is directly proportional to the pressure of the gas above the fluid. In this example, doubling the pressure of the gas above the liquid, doubled the concentration of gas within the liquid.

Figure 7.11

Boyle's Law.

Figure 7.12

This is due to a physical process.

Figure 7.13

1. O_2: 40-100 mm-Hg; CO_2: 40-45 mm-Hg
2. Water vapor, from the inventory of water in our bodies.

What a Chemist Sees

21 mg, which is equivalent to 0.021 g.

Section 7.2 Concept Check

1. Since the pressure of CO_2 in the atmosphere is so low, Henry's Law predicts that very little CO_2 will remain dissolved in the liquid over time, which is what we observe.
2. In each case, the substances are present in the blood and diffuse through the pulmonary capillaries to the alveoli of the lungs.

Know Before You Go

1. 1.0 mol NH_3.
2. 3.6 L.
3. a) 0.036 *M* NaCl; b) 0.146 *M* sucrose.

Figure 7.14

1. (a) isopropanol; (b) water
2. The rubbing alcohol and saline solution

Know Before You Go

4. 2.4% (w/w)

Figure 7.15

Blood alcohol concentrations will be systematically lower in the group of larger individuals, other things being this same, because the individuals in this group have a greater average total blood volume.

Know Before You Go

5. 20–60 ppm zinc.

Section 7.3 Concept Check

1. The number of moles of solute and the volume of the solution.
2. Acetic acid; 5% w/w.
3. 1 mg/L or 1 ppm.

Figure 7.17

(a) 0.75%; (b) 0.007%

Figure 7.19

Since the solution is so dilute we can assume the density is the same as that of water: 1 g/mL. This means 1 kg of solution has a volume of 1 L, so a concentration of 1 mg/kg is the same as 1 mg/L.

Table 7.1

0.05 mg/L exceeds the acceptable limits for arsenic, cadmium, lead, and mercury.

Figure 7.20

1.

#1	#2	#3	#4	#5
5000 ppm	500 ppm	50 ppm	5 ppm	0.5 ppm

2. a) #3; b) #2

Figure 7.22

Less concentrated.

Did You Know?

1. Thermal (or heat) energy to heat the water; and mechanical energy (to apply pressure and force the water through the membrane).
2. a) The answer depends on your location. Typically the source of the water is indicated near the beginning of the water-quality report.
   b) The formatting of the reports vary, but typical categories include bacterial, inorganic, organic, and radioactive contaminants. Sometimes disinfectant byproducts are listed as well.
   c) Parts per million; parts per billion; maximum contaminant level; none detected.

Section 7.4 Concept Check

1. About 0.3%
2. (i) Common anions include chloride (Cl^-), bicarbonate (HCO_3^-), and sulfate (SO_4^{2-}). (ii) Common cations include calcium (Ca^{2+}), magnesium (Mg^{2+}), potassium (K^+), and sodium (Na^+).

3. The Environmental Protection Agency (EPA)
4. Pressure is required to counteract and overcome the force of osmosis.

What is happening in this picture?

Hydrogen sulfide, H_2S, sulfuric acid, H_2SO_4, and calcium sulfate, $CaSO_4$. The deposits are $CaSO_4$

Chapter 8

Figure 8.2

acids: citric, ascorbic (also: acetic)
base: calcium carbonate

What a Chemist Sees

Sulfuric acid: sulfur; Nitric acid: nitrogen; Boric acid: boron; Carbonic acid: carbon

Figure 8.6

One hydronium ion is produced.

Know before you go

1. a. acid: HI; base: LiOH; products: LiI + H_2O;
   b. acid: HNO_3; base: NH_3; products: $NH_4^+ + NO_3^-$

Secion 8.1 Concept Check

1. a. red; b. blue; c. blue
2. a salt
3. Because it does not require the presence of water as solvent.

Figure 8.8

Equal amounts of hydronium and hydroxide ions are produced, so they exist in a one-to-one ratio.

Know before you go

2. a. 10^{-3} *M*
3. 0.00001 *M*; 0.00002 mol H_3O^+
5. pH = 8
6. 10^{-3} *M*

Figure 8.11

1. a. $10^{-7} \times 10^{-7} = 10^{-14}$; b. $10^{-3} \times 10^{-11} = 10^{-14}$; c. $10^{-11} \times 10^{-3} = 10^{-14}$
2. $[H_3O^+] = 10^{-9}$ *M*; pH = 9; basic

Figure 8.12

1. a. basic; b. acidic; c. acidic
2. decreases
3. a. increases 10 fold; b. increases 100 fold; c. increases 1000 fold

Figure 8.14

Answer (a), because HCl completely dissociates, whereas acetic acid does not.

Know before you go

7. a. 0.01 *M* HCl; b. 0.01 *M* acetic acid; c. 0.01 *M* HNO_3

Section 8.2 Concept Check

1. The concentrations are constants because water is in dynamic equilibrium with the ions it forms. The concentrations are equal to each other because each ionization event produce one hydronium and one hydroxide ion.
2. pH = 6
3. It is less in the case of the acetic acid solution because acetic acid only partially ionizes. It is a weak acid.

Figure 8.18

the central carbon

$$\mathrm{HO{-}\boxed{C}{-}CO_2H} \text{ with } \mathrm{CH_2{-}CO_2H} \text{ above and } \mathrm{CH_2{-}CO_2H} \text{ below}$$

Figure 8.19

magnesium chloride, $MgCl_2$

Figure 8.20

a. H_2O/H_3O^+ and c. citric acid ($C_6H_8O_7$)/citrate ion ($C_6H_7O_7^-$)

Figure 8.24

calcium ions (Ca^{2+}) and carbonate ions (CO_3^{2-})

Did You Know?

1. blue
2. Because of buffering, these solutions are expected to maintain a stable pH, which is necessary to minimize eye irritation.
3. Various correct answers apply, including copper and zinc.

Section 8.3 Concept Check

1. acidic
2. the carboxyl group, $-CO_2H$
3. Antacids provide enough base to just neutralize excess stomach acid. The stomach needs to maintain acidity for normal functioning.
4. sulfur dioxide (SO_2) and nitrogen oxides (NO_x)

What is happening in this picture?

1. carbon dioxide (CO_2)
2. sodium citrate ($NaC_6H_7O_7$)

Chapter 9

Figure 9.3

1. 4 u, due to the emission of the alpha particle, a He-4 nucleus.
2. The atomic number decreases (by two) because the alpha particle emitted contains two protons.

Figure 9.4

In beta decay, the atomic number increases by one, so iodine (I), atomic number 53, would decay to xenon (Xe), atomic number 54.

Know Before You Go

1. (a) ${}^{226}_{88}Ra \longrightarrow {}^{222}_{86}Rn + {}^{4}_{2}\alpha$

   (b) ${}^{14}_{6}C \longrightarrow {}^{14}_{7}N + {}^{0}_{-1}\beta$

Section 9.1 Concept Check

1. Becquerel initially assumed the phenomenon he observed was due to x-rays accompanying induced phosphorescence. Becquerel then happened to store sealed photographic plates and uranium in a desk drawer. He later discovered an exposure on these photographic plates even though there

was no sunlight. Without sunlight there could be no induced phosphorescence, leading him to the insight that some other radiation must be emitting from the uranium.

2. Radioactivity originates in the nucleus. Atoms are prone to radioactive decay if the nucleus has an atomic number of 84 or greater of if the ratio of neutrons to protons within the nucleus is above or below a stable range.
3. (i) α decay; (ii) β decay.
4. A β particle.

Figure 9.9

Some elements appear more than once, because different isotopes of the element are formed in the decay series. For example, thorium appears twice, as Th-234 and as Th-230, as a result of the following decay series:

$$\text{Th-234} \xrightarrow{\beta\text{ decay}} \text{Pa-234} \xrightarrow{\beta\text{ decay}} \text{U-234} \xrightarrow{\alpha\text{ decay}} \text{Th-230.}$$

Figure 9.12

Po-210 undergoes α decay to produce Pb-206, its daughter isotope.

Figure 9.13

0.625 g of C-14 would remain after 4 half-lives.
$(4 \times 5730) = 22{,}920$ years, which is about 23,000 years.

Figure 9.14

$(50{,}000 \div 5730) = 8.7$ half-lives, which rounds to 9 half-lives.

Section 9.2 Concept Check

1. Penetrating power: $\gamma > \beta > \alpha$.
   Ionizing power: $\alpha > \beta > \gamma$.
2. Radon is formed as part of the radioactive decay series of uranium. Environmental exposure to radon can cause lung cancer.
3. C-14
4. Nitrogen.

Section 9.3 Concept Check

1. The law of conservation of mass applies to conventional chemical reactions, such as burning wood, so we would not see the conversion of mass (or matter) into energy in this case.
2. Pu-239 has 94 protons, 94 electrons, and 145 neutrons.

$$\begin{aligned} 94 \text{ protons} \times 1.007 \text{ u (mass of an individual proton)} &= 94.658 \text{ u} \\ 94 \text{ electrons} \times 0.0005 \text{ u (mass of an individual electron)} &= 0.047 \text{ u} \\ 145 \text{ neutrons} \times 1.009 \text{ u (mass of an individual neutron)} &= 146.305 \text{ u} \\ \hline \text{Sum} &= 241.01 \text{ u} \end{aligned}$$

Mass defect = (241.01 − 239.05) u, which is about 2.00 u.

Figure 9.18

$${}^{1}_{0}\text{n} + {}^{235}_{92}\text{U} \longrightarrow {}^{92}_{36}\text{Kr} + {}^{141}_{56}\text{Ba} + 3\,{}^{1}_{0}\text{n}$$

Figure 9.19

$3 \times 3 \times 3 \times 3 \times 3 = 243$ neutrons

Figure 9.22

An additional 5 blue dots would have to be removed, leaving 2 blue dots remaining. The mixture will be 60% enriched in red dots when there are 3 red dots and 2 blue dots, because 3 out of a total of 5 particles represents 3/5 or 60%.

Figure 9.27

U-238 makes up the vast majority of spent nuclear fuel, because the fuel is already about 97% U-238 to begin with.

Section 9.4 Concept Check

1. A U-235 atom needs to absorb a neutron in order to initiate fission.
2. Enrichment is necessary because U-235, the desired uranium isotope for nuclear fission, is naturally present in such small quantities in uranium. Enrichment is commonly accomplished through gas centrifugation of UF_6.
3. The heat released in the nuclear reactor is used to boil water, which then drives a steam turbine to produce electricity.
4. Deep, geological storage.
5. Fusion involves the fusing together of very light nuclei, whereas fission involves the splitting of very heavy nuclei.

What is happening in this picture?

The image was take about 60 yrs following the explosion, which represent 2 half-lives. After 1 half-life, 50% of these radioisotopes remain; after 2 half-lives, 25% remain.

Chapter 10

Figure 10.2

1. The vitamin C loses electrons and is oxidized.
2. The hypochlorite ion becomes reduced.

Figure 10.3

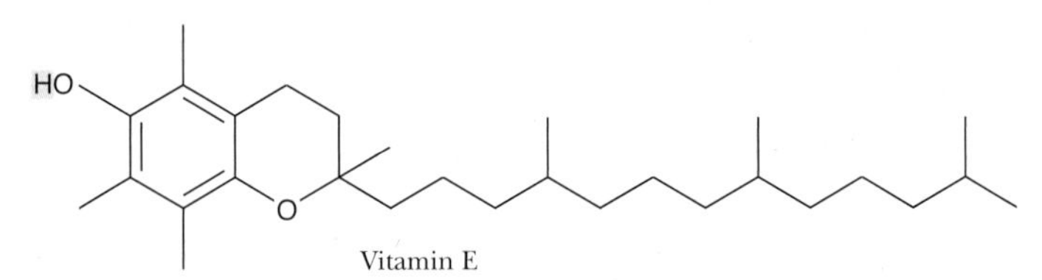

Vitamin E

Figure 10.4

A peroxide compound such as calcium peroxide.

Figure 10.6

1.

	Oxidizing agent	Reducing agent
a.	Ag^+	Cl^-
b.	Cu^{2+}	Ag

2. The silver ion has one fewer electrons than protons so bears a single positive charge. When this ion gains an electron to become a silver atom, the number of protons and number of electrons become equal so there is no electrical charge.

Figure 10.7

a. Yes, here oxidation can be described in terms of gain of sulfur and reduction can be described in terms of loss of sulfur.

b. Sulfur is in the same group (column) as oxygen, one period (row) below oxygen.

Figure 10.8

Carbon is oxidized to carbon dioxide, and the iron ions (Fe^{3+}) of iron oxide are reduced to iron metal.

What a Chemist Sees

1. It is a reduction. 2. Water is a reactant, both in the reaction shown in Step 2, as well as in the second reaction shown in Step 3.

Section 10.1 Concept Check

1. Chlorine (Cl_2) is the oxidizing agent, and sodium (Na) is the reducing agent.
2. The are all redox processes.

Figure 10.9

1. The zinc is the anode, and the copper is the cathode.

2.

	Number of protons	Number of electrons
a.	30	30
b.	30	28
c.	29	29
d.	29	27

Know Before You Go

1. Yes, the positively charged sodium cations would migrate towards the copper sulfate solution and the negatively charged chloride ions would migrate towards the zinc sulfate solution.
2. Because as the chemical species is reduced it causes another to be oxidized and hence acts as an oxidizing agent.
3. a. $F_2(g)$; b. $Li^+(aq)$
4. $Li(s)$. A reducing agent causes other chemical species to be reduced and so undergoes oxidation itself. When evaluating the reverse of the reactions shown in the table (which represent oxidations), we find the oxidation of lithium metal has the most positive oxidation potential: $Li(s) \longrightarrow Li^+(aq) + e^-$ (+3.04 V).
5. $(0.54 + -(-0.76)) = 1.30$ V; $Zn^0 + 2\,I^- \longrightarrow I_2 + Zn^{2+}$

Figure 10.12

a. zinc anode b. (−)-terminal

Figure 10.14

To balance the negative charge created by electrons flowing to the graphite electrode during recharge.

Figure 10.15

Since 2 moles of hydrogen gas (H_2) are produced for every 1 mole of oxygen gas (O_2), the right tube (which contains more gas) contains hydrogen, and the left tube contains oxygen.

Figure 10.16

1. Sodium (Na^+) and chloride (Cl^-) ions **2.** Hydroxide ion (OH^-) **3.** Bromine (Br_2), hydrogen (H_2), and sodium hydroxide (NaOH) **4.** The availability of inexpensive electricity (run by hydroelectric power) to operate these chemical factories.

Figure 10.17

1. Reduced **2.** Because of the high energy requirements of electrolysis as well as the costs of mining bauxite.

Figure 10.18

The silver electrode will decrease in size as silver atoms (Ag^0) are oxidized to silver ions (Ag^+).

Section 10.2 Concept Check

1. The oxidation state of copper reduces from +2 to 0. The oxidation state of zinc increases from 0 to +2.
2. Au^+
3. Because the circuit provides a path for electrons to flow from anode to cathode, thus coupling the two half-cell reactions and allowing the net redox reaction within the battery to occur.
4. Because chemical changes are produced by passing an electrical current through the battery.

Figure 10.19

Electrons flow from the anode (the site of oxidation) to the cathode (the site of reduction).

Did You Know?

Both devices use chemical reaction to create electrical power. In the case of batteries, the reacting chemicals are sealed within these devices. In the case of fuel cells, the reacting chemicals must be supplied externally.

Figure 10.22

Antireflective coatings on solar cells reduce the amount of light that reflects off the surfaces of these devices. Improved antireflective coatings further increase the amount of light absorbed by solar cells, thereby improving their efficiency.

Section 10.3 Concept Check

1. Water vapor.
2. To improve silicon's electrical conductivity, and specifically to prepare p-type and n-type semiconductors for use in solar cells.

What is happening in this picture?

1. Oxidizing agent
2. Lycopene. Yes, it absorbs some of the components of visible light, giving rise to a colored appearance.

Chapter 11

Figure 11.1

1. $C_{18}H_{35}NaO_2$
2.

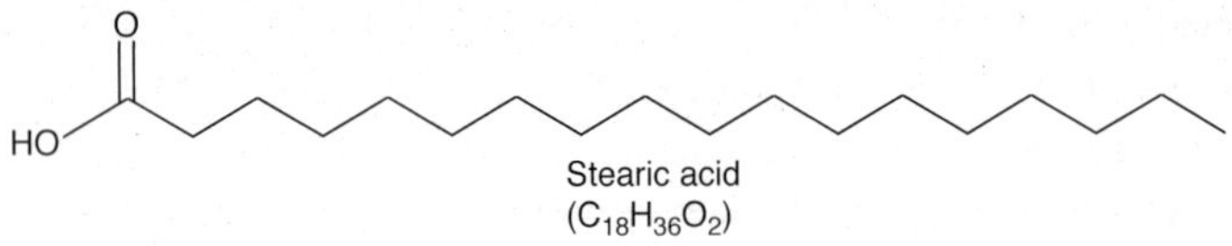

Stearic acid
($C_{18}H_{36}O_2$)

Figure 11.3

Water molecules are polar and thus form attractions to the polar, hydrophilic heads of detergent molecules but are repelled by the nonpolar, hydrophobic tails of detergent molecules.

Figure 11.5

(a) Its nonpolar, hydrophic tail. (b) Its polar, hydrophilic head.

Figure 11.6

Methanol and sodium acetate.

Figure 11.7

All triglycerides are triesters of glycerol and fatty acids (discussed in Section 5.1) so all triglycerides contain: i) a three-carbon glycerol backbone, ii) three ester linkages, and iii) three hydrocarbon chains.

Figure 11.8

Glycerol is a byproduct of saponification (as shown in Figure 11.7) and serves as a skin conditioner.

Figure 11.9

1. Yes
2. No
3. No
4. Twelve carbon atoms in each molecule.
5. Similarities: Each has a nonpolar, hydrophobic tail and a polar, hydrophilic head. Each has 12 carbon atoms. Differences: The structure of the polar, hydrophilic head is different in each. In sodium laurate the head is a CO_2^- group. In sodium laurel sulfate it is an SO_4^- group.

Figure 11.13

1.

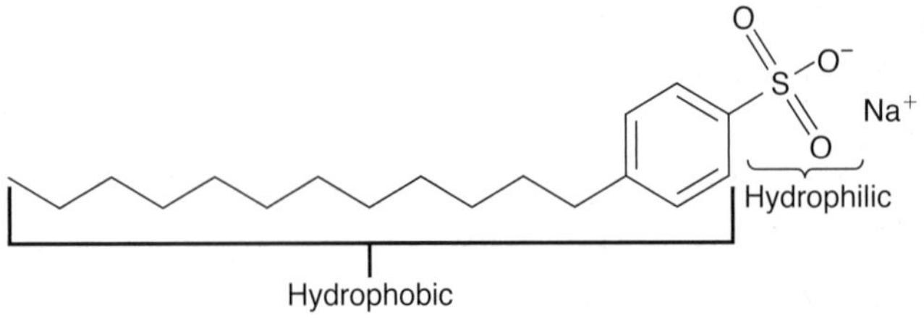

Hydrophilic

Hydrophobic

2.

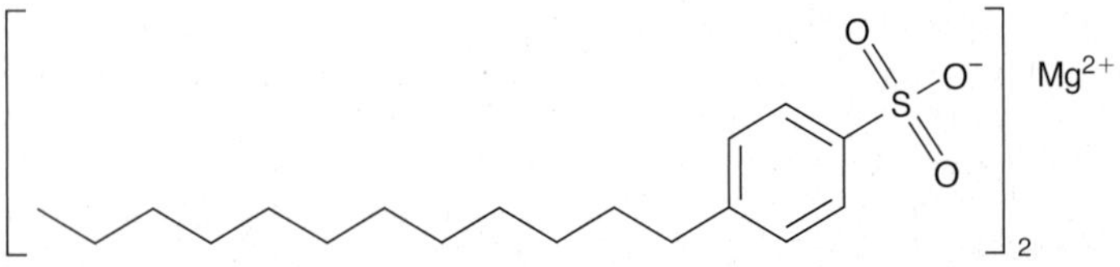

This structure should be more soluble in water as compared to the magnesium salt of a soap.

Did You Know?

Soap molecules are semisynthetic because they are produced from a chemical reaction involving naturally-occurring fats and oils (saponification).

Section 11.1 Concept Check

1. Soap molecules form a monolayer at the water's surface and thereby disrupt attractive forces between water molecules at the surface.
2. Nonpolar interactions (such as dispersion forces discussed in Section 6.1) between the hydrophobic tails of detergent molecules.
3. The molecule's hydrophilic head.
4. Hard water ions, such as Mg^{2+}, Ca^{2+}, and Fe^{2+}, are replaced by Na^+ ions.

Figure 11.16

The outermost layer of the epidermis (the stratum corneum), the remainder of the epidermis, and the dermis.

Figure 11.17

Hydrogen bonding.

What a Chemist Sees

1. Mascaras make the lashes appear longer and thicker by darkening both the ends and shaft, for more contrast and volume.
2. The system of alternating single and double bonds (conjugated system) is noted in red. This system produces color.

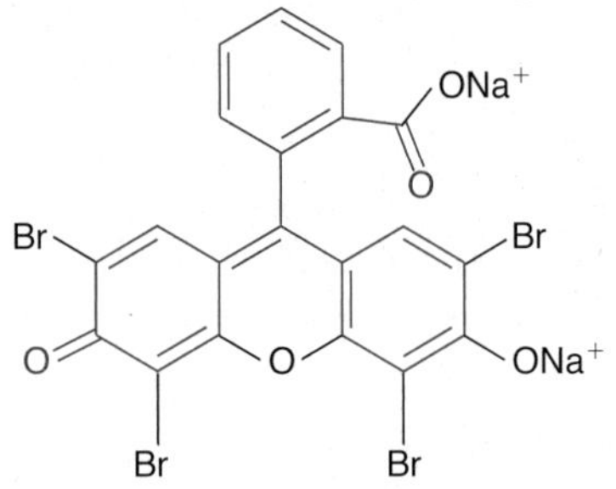

Figure 11.18

1. 2-phenylethanal: aldehyde; 2-phenylethanol: alcohol; civetone: ketone.
2. Civetone < 2-phenylethanol < 2-phenylethanal

Figure 11.20

X-rays

Figure 11.21

(a) Octisalate, octocrylene (b) Avobenzone

Figure 11.22

Ketone and alcohol functional groups.

Section 11.2 Concept Check

1. The fragrance component.
2. An oil-in-water emulsion.

Figure 11.24

Sulfur

Figure 11.26

1. Because the polar, hydrophilic head of each is negatively charged.
2. The structure of the cations.
3. Each is a nitrogen-based cation, with four bonds to nitrogen.
4. They tend to be more soluble in water and gentler on the skin as compared to soaps.

Section 11.3 Concept Check

1. To remove accumulated plaque.
2. Shampoos remove dirt but also some of the accumulated sebum (natural oils) from the hair. Conditioners replace lubricating oils lost in shampooing.

What is happening in this picture?

1. Detergents trap oil particles in micelles, which disperse within water.
2.

Hydrophobic Hydrophilic

Span 80 is a nonionic surfactant.

Chapter 12

Figure 12.1

1. A benzene ring with the following substitution pattern:

2. Because aspirin was developed by chemical modification of a natural product.

Figure 12.2

1.

Carboxylic acid group

Salicylic acid

Acetylsalicylic acid (aspirin)

2. Acetic anhydride adds an acetyl group to salicylic acid, yielding aspirin.

Figure 12.3

(a) I (b) P (c) I

Figure 12.4

1. Phenacetin and acetaminophen both contain within them the acetanilide structure:

Acetanilide

2. They both contain within them the *p*-aminophenol core structure:

p-aminophenol core structure

3. *N*-ace**tyl**-*p*-aminoph**enol**

Section 12.1 Concept Check

1. Aspirin inhibits the activity of enzymes involved in the production of prostaglandins.
2. Histamine is a signaling compound involved in allergic response. Antihistamines block the action of histamine.

Figure 12.7

Amoxycillin is an analog of penicillin because it shares the same basic structure as penicillin, but has small modifications to alter its performance.

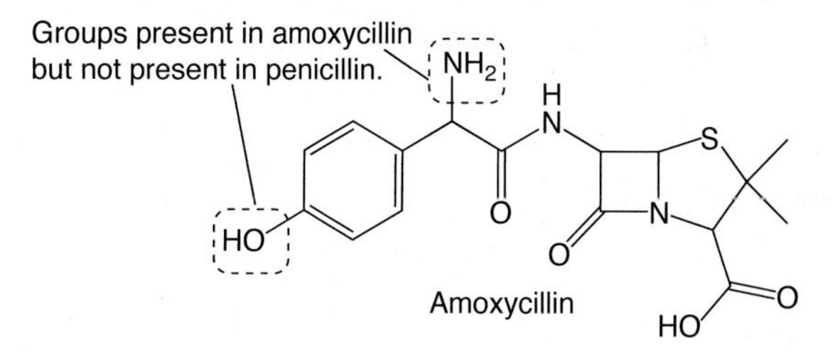

Groups present in amoxycillin but not present in penicillin.
NH2
HO
Amoxycillin

Figure 12.9

Fluorine.

Figure 12.10

1. 17
2.

Alcohol functional group

Testosterone

Figure 12.11

1. (a) Progesterone and norethindrone share the following structural features in common, shown in black.

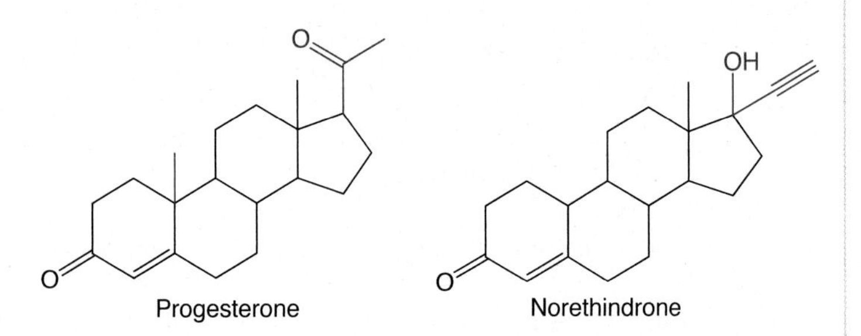

(b) The structures of progesterone and norethindrone differ in the groups shown in red.

2. Both are ketones.

Section 12.2 Concept Check

1. To determine whether the drug undergoing investigation provides true benefit above that demonstrated by the placebo.
2. British bacteriologist Alexander Fleming noticed that a mold contaminating a petri dish had killed bacteria that had been growing there.

Figure 12.13

The nonpolar hydrocarbon-rich regions of THC are noted in red.

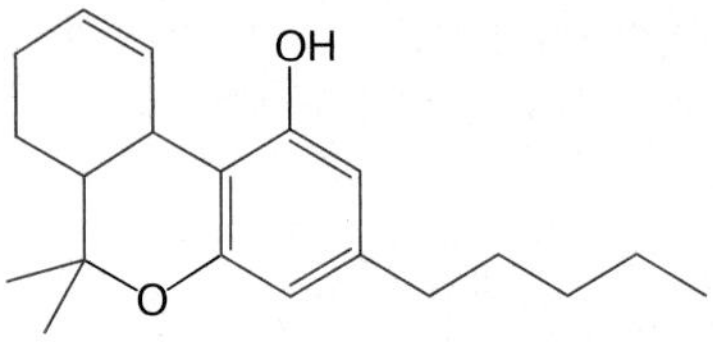

OH

O

Tetrahydrocannabinol (THC)

Figure 12.14

Both structures contain amine groups.

Figure 12.15

1. Codeine contains a methyl group (CH_3) bonded to the oxygen in red, whereas morphine has a hydrogen atom (H) bonded to this oxygen.

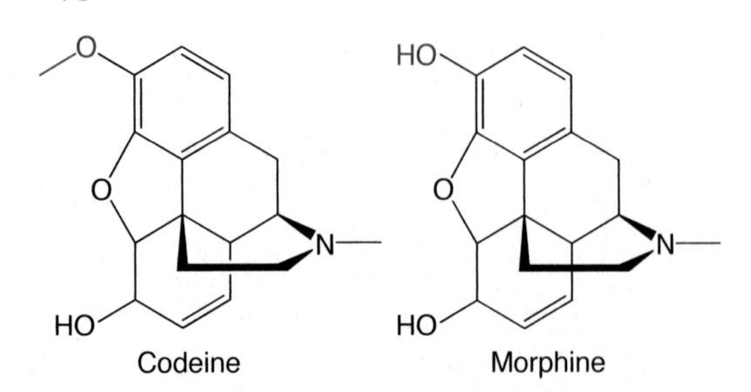

2. Both compounds contain the basic, amine group (and are derived from plants).

Figure 12.16

1. Morphine has alcohol functional groups. These groups are not present in heroin.

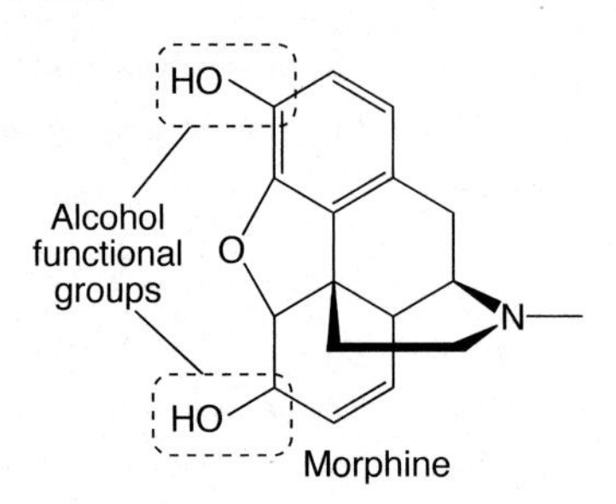

2. Because heroin is derived from the chemical modification of a natural product (morphine).

Figure 12.17

1. Oxycodone and hydrocodone differ from codeine as shown in red.

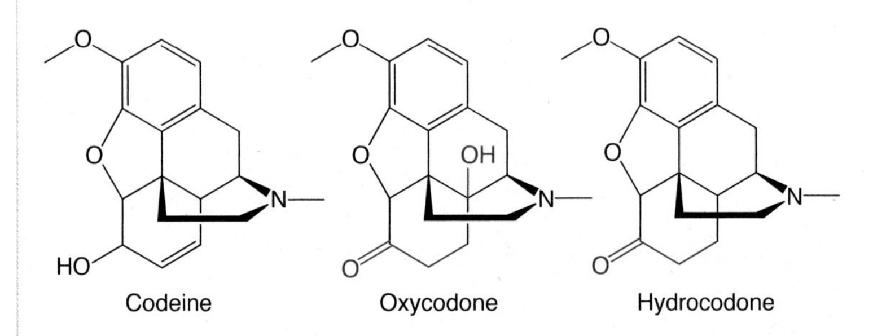

2. Oxymorphon differs from morphine as shown in red.

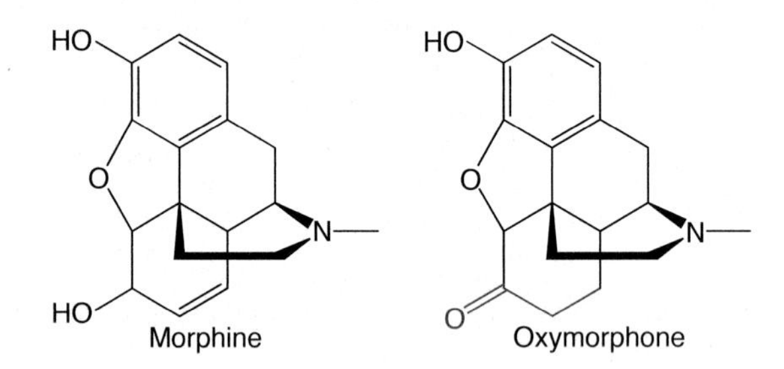

Figure 12.18

1.

HO

O

N

Cl⁻

H

HO

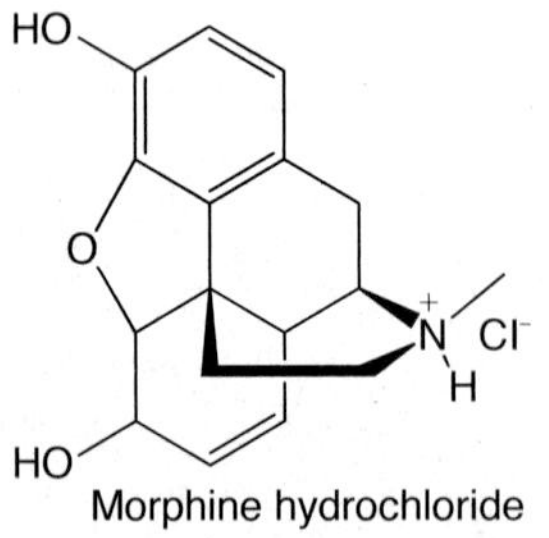

Morphine hydrochloride

2. More soluble in water since it is a salt.

Figure 12.19

1. Both LSD and mescaline contain the amine functional group.

2.

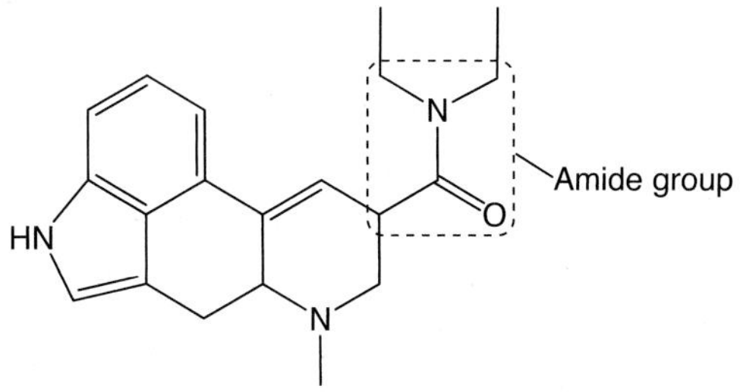

Amide group

Lysergic acid diethylamide (LSD)

A Deeper Look

Treat with a base, such as sodium bicarbonate.

Figure 12.20

1. Methamphetamine contains a methyl group (CH_3) bonded to nitrogen.
2. MDMA has an oxygen-containing ring structure that is not present in methamphetamine.

Section 12.3 Concept Check

1. Acetaldehyde.
2. An amine.
3. 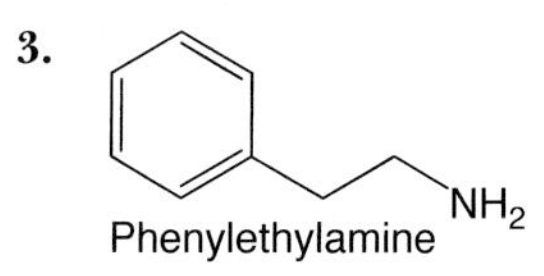

Phenylethylamine

Figure 12.23

Cytosine

Figure 12.24

1. If the strands were identical, identical bases would pair with one another. Instead, complementary bases pair with one another (A to T, G to C), so the strands are complementary.
2. CGCCTACTA

Figure 12.25

Hydrogen bonds between complementary bases.

Table 12.1

Histidine (His). CAU.

Figure 12.27

GAA: Glutamic Acid (Glu)

CAA: Glutamine (Gln)

UGA: Terminate peptide chain (also known as a Stop codon)

Figure 12.28

1. This would result in normal hemoglobin since the codon GAA also corresponds to glutamic acid (Glu).
2. The glutamic acid side chain is polar because of the polar carboxyl group (CO_2H). The valine side chain is nonpolar because of the nonpolar hydrocarbon group.

Figure 12.29

(a) Yes. (b) No.

Did You Know?

In the first case, if it is used to treat a medical condition, it would more appropriately be called a medicine, although it provides nutritional benefits as well. In the second case, if it is used prophylactically, it functions both as a food and as a (prophylactic) medicine.

Section 12.4 Concept Check

1. A gene contains the sequence of DNA bases that code for a protein (polypeptide chain). During transcription, a complementary strand of messenger RNA (mRNA) is produced from the DNA of the gene. This mRNA consists of a sequence of codons that dictate the sequence of amino acids of the forthcoming protein. In translation, each sequential amino acid of the growing protein chain is transported to its corresponding codon of mRNA by a molecule of transfer RNA (tRNA).
2. Sickle cell anemia results from a genetic mutation that replaces a polar glutamic acid unit with a nonpolar valine unit at a precise point within one of the pairs of the polypeptide chains of hemoglobin. As a result, the hydrogen bonds that form the higher structures of hemoglobin produce an abnormal molecular structure, causing hemoglobin molecules to aggregate or clump together.
3. Both processes involve nuclear-transfer, in which the nucleus of an egg cell (with its half set of chromosomes) is replaced by a nucleus extracted from a body-cell of another individual, which has a full set of chromosomes. In reproductive cloning, this egg forms the basis of the development of a new infant, whereas in therapeutic cloning, this egg forms the basis of the development of stem cells, which have therapeutic uses.

What is happening in this picture?

1. Non-coding refers to regions of DNA that do not code for proteins and hence do not represent genes.
2. DNA is present in skin cells. Yes, the sample could become cross contaminated with human DNA.
3. Yes, dog and human cells contain different numbers of chromosomes. Human somatic cells have 23 pairs of chromosomes and those of dogs contain 39 pairs.

Chapter 13

Figure 13.3

1. Ethylene
2. a. four b. two

Table 13.1

a. It is similar in that it contains a carbon-carbon double bond. b. It is different in that its carbon atoms form single bonds exclusively to hydrogen atoms. In the remainder of the monomers in the table, one or more these hydrogen atoms have been replaced by another element.

Figure 13.4

1.

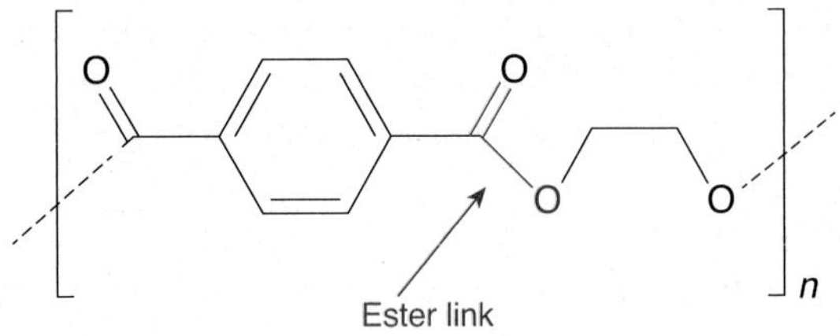

Ester link

2.

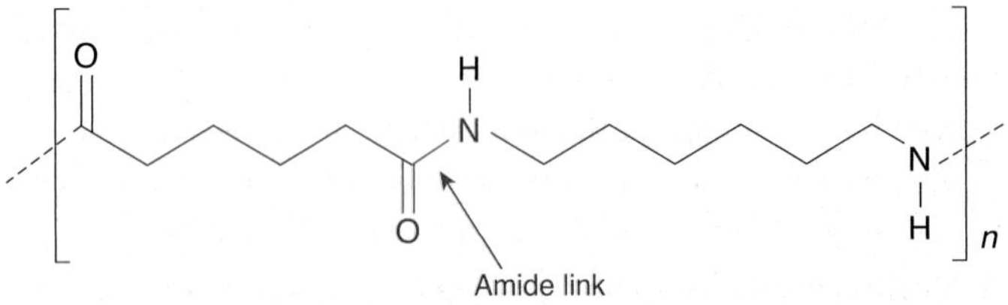

Amide link

3. These represent bonds linking the recurring unit of the polymer.

4.

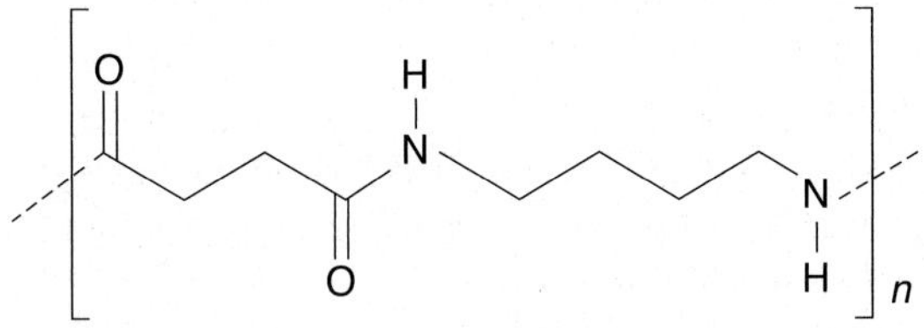

Figure 13.5

1. Because all the atoms originally present in the monomer units are retained in the resulting polymer.
2. a. 5 b. 5,000

Section 13.1 Concept Check

1. Commercial plastics are made from the polymerization of monomers and hence are polymers. However, various naturally occurring polymers, such as proteins and polysaccharides, are not plastics.
2. Addition polymerization retains all the atoms originally present in the monomer units. Condensation polymerization releases a small molecule, such as water, with the formation of each link, so not all of the atoms originally present in the monomer units are retained in the polymer.
3. Because once cured, they harden permanently so will not soften during high temperature use.

Figure 13.6

Yes, because it is made by chemical modification of a naturally occurring product.

Figure 13.7

1. Because it is made from a single type of monomer unit.
2. It has a chlorine atom in the place of a methyl group.

Figure 13.8

Formaldehyde

What a Chemist Sees

1.

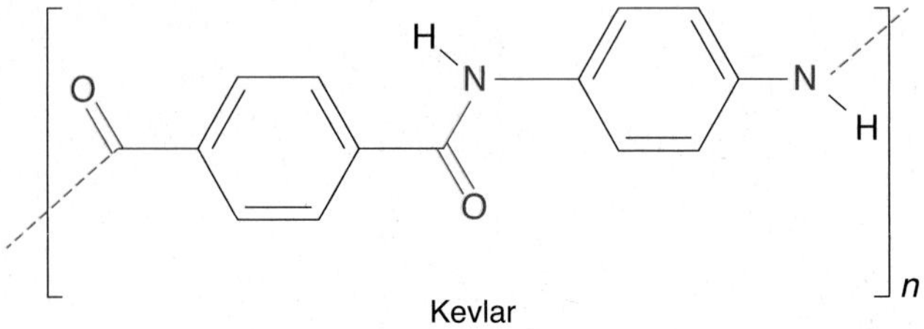

Kevlar

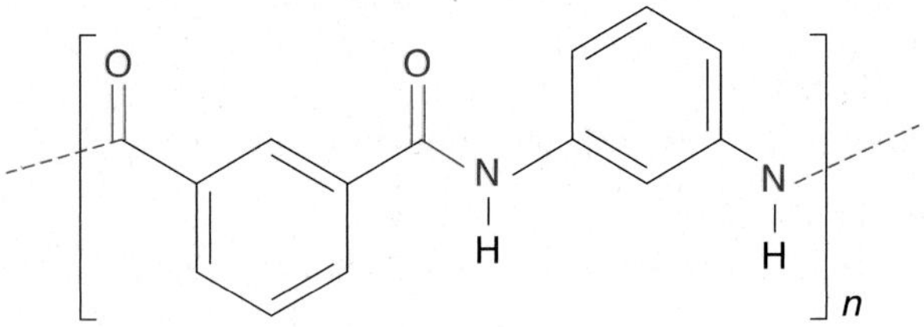

Nomex

2. a. H_2N–C_6H_4–NH_2

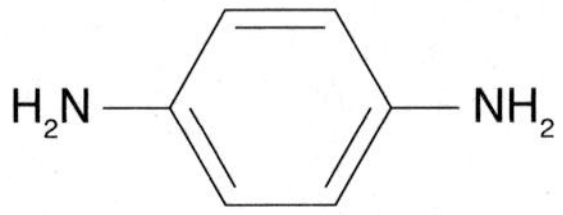

b. H_2N NH_2

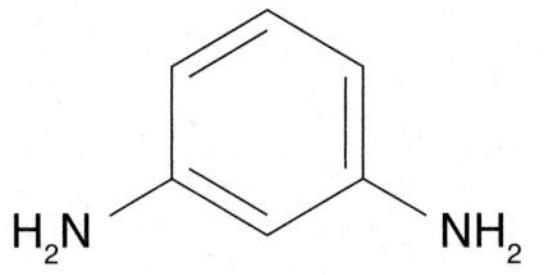

Figure 13.11

1. PLA is all three. It is a polyester because it contains repeating ester linkages. It is a homopolymer because it is made from a single type of monomer. It is a thermoplastic because it softens with heat.

Section 13.2 Concept Check

1. It is a strengthened form a rubber made by heating natural (latex) rubber with sulfur. Unlike natural rubber, vulcanized rubber retains its elasticity when heated because the sulfur cross-links between polyisoprene chains help prevent the chains from slipping past one another when stretched.
2. Resin identification codes are imprinted on consumer plastics as an aid to classify them for recycling. Numbers 1 through 6 correspond to the most common recyclable plastics, and number 7 includes all other types of plastics.

Figure 13.12

No. For example, the air can contain ozone and volatile organic compounds from natural sources; the rain contains carbonic acid produced from the absorption of carbon dioxide in the air; the pond water contains dissolved minerals and microbiological contaminants.

Figure 13.15

1. The concentration of large numbers of vehicles and their emissions of nitrogen oxides and hydrocarbons can contribute to smog.
2. $O_3 + NO \rightarrow O_2 + NO_2$

Table 13.2

a. mobile b. stationary

Figure 13.22

1. Carbon, hydrogen, and chlorine.
2. They all contain aromatic (benzene) rings bonded to one or more chlorine atoms.

Section 13.3 Concept Check

1. Because a contaminant may be considered pollution in one location but not in another. For example, ozone can be a pollutant at ground level, but not in the stratosphere, where it protects against incoming ultraviolet radiation.
2. Scrubbers that force the gases through a slurry of basic materials, such calcium carbonate, $CaCO_3$, or magnesium hydroxide, $Mg(OH)_2$. This reduces emissions of dioxide, a cause of acid rain.
3. a. These materials include paper and paperboard, plastics, metals, glass, food waste, and lawn and yard trimmings.
   b. These materials include paints, solvents, motor oils and antifreeze, lawn and garden chemicals, electronic devices, and certain types of batteries.

What is happening in this picture?

1. It means that a small molecule is released with the formation of each link in the resulting polymer.
2. Hydrochloric acid, HCl.

Chapter 14

Figure 14.1

1. Four.
2. No because it does not contain the carboxyl group, $—CO_2H$.
3. Ascorbic acid has polar hydroxyl groups, —OH, which are capable of hydrogen bonding with water molecules. In addition to these four hydroxyl groups, ascorbic acid has two additional oxygen atoms that are capable of forming hydrogen bonds with the hydrogen atoms of water molecules.

Figure 14.2

1. A hydroxyl group, —OH. Since this hydroxyl group is bonded to a carbon atom in retinol, this constitutes an alcohol functional group.
2. Both β-carotene and retinol have extensive hydrocarbon chains, which are nonpolar and which can form associations with the hydrocarbon chains of fats.

Figure 14.3

1. 7-dehydrocholesterol contains the four-ring steroidal structure (see Figure 12.10), but vitamin D_3 does not contain this structure. In addition, a CH_3 group in 7-dehydrocholesterol becomes a CH_2 group in vitamin D_3.

What a Chemist Sees

Hydrogen is located in the first period (row) and first group (column) of the periodic table. Carbon, nitrogen, and oxygen are located in the second period and in the 14th, 15th, and 16th groups, respectively, of the periodic table.

Section 14.1 Concept Check

1. Carrots are a good source of β-carotene, which is converted into retinol (vitamin A) within the body.
2. (a) Calcium (b) Iron

Figure 14.4

EDTA and oxalic acid both contain carboxylic acid functional groups. Just as EDTA can use the lone pairs on oxygen atoms to form bonds to metal ions, oxalic acid can do so as well. In binding to certain minerals, such as the 2+ ions of calcium and iron (Ca^{2+} and Fe^{2+}) oxalates interfere with the body's absorption of these minerals.

Section 14.2 Concept Check

1. The U.S. Food and Drug Administration (FDA)
2. The purpose of the GRAS list is to allow chemical substances that are generally regarded as safe to be added to foods within prescribed levels.
3. Ascorbic acid (vitamin C) added to foods functions both as a nutritional agent and as an antioxidant and antimicrobial agent.

Table 14.8

Arsenic trioxide.

Figure 14.8

Alcohol (-ol); Aldehyde (-al); Alkene (-ene); Ester (-ate); Ketone (-one)

Figure 14.10

That neither the substance in question nor its metabolites act as mutagens in this test.

Section 14.3 Concept Check

1. The test animal and the method of administration.
2. Because no substance is free from potential hazards.

What is happening in this picture?

1. Because they live at a far northern latitude and receive limited sunlight for a significant portion of the year.
2. The treatment exposes the skin to ultraviolet light, which helps to produce vitamin D.
3. Adding vitamin D supplements to the diet, such as through vitamin D-fortified milk.

Answers to End-of-Chapter Exercises

Chapter 1

1. (a) coal, natural gas, and oil; (b) biologically derived fuels (biofuels); solar, wind, geothermal, hydroelectric, and nuclear energy.
2. (a) gasoline or diesel fuel; (b) coal, natural gas, or wood
3. (a) soap; (b) they serve as cleansing agents
4. (a) aspirin; (b) acetylsalicylic acid; (c) it inhibits blood's ability to clot
5. Clothing, plastics
6. Multiplication
7. (a) meter; yard; (b) liter; quart or gallon; (c) kilogram; pound

8. mega-; kilo-; centi-; milli-; micro.
9. Paracelsus was a 16th century Swiss physician and alchemist who recognized that any substance that enters our bodies can harm us as well as benefit us, and that the amount of harm is usually proportional to the quantity. He stated this principle as "Poison is in everything, and no thing is without poison. The dosage makes it either a poison or a remedy." Today we state this more simply as, "The poison is in the dosage."
10. (a) Bisphenol-A, used in plastics and resins to render bottles and similar containers resistant to breaking or shattering; and in liners of food cans to prevent foods from changing in flavor or consistency when kept for long periods of time; (b) BPA may pass from the container or liner into the stored food or drink, and then into our bodies as we eat or drink the contaminated contents.
11. (a) fresh water, trees, food crops; these grow naturally or are renewed readily; (b) fossil fuels such as natural gas, coal and petroleum products; these do not grow nor are they renewed readily; as we consume them their natural supplies shrink steadily
12. Combustion of fossil fuels can be used to produce electricity which runs fans and air conditioning systems.
13. (a) kilogram, meter, second; (b) Because it's necessary to have absolute definitions of these important units.
14. This is a matter of personal opinion, but arguably, the metric system is more useful to society as almost all countries have adopted this system and the prefixes used in the metric system allow for more convenient conversions between units.
15. 1 gram.
16. (a) Basic research seeks to discover new knowledge for its own sake, whereas research and development has a particular application or end use in mind. (b) Discoveries from basic research have been shown to lead the way to many new products, technologies, and even new industries that have benefited society.
17. (a) A hypothesis is a tentative explanation for a relatively small set of observations. whereas a theory is a generally accepted principle based on a large set of confirmed observations. (b) In an experiment, a test or procedure is performed. An observation is simply the act of recording or carefully examining something.
18. (a) Macronutrients, such as proteins, carbohydrates, and fats and oils, are the major components of our foods that provide us with energy and the materials that form our bodies. Micronutrients are the vitamins, minerals and other minor components of our foods that our bodies need in only small amounts each day. (b) The water would contain both macro- and micronutrients, however, in small amounts.
19. Raw materials: The product could be reformulated so that fewer ingredients are used (such as the absence of dyes and fragrances) or more sustainable resources are used, such as plant-derived cleaning agents. Manufacturing: The manufacturing process could be designed to require less energy or produce less waste. Packaging: The packaging could be designed to use fewer materials, those from more sustainable resources, or those that can be recycled more easily. Use: The product could be formulated to be more concentrated so that less energy is required to transport it and less of it is used. The product could be designed to work in cold water so that less energy is required.
20. From the figure, 100 pounds is about the same as 45 kilograms, so 100/45 is roughly 2.2 pounds per kilogram.
21. (a) 1.61×10^6 millimeters (b) 3.15×10^{10} milliseconds (c) 9.07×10^{11} micrograms
22. 0.001 kg
23. 4.1×10^8 cm
24. (a) 9.46×10^{15} meters (b) 5.88×10^{12} miles

Chapter 2

Review

1. a. Ernest Rutherford found that an atom's positive charge is localized in an extremely small, dense center, the atom's nucleus.
b. Neils Bohr proposed that electrons occupy well-defined orbits, now known as quantum shells.
c. J. J. Thomson discovered electrons as negatively charged particles contained within atoms.
2. Aristotle believed that all matter is completely continuous and is not composed of atoms.
3. a. Dmitri Mendeleev devised the periodic table.
b. Erwin Schrodinger, Max Born, and Werner Heisenberg developed the modern, quantum-mechanical model of the atom, based on wave mechanics.
4. Democritus and John Dalton both contributed to our understanding that elements are composed of extremely small particles known as atoms.
5. a. An atom's atomic number represents the number of protons in the atom's nucleus.
b. At atom's mass number represents the combined number of protons and neutrons in the atom's nucleus.
c. In contrast to the mass of an individual atom, which defines the mass of that individual atom in atomic mass units, the atomic mass of an element represents the weighted average of the masses of all of the isotopes of that element.
6. a. The superscript preceding the elemental symbol represents the mass number of the atom identified by that symbol.
b. The subscript that appears to the left of an elemental symbol represents the atomic number of an atom of that element.
7. Electrons occupy an atom's quantum shells.
8. Protons and neutrons have similar masses. Protons are positively charged particles; neutrons are electrically neutral.
9. Electrons contribute least to an atom's mass.
10. Carbon, specifically the isotope C-12, is used to define the atomic mass unit.
11. The number of neutrons in their nuclei distinguishes one isotope of an element from another isotope of the same element.
12. ${}^{1}_{1}H$; ${}^{2}_{1}H$; ${}^{3}_{1}H$
13. The nucleus of ${}^{1}_{1}H$ is unique in that it contains no neutrons.
14. The atoms must differ in the number of protons present in their nuclei.

15. a. 4; b. 9; c. beryllium; $^{9}_{4}Be$

16. The colored lines result from energy emitted by electrons in higher energy shells as these electrons drop to lower energy shells.

17. The inert gases or the noble gases

18. The value of $2(n^2)$ represents the maximum number of electrons an atom's electron shell can hold, where n represents the shell's quantum number.

19. a. An object's weight is a product of the effect of gravity on a body.
b. An object's mass reflects its resistance to acceleration or to any other change in its motion, wherever in the universe it might be.
c. The body's weight will change, but not its mass. On moving from the earth to the moon, the body's resistance to acceleration or to any other change in its motion does not change, but the gravity affecting the body does change, resulting in a change in the body's weight.

20. We find the element lithium in certain kinds of batteries.

21. a. Beryllium (Be), magnesium (Mg), strontium (Sr), barium (Ba).
b. They are in the same period. Aluminum is a metal, silicon is a metalloid, and phosphorus is a nonmetal.
c. Nitrogen (N), neon (Ne), sodium (Na), nickel (Ni).

Think

22. No. Isotopes are atoms of the same atomic number – and therefore representing the same element – but with different mass numbers.

23. Atoms of two different elements can have the same *total number* of protons and neutrons, and therefore have the same mass number, but if the atoms represent different elements the atoms must have different atomic numbers.

24. We can identify various elements by using their atomic numbers. We could identify carbon as element 12, hydrogen as 1, nitrogen as 7, and oxygen as 8.

25. Each element of this column, including both hydrogen and all the alkali metals, has only one electron in its valence shell (the outermost quantum shell).

26. a. Rutherford found that an atom has an extremely small nucleus, in which all the positive charges of the atom are concentrated; Thomson believed that the entire atom is a positively charged sphere.
b. Bohr viewed an atom as having a positively charged nucleus with electrons occupying a series of concentric shells surrounding the nucleus. Thomson believed that the electrons were scattered throughout the positively charged sphere, somewhat like raisins in a raisin-pudding.

27. Electrons are at their lowest energy levels when they occupy the quantum shells that Neils Bohr postulated. It would take more energy than that provided by the attraction of opposite electrical charges to pull the electrons out of their quantum shells and into the nucleus.

28. A neutron would be produced. The negative charge of the electron and the positive charge of the proton would nullify each other, producing an electrically neutral particle, the neutron.

29. a. The two atoms have the same number of protons and the same number of electrons, and they have their electrons distributed in the same way in their quantum shells.
b. The two atoms differ in the number of neutrons they hold in their nuclei.

30. a. $^{3}_{1}H$
b. $^{3}_{2}He$

31. a. $^{13}_{6}C$
b. $^{13}_{5}B$

32. $^{4}_{2}He$

33. a. No. The cork sphere is larger.
b. Each weighs one kilogram
c. Each weighs 1/6 kilogram; each has a mass of one kilogram.

34. The second element in each pair has a lower atomic mass. In the periodic table, we generally find that as we increase atomic number from one element to the next, we see a steady increase in atomic mass, because each subsequent element has one additional proton. The pairs of elements in the question are exceptions to this trend because of anomalies in isotopic abundances. For instance, the most common isotope of Ar is relatively heavy, with a mass number of 40 (18 protons and 22 neutrons), whereas the most common isotope of K, is lighter, with a mass number of 39 (19 protons and 20 neutrons).

35. The electrons of the surface atoms of the ball and the electrons of the surface atoms of the bat repel each other.

36. The hydrogen atom has several quantum shells of different energies. As the hydrogen atoms of the entire mass of hydrogen are excited, their single electrons can move among these several quantum shells, absorbing and emitting different amounts of energy with their various transformations.

Calculate

37. $1 \text{ meter} \times 10^{-10}/10^{-15} = 1 \text{ meter} \times 10^5$
$= 100{,}000 \text{ meters}$

38.

Notation	$^{11}_{5}B$	$^{40}_{20}Ca$	$^{58}_{28}Ni$	$^{22}_{10}Ne$
Atomic number	5	20	28	10
Mass number	11	40	58	22
Number of neutrons	6	20	30	12
Number of electrons	5	20	28	10

39. $1 \text{ g} \times 1 \text{ proton} /(1.67 \times 10^{-24})\text{g}$
$= 0.599 \times 10^{24} = 5.99 \times 10^{23}$ protons

40. Let the percentage abundance of B-11 equal n. Since there are only two naturally occurring isotopes of boron:

$$(11 \text{ u} \times n) + (10 \text{ u} \times (1 - n)) = 10.8 \text{ u}$$
$$11n \text{ u} + 10 \text{ u} - 10n \text{ u} = 10.8 \text{ u}$$
$$1\, n \text{ u} + 10 \text{ u} = 10.8 \text{ u}$$
$$n \text{ u} = 0.8 \text{ u}$$
$$n = 0.8, \text{ or } 80\%$$

B-11 makes up 80% of all naturally occurring boron, and B-10 makes up 20%.

41. Oxygen: 60 kg × 0.61 = 36.6 kg
Carbon: 60 kg × 0.23 = 13.8 kg

Hydrogen: 60 kg × 0.10 = 6.0 kg
Nitrogen: 60 kg × 0.026 = 1.6 kg
(b) 60 kg × (1 u/1.67 × 10^{-27} kg) ×
(1 carbon atom/12.01 u) = 3.0 × 10^{27} carbon atoms
(c) Since hydrogen is the lightest element, it makes up a much larger percentage than 10% in terms of number of atoms.

42. Write a 1 followed by 23 zeros.
43. Protons and neutrons have essentially the same mass, 1.67 × 10^{-24} kg each.
For a cell phone with a total mass of 105.0 g, we have
105.0 g × (proton or neutron)/(1.67 × 10^{-24})g
= 62.9 × 10^{24} protons or neutrons
= 6.29 × 10^{25} protons or neutrons

Chapter 3

Review

1. Water taken from the ocean. It is more likely to contain dissolve salt (sodium chloride).
2. An atoms valence shell is its outermost quantum shell containing electrons.
3. The halogens.
4. The electrons in an atom's valence shell.
5. Two
6. Hydrogen and helium. Columns 1 and 18.
7. The alkali metals.
8. Four
9. a. In the operation of the octet rule, atoms react chemically to obtain eight electrons in their valence shells,
b. In the formation of the ionic compound sodium chloride, the sodium atom, which has one electron in its valence shell, loses that single electron and its next lower electron shell, which contains eight electrons, becomes the sodium ion's valence shell. The chloride atom, which has seven electrons in its valence shell acquires the electron lost by the sodium atom and now holds eight electrons in its valence shell.
In the formation of water from one oxygen atom and two hydrogen atoms, an electron from the valence shell of one hydrogen atom and an electron from the valence shell of the oxygen atom form a pair of electrons shared by the oxygen and the hydrogen atom. This sharing occurs as well with a second hydrogen atom. As a result of the sharing of two pairs of electrons, the oxygen atom's valence shell is now occupied by eight electrons and each of the valence shells of the hydrogen atoms is filled with two electrons.
c. The valence shells of the two hydrogens each contain two rather than eight electrons. Thus the two hydrogen atoms do not obey the octet rule.
10. a. Six
b. One for each hydrogen, for a total of two. Covalent.
c. Non-bonding electrons
11. Hydrogen
12. They are gases with little or no chemical reactivity.
13. a. Polar covalent, i; nonpolar covalent, ii
b. Polar covalent, the covalent bond binds atoms of two different element; nonpolar covalent, the covalent bond binds atoms of the same element.
c. Electronegativity
14. The reaction is reversible.
15. When HCl is dissolved in water, the molecule ionizes to H^+ and Cl^-.
16. Covalent
17. 1) The identities of the elements that compose the molecule and 2) the number of atoms of each element in the molecule.
18. O_2
19. Triple bond.
20. a. Na and Mg
b. N, O and F
c. Ne
21. a. hydrogen
b. oxygen
c. nitrogen
22. a. LiF and BeO
b. Li_2O
c. BeF_2
d. Li_3N
23. The ratio of the number of charges on the cation and the anion.
24. By multiplying the atomic weight of each element by its subscript and adding all the products of these multiplications.
25. a. Ball and stick: Advantage: Bond angles and relative sizes of atoms appear clearly. Disadvantage: Overstates distances between atoms (bond lengths).
b. Space filling: Advantage: The three-dimensional contour of the molecule can be seen clearly. Disadvantage: Bonds and bond angles are vague.
26. a. Both carbon-oxygen bonds lie on the same straight line.
b. linear
27. a. An ion consisting of two or more atoms joined by covalent bonds.
b. Epsom salt; milk of magnesia; baking soda; bleach
c. Epsom salt: magnesium sulfate, $Mg(SO_4)$; milk of magnesia: magnesium hydroxide, $Mg(OH)_2$; baking soda: sodium bicarbonate, $NaHCO_3$; bleach: sodium hypochlorite, NaOCl
28. Lithium fluoride, LiF
29. The electrically charged cations and anions of the ionic crystals attract each other more strongly than do the electrically neutral molecules of the covalent compounds
30. Magnesium oxide, MgO
31. a. Hydroxide ion, HO^-
b. i) KOH and H_2; ii) RbOH and H_2

32. a) i) propane, C_3H_8; ii) 2-methylpropane, C_4H_{10}; iii) methane, CH_4; iv) ethane, C_2H_6; v) butane, C_4H_{10}
b. ii and iv

33. a.

```
        H
      H | H
  H  H \C/ H  H
  |  |  |  |  |
H-C--C--C--C--C-H
  |  |  |  |  |
  H  H  H  H  H
```

3-methylpentane
C_6H_{14}

```
        H
      H | H
        C   H
   H    |   |    H
    \   |   |   /
   H-C--C---C--C-H
    /   |   |   \
   H    |   |    H
        C   H
      H | H
        H
```

2,2-dimethylbutane
C_6H_{14}

```
        H
      H | H
        C   H
   H    |   |    H
    \   |   |   /
   H-C--C---C--C-H
    /   |   |   \
   H    H   C    H
          H | H
            H
```

2,2-dimethylbutane
C_6H_{14}

b. All three are isomers of each other

34. 1800: Chemicals derived from substances that are alive or were once alive. Today: A covalent compound containing carbon.

Think

35. All the noble gases have filled valence shells, so they do not need to react with other atoms to acquire, lose, or share electrons in order to fill their valence shells.

36. a. Lithium has two electrons shells; sodium has three.
b. Both have a single electron in their valence shells.

37. Ionic bonds form when atoms gain or lose electrons to acquire an octet in their valence shells. Covalent bonds form when atoms share electrons to obtain a valence shell octet.

38. It is an ionic compound, with each of the X ions carrying two positive charges and each of the Y ions carrying two negative charges.

39. Removing one or more electrons from a neutral atom produces an ion with greater number of protons than electrons. The surplus of protons is better able to attract the remaining electrons towards the nucleus. Adding one or more electrons to a neutral atom produces an ion with greater number of electrons than protons. This deficit of protons is less able to attract the electrons towards the nucleus.

40. A compound formula shows the elements present in an ionic compound and the ratio of each. A molecular formula shows the elements present in a covalent compound and the ratio of each.

41. a. 10; b. 20; c. C_3H_8

42. The two nuclei would come close enough for their identical positive charges to repel each other.

43. When HCl dissolves in water it ionizes to for a proton and a chloride anion, H^+ and Cl^-.

44. No. The chemical formula of a covalent compound shows the total number of atoms of each element in each of the compound's molecules.

45.

	Na^+	Na	K^+	K
Electron configuration	(2, 8)	(2, 8, 1)	(2, 8, 8)	(2, 8, 8, 1)

Na^+ and K^+ have valence shells with filled octets and are therefore very stable. Na and K each have a single electron in their valence shells and therefore are highly reactive, undergoing electron transfer reactions with water and liberating hydrogen gas.

46. No. Calcium phosphide is made from the reaction of calcium, a metal, and phosphorus, a nonmetal, and therefore is an ionic compound.

47. Li_2F

48. The valence shells of the elements in Column 2 are not filled by the two electrons they contain. Conversely, the two electrons of helium's valence shell do fill that shell, so helium belongs with the other elements of Column 18, all of which have filled valence shells.

49. No. Some molecules represent covalent elements rather than covalent compounds. Examples include H—H, O=O and N≡N.

50. a. A hydrogen atom can accept an electron from an atom of another element to fill its valence shell with two electrons, just as a halogen atom can accept an electron to fill its valence shell, but with a different number of electrons.
b. $H:^-$

51. $:\dot{N}=\ddot{O}$ (with two more dots below the O)

52. a. covalent bonds; CBr_4; 331.6 u
b. ionic bonds; KCl; 74.55 u

53. If A is a metal and B is a nonmetal, they tend to have a large enough difference in electronegativity such that one or more electrons are transferred from A to B to form an ionic compound. If both A and B are nonmetals, their differences in electronegativities are comparatively small, so they react through sharing electrons to form a covalent compound.

Calculate

54. a. 19 protons, 18 electrons;
b. 16 protons, 18 electrons;
c. 13 protons, 10 electrons;
d. 26 protons, 24 electrons

55. a. 191.36 u; b. 101.96 u; c.78.05 u; d. 34.83 u; CdSe has the largest formula mass; Li_3N has the smallest

56. Vitamin C

57. a. SnF_2
b. 156.7 u

58. 180.16 u

59. 28×10^{19} molecules

60. a. 39.34%; b. 5.85 g; c. 1.17 teaspoon

Chapter 4

Review

1. About 80% of the world's energy is currently supplied by fossil fuels. (b) In 2010 fossil fuels accounted for 84% of the total energy consumed in the U.S.
2. (a) natural gas, petroleum (oil), and coal (b) petroleum (c) coal and natural gas
3. coal (greatest) > petroleum > natural gas (least)
4. (a) petroleum (or crude oil) (b) coal
5. (a) carbon dioxide and water (b) carbon monoxide (and unburned hydrocarbons)
6. (a) The higher the compression ratio, the more energy that can be obtained from the hydrocarbons, up to the limit of the chemical energy they contain. (b) The octane rating must be increased to inhibit knocking.
7. The volatility of a liquid is a measure of how readily the substance evaporates. Highly volatile compounds have low boiling points and low volatility compounds have high boiling points. Petroleum refining begins with fractional distillation, in which groups of compounds in petroleum are separated based on volatility.
8. i) the very top of the column; ii) the second draw-off point from the top; iii) the very bottom of the column.
9. (a) To increase the gasoline's octane rating.
   (b) Lead can poison the catalysts of an automobile's catalytic converter.
   (c) Lead acts as a poison to humans and other living things, so removing lead from gasoline makes the gasoline safer for humans and for the environment.
   (d) methyl *tert*-butyl ether (MTBE)
   (e) catalytic reforming.
10. (a) Carbon monoxide; hydrocarbons; nitrogen oxides.
    (b) incomplete combustion of the hydrocarbons of gasoline.
    (c) i) CO and C_xH_y; ii) NO_x
11. (a) unburned hydrocarbons, carbon monoxide, and nitrogen oxides
    (b) Since catalytic converters oxidize carbon monoxide and unburned hydrocarbons, carbon dioxide and water vapor emissions are slightly increased.
12. (a) An octane enhancer increases the octane rating of a gasoline. An oxygenate "improves the efficiency of hydrocarbon combustion."
    (b) Yes. Ethanol serves as both an octane enhancer and an oxygenate.
13. Both photosynthesis and cellular respiration are part of the carbon cycle.
    (b) In photosynthesis plants transform water and atmospheric carbon dioxide into organic compounds and oxygen. In cellular respiration animals combine organic compounds with atmospheric oxygen to produce carbon dioxide and water.
14. Methane, CH_4. Sources include landfills, cattle farming, rice farming, and coal mining.
    Nitrous oxide, N_2O. Sources include nitrogen-containing fertilizers in commercial agriculture.
15. This means generated by human activity, such as burning fossil fuels.
16. Infrared radiation; vibrational motions of the atoms within the covalent bonds of greenhouse gas molecule.
17. Venus atmosphere is highly dense and is made up almost almost entirely of the greenhouse gas, carbon dioxide. Mercury has practically no atmosphere to speak of.
18. Gasoline is derived from petroleum, a fossil fuel. Fossil fuel formation is infinitesimally slower as compared to the rate at which gasoline (and other fossil fuel) combustion releases CO_2 into the atmosphere.
19. (a) A catalyst (b) Reaction 2 (c) Reaction 1
20. (a) A gasoline-air mixture.
    (b) Divide the volume of the gasoline-air mixture at beginning of the compression stroke by the volume of the gasoline-air volume at the moment the spark plug fires.
    (c) Carbon dioxide and water vapor.

Think

21. Lead is harmful to living things.
22. This would increase the greenhouse effect because methane is a more effective greenhouse gas than carbon dioxide.
23. Plants absorb carbon dioxide, an atmospheric greenhouse gas, and combine it with water to produce carbohydrates. This would reduce the concentration of carbon dioxide within the atmosphere.
24. The carbon of the metabolized plankton would have been oxidized into carbon dioxide and released into the atmosphere.
25. They have identical boiling points.
26. Because fossil fuels are used to run the farm equipment, factories, and trucks used to produce and distribute ethanol.
27. To produce the electricity that electric cars use, power-plants burn fossil fuels and release their carbon into the atmosphere as carbon dioxide.
28. Fracking involves injecting high-pressure fluids underground to aid in the recovery of natural gas and oil deposits. Potential benefits include that it is often used to extract natural gas, which has a smaller carbon footprint than coal, for example. Environmental concerns include risks of groundwater contamination and disposal of waste fluids.
29. Estimates vary, but roughly 200 million gallons of oil spilled from the Deepwater Horizon offshore oil rig and roughly 11 million gallons of oil spilled from the Exxon Valdez oil tanker. Domestic oil production reduces the need for imports, but raises the risk of environmental damage as in the case of these two spills.
30. Yes, a country that consumes less energy through fossil fuels may be more reliant on coal, the fossil fuel that emits the most CO_2 for a given amount of energy produced. For example, the U.S. consumes significantly more energy through fossil fuels than China, but China emits more CO_2 because of a greater reliance on coal.
31. CO_2 has a molecular mass of 44 u: 32 u from the oxygen and 12 u from the carbon. So most of the weight of a CO_2 molecule is due to the oxygen atoms. This oxygen comes primar-

ily from the oxygen in the air, not the from the hydrocarbons in gasoline.

Calculate

32. 93

33. This would represent a 20% increase from the 1970 level.

34. (a) 18.6 mpg
(b) 150 gallons
(c) 10,700 pounds per year

35. (a) 7.3 billion barrels annually
(b) 24.3 barrels
(c) 1020 gallons

36. 45%; 9 million gallons

37. 3.1×10^7 cal; 3.1×10^4 kcal; nearly 58

Chapter 5

Review

1. (a) calorie (b) 1000 cal

2. Fats and oils

3. Short-term: glycogen Long-term: adipose tissue (body fat)

4. At room temperature, a fat is solid or semi-solid and an oil is a liquid.

5. Saturated fats and cholesterol

6. Essential fatty acids

7. Omega-9 fatty acids

8. Plant-based foods

9. It becomes more saturated (replacing C=C bonds with C—C bonds), and it can contain trans C=C bonds.

10. (a) glucose, fructose, galactose; (b) sucrose, lactose, maltose, cellobiose; (c) cellulose, glycogen, starch

11. (a) polysaccharide, (b) monosaccharide, (c) disaccharide, (d) disaccharide, (e) polysaccharide, (f) disaccharide, (g) disaccharide, (h) monosaccharide

12. (a) Hydrolysis of a polysaccharide, (b) Enzymes

13. (a) fructose and glucose, (b) glucose and galactose, (c) β-glucose, (d) α-glucose

14. I and IV apply to figure (a). II and III apply to figure (b).

15. It is a stored form of glucose, vital for quick energy.

16. Amino acids

17. (a) glycine and alanine, (b) peptide, (c) 1

18. Essential amino acids cannot be produced by the body so must come from diet. Nonessential amino acids can be produced by the body from other nutrients. See Table 5.1 for examples of each type.

19. The contain sulfur atoms.

20. All except serine

21. Arginine, cysteine, serine

22. Eggs, meat, poultry, fish, dairy products

23. They provide essential amino acids.

24. Structurally, fibrous proteins form more elongated fibrous bundles with themselves, while globular proteins are spherical and generally not bonded with other proteins. Fibrous proteins are in tissue, muscle, hair, and nails, while globular proteins are in water-based media, such as blood.

25. (a) High-density and low-density lipoproteins (HDLs and LDLs), (b) glycogen

26. (a) triglyceride, (b) didpeptide, (c) carbohydrate, (d) amino acid, (e) fatty acid

Think

27. This increases the melting point.

28. (a) Monoglyceride: a glycerol side chain linked to one fatty acid; diglyceride: a glycerol side chain linked to two fatty acids, (b) C=C bonds

29. It should decrease the iodine number because it converts C=C bonds to C—C bonds, which cannot take up iodine.

30. The baby will not be able to digest lactose present in the mother's milk.

31. (a) an enzyme, (b) galactose

32. carbon

33. HFCS is made in three steps, each with the aid of enzymes:

1. α-amylase helps break cornstarch down into shorter segments
2. glucoamylase breaks these segments into glucose
3. xylose isomerase converts glucose into a mixture of glucose and fructose. There is no single answer to this question. Some argue that it is not natural since intervention with these enzymes is required to break down corn starch. Others argue that it is, since it comes from corn and in the end contains glucose and fructose, themselves natural products.

34. Its higher-order structure is disrupted, and it loses its native shape and typically its function.

35. The proteins have undergone an irreversible denaturation.

36. These types of proteins are soluble in fluids.

37. Eggs provide complete protein (adequate amounts of essential amino acids), while peanuts are deficient in one or more essential amino acids.

38. (a) nitrogen, (b) sulfur

39. fibrous

Calculate

40. (a) coffee: 336 Cal; muffin: 359 Cal, (b) coffee: 10.7% muffin: 27.6%

41. 58 Calories

42. 1800 Calories

43. (a) stearic acid: $C_{18}H_{36}O_2$, oleic acid: $C_{18}H_{34}O_2$, (b) The number of hydrogen atoms increases by two for each C=C bond that becomes a C—C bond.

44. Canola oil

45. 10.6%

46. $C_{12}(H_2O)_{11}$

Chapter 6

Review

1. (a) The volume of the gas is inversely proportional to pressure. (b) The volume of the gas is directly proportional to the Kelvin temperature.
2. The energy is proportional to the temperature of the gas.
3. Steam > water > ice
4. Nitrogen, N_2.
5. Rotational energy resembles the twirling dancer. Translational energy resembles the runner. Vibrational energy resembles the chest expanding and contracting.
6. When steam is cooled to water, the kinetic energy lessens and attractive forces between molecules are mostly greater than the kinetic energy, but molecules still move about one another. When water cools to ice, kinetic energy lessens even more and attractive forces help the molecule maintain more fixed positions within the crystal lattice.
7. (a) Solid; (b) Hydrogen bonds; (c) Because of an attraction between the electron poor (δ^+) hydrogen atom in one molecule and the electron rich; (δ^-) oxygen atom in an adjacent molecule.
8. Intermolecular forces are stronger in acetic acid and water and weaker in ammonia and propane.
9. Decane > Heptane > Butane > Methane. The boiling points of hydrocarbons generally increase with molecular size.
10. 760 mmHg = 1 atm = 14.7 psi
11. 273°C
12. Ice floats because it is less dense than water.
13. Density is an intrinsic property of a substance and does not depend on the sample size.
14. Water molecules adopt a hexagonal packing arrangement in the crystal lattice.
15. The molar mass is the substance's atomic, molecular, or formula weight expressed in units of g/mol.
16. (a) sublimation; (b) condensation; (c) melting; (d) freezing; (e) evaporation; (f) boiling
17. Coefficients serve to balance chemical equations, so that equal numbers of atoms of a given element are represented on both sides of the reaction arrow.

Think

18. (a) Evaporation and boiling are similar in that they involve a change from liquid to gas. (b) They are different in that evaporation occurs only at the surface of the liquid and at temperatures below the boiling point, while boiling occurs throughout the bulk of the sample and at the boiling point.
19. As the balloon rises, atmospheric pressure drops which causes the balloon to expand. While temperature generally drops too, the expansion caused by lower pressure is larger than the contraction caused by lower temperature, and eventually the balloon bursts.
20. (a) The gas condenses to a liquid.
    (b) The chemical particles of real gases are not infinitely small spheres, but rather have some intrinsic volume since they are made up of atoms. Also, the chemical particles of real gases exhibit small, but real attractive forces between one another.
    (c) Charles's Law predicts the volume would shrink to nothing. This makes sense for an ideal gas, in which the particles have no volume, but is inconsistent with real gases, which condense to liquids and then to solids prior to reaching absolute zero.
21. The chemical particles of real gases have some intrinsic volume since they are made up of atoms. Also, the chemical particles of real gases exhibit attractive forces between one another.
22. Because the mass of one mole depends on the substance being measured. For example, 1 mole of H_2 has a mass of 2 g whereas 1 mole of O_2 has a mass of 32 g. 1 mole of each of these substances has a different mass, so it would not be possible to universally define the mass of 1 mole of matter.
23. Yes, you could by recognizing that the molar mass of sucrose is 342 g/mol and the molar mass of water is 18 g/mol. The ratio of these two values, 342/18 = 19. If you weigh out a mass of sugar that is 19 times the mass of water, you will have an equal number of moles of the two substances and hence an equal number of molecules.
24. Because air pressure decreases with altitude. As a result there is less oxygen available. Pressurizing the cabin of an airplane increases the level of oxygen available.
25. The condensation trails evaporate, turning from liquid to gas.

Calculate

26. (a) $12\ g \times (1\ mL/19.3\ g) = 0.62\ mL$ or $0.62\ cm^3$
    (b) $(150\ cm \times 15\ cm \times 5\ cm) \times (0.8 g/cm^3) = 9000\ g = 9.0\ kg$
27. $45.5\ carats \times (0.2\ g/carat) \times (1\ mL/3.5\ g) = 2.6\ mL$ or $2.6\ cm^3$
28. (a) $273 + 20 = 293\ K$
    (b) $20 - 273 = -253\ °C$
    (c) $520\ mmHg \times (1\ atm/760\ mmHg) = 0.68\ atm$
    (d) $29.4\ psi \times (1\ atm/14.7\ psi) = 2.0\ atm$
29. (a) The volume would decrease by half.
    (b) The volume would double.
    (c) The volume would decrease by 1/3, meaning it would be 2/3 of the original volume.
30. (a) $5\ X + 15\ Y \longrightarrow 5\ XY_3$
    (b) $X + 3\ Y \longrightarrow XY_3$
31. Since volume is directly proportional to Kelvin temperature by Charles's Law, first convert the temperature to Kelvin: 27 + 273 = 300 K. The balloon will have 80% volume at $0.8 \times 300 = 240\ K = -33\ °C$.
32. 100 K, which is −173 °C.
33. $(P_1 \times V_1)/T_1 = (P_2 \times V_2)/T_2$; since $V_1 = V_2$, $P_1/T_1 = P_2/T_2$; so $P_2 = (P_1 \times T_2)/T_1$
    $P_1 = 32\ psi$
    $T_1 = 17 + 273 = 290\ K$
    $T_2 = 37 + 273 = 310\ K$
    $P_2 = (32\ psi \times 310\ K)/290\ K = 34.2\ psi$
34. $(P_1 \times V_1)/T_1 = (P_2 \times V_2)/T_2$; so $V_2 = (P_1 \times V_1 \times T_2)/(T_1 \times P_2)$

In each case:
$V_1 = 1$ L
$P_1 = 760$ mmHg
$T_1 = 27 + 273 = 300$ K
Everest:
$P_2 = 245$ mmHg
$T_2 = -36 + 273 = 237$ K
$V_2 = (760 \text{ mmHg} \times 1 \text{ L} \times 237 \text{ K})/(300 \text{ K} \times 245 \text{ mmHg}) =$ 2.45 L
McKinley:
$P_2 = 345$ mmHg
$T_2 = -42 + 273 = 231$ K
$V_2 = (760 \text{ mmHg} \times 1 \text{ L} \times 231 \text{ K})/(300 \text{ K} \times 345 \text{ mmHg}) = 1.7$ L
Kosciuszko:
$P_2 = 590$ mmHg
$T_2 = -18 + 273 = 255$ K
$V_2 = (760 \text{ mmHg} \times 1 \text{ L} \times 255 \text{ K})/(300 \text{ K} \times 590 \text{ mmHg}) = 1.1$ L
(b) If $V_2 > V_1$ then the expansion caused by a drop in pressure is a larger effect than the contraction caused by a drop in temperature. In each case, $V_2 > V_1$ so pressure is the dominant factor.

35. $(P_1 \times V_1)/T_1 = (P_2 \times V_2)/T_2$; so $V_2 = (P_1 \times V_1 \times T_2)/(T_1 \times P_2)$
$V_1 = 1$ L
$P_1 = 760$ mmHg
$T_1 = 27 + 273 = 300$ K
$P_2 = 770$ mmHg
$T_2 = 57 + 273 = 330$ K
$V_2 = (760 \text{ mmHg} \times 1 \text{ L} \times 330 \text{ K})/(300 \text{ K} \times 770 \text{ mmHg}) = 1.09$ L

36. (a) $Cl_2 + 2\,NaBr \longrightarrow 2\,NaCl + Br_2$
(b) Balanced properly.
(c) $2\,Cu + O_2 \longrightarrow 2\,CuO$

37. (a) 7 left shoes
(b) ½ dozen yolks
(c) ½ mole of sodium cations

38. 2 moles of O atoms

39. (a) 20.2 g; (b) 215.7 g; (c) 28 g; (d) 96 g; (e) 17.7; (f) 141.8 g; (g) 1 g

40. (a) 5.7×10^{-3} mol $C_6H_8O_6$; (b) 1.8×10^{-3} mol $C_9H_8O_4$; (c) 3.3×10^{-3} mol $C_8H_9NO_2$; (d) 3.4×10^{21} $C_6H_8O_6$ molecules; 1.1×10^{21} $C_9H_8O_4$ molecules; 2.0×10^{21} $C_8H_9NO_2$ molecules

41. (a) 1.5 mol CO_2; (b) 66 g

42. (a) marshmallows; (b) 4×10^{21} kg

43. (a) 6.9 g Cl_2; (b) 0.92 g Na; (c) Yes, sodium would be left over.

44. (a) 0.1 mol SO_3; (b) 0.1 mol H_2SO_4; (c) 9.8 g

45. (a) 7.0×10^4 mol protons and neutrons combined; (b) 4.2×10^{28} protons and neutrons combined.

46. (a) $2\,CO + O_2 \longrightarrow 2\,CO_2$; (b) 2 (c) 1.75 g CO

47. (a) 10, 13, and 16 bonds, respectively.; (b) 53 kcal per mole of bonds; Yes, the energy released roughly corresponds to the number of covalent bonds; (c) 1331 kcal; 1174 kcal

48. (a) $C_2H_6O + 3\,O_2 \longrightarrow 2\,CO_2 + 3\,H_2O$; $C_5H_{12} + 8\,O_2 \longrightarrow 5\,CO_2 + 6\,H_2O$; (b) 191 g CO_2; 305 g CO_2

49. 76.3×10^9 pennies wide

Chapter 7

Review

1. Homogeneous in this case means having a single phase and uniform composition.

2. The physical state of the solvent dictates the physical state of the resulting solution.

3. Air is a homogeneous mixture of different gases.

4. An alloy.

5. Iron; carbon.

6. a) Na^+; b) Cl^-. Opposite electrical charges attract one another. Oxygen, with a partial negative charge (δ^-), associates with positively charged sodium ion. Hydrogen, with a partial positive charge (δ^+), associates with negatively charged chloride ion.

7. Colloids and suspensions have very fine particles suspended in another phase, whereas solutions are homogeneous and a single phase. The Tyndall effect is the scattering or illumination of a light beam passing through a colloidal mixture.

8. a) Blood contains cells and platelets in suspension. b) Blood contains a variety of dissolved substances, including sugars, electrolytes, enzymes, hormones, etc.

9. Nonpolar compounds are hydrophobic ("water-fearing"), because water is a polar compound.

10. Water is a polar compound. The oxygen atom of water carries a partial negative charge and the hydrogen atoms of water carry partial positive charges. a) Water forms attractions with ions in solution. The oxygen atom of water associates with cations and the hydrogen atom associates with anions. b) Water forms attractions, called hydrogen bonds, with hydroxyl groups (–OH) on compounds such as sugars and alcohols.

11. At a given temperature, the amount of gas that dissolves in a liquid is directly proportional to the pressure of the gas above the surface of the liquid. A common example is the carbonation of water in soft drinks by exposure to pressurized carbon dioxide.

12. According to Henry's law, the pressure of N_2 above the surface of the water influences its solubility. The greater the surface between water and gas, the faster it can dissolve.

13. The hydrocarbon components of oils are nonpolar and do not form attractions to water, a polar substance.

14. Emulsifiers contain both nonpolar and polar groups within the same molecule allowing them to associate with both oils and water.

15. Molarity.

16. Percentage concentrations.

17. Parts per million (ppm) or parts per billion (ppb)

18. Parts per million (ppm)

19. The Safe Water Drinking Act.

20. A serial dilution refers to series of solutions, one more dilute than the next by a constant factor.

21. Both distillation and reverse-osmosis remove solutes from water. In distillation, water is boiled and vapors are condensed and collected as water largely free of contaminants. In reverse-osmosis, pressure is applied to water, forcing it through an ultrafine membrane which blocks the passage of most solutes.

22. Nucleation.

23. The dissolved salts would crystallize as the solvent (water) completely evaporates.

24. When solvent evaporates from a saturated solution it can become supersaturated. Also, if the solubility of the solute increases with temperature, decreasing the temperature of a saturated solution can make it supersaturated.

Think

25. a) The ratio is 10:4.
b) The ratio would increase, because gases are less soluble in warmer fluids.
c) Yes, it could impact the reading by overestimating the BAC value.

26. a) According to Henry's law, more nitrogen is expected to be present in the blood and tissues at greater depths, because of the higher pressure under which the air is inhaled. As the diver ascends, the pressure decreases, and nitrogen becomes less soluble. If the ascent happens too rapidly, nitrogen can bubble out of solution within the blood causing decompression sickness.
b) Hyperbaric oxygen therapy increases oxygen levels in the blood.

27. Figures a. and c. show favorable interactions, pairing an atom with a partial positive charge with an atom bearing a partial negative charge. Figure b. shows an unfavorable interaction because each of the oxygen atoms bears a partial negative charge.

28. (a) An emulsifying agent. (b) As an example:

```
                 H  H  H  H  H  H  H  H  H  H  H  H  H  H  H
        O        |  |  |  |  |  |  |  |  |  |  |  |  |  |  |
  H      \\C—C—C—C—C—C—C—C—C—C—C—C—C—C—C—C—H
  |      /       |  |  |  |  |  |  |  |  |  |  |  |  |  |  |
H—C—O            H  H  H  H  H  H  H  H  H  H  H  H  H  H  H
  |
H—C—O                          Nonpolar tail
  |    \  H
H—C—O
  |    \  H
  H

Polar head
```

29. The compounds likely have nonpolar character.

30. There is no single correct answer. Generally, concentration limits for contaminants of this type fall within the parts per million or parts per billion range. Dosage studies using animal models, such as mice or rats, are typically done to estimate levels at which the substance poses health risks.

31. There is no single correct answer. What's considered an acceptable level can vary from individual to individual.

32. Because distilled water is essentially free of dissolved substances such as sodium. If tap water was used to make the solutions, the concentrations of sodium would be expected to be higher than those calculated.

33. (a) The EPA establishes limits for contaminants in drinking water. One could argue that water that violates any of these standards would be considered "polluted", but opinions can vary. For example, different countries can establish different standards for water quality.
(b) There is no single correct answer. One consideration is the nature of the contaminants and whether some present more of a risk than others if their MCL is exceeded.
(c) There is no single correct answer, but consider that some contaminants, such as sodium, are not regulated by the EPA. Levels of sodium could be too high for taste considerations, yet the water may not be labeled "polluted" in the conventional sense.
(d) There is no single correct answer, but consider that different contaminants pose different risks and that the amount of exposure also depends on the quantity of the "polluted" water consumed.

34. There is no single correct answer, but one could filter the water using home a filtration system or boil the water to remove microbial or volatile contaminants.

35. The answers to these questions depend on your personal opinions.

Calculate

36. 0.1 % is equivalent to 1 part per thousand.

37. (a) 0.143 parts per billion
(b) 143 parts per trillion.

38. 10 ppm N_2 and 50 ppm O_2.

39. Three 0.85-L bottles of Voss.

40. a. 300 g water; 700 g isopropyl alcohol.
b. 16.7 moles of water; 11.7 moles of isopropyl alcohol.
c. Isopropyl alcohol is solvent based on weight.
d. Isopropyl alcohol is solute based on number of moles.
e. 16.7 M
f. 11.7 M

41. $8.7 \times 10^{-4\%}$ (w/w)

42. (a) 1.0 moles
(b) 180 g glucose
(c) 7.5 L

43. (a) 615 g
(b) 0.586 L
(c) 0.044 mol sucrose
(d) 0.075M

44. Yes, the new solution meets federal drinking water standards for both selenium and barium.

45. 0.01 M

46. (a) 5.0×10^{-3} ppm
(b) 5.0 ppb

47.

#1	#2	#3	#4	#5
0.22 M	2.2×10^{-2} M	2.2×10^{-3} M	2.2×10^{-4} M	2.2×10^{-5} M

48. (a) 0.04 g ethanol
(b) 2.0 g

49. 0.25 ppm

Chapter 8

Review

1. a. A substance, such as water, that can act as either an acid or a base.
   b. A medicine that neutralizes excess stomach acid.
   c. An acid that ionizes partially in water.
   d. A carbon-bearing acid characterized by the presence of the carboxyl group, $—CO_2H$.
   e. The cation H_3O^+, produced when acids release hydrogen ions (H^+) in water.
   f. The anion OH^-, produced when bases dissolve in water.
   g. A reaction between an acid and a base to produce a salt and oftentimes water.
   h. An ionic compound that can produced by the reaction of an acid and a base. In common usage, *salt* or *table salt* refers specifically to sodium chloride (NaCl).
   i. A state in which the rate of a forward reaction (such as ionization) equals the rate of the reverse reaction (such as recombination), such that the concentrations of all chemical species remains constant.
2. a. sulfuric acid, H_2SO_4
   b. potassium hydroxide, KOH
   c. ascorbic acid, $C_6H_8O_6$
   d. sodium bicarbonate, $NaHCO_3$
   e. carbonic acid, H_2CO_3
   f. acetic acid, CH_3CO_2H
   g. citric acid, $C_6H_8O_7$
   h. sodium hydroxide, NaOH
3. a. blue; b. red; c. blue; d. red; e. red
4. a. salt, calcium carbonate, antacids
   b. acid, acetic acid, vinegar
   c. acid, sulfuric acid, car batteries
   d. acid, hydrochloric (also known as muriatic) acid, certain toilet cleaners
   e. base, sodium bicarbonate, baking soda
   f. acid, boric acid, certain eye washes
   g. base, magnesium hydroxide, milk of magnesia
   h. salt, magnesium sulfate, Epsom salts
   i. salt, potassium iodide, iodized salt
   j. base, sodium hydroxide, certain oven and drain cleaners
   k. base, ammonia, household ammonia
5. A salt is an ionic compound that can be produced by the reaction of an acid and a base.
6. a hydronium ion, H_3O^+
7. It means that water can act as an acid or a base.
8. A hydrogen ion, H^+, is a proton. A hydronium ion, H_3O^+, is formed when a proton bonds to a molecule of water.
9. carbon dioxide (CO_2)
   $NaHCO_3$ (sodium bicarbonate) + $C_6H_8O_7$ (citric acid) $\longrightarrow$ $NaC_6H_7O_7$ (sodium citrate) + CO_2 + H_2O
   $KHCO_3$ (potassium bicarbonate) + $C_6H_8O_7$ (citric acid) $\longrightarrow$ $KC_6H_7O_7$ (potassium citrate) + CO_2 + H_2O
10. HF + NaOH $\longrightarrow$ NaF (sodium fluoride, a salt) + H_2O
11. The carboxyl group, $—CO_2H$
12. Oxalic acid, citric acid (also: acetic acid)
13. A weak acid, such as acetic, boric, or carbonic acid, ionizes partially in water. A strong acid, such as hydrochloric, sulfuric, and nitric acid, ionizes completely in water.
14. A neutralization reaction. $CH_3—CO_2H$, is acetic acid. NaOH is sodium hydroxide, a base. $Na^+CH_3—CO_2^-$, is sodium acetate, a salt.
15. a. carbonic acid
    b. sulfuric and nitric acids
16. $C_6H_8O_7 + NaHCO_3 \longrightarrow NaC_6H_7O_7 + CO_2 + H_2O$
17. Water
18. This represents a test for acids. The pH of the solution is below 7.
19. Figure a depicts a solution of weak acid in water. Figure b depicts a solution of a strong acid in water.

Think

20. $2\ HCl + CaCO_3 \longrightarrow CaCl_2 + CO_2 + H_2O$. If an unknown substance is an acid, it will neutralize the calcium carbonate ($CaCO_3$) of the chalk to form CO_2, observable as consuming the chalk and producing a fizz.
21. a. yes; b. no; c. no
22. The logarithmic pH is more convenient because it spans a short range of approximately 0–14 and does not involve exponential notation.
23. 1. d. NH_3, ammonia is the only base. The rest are acids.
    2. c. Nitric acid is the only compound in the list that is not a carboxylic acid.
    3. e. Sodium hydroxide is the only compound in the list that is neither a weak acid nor a weak base.
24. Acids react with certain metals, such as zinc and iron, to produce hydrogen gas.
    Zn (from galvanized nails) + 2 CH_3CO_2H (acetic acid) $\longrightarrow$ $Zn(CH_3CO_2)_2$ (zinc acetate) + H_2 (hydrogen gas)
    We expect a similar result for the iron nails since iron reacts with acid. Lemon juice should produce a similar result as well because it contains citric acid.
25. Such indicators are acids because they release a proton.
26. $2\ NaHCO_3 + H_2SO_4 \longrightarrow Na_2SO_4 + 2\ CO_2 + 2\ H_2O$
27.

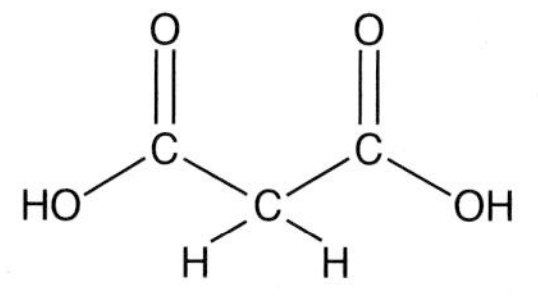

O O
C C
HO C OH
H H

28. a. $H^+ + HPO_4^{2-} \longrightarrow H_2PO_4$
    b. $OH^- + H_2PO_4^- \longrightarrow HPO_4^{2-} + H_2O$
29. When an acid donates a proton, a base accepts it, so both acid and base must be present.
30. a. Carbon dioxide (CO_2) b. Sodium bicarbonate acts a base and the source for the carbon dioxide gas.

31. Water is a Brønsted-Lowry acid because it releases a proton when it ionizes. Water also acts as a Brønsted-Lowry base because it accepts a proton to form the hydronium ion.

$$H\text{—}\ddot{O}\text{—}H + H\text{—}\ddot{O}\text{—}H \rightleftharpoons H_3O^+ + H\text{—}\ddot{O}:^-$$

acts as acid acts as base

32. A candidate's stance on any one of number of litmus test issues may sway a voter who is passionate one way or the other about the topic. Examples include, but are not limited to, abortion, gun control, the environment, and taxation.

33. a. $2\ H_3NSO_3 + CaCO_3 \longrightarrow Ca(H_2NSO_3)_2 + CO_2 + H_2O$
    b. The salt, $Ca(H_2NSO_3)_2$ (calcium sulfamate), is formed.

34. a. pH Up
    b. pH Down
    c. pH Down
    d. pH Up

35. a. citric acid
    b. acetylsalicylic acid
    c. both

Calculate

36. a. pH = 4
    b. pH = 10
    c. pH = 7

37. a. $10^{-3}\ M$
    b. $10^{-10}\ M$

38. a. 0.01 M
    b. pH = 2

39. a. 0.01 M
    b. $10^{-12}\ M$
    c. pH = 12

40. 0.01 M

41. The pH is greater than 2 and less than 7. $[F^-]$ is less than 0.01 M.

42. a. 1 liter
    b. 1 liter
    c. 0.3 liter

43. a. 0.1 liter
    b. 1 liter
    c. 0.4 liter

44. The negative of the logarithm of the hydroxide molarity.

45. pH + pOH = 14.

Chapter 9

Review

1. The discovery of radioactivity.
2. The discovery of nuclear fission.
3. (1) high atomic number (84 or greater); (2) the ratio of neutrons to protons within the nucleus is above or below a stable range.
4. a. alpha radiation; b. beta radiation; c. positron emission; d. gamma radiation
5. a. $_{-1}^{0}\beta$ b. $_{0}^{0}\gamma$ c. $_{2}^{4}\alpha$ d. $_{1}^{0}e$
6. Because these radioactive emissions can convert atoms or molecules into ions.
7. a. C-11 is used to produce medical images in positron emission tomography; b. Tc-99m produces gamma rays used in medical imaging; c. Co-60 serves as an external source of gamma rays used in destroying cancer cells.
8. Gamma rays would be most hazardous because of their penetrating power. Alpha rays would be least hazardous because of their lack of penetrating power.
9. The biological damage caused by ionizing radiation depends on several factors, including the energy of the radiation, how long we're exposed to it, its penetrating and ionizing power, and whether it's coming from outside or from within our bodies.
10. Cosmic radiation coming from outer space; internal radiation originating from within our own bodies, and terrestrial radiation produced by radioactive elements within the Earth's soil.
11. Biological harm extensive enough to cause illness or death.
12. A variety of correct answers apply, including: a. Tc-99m (6 hrs); b. Po-210 (138 days); c. C-14 (5730 yrs); d. U-238 (4.5 billion years)
13. Objects b, c, and d. Each of these is composed of organic substances and therefore contains the isotopes of carbon.
14. Tc-99m is used to generate diagnostic images because this isotope has a short half-life (6 hours) and the gamma rays it emits penetrate the body easily.
15. Positron Emission Tomography, a medical imaging technique; C-11 and F-18.
16. Cancer cells divide more rapidly than normal cells and are more active metabolically. Thus cancer cells are more susceptible to damage by ionizing radiation.
17. Cancer therapy in which a large number of low-intensity gamma ray beams are focused on a tumor, all from different angles. Each beam is too weak to damage surrounding tissue, but the intensity is powerful where all the beams come to a common focus. A gamma-ray emitter, such as Co-60, is used in this therapy.
18. A portion of the total mass is converted into the binding energy that holds the nuclear particles together.
19. The entire sequence of successive radioactive decays that converts a radioisotope into a stable isotope.
20. a. nuclear fission; b. A neutron must be absorbed by a U-235 nucleus to produce fission; neutrons must be released during the fission process so that other U-235 can absorb the released neutrons to continue the fission process.
    c. No. A critical mass must be present in order to provide a sufficient density of U-235 nuclei to absorb the released neutrons and generate a chain reaction.
21. To design and create an atomic bomb.
22. a. The first self-sustaining, controlled nuclear chain reaction took place at the University of Chicago. b. The bomb was designed and assembled in Los Alamos.

c. Hiroshima was the first city to be struck by an atomic bomb, a uranium bomb.

23. Gas centrifugation.

24. Advantage: Fission of U-235 can produce a self-sustaining chain-reaction. Fission of U-238 does not produce this. Disadvantage: Naturally existing uranium consists of less than 1% U-235.

25. Nuclear explosions require highly enriched uranium (typically greater than 80% enriched in U-235 for weapons-grade material). The uranium used as fuel in commercial fission reactor is less than 5% enriched in U-235. Accidents that can occur include the accidental overheating of the reactor core, possibly leading to meltdown and the occurrence of non-nuclear fires or explosions that might release radioactive material.

26. a. Nuclear fission; b. nuclear fusion; (i) Similar in that both reactions convert mass into energy; (ii) Different in that fission produces particles of smaller mass than the reactant; fusion produces a particle of greater mass than either of the reactants.

27. a. Movable control rods, made of neutron absorbing materials, such as boron or cadmium, slide in or out of the reactor core to control the rate of energy production. b. Nuclear reactors generate excess heat, so require large volumes of water for cooling purposes.

28. ${}^{2}_{1}H + {}^{3}_{1}H \longrightarrow {}^{4}_{2}He + {}^{1}_{0}n$

29. The positron and the electron are converted into two gamma-rays. The masses of the positron and the electron are converted into energy.

30. The iodine of our body tends to concentrate in the thyroid gland.

31. Radon formed by the decay of naturally occurring radioisotopes of Earth's soil can seep upward, enter our homes and other buildings and present a risk of producing lung cancer. Those with elevated levels of radon in their homes can take appropriate measures to reduce these levels.

Think

32. External: gamma radiation; internal: alpha radiation

33. Lead is the stable isotope formed at the end of the U-238 radioactive decay chain.

34. Disappearance through radioactive decay.

35. A well sealed house could hold within itself air containing radon gas emanating from the land it stands on. A house with a continual flow of air through it would allow the radon gas to escape.

36. Fermi created the first controlled nuclear pile, a prototype of the reactor now used in commercial nuclear power plants.

37. Since nuclear power plants are not capable of producing nuclear explosions, the most serious form of damage would be the release and dispersal of radioactive debris into the environment.

38. Your answers depends on your own thoughts about this issue. There are no correct or incorrect answers.

39. Your answers depends on your own thoughts about this issue. There are no correct or incorrect answers.

40. During each half-life 50% of the radioisotope that's present at the start of the half-life is converted into some other isotope by the end of the half-life.

Calculate

41. No; During each half-life 50% of the radioisotope that's present at the start of the half-life is converted into some other isotope by the end of the half-life. So after, 8 hours, 50% of the sample will remain; and after 16 hours, 25% will remain.

42. The isotope with a half-life of 50 years.

43. The atomic number decreases by two units; the mass number decreases by four units.

44. The atomic number increases by one unit.

45. An α particle is lost. ${}^{210}_{84}Po \longrightarrow {}^{206}_{82}Pb + {}^{4}_{2}He$

46. Nitrogen.

47. With each successive half-life (or 5730 year period), the C-12/C-14 ratio doubles due to decay of C-14 in the sample. So after 1 half-life the ratio is twice its original value. After 2 half-lives, it's four times its original value. After 3 half-lives it's eight times its original value. And after 4 half-lives (which corresponds to 4 × 5730 years, or 22,920 years) it's sixteen times its original value,. We can say that the ancient utensil was made from wood taken from a living tree about 23,000 years ago.

48. With a mass number of 56 and an atomic number of 26, an atom of Fe-56 must contain 26 protons, (56 – 26) = 30 neutrons, and 26 electrons. Thus the calculated mass of an atom of Fe-56 is

$$26 \times 1.007\ u\ \text{(mass of an individual proton)} = 26.182\ u$$
$$30 \times 1.009\ u\ \text{(mass of an individual neutron)} = 30.270\ u$$
$$26 \times 0.0005\ u\ \text{(mass of an individual electron)} = 0.013\ u$$

for a total of 56.465 u.

With an experimentally measured mass of 55.935 u, this produces a mass defect of (56.465 u – 55.935 u) = 0.530 u

49. 2 neutrons.

${}^{235}_{92}U + {}^{1}_{0}n \longrightarrow {}^{137}_{52}Te + {}^{97}_{40}Zr + 2\,{}^{1}_{0}n$

Here, the atomic numbers balance: (92 +0) = (52 + 40 + 0).

For the mass numbers to balance, 2 neutrons must be produced:

(235 + 1) = (137 + 97 + (2 × 1)).

50. The mass of the reactants is (1.01 + 235.04) u = 236.05 u

The mass of the products is (138.92 + 93.92 + 3.03) u = 235.87 u

The mass lost in the reaction is (236.05 – 235.87) u = 0.18 u

The lost mass represents (0.18/236.05) × 100 = 0.076% of the reactant mass

Chapter 10

Review

1. a. A battery's negative terminal b. a battery's positive terminal; c. a path by which electrons flow through an electrical device; d. the flow or movement of electrons through a circuit; e. the sum of the two half-cell reactions taking place in a battery; f. a process in which an electrical current generates a redox reaction; g. an electrochemical cell that produces current as long as it is supplied with fuel; h. a substance that causes another agent to be oxidized; i. a substance that causes another substance to be reduced; j. a component of a Daniell cell that permits ions to diffuse from either of the cell's beakers into the other beaker.
2. An electrical current flows through the circuit.
3. Zinc
4. Battery
5. Potassium hydroxide
6. The oxidizing agent is fluorine. The reducing agent is lithium.
7. Electrons move from reducing agents to oxidizing agents.
8. a. From the anode to the cathode; b. From the cathode to the anode.
9. A positive redox potential indicated that the redox reaction will occur spontaneously; a negative potential indicates that the redox reaction will not occur spontaneously.
10. By adding energy to the battery, as from a car's alternator or generator.
11. a. By noting the value of the standard reduction potential associated with the reduction reaction listed in Table 10.1. b. By noting the value of the reverse of the oxidation reaction listed in the table, and reversing the algebraic sign given in the table. c. By adding the standard reduction potential and the standard oxidation potential of the two half-cell reactions that compose the redox reaction.
12. Na is being oxidized and Cl is being reduced. Cl is the oxidizing agent and Na is the reducing agent. Reduction half-cell reaction: $Cl_2 + 2\,e^- \longrightarrow 2\,Cl^-$

    Oxidation half-cell reaction: $2\,Na \longrightarrow 2\,Na^+ + 2e^-$
13. a) Cl_2; b) Zn.
14. a. H_2 and O_2; b. Cl_2, H_2, and NaOH; c. Cl_2, H_2, and KOH
15. a. n-type; p-type b. As c. Ga
16. Calcium peroxide, CaO_2
17. a. The rusting of iron or steel b. The oxidation of zinc on galvanized iron or steel (or the oxidation of sacrificial blocks of aluminum, magnesium, or zinc attached to iron or steel tanks).
18. By attaching a sacrificial block of aluminum, magnesium or zinc to the outside of the tank.
19. Electric cars do not use gasoline or other petroleum derivatives and do not produce exhaust gases that can pollute the atmosphere. Electric cars suffered from disadvantages of limited driving range, low speeds, and the great weight and long recharge times of their batteries.
20. Antioxidants

Think

21. a. Because the voltage depends on the identities of the substances involved (and their underlying oxidation and reduction potentials), not on their quantities or sizes. b. Because a larger surface area allows for more atoms and ions to react concurrently thereby producing a greater current.
22. The presence of other substances that can act as either reducing or oxidizing agents.
23. Rinsing the area removes the colorless I^- produced by the reduction of the I_2 by the vitamin C. If I^- remains on the cotton, atmospheric oxygen can oxidize it back to I_2, which appears as a dark spot.
24. Household bleach, an oxidizing agent, oxidized the colorless I^- to I_2, which exhibited a color in solution.
25. The bleach oxidized the colorless Br^- to an orange-brown Br_2.
26. Pure water is a poor conductor of electricity so electrolysis proceeds slowly. In this case, water is oxidized to O_2 at the positive terminal. The addition of table salt introduces Na^+ and Cl^- to the water, which helps improve the electrical conductivity and rate of electrolysis. Here the Cl^- is oxidized to Cl_2 at the positive terminal.
27. The cell would not produce a voltage because all of the Cu^{2+} ions have been consumed.
28. No. The salt bridge is necessary so that ions in solution can diffuse from each beaker into the other beaker. Ions cannot migrate across a copper wire.
29. Vitamin C acts as an antioxidant. It is preferentially oxidized by atmospheric oxygen and thus protects other food components from oxidation.
30. It acts as a galvanic cell when internal redox reactions produce an electric current. It acts as an electrolytic cell as it is being recharged and externally applied electrical energy cause redox reactions to take place within the battery.
31. a.

    Structural features in common

    OH OH

    O

    BHA BHT

    b.

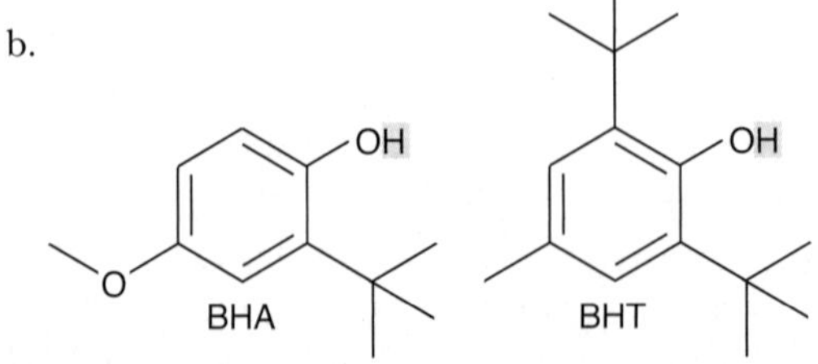

32. Yes. The electric car would not add CO_2, a greenhouse gas, to the atmosphere.
33. Through the electrolysis of water or aqueous solutions.

34. This is a matter of opinion. Advantages of electric cars over gasoline-powered cars include their lower energy and resource consumption and the absence of tailpipe emissions. Advantages of gasoline-powered cars over electric cars include their typical longer driving ranges and faster refueling times (as compared to recharging times of electric cars).

35. This is a matter of opinion.

36. Their batteries would be depleted of power too rapidly.

37. The H that becomes H^+ is losing electrons and is therefore being oxidized. The O to which this H is bonded is gaining the H's electrons and so is being reduced.

38. (The answer depends on when this question is being researched.)

Calculate

39. $(0.34 + 2.38) = 2.72$ volts

40. In the first case, Cu^{2+} ions of the solution are reduced to Cu^0 and the Zn^0 atoms of the zinc strip are oxidized to Zn^{2+}. From Table 10.1, both the reduction of Cu^{2+} to Cu^0 (0.34 V) and the oxidation of Zn^0 to Zn^{2+} (0.76 V) have positive standard potential, so the combined redox potential (1.10 V) is positive and the process is spontaneous. In the second case, both the reduction of Zn^{2+} to Zn^0 (-0.76 V) and the oxidation of Cu^0 to Cu^{2+} (-0.34 V) have negative standard potential, so the combined redox potential (-1.10 V) is negative and hence no reaction occurs.

41. Neither battery would work. In each case the oxidation reaction would simply be the reverse of the reduction reaction, with a net voltage of zero.

42. a. Yes; b. Yes; c. F_2 and Cl_2

43. By constructing the car's battery so that it consists of a set of six individual batters joined in series so that the aggregate sum of their voltages is 6×2 volts or 12 volts.

44. a. $I_2 + 2\,e^- \longrightarrow 2\,I^-$ 0.54 volts;

ascorbic acid + $I_2 \longrightarrow$ dehydroascorbic acid + 2 HI;
(0.54 volts − 0.06 volts) = 0.48 volts.

The redox reaction is spontaneous since its redox voltage is positive.

b. $ClO^- + H_2O + 2\,e^- \longrightarrow Cl^- + 2\,HO^-$
voltage = 0.84 volts;

$2\,I^- + ClO^- + H_2O \longrightarrow I_2 + Cl^- + 2\,HO^-$
(-0.54 volts + 0.84 volts) = 0.30 volts.

Oxidation of iodide by bleach is spontaneous since its redox voltage is positive.

45. a. All other values would decease by 2.87 volts, the current reduction potential for F_2
b. -2.87 volts
c. The measured voltage would be unchanged.
d. The new value for the reduction half-cell reaction, $Cu^{2+} + 2\,e^- \longrightarrow Cu^0$, would be -2.53 volts and the new value for the oxidation half-cell reaction, $Zn^0 \longrightarrow Zn^{2+} + 2\,e^-$, would be $+3.63$ volts. Together, these two potentials produce a voltage of 1.10 volts, the same as the previous value.
e. Yes
f. We would use the Table exactly as we did previously.
g. —
h. No because all other values of reduction potentials (and hence oxidation potentials) are set with respect to the standard.

Chapter 11

Review

1. a. Spherical aggregations of detergent molecules. b. Water-avoiding. c. Water-seeking. d. A natural detergent obtained by the alkaline hydrolysis of the triglycerides of fat. e. Water containing substantial concentrations of the salts of calcium, magnesium, and/or iron. f. A surface-active agent. g. A stable dispersion of fine particles of one substance in another. h. A chemical combination of a carboxylic acid and an alcohol, with the loss of a water molecule. i. A carboxylic acid containing an even-numbered, unbranched chain of 10–18 carbons. j. A triester of glycerol and fatty acids. k. Sodium hydroxide, NaOH. l. Excessive growth of algae in a body of water due to the release of phosphates, which depletes oxygen and kills fish and other aquatic animals. m. The hydrolysis of an ester by heating with water in the presence of sodium hydroxide. n. Water-insoluble precipitates of the fatty acid anions of soap molecules and the Ca^{2+}, Mg^{2+}, or Fe^{2+} cations of hard water. o. The cleavage of a molecule through the action of water. p. $HOCH_2$-CHOH-CH_2OH, a triol formed from the hydrolysis of naturally occurring fats and oils. q. A cage-like compound composed of aluminum, silicon and oxygen that may serve as a replacement for phosphates in synthetic detergents. r. A type of synthetic detergent containing a long-chain alkyl group and a sulfonate group, both bonded to a benzene ring.

2. Add some surfactant, such as soap or detergent, to the water.

3. Detergents lower the surface tension of water, allowing the water to penetrate into (or "wet") the fabric.

4. a. i. nonionic; ii. anionic; iii. cationic
b.

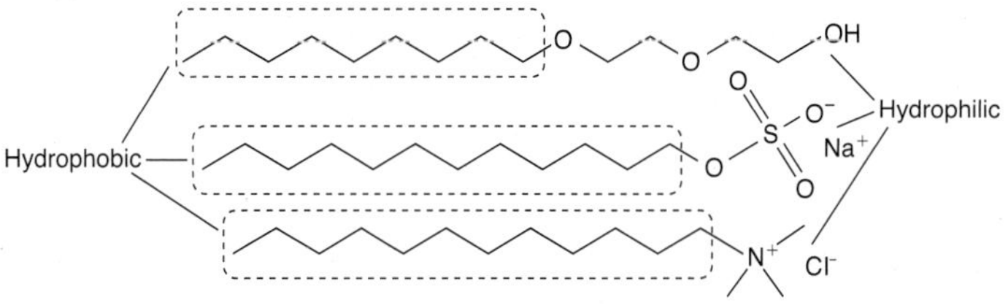

Hydrophobic
Hydrophilic

5. (a) An alcohol and a carboxylic acid. (b) An alcohol and the salt of a carboxylic acid.

6. Glycerol, also known as glycerine.

7. (a) The round head represents the carboxylate ion head ($-CO_2^-$) of the soap molecule. The zig-zag tail represents the hydrocarbon chain tail ($CH_3(CH_2)_n-$) of the soap molecule.
(b) Because the nonpolar hydrophobic tails of these molecules are repelled by water and protrude out from the water's surface. The polar, hydrophilic heads of these molecules are attracted to water and remain embedded among water molecules at the surface.
(c) They aggregate into micelles within the bulk of the fluid.

8. (a) Goat (or other animal) fat (b) Lye or extracts of wood ashes.

9. Their germicidal action.
10. Ca^{2+}, Mg^{2+} and Fe^{2+}.
11. (a) The action of hard water on soap, which lowers the soap's solubility.
(b) $2\ CH_3(CH_2)_nCO_2^-\ X^{2+}$, where X = Ca, Mg, or Fe, and n = an even number in the range of 8-18.
12.

Component	Function
Builders	Soften water and increases surfactant's efficiency.
Enzymes	Help remove stains, such as from blood and grass.
Suspension agents	Helps keep dirt from redepositing on fabric.
Color-safe bleaches	Help remove stains.
Optical whiteners	Make white clothes appear brighter.
Fragrance	Add pleasing odor to both detergent and fabrics.

13. Phosphates were added as builders to soften water and increase detergency. Phosphates were removed due to their contribution to eutrophication.
14. (a) A linear alkylbenzenesulfonate.
(b) Unlike soap, it remains effective in hard water.
15. Zeolites soften water by exchanging their sodium ions for the ions of hard water.
16. (a) The FDA defines a cosmetic as anything intended to be applied directly to the human body for cleansing, beautifying, promoting attractiveness, or altering the appearance.
(b) We can consider that anything we apply directly to our bodies to make us more attractive fits the definition of a cosmetic.
17. Surfactants
18. Sulfur
19. Antiperspirants inhibit sweating and keep the body relatively dry. Deodorants mask or eliminate odors with pleasant fragrances and also with antibacterial agents that remove the bacteria.
20. Perfumes are more concentrated than colognes. Perfumes consist of 10–25% solutions of fragrances in alcohol. The concentrations of the fragrant oils in colognes run about a tenth of those used in perfumes.
21. (a) The top note makes the first impression, the middle note is the most noticeable, and the base note is the fragrance that lingers. (b) Various correct answers exist. The chapter mentions 2-phenylethanal, 2-phenylethanol, and civetone as compounds responsible for top, middle, and base notes, respectively.
22. Ethyl acetate is an ester. Acetone is a ketone.
23. a. A protein that is the main structural component of hair. b. The central core of each hair strand. The cortex contains the hair's pigments and forms the bulk of the hair shaft. c. The small sac from which the hair grows. d. An oily substance secreted by sebaceous glands which lubricates the skin and hair. e. Sweat glands that lie in the armpits, secretions from which can be degraded by bacteria into unpleasant-smelling products. f. The lifeless, outermost layer of the skin. g. dermis h. epidermis i. A a thin, translucent, scaly sheath which envelops the hair's cortex. j. A compound that helps attract and retain moisture. k. A lotion that softens the skin. l. A colloidal dispersion of two or more liquids that are insoluble in each other. m. A substance that helps remove plaque from tooth enamel through the action of brushing.
24. a. A common detergent. b. A sulfur-containing amino acid prominent in keratin. c. A calcium-based mineral that largely makes up the enamel of teeth. e. A thin, clear, adhesive, polysaccharide film that attaches to the enamel of teeth. f. Tooth decay or cavities. g. An inorganic compound used as a scattering agent in sunscreens and used in a variety of other skin care and cosmetic products. h. A dark pigment produced by the skin to help protect against sun damage. i. A sunscreen that protects against both UV-A and UV-B radiation. j. A compound in sunless tanners.
25. (a) Frequency is inversely proportion to wavelength. For example, high frequency corresponds to short wavelength. How is the frequency of radiation related to its wavelength? (b) The energy of radiation is directly proportional to its frequency.
26. (a) i. X-rays ii. Ultraviolet region iii. Visible region iv. Infrared rays v. Microwave region (b) Radio waves (c) Gamma rays (d) Ultraviolet region (e) These represent the lower and upper boundaries, respectively, of the visible region, expressed in nanometers (nm).
27. Name each of the following compounds and describe its function in personal care products: a. Zinc oxide, used as an scattering agent in sunscreens and as a masking agent in face powders. b. Titanium dioxide, used as a scattering agent in sunscreens. c. Sodium fluoride, used a fluoridation agent used in some toothepastes to protect tooth enamel from decay d. Thioglycolic acid, used in the first step of "perm" hair treatments to disrupt disulfide links between keratin strands.
e. Lithium hydroxide, used in hair relaxers to disrupt disulfide links between keratin strands.
28. Henna
29. Fragrances
30. They tend to be more soluble in cold water and gentler on sensitive skin, and they don't dry out the hair as much.
31. The cortex forms the central core of the hair fiber and contains the hair's pigments. The cuticle is a thin, translucent, scaly sheath surrounding the cortex.
32. 1. Disulfide links—the sulfur-sulfur covalent bonds formed from thiol groups (–SH) of cysteines in neighboring polypeptide chains of keratin protein.
2. Salt bridges—the ionic bonds that form between the acidic group of one amino acid and a basic (amino) group of another amino acid located somewhere else on the same or an adjacent protein molecule;
3. Hydrogen bonds—weak, but plentiful links that form between the H of an —OH or a —NH group and the O or N of another —OH or —NH group.
33. (a) i. Epidermis ii. Dermis iii. Stratum corneum iv. Apocrine sweat gland v. Sebaceous gland vi. Eccrine sweat gland (b) The sebaceous gland (c) Stratum corneum (d) The eccrine sweat gland (e) The apocrine sweat gland
34. (a) UV-A radiation (320–400 nm) is less energetic and has longer wavelengths than UV-B radiation (290 to 320 nm).
(b) i. Premature aging of the skin (ii) Sunburn

35. As the aluminum cations (Al^{3+}) of these compounds enter skin cells, they attract water and the cells swell, causing the ducts of the eccrine glands to squeeze shut, thereby preventing the release of sweat.

36. Triclosan reduces levels of bacteria including levels those that cause odors.

37. (a) Zinc oxide scatters or blocks all the radiation, letting none through to the skin. (b) Octocrylene absorbs UV-B and shorter wavelength UV-A radiation (c) Avobenzone absorbs UV-A radiation.

38. Broad spectrum protection.

39. This nanoscale material is colorless, yet still offers protection against UV radiation.

Think

40. A surfactant can exist that is not a soap, because surfactants represent a broader class of compounds that contains within it the subclass of soaps. A soap can not exist that is not a surfactant (in other words, all soaps are surfactants) because soaps represent a subclass of compounds within the larger category of surfactants.

41. The surface tension of the cold water should be greater because surface tension results from attractions between neighboring water molecules and as the temperature increases, the greater thermal motion of water molecules interferes with attractions between them.

42. Soil. A rainwater filters through soil various minerals dissolve within it.

43. No, the high levels of dissolved salts make seawater hard and would reduce the the soap's effectiveness. Yes, because a synthetic detergent is more effective than soap in hard water.

44. The mineral deposits that can result from hard water are chiefly salts of the calcium cations (Ca^{2+}) and magnesium cations (Mg^{2+}) present in hard water and are not due to specific anions present.

45. Add an equal amount of soap solution to both samples of water, then cover the samples and shake. You should see more sudsing in the sample with softer water (or lower concentration of calcium ions).

46. The ion exchange resin would not be recharged with sodium ions and the device would not soften water.

47. You would need to install a water softener.

48. Baking soda is a base that can neutralize the acid that erodes the calcium and the phosphate from the hydroxyapatite of the teeth.

49. Both products contain surfactants that provide sudsing and cleansing action. The primary cleaning agents in toothpaste, however, are mild abrasives that help to remove plaque, whereas, the primary cleaning agents in shampoos are surfactants that help remove dirt and excess oil from the hair.

50. The instensity of UV radiation increases at higher alititudes because there is less atmosphere to absorb this radiation, so a stronger sunscreen is needed.

51. Dental plaque naturally forms on the tooth's enamel or exterior surface. Bacteria in the mouth convert this plaque to acids, which can erode the tooth enamel and lead to tooth decay.

52. In a wet wave, water molecules disrupt the hydrogen bonds that keep keratin strands aligned with each other, allowing them to shift a bit. As hair dries, the water molecules leave and the hair retains its new shape, held intact by large numbers of hydrogen bonds. "Perms" break the sulfur–sulfur bonds, reorganize the strands, and then regenerate new sulfur–sulfur bonds. Unlike a wet wave, the permanent wave keeps its shape through the newly formed covalent bonds. These remain firm whether the hair is wet or dry.

53. Primary: to inhibit perspiration. Secondary: to reduce body odor.

54. At a slightly acidic pH (4–6) the cuticle of the hair is tighter and more organized which reflects light coherently and gives hair a luster. At higher pH, the scales of the cuticle tend to swell up and fluff out, causing reflected light to scatter and making the hair look dull.

55. Cysteine.

Calculate

56. (a) 0.20 g fluoride (b) 0.64 mg fluoride

57. 225 minutes

58. (a) Decrease (b) Orange

Chapter 12

Review

1. (a) A non-steroidal anti-inflammatory drug (NSAID)
(b) Stimulant drugs, such a methamphetamine, that contain a phenylethylamine core structure.
(c) The chemical name for aspirin
(d) Plant-derived compounds, often bitter, that contain the basic, amine group.
(e) A compound that relieves pain.

2. (a) Salicin
(b) Nicotine
(c) Codeine (or morphine)
(d) Tetrahydrocannabinol
(e) Mescaline
(f) Cocaine

3. Describe the connection or distinction between the items in the following pairs.
(a) Salicylic acid is a semisynthetic precursor to aspirin. Treating salicylic acid with acetic anhydride produces aspirin.
(b) NSAIDs are drugs that inhibit the formation of prostaglandins
(c) Morphine is a semisynthetic precursor to heroin. Treating morphine with excess acetic anhydride produces heroin.
(d) Methamphetamine and MDMA are structurally related but have different psychotropic effects.
(e) Preclinical trials test drugs on lab animals. Clinical trials test them on humans.
(f) Generic drugs represent the same chemical entities and offer the same therapeutic profiles of their brand-name counterparts. A newly introduced brand-name drug is typically sold

under patent protection. When the patent expires generic drugs equivalent to the brand-name drug can be sold.

4. (a)

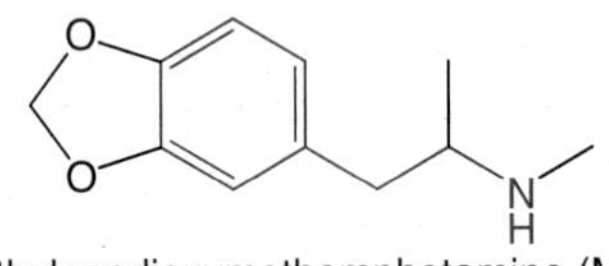

Methylenedioxymethamphetamine (MDMA)

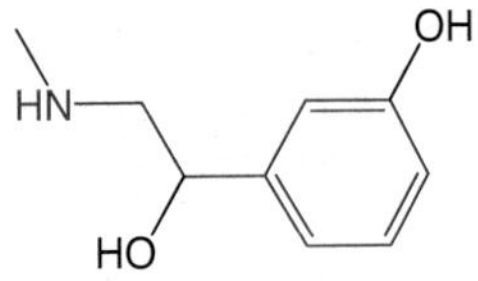

Phenyleprine

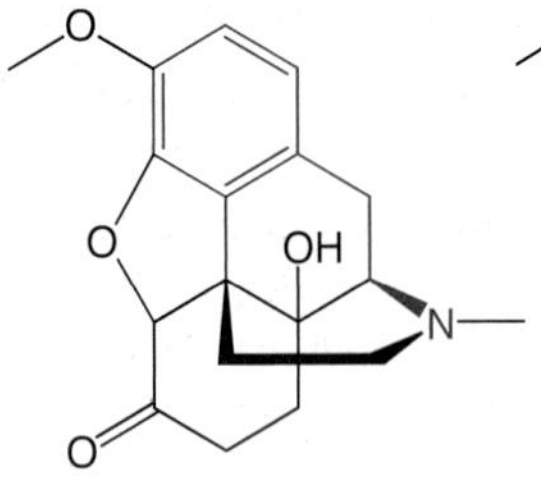

Oxycodone

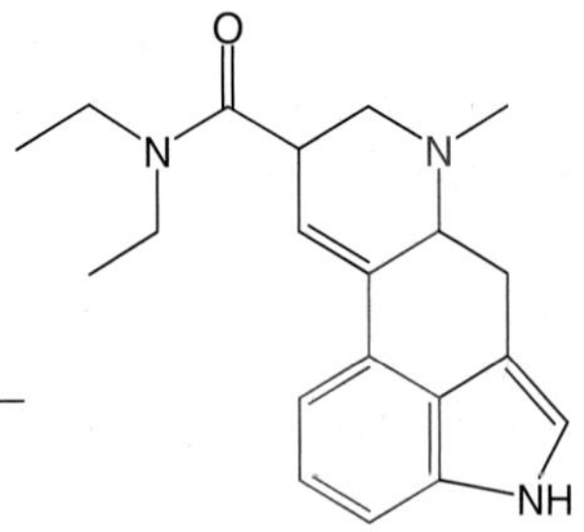

Lysergic acid diethylamide (LSD)

(b) MDMA: stimulant and hallucinogen; Phenylephrine: decongestant; Oxycodone: opioid narcotic; LSD: hallucinogen

5. Acetylation of salicylic acid produces aspirin. Acetylation of both alcohol groups of morphine produces heroin.
6. An analgesic reduces pain. An antipyretic reduces fever.
7. Symptoms of inflammation.
8. (a) Non-steroidal anti-inflammatory drug
   (b) Pain, fever, and inflammation.
9. (a)

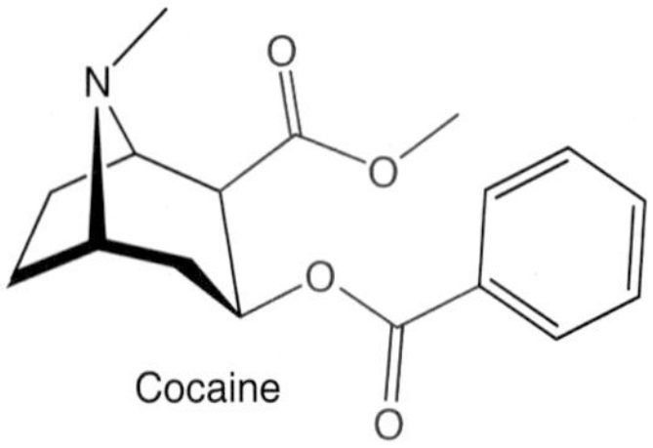

Cocaine

(b) Basic form. (c) Amine.

10. (a) Nicotine. (b) Many correct answers apply including carbon monoxide (CO).
11.

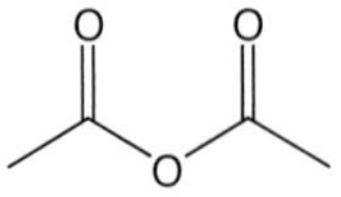

Acetic anhydride

12. (a)

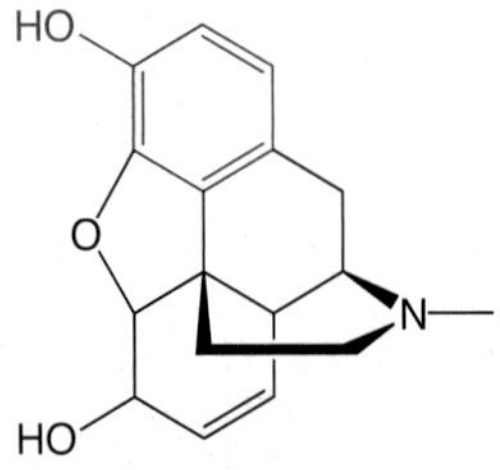

Morphine

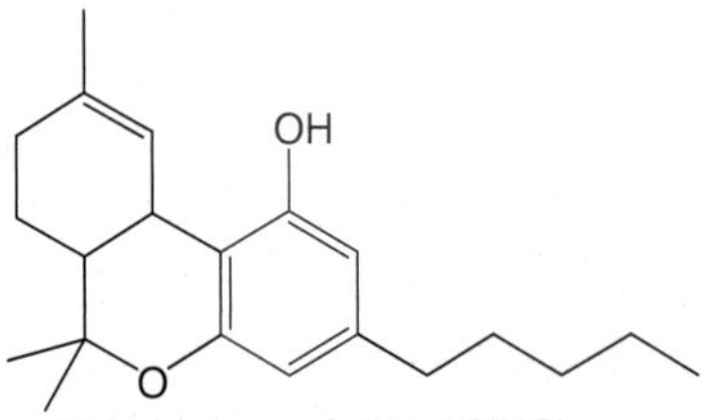

Tetrahydrocannabinol (THC)

(b) Morphine: opium poppy; tetrahydrocannabinol: cannabis.
(c) An ester.

13. Diacetylmorphine, also known as heroin, is powerfully addictive and also produces very strong withdrawal effects.
14. (a) Both are opioid narcotics offering strong analgesic effects and are addictive.
    (b) Methadone doesn't provide the euphoria or other psychological effects of heroin.
15. Caffeine and nicotine.
    (b) Caffeine is most commonly consumed in the form of caffeine-containing beverages such as coffee and tea. Nicotine is most commonly consumed in the form of tobacco.
16. (a) In mitosis, a cell divides to produce two identical copies of itself. This represents the mode by which most body cells, including those of the skin and the blood, divide.
    (b) In meiosis, the cell divides to produce two cells, each with half the chromosomes present of the parent cell. This kind of division occurs in the formation of sperm and egg cells.
17. The nuclei of cells.
18. DNA: Adenine (A); thymine (T); guanine (G); cytosine (C). RNA: Adenine (A); uracil (U); guanine (G); cytosine (C).
19. Hydrogen bonds between complementary bases.
20. (a) Adenine (left); thymine (right).

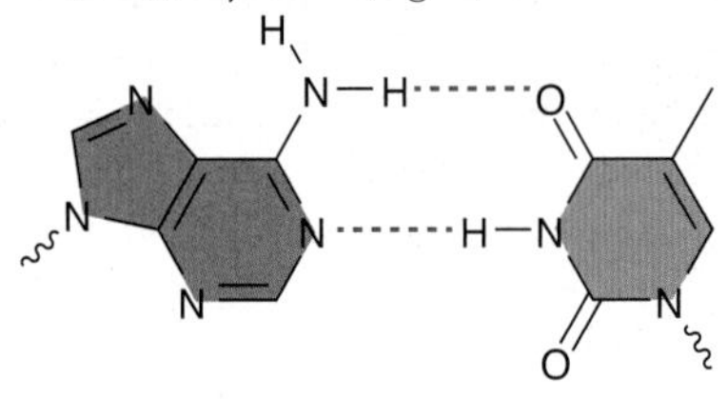

(b) Hydrogen bonds
(c) 2-deoxyribose
(d) Amines
(e) Thymine. Uracil.

21. GUA, GUG, GUC, GUU
22. The correspondence between any particular RNA codon and the specific amino acid or function it represents.
23. (a)

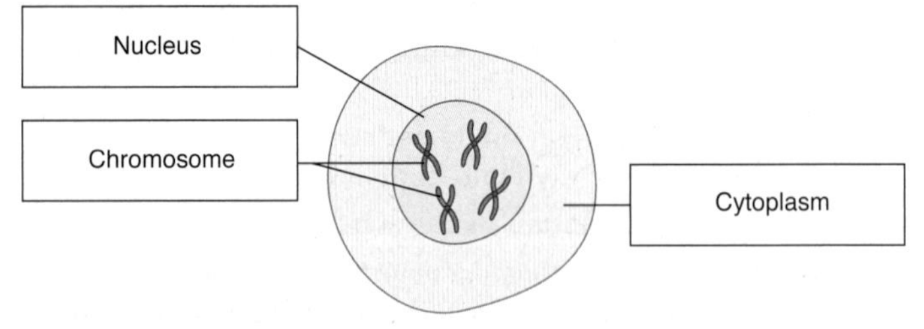

(b) Sister chromatids
(c) Yes. A chromosome holds its duplicated DNA in the form of sister chromatids joined together at the centromere.

24. (a) Phenotypes are observable traits of an organism. These phenotypes originate in genetic structures carried by a parent, called genotypes.
    (b) A gene is a segment of DNA that carries the information needed to create a specific polypeptide or protein. A genome is the complete set of genes carried by an organism.
    (c) When the cell is not dividing, its genetic material lies

diffused within the nucleus as thread-like filaments called chromatin. But as the cell prepares to divide, these filaments coalesce into more highly ordered structures called chromosomes.
(d) DNA is the molecule within the cell's nucleus that carries the genetic blueprint of the entire organism. DNA contains the sugar 2-deoxyribose and the bases adenine, thymine, guanine and cytosine. RNA carries information stored in the DNA in order to produce proteins. RNA contains the sugar ribose and the base uracil in place of thymine.
(e) A nucleotide is the repeating structural unit of DNA. Each nucleotide consists of three components: a phosphate group, the sugar 2-deoxyribose and one of four DNA bases. A DNA base is a component of a nucleotide.
(f) In transcription, mRNA (messenger RNA) carries information stored in DNA from the nucleus to the cytoplasm, where protein synthesis occurs. In translation, tRNA (transfer RNA) finds and transports each required amino acid to the site where the peptide bonds are formed under the direction of the mRNA. (g) In mitosis, a cell divides to produce two identical copies of itself. In meiosis, the cell divides to produce two cells, each with half the chromosomes present of the parent cell.
(h) A codon is a three-base sequence of mRNA that encodes for an amino acid or provides instructions to start or stop the chain-forming process of making a protein. An amino acid represents a component or one of the repeating links of a protein.
(i) An amino acid represents a component or one of the repeating links of a protein. A polypeptide, such as a protein, is a large molecule made up of a linear sequence of many linked amino acids.
(j) In reproductive cloning, the egg (which has a full set of chromosomes from a single parent) forms the basis of the development of a new infant, whereas in therapeutic cloning, this egg forms the basis of the development of stem cells.
(k) A single-gene disorder results from a mutation of a single gene passed from parent to child. A multifactorial disorder results from a variety of causes which may include both genetic and environmental factors.
(l) A single-gene disorder results from a mutation of a single gene passed from parent to child. Gene therapy offers the potential for treating single-gene disorders. In this approach, DNA containing the corrected gene is transferred into the patient's affected cells.
(m) A plasmid is a circular piece of DNA located outside the nucleus of bacterial cell. A restriction enzyme is an enzyme that cleaves DNA at a specific location along its strand.

25. Red hair is a phenotype which originates in a genotype, the carrier of the genetic information that transmits this characteristic from parent to child.

26. (a) A phosphate group and a sugar group, respectively
(b) From top to bottom, UACG.

27. (a) A plasmid is a circular piece of DNA located outside the nucleus of bacterial cell.
(b) A restriction enzyme is an enzyme that cleaves DNA at a specific location along its strand. In gene splicing, a restriction enzyme is used to remove a gene from one organism and also used to split open the bacterial plasmid so that the foreign gene can be inserted into the plasmid.
(c) Glyphosate is a common herbicide. Genetically modified crops containing a gene that confers tolerance to glyphosate allow this chemical to be sprayed on these crops so as to selectively kill weeds.
(d) A protein produced from recombinant DNA, genetic material that contains DNA sequences from more than one organism.
(e) transgenic

28. The first reported reproductive cloning of an adult mammal was the cloning of a sheep, producing a lamb named Dolly. This occurred in 1997 at the Roslin Institute in Scotland and came after 277 previous failures by the same team of scientists.

Think

29. Use aspirin is not advised for those with ulcers, those prone to bleeding or with impaired blood clotting function, those allergic to aspirin or NSAIDs, or children and adolescents recovering from chicken pox, the flu, and other viral infections, due to risk of Reye's syndrome.

30. One of the by-products of the acetylation of morphine with acetic anhydride to produce heroin is acetic acid, the major organic component of vinegar. The vinegar-like odor of heroin due to the residual acetic acid.

31. Under these conditions, the acetyl group of aspirin can slowly hydrolyze, yielding acetic acid (the major organic component of vinegar) and salicylic acid.

32.

33. This is a matter of personal opinion and may depend on the substance in question. For example, consider that the sale of alcohol in the United States was at one time prohibited.

34. Alcohol acts as a central nervous system (CNS) depressant, slowing down signaling between the brain and the rest of the body. Caffeine acts as a CNS stimulant, increasing pulse rate heightening awareness. Someone who consumes both substances at the same time is likely to perceive that they are less impaired from the alcohol than they truly are.

35. (a) Yes, this is possible if the person perceives that the substance contains caffeine.
(b) This is attributable to the placebo effect.

36. During the course of the study, only the third party maintains information on the true contents of each pill or medicine. If those administering the medicine, receiving the medicine or monitoring the clinical response to the medicine had this information, it could unwittingly bias the results of the experiment.

37. This is a matter of debate.

38. Following the introduction of antibiotics, people observed a rise in the incidence of antibiotic-resistant strains of bacteria, a concern that continues today.

39. In both cases, widespread application of a chemical designed to kill target organisms has led to a rise in the prevalence of strains of these organisms resistant to the effects of the chemical.

40. DNA is not a protein because it is not a polypeptide, but DNA can contain the genetic information to encode for a protein.
41. Inward-pointing bases allow for complementary base pairing (through hydrogen bonding), which stabilizes the double helix.
42. A plant can produce alkaloids – bitter, toxic substances that can deter insect pests from foraging on the plant.
43. (a) Meiosis produces sperm and egg cells. (b) Unlike mitosis, meiosis (followed by fertilization of the egg cell by the sperm cell), with its resultant mixture of the genetic properties of two parents, produces something new rather than more of the same.
44. Because factors other than DNA influence at least some aspects of physical appearance. These factors include random events that may occur during the development of the fetus.
45. This is a matter of personal opinion.
46. This is a matter of personal opinion.
47. This is a matter of personal opinion, but arguments for the use of recombinant DNA technology include the development of life-saving protein-based medicines as well as crops with improved attributes. Arguments opposing the use of this technology include concerns about the potential risks of the use of genetically-modified organisms.

Calculate

48.

First letter	Second Letter: A	U	C	G
A	AAA	AUA	ACA	AGA
	AAU	AUU	ACU	AGU
	AAC	AUC	ACC	AGC
	AAG	AUG	ACG	AGG
U	UAA	UUA	UCA	UGA
	UAU	UUU	UCU	UGU
	UAC	UUC	UCC	UGC
	UAG	UUG	UCG	UGG
C	CAA	CUA	CCA	CGA
	CAU	CUU	CCU	CGU
	CAC	CUC	CCC	CGC
	CAG	CUG	CCG	CGG
G	GAA	GUA	GCA	GGA
	GAU	GUU	GCU	GGU
	GAC	GUC	GCC	GGC
	GAG	GUG	GCG	GGG

49. A strand of DNA can contain four different nucleotides because each nucleotide contains a single base and there are four different bases: A, T, G, C.
50. The mRNA corresponding to this gene would contain approximately 7.7×10^5 codons.
51. Approximately 7.8 billion hydrogen bonds.
52. Based on the probabilities provided, there would be a 1 in 580 trillion chance that two randomly selected individuals share the same STR type at all 13 loci.

$1/[(0.075) \times (0.063) \times (0.036) \times (0.081) \times (0.195) \times (0.112) \times (0.158) \times (0.085) \times (0.065) \times (0.067) \times (0.039) \times (0.028) \times (0.089)] = 5.8 \times 10^{14}$

Chapter 13

Review

1. a. A monomer is a small molecule, many of which link together to form a polymer.
b. A homopolymer is made from one type of monomer. A copolymer is produced from two or more different types of monomers.
c. In an addition polymer, all atoms originally present in the monomer units are retained. By contrast, the formation of each link in a condensation polymer is accompanied by the release of a small molecule.
d. Thermoplastics soften in response to heat (and harden when cooled). By contrast, thermosets retain their hardness in heat.
e. A biopolymer is produced naturally. A semisynthetic polymer results from chemical modification of a biopolymer. Synthetic polymers are produced in a laboratory or factory from monomers typically derived from petroleum.
f. A polyamide, such as nylon, contains peptide (amide) links. A polyester, such as polyethylene terephthalate, contains ester links.
g. Primary air pollutants are emitted directly from either human or natural sources. Secondary air pollutants result from chemical reactions involving primary air pollutants.
2. b. HOM, ADD, TP, SP c. HOM, CON, TP, SP d. HOM, ADD, EL, SSP e. COP, CON, TS, SP f. HOM, CON, TP, SP g. HOM, ADD, EL, SP h. COP, CON, TP, SP
3. a. Nylon, whose uses include ropes, fishing lines, synthetic fabrics, and industrial parts.
b. Bakelite was the first synthetic polymer to be produced. Noted for its resistance to heat, it is used in many products, including the heat-resistant handles on many pots and pans.
c. Styrene-butadiene copolymer is the most common synthetic rubber, with uses that include tire treads and automotive belts and hoses.
d. Polyethylene terephthalate (PET), a clear plastic widely used in consumer packaging and in fabric form as polyester.
4. a. Uses include Mylar balloons.

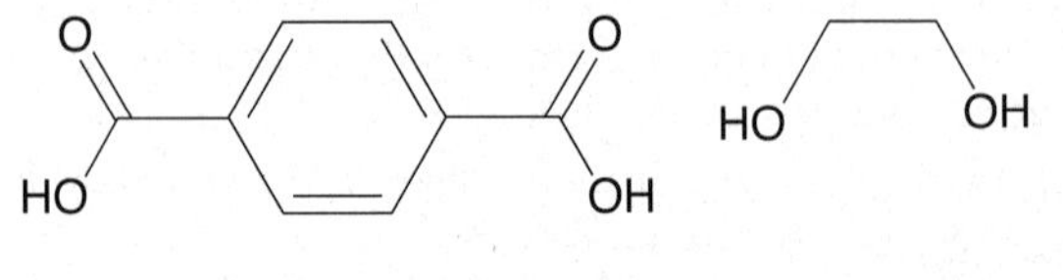

b. Uses include Teflon anti-stick coating on pots and pans.

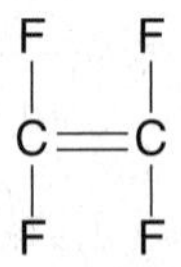

c. Uses include fire-resistant body suits.

X = OH or Cl

d. Uses include bulletproof vests and helmets.

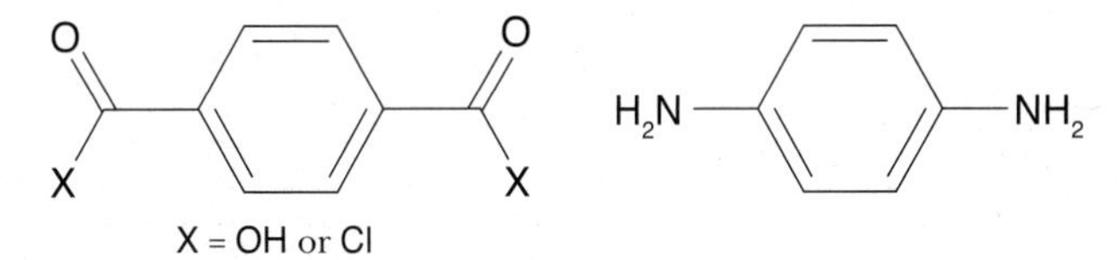

X = OH or Cl

5. a. They do not differ.
b. The chains of LDPE contain branches. By contrast, the chains of HDPE are linear, without branches.
c. The polymer chains of LDPE produce a disordered, tangled network. By contrast, the polymer chains of HDPE tend to be more ordered and produce pockets of aligned molecules, or crystallinities, within the bulk material.
d. HDPE is denser and more durable than LDPE and exhibits a higher melting point. LDPE, by contrast, is less durable, with a lower melting point.
6. Starch, cellulose, and proteins
7. Polyethylene
8. Correct answers include polyethylene terephthalate (PET), polystyrene, styrene-butadiene copolymer, Bakelite, Kevlar, and Nomex
9. a. polyamide b. polyamide c. polyester d. polyester e. polyamide f. polyamide
10. a. Most plastics are produced from fossil fuels, nonrenewable resources. As a result, manufacturing plastic from these resources is not sustainable. In addition, most plastic waste is discharged into land and sea where it accumulates because it does not generally decompose readily.
b. Methods can be developed to manufacture plastics from renewable feedstocks and plastics can be designed to more readily decompose in the environment.
11. a. Correct answers include polyvinylchloride (PVC) and neoprene
b. Teflon
c. Correct answers include nylon, Kevlar, and Nomex
d. polyvinylacetate
12. Since plastics are generally less dense the seawater, most plastic waste and debris floats. Much of this waste accumulates in rotating ocean current systems called gyres.
13. a.

1	2	3	4	5	6
Polyethylene terephthalate	High-density polyethylene	Polyvinyl chloride	Low-density polyethylene	Polypropylene	Polystyrene
PET	**HDPE**	**PVC**	**LDPE**	**PP**	**PS**

b. The number 7 includes all other types of plastics.
14. The action of microorganisms under favorable environmental conditions of moisture and heat.
15. Examples include Schoenbein's discovery of guncotton (nitrocellulose), Goodyear's discovery of vulcanized rubber, and Baekeland's discovery of Bakelite.
16. a. Coal-fired power plants, medical waste incinerators, and municipal waste incinerators
b. Volcanic releases
c. Airborne mercury can settle or deposit into bodies of water either through wet deposition, as a component of rain and snow, or through dry deposition, as a component of the air.
d. Through runoff carried by deposited rain and through migration into groundwater.
e. Volatilization refers to the evaporation of mercury. Volatile organic compounds enter the atmosphere through evaporation as well.
f. Mercury enters the food chain through its conversion into methylmercury by microorganisms in soil and bodies of water.
17. Primary: NO; NO_2. Human sources of these gases include the combustion of fossil fuels and their derivatives, such as from car exhaust.
Secondary: Correct answers include NO_2 (from oxidation of NO), HNO_2 and HNO_3 (from subsequent oxidations of NO_2), and peroxacetyl nitrate (from reactions forming smog).
18. Sulfur
19. a. FeS_2 b. NO_2 c. NO d. $CaCO_3$ e. $MgSO_3$
20. a. Scrubbers operate by passing exhaust gases through water or a slurry, often present as a fine spray. For example, a slurry of calcium carbonate, $CaCO_3$, or magnesium hydroxide, $Mg(OH)_2$, can convert the sulfur dioxide, SO_2, present in exhaust gases into a solid material, which settles within the base of the scrubber and can be removed.
b. Electrostatic precipitators remove particulate matter present in exhaust gases by passing the gas between two charged vertical plates or electrodes. The particles pick up electrons supplied by the negative electrode and move to the more positive one where they accumulate and can be removed.
21. It reduces the total amount of coal needed to produce a desired quantity of energy and also reduces the amount of carbon dioxide and sulfur dioxide emissions.
22. a. $2\ SO_2 + O_2 \longrightarrow 2\ SO_3$
b. $SO_2 + Mg(OH)_2 \longrightarrow MgSO_3 + H_2O$
c. 1. Breakdown of nitrogen dioxide: $NO_2 \longrightarrow NO + O$
2. Formation of ozone: $O_2 + O \longrightarrow O_3$
d. $SO_3 + H_2O \longrightarrow H_2SO_4$
23. Lightning can cause the formation of nitric oxide and ozone.
24. $4\ FeS_2 + 15\ O_2 + 8\ H_2O \longrightarrow 2\ Fe_2O_3 + 8\ H_2SO_4$
25. Sedimentary water pollution.
26. Chlorine gas (Cl_2) and ozone (O_3).

27.

Name	Description	Harm to aquatic life
Biological	Results from the presence of disease-causing and life-threatening microorganisms, especially in drinking water.	Pathogenic microorganisms in the water can harm aquatic as well as human life.
Thermal	Results from the transfer of heat into bodies of water thus raising the temperature.	Gases, including oxygen, are less soluble in warm than in cold water. Thus a rise in the temperature of a body of water can deprive fish and other aquatic animals of needed oxygen.
Sedimentary	Results from the accumulation of suspended particles within a body of water.	Suspended particles in water block sunlight and reduce visibility, thus interfering with photosynthesis by aquatic plants and decreasing the ability of aquatic animals to see and find food.
Chemical	Results from the release of harmful or undesirable chemicals into bodies of water.	Various harms. For example, the runoff of agricultural fertilizers can cause eutrophication, which diminishes dissolved oxygen and harms aquatic animals.

28. Nitric acid (HNO_3); sulfuric acid (H_2SO_4)

29. Through eutrophication, fertilizer runoff stimulates excessive growth of surface algae. The algae create mats that cover the surface, sealing off the rest of the water from the oxygen of the air. Deprived of dissolved oxygen, fish and other aquatic animals virtually disappear from these waters.

30. Volatile organic compounds. You could inhale gasoline vapors or the VOCs present in house paint.

31. a. (i) chlorofluorocarbons (CFCs); (ii) persistent organic pollutants (POPs)

b. CFCs were banned because they catalyze the destruction of stratospheric ozone. POPs were banned because they resist degradation in the environment and can undergo bioaccumulation in the food chain.

32. a. Methyl isocyanate was released due to an explosion at a chemical plant that manufactured pesticides.

b. Approximately 22,000 tons of hazardous waste from a local chemical company was buried in drums. Years later, a community settled on this land, but over time the drums corroded and leaked their contents, contaminating the area with hazardous waste.

Think

33. Because thermoplastics can be melted and reformed into pellets or other necessary shapes whereas thermosets can not be melted and reformed.

34. Three

35. Here we show one butadiene (BD) and three styrene (S) units forming in the following sequence from left to right: BD-S-S-S, but correct answers could also show this sequence any number of other ways, including S-BD-S-S, S-S-BD-S, or S-S-S-BD.

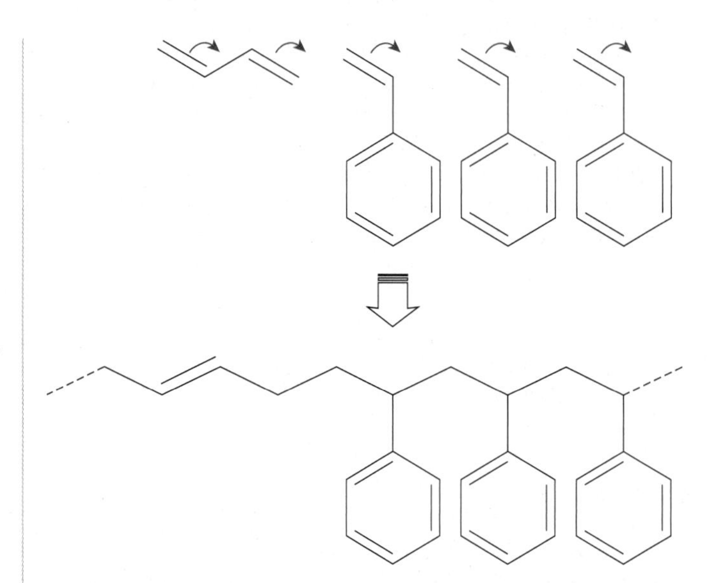

36. The term plastic means a material that can be molded and shaped. Commercial plastics are polymers, very large molecules formed by the repeated combination of much smaller molecules called monomers. However, not all polymers are plastics.

37. Vulcanization yielded a strengthened rubber that has found many valuable uses such as in automobile and truck tires, rubber gaskets and seals, shoe soles, and other products. Without vulcanization rubber would not be suited for these applications.

38. a. Starch or cellulose; b. Proteins; c. Latex rubber

39. a. Polyethylene or any of its derivatives listed in Table 13.1, as well as Neoprene.

b. Styrene-butadiene copolymer.

c. Bakelite, PET, Nylon, Kevlar, Nomex

40. Correct answers include adipic acid, used in the manufacture of nylon; and an isomer of terephthalic acid in which the carboxyl groups are attached at the 1 and 3 positions of a benzene ring (instead of at the 1 and 4 positions), used in the manufacture of Nomex.

41. Correct answers include the structure of the monomer(s) from which the polymer forms; the average length or molecular mass of the polymer chain; the architecture of the polymer chains—whether linear or branched, for example; the presence of attractive forces or bonds between the polymer chains in the material; the presence of ordered regions or crystallinities within the material; and the presence of additives, such as plasticizers, that may affect the flexibility and other characteristics of the material.

42. Polyvinyl bromide

43.

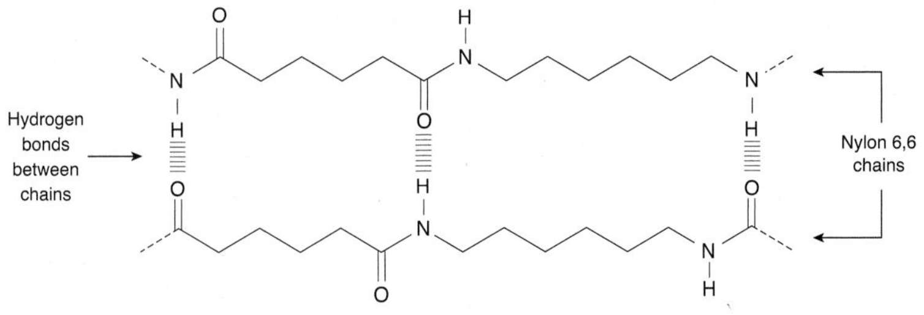

Hydrogen bonds between chains

Nylon 6,6 chains

44. a. Teflon b. Mylar c. Nylon d. Bakelite e. Neoprene

45. Because molecules of LDPE have branch points, and the shorthand form, $—(CH_2)—_n$, represents a linear or unbranched structure.

46. $\left[CH_2—\underset{\displaystyle Cl}{\underset{|}{C}}=\underset{\displaystyle H}{\underset{|}{C}}—CH_2 \right]_n$

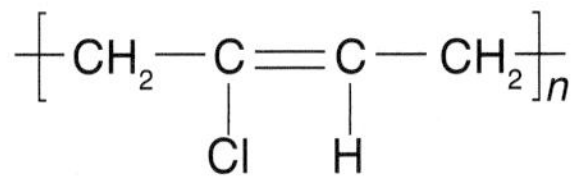

47. They are similar in that they are both polyamides and copolymers and can experience hydrogen bonding between chain segments. They are different in various ways, including that nylon is made from two monomer types, with a repeating structure $-(A\text{-}B)\text{-}_n$, whereas proteins are made from up to twenty monomer types, with a far more complex linear sequence; and protein chains can adopt a more complex higher order structure due to a greater variety of interactions between chain segments.

48. HDPE trash bags may lack the flexibility needed for that use and LDPE containers for liquid detergents may lack the rigidity and durability needed for that use.

49. A thermoplastic.

50.

Problem	Solution	Challenges of solution
Plastics derive from fossil fuels, which are nonrenewable resources.	1. Recycling reduces the need to produce as much plastic. 2. Bioplastics can been developed.	1. Not all plastics are recyclable; recycling compliance rates can be low. 2. Few bioplastics are widely available. For those that are available, few dedicated composting facilities exist and intermingling of bioplastics with recyclable plastics poses challenges.
Most plastic ends up as waste or pollution, where it lingers in the environment because it doesn't break down readily.	1. Make recycling programs more widespread and convenient so less plastic is discarded. 2. Design inexpensive plastics that readily decompose in the environment.	1. Plastics are used for many single-use, inexpensive items; it is often more convenient to discard these items than to recycle them; regardless, not all plastics are recyclable. 2. Due to various factors, including technical and economic, biodegradable plastics are not in widespread use.

51. Polyamides, such as nylon, Kevlar, and Nomex, are capable of exhibiting secondary structure due to hydrogen bonding between polymer chain segments.

52. DNA represents a copolymer because there are four different repeating units, nucleotides made from either adenine, cytosine, thymine, or guanine.

53. [Answers will vary depending on the community in which you live.]

54. The air itself, in the form of nitrogen gas, N_2.

55. The metal of the pan is first sandblasted or chemically treated to leave a microscopically rough surface. A base or primer coat of Teflon is then applied, which adheres to the surface by filling up the microscopic nooks and crannies. The properties of this primer coat allow it to also stick to subsequent layers of Teflon that may be applied to it. The pan may also be heated or baked between applications of these layers to improve bonding.

56. In principle, possibly, but in practice, unlikely because fresh water is a limited resource and the quantities needed to reduce the concentrations of contaminants to acceptable levels would be too large.

57. This is a matter of personal opinion. Since the salt present in seawater is due to natural processes, one could argue this salt does not constitute pollution; but from the vantage of using seawater as a potential source of drinking water, this salt constitutes a pollutant.

58. Energy conservation would reduce the amount of electricity we use and the amounts fossil fuels we consume for both transportation and heating. This would help reduce gaseous emissions and hence air pollution.

59. Since most CO emissions are due to the combustion of carbon-based transportation fuels, one solution would be to transition to alternative fuels such as hydrogen. Other solutions include improving the combustion efficiency of gasoline (through the use of oxygenates and engine control systems that optimize air-to-fuel combustion ratios), as well improving the performance of catalytic converters.

60. Stratospheric ozone is beneficial to life in shielding us from the sun's ultraviolet (UV) radiation. Ground-level ozone is harmful as a component of photochemical smog.

61. Although CFCs are not normally reactive, they are so when exposed to ultraviolet (UV) radiation. For instance, CFCs act as pollutants in the stratosphere where they can react with sunlight's UV radiation to release free chlorine atoms. These highly reactive chlorine atoms catalyze the destruction of ozone.

62. Over 80% of our energy needs are met by fossil fuels. Combustion of fossil fuels and their derivatives, such as gasoline, is the single greatest contributor to air pollution.

63. Natural gas generally contains lower levels of sulfur contaminants than petroleum or coal. Thus combustion of natural gas generates fewer sulfur dioxide (SO_2) emissions by comparison. In the atmosphere, SO_2 is a precursor to the formation of sulfuric acid, a component of acid rain, so fewer SO_2 emissions would contribute to lower levels of acid rain.

64. This would diminish air pollution, including CO, CO_2, NO_x, SO_2, VOCs, and particulate matter emissions. The environmental trade-off would be the increased burdens of mining uranium fuel and the handling and storage of greater quantities of radioactive waste generated by these nuclear power plants.

65. It means that these substances do not readily decompose in the environment and as a result are more likely to enter the food chain.

66. Various correct answers apply including the variety of products we use in our homes, such as plastics, paints, solvents, automotive fluids, lawn and garden chemicals, electronic devices, and certain types of batteries. If used or disposed of improperly, these substance can act as pollutants. Other examples include fertilizers and other agricultural chemicals, CFCs, and ozone.

67. This is a matter of opinion, but from the saltwater fish's perspective, fresh water could be considered to be polluted with an excess of water since it is deficient in salt.

68. Due to seasonal as well as yearly variations in ozone levels, your answer will depend on the point in time in which you research this topic.

69. Principally hydrofluorocarbons (HFCs).

70. Tetrachlorodioxin is a persistant organic pollutant. Accumulation of this and similar compounds in the body has been linked to reproductive and developmental effects, as well as certain cancers.

71. a. Chlorine; b. Sulfur

72. Cellulose, nitric acid, and sulfuric acid.

73. Plastic has a low density, meaning that it has a low mass for a given volume of material.

74. Scrubbing

75. Mercury enters the food chain through its conversion into methylmercury by microorganisms in soil and bodies of water. This methylmercury is initially taken up by small aquatic organisms such as algae and plankton. From there this contaminant moves up the aquatic food chain.

76. Predatory fish exist higher up the food chain and as a result accumulate greater levels of mercury. In addition, the form of mercury ingested is methylmercury, a nonpolar and fat soluble substance. The greater fat stores of these fish put them at higher risk of accumulating this form of mercury.

Calculate

77.

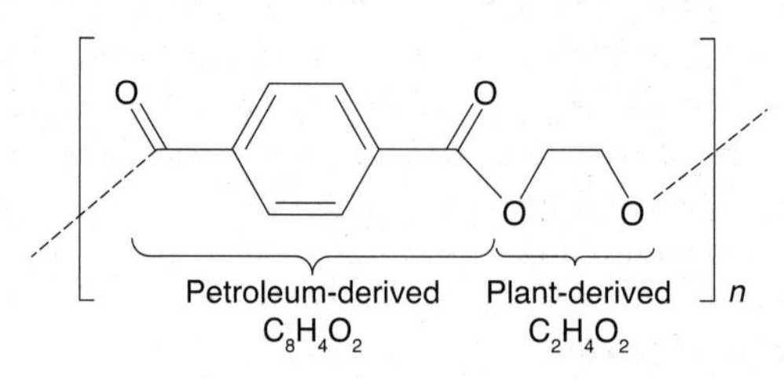

Element	Mass per atom (u) × # of atoms =		Mass contribution (u)	Mass per atom (u) × # of atoms =		Mass contribution
C	12	8	96	12	2	24
H	1	4	4	1	4	4
O	16	2	32	16	2	32
			Total = 132			Total = 60

Petroleum-derived mass contribution:

$$\frac{132}{132 + 60} \times 100\% = 68.8\%$$

Plant-derived mass contribution:

$$\frac{60}{132 + 60} \times 100\% = 31.2\%$$

78. a. Fuel combustion, stationary sources b. Fuel combustion, mobile sources; c. Industrial processes other than fuel combustion

79. a. $N_2 + 2\,O_2 \longrightarrow 2\,NO_2$

b. $NO_2 + O_2 \longrightarrow O_3 + NO$

c. To answer this, first multiply the equation from part b. by 2, to give the following:

$2 \times (NO_2 + O_2 \longrightarrow O_3 + NO) =$

$$2\,NO_2 + 2\,O_2 \longrightarrow 2\,O_3 + 2\,NO$$

Then add this equation to that for part a. (canceling common terms) to give the final answer:

From part a:	$N_2 + 2\,O_2$	$\longrightarrow$	$\cancel{2\,NO_2}$
From part b:	$\cancel{2\,NO_2} + 2\,O_2$	$\longrightarrow$	$2\,O_3 + 2\,NO$
Net equation:	$N_2 + 4\,O_2$	$\longrightarrow$	$2\,O_3 + 2\,NO$

80. a. 19.3 million tons

b. (i) 29.1 million tons; (ii) 24.1 million tons

Chapter 14

Review

1. vitamins; micronutrients; minerals; calcium; bones and teeth; potassium; chloride; fat soluble; water soluble; ascorbic acid; scurvy; rickets; night blindness; β-carotenes

2. food additive; preserve it or make it more nutritious or appealing; more appealing to the senses; monosodium glutamate; food sweeteners; titanium dioxide; color; nutritional value; potassium iodide; BHA and BHT; EDTA; metal ions; catalyze; mono- and diglycerides; emulsify.

3. a. The chemical name for vitamin D_3 b. Chemical substances added to foods that are generally regarded as safe c. A fibrous protein and component of our connective tissues that is susceptible to weakening due to vitamin C deficiency. d. The chemical name of vitamin A. e. Retinol Equivalents represent

the total amount of vitamin A and all its physiological equivalents that can be converted into or used as the vitamin. f. The Food and Drug Administration, the agency that ensures the safety of the chemicals added to foods and the drugs sold to the public. g. An antimicrobial agent added to certain packaged fruit drinks, soft drinks, and other processed foods. h. Enlargement of the thyroid gland due to a deficiency of iodine. i. Minerals required in the diet in amounts of less than 100 mg per day.

4. Macronutrients are nutrients such as proteins, carbohydrates, and fats, required to sustain life. Micronutrients are dietary substances such as vitamins and minerals, needed in trace amounts for proper health.
5. a. It forms compounds with calcium to give strength to bones and teeth. b. It forms the dominant cation (as Na^+) outside of the cells, keeping the water content of the intracellular and extracellular fluids in balance and thereby helps maintain proper fluid levels in the body. c. It helps harden the enamel of teeth, making them resistant to decay. d. In combination with protons, it provides the hydrochloric acid in gastric juices. e. It serves as a constituent of vitamin B_{12}.
6. a. vitamin C b. vitamin D
7. a. Deeply colored vegetables and fruits, such as carrots, cantaloupe, pumpkin, spinach, kale, apricots, mangos, and papayas; liver and liver oils. b. Fruits, especially citrus; many vegetables. c. Fortified milk; liver and liver oils.
8. IU and RE stand for International Units and Retinol Equivalents, respectively. The advantage of describing the recommended daily dosage of vitamin A in terms of its IUs and REs is that these terms include not only the quantity of the vitamin itself, but also the total amount of related structures that can be converted into or used as the vitamin.
9. a. EDTA b. six c. That of a metal d. It serves as an antioxidant and prevents food spoilage by sequestering metal ions that may catalyze the oxidation of components of foods.
10. a. It acts as a humectant, meaning that it helps retain moisture. b. Through hydrogen bonding to water molecules.
11. a. Thc major mincrals arc Na, K, Mg, Ca, P, S, and Cl. Thc trace minerals are the remainder. b. (i) cations (ii) anions
12. The gums are rich in collagen and are normally susceptible to wear and abrasion. Vitamin C deficiency causes a deterioration of collagen and hence early signs of this deficiency can be seen in bleeding gums.
13. To increase their appeal; to improve nutrition; to preserve their freshness and prevent spoilage; and to make them easier to process and keep them stable during storage.
14. a. Prevents goiter b. Act as colorants c. Inhibits growth of molds and other microorganisms d. Emulsify e. Inhibits growth of microorganisms in processed meats
15. Substances a. and b. improve nutrition; Substances c., d., and e. improve food appeal.
16. A humectant helps retain moisture. Humectants include glycerine, glyceryl monostearate, and propylene glycol. Some brands of coconut flakes contain propylene glycol.
17. It serves to improve nutrition (as a source of vitamin C) and to prevent oxidation of foods.
18. The lack of good methods for preserving foods led to the use of spices to mask the tastes and odors of decay in food. The opening of trade routes to the Orient and to the Western Hemisphere from Europe was at least partly due to the search for foreign spices.
19. Because they are listed as one of the exemptions under Chapter IV of the FD&C Act, the legislation which covers food additives.
20. a. He defined safety as the acceptability of risk. b. He developed a widely used test that determines the mutagenic potential of a chemical substance. c. He headed the first group within a federal agency to examine both the usefulness and the dangers of food additives.
21. a. Dairy products, such as milk and cheese b. Licorice root c. Various substances, including black pepper, carrots, celery, dill, mace, nutmeg, and parsley d. spinach and rhubarb
22. a. The U.S. Public Health Service (USPHS) b. The Bureau of Alcohol, Tobacco, Firearms, and Explosives (ATFE) c. The U.S. Environmental Protection Agency (EPA) d. The Occupational Safety and Health Administration (OSHA) e. The U.S. Food and Drug Administration (FDA)
23. A mutagen produces a genetic mutation in an organism. A carcinogen causes cancer.
24. Its chemical characteristics; how much of it we use or are exposed to; how we use it and the route of exposure; how susceptible to its hazards we are as humans as well as individuals.
25. a. b. coffee or tea c. comfrey d. oil of sassafras
26. Its mutagenic potential

Think

27. Proteins because they are the only macronutrients that contain sulfur.
28. a. Each unit of β-carotene supplies two units of retinol. b. Because various chemical substances can be converted into or used as the vitamin.
29. a. Yes b. No

    Minerals can't be created by living things, but must be absorbed from the earth in which they exist. Vitamins are synthesized by the living things, so any vitamins in the vegetable would be produced by the plants themselves and not absorbed from the soil.
30. The quality of the vitamin C is identical in both cases because vitamin C (ascorbic acid) represents a single chemical entity.
31. There is no difference since vitamin C (ascorbic acid) represents a single chemical entity regardless of how it was made, whether naturally or synthetically.
32. Cattle, like many plants and animals (but unlike humans), can synthesize vitamin C and so do not require it in their diet.
33. Through consumption of fish liver oils and fatty fish, if available, and through skin exposure to sunlight during warmer months.
34. Because water-soluble vitamins, such vitamins B, C, and folic acid, can leach into the cooking water, thus reducing the nutritional value of the vegetables. Some minerals may dissolve in the water as well.

35. a. One of the alcohol functional groups (—COH) in panthenol has been oxidized to a carboxylic acid functional group (—CO_2H) to give pantothenic acid. b. β-carotene is converted to retinol (vitamin A) by reactions within the body; 7-dehydrocholesterol is converted into cholecalciferol (vitamin D_3) through exposure of the skin to ultraviolet radiation.
36. The polar hydroxyl groups (—OH) of both molecules are capable of hydrogen bonding to water thus helping to retain moisture within the food. The polar hydroxyl group and nonpolar hydrocarbon chain within each molecule allow them to associate with both aqueous and oily components of foods thus helping to form a stable mixture or emulsion.
37. Emulsifiers must contain both polar and nonpolar components. Mono- and diglycerides both contain at least one (polar) hydroxyl group and at least one (nonpolar) hydrocarbon chain, allowing them to serve as emulsifiers. Triglycerides contain nonpolar groups in the form of hydrocarbon chains but lack polar groups.
38. Because methods of preserving foods were limited and spices aided in masking off-tastes of spoiled foods.
39. a. It is more useful than the legal definition in that it is practical and non-restrictive. b. It is less useful than the legal definition in that it would pose challenges to regulate and to ensure the safety of food additives.
40. This is a matter of personal opinion, but many would argue that the lack of nutritive value of refined sugar would not be offset by the vitamin- and mineral-fortification. It would be much better to obtain these micronutrients through a healthy diet.
41. They are generally regarded as safe (GRAS) substances.
42. This is a matter of personal opinion.
43. This is a matter of personal opinion and depends on your own tolerance to risk and your perception of the level of risk presented by any given activity.
44. The LD_{50} for nicotine administered to mice is 230 mg/kg orally, 9.5 mg/kg intraperitoneally, and just 0.3 mg/kg intravenously.
45. This is a matter of personal opinion. If you are inclined to think it is safe, consider that no substance is entirely free of hazards, and though the chemical appears to pose little risk to the animals tested, its toxicity in humans or long-term health effects may not be known. If you are inclined to think it is not safe, consider that if the substance truly produces no effect on the animal tested, regardless of the dosage, it would appear to be highly safe.
46. It means that any medicine (just like any substance we consume) can potentially poison if taken in excess.
47. a. and b., but not c., although cigarette packaging states dangers associated with smoking.
48. This is a matter of personal opinion. Each of these substances offer potential appeal (to some more than others), yet the use of each poses hazards.
49. Since the toxicity of a substance is related to its dosage, an argument can be made that there exists a low enough exposure to any given substance so as to constitute an acceptable risk.
50. You could conclude that the substance itself, but not necessarily its metabolites, is mutagenic. You could conclude that only the substance itself, but not necessarily its metabolites, is not a mutagen under the conditions of the test.
51. Because individuals within a population vary in their tolerance to a poison. An LD_{100} would underestimate the toxic potential of substance for a large portion of the population.
52. These vitamins are not water soluble.

Calculate

53. $$\frac{1\ \cancel{\text{g calcium}}}{\sim 0.5\ \cancel{\text{g phosphorus}}} \times \frac{1\text{ mole calcium atoms}}{\sim 40\ \cancel{\text{g calcium}}} \times \frac{\sim 31\ \cancel{\text{g phosphorus}}}{1\text{ mole phosphorus atoms}} = \frac{1.55\text{ mole calcium atoms}}{1\text{ mole phosphorus atoms}} \cong \frac{3\text{ calcium atoms}}{2\text{ phosphorus atoms}}$$

54. $$\frac{8000\ \cancel{\text{IUs}}}{72\text{ g carrot}} \times \frac{0.3\ \cancel{\mu\text{g retinol}}}{1\ \cancel{\text{IU}}} \times \frac{1\text{ g retinol}}{10^6\ \cancel{\mu\text{g retinol}}} \times 100\% = 3.33 \times 10^{-3}\%$$

55. 1.50 REs; 3.33 IUs
56. a. arsenic trioxide b. aspirin c. caffeine d. acetaminophen
57. arsenic trioxide, caffeine, nicotine; ethyl alcohol, thiamine hydrochloride

58. $$70\ \cancel{\text{kg body weight}} \times \frac{1.5\ \cancel{\text{g aspirin}}}{\cancel{\text{kg body weight}}} \times \frac{10^3\ \cancel{\text{mg aspirin}}}{1\ \cancel{\text{g aspirin}}} \times \frac{1\text{ aspirin tablet}}{325\ \cancel{\text{mg aspirin}}} = 323\text{ aspirin tablets}$$

59. $$70\ \cancel{\text{kg body weight}} \times \frac{4\ \cancel{\text{g sodium benzoate}}}{\cancel{\text{kg body weight}}} \times \frac{100\ \cancel{\text{g drink}}}{0.1\ \cancel{\text{g sodium benzoate}}} \times \frac{1\ \cancel{\text{mL drink}}}{1\ \cancel{\text{g drink}}} \times \frac{1\text{L drink}}{1000\ \cancel{\text{mL drink}}} = 280\text{ L drink}$$

60. sodium cyanide > aspirin > sodium chloride > water

Glossary

absolute zero The coldest possible temperature, corresponding to zero Kelvin or −273°C.
acid A chemical compound that turns litmus paper red and reacts with bases to form salts.
acid rain Rain with a pH lower than 5.6, typically due to sulfuric and nitric acid contamination.
addition polymer A polymer in which all atoms originally present in the monomer units are retained.
alkali metals The elements in the first column of the periodic table, with the exception of hydrogen.
alkaline earth Any element in the second column of the periodic table.
alkaloid A class of plant compounds, often bitter, that contain the basic, amine group.
alkane A hydrocarbon molecule containing only single bonds.
alloy A solid metal solution containing two or more elements.
Ames test A test that determines the mutagenic potential of a chemical substance.
amino acid A small organic molecule containing both **amino** and **carboxylic acid** functional groups that serves as a structural unit for proteins.
ampere A measure of the rate of flow of an electrical current.
amphoteric Able to act as either an acid or a base.
anion A negatively charged ion.
anode A battery's negative terminal.
antibiotics Substances, originally derived from certain molds and fungi, that kill other microbes.
antioxidant A substance that readily undergoes oxidation, thereby preventing other compounds from being oxidized.
aqueous Water-based or having water as solvent.
atom A fundamental particle of matter.
atomic mass The weighted average of the masses of all of the naturally occurring isotopes of an element.
atomic number The number of protons in the nucleus of an atom.
atomic orbitals Lobes or regions of space outside the atom's nucleus where there is a high likelihood of finding the atom's electrons. A maximum of two electrons may occupy each atomic orbital.
atomic structure The combination of all the particles that compose an atom, their relationships to one another, and their locations within the atom.
α particle A positively charged particle consisting of two protons and two neutrons.
β particle A high-speed electron emitted during (β) radioactive decay.
background radiation Ionizing radiation emanating from within us as well as from our environment and outer space.
balanced chemical equation An equation showing reactants, a reaction arrow, and products, with the same number of atoms of a given element on each side of the arrow.
basal metabolism The body's energy expenditure while at rest.
base A chemical compound that turns litmus paper blue and reacts with acids to form salts.
basic research Fundamental research that increases our understanding of the world.
binding energy The energy required to hold the protons and neutrons together in the nucleus of an atom.
bioaccumulation A process whereby pollutants accumulate in the tissues of animals and plants.
biofuels Fuels derived from living matter or biomass, a renewable resource.
blood alcohol content The concentration of ethanol in the blood, typically measured in grams of ethanol per 100 mL blood.
boiling point The temperature at which a liquid is transformed into a gas.
BPA An abbreviation for bisphenol-A, a chemical added to plastics and resins that improves and strengthens their physical properties.
buffer A combination of a weak acid and its conjugate base that resists changes in pH.
Calorie A dietary unit of energy defined as 1000 calories, or 1 kilocalorie.
calorie The amount of heat (or energy) needed to raise the temperature of 1 gram of water 1°C.
calorimetry The measurement of heat released or absorbed by a chemical or physical process.
carbohydrates A class of macronutrients including sugars and starches that is an important source of energy for organisms.
carbon cap and trade A system whereby total CO_2 emissions are limited or capped by a government body. Participants in the system, such as electric utilities, buy permits allowing them to emit defined amounts of CO_2. Those who emit less than their quota can sell their excess allowances to others in the system.
carbon cycle A global cycle in which carbon is exchanged among the atmosphere, the oceans, geological systems, and living things.
carbon taxes An alternative to carbon cap and trade, whereby a tax is levied on the use of fossil fuels.
carboxylic acid A carbon-bearing acid characterized by the presence of the carboxyl group, $—CO_2H$.
carcinogen A cancer-causing substance.
catalyst A substance, often a specialized metal, that speeds up the rate of a chemical reaction without itself being consumed.
catalytic converter A device built into an automobile's exhaust system that uses catalysts to reduce the levels of hydrocarbons, carbon monoxide, and other pollutants emitted.
catalytic cracking A petroleum refining process that uses a catalyst to break down (crack) higher-boiling, higher-molecular-weight hydrocarbon molecules into lighter molecules.
catalytic hydrogenation A process by which hydrogen is added to unsaturated chemical compounds in the presence of a catalyst.
catalytic reforming A petroleum refining process in which a catalyst converts low-octane- rated compounds into those more suitable for gasoline.
cathode A battery's positive terminal.
cation A positively charged ion.
chemical change A process that produces substances with new chemical compositions.
chemical mutagen A chemical substance that produces a genetic mutation in an organism.
chemistry The study of matter—its composition, its properties, and the changes it undergoes.
coal A solid, carbon-rich fuel with a widely varying composition that depends on its source. This relatively plentiful fuel is used chiefly for generating electricity.
colloid A stable dispersion of fine particles of one substance in another.
combustion The reaction of a fuel with oxygen to produce heat and light.
compound A pure substance formed by the chemical combination of two or more elements in a specific ratio.
compound formula A chemical formula showing the elements present in an ionic compound and the ratio of each.
compression ratio A measure of the extent to which the fuel–air mixture is compressed during the compression stroke.
concentration A measure of the amount of solute dissolved within a given quantity of solvent or solution.
condensation polymer A polymer in which the formation of each link is accompanied by the release of a small molecule.
conservation of mass Matter can neither be created nor destroyed as a result of a conventional (nonnuclear) chemical reaction.
covalent bond A bond consisting of a pair of electrons shared by two atoms.
covalent compound A substance consisting of discrete molecules, each containing atoms of different elements held together by covalent bonds.

critical mass A mass of fissionable material large enough to ensure that released neutrons are absorbed by other fissionable nuclei, thereby sustaining a nuclear chain reaction.

crystal lattice Orderly, three-dimensional arrangement of the chemical particles that form a crystal.

denaturation A process by which a protein's higher-order structure is disrupted due to heat or exposure to chemical compounds.

density A measure of how much mass is in a given volume of a substance.

dentifrice A substance, typically a paste, for cleaning teeth.

deoxyribonucleic acid (DNA) The molecule within the cell's nucleus that carries the genetic blueprint of the entire organism.

detergent A cleansing agent consisting of molecules that contain hydrophilic heads and hydrophobic tails.

dispersion force A weak attractive force between nonpolar molecules in close proximity.

double bond Two pairs of shared electrons serving as two covalent bonds between two atoms.

dynamic equilibrium A state in which the rate of a forward reaction (such as ionization) equals the rate of the reverse reaction (such as recombination) so that the concentrations of all chemical species remain constant.

electrolysis Chemical changes caused by passing an electric current through a substance.

electrolyte A substance that ionizes efficiently in solution and conducts electricity.

electrolytes The cations and anions that constitute the ionic compounds of our blood and other bodily fluids.

electromagnetic spectrum The full range of frequencies that characterizes electromagnetic radiation.

electron configuration The arrangement of an atom's electrons in its quantum shells.

electronegativity A measure of the ability of an atom to attract bonding electrons.

electrons Small, negatively charged particles located in shells surrounding an atom's nucleus.

element A substance, all of whose atoms have the same atomic number.

emulsion A colloidal dispersion of two or more liquids that are insoluble in each other.

endothermic Any process that absorbs heat.

energy The capacity to perform work.

entropy A measure of the disorder or randomness of the positions of a collection of atoms, ions, or molecules.

enzyme A biological molecule, typically a protein, that acts as a catalyst.

enzyme inhibition Suppressing an enzyme's normal function.

evaporation A process by which a portion of a liquid at its surface turns into gas.

exothermic Any process that releases heat.

food additives Chemicals added to foods to help preserve them and to make them more nutritious and more appealing.

formula mass Sum of the atomic masses of all the atoms in the formula of an ionic compound.

fossil fuels Fuels such as natural gas, petroleum (oil), and coal, derived from decaying plant and animal matter.

fractional distillation A process by which a liquid mixture of compounds is separated into fractions based on boiling points.

free radical A chemical species that has an odd number of electrons.

fuel cell An electrochemical cell that produces current as long as it is supplied with fuel.

galvanic cell A device that creates electrical energy from a chemical reaction.

galvanization Applying a thin coating of zinc to a base metal, such as iron or steel, in order to inhibit corrosion.

gamma (γ) ray Very high-energy (short wavelength) radiation emitted by excited-state nuclei.

gene A segment of DNA that carries the information needed to create a specific polypeptide or protein.

generic drugs Drugs that are chemically equivalent to patented drugs.

genetically modified organism (GMO) An organism whose genetic material has been altered through genetic engineering.

glucose A simple carbohydrate that serves as the main source of energy for the functioning of our bodies.

GRAS list A list of chemical substances generally regarded as safe by the FDA.

green chemistry Chemical practices that aim to conserve resources and reduce the generation of waste and toxic substances.

greenhouse effect A process by which infrared radiation is trapped by certain atmospheric gases, thereby warming Earth's surface and lower atmosphere.

greenhouse gases Gases that trap infrared radiation in the atmosphere.

groups Columns in the periodic table.

half-cell reaction Either the reduction or oxidation reaction within an electrochemical cell.

half-life The length of time it takes for one-half of a given quantity of a radioactive isotope to decay. Every radioisotope has its own characteristic half-life.

halogens Elements in column 17, the next to last column of the periodic table.

hazardous waste Chemical waste that may endanger public health or wildlife.

heat The energy that flows from a warmer body to a cooler one.

hydrocarbons A class of chemical compounds composed of carbon and hydrogen. The general molecular formula C_xH_y is used to represent any hydrocarbon molecule, where *x* and *y* denote the number of carbon and hydrogen atoms, respectively.

hydrogen bond An attractive force between an oxygen, nitrogen, or fluorine atom and a nearby hydrogen atom bonded to an oxygen, nitrogen, or fluorine atom.

hydrolysis The process of breaking down a chemical compound through reaction with water.

hydronium ion The cation H_3O^+, which is produced when acids release hydrogen ions (H^+) in water.

hydroxide ion The anion OH^-, which is produced when bases dissolve in water.

hypothesis A tentative explanation for a relatively small set of observations.

ideal gas A gas that behaves as if it were made up of molecules or atoms that have no volumes of their own and bounce off each other without losing any energy.

inert or **noble gases** Elements in column 18, the last column of the periodic table.

infrared radiation A type of radiation that we sense as heat.

intermolecular forces Attractive forces that exist between molecules in close proximity.

ion An atom or a group of atoms that carries an electrical charge.

ionic bond A chemical bond resulting from the mutual attraction of oppositely charged ions.

ionization Conversion of a covalent molecule into ions.

ionizing power A measure of how efficiently ionizing radiation causes ionization as it penetrates matter.

ionizing radiation Radiation (such as α, β, γ, x-, and cosmic rays) capable of ionizing atoms or molecules in its path.

isomers Two or more compounds that share the same molecular formula but differ in structure.

isotopes Atoms of the same element that differ in the number of neutrons and therefore in mass number.

isotopic enrichment A process of increasing the abundance of a desired isotope in a mixture by removing the undesired isotope.
joule The standard unit of energy in the metric system, abbreviated J, equivalent to 0.24 calories. One kilojoule (kJ) is 1000 joules.
kinetic energy The energy of motion.
knocking A metallic pinging sound sometimes heard from automobile engines when the air–fuel mixture combusts erratically in pockets, instead of in one smooth wave emanating from the spark plug.
law of conservation of energy A natural law recognizing that energy can neither be created nor destroyed but can only be converted from one form to another.
LD_{50} The dose of a substance that kills half of the animals tested.
Lewis or **electron dot structures** Notations that show only the valence electrons of an atom, arranged as dots around the element's symbol.
life-cycle assessment Considering a product's full environmental impact, from the process of obtaining its raw materials to the disposal or recycling of the exhausted product.
litmus paper An absorbent paper strip containing a natural dye that turns red in acid solution and blue in basic solution.
lone pairs A pair of valence electrons not involved in bonding.
macronutrients The major components of our foods that provide us with energy and the materials that form our bodies.
major minerals Minerals required in the diet in amounts greater than 100 mg per day or present in the body in amounts greater than 0.01% of body mass.
mass A measure of a body's resistance to acceleration.
mass defect The difference between the mass of an atomic nucleus and the (larger) sum of the masses of its constituent particles.
mass number The combined number of protons and neutrons in the nucleus of an atom.
meltdown Accidental overheating of the fuel in a nuclear reactor, leading to a melting of the reactor core.
melting point The temperature at which a solid is transformed into a liquid.
metabolism All the chemical reactions that take place in a living organism to support life.
micelles Spherical aggregations of detergent molecules.
micronutrients Dietary substances needed in trace amounts for proper health.
minerals Elements needed by the body in small amounts for proper health.
molarity A measure of solution concentration defined as the number of moles of solute per liter of solution.
mole An amount of substance that contains 6.02×10^{23} chemical particles.
molecular formula The chemical formula of a covalent compound.
molecular mass Sum of the atomic masses of the atoms in a molecule.
molecular structure The arrangement in space of the atoms that make up a molecule.
molecule An electrically neutral assembly of atoms held together by covalent bonds.
monomer A small molecule, many of which link together to form a polymer.
municipal solid waste Solid materials discarded by homes, businesses, and public institutions.
nanotechnology The science of developing materials with dimensions on the nanoscale (less than 100 nm in size) and often exhibiting novel properties.
natural gas A mixture of hydrocarbon gases, chiefly methane, often found associated with petroleum deposits.
natural product A chemical compound produced by a living organism.
neutralization A reaction between an acid and a base to produce a salt and water.
neutron A nuclear particle that carries no electric charge.
nitrogen balance The balance between the amount of nitrogen consumed in the diet (in the form of protein) and the amount excreted.
nonpolar covalent bond A covalent bond in which the shared pair of electrons lies equidistant from both bonded atoms.
nonpolar Lacking an overall separation of charge within a molecule.
nonrenewable resources Natural resources that are not replenished readily by natural processes and become depleted as they are used.
NSAID A non-steroidal anti-inflammatory drug, typically used to treat pain, fever, or inflammation.
nuclear fission A reaction that splits a relatively massive nucleus into two or more sizable fragments, releasing energy in the process.
nuclear fusion A nuclear reaction in which light atomic nuclei fuse together with the release of energy.
nucleation A process that initiates the formation of a distinct state of matter within a solution, such as bubbles (gas) or crystals (solid).
nucleotide The repeating structural unit of DNA.
nucleus The extremely small, dense center of the atom in which the atom's positive charge is localized.
nuclide The nucleus of an isotope, characterized by the number of protons and neutrons it contains.
ocean acidification A decrease in the pH of the oceans due to the absorption of atmospheric CO_2.
octane rating A measure of the antiknock properties of a fuel—a gasoline with a higher octane number is less prone to knocking than a gasoline with a lower octane number.
octet rule Atoms often react to obtain exactly eight electrons in their valence shell.
omega-3 fatty acid An unsaturated fatty acid with a carbon-carbon double bond occurring between the third and fourth carbons from the end of the chain. (The last carbon of the fatty-acid chain is designated as the omega carbon. Omega is the last letter of the Greek alphabet.)
organic compound A covalent compound containing carbon.
osmosis The tendency for a solvent, such as water, to migrate through a semipermeable membrane from a dilute solution to a more concentrated solution.
oxdiation Loss of electrons.
oxidation state A number describing the degree of oxidation of an atom.
oxidizing agent A substance that causes another to be oxidized.
oxygenate An oxygen-containing compound added to gasoline to improve oxidation of fuel and decrease harmful emissions.
part per million (ppm) A solution concentration equal to 1 mg of solute in 1 kg of solvent.
penetrating power The ability of ionizing radiation to penetrate matter.
peptide A compound made up of linked amino acids.
periodic table An orderly arrangement of all chemical elements into rows (called periods) and columns (called groups or families).
periods Rows in the periodic table.
petroleum An oily, usually dark, flammable liquid, consisting of a complex mixture of hundreds of hydrocarbons and other minor components.
petroleum refining A process by which petroleum is separated into its different components.
pH A measure of acidity, defined as the negative logarithm of the hydronium ion concentration: $pH = -\log [H_3O^+]$.
photochemical smog A complex combination of pollutants resulting from the interaction of sunlight with nitrogen dioxide and subsequent reactions involving atmospheric oxygen and hydrocarbon pollutants.
photovoltaic cell A cell that converts solar energy into electrical energy.

physical change A transformation of matter that occurs without any change in chemical composition.
placebo An inert substance used as a scientific control in tests of an actual medication.
poison A chemical that enters our body and causes illness or death.
polar covalent bond A covalent bond in which the shared pair of electrons lies closer to one of the bonded atoms.
polar Having a separation of charge due to differences of electronegativities of component atoms or ions.
polyatomic ion An ion made up of more than one atom.
polymer A very large molecule formed by the repeated combination of much smaller molecules.
positron A positively charged analog of an electron.
positron emitters Radioisotopes that decay with positron emission, some of which are used in medical imaging.
potential energy Stored energy due to an object's position in space or due to the composition of a substance.
pressurized water reactor A reactor in which circulating water is kept at high pressure to prevent it from boiling as it's heated by an atomic pile. Steam is then generated in a secondary loop.
primary air pollutant An air pollutant generated as the direct result of a specific activity.
protein A polymer made up of a long sequence of amino acids.
proton A positively charged particle found within the nucleus of an atom.
quantum shells The orbits of electrons at fixed distances from the nucleus, as described in the Bohr model of the atom.
radioactive fallout Airborne radioactive particles, from a nuclear explosion or accident, which settle to the ground. Depending on the severity of the incident and weather patterns, fallout can potentially distribute across the globe.
radioactivity The emission of radiation by the spontaneous decay of an unstable atomic nucleus.
radiocarbon dating A technique, based on the characteristic decay rate of carbon-14, for determining the age of an artifact of biological origin.
recombinant DNA Genetic material produced by introducing DNA sequences of one organism into the DNA of another organism.
recommended daily allowance (RDA) The level of daily consumption of a given macronutrient recommended for maintaining proper health.
reducing agent A substance that causes another to be reduced.
reduction Gain of electrons.
reference standards Precise quantities upon which SI base units are defined.
rem A unit of ionizing radiation representing an equivalent dose to humans, regardless of the type of radiation.
renewable resources Natural resources that can be renewed or replenished readily by natural processes.
research and development Gathering knowledge with the goal of creating new products or improving existing ones.
restriction enzyme An enzyme that cleaves DNA at a specific location along its strand.
ribonucleic acid (RNA) A type of nucleic acid that acts as an intermediary in the production of proteins within an organism.
salt An ionic compound produced by the reaction of an acid and a base. In common usage, **salt** or **table salt** refers specifically to sodium chloride (NaCl).
saponification Heating fats or oils in the presence of lye to produce soap and glycerol.
saturated fatty acid A fatty acid with no carbon-carbon double bonds.
scientific method The process by which science operates, involving the development of explanations for observations of the universe.
secondary air pollutant A pollutant formed by the further reaction of a primary air pollutant.
semisynthetic A substance made by chemical modification of a natural product.
SI base units Fundamental units of the SI system, such as the meter, kilogram, and second.
SI prefixes Prefixes that scale an SI unit either larger or smaller by some factor of ten.
single bond One pair of shared electrons serving as a covalent bond between two atoms.
smelting A process for producing metals from their ores.
soap The sodium salt of a long-chain carboxylic acid.
solubility A measure of the degree to which a solute dissolves in a solvent.
solute A substance present in lesser amount in a solution.
solution A homogeneous mixture of two or more substances.
solvent A substance present in greater amount in a solution.
standard reduction potential A standard measure of how readily a chemical particle accepts electrons.
strong acid An acid that ionizes completely in water to form an equivalent amount of hydronium ions.
strong base A base that dissociates completely in water to form an equivalent amount of hydroxide ions.
sublimation Changing directly from a solid to a gas, without passing through a liquid state.
sun protection factor (SPF) A measure of protection against UV-B radiation.
surface tension A property of liquids that causes them to behave as if the surface is covered by a thin membrane.
surfactant Shortened form of surface-active agent; a chemical that accumulates at a liquid's surface and changes the properties of that surface.
suspension A dispersion of solid particles in a fluid that tends to settle out over time.
theory A generally accepted principle based on a large set of confirmed observations.
thermic effect of food (TEF) Energy the body expends in digesting and metabolizing food.
toxin A poison of biological origin.
trace minerals Minerals required in the diet in amounts of less than 100 mg per day.
trans fatty acid An unsaturated fatty acid containing one or more bonds with a *trans* orientation.
triglycerides Molecules composed of glycerol and three fatty acid chains.
triple bond Three pairs of shared electrons serving as three covalent bonds between two atoms.
Tyndall effect The scattering or illumination of a light beam passing through a colloidal mixture.
unit cancellation A method of converting from one unit to another by multiplying by one or more equivalences.
unsaturated fatty acid A fatty acid with one or more carbon-carbon double bonds.
valence electrons The electrons contained in an atom's valence shell.
valence shell The outermost quantum shell of an atom.
vitamins Organic compounds required in small amounts for proper health.
volatile organic compounds (VOCs) Organic (carbon-based) compounds with relatively low boiling points ($<250°C$) and a propensity to evaporate.
volatility A measure of how readily a substance evaporates. Highly volatile compounds have low boiling points.
volt A measure of electrical potential.
weak acid An acid that ionizes partially in water.
weak base A base that ionizes partially in water.
weight A force due to the pull of gravity on an object.

Index

Note: Entries followed by "t" appear within tables.
Entries followed by "f" appear within figures.

B

C

D

E

F

H

I

J

K

L

M

N

Q

R

T

U

V

W

X

Z